Handbook of Research on Open Source Software:
Technological, Economic, and Social Perspectives

Kirk St.Amant
Texas Tech University, USA

Brian Still
Texas Tech University, USA

INFORMATION SCIENCE REFERENCE

Hershey · New York

Acquisitions Editor:	Kristin Klinger
Development Editor:	Kristin Roth
Senior Managing Editor:	Jennifer Neidig
Managing Editor:	Sara Reed
Assistant Managing Editor:	Diane Huskinson
Copy Editor:	Nicole Dean, Shanelle Ramelb, Ann Shaver, and Sue VanderHook
Typesetter:	Diane Huskinson
Cover Design:	Lisa Tosheff
Printed at:	Yurchak Printing Inc.

Published in the United States of America by
    Information Science Reference (an imprint of Idea Group Inc.)
    701 E. Chocolate Avenue, Suite 200
    Hershey PA 17033
    Tel: 717-533-8845
    Fax: 717-533-8661
    E-mail: cust@idea-group.com
    Web site: http://www.info-sci-ref.com

and in the United Kingdom by
    Information Science Reference (an imprint of Idea Group Inc.)
    3 Henrietta Street
    Covent Garden
    London WC2E 8LU
    Tel: 44 20 7240 0856
    Fax: 44 20 7379 0609
    Web site: http://www.eurospanonline.com

Library of Congress Cataloging-in-Publication Data

Handbook of research on open source software : technological, economic and social perspectives / Kirk St.Amant and Brian Still, editors.
  p. cm.
  Summary: "This book examines how use of open source software (OSS) is affecting society, business, government, education, and law,including an overview of the culture from which OSS emerged and the process though which OSS is created and modified. Readers will gain an understanding of the complexities and the nuances related to OSS adoption and the range of its applications"--Provided by publisher.
  Includes bibliographical references and index.
  ISBN 978-1-59140-999-1 (hardcover) -- ISBN 978-1-59140-892-5 (ebook)
  1. Shareware (Computer software)--Handbooks, manuals, etc. 2. Open source software--Handbooks, manuals, etc. I. St. Amant, Kirk, 1970- II. Still, Brian, 1968-
  QA76.76.S46H35 2007
  005.3--dc22
                        2006039844

British Cataloguing in Publication Data
A Cataloguing in Publication record for this book is available from the British Library.

All work contributed to this book set is new, previously-unpublished material. The views expressed in this book are those of the authors, but not necessarily of the publisher.

Editorial Advisory Board

List of Contributors

Table of Contents

Section I
Culture, Society, and Open Source Software

Section IV
Laws and Licensing Practices Affecting Open Source Software Uses

Section VII
Educational Perspectives and Practices Related to Open Source Software

Detailed Table of Contents

Section 1
Culture, Society, and Open Source Software

The chapter introduces and explains some of the most relevant features of the free software philosophy formulated by Richard M. Stallman in the 1980s. The free software philosophy and the free software movement built on it historically precede the open source movement by a decade and provide some of the key technological, legal and ideological foundations of the open source movement. Thus, in order to study the ideology of open source and its differences with regard to other modes of software production, it is important to understand the reasoning and the presuppositions included in Stallman's free software philosophy.

This chapter introduces Greasemonkey, a new extension for the Firefox browser, which enables users to alter the behavior and appearance of Web pages as the pages load. The chapter claims that Greasemonkey is forcing a reevaluation of what it means to be an author in digital environments. Using Michel Foucault's original question, "What is an author?" the chapter argues that creators of Greasemonkey scripts take on the additional roles of designer and programmer.

This chapter analyzes the differences between the philosophy of the Free Software Foundation as described by Richard Stallman and the open source movement as described in the writings of Eric Raymond. It argues that free software bases its activity on the argument that sharing code is a moral obligation and open source bases its activity on a pragmatic argument that sharing code produces better software. By examining the differences between these two related software movements, this chapter enables readers to consider the implications of these differences and make more informed decisions about software use and involvement in various software development efforts.

Chapter IV

This conceptual chapter aims to contribute to our understanding of the FLOSS innovation and how it is shaped by and also shapes various perceptions on and practices of hacker culture. The author argues that hacker culture has been continuously defined and re-defined, situated and re-situated with the ongoing development and growing implementation of FLOSS. The story on the development of *EMACSen* illustrates the consequence of different interpretations and practices of hacker culture clash.

Chapter V

This chapter investigates the premise that software is culture. It explores this proposition through the lens of peer production of knowledge-based goods circulating in the electronic space of a digital commons, and the material space of free media labs. Computing history reveals that technological development has typically been influenced by external socio-political forces. However, with the advent of the Internet and the free software movement, such development is no longer solely shaped by an elite class.

Chapter VI

A framework is proposed that creates, uses, communicates and distributes information whose organizational dynamics allow it to perform a distributed cooperative enterprise also in public environments over open source systems. The approach assumes the Web services as the enacting paradigm, possibly over a grid, to formalize interaction as cooperative services on various computational nodes of a network. By discussing a case study, the chapter details how specific classes of interactions can be mapped into a service-oriented model whose implementation is carried out in a prototypical public environment.

Chapter VII

Development organizations and international non-governmental organizations have been emphasizing the high potential of free and open source software for the less developed countries. Cost reduction, less vendor dependency and increased potential for local capacity development have been their main arguments. In spite of its advantages, free and open source software is not widely adopted on the African continent. In this chapter the experiences of one of the largest free and open source software migrations in Africa is evaluated. The purpose of the evaluation is to make an on-the-ground assessment of the claims about the development potential of FOSS and draw up a research agenda for FOSS community concerned with the less developed countries.

This chapter presents the benefits of FLOSS including its superior quality and stability. Challenges to FLOSS use particularly for developing countries are described. It indicates that despite the greater benefits to developing countries of technology transfer of software development skills and the fostering of ICT innovation, the initial cost of acquiring FLOSS has been the key motivation for many developing countries adopting FLOSS solutions. It illustrates this by looking at the experience of a university in a developing country, The University of the West Indies, St. Augustine Campus in Trinidad and Tobago.

Computing practices in developing countries can be complex. At the same time, open source software impacts developing countries in various ways. This chapter examines the social and economic impacts of open source software (OSS) on three such nations: China, South Korea, and India. In so doing, the chapter discusses and analyzes benefits as well as downsides of the social/political and financial impacts on these developing countries. Topics covered in this chapter are piracy, software licensing, software initiatives, social and political components involved in OSS implementation and software compatibility issues.

Section II
Development Models and Methods for Open Source Software Production

This chapter discusses the issues of dependencies, distances, and priorities in open source project networks, from the standpoint of both technological and social networks. Thus, a multidisciplinary approach to the phenomenon of open source software development is offered. There is a strong empirical focus maintained, since the aim of the chapter is to analyze open source software network characteristics through an in-depth, qualitative case study of one specific open source community: the Open Source Eclipse plugin project Laika.

This chapter explores the concept of patchwork prototyping—the combining of open source software applications to rapidly create a rudimentary but fully functional prototype that can be used and hence evaluated in real life situations. The use of a working prototype enables the capture of more realistic and informed requirements than traditional methods that rely on users trying to imagine how they might use the envisaged system in their work, and even more problematic, how that system in use may change how they work. Experiences with the use of the method in the development of two different collaborative applications are described.

Chapter XII

Open source software (OSS) development has been a trend parallel to that of agile software development, which is the highly iterative development model following conventional software engineering principles. Striking similarities exist between the two development processes as they seem to follow the same generic phases of software development. This chapter expounds on this connection by adopting an agile perspective on OSS development to emphasize the similarities and dissimilarities between the two models.

Chapter XIII

Although open source software (OSS) has been widely implemented in the server environment, it is still not as widely adopted on the desktop. This chapter presents a migration model for moving from an existing proprietary desktop platform (such as MS-Office on a MS-Windows environment) to an open-source desktop such as OpenOffice on Linux using the Gnome graphical desktop. The model was inspired by an analysis of the critical success factors in three detailed case studies of South African OSS-on-the-desktop migrations.

Chapter XIV

This chapter contributes to the sociological understanding of open source software (OSS) production by identifying the social mechanism that creates social order in OSS communities. OSS communities are identified as production communities whose mode of production employs autonomous decentralized decision making on contributions and autonomous production of contributions while maintaining the necessary order by adjustment to the common subject matter of work. Thu, OSS communities belong to the same type of collective production system as scientific communities.

Section III
Evaluating Open Source Software Products and Uses

Chapter XV

The chapter will present a detailed definition of open source software, its philosophy, operating principles and rules, and strengths and weaknesses in comparison to proprietary software. A better understanding of the philosophy underlying open source software will motivate programmers to utilize the opportunities it offers and implement it appropriately.

This chapter provides an insight into open source software and its development to those who wish to evaluate open source software. Using existing literature on open source software evaluation, a list of nine evaluation criteria is derived including community, security, license and documentation. In the second section these criteria and their relevance for open source software evaluation are explained. Finally the future of open source software evaluation is discussed.

Open source software is required to be widely available to the user community. To help developers to fulfill this requirement, Web portals provide a way to make open source projects public, so that the user community has access to their source code, can contribute to their development and can interact with the developers team. However, choosing a Web portal is not an easy task. There are several options available, each of them offering a set of tools and features to its users. The goal of this chapter is to analyze a set of existing Web portals (SourceForge.net, Apache, Tigris, ObjectWeb and Savannah), hoping that this will help users to choose a hosting site to their projects.

The aim of this chapter is to explore the differences and commonalities between open source software and other cases of open technology. The concept of open technology is used here to indicate various models of innovation based on the participation of a wide range of different actors who freely share the innovations they have produced.

This chapter introduces the hybrid GLW information infrastructure as an alternative to proprietary-only information infrastructures with lower costs. The author argues that the use of FLOSS servers in a client-server infrastructure reduces the transaction costs relative to the data processing and the contract management that organizations have to support, preserving the investment already made with the installed base of clients in comparison to the use of proprietary managed servers.

This chapter examines the main issues that have to be considered when selecting an open source content management system. It involves a discussion of literature and the experiences of the authors after installing and testing four widely used open source CMSs (Moodle, Drupal, Xoops and Mambo) on a stand-alone desk-top computer. It takes into consideration Arnold's (2003) and Han's (2004) suggestions for the development of CMSs, and identifies six criteria that need to be considered when selecting an open source CMS for use.

Chapter XXI

This chapter introduces a prototyping approach to evaluate OSS components. The prototyping approach provides decision makers with context-specific evaluation results and a prototype for demonstration purposes. The approach can be used by industrial organizations to decide on the feasibility of OSS components in their concrete business cases.

Chapter XXII

This chapter first explores why there is a need for data on free/libre and open source software (FLOSS) projects. Then the chapter outlines the current state-of-the art in collecting and using quantitative data about FLOSS project, focusing especially on the three main types of FLOSS data that have been gathered to date: data from large forges, data from small project sets, and survey data. Finally, the chapter will describe some possible areas for improvement and recommendations for the future of FLOSS data collection.

Chapter XXIII

This chapter attempts to bring to light the field of one of the less popular branches of the open source software family, which is the open source database management systems branch. The main system representatives of both open source and commercial origins will be compared in relation to this model, leading the reader to an understanding that the gap between open and closed source database management systems has been significantly narrowed, thus demystifying the respective commercial products.

Chapter XXIV

The chapter discusses the adoption and assimilation process of open source software as a new form of information technology. Specifically, although it reports a generally positive attitude towards OpenOffice. org, a widely used open source suite, it first shows the difficulties of the first early adopters to lead the innovation process and push other users. Different usage patterns, interoperability issues and in general the reduction in personal productivity typical of the early phases of adoption are also remarked.

Section IV
Laws and Licensing Practices Affecting Open Source Software Uses

This chapter discusses legal and economic rationale in regards to open source software protection. It examines copyright and patent issues with regard to software in the United States and Europe. Ultimately, it argues that there is a need to rethink approaches to property law so as to allow for viable software packaging in both models.

This chapter provides an anecdotal case study of the adoption of open-source software by government-funded nonprofit organizations in the legal services community. It focuses on the open source template, a Website system that provides information to the public on civil legal matters, and collaborative tools for legal aid providers and pro bono attorneys. It is hoped that this chapter will assist those considering the adoption of open source software by identifying the specific factors that have contributed to the success within the legal services arena and the real-world benefits and challenges experienced by the members of that community.

This chapter has two distinct objectives. Firstly to survey the political economic foundation of copyleft as it applies to open source computer software, and secondly, to provide some preliminary legal analysis in relation to the General Public License (GPL) which legally embodies copyleft principles. The chapter begins its philosophical exploration by giving a brief overview of copyright as it applies to the language of computer software, specifically source code. This is followed by a discussion that contrasts closed source and open source software development.

This chapter describes the background and spirit of the GPL and as well as discusses its importance. The chapter also examines certain socio-technical developments that challenge the effectiveness of existing licensing practices and describes the process of moving from GPL version 2 to version 3—a move intended to meet these challenges. This approach helps readers understand the importance of the GPL and understand how it creates a regulatory instrument to meet new challenges while maintaining its ability to offer the freedoms the license entails.

This chapter examines the free access to law movement—a set of international projects that share a common vision to promote and facilitate open access to public legal information. The project creates synergies between the notion of freeing the law by providing an alternative to commercial systems and the ideals that underpin open source software. To examine these factors, this chapter outlines the free access to law movement and examines the philosophies and principles behind it. The chapter also reviews the role open source software has played in the movement's development. The chapter then concludes with an assessment of what has been achieved and of the similarities between the free access to law and open source software movements.

This chapter presents a novel perspective of using the Creative Commons (CC) licensing model to compare 10 commonly used OSS licenses. In the chapter, the authors present a license compatibility table that shows whether it is possible to combine OSS with CC-licensed open content in a creative work. By presenting this information, the authors hope to help people understand if individuals can re-license a work under a specific CC license. Through such an increased understanding, readers will be able to make a better decision on license selection related to different projects.

This chapter examines certain issues related to what users of free/libre open source software (FLOSS) licenses are attempting to address through such uses. The chapter begins with a discussion of legal terms applicable to intellectual property and FLOSS. The chapter then examines software terms and their definitions as part of software development and engineering. The author then presents a taxonomy of FLOSS licenses. The chapter concludes with a brief discussion of how the perspectives of FLOSS users may change the need for a type of license.

Section V
Public Policy, the Public Sector, and Government Perspectives on Open Source Software

This chapter presents a critical review of the main arguments for and against public intervention supporting F/OS. The authors also provide empirical evidence related to public interventions taking place in Europe. The authors begin by providing a general analytical framework for examining public interventions. They then present evidence concerning the main public OSS initiatives in Europe. The chapter then concludes with a discussion of how to integrate theoretical perspectives with empirical analysis.

This chapter examines the integration of OSS in local and territorial e-administration in France. The policies defined in France and promoted by initiatives from the European Union are leading to the definition of a normative framework intended to promote interoperability between information systems, the use of free software and open standards, public-private partnerships, and the development of certain abilities. These policies are applicable to state agencies but are not required for local and regional collectives because of the constitutional principle of administrative freedom. To examine such issues, the authors of this chapter discuss how the integration of all administrative levels can be achieved through an e-administration OSS-based framework that coexists with proprietary software use.

This chapter introduces the L-PEST model as a tool for better understanding the motivations governments in their adaptation of FLOSS. The primary objective of this chapter is to identify and describe the actors associated with the use of FLOSS in the public sector, and in so doing, addresses a gap in the research on this topic. It is hoped the analytical model proposed in this chapter will help clarify the intricate relationship between relevant factors affecting FLOSS adoption and use by governments.

This chapter examines the economic and temporal/labor demands of creating free/libre and open source software (FLOSS). By examining the symbiotic relationship individuals have with commercial or closed software development, the author presents a new way to understand such interactions. This perspective is coupled with an examination of how this economic structure could conceivably be exploited for increased economic gain at the expense of those individuals actually involved in the creation of the software. The chapter then concludes with a discussion of possible ways in which FLOSS software could be opened up more broadly to non-technical software users.

This chapter contextualizes open source development and deployment in the nonprofit sector and discusses issues of ideology that often accompany such development and deployment. The chapter separates and defines the ideologies of application development, selection, and use by describing the different issues each creates in the nonprofit context. The author's purpose in presenting such informaiton is to clearly articulate the unique dynamics of application development and deployment in the nonprofit, or social value, context and where to apply related ideological considerations for best effect.

This chapter examines the critical task of governing the open source environment with an open source repository. As organizations move to higher levels of maturity, the ability to manage and understand the open source environment is one of the most critical aspects of the architecture. Successful open source governance requires a comprehensive strategy and framework which this chapter presents through historical, current-state, and future perspectives. By understanding the role of open source metadata and repositories, researchers will continue to expand the body of knowledge around asset management and overall architecture governance.

Section VI
Business Approaches and Applications Involving Open Source Software

Surprisingly little empirical research has been performed to understand firms' participation in OSS communities. This chapter aims to fill this gap in state-of-the-art research on OSS by discussing the results of a survey involving 125 Dutch high-technology firms that are active in the market for OSS products and services. The results presented in this chapter contribute to research on OSS by providing a new model for perceiving the concept of community in relation to OSS. The results also suggest that firms view their internal investments in R&D as a complement to their external product-development activities in OSS communities.

This chapter discusses the role of the project/product community in the open source product life cycle. The chapter outlines how a community-driven approach affects not only the development process, but also the marketing/sales process, the deployment, the operation, and the resulting software product. Participation in the community is essential for any organization using the product, leading to the concept of a *community customer*. For this reason, the chapter presents specific community participation guidelines that organizations and individuals can use to further develop OSS products or to offer lifetime services on those products.

Chapter XL

This chapter focuses on the economics of open source strategies. From a strategic perspective, the concept of open source falls into a category of business models that generate advantages based on customer and user involvement (CUI). While open source has been a novel strategy in the software business, CUI-based strategies have been used elsewhere before. This chapter presents a review of CUI-based competition, clearly delineates CUI antecedents and business value consequences, and concludes with a synopsis of managerial implications and a specific focus on open source.

Chapter XLI

This chapter studies how venture capitalists invest in open source-based companies. The evaluation of knowledge-intensive companies is a challenge to investors, and the rise of open source companies with new value propositions brings new complexity to deal-making. This chapter highlights some experiences some venture capitalists have had with open source companies. The authors hope that the overview of venture capital processes, the methodology, and the two case examples in this chapter provide both researchers and entrepreneurs with new insights on how venture capitalists work and make investments in open source companies.

Chapter XLII

Profit-oriented business behavior has increased within the open source software movement, yet it has proven to be a challenging and complex issue. This chapter explores considerations in designing profitable revenue models for businesses based on open source software. The authors approach the issue through two business cases: Red Hat and MySQL, both of which illustrate the complexity and heterogeneity of solutions and options in the field of OSS. The authors focus on the managerial implications derived from the cases and discuss how different business elements should be managed when doing business with open source software.

Chapter XLIII

This chapter introduces open source software (OSS) for accounting and enterprise information systems. It covers the background, functions, maturity models, adoption issues, strategic considerations, and future trends for small accounting systems as well as large scale enterprise systems. The authors hope that understanding OSS for financial applications will not only inform readers of how to better analyze accounting and enterprise information systems, but also assist in the understanding of the relationships between the different functions of these systems.

Chapter XLIV

This chapter discusses different ways the open source software (OSS) methods of software development interact with the corporate world. The success achieved by many OSS products has produced a range of effects on the corporate world, and OSS has presented the corporate world with opportunities and ideas, prompting some companies to implement components from OSS business models. The consumer of software is sometimes baffled by the differences in the two, often lacking understanding about two models and how they interact. This chapter clarifies common misconceptions about the relationship between OSS and the corporate world and explains facets of the business models of software design to better inform potential consumers.

Chapter XLV

This chapter explores how use of business models enables value creation in open source software (OSS) environments. In the chapter, the authors argue that this value can be attained by analyzing the value creation logic and the elements of potential business models emerging in the OSS environment, as profitable business is all about creating value and capturing it properly. Open Source offers one possibility for firms that are continuously finding new opportunities to organize their business activities and increase the amount of value they appropriate via their capabilities. Furthermore, the concept of a business model is considered as a tool for exploring new business ideas and capturing the essential elements of each alternative. The authors, therefore, propose that a general business model that is also applicable in the context of OSS, and they provide a list of questions that may help managers with their uses of OSS in business.

Chapter XLVI

This chapter presents an open source business strategy that is a feasible and profitable option for organizations to consider. In examining such a strategy, the author uses a practical case study involving Novell, Inc.—a company that is successfully implementing a free/libre open source software (FLOSS) business strategy. In so doing, the author addresses the concern that there is no substantial evidence on whether OSS processes and practices are effective within the business environment. The author also taps an emerging interest in finding ways to use OSS as an effective part of an organization's business strategy.

Section VII
Educational Perspectives and Practices Related to Open Source Software

Chapter XLVII

This chapter examines the use of communities of practice in the process of disseminating open source software (OSS) in the University of São Paulo, Brazil. In this particular case, the change management process included establishing an OSS support service and developing a skills-building training program for its professional IT staff that was supplemented by a community of practice supported by an Internet-based discussion list. After relying on these resources extensively during the early phases of the adoption process, users changed their participation in this local community by a mostly peripheral involvement in global OSS communities of practice. As a result of growing knowledge and experience with OSS, in this context, users' beliefs and attitudes toward this technology became more favorable. These results, consistent with the theory of planned behavior constructs, provide a useful guide for managing the change process.

Chapter XLVIII

This chapter examines the use of computing technologies in educational settings. The author explains that different countries are at different stages in this process, but in general, the deployment of technologies is moving from individual, stand alone computers to integrated technologies that are networked. Since models of open source software development are based around contributing to the public good through online networked activities, the paradigm shift away from personal to networked computers linked to the Internet makes open source software viable both technically and philosophically for the education sector in many regions. To help others better understand such factors, the author explores some of the technical and philosophical contributions open source software can make to education with the objective of helping readers develop criteria for identifying suitable open source software for use in schools.

Chapter XLIX

This is chapter presents evaluation criteria used by a higher education institution to evaluate an open source e-learning system. The idea the author explores is that e-learning applications are becoming commonplace in most higher education institutions, and some institutions have implemented open source applications such as course management systems and electronic portfolios. These e-learning applications are the first step towards moving away from proprietary software such as Blackboard and WEBCT and toward open source products. To examine such shifts, the author presents aspects educators need to consider in relation to such systems.

This chapter discusses the rapid transition from paper to electronic distribution of scholarly journals and how this has led to open access journals that make their content freely available over the Internet. The chapter also presents the practical and ethical arguments for providing open access to publicly funded research and scholarship and outlines a variety of economic models for operating these journals. In so doing, the chapter examines the practical and ethical arguments for open access to research and scholarship and explores alternative models for funding the dissemination of scholarship and the key role open source software can play in facilitating open access to scholarship. The chapter then concludes with a discussion of future trends in the organization and funding of scholarly publication.

This chapter presents a case study of a migration to OSS in a South African school. The innovative aspect of the case study lies in how the entire implementation was motivated by the collapse of the school's public address system. It was found that an OSS-based message provided a more cost-effective replacement option whereby the speakers in the school were replaced with low-cost workstations (i.e., legacy systems) in each classroom. (Interestingly, this OSS implementation happened despite the fact that in South Africa, Microsoft Windows and MS-Office are available free of charge to schools under

Microsoft's Academic Alliance initiative.) The chapter also analyzes some critical themes for adoption of OSS in the educational environment.

Chapter LII

This chapter examines how leading-edge, industrial-strength software can be introduced into the university classroom by using open source software, open standards, distance learning, and infrastructure shared among cooperating universities. In addressing this topic, the authors describe the evolution of software development during the twentieth century, the paradigm change at the beginning of the twenty-first century, and the problems with the existing university information technology education. The authors then describe a shared software infrastructure program (SSIP) to rapidly introduce leading edge industrial software solutions into university classrooms at no cost to SSIP Member Universities.

Chapter LIII

This chapter explores the role of technology in learning in relation to the use of wikis in educational environments. In examining this topic, the author first overviews the use of technology in educational environments. The author next examines wikis and their development and then reviews the emerging literature on the application of wikis to education. Through this approach, the author uses wikis as an exemplary model of open source learning that has the potential to transform the use of information communication technologies in education.

Chapter LIV

This chapter provides a model for using open source software within engineering education. The author begins the chapter with an overview of OSS and his position on the use of OSS in educational settings. The author next presents a treatment of central software engineering practices as they relate to OSS perspectives and uses. The author next discusses the use of OSS in software engineering education (SEE). The chapter concludes with a discussion of both the challenges related to the use of OSS in such educational settings and directions for future research on the topic.

Foreword

THE MANY QUESTIONS FROM OPEN SOURCE SOFTWARE...

Open source software has for some years been one of the most hotly debated topics, both in research and in practice. One reason for this is that several open source products like GNU/Linux or Apache have now for years been in the spotlight as leaders in their respective application areas, and continue to be so, while others like MySQL or even ERP packages come to the front in new application areas not traditionally associated with open source software. This fact has demonstrated one thing, to people from academia, industry and public organizations: Open source projects can indeed lead to software systems that exhibit and maintain high functionality and quality.

Nevertheless, there are numerous questions remaining, of interest to different groups. While researchers want to uncover the mechanisms and reasons for this development model to work, management wants to know how to use open source software to its fullest advantage or how to base a business on it, and public organizations both on the national and international level struggle with the question of how to deal with this phenomenon. This handbook succeeds in bringing together papers addressing the whole range of topics in the area of open source software. Given the diversity of this field, this is not an easy task, but researchers, managers and policy-makers will all find interesting answers and even more interesting new questions within the pages of this handbook.

Scanning the different entries gives a great impression of what different subjects currently garner the most attention.

OSS Evaluation and Adoption

Given the amount of different projects, even within a set application area, this is a growing concern, especially with practitioners. For example, van den Berg gives an overview of approaches to evaluating open source software in "Open Source Software Evaluation," while Carbon, Ciolkowski, Heidrich, John, and Muthig present a new method of evaluation through prototype development in "Evaluating Open Source Software through Prototype Development." Also some special cases (the IT in schools by Moyle in "Selecting Open Source Software for Use in Schools") and concrete application areas like content management systems ("Issues to Consider when Choosing Open Source Content Management Systems (CMSs)" by Boateng and Boateng), database management systems ("A Generalized Comparison of Open Source and Commercial Database Management Systems" by Evdoridis and Tzouramanis) or business functions ("Open Source for Accounting and Enterprise Systems" by Tribunella and Baroody) are explored in detail. But evaluating and choosing an optimal open source software does not end the process, adoption does not depend on software functionality and quality alone. For example, Rossi, Russo, and Giancarlo Succi detail migrations in public administrations ("Evaluation of a Migration to Open Source

Software"), and Brink, Roos, Van Belle, and Weller propose a model for desktop migration ("A Model for the Successful Migration to Desktop OSS"), backed up by a case study ("An Innovative Desktop OSS Implementation in a School"). Of special interest is the entry by Humes titled "Communities of Practice for Open Source Software," in which concepts from theory of planned behavior are used in a case study showing the positive effects of establishing communities of practice for adoption and diffusion. Following the increased adoption rates, most IT architectures today tend to become hybrid incorporating both proprietary and open source software, with Vinicius acknowledging this fact and exploring the concept of transaction costs in the context of information infrastructures ("Reducing Transaction Costs with GLW Infrastructure"). Lastly, Stephens takes a look beyond a single adoption or migration project and proposes establishing a centralized repository for downloading certified open source products to ensure good governance ("Governance and the Open Source Repository").

Areas of Special Interest: Science and Education

There are also a few areas that are of special interest regarding the adoption of open source software, for example the scientific process itself, which has often been compared to open source development. Bosin et al. propose an architecture for cooperative scientific experiments ("ALBA Architecture as a Proposal for OSS Collaborative Science"), and Solomon details "The Role of Open Source Software in Open Access Publishing." The other area is education, where beyond the entries already mentioned above, two more chapters highlight the importance of open source software in this context ("A Perspective on Software Engineering Education with Open Source Software" by Kamthan and "Rapid Insertion of Leading Edge Industrial Strength Software into University Classrooms" by Simmons, Lively, Nelson, and Urban).

OSS in Public or Nonprofit Organizations

Also of high interest, and in some areas overlapping with choosing and adopting open source software, is the relationship with public or nonprofit organizations. The interactions between open source and public organizations can be broadly grouped as adopting open source software, becoming co-developers or sponsors in projects and finally acting as regulatory authorities, most notably regarding software patents. The issue of adoption has already been touched on by some entries also in the context of public organizations; for example Favier, Mekhantar, and Terrasse delve into more detail in "Use of OSS by Local E-Administration: The French Situation," or Agostinelli in "OSS Adoption in the Legal Services Community." Moving from passive use to development or sponsoring, Laszlo provides an inductive general conceptual model of various public sector and government initiatives for promoting or using open source ("Issues and Aspects of Open Source Software Usage and Adoption in the Public Sector"), while Peizer explicitly contextualizes open source development and deployment in the nonprofit sector and discusses issues of ideology that often accompany its use ("Open Source Technology and Ideology in the Nonprofit Context"). Public policies in the European context also form the basis for yet another entry ("On the Role of Public Policies Supporting Free/Open Source Software" by Comino, Manenti, and Rossi). Finally, the role open source might play for developing countries is the topic of a chapter by Dudley-Sponaugle, Hong, and Wang, titled "The Social and Economical Impact of Open Source Software in Developing Countries."

OSS Business Models

These topics relevant for public organizations distinctly differ from private firms, which, beyond adopting open source software, increasingly participate in projects or explore related business model. In this handbook, business models feature in several entries: Seppänen, Helander, and Makinen give an introduction to this topic with "Business Models in Open Source Software Value Creation" and the basic message of this chapter is that the elements of a business model remain the same regardless of industry. Rajala, Nissilä, and Westerlund also take up this topic, and discuss revenue models, based on case studies of Red Hat and MySQL ("Revenue Models in the Open Source Software Business"). This is complemented by yet another famous and successful case study, Novell, by du Preez ("Novell's Open Source Evolution"). An interesting new viewpoint is introduced by Puhakka, Jungman, and Seppänen in their chapter "Investing in Open Source Software Companies: Deal Making from a Venture Capitalist's Perspective," in which they conclude that venture capitalists do not seem to put special value to open source companies, but some recognize different elements in evaluating those companies. Finally, Stam and van Wendel de Joode (Analyzing Firm Participation in Open Source Communities") explore the participation of firms in open source projects based on a survey. They distinguish between technical and social activities, and highlight factors leading to different types and levels of engagement. One important result concerns the finding that firms seem to view their internal investments in R&D as a complement to their external product-development activities in OSS communities.

OSS Theory

For the researcher, the reasons and workings behind open source software and its development are key topics. A number of entries in this handbook reflect this, which deal with the theoretic underpinnings of this movement. The discussion around the protection of software programs and open source licenses are manifold, and, for example, de Vuyst and Fairchild highlight this in "Legal and Economic Justification for Software Protection." Also related is an entry by Cunningham titled "The Road of Computer Code Featuring the Political Economy of Copyleft and Legal Analysis of the General Public License." Both chapters go beyond a strictly legal discussion as provided by Lin, Lin, and Ko in "Examining Open Source Software Licenses through the Creative Commons Licensing Model" and incorporate a political, social and economic perspective. Ballentine in a highly interesting chapter challenges the underlying notion of authorship itself ("Greasemonkey and Challenges to Authorship").

But open source software is not only based on its licenses, but also a different ideology or culture. These are the topics in three different entries: "Free Software Philosophy and Open Source" (Vainio and Vadén), "Morality and Pragmatism in Free Software and Open Source" (Yeats), and "Hacker Culture and the FLOSS Innovation" (Lin), a chapter acknowledging the importance of the continuously evolving hacker culture for open source, while discussing its changing mainstream perception. O'Donnell, in the chapter, "The Labor Politics of Scratching an Itch" also highlights the base for open source development by examining the relationships with educational, employment and work compensation and the results on the overall demographics of this movement.

OSS Development and Community

Lastly, open source software is also about software development and communities. The issue of whether this constitutes a new or more efficient way of production is one of the main questions surrounding this phenomenon. Gläser defines open source communities as production communities that apply a distinct mode of production of decentralized task definition ordered by the common subject matter of work ("The Social Order of Open Source Software Production"). Within this process, Hoppenbrouwers identifies "Community Customers," individuals or organizations who want to deploy an open source product, without having a direct aim to further develop the product, and who actively engage in the community to assure future suitability of the product, and discusses their role. In research, many theories can be developed or discussed, but ultimately need to withstand empirical validation. Empirical research into different aspects of open source software and its production has therefore been performed for some years now, and Conklin gives an overview of methods and results ("Motives and Methods for Quantitative FLOSS Research"). As an example, Järvensivu, Helander, and Mikkonen present an empirical case study on an open source project, where both the underlying technological and social networks, both internal and external, are explored. Finally, the relationships between open source software development and traditional software engineering techniques have often been discussed. Sahraoui, Al-Nahas, and Suleiman put this model in the context of agile software development practices, and uncover striking similarities ("An Agile Perspective on Open Source Software Engineering"), while Jones, Floyd, and Twidale propose a rapid prototyping-based approach to requirements gathering using open source software ("Patchwork Prototyping with Open Source Software").

This comprehensive handbook successfully demonstrates the diversity of subjects surrounding the deceptively simple term of *open source software*. This fact alone will ensure that in the future, open source software will certainly continue to be an issue for researchers and practitioners. We cannot yet foresee where this trend will go, or where it will take us, but one thing is certain: Open source software is here to stay.

Stefan Koch
Department of Information Systems and Operations
Vienna University of Economics and Business Administration
Vienna, Austria

Stefan Koch is an associate professor of information business at the Vienna University of Economics and Business Administration, Austria. He received an MBA in management information systems from Vienna University and Vienna Technical University, and a PhD from Vienna University of Economics and Business Administration. Currently, he is involved in the undergraduate and graduate teaching program, especially in software project management and ERP packages. His research interests include open source software development, cost estimation for software projects, IT investment analysis and governance, the evaluation of benefits from information systems, applications of data envelopment analysis, and ERP systems. He has edited a book titled *Free/Open Source Software Development* for an international publisher in 2004, and acted as guest editor for Upgrade for a special issue on libre software. He has published 10 papers in peer-reviewed journals, and over 25 in international conference proceedings and book collections. He has given talks at more than 20 international and national conferences and events, acted as program committee member for 8 conferences, including the Second Conference on Open Source Systems, and served as a reviewer for several journals including *Management Science, CACM, Journal of Systems and Software, IEEE Transactions on Software Engineering, Research Policy,* and *Journal of Management Information Systems*, as well as numerous conferences including HICSS and ICIS.

Preface

SOFTWARE IN THE MODERN CONTEXT

In many ways, software has become the life's blood of the modern world. Software allows businesses to compile and share data—literally—at the speed of light. Software also permits governments and other public sector organizations to oversee numerous activities, administer vast territories, and analyze developing situations. So powerful is the hold software has over the modern world that the fear of a global software crash in the form of the *millennium bug* led many to associate widespread software glitches with the end of civilization as we know it.

While the turn of the millennium passed without major incident, the importance of software in society continues to grow. This time, however, individuals and organizations are increasingly turning their attention from software use in industrialized nations to computing practices and online access in the developing world. As a result, concerns over the millennium bug have given way to growing interest in the global digital divide. And public and private organizations alike are increasingly examining how computers, online access, and software might help developing nations advance economically, politically, and socially.

The expanding and interlinked nature of global software use presents new situations and raises new questions for organizations and individuals alike. For example, what kinds of software should be used when and how? What are the economic, social, and political ramifications of deciding to use one kind of software instead of another? Likewise, choices based on responses to these questions can affect interaction and integration on both local and global scales. For these reasons, it is now more important than ever that individuals have access to the information needed to make informed choices about software adoption and use.

This edited collection is an initial step toward providing some of the information needed to make such informed choices. By presenting readers with a range of perspectives on a particular kind of software—open source software (OSS)—the editors believe they can shed light on some of the issues affecting the complex nature of software use in different contexts. Moreover, by bringing together the ideas and opinions of researchers, scholars, businesspersons, and programmers from six continents and 20 nations, the editors believe this collection can help readers appreciate the effects of software-related decisions in today's interconnected global environment.

OPEN SOURCE SOFTWARE (OSS): AN OVERVIEW

Software is essentially programming code—or *source code*—that provides a computer's operating system with instructions on how to perform certain tasks. The source code of a word processing program, for example, provides an operating system with information on how to translate certain keystrokes into specific text that appears on digital (and later print) pages. In essence, if a person knows the source code needed to make a computer perform a particular operation, then that individual can simply enter such source code into his or her operating system, and the computer will respond as desired. Such a response, moreover, would be the same one the computer would provide in relation to the original software product.

Within this framework, if an individual can access the underlying source code for a software product, then that person can just copy the code and not need to purchase the original program. For this reason, many software companies *close off* access to the source coding that allows their programs to work. As a result, users cannot readily access and copy the *mechanism* that makes the product work (and thus, gives it value). Instead, individuals must use an interface that allows them to activate certain commands indirectly within a software program's underlying source code. Such closed programs are know as *proprietary software*, for only the creator/copyright holder of that software is allowed to open or to see and to copy or manipulate the underlying source code.

Open source software (OSS), however, represents a completely opposite perspective in terms of access to source code. OSS products are, in essence, created in such a way that access to the underlying source code is *open* and available for others to access and review. A very basic yet common example of such open access to coding is HTML, which allows browsers to display Web pages. The coding of these pages is generally open for anyone to review and replicate. All one needs to do is access a page's underlying coding by using the "View Source"—or other related option—in his or her browser.

Such openness means individuals do not need to buy open source software in order to use it. Rather, they can review and copy a program's underlying source code and thus create a "free" version of the related software. Such openness also means individuals can modify source code and thus alter or enhance the related program. So, in theory, the foundational source code of one software product could be modified to perform different functions—or even become an entirely new program. Updating software, in turn, becomes a matter of copying new or modified code vs. purchasing the newest version of a product.

From both a local and an international perspective, OSS can provide individuals with access to affordable software that allows them both to engage in computing activities and to access others via online media. Moreover, the flexibility of OSS means individuals can modify the software they use to perform a wide variety of tasks. Doing so reduces the need for buying different programs in order to perform different activities. Thus, it is perhaps no surprise that the use of OSS is growing rapidly in many of the world's developing nations where the costs of proprietary software is often prohibitive for many citizens.

LIMITATIONS OF OPEN SOURCE SOFTWARE: A BASIC PERSPECTIVE

The easily accessible and flexible nature of open source software makes it ideal to use in a variety of contexts. OSS, however, also brings with it certain limitations that could affect interactions among individuals. Many of these limitations, in turn, arise from the fact that OSS is often developed and supported by communities of *volunteer* programmers who create and modify items in their free time.

First, and perhaps foremost, because OSS is open for the user to modify as desired, it is easy for each individual to use the same programming foundation/source code to develop different non-compatible softwares. Such divergence is often referred to as *forking code*. In such cases, each programmer working on the development of an OSS item can take a different "fork" in the programming *road*, and with each different fork, two programs that were once identical become increasingly different from one another.

Such forking code, moreover, has long been considered a major problem in OSS development.

These prospects for divergence mean OSS use is open to a variety of problems involving compatibility. Such problems include:

- Individuals generating software that others cannot use due to compatibility issues
- Software that does not work as desired or work in unexpected ways
- Parts of distributed programming projects not working as intended or not working at all
- Users becoming frustrated with and abandoning software they consider too time-consuming or cumbersome to operate

Thus, the freedom that allows one individual to operate software might prevent others from making use of such materials. As a result, access to and exchanges among others—perhaps the central focus of many of today's software packages—are impeded by the software itself.

Some companies, such as Linux, have addressed the problem of forking code and compatibility through focused oversight processes that govern programming practices. The result has been successful and relatively stable software products that work effectively with other systems. The same kind of management, oversight, and standardization, however, becomes more complicated in most OSS development/production situations where the standard approach often involves a group of globally dispersed, unpaid individuals creating OSS products in their spare time.

A second major problem area for OSS involves the technical support available to users of such software. Because it is often the case that no individual or organization really *owns* an open source software product, there is often no formal or standard mechanism for providing technical support to OSS users. Such support instead tends to come from loose networks of OSS developers and aficionados who interact informally in online contexts such as chat rooms or listserevs. Within this context, *technical support* generally means a user who is experiencing difficulty posts a query to an online OSS forum and then waits for a member of that forum to read the posting and reply.

One limitation of such an informal support system is that answers are not readily available. Instead, individuals could find themselves waiting for anywhere from seconds to days from some random community member to respond. Such delays could, in turn, have a major effect on the usability and the desirability of OSS products—not to mention the successes with which individuals can use such products to interact. Equally problematic is that such technical support systems are open for anyone to participate in and provide advice or solutions—regardless of the technical skills of the individual. Thus, the quality of the advice provide by OSS support systems can be haphazard, inconsistent, or even incorrect.

While these are but two problem areas, they illustrate the complexities bound up in selecting and using software effectively. Moreover, the use of OSS vs. proprietary software becomes increasingly intertwined with different social and political perspectives related to computing use. As a result, software choices can be as much a matter of socio-political ideology as they can be about using a product to perform a task.

SOCIAL PERSPECTIVES ON OSS: OWNERSHIP AND ECONOMICS

Much of the software we use today is proprietary. In other words, the source code that makes it work is not accessible for modification by those that purchase it. Notable opponents of proprietary software, such as the creator of the GNU operating system, Richard Stallman, led efforts to develop and distribute software and its source code freely. These efforts became known as the free software movement (FSF) and enabled software developers to both use and modify FSF items. The only stipulation these initial programmers imposed was that individuals who used FSF as a foundation for developing other items needed to make such modified code freely available. Perhaps the most successful example of FSF is the GNU/Linux operating system, which continues to increase its market share in direct competition with Microsoft's proprietary systems.

As attractive as Stallman's free software approach was to many, it also alienated others who believed that such a revolutionary and also intransigent stand—one that insisted all software code be made freely available—was not feasible. Additionally, this socio-political stand to programming was often confusing—particularly to individuals from outside of this community of programmers, for many of them interpreted the word *free* to mean nothing could be charged for and no profit could be made from developing such software. As a result of this perception, many entrepreneurial developers and software companies refrained from participating more actively in supporting Linux and other FSF products. Because of this view, a number of FSF developers, including Eric Raymond and Bruce Perens, met in 1998 to put a more business-friendly face on free software. What they eventually developed was the concept of open source software (OSS). Unlike free software, OSS was more flexible and even offered additional licensing possibilities that allow individuals to mix proprietary and open software. One example of this "hybrid" approach is the Berkeley Distribution License, which allows software developers to modify source code and then take their modifications private, even selling them for a profit, rather than having to make them freely available to others, including competitors.

PRAGMATIC APPLICATIONS: EXPLORING OPTIONS FOR OSS USE

For all of the inroads OSS has made in encouraging businesses to adopt it, it still has its detractors who rightly point out that—as noted earlier—most OSS is produced by volunteer hobbyists with no financial incentive to contribute to or continue supporting products they've produced. OSS also suffers from customer service problems since no one company necessarily exists to stand behind a product. In addition, with developers free to modify source code and generally distribute it however they wish, many different derivatives of the same basic software (*forking code*) can exist, leading to confusion and incompatibility. Finally, because of its ever-increasing popularity, OSS is something more businesses and more developers are interested in leveraging or contributing to, resulting in a market that some argue supports too many products and too few developers to go around supporting them all. For example, although there were only a handful of OSS content management systems just a few years ago, there are now scores of such systems—a situation that makes it difficult for consumers to decide which one to use.

Despite such negatives, OSS is not a passing fancy. In fact, many organizations have followed the lead of RedHat (a distributor and service supporter of Linux) in exploring ways to develop business models that maximize the advantages of OSS while maintaining the openness and flexibility of products. IBM, for example, commits a large part of its core group of developers to building and/or enhancing OSS. In addition, organizations and even governmental bodies, ranging from small not-for-profits to the European Union, have actually adopted OSS for use, with even more exploring how OSS can contribute to their core operations. Understanding how to do this, considering that OSS is a relatively new player, is challenging. It requires knowledge not only of the origins and the operating principles of OSS, but also knowledge of the social, legal, and economic factors that affect the use of OSS products.

SEARCHING FOR UNDERSTANDING WITHIN THE COMPLEXITY: OBJECTIVE AND ORGANIZATION OF THIS HANDBOOK

The decision to purchase or to use a particular software product is an important one that can contribute to the success or the failure of the related organization. For this reason, decision makers at different levels and in a variety of fields need to familiarize themselves with the various factors that contribute to the successful adoption and use of software products. Similarly, individuals need to make better-informed choices about what software to select or personal use and why. In the case of open source software, such decisions are further complicated by the social agendas and economic goals many developers and users attach to the use of OSS materials.

The objective of this handbook is to provide readers with a foundational understanding of the origins, operating principles, legalities, social factors, and economic forces that affect the uses of open source software. To achieve this objective, this handbook is divided into seven major sections, each of which examines a different factor related to or affecting OSS development, adoption, and use. The topic of each major section, in turn, is examined by 7-10 authors from different cultures and backgrounds—authors who provide a broad range of perspectives on the related topic. As a result, each major section provides readers with both a more holistic treatment of each subject and a broad base of information upon which more informed decisions can be made.

The seven major sections of this handbook are as follows:

- **Section I: Culture, Society, and Open Source Software:** The entries in this section overview the internal culture of the individuals who create OSS products as well as examine both social perspectives on OSS use ant the potential OSS has to change societies.
- **Section II: Development Models and Methods for Open Source Software Production:** These chapters explore different methods for creating OSS products and discuss the benefits and the limitations of such methods as well as consider approaches for maximizing the more successful elements of such methods.
- **Section III: Evaluating Open Source Software Products and Uses:** Authors in this section both present models for assessing the effective uses of various OSS products and provide opinions on what makes some OSS items successful while others are not.
- **Section IV: Laws and Licensing Practices Affecting Open Source Software Uses:** In this section, chapters examine how legal factors and licensing strategies try to shape OSS development and use and also explore the new legal situations created by OSS products.
- **Section V: Public Policy, the Public Sector, and Government Perspectives on Open Source Software:** The chapters provide both examples of how government agencies and other non-profit organizations have adopted or adapted OSS use to meet programming needs; they also present ideas for how such public-sector entities should view OSS within the context of their activities.
- **Section VI: Business Approaches and Applications Involving Open Source Software:** This section's authors present models and cases for OSS development approaches and uses in for-profit endeavors as well as explore how business can address some of the more problematic aspects of OSS adoption and use.
- **Section VII: Educational Perspectives and Practices Related to Open Source Software:** Entries in this concluding section employ a range of perspectives and approaches to examine how OSS products can be integrated into educational activities in different contexts within and across societies.

While this collection of chapters provides readers with a wealth of OSS-related information, this text only begins to explore the complex environment in which software is operated. The foundation provided by the essays in this handbook, however, is an essential one for helping readers understand key concepts and ask the right questions when exploring software adoption and use. By using this information and building upon these ideas and perspectives, readers can enhance their views of software use in society while also shaping policies and practices related to software.

Kirk St.Amant and Brian Still
Lubbock, TX, USA
April 2007

Acknowledgment

The editors would like to thank all involved in the creation of this collection including the reviewers, who dedicated their time and expertise to this project, and the editorial staff at IGI Global, Kristin Roth in particular, for their assistance and professionalism throughout this project.

Our thanks also go out to all of the individuals who contributed to this collection; their intelligence, insights, and commitment to examining how open source software affects different aspects of our daily lives are all greatly appreciated.

Finally, Kirk wishes to thank his daughter, Lily Catherine St.Amant, for being a continual source of inspiration in all that he does, his wife, Dori St.Amant, for her unwavering patience and understanding during this project, and his parents, Richard and Joan St.Amant, for encouraging his early interests in technology and communication. Brian wishes to thank his family—Amy, Jack, and Olivia—for their love and support during this project.

Kirk St.Amant and Brian Still
Lubbock, TX, USA
April 2007

Section I
Culture, Society, and Open Source Software

Chapter I
Free Software Philosophy and Open Source

Niklas Vainio
University of Tampere, Finland

Tere Vadén
University of Tampere, Finland

ABSTRACT

This chapter introduces and explains some of the most relevant features of the free software philosophy formulated by Richard M. Stallman in the 1980s. The free software philosophy and the free software movement built on it historically preceded the open source movement by a decade and provided some of the key technological, legal and ideological foundations of the open source movement. Thus, in order to study the ideology of open source and its differences with regard to other modes of software production, it is important to understand the reasoning and the presuppositions included in Stallman's free software philosophy.

INTRODUCTION

The free software (FS) movement is the key predecessor of the open source (OS) community. The FS movement, in turn, is based on arguments developed by Richard M. Stallman. In crucial ways, Stallman's social philosophy creates the background for the co-operation, co-existence and differences between the two communities. Stallman started the FS movement and the GNU project prompted by his experiences of the early hacker culture and subsequent events at the MIT artificial intelligence lab in the 1980s. The project was founded on a philosophy of software freedom, and the related views on copyright or the concept of *copyleft*. After the creation of the open source movement in 1998, debates between the two movements have erupted at regular intervals. These debates are grounded in the different ideological perspectives and sociopsychological motivations of the movements. The FS movement has laid technological, legal and ideological cornerstones that still exist as part of the open source movement.

THE SOCIOHISTORICAL BACKGROUND OF THE FREE SOFTWARE PHILOSOPHY

The first computer systems were built in the 1940s and 1950s mainly for military and scientific purposes. One of the earliest research institutes to use and study computers was the Massachusetts Institute of Technology (MIT). The artificial intelligence (AI) lab at MIT was founded in 1958 and became one of the birthplaces of computer science and computer culture.

In *Hackers* (1984), Steven Levy describes the subculture around the AI lab computers in the 1960s. Young male electronics hobbyists devoted their time to programming and studying these machines. They called themselves *hackers*, a word denoting a person who enjoys exploring computer systems, being in control of the systems, and facing the challenges they present. For a hacker, a computer is not just a tool, it is also an end in itself. The computer is something to be respected and programming has an aesthetics of its own (Hafner & Lyon, 1996; Levy, 1984; Turkle, 1982).

A subculture was created among the MIT hackers with traditions and social norms of its own. Important values for the community were freedom, intelligence, technical skills, and interest in the possibilities of computers while bureaucracy, secrecy, and lack of mathematical skills were looked down on. The six rules of this *hacker ethic* as later codified by Levy were:

1. *Access to computers—and anything which might teach you something about the way the world works—should be unlimited and total. Always yield to the hands-on imperative!*
2. *All information should be free.*
3. *Mistrust authority—promote decentralization.*
4. *Hackers should be judged by their hacking, not bogus criteria such as degrees, age, race, or position.*
5. *You can create art and beauty on a computer.*
6. *Computers can change your life for the better.* (Levy, 1984, pp. 40-45)[1]

Computer programs were treated like any information created by the scientific community: Software was free for everyone to use, study, and enhance. Building on programs created by other programmers was not only allowed, but encouraged. On one hand, nobody owned the programs, and on the other, they were common property of the community.

In the early 1980s, a conflict arose in the AI lab when some of the hackers formed a company called Symbolics to sell computers based on technology originally developed in the lab. Symbolics hired most of the hackers, leaving the lab empty. This, together with the fact that the software on Symbolics machines was considered a trade secret, caused a crisis. The community and its way of life had been destroyed and Stallman later described himself as "the last survivor of a dead culture" (Levy, 1984, p. 427; see also Williams, 2002).

Stallman saw an ethical problem in the growing trend of treating software in terms of property. In the AI lab, there was a strong spirit of co-operation and sharing, making the code, in a way, a medium for social interaction. Thus restrictions in the access to code were also limitations on how people could help each other.

In 1984, Stallman published *The GNU Manifesto* announcing his intention to develop a freely available implementation of the Unix operating system. He explained his reasons in a section titled *Why I Must Write GNU*:

I consider that the golden rule requires that if I like a program I must share it with other people who like it. Software sellers want to divide the users and conquer them, making each user agree not to share with others. I refuse to break solidarity with other users in this way. I cannot in good conscience sign a nondisclosure agreement or a software license agreement…So that I can continue to use computers without dishonor, I have decided to put together a sufficient body of free software so that I will be able to get along without any software that is not free. (Stallman, 2002d, p. 32)

The project gained interest and Stallman started receiving code contributions from developers. During the 1980s, major components of an operating system were developed, including a system library, shell, C compiler, and a text editor. However, a core component, the kernel, was still missing until Linus Torvalds began to work on the Linux kernel in 1991. During the 1990s, free software systems based on the Linux kernel gained in popularity, media hype, and venture capital investments.

STALLMAN'S ARGUMENTS IN *THE GNU MANIFESTO* AND THE *FREE SOFTWARE DEFINITION*

Stallman's main argument in *The GNU Manifesto* (1984) is the "golden rule" quoted previously: A useful program should be shared with others who need it. Stallman started GNU in order to "give it away free to everyone who can use it" (Stallman, 2002d, p. 31) in the spirit of co-operation, sharing and solidarity. He criticizes proprietary software sellers for wanting to "divide the users and conquer them" (Stallman, 2002d, p. 32). Stallman's intention here is not anti-capitalist or anti-business. He gives suggestions on how software businesses can operate with free software. The fundamental ethical problem Stallman sees in proprietary software is the effect it has on community and co-operation. For Stallman, himself a master programmer, the "fundamental act of friendship among programmers is the sharing of programs" (Stallman, 2002d, p. 32). Restrictions on sharing would require programmers to "feel in conflict" with other programmers rather than feel as "comrades" (Stallman, 2002d, pp. 32-33).

Stallman suggests that software businesses and users could change the way they produce and use software. Instead of selling and buying software like any other commodity, it could be produced in co-operation between users and companies. Although the software would be free, users would need support, modifications and other related services which companies could sell. Stallman argues this would increase productivity by reducing wasteful duplication of programming work. Also it would make operating systems a shared resource for all businesses. If the business model of a company is not selling software, this would benefit the company. Being able to study source code and copying parts of it would increase the productivity of the programmer.

An important goal in the manifesto is increasing the users' independence from software sellers. When software is free, users are no longer at the mercy of one programmer. Because anyone can modify a free program, a business can hire anyone to fix the problem. There can be multiple service companies to choose from.

For Stallman, the main reason for rejecting software ownership is good civil spirit, but he also argues against the concept of copyright and authorship: "Control over the use of one's ideas' really constitutes control over other people's lives; and it is usually to make their lives more difficult," Stallman (2002d, p. 37) notes. He denies the idea of copyright as a natural, intrinsic right and reminds us that the copyright system was created to encourage literary authorship at a time when a printing press was needed to make copies of a book. At the time, copyright restrictions did little harm, because so few could invest in the equipment required to make a copy. Today, when copies of digital works can be made at practically zero cost, copyright restrictions cause harm because they put limits on the way the works can benefit society. Stallman (2002d, p. 37) notes that copyright licensing is an easy way to make money but is "harming society as a whole both materially and spiritually." He maintains that even if there was no copyright, creative works would be created because people write books and computer programs on other grounds: fame, self-realization, and the joy of being creative.

Depending on the context, four different meanings of the term *community* can be found in Stallman's argument. The first one is that of a *hacker community* like the one at MIT's AI lab. The second is the *computer-using community* interested in business, wasted resources, and independence from software sellers. The third community is the *society* that will benefit from co-operation and face costs from complicated copyright and licensing mechanisms and enforcement. The fourth level of community Stallman mentions is *humanity*. He argues that because of the copyright restrictions on computer programs, the "amount of wealth that humanity derives" is reduced (Stallman, 2002d, p. 36). In these four meanings of the term, we can see community grow from a small group of friends to an interest group, then to society and finally to humanity as a whole. As the communities grow in size, the temporal perspective is expanded: for hacker friends, the benefits are direct and immediate whereas in the case of humanity change may require decades.

In *The GNU Manifesto*, Stallman mentions that "everyone will be permitted to modify and redistribute GNU, but no distributor will be allowed to restrict its further redistribution. That is to say, proprietary modifications will not be allowed" (Stallman, 2002d, p. 32). In *Free Software Definition* (Stallman, 2002a), he lists the *four freedoms* which a piece of software must meet in order to be free software. The freedoms are:

- **Freedom 0:** The freedom to run the program, for any purpose
- **Freedom 1:** The freedom to study how the program works, and adapt it to your needs; access to the source code is a precondition for this

- **Freedom 2:** The freedom to redistribute copies so you can help your neighbor
- **Freedom 3:** The freedom to improve the program, and release your improvements to the public, so that the whole community benefits; access to the source code is a precondition for this (Stallman, 2002a, p. 41)

Freedom of software is defined by referring to the rights of the computer user, who may run the program for any purpose, good or evil, study and adapt the software, and distribute copies of the program, modified or original. It should be noted that the definition assumes sharing is always beneficial and desired. It does not matter if the neighbor or the community has any use for the software or the skills to use it.

For a piece of software to be free, it would not be enough to abolish the copyright system. Because a user needs the source code in order to effectively exercise freedom 3, the author must actively promote software freedom by releasing the source code. Therefore, a co-operative community is already needed for software freedom.

Stallman makes an important distinction between *free as in free speech* and *free as in zero price*. The concept of free software is not against selling software, it is against restrictions put on the users. Free software can be sold but the seller may not forbid the users to share or modify it.

COPYLEFT: THE GPL AS LEGAL AND SOCIAL DEVICE

Because Stallman was the copyright holder of the GNU programs that he wrote, he could have handed the programs to the public domain. Thus the programs would have been free. However, releasing the programs to the public domain would have meant that people would have been able to distribute the programs in ways which would have restricted the freedom of users, for instance, by distributing them without the source code. A free program would have become non-free. Stallman wanted the distribution of his programs or any other free software to stay free forever, and together with Free Software Foundation (FSF) legal counsel Eben Moglen, they devised the GNU General Public License (GPL) for this purpose (Stallman, 1989; Stallman, 2002b). The main idea of the GPL is that anyone is free to use, modify and redistribute a GPLed program on the condition that the same freedom of use, modification, and redistribution is also given to the modified and redistributed program. The easiest way to fulfill the condition is to release the redistributed and modified program under the GPL. The GPL is in this sense "viral": a GPLed program can be unified with other code only if the added code is compatible with the GPL. The purpose of the GPL is to keep free software free and to stop it ever becoming a part of pro-prietary software (Stallman, 2002a, pp. 89-90; Stallman, 2002c, pp. 20-21). The GPL is called a *copyleft* license, because in a sense it turns around the copyright by giving the user, not only the author, the freedom to use and to continue to build on the copylefted work. In this sense, copyright law and the GPL license built on it are the artifices that make the free software move-ment possible. There is some irony to the fact that the movement in this sense needs the copyright law in order to function. This is also the reason why it is not correct to describe the movement as being against copyright. Consequently, the GPL has to function well. The original GPL version 1 has been modified into version 2, under which, for instance, the Linux kernel is released. Cur-rently, in 2006, a new version, GPLv3, is being prepared by Stallman and the FSF. The somewhat unorthodox twist that GPL gives to copyright law has sometimes aroused suspicion over whether the GPL is a valid and enforceable license. As Moglen (2001) notes, most often GPL violations are settled without much publicity in negotiations between the FSF and the violator. As the FSF seeks only the freedom of software, a violator can easily rectify the situation by starting to comply with the GPL. It is sometimes argued that the fact that code under GPL can not lose the property of being free does not give the user maximum freedom with the code: the user is not permitted to "close" the code and release it under a proprietary software license. For instance, a typical Berkeley Software Distribution (BSD) license does not require that modifications or derivative works be free. Proponents of BSD see this as a good thing, maybe even as a benefit over the GPL, because the BSD license gives the developer more possibilities. However, for Stallman this is not desired, as closing the source tramples on the possible future uses of the code: "It is absurd to speak of the 'freedom to take away others' freedom'" (Stallman cited in Butler, 2005).

FREE SOFTWARE AS A POLITICAL PHILOSOPHY

As described above, Stallman's free software philosophy goes beyond the freedom and needs of an individual programmer. In Stallman's work, we find a political philosophy that has roots both in the liberalist and the communitarian traditions but accepts neither as such.

Stallman's ideas on user's freedom have roots in the liberalist political philosophy of Thomas Hobbes, John Locke, John Stuart Mill and oth-ers. In the *Second Treatise of Government* (1690), Locke argued that societies are built on a social contract in which people agree to give away some of their personal liberty to escape the cruel reality of the "state of nature" and receive protection for the fundamental rights which are life, liberty, and property. Locke's influence on political philosophy can be seen, for example, in the formulation of the U.S. Declaration of Independence and in the Constitution.

Stallman describes his relation to the liberalist tradition as follows:

The philosophy of freedom that the United States is based on has been a major influence for me. I love what my country used to stand for. ... Science is also an important influence. Other campaigns for freedom, including the French and Russian revolutions, are also inspiring despite the ways they went astray. (Stallman, 2004)

The four freedoms of free software were named after the influential speech given by the U.S. President Franklin D. Roosevelt during the Second World War in 1941, called *The Four Freedoms* (Roosevelt, 1941).

For Locke, freedom means "to be free from restraint and violence from others" (Locke, 1690, para. 57). For Stallman, software freedom means the freedom to run, study, modify, and distribute the program Locke described the time before organized society as a natural state where everybody had complete freedom but had to live in constant danger. Stallman has described the American society as a "dog-eat-dog jungle" where antisocial behavior like competition, greed and exclusion is rewarded instead of co-operation (Levy, 1984, p. 416; Stallman, 2002e). Because sharing of software is forbidden, freedom is restricted in such a society.

The tension between individualism and communitarianism is constant in Stallman's philosophy. He started the GNU project because of his own moral dilemma, but he also argues for it on a collectivist basis. In the first announcement of the GNU project (Stallman, 1983), the perspective was individualist: "So that I can continue to use computers without violating my principles, I have decided to put together a sufficient body of free software so that I will be able to get along without any software that is not free." In *The GNU Manifesto* (Stallman, 2002d), the words "violating my principles" were replaced with the word "dishonor," indicating a move towards a more communal view. The tension also arises if we ask for what and for whom software freedom is intended. Isaiah Berlin (1969) has introduced

a distinction between the notions of negative and positive freedom: negative freedom means freedom from obstacles and restrictions while positive freedom means control over one's life and positive opportunities to fulfill a goal. Both the liberalist tradition and Stallman mainly use the negative concept of freedom, but in his emphasis on community we can also see aspects of positive freedom.

Freedom 0, the freedom to run the program, is a pure example of the negative concept of freedom. The user has the right to use the software, whatever the purpose might be. Freedom 1 has two components: having permission to study the program and having the source code. In this sense freedom 0 is not only about absence of restraints, it is also about presence of the source code and in this sense a positive freedom. Likewise, freedom 2 is not only an individualist or negative freedom: the freedom to redistribute copies is necessary to help the neighbour. Freedom 3 to improve the program and release the improvements to the community is also of a positive nature: It is required to build a community.

For a programmer, freedom of software is a fundamental issue related to a way of life, to the identity of a hacker. Is the freedom relevant only for programmers? Bradley Kuhn and Richard Stallman reply:

We formulated our views by looking at what freedoms are necessary for a good way of life, and permit useful programs to foster a community of goodwill, cooperation, and collaboration. Our criteria for Free Software specify the freedoms that a program's users need so that they can cooperate in a community. We stand for freedom for programmers as well as for other users. Most of us are programmers, and we want freedom for ourselves as well as for you. But each of us uses software written by others, and we want freedom when using that software, not just when using our own code. We stand for freedom for all users, whether they program often, occasionally, or not at all. (Kuhn & Stallman, 2001)

This freedom is for everyone, whether they need it, use it, or not, just like freedom of speech. But freedom of software is just a means to a more important end, which is a co-operative, free society. Stallman wants to contribute to a society that is built on solidarity and co-operation, not exclusion and greed. In a communitarian way, the argument sees morality and the good of the individual co-dependent on the good of the community.

POLITICAL MOVEMENT OR DEVELOPMENT MODEL? A COMPARISON OF FS AND OS IDEOLOGIES

One of the motivations for launching the Open Source Initiative (OSI) was the perception that the ideology and concepts used by the FS movement, in general, and Richard Stallman, in particular, were putting off potential collaborators, especially business partners. Eric S. Raymond explains his motivations as tactical, rather than principal:

The real disagreement between the OSI and the FSF, the real axis of discord between those who speak of "open source" and "free software," is not over principles. It's over tactics and rhetoric. The open source movement is largely composed not of people who reject Stallman's ideals, but rather of people who reject his rhetoric. (Raymond, 1999)

Thus, the aim of the term *open source* is to emphasize the practical benefits of the OS development model instead of the moral philosophy behind the free software ideal. For the actors in the OS movement, the creation of OS software is an utilitaristic venture of collaboration, based on individual needs. According to Eric S. Raymond, "Every good work of software starts by scratching a developer's personal itch" (Raymond, 1999). This is in clear contrast with the intentional, systematic, and collective effort described by Stallman: "essential pieces of GNU software were developed in order to have a complete free operating system. They come from a vision and a plan, not from impulse" (Stallman, 2002c, p. 24).

The main ideological shift was in the professed motivation for writing code. The software itself often stayed the same: by definition, free software is a subset of open source software. For the outside world this ideological shift may present itself as relatively minor, so that in the name of simplification a common name such as FOSS (free/open source software) or FLOSS (free/libre and open source software) is often used. Initially the two communities also overlapped to a large degree, but lately some polarization has been in evidence. For instance, in a recent survey a large majority of Eclipse developers reported that they identify with the OS movement, while a clear majority of Debian developers reported identification with the FS movement (see Mikkonen, Vainio, & Vadén, 2006). This development may be expected to continue, as companies are increasingly taking part and employing programmers in OS development.

A crucial difference between OS and FS has to do with the political economy of software production. However, this distinction is best described as the difference between *business friendly* open source and *ideological/political* free software, or *capitalist* open source and *communist* free software. These are not the correct levels of abstraction. For instance, sometimes the GPL license is more *business friendly* than a given non-GPL-compatible open source license. The fact that the OS community treats code as a *public good* might be perceived as odd in certain types of market economies, while in others such public goods are seen as necessary drivers of capitalism. By making software a non-scarce resource, OS has an effect on where and how a revenue stream is created. However, this merely reorganizes production and labour, instead of changing their mode.

Schematically put, FS is a social movement, while OS is a method for developing software. Whatever the definitions of systems of economical production—such as capitalism, communism, market economy, and so on—may be, OS is non-committal with regard to the current issues of political economy, such as copyright, intellectual property rights and so on. Individual members of the OS community may or may not have strong views on the issues, but as a community OS is mostly interested in the benefits of openness as a development model. This attitude is well exemplified in the views expressed by Linus Torvalds: "I can't totally avoid all political issues, but I try my best to minimize them. When I do make a statement, I try to be fairly neutral. Again, that comes from me caring a lot more about the technology than about the politics, and that usually means that my opinions are colored mostly by what I think is the right thing to do technically rather than for some nebulous good" (quoted in Diamond, 2003). This pragmatic or "engineering" view on FOSS is intended to work better than ideological zealotry in advancing the quality and quantity of code. In contrast, in order to change a political system one needs a social movement. As noted previously, the FS movement is a social movement based on shared values. While these values are close to the loosely interconnected values of the anti-globalization movement (see Stallman, 2005, 2002f), they are not the defining values of socialist or communist parties or movements. For instance, the FS movement does not have a stand on class relations or on how to treat physical property, and so on. In this sense the FS movement as a social movement is a specialized, one-cause movement like many other post-modern social movements. Again, here lies a crucial distinction: the ethical principles of FS concern only information, and only information that is a tool for something. Typically, a socialist or communist set of values

would emphasize the importance of material (not immaterial) things and their organization.

Ideologically proximate groups often behave in a hostile manner towards each other in order to distinguish themselves; the public controversies between the FS and OS communities are a good example. Extra heat is created by the different perspectives on the politics of freedom. The Torvaldsian view of "no politics" is tenable only under the precondition that engineering can be separated from politics and that focusing on the engineering part is a non-political act. Stallman, for one, has consistently rejected this precondition, and claims that the allegedly non-political focus on the engineering perspective is, indeed, a political act that threatens the vigilance needed for reaching freedom.

A good example of these controversies is the one over the name (Linux or GNU/Linux) of the best known FOSS operating system. Since the mid 1990s, Stallman and the FSF have suggested that developers use the name GNU/Linux, arguing that "calling the system GNU/Linux recognizes the role that our idealism played in building our community, and helps the public recognize the practical importance of these ideals" (FSF, 2001). However, true to their pragmatical bent, OS leaders such as Raymond and Torvalds have replied that the name Linux has already stuck, and changing it would create unnecessary inconvenience. Some of the distributions, such as Debian, have adopted the naming convention suggested by the FSF.

CONCLUSION: FS AS A HISTORICAL BACKDROP OF OS

The FS movement initiated by Stallman predates the OS movement by over a decade and the latter was explicitly formed as an offshoot of the former. Consequently, the definition of OS software was developed in the context of an ongoing battle between the FS and proprietary software models. Arguments presented by Stallman in the early 1980s still form some of the most lucid and coherent positions on the social and political implications of software development. Most importantly, the polarization of the FOSS community into the FS and OS camps has been only partial. All of these facts point out how FS has acted as a necessary background for OS. This background can roughly be divided in technological, legal and ideological parts.

On the technological side, FS code often forms a basis and ancestry for OS projects. The formation of the operating system Linux or GNU/Linux is one of the examples where the functions of the FS movement form an essential cornerstone of existing OS software. Typically Linux distributions include major technological components (such as glibc [the GNU C Library], Coreutils, and gcc) from the GNU project.[2] It is uncontroverted that without the systematic and prolonged effort by the FSF the development and adoption of Linux (the operating system) would not have been as rapid or widespread as it has been. However, it is equally clear that several key OS projects, such as Apache or Eclipse, are not technologically dependent on GNU.

The legal cornerstone provided to the OS community by the FSF and Stallman is the GPL license, under which Linux (the kernel) and several other key OS projects were developed. The GPL is concurrently clearly the leading FOSS license, comprising over 50% of the code in projects maintained at SourceForge and of major GNU/Linux distributions (Wheeler, 2002). The GPL as a license and the ideal of freedom that it embodies are the legal lifeblood of both the FS and the OS communities, even though several other families of licenses are crucially important.

The ideological foundation provided by the FS movement is difficult to gauge quantitatively. Suffice it to say the OS movement is, according to its own self-image, a tactical offshoot of the FS movement. Many of the sociocultural arguments (openness for reliability, longevity of

code, and user control) and ways of functioning (collaborative development based on the GPL) that the OS community uses were spearheaded by the FS community. Moreover, now that OS is moving outside its niche in software production and gaining ground as a *modus operandi* in other fields (such as open content, open medicine, open education, open data, and so on), the OS movement finds itself again in closer proximity to the ideals expressed by the FS movement. However, there are also trends that tend to emphasize the neutral, engineering point of view that created the need for the separation of OS from FS in the first place: as OS software becomes more commonplace and even omnipresent, the ideological underpinnings are often overlooked with or without purpose.

REFERENCES

Berlin, I. (1969). Two concepts of liberty. In I. Berlin, *Four essays on liberty*. Oxford, UK: Oxford University Press.

Butler, T. (2005, March 31). Stallman on the state of GNU/Linux. *Open for Business*. Retrieved February 20, 2006, from http://www.ofb.biz/modules.php?name=News&file=article&sid=353

Diamond, D. (2003, July 11). The Peacemaker: How Linus Torvalds, the man behind Linux, keeps the revolution from becoming a jihad. *Wired*. Retrieved February 20, 2006, from http://www.wired.com/wired/archive/11.07/40torvalds.html

FSF. (2001). *GNU/Linux FAQ*. Retrieved February 20, 2006, from http://www.gnu.org/gnu/gnu-linux-faq.html

Hafner, K., & Lyon, M. (1996). *Where wizards stay up late: The origins of the Internet*. New York: Touchstone.

Himanen, P. (2001). *The hacker ethic and the spirit of the information age*. New York: Random House.

Kuhn, B., & Stallman, R. (2001). *Freedom or power?* Retrieved February 20, 2006, from http://www.gnu.org/philosophy/freedom-or-power.html

Levy, S. (1984). *Hackers: Heroes of the computer revolution.* London: Penguin.

Locke, J. (1690). *Second treatise of government.* Indianapolis: Hackett.

Mikkonen, T., Vainio, N., & Vadén, T. (2006). Survey on four OSS communities: Description, analysis and typology. In N. Helander & M. Mäntymäki (Eds.), *Empirical insights on open source business.* Tampere: Tampere University of Technology and University of Tampere. Retrieved June 27, 2006, from http://ossi.coss.fi/ossi/fileadmin/user_upload/Publications/Ossi_Report_0606.pdf

Moglen, E. (2001). *Enforcing the GNU GPL.* Retrieved February 20, 2006, from http://www.gnu.org/philosophy/enforcing-gpl.html

Raymond, E. S. (1999). *Shut up and show them the code.* Retrieved February 20, 2006, from http://www.catb.org/~esr/writings/shut-up-and-show-them.html

Raymond, E. S. (2003). *The jargon file.* Retrieved June 6, 2006, from http://www.catb.org/jargon/

Roosevelt, F. D. (1941). *The four freedoms.* Retrieved February 20, 2006, from http://www.libertynet.org/~edcivic/fdr.html (May 27, 2004)

Stallman, R. (1983, September 27). *New UNIX implementation. Post on the newsgroup net.unix-wizards.* Retrieved February 20, 2006, from http://groups.google.com/groups?selm=771%40mit-eddie.UUCP

Stallman, R. (1989). *GNU General Public License version 1.* Retrieved February 20, 2006, from http://www.gnu.org/copyleft/copying-1.0.html

Stallman, R. (2002a). Free software definition. In J. Gay (Ed.), *Free software, free society: Selected essays of Richard M. Stallman* (pp. 41-43). Boston: GNU Press.

Stallman, R. (2002b). GNU General Public License version 2. In J. Gay (Ed.), *Free software, free society: Selected essays of Richard M. Stallman* (pp. 195-202). Boston: GNU Press.

Stallman, R. (2002c). The GNU project. In J. Gay (Ed.), *Free software, free society: Selected essays of Richard M. Stallman* (pp. 15-30). Boston: GNU Press.

Stallman, R. (2002d). The GNU manifesto. In J. Gay (Ed.), *Free software, free society: Selected essays of Richard M. Stallman* (pp. 31-39). Boston: GNU Press.

Stallman, R. (2002e). Why software should be free. In J. Gay (Ed.), *Free software, free society: Selected essays of Richard M. Stallman* (pp. 119-132). Boston: GNU Press.

Stallman, R. (2002f). *The hacker community and ethics: An interview with Richard M. Stallman.* Retrieved February 20, 2006, from http://www.uta.fi/~fiteva/rms_int_en.html

Stallman, R. (2004, January 23). *A Q&A session with Richard M. Stallman.* Retrieved February 20, 2006, from http://puggy.symonds.net/~fsug-kochi/rms-interview.html

Stallman, R. (2005, December 18). Free software as a social movement. *ZNet.* Retrieved February 20, 2006, from http://www.zmag.org/content/showarticle.cfm?SectionID=13&ItemID=9350

Turkle, S. (1982). The subjective computer: A study in the psychology of personal computation. *Social Studies of Science, 12*(2), 173-205.

Wheeler, D. (2001). *More than a gigabuck. Estimating GNU/Linux's size.* Retrieved February 20, 2006, from http://www.dwheeler.com/sloc/redhat71-v1/redhat71sloc.html

Wheeler, D. (2002). *Make your open source software GPL-compatible: Or else.* Retrieved

February 20, 2006, from http://www.dwheeler.com/essays/gpl-compatible.html

Williams, S. (2002). *Free as in freedom: Richard Stallman's crusade for free software*. Sebastopol, CA: O'Reilly.

KEY TERMS

Communitarianism: A philosophical view holding that the primary political goal is the good life of the community.

Copyleft: The practice of using copyright law in order to remove restrictions on the distribution of copies and modified versions of a work for others and require the same freedoms be preserved in modified versions.

Free Software (FS): Software that can be used, copied, studied, modified, and redistributed without restriction.

General Public License (GPL): A widely used free software license, originally written by Richard M. Stallman for the GNU project.

Hacker Community: A community of more or less likeminded computer enthusiasts that developed in the 1960s among programmers working on early computers in academic institutions, notably the Massachusetts Institute of Technology. Since then, the community has spread throughout the world with the help of personal computers and the Internet.

Liberalism: A philosophical view holding that the primary political goal is (individual) liberty.

ENDNOTES

¹ For alternative formulations of the hacker ethos, see the entry "hacker ethic" in *The Jargon File*, edited by Raymond (2003) and *The Hacker Ethic* by Himanen (2001), who gives the concept a more abstract scope.

² For a view of the complexity of a GNU/Linux distribution (see Wheeler, 2001).

Chapter II
Greasemonkey and a Challenge to Notions of Authorship

Brian D. Ballentine
West Virginia University, USA

ABSTRACT

This chapter introduces Greasemonkey, a new extension for the Firefox browser, which enables users to alter the behavior and appearance of Web pages as the pages load. The chapter claims that Grease-monkey is forcing a reevaluation of what it means to be an author in digital environments. Using Michel Foucault's (1979) original question, "What is an author?" the chapter argues that creators of Grease-monkey scripts take on the additional roles of designer and programmer. Also, the chapter cautions that since Greasemonkey scripts have the ability to alter the layout, navigation, and advertising on a Web page, there may be legal ramifications in the future for this open source project.

INTRODUCTION

The question "What is an author?" has been the source of much scholarship in the humanities at least since the publication of Michel Foucault's (1979) famous essay, and recent developments in computer generated texts have only made it more pressing that scholars grapple with this fundamental question. Currently, there is increased recognition that the very idea of the "author-function" since the rise of print culture and intellectual property rights cannot be comprehensively understood without taking into account the complementary idea of a "designer," especially with respect to the production of digital texts. Consequently, hypertextual and digital theorists have adopted the twin notions of author and designer to account for the assembly of interactive texts. While the addition of a designer has certainly deepened our understanding of how text gets produced, assembled, and disseminated, and thus represents a significant advance in the study of authorship and digital writing, current scholarship has yet to account for the role of the programmer as a distinct aspect of the author-function. The open source community and the technologies the community produces, present the opportunity to examine and question a programmer's status as an author. This chapter will assess hypertext and digital theories as they pertain to authors and designers and then show how the addition of the programmer to the theoretical nomenclature will advance our understanding of the author-function in digital environments. While there are many innovative projects under development within the open source community, this chapter focuses on a new technology called Greasemonkey and the freedoms (and risks) it provides an author/designer/programmer.

GREASEMONKEY BACKGROUND

Greasemonkey is an extension for the Firefox browser that enables users to install client-side "user scripts" that alter the behavior and appearance of Web pages. The alterations occur as a Web page is downloaded and rendered in a user's Web browser. The alterations occur without the consent of the site owners. Traditionally, Web pages are fixed offerings developed for an audience to use but not to alter *and* use. All major Web browsers are equipped with the option to "view source" which reveals the source code responsible for a particular Web page. The source code can be copied, saved, and edited by an end-user. Greasemonkey is vastly different from simply acquiring the code in that edits occur as the page loads in Firefox allowing a user to continue to interact with a company's Web page even after edits are complete. Greasemonkey's functionality, therefore, enables an examination of the roles of authors, designers, and programmers as these figures write scripts that actively manipulate Web pages.

For example, a Greasemonkey script titled "Book Burro" enables users to simultaneously see competitive prices from other bookstores while searching Amazon.com. The script also searches for a book's availability in local and national libraries. Web sites, especially large retail sites such as Amazon, are strategically designed, programmed, and "authored" to be effective marketing and sales tools. Greasemonkey enables users to reclaim the roles of author, designer, and programmer and recalibrate, edit, or "remix" Amazon's strategies.

Figure 1. Greasemonkey screen capture showing the Book Burro script

This phenomenon is known as "active browsing." While anyone may program a Greasemonkey script on their own, there are hundreds of scripts posted on sites dedicated to Greasemonkey such as userscripts.org.[1] The example in Figure 1 is the Book Burro script that displays competitive prices and library availability for a sought-after book. The information is displayed in a new menu in the left-hand corner of the browser window. Users may add or delete online stores or libraries from the display.[2]

MAIN FOCUS OF THE CHAPTER

Author/Designer/Programmer: The Author-Function

Is using Greasemonkey to create scripts such as Book Burro "writing" and worthy of the "author" distinction? Academics have had a relatively short but complex relationship with digital writing and digital texts. Throughout the 1990s, scholars tackled the complicated similarities between digital writing and popular critical theory movements such as post-structuralism and deconstructionism.

The importance of the reader's interpretation of a text, the text's relationship to other text, and a text's inability to create true closure are theoretical attributes adopted by hypertext theorists. As Jay David Bolter made clear: "[E]ven the most radical theorists (Barthes, de Man, Derrida, and their American followers) speak a language that is strikingly appropriate to electronic writing" (Bolter, 1990, p. 161). Indeed, Jacques Derrida conveniently broadened the scope of "writing" in *Of Grammatology* in terms that are easy to envision a home for digital writing:

And thus we say "writing" for all that gives rise to an inscription in general, whether it is literal or not and even if what it distributes in space is alien to the order of the voice: cinematography, choreography, of course, but also pictorial, musical, sculptural "writing." (Derrida, 1974, p. 9)

Foucault made similar inclusive adjustments for his definition of the author-function: "Up to this point I have unjustifiably limited my subject. Certainly the author-function in painting, music, and other arts should have been discussed" (Foucault, 1979, p. 153). Hypertext theorists have felt compelled to add digital writing to the list. But to Bolter's credit, he recognized that the theories do not line up as neatly as proposed. "Electronic writing takes us beyond the paradox of deconstruction, because it accepts as strengths the very qualities—the play of signs, intertextuality, the lack of closure—that deconstruction poses as the ultimate limitations of literature and language" (Bolter, 1990, p. 166). Ultimately, Bolter determines that these theories only serve to inform us "what electronic writing is not. We still need a new literary theory to achieve a positive understanding of electronic writing" (Bolter, 1990, p. 166).

In 2001, Bolter offered a vision for a new theory that might properly encompass writing a digital space. According to Bolter:

[T]he work of remediation in any medium relies on two apparently opposite strategies. Sometimes the artist tries to erase the traces of the prior medium in her work and seeks to convince us that her work in the new medium represents the world directly. At other times, she accepts and even foregrounds the older medium. We call the first strategy "transparent immediacy" and the second "hypermediacy." In its remediation of print, hypertext adopts both of these strategies. When the author elects to leave the reader alone with an episode of conventional prose, she is relying on the power of traditional narrative prose to be transparent. When she emphasizes the reader's choice through the process of linking, she is evoking a strategy of hypermediacy. (Bolter, 2001, p. 1856)

Bolter's theory can be transposed to incorporate any medium from television to cinema to software. However, his discussion is an oversimplification of the current capabilities of writing in a digital space. Even if an author has chosen to "leave the reader alone" with a text in a digital environment there is much to be considered in terms of *how* the author was able to make that choice. The example Greasemonkey script demonstrates that authoring in a digital environment has changed now that a reader can actively edit and alter an author's text. Greasemonkey is complicating digital writing now that the *reader* has an option to be left alone or actively engage with and alter a text.

However, critics have become frustrated with the prominence of digital writing and rather than work with the existing theories (or develop new ones like Bolter) are driven to discredit it altogether. There are those that rail against hypertext's "fashionable tale" as Richard Grusin (1996, p. 39) phrased it. Grusin's work begins, appropriately enough, by quoting Foucault and a claim that proponents of hypertext readily embrace:

The author—or what I have called the "author-function"—is undoubtedly only one of the possible specifications of the subject and, considering past historical transformations, it appears that the form, the complexity, and even the existence of this function are far from immutable. We can easily imagine a culture where discourse would circulate without any need for an author. (Grusin, 1996, p. 39)

Grusin used Foucault's ideas to advance a theory that hypertext criticism is technologically determined and in fact supports a larger techno-logical fallacy. "This fallacy," according to Grusin "most often manifests itself in propositional state-ments that ascribe agency to technology itself, statements in which the technologies of electronic writing are described as actors" (Grusin, 1996, p. 41). Grusin examines the work of Bolter, George Landow, and Richard Lanham filtering out evi-dence including Lanham's famous remark that "[t]he electronic word democratizes the world of arts and letters" (Grusin, 1996, p. 41).

Is Lanham's remark so problematic now that we are many years removed from Grusin's com-plaint? Recently, Nicholas Rombes described what technologies such as "personal websites and blogs" have done for the resurrection of the author:

[T]he author has grown and multiplied in direct proportion to academic dismissals and denun-ciations of her presence; the more roundly and confidently the author has been dismissed as a myth, a construction, an act of bad faith, the more strongly she has emerged. The recent surge in personal websites and blogs—rather than diluting the author concept—has helped create a tyrannical authorship presence ... (Rombes, 2005)[3]

Such a claim for technology is exactly Grusin's complaint. What is missing is an examination of those wielding the technology. To argue instead that the choices made by the designers and the programmers in digital spaces have created an opportunity to reevaluate the author-function switches the agency from the technology to the users of that technology.

Author/Designer/Programmer: The Designer-Function

The Book Burro script succeeds because it meets the difficult task of integrating into a page dense with information and navigation options. The soft, light-orange background compliments Amazon's color scheme instead of competing with it. The semi-transparent menu allows the user to see what the Book Burro script is covering over so there is no sense that Amazon's functionality has been obstructed or lost. The icons at the top of the Book Burro menu allow the user to hide or "window shade" the contents of the menu leaving only the small, thin toolbar. A user may also close Book Burro altogether. All of this is to say that Book Burro creates a skillful unity with Amazon's own design by adhering to gestalt grouping principles such as similarity, proximity, closure, and con-tinuation. To ignore these design principles would create tension in the remixed Amazon page. "Ten-sions tend to tear the visual message apart into separate and competing elements for attracting a viewer's attention, conveying sense of chaos not choreography" (Dake, 2005, p. 26). The design of the Book Burro script avoids tension by bonding with the existing design on Amazon's page.

Even though the term "designer" has made its way into the work of theorists such as Landow, Bolter, Slatin, and Hayles, this inclusion has not produced a universal definition or understand-ing of the term as it relates to the production of a "text."[4] The rapid advancements of Internet technologies along with the transitions between print and digital media contribute to the discrep-ancies in definition. When Landow's *Hypertext* first appeared in 1992, graphically intensive user-interfaces existed on a much smaller scale due to hardware, software, and network limitations. Simply put, the designer was technologically and economically limited from playing a central part in the creation of digital text.

Similarly, cost and technology have been limiting factors for printed text. But, as demonstrated by Hayles, as those costs decrease there is much to be gained by taking advantage of a designer's skills. Hayles brought the designer function to the foreground with her book *Writing Machines* in which she explored the importance of the materiality of a text whether that text is print or electronic. She credits Anne Burdick (who is given the label "designer" on the title page) for a part in the success of the project. In a closing section of the book titled "Designer's Notes," Burdick applauds Hayles for breaking the "usual chain of command" by having her work on the project from the beginning and "*with* words" instead of "*after* words" (Hayles, 2002, p. 140). That is, she took part in the authoring of the text and Hayles gives her credit for that, stating: "Also important were Anne Burdick's insights. More than any of us, Anne remained alert to the material qualities of the texts I am discussing and producing, pointing out places where descriptions needed to be more concrete, engaging, and specific in their attention to materiality" (Hayles, 2002, p. 143).

Crediting and distinguishing the designer role as being part and parcel to the authoring of a text is not common practice. Ironically, early perceptions of an author were even less generous than those afforded to the contemporary designer. In her essay, *The Genius and the Copyright,* Martha Woodmansee (1984) begins by providing an overview of the concept of an author up until the mid-eighteenth century. According to Woodmansee, the author was "first and foremost a craftsman" (Woodmansee, 1984, p. 426). She continues by describing the "craftsman" as a "skilled manipulator of predefined strategies for achieving goals dictated by his audience" (Woodmansee, 1984, p. 427). Likewise, early Web designers worked under technological constraints that kept their contributions to a text at a minimum. Quite often the designers of digital text did little more than plug

content into existing templates. It is not surprising that their contribution was viewed as administrative ("craftsman") and not in terms of the inspired, "original genius" that Woodmansee develops in her essay (Woodmansee, 1984, p. 427).

In *Bodies of Type: The Work of Textual Production in English Printers' Manuals*, Lisa Maruca (2003) offers a different perspective by focusing on the act of making books and claiming that those responsible for the physical production of a text share in the author-function. In her analysis of Joseph Moxon's 1683 *Mechanick Exercises* she finds evidence that:

the body of print emerges as a working body, a laborer whose physical construction of print is every bit as, if not more, important than the writer who supplies the text. Indeed, the print worker is understood as a collaborator in the construction of the meaning of print text. (Maruca, 2003, p. 324)

It is a bold claim to attach so much weight to a process often dismissed as mere mechanistic output. The validation, Maruca insists, is found, "by looking more closely at the multiple possible and actual uses of a machine in the hands of variously ideologically situated owners and workers" (Maruca, 2003, p. 324). The materials they produce, their text, should not be considered separately from "the metaphysical text" (Maruca, 2003, p. 324). Instead, these materials are "in fact always ultimately textual" (Maruca, 2003, p. 323).

Over three hundred years later, technological advancements have put new tools in the hands of those ready to influence a digital text's physical form. Tools, such as Greasemonkey, are challenging the stigma of craftsman and promoting the type of authorial collaboration discussed by Maruca. Many of the available Greasemonkey scripts were developed in response to what an author/designer/programmer deemed to be poor or underdeveloped design. The "usability" category on userscripts.org now has 120 scripts that work to edit and advance functionality for sites such as Amazon, Gmail, Flickr, and A9.[5] Greasemonkey authors/designers/programmers are experimenting with new concepts in navigation and usability and making their work available online for anyone to use, edit, and improve.

But now that technology has advanced to the degree that design can be implicated in the authoring of a text, the task of keeping a text "engaging" as Hayles required is still vague and open-ended. Yet it is the bottom-line for anyone involved with the design of digital texts. Someone interested in a career in Web design will face this vague requirement of "engagement" and can expect a high degree of ambiguity in their job search. A visit to a major employment site such as Monster.com or ComputerJobs.com and a search for "Web designer" will yield not only an abundance of opportunities but broadly defined job postings.

For example, a company advertising for a "Web Design Guru" began their job description with a series of questions: "Would you like the opportunity to prove that usability need not be ugly? Do you have an extreme imagination and the intelligence to back up your great designs?" (Propeller, 2006). The requirement to produce "great designs" with "extreme imagination" is approximately the equivalent of asking an author to make a story "interesting." But, the primary reason these descriptions do not and *should not* get more specific is because as with any creative authoring process, engaging material is developed on an individual case basis. Successful collaborations in the authoring of a digital text require broad definitions for job descriptions to keep the designer from serving as just a craftsman. If not, the result is to keep with the "usual chain of command" and bring in the designer after the fact. While treating the design role as a secondary function may still be the norm for the print medium it is no longer the case for digital text.

However, the "great design" found in a Greasemonkey script such as Book Burro, is possible because the author also has the ability to write code. A successfully "authored" Greasemonkey script will be a blend of innovative design and programming. It is this blending of responsibilities and roles that further complicates what it means to author a digital text. While notable hypertext and digital theorists have made use of the term "designer" in regard to the production of digital materials, perhaps it is the blending and blurring that continues to prohibit a solidified agreement or understanding of the roles the designer and programmer play.

Author/Designer/Programmer: The Programmer-Function

Based on Woodmansee's definition of the "craftsman," it is alarming that Bolter remarked: "No longer an intimidating figure, the electronic author assumes the role of a craftsperson, working with prescribed materials and goals. She works within the limitations of a computer system ..." (Bolter, 2001, p. 168). Equally problematic is that it is difficult to discern to what degree, if any, an electronic author programs. Indeed, substantial discussion of the programmer is difficult to find in hypertext theory. Slatin (1991) touched on the designer function and tangentially introduces the programmer in an early essay on hypertext. He wrote: "'Writing,' in the hypertext environment, becomes the more comprehensive activity called 'authoring'" (Slatin, 1991, p. 160). This "authoring," he noted, might involve "a certain amount of programming" (Slatin, 1991, p. 160). The reference to "a certain amount of programming" is of course vague and Slatin's quote only scratches the surface of the programming as it relates to the author-function.

Hayles recognized:

an unfortunate divide between computer science folks, who knew how the programs and hardware worked but often had little interest in artistic practices, and literary critics, who too often dealt only with surface effects and not with the underlying processes of the hardware and software. (Hayles, 2002, p. 66)

But whether they are interested or not, critics will need to begin bridging this divide if their scholarship is to be applicable. Hayles' agenda in *Writing Machines* was to deal with the materiality of a text. However, the materiality of a digital text is dependent on these "computer science folks" and their skills.

Along with the absence of a substantive discussion of the programmer, an examination of early hypertext theory shows a reverence for the programmer's ability to network and link documents. The nonlinear and poly-vocal attributes of digital writing captivated theorists during much of the 1990s. The endless potential of links between sites and pages was treated as a victory for the reader against the tyranny of the printed line and the oppression of the author. In *From Text to Hypertext*, Silvio Gaggi (1997) explains: "In electronic networks no single author addresses any single reader, or, if one does, their exchange emerges from and immediately reenters a broader context of multiple speakers and listeners. There is a polyphony of voices" (Gaggi, 1997, p. 111). Likewise, design specialist Jakob Nielson (2000) wrote: "Hypertext basically destroys the authority of the author to determine how readers should be introduced to a topic" (Nielson, 2000, p. 171). In Nielson's section, *The Authority of the Author*, he explains: "Authoring takes on an entirely new dimension when your job is changed to one of providing opportunities for readers rather than ordering them around" (Nielson, 2000, p. 171). These critics were clearly taken by what they perceived to be "transference of authorial power" (Landow, 1997, p. 90).

However, those so-called choices and links have an author, or rather, a programmer. Prior to the introduction of Greasemonkey, a reader could not create and follow a link that was not there. Of course, closing a browser is no more empowering than closing one book and picking up another and this option does not uniquely strengthen the reader position in a digital environment. The reader is only using avenues put in place by the "electronic author." Greasemonkey, on the contrary, has the potential to fulfill the early promises of the Internet to re-empower the reader and return the authoring capabilities to the audience.

The question then becomes whether or not writing code necessitates the same "extreme imagination" required of designer and authors? Fortunately, scholars have begun studying the parallels between "writing" and "code." This research provides some insight for determining whether or not the programmer should be relieved of the craftsman label. Such an examination requires a definition, or at least an attempt to define "code." In his introduction to *Codework*, Alan Sondheim writes:

In a narrower sense, code refers to a translation from natural language to an artificial, strictly defined one; the syntax of Morse code, for example, has no room for anomalies or fuzziness. Computer code generally requires strictly defined codes that stand in for operations that occur "deeper" in the machine. Most users work on or within graphic surfaces that are intricately connected to the programming "beneath"; they have little idea how or why their machines work. (Sondheim, 2001)

While Sondheim could easily back up his claim about "most users" and their levels of engagement with what goes on below the surface of a user-interface, the open source community and programs such as Greasemonkey are changing the level of engagement users have with code "'deeper' in the machine."

In this instance, it will be demonstrated that with the advancements of code, programmers have choices that require not just craftsman-like skill but imagination. These choices challenge programmers to create and "author" because intentionality and predetermined outcomes are not built into their computer languages. The argument posed here is that computer languages have become so robust and complex that programmers are using code to author pages in ways that were never conceived of when the languages were developed. Perhaps the best way to prove that imagination is required of programmers is to examine programmers discussing code.[6]

The following discussion is taken from a JavaScript message board involving myself and two other programmers. The programmer posting the problem wishes to allow images on his Web site to overlap at will. His first instinct is to use the HTML <LAYER> tag which, among its properties, does allow overlapping. Unfortunately, his code fails.

Subject: [javascript] I need to share a Layer between two Frames.

I have a web site wich I use Frames. But I need to display an image qich its gone take an area biguer than the original Frame. Now when I try to do this part of the Layer its hide under the other frame. I want to open this layer and need to be displayed between two frames. Some Ideas. Thank You.[7]

Another programmer responded to the message with bad news. After explaining the limitation of frames as a means to construct a site, he recommends that the programmer either recreate his entire site without frames or deal with the limitations of the <LAYER> tag.

You can't share a layer between two frames.

Each frame consists of a single page. This means for your two frames you must have two separate pages being viewed in them. A layer, though it can

be positioned anywhere on a page, can't go any further than the constraints of this page, which means it can't be on two pages at once.

Either change the way your sites set up and don't use frames or compromise on what you want to achieve with this layer—sorry.

Even though the <LAYER> tag was the understood tactic for stacking or layering images, if another solution could be found there was no reason to implement the <LAYER> tag. The solution involved the use of a newer tag called an in-line frame or <IFRAME>. This tag is generally used for an easy means to format menus and other navigational systems on Web sites. In-line frames are thought of as "useful for sidebar and other information of interest that isn't directly relevant to the main body of content" (Tuck, 2003). However, a closer examination of the tag's properties shows that with imaginative manipulation, it does allow layering. It is an unconventional use of a tag beyond its designated and pre-concieved task that solves the problem.

... Start by creating a new page that will serve as your new index page. In it, place 2 IFrames in the BODY tag. Set the source of the first IFrame to your old index page. Make sure you set the STYLE to something like this {z-index:1; position: static; top:0; left:0; width:100%; height:100%} Set the source of the second IFrame to the page that has your Layer and image in it. Make sure "AllowTransparency" is set to TRUE and then set the IFrame's STYLE to something like {z-index:0; position:absolute; top:0; left:0; width:100%; height:100%} Note that the second IFrame's z-Index is lower than the first. Give each IFrame an ID and using JavaScripts you can control when you want the lower or second IFrame to show up over the first. I would call the second IFrame something like ID="glassIFrame" because that is essentially what it is. A clear layer that you will

use to overlay images but will allow your original frames to show through.

Hope this helps,

Brian

Shortly after this was posted, the programmer responded simply: "Thank You. This its working to me [sic]." This example serves to demonstrate that the boundaries of code can be pushed and stretched by the imagination of the programmer. Such creative practice can directly affect "writing" or "authoring" in a digital environment. That is, the programmer must be included in the theoretical nomenclature for analyzing and discussing digital text. The "computer science folks" often dismissed as craftspeople, are not outliers in the production of these works. Greasemonkey is an excellent example of the designer and programmer roles working in the creation of scripts such as Book Burro.

FUTURE TRENDS AND THE POTENTIAL "CATASTROPHE" OF GREASEMONKEY

In his *The Code is not the Text (Unless It is the Text)*, John Cayley is concerned with, among other things, the relationship between code and literature. Literature, as Cayley writes, is subject to the transformative forces inherent in electronic spaces:

[M]utation is indeed a generative catastrophe for "literature' in the sense of immutable, authoritative corpus. As writing in networked and programmable media, language and literature mutate over time and as time-based art, according to programs of coded texts which are embedded and concealed in their structures of flickering signification. (Cayley, 2002)

Greasemonkey enables, even promotes, the mutation of the otherwise unalterable offerings of companies such as Amazon. The previous section demonstrated the importance of the programmer role or the creative possibilities for the person who controls these "structures." Greasemonkey is perhaps one of the most innovative open source technologies that effectively enables readers to assume the multi-faceted roles of a digital author including designer and programmer. However, as the popularity of Greasemonkey grows, many authors/designers/programmers are turning their attention to undoing or mutating corporate business strategies. Book Burro does facilitate the purchase of a competitor's book while a user researches materials with Amazon. Other scripts simply eliminate advertising. For example, Slashdot is a Web site that supports open source projects. It is subtitled, *News for Nerds and Stuff that Matters.*[8] At the top of Slashdot's homepage is a prominent advertisement section. The advertisers are often major computer or software companies that are well-aware of Slashdot's readership. With the installation of the Greasemonkey script *Slashdot Ad Removal*, Firefox users no longer have to view the advertisements.[9] Who will continue to advertise on sites that have known Greasemonkey scripts? Perhaps more alarming is the understanding that technology that can be demonstrated to have a negative effect on financial gains may be subject to litigation. The Napster case demonstrated what happens if a company was found to willfully facilitate the sharing of copyrighted material. It is not stretching the imagination to see a day when lawsuits are filed to prevent the willful defacement of paid advertising. Consequently, as we add the programmer to the theoretical nomenclature for authorship studies, we see not just the roles the programmer plays but the power and significance of this role.

REFERENCES

Bolter, J. D. (1990). *Writing space: The computer in the history of literacy.* Hillsdale, NJ: Lawrence Erlbaum Associates.

Bolter, J. D. (2001). *Writing space: Computers, hypertext, and the remediation of print* (2nd ed.). NJ: Lawrence Erlbaum Associates.

Cayley, J. (2002). The code is not the text (unless it is the text). *Electronic Book Review.* Retrieved January 25, 2006, from http://www.electronic-bookreview.com/thread/electropoetics/literal

Dake, D. (2005). Creative visualization. In K. Smith, S. Moriarty, G. Barbatsis, & K. Kenny (Eds.), *Handbook of visual communication: Theory, methods, and media* (pp. 23-42). Mahwah, NJ: Lawrence Erlbaum Associates.

Derrida, J. (1974). *Of grammatology* (G. Spivak, Trans.). Baltimore: The Johns Hopkins University Press.

Foucault, M. (1979). What is an author? In J. V. Harari (Ed.), *Textual strategies: Perspectives in post-structuralist criticism* (pp. 141-160). Ithaca: Cornell University Press.

Gaggi, S. (1997). *From text to hypertext: Decentering the subject in fiction, film, the visual arts and electronic media.* Philadelphia: University of Pennsylvania Press.

Grusin, R. (1996). What is an electronic author? In R. Markley (Ed.), *Virtual realities and their discontents* (pp. 3953). Baltimore: The Johns Hopkins University Press.

Hayles, N. K. (2002). *Writing machines.* Cambridge, MA: MIT Press.

Landow, G. (1997). *Hypertext 2.0: The convergence of contemporary critical theory and technology.* Baltimore: The Johns Hopkins University Press.

Maruca, L. (2003). Bodies of type: The work of textual production in English printers' manuals. *Eighteenth-Century Studies, 36*(3), 321343.

Nielson, J. (2000). *Designing web usability: The practice of simplicity.* NewYork: New Riders Publishing.

Propeller. (2006). Web Design Guru. *Monster.com.* Retrieved February 4, 2006, from http://job-search.monster.com/getjob.asp?JobID=39246359&AVSDM=2006%2D02%2D03+13%3A21%3A54&Logo=1&q=Web+design&fn=554&cy=us

Rombes, N. (2005). The rebirth of the author. In A. Kroker & M. Kroker (Eds.), *CTHEORY.* Retrieved January 25, 2006, from http://www.ctheory.net/articles.aspx?id=480

Slatin, R. (1991). Reading hypertext: Order and coherence in a new medium. In P. Delany & G. Landow (Eds.), *Hypermedia and literary studies* (pp. 153-169). Cambridge, MA: The MIT Press.

Sondheim, A. (2001). Introduction: Codework. *American Book Review, 22*(6). Retrieved January 25, 2006, from http://www.litline.org/ABR/issues/Volume22/Issue6/abr226.html

Tuck, M. (2003). Practical Web design: Frames and frame usage explained. *SitePointe.com.* Retrieved February 6, 2006, from http://www.sitepoint.com/article/frames-frame-usage-explained/5

Woodmansee, M. (1984). The genius and the copyright: Economic and legal conditions of the emergence of the "author." *Eighteenth-Century Studies, 17*(4), 425-448.

KEY TERMS

Active Browsing: Using Greasemonkey scripts allows individuals browsing a Web site to take control and alter that site's appearance and even functionality. The term active browsing is used in contrast to what Greasemonkey users deem the traditional, passive approach to Web browsing.

Author-Function: Michel Foucault developed this term to call into question our ideas about what it means currently to create a work or a text. His definition of a text extends beyond traditional printed works and into other media. The author-function is not a direct analog for the person or individual we call the author. Rather, it is our understanding of how text is produced, distributed, and consumed. Foucault states that the idea of an author has not always existed and that there may be a time when a text is produced, distributed, and consumed without the individual we call the author.

Book Burro: A Greasemonkey script that works with Amazon.com's Web site. The script displays competing pricing as well as library availability of a book found on Amazon's site.

Greasemonkey: An extension for the Mozilla Firefox browser that enables users to install client-side user scripts that alter the behavior and appearance of Web pages. The alterations occur as a Web page is downloaded and rendered in a user's Web browser. The alterations occur without the consent of the site owners.

JavaScript: An object-oriented, cross-platform, Web scripting language originally developed by Netscape Communications, JavaScript is most commonly used for client side applications.

Napster: The first peer-to-peer file-sharing service that by the middle of 2000 had millions of users. Record industries sued Napster for facilitating the free exchange of copyrighted material. By July of 2001, Napster was shut down.

User Scripts: Computer programming that can be activated in order to alter the appearance of a Web page.

ENDNOTES

[1] For more information on Greasemonkey or to download and use it, please see: http://greasemonkey.mozdev.org/

[2] See http://userscripts.org/scripts/show/1859 to download and install the Book Burro script

[3] I would add the growth of "wikis" to Rombes' discussion of technologies that push the author-function.

[4] Here digital writing and media - even Greasemonkey scripts are treated as "text."

[5] See http://userscripts.org/tag/usability for a list of usability scripts

[6] There are examples of scholars debating the roles code and computers play in sharing knowledge and facilitating invention. Douglas Hofstadter contemplated a similar issue back in 1979 in a section from his *Gödel, Escher, Bach: An Eternal Braid* titled "Are Computers Super-Flexible or Super-Rigid?" Code is also associated with control in digital environments. More recently, Lawrence Lessig published *Code and other Laws of Cyberspace*. Likewise, John Cayley's *The Code is not the Text (unless it is the Text)* has been influential for academics working with unraveling the relationships among writers, digital media, and code. The example that follows is more practical than theoretical. That is, it was necessary instead to show real programmers working with a real issue.

[7] Rather than edit for spelling and grammar, the original posting has been preserved.

[8] See http://slashdot.org

[9] See http://userscripts.org/scripts/show/604

Chapter III
Morality and Pragmatism in Free Software and Open Source

Dave Yeats
Auburn University, USA

ABSTRACT

This chapter analyzes the differences between the philosophy of the Free Software Foundation (FSF) as described by Richard Stallman and the open source movement as described in the writings of Eric Raymond. It argues that free software bases its activity on the argument that sharing code is a moral obligation and open source bases its activity on a pragmatic argument that sharing code produces better software. By examining the differences between these two related software movements, this chapter enables readers to consider the implications of these differences and make more informed decisions about software use and involvement in various software development efforts.

INTRODUCTION

As governments around the world search for an alternative to Microsoft software, the open source operating system Linux finds itself in a perfect position to take market share from Microsoft Windows. Governments in France, Germany, The Netherlands, Italy, Spain, and the United Kingdom use Linux to encourage open standards, promote decentralized software development, provide improved security, and reduce software costs (Bloor, 2003). The Chinese government strongly supports Linux as its operating system of choice because Chinese experts have complete access to the source code and can examine it for security flaws (Andrews, 2003). In Brazil, leftist activ-

ists gathered to promote the use of open source software (OSS) (Clendenning, 2005).

There is a connection between the technological reasons for choosing open source software and the political ones. Many governments see open source as a way to promote a socialistic agenda in their choices of technology. Open source advocates, however, do not necessarily make these connections between the software development methods involved in open source and political movements of governments. There is evidence, however, that leaders in the open source movement have expressed their rationale for advocating opening the source code of software.

The open source movement can trace its roots back to an alternate, still very active, software movement known as free software. While open source and free software can (and do) coexist in many ways, there are some essential differences that distinguish the two groups from one another. Perhaps most notably, the free software movement is based on a belief in a moral or ethical approach to software development, while open source takes a much more pragmatic view. While both groups argue for the open sharing of source code, each has its own reason for doing so. Understanding the differences between open source and free software can help open source researchers use more precise terminology and preserve the intent of each of these groups rather than assuming that they are interchangeable.

The following chapter begins with a brief historical overview of the free software and open source movements and highlights some of the main beliefs of each. The chapter then offers an examination of both the moral and pragmatic aspects of open source software. The conclusion invites readers to consider the implications of the differences between the two viewpoints and suggests ways for readers to apply this information when making choices about software.

BACKGROUND

The open source movement grew out of the software development practices in academic settings during the 1970s. During those early years of software development, computer scientists at colleges and universities worked on corporate-sponsored projects. The software developed for these projects was freely shared between universities, fostering an open, collaborative environment in which many developers were involved in creating, maintaining, and evaluating code (Raymond, 1999).

In his *A Brief History of Open Source* article, Charlie Lowe (2001) describes the end of open and collaborative methods of developing computer software in the 1980s when the corporate sponsors of academic software projects began to copyright

the code developed for them. Corporations claimed that the university-run projects created valuable intellectual property that should be protected under law. This, of course, was just one of the signs of the shift from the commodity-based economy in the U.S. to a knowledge-based one. The wave of copyrights threatened to end the collaboration between computer scientists and slow the evolution of important projects. It looked as if the computer scientists would be required to work in smaller groups on proprietary projects.

Richard Stallman (1999) reports that he created the GNU General Public License (GPL) to maintain the ability to collaborate with other computer scientists on software projects, without restriction. The name GNU is a self-reflexive acronym meaning "GNU's Not UNIX," a play on words that pays homage to and differentiates itself from the UNIX legacy.[1] Stallman was concerned that the UNIX operating system, created during the collaborative era of the 1970s, would no longer be supported by new programs that used its stable and robust architecture when access to the source code was cut off. Stallman started the GNU initiative (which enabled the establishment of the Free Software Foundation [FSF]) to ensure that new software would be freely available.

The GNU GPL gave programmers the freedom to create new applications and license them to be freely distributable. Specifically, the GNU GPL gives anyone the right to modify, copy, and redistribute source code with one important restriction: Any new version or copy must also be published under the GNU GPL to insure that the improved code continues to be freely available. Many programmers (both those accustomed to the academic practices of the 1970s and new computer enthusiasts) adopted the GNU GPL and continued to work in open, collaborative systems.

Arguably the most important piece of software developed under the GNU GPL is the Linux operating system. Linus Torvalds, while still a student at the University of Helsinki in 1991, created a new operating system based on the ideas found in the UNIX operating system. This new piece of software, Linux, not only proved the success of GNU GPL, but it also represented a further shift toward a widely cooperative effort in software development. According to Eric Raymond, Linux debunked the myth that there were software projects with an inherent "critical complexity" that necessitated a "centralized, *a priori* approach" (Raymond, 2001, p. 21). With wide adoption among software developers and computer scientists, Linux proved to be a stable and powerful system despite its complexity and rapid development schedule.

In 1998, when Netscape decided to make its source code public as part of the Mozilla project, Eric Raymond and Bruce Perens suggested the use of the term "open source" in response to confusion over the term "free." Stallman, in many cases, found himself explaining that he was using the term "free" in the sense of "freedom" or "liberty" rather than "without monetary cost." Raymond and Perens founded the Open Source Initiative (OSI) to differentiate the new group from the FSF.

While there are many other active voices in the free software and open source movements such as Linus Torvalds (originator of the Linux operating system) and Robert Young (co-founder and CEO of Red Hat), Richard Stallman and Eric Raymond continue to be the most influential and widely cited. While Stallman and Raymond do agree at some level that software development benefits from the free distribution of source code, they see this free distribution in two completely different ways.

THE DEBATE

Many people have written about the debate between Eric Raymond and Richard Stallman. It is widely reported (Williams, 2002) that Stallman disagrees with Raymond's pragmatic reasons for promoting the term "open source" over "free

software." In fact, Raymond's *Shut Up and Show Them the Code* and Stallman's *Why "Free Software" is Better Than "Open Source"* are two examples of the heated exchange between the two writers, each defending his own position on the issue of freely available source code. Bruce Perens reports that it is "popular to type-case the two as adversaries" (Perens, 1999, p. 174). While most studies emphasize that the term "open source" was adopted simply to avoid confusing the word "free" in "software" (DiBona, Ockman, & Stone, 1999; Feller & Fitzgerald, 2002; Fink, 2003), others are careful to point out that the shift in terminology really signaled a shift in strategy for open source advocates.

Perhaps the best work done on the differences between these groups is David M. Berry's (2004) work, *The Contestation of Code: A Preliminary Investigation into the Discourse of the Free/Libre and Open Source Movements*. By analyzing the discourse of the two movements (in the words of Stallman and Raymond), Berry concludes that the discourse of the free software movement more closely identifies with the user, is more utopian, and advocates a communal, socialist approach. The discourse of the open source movement, on the other hand, advocates a more individualistic approach that identifies with the "owners" or creators of software, resulting in a more libertarian emphasis.

In any case, the rift between those who choose to use the term free software and those who choose to use the term open source has resulted in some scholars choosing sides on the issue. Lawrence Lessig (2004), an important scholar in the area of intellectual property law, discusses open source at great length in his work, *Free Culture*. However, he quotes only Stallman's writings, not Raymond's. To Lessig, at least, the open source movement is more about Stallman's rhetoric of freedom than Raymond's pragmatism. Understanding the rationale behind such choices is important in understanding the impact of open source software outside of the software industry.

OPEN SOURCE AND FREE SOFTWARE

In *The GNU Operating System and the Free Software Movement*, Stallman suggests that the two terms "describe the same category of software … but say different things about the software, and about values" (Stallman, 1999, p. 70). The following sections examine the work of Richard Stallman and Eric Raymond to investigate the different philosophical approaches to software development espoused by each.

Free Software: The Works of Richard Stallman

Stallman's (Stallman, 2002c) theorizing about software rests on what he identifies as the four main "freedoms" of his "Free Software Definition." According to Stallman (2002c, p. 18), these freedoms are:

- **Freedom 0:** The freedom to run the program, for any purpose
- **Freedom 1:** The freedom to study how the program works, and adapt it to your needs (access to the source code is a precondition to this)
- **Freedom 2:** The freedom to redistribute copies so you can help your neighbor
- **Freedom 3:** The freedom to improve the program, and release your improvements to the public, so that the whole community benefits (access to the source code is a precondition to this)

In other words, Stallman directly relates his views about software development to a set of freedoms for users of that software. In the rhetoric of these main "freedoms," at least, Stallman is concerned more with the users of a software program than with the program itself.

In *The GNU Manifesto*, Richard Stallman makes impassioned arguments about his stance toward software development. "I consider that the golden rule requires that if I like a program I must share it with other people who like it" (Stallman, 2002a, p. 32), he writes. "So that I can continue to use computers without dishonor, I have decided to put together a sufficient body of free software so that I will be able to get along without any software that is not free" (Stallman, 2002a, p. 32). He constructs his call for radical change in the way software development occurs with several ideological claims. Specifically, Stallman claims that sharing is fundamental and that free software offers the only ethical alternative for software programmers.

Stallman's rationale for calling programmers to work on an alternative to proprietary software is based on what he calls the "fundamental act of friendship of programmers": the "sharing of programs" (Stallman, 2002a, p. 33). Stallman suggests that "[m]any programmers are unhappy about the commercialization of system software. It may enable them to make more money, but it requires them to feel in conflict with other programmers in general rather than feel as comrades" (Stallman, 2002a, p. 32-33). More than simply suggesting that the sharing of programs is ideal or simply important, Stallman argues that it is a fundamental imperative and a source of conflict. He goes so far as to suggest that programmers "must choose between friendship and obeying the law" (Stallman, 2002a, p. 33), implying that the law, on the issue of software availability, is in error.

The metaphors Stallman uses to expand on the idea of the centrality of sharing among developers makes it sound as if restricting software use is against nature itself. He writes: "Copying … is as natural to a programmer as breathing, and as productive. It ought to be as free" (Stallman, 2002a, p. 34). He goes on to equate software with air itself: "Once GNU is written, everyone will be able to obtain good system software free, just like air" (Stallman, 2002a, p. 34). Denying people the right to free software, in other words, would be like trying to regulate and restrict breathing itself. According to Stallman, restricting software use results in a kind of "police state" employing "cumbersome mechanisms" (Stallman, 2002a, p. 34) in the enforcement of copyright law.

While Stallman characterizes a software development community that shares all of its resources as a utopian society, he harshly criticizes proprietary software. He claims that restricting use of a program through intellectual property law constitutes "deliberate destruction" (Stallman, 2002a, p. 36) and a failure to be a good citizen. Stallman's rhetoric sets a scene with only two alternatives: free software based on the ideas of camaraderie, friendship, freedom, good citizenship, community spirit, and sharing of proprietary software based on the ideas of restriction, destruction, commercialization, and materialism. Clearly, Stallman's purpose is to set up a binary in which the only good is free software and the only evil is proprietary software. Any programmer who chooses to develop software in the capitalist proprietary software environment is choosing to be less moral than his or her free software counterparts.

Open Source Software: Raymond's Cathedral and Bazaar

Eric Raymond's *The Cathedral and the Bazaar* (2001), promotes open source software using two pragmatic claims that compliment each other: the promotion of the individual and the conscription of others. Unlike Stallman's emphasis on sharing and morality, Raymond emphasizes the practical aspects of open source that leads to its technical superiority. Specifically, Raymond describes the importance of the lead developers of projects while at the same time emphasizing the necessity of using others to complete work on projects. In neither case does Raymond express a belief in the moral superiority of open source development. Instead, all of the benefits of open source are described in terms of the development of a superior technological artifact.

Throughout *The Cathedral and the Bazaar*, Raymond (2001) promotes an egocentric view of technological development that emphasizes the role of the individual in the process. This egoistic approach is revealed in many ways—from Raymond's own self-congratulation, to his description of how developers find incentive to volunteer to participate in projects. Raymond's tendency to promote individuals over the group begins with his own tendency to describe himself as a gifted individual. While some of his claims may be true, it is unusual for a person to sing their own praises quite as blatantly as Raymond does. Usually, modesty does not allow for such open self-congratulation. In describing the personality traits common to good leaders for open source software projects, Raymond points to his own abilities. "It's no coincidence that I'm an energetic extrovert who enjoys working a crowd and has some of the delivery and instincts of a stand-up comic" (Raymond, 2001, p. 49). His infatuation with his own charming personality illustrates how much he values the individual over the group. More than once, Raymond cites instances where his superior programming skills enabled him to make extraordinarily wise decisions that a lesser programmer might miss. For his programming and writing skills, Raymond mentions that he got "fan mail" (Raymond, 2001, p. 38) and "help[ed] make history" (Raymond, 2001, p. 61). Clearly, Raymond's focus on the individual begins with himself.

Raymond goes beyond his own egoism, however, when he generalizes about what constitutes a good open source software project. According to Raymond, "every good software project starts by scratching a developer's personal itch" (Raymond, 2001, p. 23). This aphorism is the first of the 19 rules of open source software development. It is interesting that Raymond recognizes that the motivation behind good software comes not from a need in the community but rather from a personal interest or desire. Raymond reiterates this emphasis on the individual developer's primacy in starting a project in tenet 18: "To solve an interesting problem, start by finding a problem that is interesting to you."(Raymond, 2001, p. 49) Again, nowhere in Raymond's writing does he refer to moral behavior or developers and a need to share. Instead, he believes that the curiosity of an individual developer is enough to justify work in the open source model.

The natural conclusion to a system that encourages individuals to involve themselves in only those projects which they find personally interesting is a hierarchical system that promotes these individuals. Open source developers who choose to take on a particular software problem promote themselves to the role that Raymond calls the "core developer" (Raymond, 2001, p. 34). This role bestows the leadership upon a single individual who is, in turn, supported by a "halo of beta-testers" who exist to serve the needs of the leader (Raymond, 2001, p. 34). Naturally, this leader wields considerable power over his or her user community. And, according to Raymond, not every developer possesses the skills to be a good project leader. Raymond presupposes that any good project leader has superior technical abilities that are generally recognized in the open source community. Further, he suggests that the core developers have skills "not normally associated with software development"—people skills (Raymond, 2001, p. 48).

What Raymond calls people skills is actually an ability to provide incentive to other developers to enlist their help with a project. Raymond posits that the success of the Linux project came largely from core developer Linus Torvalds' ability to keep his volunteer developers "stimulated and rewarded" by giving them "an ego-satisfying piece of the action" (Raymond, 2001, p. 30). In his own fetchmail project, Raymond says he made a habit of "stroking [users] whenever they sent in patches and feedback" (Raymond, 2001, p. 38). In his analysis of how to encourage participation in members of the open source community, Raymond asserts that hackers find rewards in

"the intangible of their own ego satisfaction and reputation among other hackers" (Raymond, 2001, p. 53). A project leader must "connect the selfishness of individual hackers as firmly as possible to the difficult ends" involved in software development (Raymond, 2001, p. 53). Rather than provide monetary incentive, then, Raymond encourages an approach that enables project leaders to conscript users' assistance through a coercive appeal to their egoistic, selfish desire for glory. This approach simultaneously reinforces the leader's domination over other developers and de-emphasizes any development practice based on goals related to benefit to the community.

Raymond discusses how the paradigm of encouraging egoistic behavior of volunteer developers affects the individual reputation in the leader in the following passage:

Interestingly enough, you will quickly find that if you are completely and self-deprecatingly truthful about how much you owe other people, the world at large will treat you as though you did every bit of the invention yourself and are just being becomingly modest about your innate genius. (Raymond, 2001, p. 40)

It is difficult to believe that Raymond would ever be mistaken as "becomingly modest." Even when he encourages leaders to give credit to those that assist with the project, he reveals the underlying motive of additional glory and recognition in the open source community.

The dominating force that goes hand-in-hand with Raymond's suggestion that project leaders should appeal to a volunteers' selfishness is the idea that these users must be recruited and conscripted in order to create a successful open source project. Raymond quotes Linus Torvalds as saying, "I'm basically a very lazy person who likes to take credit for things that other people actually do" (Raymond, 2001, p. 27). While Torvalds is obviously speaking tongue-in-cheek here, it reveals a common theme that Raymond continues to espouse. Two of the 19 development practices include, "If you treat your beta-testers as if they're your most valuable resource, they will respond by becoming your most valuable resource" (Raymond, 2001, p. 38), and "The next best thing to having good ideas is recognizing good ideas from your users" (Raymond, 2001, p. 40). Both of these statements imply that the volunteer developers belong to and work for the project leader. In addition, the project leader can use these volunteers for his or her own purposes like a natural resource.

The idea of individual ownership extends beyond the volunteers on a particular project to the software itself. Despite the fact that projects are "co-developed" by many individual developers, the lead project coordinator actually "owns" the technology. This idea is present in another one of Raymond's main tenets: "When you lose interest in a program, your last duty to it is to hand it off to a competent successor" (Raymond, 2001, p. 26). Therefore, the technology can be bequeathed and inherited much like a traditional patriarchal succession of ownership. And when a software project is passed down to the next generation of leadership, the volunteer user base comes with it.

Speaking of this volunteer user base, Raymond suggests that "[p]roperly cultivated, they can become co-developers" (Raymond, 2001, p. 26). In addition to cultivating, Raymond suggests that users can be "harnessed" (Raymond, 2001, p. 50) to do work for the lead developer. Essentially, Raymond espouses conscripting volunteers to do the work of the lead developer. Tenet 6 summarizes his position: "Treating your users as co-developers is your least-hassle route to rapid code improvement and effective debugging" (Raymond, 2001, p. 27). The implication of that statement is not that users *really are* co-developers but rather that users should be treated *as if they were* co-developers in order to ensure that they will do work for the improvement of the system. Raymond seems to believe that core developers could build open source software projects on their own, but enlisting the help of users provides a less difficult way to achieve the goal of creating a powerful system. Conspicuously absent in this method of project management is the idea that these volunteer users are better served by participating in the development process. Instead, Raymond's main concern is with the system itself.

According to Raymond, the true benefit of this conscription model of development comes from the advantages of using a large body of volunteers to detect and fix bugs in the system. Tenet 7 is, "Release early. Release often. And listen to your customers" (Raymond, 2001, p. 29). However, Raymond's description of the value of this rule does not include a plea for technology that is sensitive to users' needs. Instead, he asserts that this innovation is simply an effective way to test the software for technological bugs, not usability problems. The goal is to "maximize the number of person-hours thrown at debugging and development, even at the possible cost of instability in the code" (Raymond, 2001, p. 30). Raymond suggests that a "program doesn't have to work particularly well. It can be crude, buggy, incomplete, and poorly documented" (p. 47). Therefore, he promotes systems exposed to "a thousand eager co-developers pounding on every single new release. Accordingly you release often in order to get more corrections, and as a beneficial side effect you have less to lose if an occasional botch gets out the door" (Raymond, 2001, p. 31).

His suggestion that less-than-usable software can be released shows that his interest is not in the value of the software to users. His interest is in the value of the users to the software.

MORALITY AND PRAGMATISM

While Stallman's emphasis in advocating for the free software movement is clearly one of moral behavior and obligation, Raymond's characterization of the open source movement emphasizes the technological superiority of a decentralized development process. Stallman's argument sets up free software as a superior and more ethical alternative to proprietary software development that focuses on the rights and freedoms of users and developers. Nowhere in Raymond's writings does he suggest that proprietary software is less ethical. That is not to say Raymond isn't critical of proprietary software. However, his main concern is always the technological implications of software rather than the moral.

Table 1 outlines a few of the more important differences between the two movements:

Faced with two very different value systems surrounding these related movements, open source software users should pay careful attention to the software they choose to use or the software communities in which they participate. While the two approaches to software development adopt similar practices, they represent two different viewpoints that are often at odds with one another. If a user is making a choice to use open source software because of a belief that it is more moral to support open intellectual property policies, they may want to seek out like-minded projects that use the Stallman approach. If a user is more concerned about the technological superiority of open source software even if that superiority comes at the cost of an emphasis on equality among users, then they may want to seek out projects that are run by maintainers that use the Raymond style.

In either case, users should be aware that the choices they make in their affiliations also signal to others that they adopt the worldview represented in those choices. While there may be many instances of individual developers and software projects that blend the ideas and beliefs of both Raymond and Stallman, it is still important to understand that these philosophies often result in development approaches at odds with each other. Though the practices are admittedly similar, a difference in *why* one would choose to develop open source software can affect *how* one carries out that choice.

Perhaps the most important thing to realize, however, is that neither the Raymond nor the Stallman approach is inherently superior for all users. Instead, the choice to adopt one of the approaches over the other rests entirely upon the needs and situation of the individual user. While the rhetoric of both Stallman and Raymond suggest that their understanding of software development represents a truly enlightened and superior approach, neither one can be said to offer the final word.

FUTURE TRENDS

The most inclusive and technically accurate description of software with freely available source code is free/libre open source software (F/LOSS or FLOSS) because it accurately maintains portions of each of the various movements in the software

Table 1. The morality and pragmatism of the free software and open source movements

Morality *Free Software/Stallman*	Pragmatism *Open Source/Raymond*
Defines the benefit of free software as a superior moral choice.	Defines the benefit of open source as a pragmatic way to develop superior software.
Emphasizes developers' moral obligation to share with others.	Emphasizes satisfying developers' personal and individual desires.
Understands the development process as a shared, communal, group effort based on socialistic principles.	Understands the development process as one driven by one or a small group of leaders who conscript volunteers to assist with the project.

community. However, the term *open source* has proven to be the most popular, partly because of the deliberate attempt by open source advocates to make the licensing structures more business-friendly. Apart from its popularity, many choose the term open source almost exclusively due to its influence on the broader culture; open source has been used as a descriptor for everything from yoga to t-shirt designs.[2] When these non-technological instances of open source are used, the suggestion is that the development of these particular creative endeavors is open to many. More than FLOSS, open source represents both a software development phenomenon and a cultural one.

However, when popular culture adopts open source to mean an open, sharing community of creative invention, it misses the main emphasis of the movement. According to Raymond, the movement is less about the moral imperative to share with others and more about the benefit of harnessing the creative energy of individuals who are free to choose their own work. Rather than seeing open source as good for the public, Raymond emphasizes the benefit of the process for the technology. In other words, Raymond is more interested in product than people.

Unfortunately, it is likely too late to correct the trend in popular culture to equate the term open source with "sharing intellectual property" even though open source refers to a process and value system much more complicated than simply sharing. While the term open source government" certainly carries with it a grandiose image of participatory, shared government in which each member of a community has a voice, but it is unclear how open source government is different from a healthy democracy.

Members of the open source community can contribute to the increased use of the term open source by helping others understand where new uses of the term open source resonate with similarities in the software movement and where they miss the mark. Very few creative endeavors have

anything akin to the source code found in software development, so making the source code freely available, the essential meaning of the phrase open source, cannot be replicated in other fields. However, the idea that creative work should be shared and that sharing can be protected with creative licensing is a contribution from the open source movement that can be adopted by others. Rather than adopting open source, other communities may benefit by using more precise and applicable language such as Creative Commons to refer to the sharing of intellectual property.

CONCLUSION

While many researchers and developers of open source software (myself included) typically lump free software and open source software together, both Stallman and Raymond adamantly insist that there are fundamental differences between the two groups. In his essay *Why "Free Software" is Better than "Open Source,"* Stallman explains that the two groups "disagree on the basic principles, but agree more or less on the practical recommendations" (Stallman, 2002b, p. 55). In other words, free software and open source software are essentially the same in *practice*, but not in *principle*.

Because open source and free software appear to operate the same way to outside observers, there is very little insight into when a piece of software should be labeled open source and when it should be labeled free software. Often, both use the same licensing structures. The essential difference is not a technological one; it is one of philosophies. Only the developers themselves can attest to the reasons they choose to develop software with freely available source code. However, it is useful for outside observers to be more precise in their allegiances. It could mean the difference between freedom and pragmatism.

REFERENCES

Andrews, P. (2003, November 24). Courting China. *U. S. News and World Report*, p. 44-45.

Berry, D. M. (2004). The contestation of code: A preliminary investigation into the discourse of the free/libre and open source movements. *Critical Discourse Studies, 1*, 65-89.

Bloor, R. (2003, June 16). Linux in Europe. *IT-Director.com*. Retrieved February 7, 2005, from http://www.it-director.com/article.php?articleid=10929

Clendenning, A. (2005, January 30). Activists urge free open-source software. *World Social Forum*. Retrieved February 7, 2005, from http://www.commondreams.org/headlines05/0130-03.htm

DiBona, C., Ockman, S., & Stone, M. (1999). Introduction. In C. DiBona, S. Ockman, & M. Stone (Eds.), *Open sources: Voices from the open source revolution* (pp. 1-17). Sebastopol, CA: O'Reilly & Associates.

Feller, J., & Fitzgerald, B. (2002). *Understanding open source software development*. London: Addison-Wesley.

Fink, M. (2003). *The business and economics of Linux and open source*. Upper Saddle River, NJ: Prentice Hall PTR.

Lessig, L. (2004). *Free culture: How big media uses technology and the law to lock down culture and control creativity*. New York: Penguin Press.

Lowe, C. (2001). A brief history of open source: Working to make knowledge free. *Kairos: A Journal for Teachers of Writing and Webbed Environments, 6*(2). Retrieved October 25, 2005, from http://english.ttu.edu/KAIROS/6.2/news/opensource.htm

Luman, S. (2005, June). Open source softwear. *Wired 13.06*.

Pallatto, J. (2005, May 13). Yoga suit settlement beggars open source ideals. *eWeek*. Retrieved October 25, 2005, from http://www.eweek.com/article2/0,1759,181,5971,00.asp

Perens, B. (1999). The open source definition. In C. DiBona, S. Ockman, & M. Stone (Eds.), *Open sources: Voices from the open source revolution* (pp. 171-188). Sebastopol, CA: O'Reilly & Associates.

Raymond, E. (1999). A brief history of hackerdom. In C. DiBona, S. Ockman, & M. Stone (Eds.), *Open sources: Voices from the open source revolution* (pp. 19-30). Sebastopol, CA: O'Reilly & Associates.

Raymond, E. (2001). The cathedral and the bazaar. *The cathedral and the bazaar: Musings on Linux and open source by an accidental revolutionary* (rev. ed., pp. 19-64). Sebastopol, CA: O'Reilly & Associates.

Stallman, R. (1999). The GNU operating system and the free software movement. In C. DiBona, S. Ockman, & M. Stone (Eds.), *Open sources: Voices from the open source revolution* (pp. 53-70). Sebastopol, CA: O'Reilly & Associates.

Stallman, R. (2002a). The GNU Manifesto. In J. Gay (Ed.), *Free software free society: Selected essays of Richard M. Stallman* (pp. 31-39). Boston: Free Software Foundation.

Stallman, R. (2002b). Why "free software" is better than "open source." In J. Gay (Ed.), *Free software free society: Selected essays of Richard M. Stallman* (pp. 55-60). Boston: Free Software Foundation.

Stallman, R. (2002c). The GNU Project. In J. Gay (Ed.), *Free software free society: Selected essays of Richard M. Stallman* (pp. 15-30). Boston: Free Software Foundation.

Williams, S. (2002). *Free as in freedom: Richard Stallman's crusade for free software*. Sebastopol, CA: O'Reilly & Associates.

KEY TERMS

Free/Libre and Open Source Software (FLOSS): A more inclusive term for all software with freely available source code.

Free Software (FS): Software with freely available source code developed in the tradition of the Free Software Foundation and influenced by the writings of Richard Stallman.

Morality: An appeal to the fundamental goodness of an act; primary rationale behind the free software movement.

Open Source Software (OSS): Software with freely available source code developed in the tradition of the Open Source Initiative (OSI) and influenced by the ideas of Eric Raymond and Bruce Perens.

Proprietary Software (PS): Software without publicly available source code, commonly seen as the opposite of free and open source software.

Pragmatism: An appeal to the usefulness of an act; primary rationale behind the open source movement.

ENDNOTES

[1] It is common for developers to use reflexive acronyms, partly as a tongue-in-cheek recognition of the overuse of acronyms in technology. Other examples include PHP (PHP hypertext protocol) and WINE (WINE Is Not an Emulator).

[2] For more information about some of the ways open source is being used outside of software, see Stuart Luman's article "Open Source Softwear" (2005, *Wired 13*[06]) and John Pallatto's article "Yoga Suit Settlement Beggars Open Source Ideals" (2005, *eWeek*, May 13).

Chapter IV
Hacker Culture and the FLOSS Innovation

Yu-Wei Lin
University of Manchester, UK

ABSTRACT

This chapter aims to contribute to our understanding of the free/libre open source software (FLOSS) innovation and how it is shaped by and also shapes various perceptions on and practices of hacker culture. Unlike existing literature that usually normalises, radicalises, marginalises, or criminalises hacker culture, I confront such deterministic views that ignore the contingency and heterogeneity of hacker culture, which evolve over time in correspondence with different settings where diverse actors locate. I argue that hacker culture has been continuously defined and redefined, situated and resituated with the ongoing development and growing implementation of FLOSS. The story on the development of EMACSen (plural form of EMACS—Editing MACroS) illustrates the consequence when different interpretations and practices of hacker culture clash. I conclude that stepping away from a fixed and rigid typology of hackers will allow us to view the FLOSS innovation from a more ecological view. This will also help us to value and embrace different contributions from diverse actors including end-users and minority groups.

INTRODUCTION

Free/libre open source software (FLOSS) has emerged as an important phenomenon in the information and communication technology (ICT) sector as well as in the wider public domain. A new research strand has attracted scholars and practitioners to analyse the development of FLOSS from many perspectives. While the FLOSS community continues to grow, diverse actors (e.g., developers, firms, end-users, organisations, governments, etc., just to name a few) are brought into play. Meanwhile, a variety of apparatus and inscriptions (e.g., technical ones such as software and hardware tools, socioeconomic ones such as licences, educational ones such as certificates, and sociocultural ones such as online/off line discussion forums) are developed and employed to maintain the practice. The complex composition of the FLOSS community entails a heterogeneous field where innovation is *sociotechnically constructed*. Practices and values in the FLOSS community are interpreted differently in support of individual and organisational demands (social, economic, political, and technical) of the actors. Such a heterogeneous world resembles an ecological system that contains diversity while resources (information, knowledge, and tools) are commonly shared amongst actors.

Technically speaking, current research on FLOSS, across academic disciplines and industry fields, mainly focuses on measuring the efficiency and productivity in terms of code reuse, density of bugs, and complexity of code or frequency of release, usage, and adoption in the software engineering approach of productivity cycles. A prominent example with regard to determining the benefits of the FLOSS development model is improving security. Given the nature of software technologies, it is generally agreed that "given enough eyeballs, all bugs are shallow" (Raymond, 1999). Moreover, FLOSS also contributes to open standards and interoperability because the availability of source code increases the transparency of software and eases the development of compatible complementary software (DiBona, Ockman, & Stone, 1999; Feller & Fitzgerald, 2001).

While these studies focus on a technologically deterministic perspective of the FLOSS innovation, the intense interactions between people all over the globe in the FLOSS community indicate the importance of mutual shaping between all economic, sociocultural, and technical factors in the FLOSS innovation process. One of the key factors that shape the FLOSS innovation is said to be the hacker culture (Himanen, 2001; Levy, 1984; Moody, 2001; Raymond, 1999; Williams, 2002). Much of the existing literature dedicated to understanding the motivations of those participating in the FLOSS development have treated hacker culture as an incentive that drives programmers to compete or collaborate with each other. A collaboration-oriented argument highlights the features of gift culture, community-forming, knowledge-sharing, and social networking in the FLOSS innovation, whilst a competition-oriented argument emphasises the mutual challenging and self-exploring aspects in a reputation-reward

system. Either account, nonetheless, repeatedly overstates "the hackers" as such a homogeneous group that "fails to account for the plasticity of human motivations and ethical perceptions" (Coleman, 2005, chap. 5). As MacKenzie comments on Himanen's work:

Its focus on hacker heroes and their individual ethical values as the core of hacker culture largely ignores the complicated practices of software development for the sake of what I can only read as an uncritical individualism centred on passion: "hackers want to realize their passions." (MacKenzie, 2001, p. 544)

In line with MacKenzie, I argue that sociological research on FLOSS communities should go beyond the idealised and self-serving versions of FLOSS projects towards understanding the FLOSS development as a sociological phenomenon. It is important to analyse material practices and mechanisms as well as social practices that "developers commit themselves to an ethical vision *through*, rather than prior, to their participation in a FLOSS project" (Coleman, 2005, chap. 5). That said, hacker culture shall not be seen as a preexisting norm in the FLOSS social world; it is negotiated semantically and contextually practised to embody different voices towards hacker culture. Thereby, FLOSS should be better treated as socially-informed algorithms where hacker culture is defined, annotated, practised, situated, and redefined by a diverse range of actors.

BACKGROUND

As said, a *hacker-driven* innovation has been proposed to denote the FLOSS development and this idea has been appropriated widely by researchers and practitioners in this field. It is generally recognised that FLOSS was originated from the hacker culture of the 1960s and 1970s, when hackers defined themselves as "clever software programmers who push the limits of the doable" (Rosenberg, 2000, p. 6). Existing studies on participants' motivations of sharing source code usually presume a firm open source "hacker" culture that is widely shared amongst members in FLOSS communities and drives them to voluntarily participate in the FLOSS development and share their work (e.g., Hannemyr, 1999; Himanen, 2001; von Hippel & von Krogh, 2003; Weber, 2004).

But the definition of hackers is so ambiguous that it is very difficult to identify the object even if a variety of writings have been dedicated to this goal.

The first text systematically introducing computer hackers appeared in 1984 when Levy compiled a detailed chronology of hackers in his book entitled *Hackers: Heroes of the Computer Revolution*. This book revealed an unknown world where technical innovation was developing at a high speed. Levy described how the activities of hackers influenced and pushed the computer revolution forward. In Levy's account, the era of hacking had commenced in the 1960s in university computer science departments where highly skilled students worked and shared information through computer networks. Members of this world tried to mobilise the power of computing in entirely novel ways. They communicated with each other through computer networks in source code. Because this world was so different from wider social life, its members were regarded with suspicion and often seen as deviant. Levy classified hackers into three generations from the 1950s to 1980s according to their various actions and beliefs "associated with hacking's original connotation of playful ingenuity" (Taylor 1999, p. 36). According to Levy, the earliest hackers, the pioneering computer enthusiasts at MIT's laboratories in the 1950s and 1960s, were the first generation of hackers, who were involved in the development of the earliest computer programming techniques. Then there was the second generation of hackers who were engaged

in computer hardware production and the advent of the PC; the third generation of hackers (who were also fanatic game players), devoted their time to writing scripts and programmes for game architecture.

Not surprisingly, perhaps, with the popularisation of PCs, some hackers from the 1980s gained great success in computer businesses such as Apple and Hewlett Packard. Apart from working on making hardware, some hackers created software applications and programs for PCs. Bill Gates' Microsoft was started at this time. It seemed that their business success was so marked that their identity as "hackers" *per se* was downplayed. To add to Levy's categories, I have also observed that with the growth of Internet technologies, an unbalanced global software market dominated by Microsoft, and a wider political milieu suffering from all sorts of anti-terrorist discourses, the contemporary hacker generation is engaging with new ".net" issues such as licensing, patents, security, and privacy. In addition to developing software technologies, hackers at the age of Web 2.0 also have to deal with more social and political issues than before.

Levy's chronological categories of hackers was soon overtaken by scholarly studies in the 1990s investigating the hacker world and understanding the key role that computer hackers play in the ICT network society. However, a thorough picture has never been mapped. Researchers invariably situate hackers in the field of computer network security and can hardly avoid dichotomizing hackers into black hat or white hat. The sensational coverage of computer crime in mainstream media leads many scholars to place hackers in the context of deviance, crime or the expression of an obsessed user subculture with a gang mentality. Chantler (1996) observed hackers since 1989 and finally brought all the materials together in a thesis in 1996 titled *Risk: The Profile of the Computer Hacker*, which mainly introduces the biographical life of hackers and their activities. Meyer (1989), a criminologist, studied the social organization of the computer underground from a postmodernist view. Taylor's (1999) book titled *Hackers: Crime in the Digital Sublime*, which tries to explore the hacker subculture from a more open perspective, nevertheless, still locates it in the context of digital crime, as the book title suggests. Thomas (2002) discusses the relationship between hackers and technology and portrays hacker culture in terms of their perception of technology, and human relationships (Thomas, 2002). In this sense, hacker culture was seen as being formed through interaction with technology, culture, and subculture. Thomas concludes his analysis of hacker culture with an account of the two controversial hacker figures, Kevin Mitnick and Chris Lamprecht, both used to conducting unlawful network penetration activities. Skibell's work that demonstrates that "the computer hacker that society assumes is the principal threat is nothing more than a mirage, and that a revaluation of the dangers to computer security needs to be undertaken before sensible policy can emerge" (Skibell, 2002, p. 337). He comes to the conclusion that "the majority of computer intruders are neither dangerous nor highly skilled, and thus nothing like the mythical hacker" (Skibell, 2002, p. 336). In stating that "the hacker only exists in the social consciousness," (Skibell, 2002). Skibell's work points out that perceptions of hackers are socially constructed. This association with computer network security has been widely represented by mass media and internalised by the public. Whenever there is an incident involving computer network security, "hackers" are blamed.

In contrast to the aforementioned literature, which was largely inspired by the sensational media coverage about the huge damage to companies from the attacks of malicious "hackers," and their portrayal as negative factors in the development of ICT, a hacker, in the tradition of FLOSS literature, is regarded as a creative and enthusiastic programmer for some groups of actors (e.g., Raymond, 1999). These hackers more or less resonate Levy's first second-generation

hackers. The difference lies in their understanding of hacker ethics, a manifesto of freedom of information. Their acts, no matter if they are coding, writing, or other performances, pursue a meaning of liberating information and challenging authority. Even in the context of a system attack, hacking is seen as a technical activity deploying arbitrary codes to free information, to challenge the weakness of software, database, or firewall. Such codes include viruses and scripts that are both programmes. The operation of these codes might raise people's vigilance towards network security. Under these circumstances, codes are written to improve software quality or reliability in a way. Most of the time, these hacking tools are available on the Internet. Whilst this situation is said to allow "script kiddies" to perform malicious acts on the Web (e.g., to deface Web pages or send viruses), their activities can be seen as an alternative form of self-expression as well that demonstrates trial-and-error mindset. It is possible that the existing tools can be improved or a new tool can be created to conduct these actions. In light of this, Ross (1991) summarises a variety of narratives found within the hacker community that express their behaviour:

- *Hacking performs a benign industrial service of uncovering security deficiencies and design flaws.*
- *Hacking, as an experimental, free-form research activity, has been responsible for many of the most progressive developments in software development.*
- *Hacking, when not purely recreational, is [a sophisticated] educational practice that reflects the ways in which the development of high technology has outpaced orthodox forms of institutional education.*
- *Hacking is an important form of watchdog[, countering] to the use of surveillance technology and data-gathering by the state, and to the increasingly monolithic communications power of giant corporations.*

• *Hacking, as guerrilla know-how, is essential to the task of maintaining fronts of cultural resistance and stocks of oppositional knowledge as a hedge against a technofascist future.* (pp. 81-82)

Hannemyr (1999) shares a similar view with Ross and sees hacking as a positive method adopted by hackers for creating information systems and software artifacts that offer higher flexibility, tailorability, modularity, and open-endedness compared with methods used by industrial programmers. Their interpretations of hacking echoes the hacker ethics defined in the *New Hacker's Dictionary* or *The Jargon File* (Raymond & Steele, 1991). While the majority of the public still regards the hacker as hostile, for Raymond and Steele, in the hacker community, being a hacker does not necessarily mean being exactly good or bad; rather, being a hacker means being creative and innovative. Their intention of differentiating hackers from "crackers" nonetheless prescribes an elite hacker class. Instead of criminalising hackers, they normalise hacker culture and expect people to follow the already determined ethics.

Although a good number of practitioners do refer their hacker identity to this version of hackers, a single and stable definition of the "hacker" is hard to give. "Hacker" remains an obscure term. Having read these writings that mainly assign hackers into either the field of computer network security or the UNIX programming world, my point is that instead of seeking a universal definition of hacker, we should treat hacker as an umbrella concept that is defined and redefined by different people, situated and resituated in different contexts. There are so many different expressions of hacker identity. It is inadequate to focus the analysis on either stigmatised hacking or UNIX geek programming life alone. It appears to me that previous literature, few of which express the diversity of the hacker in modern society, is of limited value in understanding the hacking practices and their

relationship with the ICT innovation system. It presents a reductionist notion, which appears to be, from my point of view, very problematic. In this chapter, I do not wish to begin with a proposition that categorises hackers as deviant or marginal actors, nor do I wish to portray hackers simply in a positive light.

A motivation for doing so has to do with a methodological challenge. As Taylor (1999) explains, the reason for such an ambiguous hacker identity is because of loose social ties, and an attempt to analyse the computer underground is therefore "inherently difficult" (Taylor, 1999). When Taylor studies the relationship between the computer underground and the computer security industry, it turned out to be difficult to pursue because:

Both groups are far from being coherent and well-established given the relative youth of computing and its hectic evolutionary pace. [Moreover,] the boundaries between groups are unusually fluid and there is no established notion of expert knowledge. ... It is thus at times problematic, in choosing interview subjects and source materials, to fall back on conventional notions of what constitutes an expert or even a member of a subculture. (Taylor, 1999)

Ross (1991) also gives a similar explanation:

While only a small number of computer users would categorise themselves as "hackers," there are defensible reasons for extending the restricted definition of hacking down and across the caste hierarchy of systems analysts, designers, programmers, and operators to include all high-tech workers—no matter how inexpert—who can interrupt, upset, and redirect the smooth flow of structured communications that dictates their position in the social networks of exchange and determines the pace of their work schedules. To put it in these terms, however, is not to offer any universal definition of hacker agency. There are many social agents. ... All [these social agents], then, fall under a broad understanding of the politics involved in any extended description of hacker activities. (pp. 92-93)

Given these methodological and ontological challenges, it is unwise then to characterise hackers as a homogeneous community and hierarchy akin to a gang organisation. There is no clearly bounded constituency of hackers. As Nissenbaum (2004) argues, the transformation in our conception of hacking over the past few decades is more a function of contextual shifts than of changes in hacking itself. This has been achieved not through direct public debate over conflicting ideals and interests, but through an ontological shift mediated by supportive agents of key societal institutions: legislative bodies, the courts, and the popular media. In a similar vein, my PhD dissertation (Lin, 2004) employed the *social worlds theory* (e.g., Clarke 1991) and other methodologies inspired in the field of science and technology studies (STS) (e.g., Jasanoff, Petersen, Pinch, & Markle, 1995; Sismondo, 2004) is to pursue this end. The thesis, instead of presuming hackers as a specific and relatively closed social group, treats the term "hacker" as flexibly as possible. The notion "hacker" is interpreted differently to demonstrate one's identity and self-expression. Since the notion is not predefined, it allows heterogeneous readings and performances. I also suggest to contextualise hacker culture in everyday computing practices. A definition of hacker is identified and situated in a local context where an individual or a group of actors practise what and how they understand hacker culture. A hacker identity is constructed through performing some tasks in everyday computing. These tasks, usually pertaining to coding and programming, define and situate a stream of the FLOSS innovation in a local context where the performers inhabit.

To overcome the methodological and ontological challenges, in the following, I will take a practice-based perspective to look at how hacker culture is embodied and embedded in everyday computing world and performed by individuals or groups who either share collective ideas and/or practices of hacking or demonstrate their unique understanding and performances of hacker culture. As a consequence, a hacker-driven product, an editor programme, forks into various versions whose developments differ in contexts. When one version derived from one sense of hacking gets more apparent, it would drift away from others and find another stage of performing their understanding of hacker culture. Thereby, I will conclude that a practice-based perspective is needed in order to capture the emerging, contingent, and dynamic hacker culture and its relationship with the FLOSS innovation system and the wider computing world. If there existed a universal definition of "hacker," the FLOSS innovation would not be as burgeoning as it is now.

MAIN FOCUS OF THE CHAPTER

Hacker Culture and the FLOSS Innovation

Given the critical review of existing literature on hackers and hacker culture, it is obvious that one should stay away from simplicity and stereotypes of hackers. Parallel to Coleman's anthropological research on hacker groups that contributes to our understanding of "how hacker valuations, motivations and commitments are transformed by the lived experiences that unfold in FLOSS projects and institutional that are mediated through project charters and organizational procedures," I suggest to take a practice-based perspective on the FLOSS development to strengthen the heterogeneity and diversity in the hacker "social world," where local culture and situated knowledge derive from identities and commitments largely developed through prolonged interaction toward shared, yet continually emergent, goals (Lin, 2004). In other words, I highlight multiple visions and means of achieving them by attempting empirically to view the world in the actors' own terms. In so doing, I show that hacker culture is embedded and embodied in everyday software engineering practices linked with the FLOSS development.

The development of EMACSen (plural form of EMACS—editing macros) can serve as a good illustration here. EMACS is one of the first programmes written by Richard Stallman, the founder of the Free Software Foundation (FSF) and released under the General Public License (GPL), the most popular FLOSS license. Its historical position as a classic programme allows us to see how the development of a project was mutually shaped by the act of hacking which is situated in everyday practices (e.g., programming for programmers) and by different understandings of hacker culture in various contexts.

EMACSen

According to the document *GNU EMACS FAQ*, EMACS refers to a class of text editors, possessing an extensive set of features, that are popular with computer programmers and other technically proficient computer users. The original EMACS was a set of macros written by Richard Stallman in 1976 for the TECO (Text Editor and COrrector) editor under Incompatible Timesharing System (ITS) on a PDP-10 machine. EMACS was initiated by Guy Steele as a project to unify the many divergent TECO command sets and key bindings at MIT, and completed by Stallman. It was inspired by the ideas of TECMAC and TMACS, a pair of TECO-macro editors written by Guy Steele, Dave Moon, Richard Greenblatt, Charles Frankston, and others. Many versions of EMACS have appeared over the years, but now there are two that are commonly used: GNU EMACS, started by Richard Stallman in 1984 and still maintained by him, and XEmacs, a fork of GNU EMACS which was started in 1991 and has remained mostly compatible. Both use a powerful extension language, EMACS Lisp, that allows them to handle tasks

ranging from writing and compiling computer programs to browsing the Web.

Being a popular editor programme, EMACS was able to meet the requirements of many users by being flexible (allowing users to define their own control keys). This feature of flexibility reflects EMACS's affordance and enables more actors to move into the innovation process through adopting, using, testing, and improving the software. Unlike its predecessors TECMAC and TMACS which took programmers a long time to understand each other's definitions of commands before they could bring new order to the programme, EMACS won the hearts of its users with a standard set of commands. For the sake of durable efficiency, one of the developers, Stallman, came up with an idea of sharing newly defined commands for the sake of doubling. Hence, he wrote the terms of use for EMACS to request reports of new modifications to him. He released the EMACS source code with a condition of use that requested feedback about any modification to the source code, while also allowing its redistribution at the same time. In so doing, Stallman actually redefined and broadened the boundary of the developing team and made the EMACS innovation more accessible. In issuing this social contract, on the one hand Stallman drew users' attention to the extensibility of EMACS, and on the other hand fulfilled his belief in the freedom of information, and his understanding of hacker culture. The condition he put on source code distribution therefore acted equally to engage users with a practical attitude as well as to promote his philosophy and to sustain the culture that he was used to living within the MIT AI lab, a locale that shaped his perception and behaviour.

The development of EMACS and the initiative of Stallman's social contract (which inspires the advent of the infamous GPL) both show the co-production of sociocultural milieu and technical artifacts. The technical tools such as different programming languages, the work atmosphere at the MIT AI lab in the 1970s and 1980s, and the culture of sharing knowledge embedded in the programming practice in the 1970s, all contribute to the innovation of EMACS. The original version of EMACS was created under a condition situated in a specific physical and social space (i.e., the MIT AI lab), and programmed by a specific language (i.e., Lisp). That said, the daily environment and programming tool co-construct the development of EMACS. If "culture" is defined as a way of living, which is invisible and embedded in our daily lives, the original version of EMACS that was created in a specific environment and programming culture that Stallman and others located is undoubtedly a socially-informed algorithm. It is also an embodiment of a version of hacker culture situated in the above mentioned environment, embedded in everyday programming practices. Stallman's understanding of hacker culture constantly appears in his writings and speeches, and this serves to explain the situation in which he designed and maintained EMACS and other software.

Nevertheless, while the social norm established in the original version of EMACS linked to innovation gained greater political weight, given Stallman's later act of advocating "free software," some users were reluctant to conform to the social obligations. This is one of the reasons why Stallman's GNU EMACS is labelled as a *moral* product regardless of its technical utility. People who did not share Stallman's vision went on creating other editor programmes. Furthermore, new versions of EMACS were created through yet other problem-solving process (e.g., EINE, ZWEI, Xemacs, etc., are derived from the need of porting EMACS with other programming languages). These forked versions of EMACS were created because their creators situated their hacking in different programming tools that they used everyday. In fact, as documented (e.g., Moody, 2001), the reasons for the divergences vary in social scope (e.g., disagreement with Stallman's social contract), and technical scope (e.g., the original version of EMACS did not run on other

programming languages or support certain type of machines). For instance, the versions EINE and ZWEI for Lisp, developed by Weinreb, could be considered as some hacked version of EMACS situated in some specific programming environments. While these motivations of forking all link with programmers' mundane and situated practices and influenced by complicated sociotechnical factors, the forked products represent as disagreements on Stallman's interpretation and articulation of hacker culture.

An Evolving Hacker Culture and the EMACS Development

The development of EMACS leads to several major the FLOSS innovations in roughly three aspects: technical, sociocultural, and legal. The technical innovation refers to various versions of software programmes based on the original version of EMACS initiated by Steele. The sociocultural innovation refers to a community-based type of collaboration to develop software motivated by Stallman's social contract. The legal innovation refers to the advent of GPL, which inspires the emergences of many different software licences. The hacker culture defined by the early developers such as Stallman and Steele at the time and space they were situated in has been embodied and embedded in the early development of EMACS.

However, over time, Stallman's way of hacking has been challenged, negotiated, refined, and resituated with the emergence of other versions of EMACS. Rather than being a question of which version of EMACS is technically better, which way of hacking is more efficient, which way of licensing is more politically correct, the question from a sociological perspective would be how different understandings of an intermingling of social, technical, economic, and legal factors were taken into account and taken actual form in the EMACS development. The story above thus shows that there is no one dominant or homogeneous notion of hacker culture. If a universally defined hacker culture (say, Stallman's version of hacker culture) existed and mandated all hackers' behaviours, there would not be so many versions of EMACS, different editors, and software licences to date. This also echoes my view that hacker culture needs to be understood in a practice-based sense concerning how actors perform their understandings of hacker culture, and how various the FLOSS innovations are initiated, developed, and accepted in different contexts.

Parallel to EMACS, many other FLOSS projects have been witnessing similar technical, sociocultural, and legal innovations. In terms of a practice-based view, each project and forked subproject is an embodiment of a definition and performance of hacker culture, whether practitioners explicitly or implicitly identify themselves as hackers. Some practices emerging from Stallman's way of defining and performing hacker culture have been institutionalised in ways such as open sourcing software under GPL-like licences and sharing information across various medias (e.g., mailing lists, IRC channels, Wikis, blogs, and Web sites). Although many FLOSS communities are conducting these collective practices derived from Stallman's hacker culture, it does not necessarily mean that there is a single philosophy and ontology of hacker culture indifferently shared amongst all members. A hacker social world (Lin, 2004) *de facto* accommodates heterogeneous "hacker groups" and hackers who assign different meanings to the umbrella concept "hacker." These social groups mutually engage in, interact, communicate, and negotiate with one another. In other words, if there were a hacker community, it is a social world that incorporates heterogeneity through engaging actors on a constellation of collective practices, the practices of experimenting or challenging existing knowledge paradigms and of sharing information and knowledge. Over time software technologies, the orbit within which hacking practices are found has been extended, and is much wider than was the case for those stud-

ies conducted, notably in relation to the FLOSS development. For instance, the innovation based on a community-based collaborative culture is recognised in the wider computing world such as blogging and the Wiki phenomenon.

The collective hacking practices appear to be important factors in the emergence of FLOSS. If there is a norm existing in "the hacker community," it should be in the sense of Robert Cover (cited in Coleman, 2005) who argues that "the production and stabilization of inhabited normative meanings requires ongoing and sometimes conflicting interpretation of codified textual norms." Such continual acts of reinterpretation and commitment are exactly because of the heterogeneity in the hacker social world. Heterogeneity, on the one hand, becomes the resource that helps mobilise the FLOSS innovation, and on the other hand, drives diverse actors to redefine and practise the hacker culture they perceive differently. Analysing hacker culture and understanding how collective (hacking) norms and practices are interpreted, articulated, and performed differently by different people, in this regard, provides a culturally contextualised understanding of the FLOSS innovation.

FUTURE TRENDS

Apart from valuing contributions from minority hacker groups and their contributions to the FLOSS development, future studies should also center on how different hacker groups define their territory, how different hacker groups interact with each other? In what way? Do they cooperate, or do they draw a line between each other? These sociological issues are critical to our understanding of the dynamics both in the hacker social world and the FLOSS innovation system where geeks and activist cultures are brought together.

CONCLUSION

This chapter begins with a review on the existing research into hacker culture and its relationship with the FLOSS development, and discusses different articulations and interpretations on the concepts of hackers and hacker culture. Looking at such a variety of materials, I argue that the evolution of FLOSS involves continuous negotiations of meanings and consequently a practice-based and sociological perspective is needed to help us better understand the dynamics in FLOSS evolution: the changing roles and work practices of FLOSS developers and how their cultures and identities are formed through interacting with each other and with the artefacts in their everyday environments through committing to the collective open source practices. Based on the story of the evolution of EMACS and a plurality of forked versions, I delineate how this diversity of EMACSen embodies and symbolises different practices and articulations towards "hacker culture." Unlike most of the previous research on hacker movement, I take a practice-based perspective to document the various voices on hacker movement and the evolution of FLOSS, and their interactions (conflicts and negotiations) and consequent impact. A shift from a rigid and fixed typology of hacker culture to a practice-based perspective on hacker culture would allow us to look at how the collective production of FLOSS skills and practices emerge from negotiating the meanings and interpretations of hackers. It also offers a holistic but critical view to study various performances of hacker culture and their relationships with the FLOSS development, such as the hacktivism referred to by Jordan and Taylor (2004), different hacker ethics (Coleman, 2005), Indymedia and Wikipedia's mediactivism, and other forms of digital struggles and geek fights (e.g., software patents, repression of peer to peer file-sharing, IP and data retention laws are attacking digital freedom daily). It implicates

that the FLOSS innovation system serves as a sociotechnically efficient platform to enroll wider sociotechnical resources from the public as well as the private sectors to provide for greater innovation capacity in software development because this platform allows a free evolution of hacker culture that is constantly redefined, reinterpreted, and resituated (Lin, 2004).

Having said that, once a static and normative definition of hacker culture is tackled, and the emphasis is placed on different understandings and performances of the concept, it indicates the importance of integrating end-users and minorities in this dynamic world (e.g., women, the vision-impaired, and people from developing countries) groups in the FLOSS innovation process (e.g., Lin, 2006a, 2006b). So far, the term "hacker" is either claimed by advantaged groups in software engineering (e.g., male, white) and acclaimed in their networks, or declaimed by more mainstream media as deviants. Both discourses are voiced from the positions of the advantaged that ignores other ways and interpretations of hacking. They also contributed to many of the inequalities in the FLOSS social world. These minority groups do not usually fit into the mainstream hacker culture loudly advocated by mainly an advantaged group in software engineering or stigmatised by mainstream media. A practice-based view on hacking is to distribute the right of interpreting and performing hacker culture to a wider and more diverse range of actors involved in the FLOSS development. It is less interesting for me to group who are the hackers. It is more important for me to make sure that people are equally entitled to the title "hacker" and equally allowed to practise what they believe is hacking. In so doing, I value everyday technologies, tacit knowledge, and local culture in hacking and in the FLOSS innovation. What I would like to present here is a contextualised perspective on hacking.

REFERENCES

Chantler, N. (1996). *Risk: The profile of the computer hacker*. Doctorol dissertation, Curtin University of Technology, Perth, Western Australia.

Clarke, A. E. (1991). Social worlds/arenas theory as organizational theory. In D. Maines (Ed.), *Social organization and social processes: Essays in honour of Andelm L. Strauss*. NY: Aldine Gruyter.

Coleman, E. G. (2005). *The social construction of freedom in free and open source software: Hackers, ethics and the liberal tradition*. Doctoral dissertation, Department of Anthropology, University of Chicago.

Cover, R. (1992). Nomos and narrative. In M. Minow, M. Ryan, & A. Sarat (Eds.), *Narrative, violence, and the law: The essays of Robert Cover*. Ann Arbor: The University of Michigan Press.

DiBona, C., Ockman, S., & Stone, M. (Eds.). (1999). *Open sources: Voices from the open source revolution*. Sebastopol, CA: O'Reilly.

Feller, J., & Fitzgerald, B. (2001). *Understanding open source software development*. London: Addison-Wesley.

GNU EMACS FAQ. (n.d.) Retrieved from http://www.gnu.org/software/emacs/emacs-faq.text

Hannemyr, G. (1999). Technology and pleasure: Considering hacking constructive. *First Monday* 4(2). Retrieved July 11, 2006, from http://www.firstmonday.org/issues/issue4_2/gisle/

Himanen, P. (2001). *The hacker ethic and the spirit of the information age*. London: Secker & Warburg.

Jasanoff, S., Petersen, J. C., Pinch, T., & Markle, G. E. (1995). *Handbook of science and technology studies*. London: Sage.

Levy, S. (1984). *Hackers: Heroes of the computer revolution*. Garden City, NY: Anchor Press/Doubleday.

Jordan, T., & Taylor, P. (2004). *Hacktivism and cyberwars: Rebels with a cause?* Routledge.

Lin, Y.-W. (2004). *Hacking practices and software development: A social worlds analysis of ICT innovation and the role of open source software*. Unpublished doctoral dissertation, SATSU, University of York, UK.

Lin, Y.-W. (2006a). Women in the free/libre open source software development. In E. M. Trauth (Ed.), *Encyclopedia of gender and information technology* (pp. 1286-1291). Hershey, PA: Idea Group Reference.

Lin, Y.-W. (2006b). Techno-feminist view on the open source software development. In E. M. Trauth (Ed.), *Encyclopedia of gender and information technology* (pp. 1148-1153). Hershey, PA: Idea Group Reference.

Mackenzie, A. (2001). Open source software: When is a tool? What is a commodity? *Science as Culture, 10*(4), 541-552.

Meyer, G. R. (1989). *The social organization of the computer underground*. Master's thesis, Northern Illinois University.

Moody, G. (2001). *Rebel code: Inside Linux and the open source revolution*. Cambridge, MA: Perseus Publishing.

Nissenbaum, H. (2004). Hackers and the contested ontology of cyberspace. *New Media & Society, 6*(2), 195-217.

Raymond, E. (1999). *The cathedral & the cazaar: Musing on Linux and open source by an accidental revolutionary*. Sebastopol, CA: O'Reilly. Retrieved July 7, 2006, from http://www.catb.org/~esr/writings/cathedral-bazaar/

Raymond, E. S., & Steele, G. L. (1991). *The new hacker's dictionary.* Cambridge, MA: MIT Press.

Rosenberg, D. K. (2000). *Open source: The unauthorized white papers.* Hoboken, NJ: John Wiley & Sons.

Ross, A. (1991). *Strange weather: Culture, science and technology in the age of limits.* London: Verso.

Sismondo, S. (2004). *Introduction to science and technology studies.* Oxford: Blackwell.

Skibell, R. (2002). The myth of the computer hacker. *Information, Communication & Society, 5*(3), 336356.

Taylor, P. A. (1999). *Hackers: Crime in the digital sublime.* London: Routledge.

Thomas, D. (2002). *Hacker culture.* University of Minnesota Press.

von Hippel, E., & von Krogh, G. F. (2003). Open source software and the "private-collective" innovation model: Issues for organization science. *Organization Science, 14,* 209-223.

Weber, S. (2004). *The success of open source.* Cambridge, MA: Harvard University Press.

Williams, S. 2002. *Free as in freedom: Richard Stallman's crusade for free software.* Retrieved July 10, 2006, from http://www.oreill.com/open-book/freedom/index.html

KEY TERMS

Editing Macros (EMACS): EMACS refers to a class of text editors, possessing an extensive set of features, that are popular with computer programmers and other technically proficient computer users. The original EMACS, a set of *Editor MACroS* for the TECO editor, was written in 1975 by Richard Stallman, and initially put together with Guy Steele. Many versions of EMACS have appeared over the years, but now there are two that are commonly used: GNU EMACS, started by Richard Stallman in 1984 and still maintained by him, and XEmacs, a fork of GNU EMACS which was started in 1991 and has remained mostly compatible. In this chapter, the development of EMACS is used to illustrate how hacker culture and the FLOSS innovation co-evolved over the development process.

Forks: In software engineering, a project fork or branch happens when a developer (or a group of them) takes a copy of source code from one software package and starts to independently develop a new package. The term is also used more loosely to represent a similar branching of any work, particularly with FLOSS. Associated with hacker culture, this chapter argues that forking usually happens because people improve the software based on their local needs which implicitly entails different interpretations and practices of what a hacker is and how to become a hacker alternatively.

Free/Libre Open Source Software (FLOSS): Generically indicates non-proprietary software that allows users to have freedom to run, copy, distribute, study, change, and improve the software.

General Public License (GPL): A free software licence that guarantees the freedom of users to share and change free software. It has been the most popular free software license since its creation in 1991 by Richard Stallman.

Hacker: According to Wikipedia, a hacker is a person who creates and modifies computer software and computer hardware including computer programming, administration, and security-related items. The term usually bears strong connotations, but may be either positive or negative depending on cultural context. However, this chapter challenges a fixed definition of hacker and suggests a look at different interpretations of hackers and practices of becoming hacker.

Hacker Culture: According to Wikipedia, hacker culture is a subculture established around hackers (see Hacker). Wikipedia lists two mainstream subcultures within the larger hacker subculture: the academic hacker and the hobby and network hacker. However, this chapter suggests that hacker culture evolves over time and new definitions always emerge through the negotiations of different interpretations of a hacker and practices of becoming a hacker in spatiality and temporariness.

Socially-Informed Algorithm: A socially-informed algorithm is a piece of algorithm that is designed and developed dependent of social, cultural, and organisational contexts. Broadly speaking, each written algorithm is both technically and socially informed because it is always shaped by the social environment where the developers situate and the technical tools are known and made available to the developers and users.

Chapter V
Social Technologies and the Digital Commons

Francesca da Rimini
University of Technology, Sydney, Australia

ABSTRACT

This chapter investigates the premise that software is culture. It explores this proposition through the lens of peer production, of knowledge-based goods circulating in the electronic space of a digital commons, and the material space of free media labs. Computing history reveals that technological development has typically been influenced by external sociopolitical forces. However, with the advent of the Internet and the free software movement, such development is no longer solely shaped by an elite class. Dyne:bolic, Streamtime and the Container Project are three autonomously-managed projects that combine social technologies and cooperative labour with cultural activism. Innovative digital staging platforms enable creative expression by marginalised communities, and assist movements for social change. The author flags new social relations and shared social imaginaries generated in the nexus between open code and democratic media. In so doing the author aims to contribute tangible, inspiring examples to the emerging interdisciplinary field of software studies. "Humanity's capacity to generate new ideas and knowledge is its greatest asset. It is the source of art, science, innovation and economic development. Without it, individuals and societies stagnate. This creative imagination requires access to the ideas, learning and culture of others, past and present" (Boyle, Brindley, Cornish, Correa, Cuplinskas, Deere, et al., 2005)

INTRODUCTION

Software—sets of programmed instructions which calculate, control, manipulate, model, and display data on computing machines and over digital networks—is culturally loaded. Whenever we load programs, we also load messy clusters of cultural norms and economic imperatives, social biases and aesthetic choices, into machines and networks whose own histories are linked to larger sociopolitical forces. Increasingly instrumental in facilitating new forms of cultural expression and social activism, software is used to connect and mobilise diverse communities, interest groups, and audiences; spanning local, regional and global levels.

New social assemblages, and new social relations, are thus arising out of software-assisted communication, collaborative production and the exchange of creative, intellectual artifacts[1] This model of autonomously-managed generative activity is termed "peer production." The knowledge-based outcomes of peer production are framed as contributing to a global "Digital Commons."[2] Just as the concept of the earthly commons centres around communally shared and managed material resources—land, trees, water, air, and so on—the Digital Commons can be imagined as shared immaterial resources. These are wildly proliferating nodes of electronic spaces, social technologies, intellectual goods, and cooperative labour processes enabled by, and manifested through, the Internet. The voluntary labour driving this phenomenon is occurring on an unprecedented scale, generating demonstrable effects on both knowledge generation and social organisation.

Chronicles of software as corporate culture abound, revealing the light and shadow of the giants, from IBM to Amazon to Google. Similarly, the rise of the free software movement, the open source software (OSS) participatory programming model, and the evolution of the Internet and then the World Wide Web, are well documented.[3] Less visible are the histories of the pixies, those nimble social technologies arising from the nexus of the free software movement, cultural activism, and new hybrid forms of peer production. Where documentation does exist, it is more likely to be within the fields of new media art, tactical media, and the emerging academic interdisciplinary field of software studies, or in project Wikis and blogs.[4]

This chapter places collaborative software development within the context of software as culture. Specifically, I examine some instances of software-assisted peer production in the cultural expression of social activism. The first part of the chapter draws attention to some sociopolitical factors that shaped the development of computing, giving an historical context to my proposition that software and culture are intrinsically interconnected. This is followed by a brief sketch of current theoretical propositions about some relationships between capitalism, computing technologies, knowledge-based labour, and network society.

In the second part of this chapter, I will identify distinguishing features of the Digital Commons, outlining the cooperative processes which enliven it. Moving from theory to practice, I will highlight three exemplary projects to illustrate the kinds of content, processes, and social relations contributing to the Digital Commons. I will introduce the *Dyne:bolic* distribution of the GNU/Linux operating system, and the *Streamtime* network for producing content in crisis areas. The *Container Project*, an open access digital media hub in Jamaica, will then be introduced. Speculation on future trends will signpost efforts to contain the circulation of knowledge and cultural material via systems of "digital enclosures." I will conclude by speculating on possible directions for social technologies, as network nodes proliferate globally, thereby increasing public spaces for creative cooperation. Increased peer participation and cultural diversification give rise to a concept of a multitude of interlinked Digital Commons. Such networked imaginative productive spaces not only could meet the challenges thrown down by the socially elite proponents of the new digital enclosures, but also prefigure possibilities for new global democratic sociopolitical forms.

BACKGROUND

The evolution of computing is woven through with histories of power, capital, and social control. Each major innovation benefited from a rich accretion of ideas and inventions, sometimes spanning centuries, cultures, and continents. Specific political imperatives (serving national or imperial interests) and wider societal forces shaped the development pathways of computing. From cog to code, information technologies have never been neutral.

The Politics of Invention

Joseph-Marie Jacquard's construction of the automated punch card loom, a proto-information technology, illustrates both strategic government patronage, and the collective, cumulative processes of invention.[5] The loom benefited from government intervention and financial support, as Napoleon recognised that the new loom could play a crucial role in achieving post-revolutionary France's economic goal to rival the industrial giant Britain. This same Jacquard loom directly inspired the English inventor Charles Babbage (himself assisted by the visionary mathematician Ada Lovelace), who made a series of conceptual and engineering breakthroughs in two mechanical systems for automated arithmetic calculation, the difference engine and the analytical engine.[6] Babbage was influenced by the ideas of the 18th century moral philosopher Adam Smith, the Scottish anti-mercantile proponent of laissez-faire economic liberalism, who proposed the idea of the systematic division of labour. Babbage envisaged his mechanical cog-and-card machines as furthering Britain's national economic interests, as trade and taxation would benefit from mathematical precision and reduced labour costs. Punch cards reappeared in the electro-mechanical binary punch card calculating machines developed by engineer Herman Hollerith in the late nineteenth century in the United States. The role of IBM in the programming of punch cards for customised demographic data collection by the Nazi regime throughout the 1930s-1940s, demonstrates what Christian Parenti (2003) terms the "informatics of genocide."[7] In the twentieth century, information technology played a dominant role in determining material and ideological power, within and between nations.

On the eve of World War II both Axis and Allies were thirsting for new mathematical engines.

The English and French needed information to decrypt the codes of Germany's Enigma machine;[8] the Americans needed information in the form of ballistic firing tables in order to accurately instruct their gunners which way to point their guns in their new generation of fast war planes;[9] and the Germans needed a machine which could rapidly process stacks of simultaneous equations to ensure that the frameworks of *their* new planes could withstand the stress of increased speed.[10] Each of these national objectives was answered by the injection of substantial government and corporate support for the boffins in the engine rooms; technological innovation in computing was sculpted by powerful external influences.

Network Society and Immaterial Labour

How did humanity reach what historian Paul Ceruzzi (2003) describes as "an age transformed by computing"?[11] Attempts to commercialise computers were made in the late 1940s; later the creation of small systems in the 1960s was followed by personal computing in the 1970s, and the rise of networked systems in the mid 1980s. The "deep recession" of the 1970s consolidated socioindustrial changes in the West so profound that they constituted a new regime of accumulation, termed late capitalism.[12] The markers were privatisation, deregulation, the growing power of transnational corporations, and globalisation—of markets, labour, finance, and communications. Globalisation itself required specific technological developments, including automation and computerisation of production processes, and the growth of networked communications (Castells, 2000; Webster, 2000).

In his three-volume opus *The Information Age: Economy, Society and Culture*, Manuel Castells (2000) describes the emergence of a network society around the end of the 20th century, characterised by the centrality of information and knowledge to the economy, and the rise of communication networks.[13] *The Rise of the Network Society* proposes that a "new social order" arises from a global system of "informational capitalism" (Castells, 2000, pp. 409, 508). The "revolutionary" era's distinguishing feature is the "action of knowledge upon knowledge itself as the main source of productivity" (Castells, 2000, p. 17), creating a "cumulative feedback loop between innovation and the uses of innovation," with "the human mind [as] a direct productive force" (Castells, 2000, p. 31). "Critical cultural battles for the new society" are played out in this "new historical environment" (Castells, 2000, p. 405). The central role played by "immaterial labour" within this network society was first articulated by Italian theorists.[14] In the essay *Immaterial Labour,* sociologist Maurizio Lazzarato (1996) describes a "great transformation" (Lazzarato, 1996) starting in the 1970s, which blurred the manual and mental binary framing of labour. He defines immaterial labour as "the labour that produces the informational and cultural content of the commodity" (Lazzarato, 1996). The commodity's informational content indicates "the changes taking place in workers' labour processes ..., where the skills [increasingly] involve ... cybernetics and computer control (and horizontal and vertical communication)" (Lazzarato, 1996). For "the activity that produces the 'cultural content' of the commodity, immaterial labour involves activities ... not normally recognized as 'work' [...] activities involved in defining and fixing cultural and artistic standards, fashions, tastes, consumer norms, and, more strategically, public opinion" (Lazzarato, 1996). No longer the privilege of a social elite, these activities have "become the domain of what we have come to define as 'mass intellectuality'" (Lazzarato, 1996). Immaterial labour is constituted "in forms *that are immediately* collective," (Lazzarato, 1996) existing "only in the form of networks and flows" (Lazzarato, 1996). In *Network Culture: Politics for the Information Age,* Tiziana Terranova (2004) takes the idea of flows to examine the productive relations flowing between the "thriving and hyperactive" Internet,

an "outernet" of social, cultural and economic networks, the "digital economy," and "free" labour. Terranova focuses on the critical role of the Internet, arguing that it "functions as a channel through which 'human intelligence' renews its capacity to produce" (Terranova, 2004, pp. 73-79). The Internet "highlights the existence of networks of immaterial labour and speeds up their accretion into a collective entity." Commodities become "increasingly ephemeral" and turn into "translucent objects," a transparency which reveals their "reliance on the labour which produces and sustains them"; it is this "spectacle of labour"—"creative, continuous, innovative"—that attracts users/consumers of these commodities (Terranova, 2004, p. 90).

MAIN FOCUS OF THE CHAPTER: KNOWLEDGE, CREATIVITY, AND SOCIAL CHANGE ON THE DIGITAL COMMONS

The fields of sociology and cultural theory are not alone in advancing theories about the social relations of information technology. Perspectives from the free software movement, media arts, the sciences, and the law are also contributing to new notions of the commons.[15] In his essay *Three Proposals for a Real Democracy: Information-Sharing to a Different Tune,* Brian Holmes (2005, p. 218) proposes:

the constitution of a cultural and informational commons, whose contents are freely usable and protected from privatization, using forms such as the General Public License for software (copyleft), the Creative Commons license for artistic and literary works, and the open-access journals for scientific and scholarly publications. This cultural and informational commons would run directly counter to WIPO/WTO treaties on intellectual property and would represent a clear alternative to the paradigm of cognitive capitalism, by conceiving human knowledge and expression as something essentially common, to be shared and made available as a virtual resource for future creation, both semiotic and embodied, material and immaterial.

In *Piratology: The Deep Seas of Open Code and Free Culture*, theorist Armin Medosch (2003) unpicks the labour processes creating the commons. He explains that the point-to-point communications principle on which the Internet is based "aids the creation of new transversal structures—communities, movements, interest groups, campaigns, discussion boards, file-sharing communities ... " (Medosch, 2003, p. 13). These autonomous groupings produce a "social dynamism, based on new types of technologically-supported collectivisations" (Medosch, 2003, p. 13). Medosch describes "commons-based peer production," which he defines as as "the production of goods and services based on resources that are held in a commons and organised by peers," as now having reached a "critical mass" (Medosch, 2003, p. 15).

Crucially, this has occurred "right in the centre of Western societies, within the most advanced areas of production" (Medosch, 2003, p. 15). As evidenced by pan-continental gatherings; the activity on free software lists in Latin America, India, Asia, and Africa; the blogging movements in Iran and Iraq; and the adoption of free software by various governments, I would argue that concurrent swells of participatory media are also forming in non-Western societies spanning various stages of industrialisation.[16] Medosch proposes that "without explicitly formulating itself as oppositional, this nondescript movement of movements slowly but inevitably changes society from within" (Medosch, 2003, p. 15).

The historical importance of this trend is echoed by other commentators. James Boyle (2002) describes the Internet as "one big experiment in distributed cultural production." For Free Software Foundation legal counsel Eben Moglen (2003), "the movement for free information announces the arrival of a new social structure,

born of the transformation of bourgeois industrial society by the digital technology of its own invention." Castells (2000) views the technological transformation of media participation as being of "historic dimensions," likening it to the "new alphabetic order" of the ancient Greeks, which "provided the mental infrastructure for cumulative, knowledge-based communication." Hypertext and a "meta-language" integrate oral, textual, aural and visual modalities into one system of communication, which reunites the human spirit in "a new interaction between the two sides of the brain, machines, and social contexts" (Castells, 2000, pp. 355-356).

Knowledge work "is inherently collective, it is always the result of a collective and social production of knowledge," according to Terranova (2004, p. 80). The General Public License (GPL) conceived by Richard Stallman[17] and taken up widely by free/libre open source software (FLOSS) developers, is a legal mechanism ensuring that information about software source codes remains open and unprivatised. The Free Software Foundation explains that:

The GPL builds upon the ethical and scientific principle of free, open and collaborative improvement of human knowledge, which was central to the rapid evolution of areas like mathematics, physics, or biology, and adapts it to the area of information technology.[18]

The GPL was later applied to other kinds of cultural goods, providing a framework for discussions around the role of knowledge in information society. It also inspired the open content licensing system, Creative Commons (CC).[19] Creative Commons offers a spectrum of copyright or copyleft protections which can be assigned to a wide range of content types such as film, music, and texts before they enter the public realm.[20]

Internet-assisted systems of knowledge exchange recall the ideas of educator Ivan Illich (1971) in his prescient book *Deschooling*

Society. Decades before the Internet became a popular medium, Illich proposed that the "un-hampered participation" of individual active subjects informing and empowering themselves "in a meaningful setting" via mechanisms such as autonomously organised learning webs, skill exchanges and networks of peers was fundamental to societal transformation (Illich, 1996, p. 39, pp. 76-97). Twenty years later, by conceiving and then crafting the ingenious marriage of hyper-text and computer networks, Tim Berners-Lee created the World Wide Web, thereby gifting the Internet with a user-friendly means of cre-ating self-managed electronic learning webs.[21] Scalability—the virtually infinite capacity of the Internet to add interconnected nodes of communicable content to itself—means that the Digital Commons is potentially boundless. It is constrained mainly by technical issues, such as bandwidth availability, and economic factors such as access costs. Limits to the constitution of the commons are more likely to be social in nature. Common land is bounded by hedges, fences, or historical memory, and its resources cooperatively accessed and managed by agreed upon rules. Similarly, the Digital Commons is a self-managed web of systems that follows protocols "defined by the shared values of the community sharing these resources" (Kluitenberg, 2003).[22]

Kluitenberg (2003, p. 50) stresses the hybrid and fluid qualities of the democratic media sys-tems created by "artistic and subversive media producers." According to him, the:

Successful mediator needs to be platform inde-pendent, ... able to switch between media forms, cross-connect and rewire all platforms to find new communication spaces ... they become tools to break out of the marginalised ghetto of seldomly visited websites and unnoticeable live streams.

An example of this approach is the Media Shed project by the acclaimed art group Mongrel.[23] Operating from a light industrial shed in the economically impoverished city of Southend, Eng-land, Mongrel collaborated with the local Linux Users Group to run software training sessions, assist community-generated digital art projects, and establish an Internet radio station. All projects used recycled electronic hardware, free and artist-made multimedia, and social softwares. The Me-dia Shed charter reflects Mongrel's long history of making "socially engaged culture," and resonates with the ideals expressed by similar hybridised collaborations on the Digital Commons. It aims:

To research, create and promote communica-tion through free-media technologies outside the monetary and licensing control of proprietary systems, to assist the free flow of information, self education and opinions, to encourage creative expression and to contribute to and explore the issues that are important to the everyday lives of individuals, communities and diverse cultures in a pluralist society. (Mongrel 2006)

Some discourses foreground the radical cul-tural potential of the Digital Commons, and the social agency of its "immaterial labourers." *The Delhi Declaration of a New Context for New Media* (World Information Cities, 2005) speaks of a "vigorous cluster of practices of ongoing cultural transaction within and outside formal commodity relations" which guarantees cultural diversity. Medosch (2003a) depicts artist/coders as being "at the heart of a cultural struggle" be-cause they "carry forward the cultural politics of code by supporting the foundations for the preservation and renewal of culture" (Medosch, 2003a, p. 16). With the digital tools they make, "the artist/coders liberate culture from the grips of the culture industries ... creat[ing] platforms for social experimentation" (Medosch, 2003a, p. 16). A related set of practices can be grouped un-der the umbrella of electronic civil disobedience. Jordan and Taylor (2004) describe practitioners of "hacktivism" as seeking to "re-engineer systems in order to ... confront the overarching institutions of twenty-first-century societies" (Jordan & Taylor, 2004, p. 29).[24]

Brian Holmes (2003) identifies the progressive re-engineering of public knowledge and the social imaginary in his text *Cartography of Excess*, referencing Internet-based mapping projects such as *TheyRule*, a detailed investigation into American corporate boardroom power relations.[25] Holmes opines that:

Far beyond the computer logic of open source, the great alternative project of the last decade has been mapping the transnational space invested primarily by corporations, and distributing that knowledge for free. This is the real power of "spontaneous cooperation" in a global information project like Indymedia. (p. 65)

Such projects are valuable because they make the rules of the neoliberal economy visible to a point where "we can start imagining—or exploring—a radically different map of the planet again." (Holmes, 2003, p. 65)

Social Softwares as Social Technologies: Dyne:bolic, Streamtime, and the Container Project

Creating the material circumstances to enable the democratic exchange of imagination and information is a driving factor in numerous projects on the Digital Commons. Dyne:bolic, Streamtime and the Container Project are three such examples, employing free and social softwares as tools for creative expression, social activism, and cultural transformation. If we consider the Digital Commons to be the macrostructure, then social software can be thought of a set of microsystems within this framework.[26] Matthew Fuller describes social software as:

Primarily ... built by and for those of us locked out of the narrowly engineered subjectivity of mainstream software. It is software which asks itself

what kind of currents, what kinds of machine, numerical, social, and other dynamics, it feeds in and out of, and what others can be brought into being. ... It is ... directly born, changed, and developed as the result of an ongoing sociability between users and programmers in which demands are made on the practices of coding that exceed their easy fit into standardised social relations. (Fuller, 2003, p. 24)

Dyne:bolic, a live bootable distribution of the GNU/Linux operating system, is a good example of Fuller's model of social software. Released under the GPL, it has the bonus of "a vast range of software for multimedia production [...] ready for being employed at home, in classrooms and in media centers" which have been made by "hundreds of programmers all around the world" (Jaromil 2006).[27] In order to ensure the widest spectrum of people and machines can access Dyne:bolic, it has been optimised to run on older machines. Compare this with the OS releases from the proprietary vendors—could an Apple SE circa 1995 run OS10 for example? Completely rewritten in 2005 as the "DHORUBA" release, lead developer Jaromil (2006) announces that the project is already planning its next stage, which will be "a cross-platform build environment to cover all kinds of hardware around."

In an undated text entitled *This is Rasta Software*, Jaromil links Dyne:bolic with revolutionary social movements, proclaiming:

This software is for all those who cannot afford to have the latest expensive hardware to speak out their words of consciousness and good will. This software has a full range of applications for the production and not only the fruition of information, it's a full multimedia studio ... because freedom and shar[ing] of knowledge are solid principles for evolution and that's where this software comes from ... This software is free as of speech and is one step in the struggle for Redemption and Freedom. ... (Jaromil, 2005)

Dyne:bolic and many other free software has arisen out of the Italian autonomous hackmeeting and hack labs scene.[28] Whilst "trying to recover the essence of the first hackers from MIT," these are outcomes of a significantly different cultural context to that of the liberal and libertarian interpretations of freedom characterised by American discourse (Nmada & Boix, 2003). The *social and communal* end uses and empowering possibilities of the software are valorised, more than the *individual's* right to examine and share source code. This is cooperatively-made software to "let people be Free to Create" (Jaromil 2006).[29]

The Streamtime project, a collaboration between Radio Reedflute and Rastasoft, applies this principle, gathering up free software such as Dyne:bolic from the Digital Commons, to assist the building of "autonomous networks in extreme conditions."[30] Streamtime describes itself as "a handshake in cyberspace, a hanging garden for dialogue and cooperation, generated by a sense of solidarity, hospitality and a desire to communicate and relate." An initiative of Dutch media activist Jo van der Spek, the communication platform enables self-production of media, such as low-tech wireless radio networks to stream local content. It hosts a meta-blog linking to multi-lingual chronicles of life in wartime situations in Iraq and Lebanon, audio archives (poetry and interviews), and links to other DIY media resources. Streamtime's Mission Statement explains:

Streamtime uses old and new media for the production of content and networks in the fields of media, arts, culture and activism in crisis areas, like Iraq. Streamtime offers a diffuse environment for developing do-it-yourself media. We focus on a cultural sense of finding your own way in the quagmire that is Iraq, and its representation in the global media. We should not try to change politics in order to foster cultural change; we should support cultural manifestation in order to force political change.

The Container Project, initiated by Jamaican artist Mervin Jarman, is a more localised example of cultural intervention using social technologies.[31] Mervin wanted to take "creative computer technology to ghetto people and deep rural communities in the Caribbean," so that "kids growing up in the ghettos of Jamaica [could] realize they can 'fulfill their wildest dreams'" (de Silva). The project of "technological repatriation" was inspired by Mervin's own youthful experiences of poverty, his later journeys into video making, and his participation in a digital media training program for the socially disadvantaged at ARTEC in London. He sees the Container as a "revolutionary project" that challenges the existing social order of endemic poverty, by using under-recognised rich local cultural traditions and talent to generate new entrepreneurial systems of reciprocal exchange and opportunity (Fuller, 1999).

In 2003 a volunteer team came to the Jamaican village of Palmers Cross to help realise Mervin's vision. The group converted a shipping container into a media lab housing sixteen networked computers running three operating systems (GNU/Linux, Mac, and Windows), all connected to a Linux server. The lab included a purpose-designed dedicated multimedia suite, and machines hosting a mix of proprietary and free software programs, including artist-made social softwares. Mervin used his intimate knowledge of his community's dynamics when designing the Container's architecture. The bright yellow structure was opened up with large kiosk-style windows, inviting people to get to know what was happening at their own pace. The Container Project fulfilled a community need for a social hub. Originally envisaged as a mobile lab, the local people have been reluctant to let it leave their village. Its educational and cultural exchange programs addressed a range of needs, from general computer skills to the sharing of creative talents with a world audience. Mervin views this community empowerment as a global issue, explaining:

That's why I think the Container is such an incredible and revolutionary project because it allows street-level emergence into what would be an otherwise unchallenged consortium of global culturalisation and then where would we be? What would happen to our dynamics as it relates to production, be that in the Music, Art and Craft, in the way we conduct businesses, and develop our own customized software to satisfy our specifics? ... No system should impose its will and/or cultural identity on another, the only way for software and technology to be truly dynamic is to decentralize the decision making process, open up the formats to customization on a more trans-culture and gender context. (Fuller, 1999)

In 2004, a broadband connection linked the Container Project to the wider world, and its electronic Post Office Box (e-POB) was an immediate success, tapping into fundamental communication needs. In June 2005 the young musicians and singers of the village participated in the SkintStream project, a network connecting "audiences and cultural spaces previously separated by economic, geographical, and political factors." A temporary "Poor-to-Poor" streaming radio channel was established, linking creative communities in Palmers Cross, a shanty town in Johannesburg, a diaspora Congolese community in London, a public housing community in Toronto, and young musicians in Southend.[32] It was the first time that most of the participants had performed their creative works to outside audiences, and the level of excitement with the experience was reportedly very high. SkintStream embodies one of the goals around cultural empowerment stated on the Container Project Web site—to demonstrate to people in remote and isolated communities that they too "can contribute to the future, that they will have a place in the millennium."

In March 2006, the Container Project hosted a Community Multimedia Centre Management Workshop.[33] The three week event included a Digital Storytelling Workshop, and the creation of a temporary recording studio. Based on a knowledge-sharing model, guest artists and teachers passed on technical, creative, leadership, and training skills to ten workshop participants, giving the students the ability to replicate the program themselves in other communities. The Container Project team are now working closely with local organisation ICT4D Jamaica to deliver workshops under the "community without borders" concept.[34] As Mervin explained, this "fulfills the Container mobility needs, only we move people into Palmers Cross so they get the whole ambient of what it feels like to be part of the Container family."[35] Two projects in the planning stage are the creation of a community Internet radio portal for the region, and mongrelStreet lab, a portable lab made out of wheelie bins.

Like Dyne:bolic and Streamtime, the Container Project harnesses social technologies with creative expertise to create a platform for cultural expression and exchange for disenfranchised communities. These are just three of a multitude of similar projects occurring around the world. Visible social change is happening on grassroots local levels, and ideas and project-generated content are feeding back into a multiplicity of interconnected Digital Commons. This emergent phenomenon could herald widespread social change based on the new shared social imaginaries which are being generated.

FUTURE TRENDS

Mongrel (2004) proposes that when "new common cultural spaces open up in the public domain as they did with the Internet in the 1990s, those with the proprietary right or economic might, usually attempt enclosure." Commodification and privatisation of natural and public resources and spaces present a significant challenge to the commons, earthly and electronic.[36] The various processes through which attempts are made to privatise the Digital Commons are termed the "digital enclosures." In response, new alliances of free software developers, legal and cultural activists are gathering to protect, and extend, the freedom of the commons. Two recent examples of the digital enclosures include the failed legislative bid by the European Parliament to impose software patents,[37] and the impositions of the United States' Digital Millennium Copyright Act (DMCA).[38] Battles on the contested ground of intellectual property are intensifying as the United States pressures its trading partners to "adopt laws modelled on the DMCA as part of bilateral and multilateral trade agreements" (von Lohmann, 2004).

James Boyle (2002) warns that intellectual property rights (IPR) threaten the "commons of mind," stating that "things ... formerly thought to be uncommodifiable, essentially common, or outside the market altogether are being turned into private possessions under a new kind of property regime. But this time the property ... is intangible, existing in databases, business methods, and gene sequences." He notes that, unlike a common tract of land which can be overexploited,[39] the "digitized and networked *commons of the mind*" is not depleted or destroyed by being mutually shared. Due to the fragmentary nature of information products, all fragments "can function as raw material for future innovation" (Boyle, 2002).

Despite the threat of the enclosures the Digital Commons is expanding, as peer production of democratic media projects, cultural activism, and art proliferate. The Internet is the key enabling technology underpinning the commons, and all figures point to the exponential growth of the net, especially in the global South.[40] This creates a more culturally-diverse, socially inclusive, and globalised network society, and it is unlikely that the new swarms of activity will recede or wither. These nonlinear clusters of social technologies and projects resonate with fundamental human desires to communicate, to create, to work cooperatively and collectively, and to exchange elements of ourselves and our cultures.

Empirical research is needed to analyse these new phenomena. Comprehensive documentation of a spectrum of projects energising the Digital Commons will contribute to an understanding of what is common (and different) about these projects' cooperative labour processes, their technological innovation, the new systems of cultural and social exchange developing, and the challenges faced by participants. Multiple-language translations of project documentation and case studies would offer important cross-cultural perspectives. Qualitative research would ground more speculative work, such as considerations about the shifts in social imaginaries resulting from these experiments in production and social relations. Indeed, learning how such imaginative shifts are being played out in material projects and networks could reveal unfolding global patterns and flows.

CONCLUSION

The idea that all humanity is living in a global age of advanced neoliberal capitalism, with its interconnected communicative flows of data, finances and labour is no longer new; Marshall McLuhan and others were channelling the information revolution spirits some 40 years ago.[41] In contrast, discourses around network society, knowledge work, immaterial labour, and software as culture, are still in their infancy, and the language is sometimes esoteric, or opaque. Fortunately practice outstrips theory on the Digital Commons, as new hybrid collaborations of peer production and social activism are creating democratic public spaces for communication and creativity, and generating new systems of exchange. In these contexts, far away from the Google campus, cooperation displaces competition, and the creation of shared frameworks of knowledge and action provides traction for local, regional, and transnational social change.

There is no unitary or abstract Digital Commons, but rather a multiplicity of Digital Commons across the North-South power axis. In this new millennium voices from the "Fourth World" or

"Global South" are entering the network flows, forming new autonomous networks and creative laboratories, further transforming the praxis. Their discourses emphasise software freedom as being intrinsically related to free culture,[42] community empowerment, traditional indigenous knowledge[43] and social rights. The decision by the Brazilian government to use only open source software, and to establish 1,000 free software and free culture centres in the poorest parts of the country, is directly linked to a radical social vision which is challenging knowledge ownership laws from pharmaceutical patents to file sharing. In the words of Brazilian Minister of Culture and acclaimed musician Gilberto Gil, "if law doesn't fit reality anymore, law has to be changed. ... That's civilisation as usual" (Burkeman, 2005).

And just beneath civilisation lies the unknown, the realm of spectres and magic and transformation. What is a spell if not a set of programmed instructions to create change? Open code is transforming society subtly, as social technologies are being cooperatively built, shared, and used in a deeply networked, informatised, immaterial, cultural space—the "collective subjectivity" of the Digital Commons (Dafermos, 2005).[44] The Free Software Movement has provided the impetus for the evolution of numerous thriving ecosystems, and rich hybridised sites of cultural production. The enthusiastic embrace by the "Fourth World" of free software is one sign, amongst many others, that social change on an unprecedented scale is afoot. The immaterial spaces created by networked imaginations could offer us all vital keys to comprehending such change.

REFERENCES

Agar, J. (2001). *Turing and the universal machine: The making of the modern computer.* Cambridge: Icon Books UK.

Berners-Lee, T., & Fischetti, M. (1999). *Weaving the Web.* London: Orion Business Books.

Black, E. (2001). *IBM and the holocaust: The strategic alliance between Nazi Germany and America's most Powerful corporation.* New York: Crown Publishers.

Bollier, D. (2002). *Silent theft: The private plunder of our common wealth.* New York: Routledge.

Bosma, J., Van Mourik Broekman, P., Byfield, T., Fuller, M., Lovink, G., McCarty, D., et al. (Eds.), (1999). *Readme! filtered by Nettime: ASCII culture and the revenge of knowledge.* New York: Autonomedia.

Boyle, J. (2002). *Fencing off ideas: Enclosure and the disappearance of the public domain.* Retrieved August 18, 2005, from http://centomag.org/essays/boyle

Boyle, J., Brindley, L., Cornish, W., Correa, C., Cuplinskas, D., Deere, C., et al. (2005). *Adelphi Charter on creativity, innovation and intellectual property.* Retrieved November 8, 2005, from http://www. adelphicharter.org

Burkeman, O. (2005, October 14). Minister of counterculture. *The Guardian.* Retrieved March 28, 2007, from http://technology.guardian.co.uk/news/story/0,16559,00.html

Castells, M. (1998). *End of millenium* (2nd ed., Vol. 3). Oxford: Blackwell.

Castells, M. (2000). *The rise of the network society* (2nd ed., Vol. 1). Oxford: Blackwell.

Ceruzzi, P. E. (2003). *A history of modern computing* (2nd ed.). Cambridge, MA: MIT Press.

Critical Art Ensemble. (1994). *The electronic disturbance.* New York: Autonomedia.

da Rimini, F. (2005). *Grazing the Digital Commons: Artist-made social softwares, politicised technologies and the creation of new generative realms.* Unpublished master's thesis, University of Technology, Sydney, Australia.

Dafermos, G. N. (2005). *Five theses on informational—Cognitive capitalism.* Retrieved November 28, 2005, from http://www.nettime.org/Lists-Archives/nettime-l-0511/msg00103.html

Davis, M. (2000). *The universal computer: The road from Leibniz to Turing.* New York: W. W. Norton.

de Silva, S. (n.d.). *Desperately seeking Mervin.* Retrieved March 14, 2005, from http://www.the-paper.org.au/024/024desperatelyseekingmervin.html

Essinger, J. (2004). *Jacquard's Web: How a hand loom led to the birth of the information age.* Oxford, UK: Oxford University Press.

Fitzpatrick, A. (1998). Teller's technical nemeses: The American hydrogen bomb and its development within a technological infrastructure. *Society for Philosophy and Technology, 3*(3).

Fuller, M. (1999). *Mervin Jarman - The Container.* Retrieved January 2005, from http://www.nettime.org/Lists-Archives/nettime-l-9906/msg00138.html

Fuller, M. (2003). *Behind the blip: Essays on the culture of software.* New York: Autonomedia.

Grattan-Guinness, I. (1990). Work for the hairdressers: The production of de Prony's logarithmic and trigonometric tables. *Annals of the History of Computing, 12*(3), 177-185.

Hauben, R. (n.d.). *History of UNIX: On the evolution of Unix and the automation of telephone support operations (i.e., of computer automation).* Retrieved November 7, 2005, from http://www.dei.isep.ipp.pt/docs/unix.html

Hauben, M., & Hauben, R. (1995). *Netizens: On the history and impact of the Net.* Retrieved November 7, 2005, from http://www.columbia.edu/~hauben/netbook/

Holmes, B. (2003). Cartography of excess. In T. Comiotto, E. Kluitenberg, D. Garcia, & M. Grootveld (Eds.), *Reader of the 4th edition of next 5 minutes* (pp. 63-68). Amsterdam: Next 5 Minutes.

Holmes, B. (2005). Three proposals for a real democracy: Information-sharing to a different tune. In M. Narula, S. Sengupta, J. Bagchi, & G. Lovink (Eds.), *Sarai Reader 2005: Bare acts.* Delhi: Sarai.

Illich, I. (1971, 1996). *Deschooling society.* London: Marion Boyars.

Jaromil. (2005) *This is rasta software.* Retrieved November 13, 2005, from http://dynebolic.org/manual-in-development/dynebolic-x44.en.html

Jaromil. (2006). *Dyne:bolic 2.1 codename DHORUBA.* Retrieved July 13, 2006, from http://nettime.org

Jordan, T., & Taylor, P. A. (2004). *Hacktivism and cyberwars: Rebels with a cause?* London: Routledge.

Kluitenberg, E. (2003). Constructing the Digital Commons. In T. Comiotto, E. Kluitenberg, D. Garcia, & M. Grootveld (Eds.), *Reader of the 4th edition of next 5 minutes* (pp. 46-53). Amsterdam: Next 5 Minutes.

Lazzarato, M. (1996). *Immaterial labor.* Retrieved August 11, 2005, from http://www.generation-online.org/c/fcimmateriallabour3.htm

Lessig, L. (2004). *Free culture: How big media uses technology and the law to lock down culture and control creativity.* London: Penguin.

Liang, L. (2004). *Guide to open content licenses.* Rotterdam: Piet Zwart Institute.

Linebaugh, P., & Rediker, M. (2001). *The many-headed hydra: Sailors, slaves, commoners, and the hidden history of the revolutionary Atlantic.* Boston: Beacon Press.

Lovink, G. (2002). *Dark fiber.* Cambridge, MA: The MIT Press.

Mantoux, P. (1905, 1983). *The industrial revolution in the eighteenth century: An outline of the beginning of the modern factory system in*

England (rev. ed.). Chicago; London: University of Chicago Press.

Medosch, A. (2003). Piratology: The deep seas of open code and free culture. In A. Medosch (Ed.), *Dive*. London: Fact.

Medosch, A. (2005). Roots culture: Free software vibrations "inna Babylon." In M. Narula, S. Sengupta, J. Bagchi, & G. Lovink (Eds.), *Sarai Reader 2005: Bare acts*. Delhi: Sarai.

Meikle, G. (2002). *Future active: Media activism and the Internet*. Sydney, Australia: Pluto Press.

Midnight Notes Collective. (1990). The new enclosures. *Midnight Notes, 10*.

Moglen, E. (1999). Anarchism triumphant: Free software and the death of copyright. *First Monday, 4*(8).

Moglen, E. (2003). *The dotCommunist Manifesto*. Retrieved June, 2005, from http://moglen.law.columbia.edu/

Mongrel. (2004a). *BIT_COMMON <=> CODE_OF_WAR*. Retrieved November 30, 2005, from http://www.scotoma.org/notes/index.cgi?MonsterUpdate3

Mongrel. (2004b). *About Mongrel*. Retrieved August 15, 2005, from http://www.mongrelx.org/home/index.cgi?About

Mongrel. (2006). *Free-media*. Retrieved July 22, 2006, from http://dev.mediashed.org/?q=freemedia

Moody, G. (2001). *Rebel code: Linux and the open source revolution*. London: Allen Lane, The Penguin Press.

Nmada, & Boix, M. (2003). *Hacklabs, from digital to analog*. Retrieved February 2, 2006, from http://wiki.hacklab.org.uk/index.php/Hacklabs_from_digital_to_analog

Parenti, C. (2003). *The soft cage: Surveillance in America from slave passes to the war on terror*. New York: Basic Books.

Plant, S. (1997). *Zeroes + Ones: Digital women + the new technoculture*. London: Fourth Estate.

Rheingold, H. (2000). *Tools for thought: The history and future of mind-expanding technology* (2nd ed.). Cambridge, MA: The MIT Press.

Stallman, R. (2005, June 20). Patent absurdity. *The Guardian*.

Swade, D. (2000). *The cogwheel brain: Charles Babbage and the quest to build the first computer*. London: Little, Brown and Company.

Terranova, T. (2004). *Network culture: Politics for the information age*. London: Pluto Press.

Toner, A. (2003). *The problem with WSIS*. Retrieved November 8, 2005, from http://world-information.org/wio/readme/992006691/1078414568/print

Toole, B. A. (Ed.). (1992). *Ada, the enchantress of numbers: A selection from the letters of Lord Byron's daughter and her description of the first computer*. Mill Valley, CA: Strawberry Press.

von Lohmann, F. (2004). Measuring the Digital Millennium Copyright Act against the Darknet: Implications for the regulation of technological protection measures. *Loyola of Los Angeles Entertainment Law Review, 24*, 635-650.

Weber, S. (2004). *The success of open source*. Cambridge, MA: Harvard University Press.

Webster, F. (2002). *Theories of the information society* (2nd ed.). London: Routledge.

Williams, S. (2002). *Free as in freedom: Richard Stallman's crusade for free software*. Sebastopol, CA: O'Reilly & Associates.

World Information Cities. (2005). The Delhi declaration of a new context for new media. In *IP and the City: Restricted Lifescapes and the Wealth of the Commons* (p. 15). Vienna: World-Information City.

KEY TERMS

Digital Commons: A conceptual framework for considering the common wealth of intellectual goods, knowledge products, creative works, free software tools, shared ideas, information, and so on which are freely and democratically shared, and possibly further developed, via the Internet

Free/Libre Open Source Software (FLOSS): A convenient acronym for "free libre open source software." It neatly bundles the revolutionary associations of "free (libré) as in freedom" together with the more technical and neutral connotations of "open source." The term implicitly acknowledges that differences between the two camps exist, but they are operational in the same field.

Free Software (FS): Software in which the underlying code is available to be inspected, modified, shared, with the proviso that it remains open, even following modification. To ensure it remains open, free software is distributed under the General Public License (GPL) or similar legal agreements.

Free Software Movement: The philosophical and political context underpinning the creation of free software, and the subjective sense of community shared by developers and users.

Immaterial Labour: A theoretical framing of knowledge work, labour processes, and social relations in information society, initially articulated by Italian theorists including Maurizio Lazzarato and Christian Marazzi.

Open Source Software (OSS): A strategic business-friendly "rebranding" of free software emphasising the practical benefits of the model of participatory software development and open code, and downplaying the original ideological and philosophical positions.

Peer Production: A horizontal, distributed method of cooperative, creative labour, gener-

ally facilitated by high levels of communication, information, and file sharing via the Internet.

Social Software: The term came out of the nexus between cultural and social activism, art and tactical media, and was originally used to designate software that came into being through an extended dialogue between programmers and communities of users, ensuring that the software was responsive to user needs. The phrase no longer carries the same import, as it is now applied to software-assisted social networking platforms such as MySpace.

Social Technologies: An umbrella term which could include free software, social software, recycled electronic equipment in free media labs, and so on. Technology put to use by the people, for the people.

ENDNOTES

[1] In his seminal book *Behind the Blip: Essays on the Culture of Software,* Matthew Fuller (2003) proposed that computers are "assemblages," combining technical, mathematical, conceptual and social layers. Through a process of critical examination we can better understand "the wider assemblages which they form and are formed by" (Fuller, 2003, p. 21). According to Fuller, software creates sensoriums, "ways of seeing, knowing and doing in the world that at once contain a model of that part of the world it ostensibly pertains to, and that also shape it every time it is used" (Fuller, 2003, p. 19).

[2] The Digital Commons is often discussed with reference to the changing of common land usage since medieval times. For example, eighteenth century England was "marked by the co-existence and close association between small agricultural production and small industrial production," and "the commons" referred to bounded parcels of

land which were available to be used by the local yeomanry and tenants (gleaned and gathered, cultivated, hunted, and traversed for reaching other destinations) under agreed upon protocols (Mantoux, 1983, pp. 137-139; Linebaugh & Rediker, 2000, pp. 22-26). Collective ownership and usage rights of land underlies "the clachan, the sept, the rundale, the West African village, and the indigenous tradition of long-fallow agriculture of Native Americans—in other words, it encompassed all those parts of the Earth that remained unprivatised, unenclosed, a noncommodity, a support for the manifold human values of mutuality" (Linebaugh & Rediker, 2000, p. 26).

3 The emergence of unwelcome proprietorial directives at MIT in the early 1980s inspired hacker Richard Stallman to begin work on a system enabling the free circulation of technical knowledge in the field of software. Thus began the GNU (a recursive shortening of "Gnu's Not Unix") project, which eventually resulted in the GNU/Linux operating system. The subjective sense of belonging to a global programming community which grew up around the various free software projects was fostered by an early social software—the newsgroup medium, a free, bandwidth-light, subject-based communication environment. The participatory programming method that benefited the GNU/Linux development model was enabled by the Internet, a medium in which everyone could communicate, and exchange software modules, with no geographical or timezone barriers. A comprehensive history of FLOSS (free, libré open source software) has been documented by Glyn Moody (2001) in *Rebel Code: Linux and the Open Source Revolution*. Sam Williams (2002) provides a detailed account the birth of the Free Software Movement in *Free as in Freedom: Richard Stallman's Crusade for Free Soft-*

ware. Steven Weber's (2005). *The Success of Open Source* posits open source as a "political economy," and provides perspectives on how the phenomenon functions on micro and macro levels. The website and community around www.slashdot.org is a central Anglophone forum for technically-focused discussion. *FirstMonday* is a refereed online journal focusing on FLOSS and associated cultural issues www.firstmonday.org.

4 Documentation and critique of more culturally focused software projects can be found in anthologies such as *Readme!* (1999), *Dark Fiber* (2002), *Anarchitexts: Voices from the Global Digital Resistance* (2003) and the annual *Sarai Reader* (2001-2005); and in mailing lists such as www.nettime.org. See also Fuller, 2003; Medosch 2005; da Rimini, 2005.

5 The punch card was the "software," a self-feeding set of pattern instructions, which was fed into, and controlled, the fixed loom "hardware." Different sets of punch cards could be fed into the same loom, resulting in different "outputs," patterned lengths of material. The automation of weaving processes caused the disappearance of certain jobs; specifically, the punch card completely replaced the work of the draw boy. See Essinger's (2004) fascinating account.

6 For various reasons Babbage's machines were never built beyond prototype stage in his lifetime. Illuminating histories of Babbage, Lovelace, and the Engines are to be found in Toole, 1992; Plant, 1997; Swade, 2000; and Essinger, 2004. Swade also documents the recent building of a Babbage engine from original plans.

7 See the authoritative account by Edwin Black (2004).

8 In 1937, the young English mathematics student, Alan Turing, "imagined a machine that could be used to generate complex numbers ... a *symbol-manipulating machine*" (Agar,

2001, pp. 8889, italics in original). These thought experiments generated the concept of a Universal Turing Machine, that is, "any stored-program computer [which] can be programmed to act as if it were another" (Ceruzzi, 2003, p. 149). See *Computing Machinery and Intelligence* (Turing, 1950) at www.cse.msu.edu/~cse841/papers/Turing.html. During World War II, Turing worked as a code-breaker at the Code and Cypher School at Bletchley Park, the centre of the Allies' efforts to decrypt Germany's Enigma machines. Later Turing worked with the first general purpose electronic computer, the "experimental monster" nicknamed the "Blue Pig," built in 1948 at Manchester University. The Atlas, a later version built in 1962, used a "hierarchy of memories, each slower but larger than the one below it," that "gave the user the illusion of a single-level fast memory of large capacity." This beast was "one of the most influential on successive generations" of computers (Davis, 2000, pp. 177-197; Agar, 2001, pp. 120-122; Ceruzzi, 2003, p. 245).

[9] In the build-up to the United States' entry to World War II, American mathematician Howard Aitken was funded by the U.S. Navy, and supported by IBM's machines and expertise, to construct a modern version of Babbage's Difference Engine. The Automatic Sequence Controlled Calculator, renamed Harvard Mark 1, "churn[ed] out numbers for naval weapon design." Simultaneously, "a second monster was under construction ... the Electronic Numerical Integrator and Computer—the ENIAC ... also born of speed and conflict." ENIAC's creators, physicist John W. Mauchly and J. Presper Eckert, were funded by the U.S. Army to build a "monster calculator." The army "was desperate" for a machine which would be able to rapidly process the complex simultaneous equations needed to produce ballistic tables for the new anti-aircraft guns. Finished in 1945 the ENIAC missed the war, but was soon employed for other military tasks, including thermonuclear bomb calculations for the nascent science of nuclear physics (Fitzpatrick, 1998; Agar, 2001, pp. 53-61).

[10] Engineer Konrad Zuse was employed by the Henschel aircraft company during the rearmament of Germany in the mid 1930s. Pressured to hasten production of its new, fast military planes, Henschel was hampered by the time needed for vital mathematical calculations to ensure fuselage and wing stability. Because there were up to thirty unknowns in these calculations, they were best solved by simultaneous equations, taking a team of mathematicians weeks of labour. Zuse realised that these complex processes could be mechanised, if there was a calculator which could read a "plan" or script giving the order of the equations to be sequentially calculated. Zuse's great intellectual contribution was to conceive of using binary numbers for the plan, machinic memory and calculations. In 1938 Zuse built a prototype, the Z3, at home with the help of friends, including Helmut Schreyer, a Nazi and hobbyist projectionist. The binary plan was punched into celluloid film reels. (Rheingold, 2000; Agar, 2001, pp. 41-52; Ceruzzi, 2003, pp. 83-84). See also *The Life and Work of Konrad Zuse*, by Professor Horst Zuse, online at www.epemag.com/zuse.

[11] This phrase is borrowed from Paul Ceruzzi's meticulous account of computing in the United States between 19452001 in *A History of Modern Computing* (2003, p. 2).

[12] This phase of capitalism is also framed as "post-Fordism," "late capitalism," and most commonly, "neoliberalism." The policies of the triumvirate of the World Bank, the International Monetary Fund, and the World Trade Organisation, are acknowledged as

determining the way this stage of capitalism is manifested in the Global North and Global South.

[13] Manuel Castells is a leading theorist on the relationships between information and society (Webster, 2001, p. 97). In *The Information Age: Economy, Society and Culture,* Castells (1998, 2000) combines empirical evidence with personal cross-cultural research to analyse the material features of informational societies, social movements arising out of network society, macropolitical affairs, and processes of social transformation.

[14] Notable theorists include sociologist Maurizio Lazzarato, the economist Christian Marazzi, Paolo Virno, and philosopher Antonio Negri. With many texts now translated into English, the concept permeates debates from free software to "precarious labour." Quotations in this paragraph are drawn from Lazzarato's 1996 essay, *Immaterial Labour,* using the English translation by *Paul Colilli and Ed Emory* at www.generation-online.org/c/fcimmateriallabour3.htm. A version of the essay is in Hardt, M. & Virno, P. (Eds.), *Radical Thought in Italy: A Potential Politics,* University of Minnesota Press, Minneapolis (pp. 133-147).

[15] Interrelated concepts of a knowledge commons, science commons, genetic commons, and creative commons are emerging from these dialogues. The Digital Library of the Commons (DLC) is a portal into an extensive collection of literature at dlc.dlib.indiana.edu. Other resources include: onthecommons.org; science.creativecommons.org; creativecommons.org; www.ukabc.org/genetic_commons_treaty.htm.

[16] Some representative examples follow. The Free Software Foundation Latin America (FSFLA) was founded in 2005. See http://mail.fsfeurope.org/mailman/listinfo/fsfla-anuncio. Africa Linux Chix is a lively Pan-African mailing list launched in 2004, active in promoting the benefits of FLOSS via conferences, networking and workshops. Blogging has driven the democratic media movement in the Middle East. Bloggers with the nicks of Salaam Pax, Raed, and Riverbend provided unique perspectives from Baghdad on the 2003 invasion of Iraq, with two collections of these chronicles later published in book form. See Pax, S. (2003), *Baghdad Blogger,* Penguin, London, and, Riverbend (2005), *Baghdad Burning: Girl Blog From Iraq,* Marion Boyars, London. See also Alavi N. (2005), *We Are Iran: The Persian Blogs,* Soft Skull Press, New York. Complementing bloggers' personal accounts are two independently-produced major websites, electroniciraq.net and www.iraqbodycount.org, providing information to English-speaking audiences. In East Asia the Interlocals project formed in 2006 as "a platform for facilitating cross-border dialogue on critical issues related to culture, gender, environment, social justice, peace, global/local politics, media movement, social movement and transformation, etc." Currently hosted by In-Media Hong Kong, content is created by a community of media activists around East Asia. See www.interlocals.net. In South Asia the Bytes for All initiative of Frederick Norhona and Parha Pratim Sarker is a platform showcasing innovative "IT for social changes practices." The Web site, e-zine, and mailing lists cover projects ranging from knowledge pipelines to rural areas to net portals for "slum-kids" to GNU/Linux rewritten in local languages. See bytesforall.org.

[17] Computing histories generally agree that the Free Software Movement—as a social movement—was initiated and steered by one individual, Richard M. Stallman (Moody, 2001; Williams, 2002; Ceruzzi, 2003). His achievements include the seminal GNU

Social Technologies and the Digital Commons

Project (Gnu's Not Unix, the heart of what became the GNU/Linux free operating system), the GPL (General Public License), and the establishment of the Free Software Foundation (FSF).

[18] Source: mail.fsfeurope.org/pipermail/press-release/2005q3/000116.html. The General Public License (GPL) is online at www.gnu.org/copyleft/gpl.html

[19] The organization was founded in 2001, with the first set of CC licenses released in December 2002. See creativecommons.org.

[20] In *Guide to Open Content Licenses*, researcher Lawrence Liang (2004) argues that the open content paradigm is a serious alternative to traditional copyright regimes that typically favour the interests of giant media conglomerates over both independent creators and the public.

[21] The World Wide Web, or WWW, is a cluster of communication protocols (HTTP), a programming language (HTML), and a universal addressing system (URL), that facilitates the exchange and display of documents on the Internet (via browser software), regardless of hardware platforms and operating systems. Developed by Tim Berners-Lee, the WWW was launched in March 1991 at the CERN facility in Switzerland (Berners-Lee & Fischetti, 1999). Berners-Lee had envisaged a "single, global information space" in 1980, unaware of key earlier projects. Vannevar Bush in the 1940s, and Ted Nelson, and Doug Engelbart, in the 1960s, are visionaries who made conceptual leaps in software, hardware interface and connectivity.

[22] Such self-management is explicit in the "softwiring" of collaborative authoring systems like WIKI. An example of "trust-based" software, WIKI is an open source database software for the shared authoring and "open editing" of Web pages. In the mid 1990s Ward Cunningham coded "the

simplest online database that could possibly work." The WIKI developers state that "allowing everyday users to create and edit any page in a Web site is exciting in that it encourages democratic use of the Web and promotes content composition by non-technical users" (Source: wiki.org). Content Management Systems (CMS) like WIKI, Dada, and Drupal offer features such as reversion to earlier instances of a document (useful when social boundaries have been transgressed by troublemaking "trolls"). These social softwares are designed with an awareness of human use (and abuse) of public space.

[23] Mongrel is an art group and a network, which formed in London in 199596. The original group comprised Graham Harwood, Matsuko Yokokoji, Mervin Jarman and Richard Pierre-Davis. Documentation of Mongrel's many acclaimed software art projects can be found at mongrelx.org. Mongrel describe themselves as: "... a mixed bunch of people, machines and intelligences working to celebrate the methods of a motley culture. We make socially engaged culture, which sometimes means making art, sometimes software, sometimes setting up workshops, or helping other mongrels to set things up. We do this by employing any and all technological advantage that we can lay our hands on. Some of us have dedicated ourselves to learning technological methods of engagement, which means we pride ourselves on our ability to programme, engineer and build our own software, while others of us have dedicated ourselves to learning how to work with people" (Mongrel, 2004b).

[24] The neologism "hacktivism" (reportedly coined by a journalist) denotes "electronic civil disobedience" or "ECD," a concept first enunciated by Critical Art Ensemble (CAE) in 1994. ECD employs tools developed by programmers and cultural activists. In their

book *Hacktivism and Cyberwars: Rebels with a Cause?* Jordan and Taylor (2004) describe hacktivism as "the emergence of popular political action ... in cyberspace [...] a combination of grassroots political protest with computer hacking" (Jordan & Taylor, 2004, p. 1). An example is the Floodnet program which enables non-destructive virtual sit-ins on government or corporate websites to draw attention to social issues (see analyses in Meikle, 2002; Jordan & Taylor, 2004).

25   *TheyRule*, an award-winning research project in the form of a dynamic website mapping the tangled web of U.S. corporate power relations, was created by Josh On and Futurefarmers at www.theyrule.net. Other projects mentioned by Holmes include the influential diagrammatic work by the late Mark Lombardi piecing together various banking and other scandals; and Bureau d'etudes *Planet of the Apes,* "a synoptic view of the world money game." See related texts at ut.yt.t0.or.at/site.

26   Due to the enormous take-up of web-based social networking platforms such as Friendster, MySpace and online dating sites the term "social software" has lost its original political edge. However, it remains a useful way of framing the social relations of software created by programmers and cultural activists.

27   First released in 2001, according to its makers Dyne:bolic was the first CD distribution of GNU/Linux operating system which did not require the user to install it permanently on their computer. Instead, the user would load the CD and it would open up into a user-friendly temporary GNU/Linux system, with additional media-making tools. See: dyne.org and dynebolic.org/manual-in-development/dynebolic-x44.en.html

28   As Dyne:bolic grew out of the Italian "Hackmeeting" movement, it is linked closely to the praxis of auto-gestation, or radical Do-It-Yourself (DIY). Many socially-driven cultural projects have arisen from the large Italian network of *centri sociali* or squatted social centres. See a history of Hackmeetings at wiki.hacklab.org.uk/index. php/Hacklabs_from_digital_to_analog.

29   Dyne:bolic belongs to a vision of integrated software and communication systems. For example, videos made with the free software tools on Dyne:bolic can then be distributed via online archives like New Global Vision, entering the Digital Commons. International video archives maintained by cultural activists include ngvision.org originating in Italy, and the video syndication network v2v. cc/ from Germany. The Indymedia video portal at www.indymedia.org/projects.php3 focuses on documentary material. A mammoth cultural archiving project is archive. org.

30   Quote from the Streamtime portal at streamtime.org. Interviews with key project facilitators online at wiki.whatthehack.org/index. php/Streamtime_and_Iraqi_Linux.

31   The Container Project Web site is a repository of material documenting the history of the project and links to its current activities. www.container-project.net/. Photo documentation of the process of converting the Container is online at www.container-project.net/C-Document/Album/page1.html.

32   Skint Stream was an initiative of ICA Cape Town, Mongrel and radioqualia. Find details of Skint Stream, and the participating communities, at www.jelliedeel.org/skinstream.

33   See workshop reports at www.cnh.on.ca/container.html, www.cyear01.com/containerproject/archives/blog.html and www.ict4djamaica.org/content/home/detail. asp?iData=504&iCat=292&iChannel=2&nChannel=Articles.

[34]   See www.ict4djamaica.org/content/home/index.htm.

[35]   Mervin Jarman, personal communication, September 12, 2006.

[36]   The "Old Enclosures" in England were carried out by thousands of Acts of Parliament between 1702 and 1810. Hunger and terror for the dispossessed multitudes accompanied the old enclosures, as capital wealth piled up for a minority. Expropriated peasants, day-labourers, and artisans throughout Europe did not capitulate meekly to the new rule of waged work, with fierce resistance during feudal times and throughout the Middle Ages (Federici 2004, pp. 133138). Silvia Federici argues that a new set of "enclosures"—from thefts of agricultural land through government expropriation, to the creation of vast masses of criminalised poor from the newly or generationally dispossessed—are accompanying "the new global expansion of capitalist relations" (Federici, 2004, p. 11). David Bollier (2002) documents the enclosures of various contemporary commons, including the Internet, in *Silent Theft: The Private Plunder of Our Common Wealth*.

[37]   See www.nosoftwarepatents.com/en/m/intro/index.html and www.ffii.org/ for summaries of this battle, and lpf.ai.mit.edu/Patents/patents.html for historical background on earlier bids to impose patents on software.

[38]   The controversial and "questionably constitutional" *Digital Millennium Copyright Act (DMCA)* was signed into United States law on October 28, 1998. The main objections to this law are that it is unreasonably weighted in favour of the top end of town in terms of copyright holders (the record, film, and publishing industries), criminalises very widespread social applications of communications technologies, and stifles innovation by small players. It also holds Internet Service Providers liable for the actions of their clients, which is similar to holding the postal service liable for the contents of a private letter. The law focuses on technological aspects of copy protection instead of the actual works themselves. For example, the law "creates two new prohibitions in Title 17 of the U.S. Code—one on circumvention of technological measures used by copyright owners to protect their works and one on tampering with copyright management information—and adds civil remedies and criminal penalties for violating the prohibitions" www.copyright.gov/legislation/dmca.pdf. A number of prosecutions have ensued, often targeting young users of peer-to-peer file sharing programs. Also prosecuted was the developer of a program that can "crack" video compression software (making it easier for people to potentially watch downloaded movies). Under this law even makers of DVD copying software have been prosecuted. The Electronic Frontier Foundation's *Digital Millennium Copyright Act (DMCA) Archive* contains a listing of many of the cases brought to trial or underway, and counter suits by lobby groups challenging the validity of the law. See www.eff.org/IP/DMCA/ and www.eff.org/IP/DMCA/DMCA_against_the_darknet.pdf.

[39]   Over-exploitation supposedly leads to what ecologist Garrett Hardin depicted as the "tragedy of the commons" in his classic text of the same name published in *Science* in 1968. One of the arguments supporting privatisation proposes that the "commons-ers" will always ruin the land through over use. See essay and responses online at www.sciencemag.org/sciext/sotp/commons.shtml. Paul Ceruzzi asserts that by "strict technical measures, the Internet has not come close to approaching this point of overpopulation ... [passing through] challenges like the 1988 worm, viruses, the Y2K crisis, the dot.com

collapse, and the terrorists' attacks of September 11, 2001, with hardly a hiccup. It is based on robust design. As for the content and quality of information that the Internet conveys, however, it has indeed been tragic" (Ceruzzi, 2003, p. 330).

40 Statistics breaking down internet usage on a continental basis at www.internetworldstats.com/stats.html point to the enormous take up on the net in Africa (424%), the Middle East (454%), and Latin America (353%), in the period 2000-2005. In contrast, North America had the lowest take up (110%). Detailed internet statistics are available at leading research Nielson Net Ratings at www.nielson-netratings.com.

41 See McLuhan, M. (1967). *The Medium is the Massage*. London: Penguin Books.

42 See, for example, Lawrence Lessig's blog describing the poetry slam on free culture by Brazilian Minister of Culture, Gilberto Gil. Lessig also notes the visionary "*Thousand points of culture project*—to build a thousand places around Brazil where free software tools exist for people to make, and remix, culture" (Source: www.lessig.org/blog/archives/2005_01.shtml).

43 In *The Problem with WSIS*, Alan Toner (2003) critiques the colonial relations between "information society" and "intellectual property" with reference to the World Intellectual Property Organisation (WIPO). It could be argued that this new form of colonial domination is strengthening the political resolve in Latin America, the Caribbean and Africa to use free software as a central platform for social transformation. "Where once corpses accumulated to the advance of colonialism or the indifference of commodity capital, now they hang in the profit and loss scales of Big Pharma, actuarially accounted for and calculated against

licensing and royalty revenue. With the aid of stringent IP law, companies are able to exercise a biopolitical control that takes to new extremes the tendency to liberate capital by restricting individual and collective freedoms and rights even the right to life itself" (Toner, 2003, para. 1). "In 1986, with the Uruguay Round of the GATT negotiations on the horizon the Intellectual Property Committee (IPC) determined to ensure that corporate IP concerns be inserted into the negotiation agenda and fully integrated into any ultimate agreement. It was the IPC's efforts to orchestrate business lobbying efforts on a global basis which culminated in TRIPS, now administered by the WTO. TRIPS will transfer an estimated 40 billion dollars from the poorest states over the next ten years, according to the World Bank, via patented medicines and seeds, and net rent transfers through royalties and licenses" (Toner, 2003, para. 10).

44 In *Five Theses on Informational-Cognitive Capitalism*, George N. Dafermos (2005) states: "The realm of such networks of co-operative development is underpinned by the pleasure principle ... they re-discover the joy ... that accompanies creative work ... collective subjectivity is impregnated with the sperm of radicality, as people are suddenly becoming aware of the reversal of perspective that lies in the shadows: a production setting ... [which] exposes the poverty of production effectuated for the sake of profit. A direct confrontation stretching from the terrain of ideas to the very institutional nucleus of capitalist society is underway. On the one side stands the beast of living labour organised independently of the capitalist demand, and, [on the other], the imaginary of intellectual property law ..."

Chapter VI
ALBA Architecture as Proposal for OSS Collaborative Science

Andrea Bosin
Università degli Studi di Cagliari, Italy

Nicoletta Dessì
Università degli Studi di Cagliari, Italy

Maria Grazia Fugini
Politecnico di Milano, Italy

Diego Liberati
Italian National Research Council, Italy

Barbara Pes
Università degli Studi di Cagliari, Italy

ABSTRACT

A framework is proposed that would create, use, communicate, and distribute information whose organizational dynamics allow it to perform a distributed cooperative enterprise also in public environments over open source systems. The approach assumes the Web services as the enacting paradigm, possibly over a grid, to formalize interaction as cooperative services on various computational nodes of a network. A framework is thus proposed that defines the responsibility of e-nodes in offering services and the set of rules under which each service can be accessed by e-nodes through service invocation. By discussing a case study, this chapter details how specific classes of interactions can be mapped into a service-oriented model whose implementation is carried out in a prototypical public environment.

INTRODUCTION

Open source software (OSS) for e-e-science should make reference to the paradigm of a distributed infrastructure over a multi-system grid, allowing data exchange through services, according to standard proposals in the areas of grid computing (Pollock & Hodgson, 2004) and service-oriented computing (SOC). In fact, biologists, medical doctors, and scientists in general are often involved in time consuming experiments and are aware of the degree of difficulty in validating or rejecting a given hypothesis by lab experiments.

Lab experiments are often still developed in isolation and tend to be small scale and specialized for ad hoc applications; there is limited potential for integration with broader reuse. One of the reasons for this lack of integration capability is that researchers need to be inter-networked in a cooperative enterprise style, although sharing data, programs, and resources in a nonprofit network of collaboration. Cooperative OSS environments can be a feasible solution for interconnection, integration, and large information sources sharing during experiment planning and execution. It is a common situation that information source owners, even members of a coalition, are not keen to delegate control over their resources to any common server. However, as long as ICT models, techniques, and tools are rapidly developing, there is a true hope to move towards the realisation of effective distributed and cooperative scientific laboratories. In fact, the concept of "what an experiment is" is rapidly changing in an ICT-oriented environment, moving from the idea of local laboratory activity towards a computer and network supported application including the integration of:

- A variety of information and data sources
- The interaction with physical devices
- The use of existing software systems allowing the potential deviation from a predetermined

sequence of actions as well as the verifiability of research work and accomplishments
- The peculiar and distributed expertise of the involved scientists

In general, scientific experiments are supported by activities that create, use, communicate, and distribute information whose organizational dynamics are similar to processes performed by distributed cooperative enterprise units.

According to the frame discussed in Bosin, Dessi, Fugini, Liberati, and Pes (2005), in this chapter we stress the benefits of OSS for e-science considering that as many operating nodes as possible can work cooperatively sharing data, resources, and software, thus avoiding the bottleneck of licences for distributed use of tools needed to perform *cooperative scientific experiments*. In particular, this chapter presents an architecture based on nodes equipped with a grid and with Web services in order to access OSS, showing how scientific experiments can be enacted through the use of cooperation among OSS sites. Such a choice, besides reducing the cost of the experiments, would support distributed introduction of OSS among other actors of the dynamical networks, thus supporting awareness of OSS and their diffusion.

Specifically, this chapter describes the ALBA (Advanced Labs for Bioinformatics Agencies) environment aimed at developing cooperative OSS models and processes for executing cooperative scientific experiments (e-experiments). Cooperative processes, e-services, and grid computing are the basic paradigms used in ALBA, which can effectively support, through OSS, the distributed execution of different classes of experiments, from visualization to model identification through clustering and rules generation, in various application fields, such as bioinformatics, neuro-informatics, telemonitoring, or drug discovery. By applying Web services (Alonso, Casati, Kuno, & Machiraju, 2004) and grid computing, an experiment or a simulation can be executed in a cooperative way on various computation nodes of a network equipped with OSS, allowing data exchange among researchers.

This approach allows for a correct design and set up of the experiment workflow, methods, and tools that are essential for cooperating organizations which perform joint experiments requiring specialized tools or computational power only available at specific nodes. In this sense, the overall structure is presented as an OSS environment, in that each node resource becomes an OSS ALBA site available for the inter-networked nodes.

The ALBA environment uses the SOC paradigm and the GRID structure to formalize experiments as *cooperative services* on various computational nodes of a grid network. Specifically, basic elements of the ALBA environment are models, languages, and support tools for realizing a virtual network that defines the organisational responsibility of the global experiments according to a set of rules under which each node can execute local services to be accessed by other nodes in order to achieve the whole experiment's results. From the researcher viewpoint, the ALBA environment is a knowledge network enabling data and service sharing, as well as expertise and competences to allow a team of scientists to discuss representative cases or data. The environment allows specific classes of experiments, such as drug discovery, micro array data analysis, or molecular docking, to be mapped into a service-oriented model whose implementation is carried out in a prototypical scientific environment.

As a case study, this chapter proposes a reference model for cooperative experiments, executed as e-applications, including a grid infrastructure, distributed workflows, and experimental knowledge repositories.

BACKGROUND

A key factor to promoting research intensive products is the vision of a large scale scientific exploration carried out in a networked cooperative environment in the style of cooperative information systems (COOPIS, 2005), with a high performance computing infrastructure, for example, of grid type (Berman, Fox, & Hey, 2003), that supports flexible collaborations (Hendler & De Roure, 2004), OSS, and computation on a global scale. The availability of such an open virtual cooperative environment should lower barriers among researchers taking advantage of individual innovation and allowing the development of collaborative scientific experiments (Gentleman, Carey, Bates, Bolstad, Dettling, Dudoit, et al., 2004). Up to now, however, rarely are technologies developed specifically for the research community, and ICT developments are harnessed to support scientific applications varying in scale and purpose and encompassing a full range of engagement points, from single purpose built experiments to complex software environments.

The range of accessible technologies and services useful to scientific experiments can be classified broadly into three categories:

- Toolkits specifically aimed at supporting experiments
- General purpose software tools still essential in enabling the experiments of interest (e.g., computation, data mining tools, data warehousing)
- More widely deployed infrastructures that may be useful in scientific experiments, such as Web services or grid computing

This scenario is similar to that of enterprise environments, whose progress requires large scale collaboration and efficient access to very large data collections and computing resources. Although sustainable interoperability models are emerging for market players (such as service providers, stakeholders, policy makers, and market regulators), they are currently deployed mostly in areas where high computing power and storage capabilities, usually needed by scientific environments, are not mission-critical.

Recently, emerging technologies, such as Web services (W3C, 2004) and the grid (Foster, Kesselman, Nick, & Tuecke, 2003), have enabled new types of scientific applications consisting of a set of services to be invoked by researchers. E-science is the term usually applied to the use of advanced computing technologies to support scientists. Because of their need for high-performance computing resources, as well as cooperative ICT technologies, for example, of Web style, many scientists are drawn to grid computing and the Web as the infrastructure to support data management and analysis across organizations. High-performance computing and communication technologies are enabling computational scientists, or e-scientists, to study and better understand complex systems. These technologies allow for new forms of collaboration over large distances together with the ability to process, disseminate, and share information. Global-scale experimental networking initiatives have been developed in the last few years: the aim is to advance cyber infrastructure for e-scientists through the collaborative development of networking tools and advanced grid services. Grids provide basic facilities for robust computation, efficient resource management, transfer, and sharing, and they support distributed computation. Moreover, coming from a different direction, the Semantic Web vision also was motivated by the need to support scientific collaboration. By enabling transparent document sharing, metadata annotations, and semantic integration, it addresses multidisciplinary distributed science research at the end-user level. Since both grid computing and Semantic Web deal with interoperability, from the e-science perspective they would both be necessary. Neither technology on its own would be able to achieve the full e-science vision. This integration, called Semantic Grid, would serve as the infrastructure for this vision.

MAIN FOCUS OF THE CHAPTER

OSS for Virtual Laboratories

An enabling factor of the ALBA environment for researchers is the possibility to remotely access shared resources and catalogues on an open source basis, in order to execute their own experiments and also to have continuous education on protocols, experiments, and discoveries in advance and with the possibility to consult other colleagues with limited need to travel and quick and effective access to online experiment documentation.

Thus, the ALBA environment is envisioned as an OSS-based environment supporting the execution of cooperative scientific experiments. E-services and the grid are the enabling technologies considered by the project to support the simulation/execution of different classes of experiments, in bioinformatics such as drug discovery, microarray data analysis, or molecular docking . Each experiment ranges from visualization (browsing and search interfaces), to model identification through clustering and rules generation, and requires tools for correct design and set up of the experiment workflow and for information retrieval (e.g., for searching similar protocols, or descriptive datasheets for chemical reactors). In addition, cooperating scientists who perform joint experiments may require specialized tools (e.g., data mining, or database tools) or computational power (e.g., long calculi for protein analysis based on their forms, or for discarding the irrelevant experiments in drug study) available only at specific nodes. The visualization part is given special attention, considering friendly interfaces and graphical simulations enabling an improved comprehension of currently textual explanations. Also privacy and data security are a major concern in the ALBA environment, considering both methods to select trusted nodes within the cooperation network, and/or to obscure or encrypt the transmitted and stored data, to preserve their sensitivity, according to user-formulated security requirements.

The ALBA environment aims to go beyond the existing virtual laboratory platforms that essentially enable information sharing and distributed computations—by offering to the researchers more complex and, possibly, semi-automated ways of conducting experiments, by exploiting and composing services offered by different institutions. For the ICT infrastructure, the ALBA environment assumes experiments formalization in terms of sets of tasks to be executed on various computational nodes of a network of labs. The environment allows developers to specify models, languages, and support tools enabling a public community of research centres, labs, and nonprofit organizations to realize a network infrastructure that defines the organisational responsibility of the global experiments, the framework of nodes, and the set of rules under which each node can execute local services to be accessed by other nodes in order to achieve the whole experiments' results. A knowledge network enables data and service sharing, as well as expertise and competences to allow the team of scientists to discuss representative cases or data. The outcome of the environment provides information regarding, for example, the efficiency of the machine learning techniques in discovering patterns related to genetic disorders, and also allow the identification of relevant types of gene expressions. These could possibly be abnormal expression rates for a particular gene, the presence or the absence of a particular gene or sequence of genes, or a pattern of unusual expression across a gene subset. It is envisioned that this would thereby provide help to guide physicians in determining the best treatment for a patient, for example, regarding the aggressiveness of a course of treatment on which to place a patient.

The ALBA environment only supports experiments defined as numerical evaluations carried out on selected data sets according to available methodological approaches. The experiment execution platform is composed of: (1) a workflow; (2) the distribution thereof; (3) the involved nodes and their relative roles in the experiment; (4) the set of involved resources, such as data areas, data repositories and e-services. Four main classes of methodological approaches to the experiments are supported:

1. Process simulation and visualization on the already available information sources
2. Supervised or unsupervised classification of observed events without inferring any correlation nor causality, such as in clustering, and neural networks (Liberati, Bittanti, & Garatti, 2005)
3. Machine learning: rule generation (Muselli & Liberati, 2002) and Bayesian networks (Bosin, Dessì, Liberati, & Pes, 2006) able to select and to link salient involved variables in order to understand relationships and to extract knowledge on the reliability and possibly causal relationships among related co-factors via tools like logical networks and Cart-models
4. Identification of the process dynamics (Ferrari-Trecate, Muselli, Liberati, & Morari, 2003)

Such classes, listed in increasing order of logical complexity, might have an impact on the design of the experiment and of its execution modality in terms of execution resources either on a single specialized node or in a grid structure.

Experiments of these classes have some portions, both of processes and of data or knowledge, that can be shared in a collaborative environment. One of the reasons for executing an experiment in a distributed way might be that one organization would need to process data under a specific costly product available on a node because of its lack of skill for developing or using open source equivalent; rather than acquiring the product, the organization might invoke a remote service as OSS available on the remote node. Another reason is that some cooperating organizations might want to inspect data dispersed on their databases, with no changes to their local computational environment.

In ALBA, according to the current laboratory practice, a researcher begins with the assertion of a high level goal needed to test a scientific hypothesis or to obtain some additional knowledge on a previous experiment. This goal has to be decomposed into a set of tasks (the *experiment life cycle*) each accomplished by an appropriate class of freely available services published in a UDDI registry exposing the public services available to the *community*. ALBA operates under the closed world assumption, that is, considering that all the nodes agree on OSS on the basis of pre-negotiation and software exchange contracts, for example, according to standard languages-based negotiation and contracting as described in (Callea, Campagna, Fugini, & Plebani, 2004).

From a methodological point of view, heterogeneous services can provide similar capabilities, but the researcher is in charge of choosing the most suitable methods to accomplish each task, that is, the researcher is in charge of designing the workflow of the scientific experiment. In particular, if the researcher wants to rerun an experiment, the workflow must take into account the changes in the choice of methods as well as in the availability of services and tools.

In ALBA the researcher interacts and chooses services, workflows, and data within an experimental environment whose cooperative framework has been defined to extend the integration of scientific experiments to a level of scenario-based interaction. This scenario is profitable for many reasons, like exchanging scientific data and processing tools which results in a reduced number of software acquisitions, load balancing work between specialized researchers, and so on.

Specifically, the researcher defines the experiment life cycle that consists in two basic processes: the modelling process and the implementation process.

The *modeling* process is organized in three steps:

1. The experiment is decomposed into a set of basic tasks orchestrated in a workflow of Web services.
2. A choreography model is defined that specifies the interactions among the tasks, according to the *template* of the experiment.

The implementation is based on the J2EE (Armstrong, 2004) and Oracle (Oracle, Oracle2) platforms, but the use of standard technologies (HTTP, XML, SOAP, WSDL) and languages (Java, SQL) makes it flexible and easily expandable.

Services are organized according to the multi-tier architecture shown in Figure 1, whose abstract layers are: exposed (public) Web services (WS), experiment workflows, service orchestration engines, data mining, database infrastructures, data repository, and result visualization tools. This last layer interacts with the researcher's ALBA client, which can also interact with experiment workflow definition tools and with the scientific tool selector, allowing one to declare which tools need to be used within an experiment.

ALBA nodes contain both proprietary software and specific OSS packages that the node decides to make public and expose as a Web service through the WS layer. While the tier separation can be purely logical, our prototype allows the physical separation of tiers, where each one is located on a separated and networked hardware resource.

A client tier represents the consumers of services. Scientists located across the scientific net (i.e., on the nodes that have contracted to share OSS tools for experiments based on service invocation) invoke the services provided by the service layer and orchestrated by the orchestration engine.[1]

The ALBA node is endowed with an infrastructure which varies in the number and type of tools. Accordingly, the ALBA nodes can mount different layers, depending on their availability to execute distributed experiments. For example,

Figure 1. Overall ALBA architectural scenario

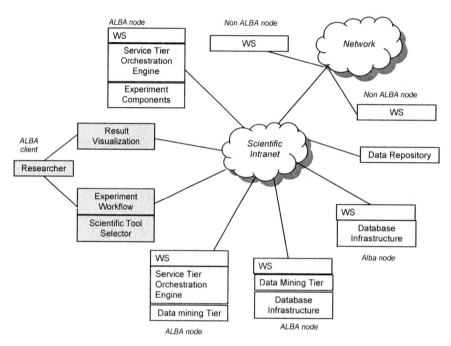

some of them can be equipped with various Experiment components (also reused from previous experiments) to be composed and orchestrated for an experiment.

Other nodes are equipped with data mining tools, others with database tools. The scientific tool selector is the (visual) layer in charge of deciding what tools of the node infrastructure should be invoked to perform a given part of the experiment, according to the specifications provided by the scientist. Outside the Scientific Net, other nodes can be accessed cooperatively, however under specific software access policies, that might include OSS.

A Sample Experiment

A sample scenario of an experiment has been implemented using the Grid OSS Taverna[2] tool to define and execute a distributed experiment sharing services, data, and computation power.

The experiment is the execution of two classification techniques of leukaemia through analysis of data from DNA-microarrays. DNA-microarrays are a powerful tool to explore biology, to make the diagnosis of a disease, to develop drugs and therapies ad hoc for patients. On the basis of such data it is then possible to classify the pathology for diagnosis purposes (e.g., to distinguish acute myeloid leukaemia from lymphatic leukaemia) or for prognostic purposes.

The analysis of the huge quantity of available data may offer a highly precise classification of the disease and is performed using methodologies that group the examined subjects into sub-groups (clusters) sharing similar values of expressions of the most relevant genes.

Classification techniques belong to two categories:

- **Supervised:** Classification is given as known for a data subset; based on such information, one searches a priori to train a classification algorithm

- **Unsupervised:** No information is a priori available and classification is performed based only on some notion of sample distance.

The experiment uses both techniques, starting in parallel an unsupervised technique based on principal component divisive partitioning and *k-means* (Liberati et al., 2005), and a supervised technique based on Bayesian networks and minimum description length (Bosin et al., 2006). Results can thus be compared, also evaluating similarities and differences between the two complementary approaches.

The architecture of the experiment is shown in Figure 2, where the use of the Taverna tool is exemplified.

- **On the site of Organization A:** The workflow is defined specifying the distributed experiment in terms of selection and choreography of resources that are available on the network. The employed tool is *Taverna*. The actor defining the experiment is a researcher or bio-informatic expert. On this site, the global results coming from

the experiment are loaded. The results are accessible both to a human and a system; both have to perform an authentication and storage of credentials for subsequent (re)use on a local or on a remote system.

- **The site of Organization B:** Offers an Unsupervised Clustering service of micro-arrays data. The service was originally a *Matlab*[3] elaboration non-usable from remote. Hence, a service has been designed and executed allowing one to use it from remote and from whatever platform. The services can be loaded with local data and with data residing on external sites and belonging to other organizations. This service has been implemented as a Java WebService (exposed as WSDL), developed under *Apache Axis.*[4]
- **The site of Organization C:** Offers a Supervised Clustering service based on Bayesian networks made available on the network.
- **The site of Organization D:** Offers data feeding the two Clustering services. Optionally, data can also be taken from more sites of different organizations, so that, for example, multiple instances of an experiment can be launched in parallel.

Figure 2. Sample cooperative e-experiment

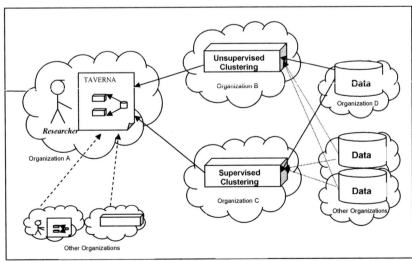
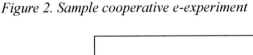

Figure 3. Snapshot of Taverna grid support tools

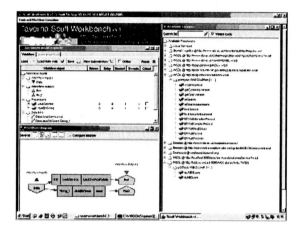

The experiment is exposed as a service and can be invoked as a whole from external organizations.

Figure 3 shows a screenshot of the use of Taverna, with reference to the phases of definition of the experiment.

FUTURE TRENDS

In future works, it will be straightforward to extend the classes of experiments of interest, to implement the proposed structure in specific application fields with needs of primary relevance, and of course to detail the sequences of interaction among actors in the specific use cases.

What remains open are many questions about the evaluation of such an approach and the actual design and implementation of the Semantic Grid panorama. Since the realisation of a distributed general purpose scientific environment is not immediate, the evaluation effort described here involves a prototypical environment based upon emerging Web service technology and applied to the previously mentioned four classes of experiments.

CONCLUSION

We have illustrated how e-science can benefit from OSS for modelling and executing scientific experiments as cooperative services executed on a grid of OSS tools. The proposed ALBA cooperative framework for distributed experiments is quite general and flexible, being adaptable to different contexts. Given the challenge of evaluating the effects of applying Web service technology to the scientific community, the evaluation performed up to now takes a flexible and multi-faceted approach: It aims at assessing task-user-system functionality and can be extended incrementally according to the continuous evolution of the scientific cooperative environment.

The first outcomes of ALBA are in terms of designed e-services supporting the simulation/execution of different classes of experiments, from visualization (browsing and search interfaces), to model identification through clustering and rules generation, in application fields, such as drug discover, microarray data analysis, or molecular docking. By applying e-services and the grid and experiment or a simulation can be executed in a cooperative way on various computation nodes of a network, also allowing knowledge exchange among researchers. A correct design and set up of the experiment workflow, visualization methods, and information retrieval tools (e.g., for searching similar protocols, or descriptive datasheets for chemical reactors) is studied to support cooperating scientists who perform joint experiments, for example requiring specialized tools (e.g., data mining or database tools) or computational power (e.g., long calculi for protein analysis based on their forms, or for discarding the irrelevant experiments in drug studies) available only at specific nodes. The visualization part is considered with special care, taking into account friendly interfaces and graphical simulations enabling an improved comprehension of currently textual explanations. Also privacy and data security are a major concern in the project, considering both methods to select trusted nodes within the cooperation network and to obscure or encrypt the transmitted and stored data, to preserve their sensitivity, according to user-formulated security requirements. Specific aims and expected results of the ALBA project aims at going beyond the existing virtual laboratory platforms that essentially enable information sharing and distributed computations by offering to the researchers more complex and, possibly, semi-automated ways of conducting experiments, by exploiting and composing services offered by different institutions. One of the interesting properties of the ALBA platform is that it is, in principle, technically open to every other actor even beyond the core of the founding institutions, in both senses of contributing to the distributed team and/or to simply exploit growing knowledge. It will be easy to implement an access policy stating, for instance, both free reading access to nonprofit institutions about the already consolidated results and an involvement subject to mutual agreement by third bodies willing to enter the platform.

REFERENCES

Alonso, G., Casati, F., Kuno, H., & Machiraju, V. (2004). Web services—Concepts, architectures, and applications. Springer Verlag.

Armstrong, E., Ball, J., Bodoff, S., Bode Carson, D., et al. (2004). *The J2EE 1.4 Tutorial*, December 16.

Berman, F., Fox, G., & Hey, T., (Eds.). (2003). *Grid computing: Making the global infrastructure a reality*. New York: John Wiley and Sons, Inc.

Bosin, A., Dessì, N., Fugini, M. G., Liberati, D., & Pes, B. (2005, September 5). Applying enterprise models to design cooperative scientific environments. *International Workshop on Enterprise and Networked Enterprises Interoperability*, Nancy (LNCS), Springer.

Bosin, A., Dessì, N., Liberati, D., & Pes, B. (2006). Learning Bayesian Classifiers from gene-expression microarray data. In I. Bloch, A. Petrosino, & A. G. B. Tettamanzi (Eds.), *WILF 2005* (LNAI 3849, pp. 297-304). Berlin; Heidelberg: Springer-Verlag.

Callea, M., Campagna, L., Fugini, M. G., & Plebani, P. (2004, September). Contracts for defining QoS levels in a multichannel adaptive information systems. In *Proceedings of IFIP Workshop on Mobile Systems, Oslo*.

COOPIS. (2005, October). *Proceedings 6th International Conference on Cooperative Information Systems (CoopIS)*, Larnaca, Cyprus (pp. 25-29). Springer Verlag.

Hendler, J. & De Roure, D. (2004). E-science: The grid and the Semantic Web. *IEEE Intelligent Systems*, 19(1), 65-71.

Ferrari–Trecate, G., Muselli, M., Liberati, D., & Morari, M. (2003). A clustering technique for the identification of piecewise affine systems. *Automatica*, 39, 205-217.

Foster, I., Kesselman, C., Nick, J. M. & Tuecke, S. (2003). *The physiology of the grid; An open grid service architecture for distributed system integration*. The Globus Project.

Gentleman, R. C., Carey, V. J., Bates, D. M., Bolstad, B., Dettling, M., Dudoit, S., et al. (2004, September 15). Bioconductor: Open software development for computational biology and bioinformatics. *Genome Biology*, 5(10), R80.

Liberati, D., Bittanti, S., & Garatti, S. (2005). Unsupervised mining of genes classifying Leukemia. In J. Wang (Ed.), *Encyclopedia of data warehousing and data mining*. Hershey, PA: Idea Group Publishing.

Muselli, M., & Liberati, D. (2002). Binary rule generation via hamming clustering. *IEEE Transactions on Knowledge and Data Engineering*, 14(6) 1258-1268.

Pollock, J. T., & Hodgson, R. (2004). *Adaptive information: Improving business through semantic interoperability, grid computing, and enterprise integration.* Wiley Series in Systems Engineering and Management. Wiley-Interscience.

Oracle. (n.d.). Retrieved from http://www.oracle.com

Oracle2 (n.d.). http://www.oracle.com/database/Enterprise_Edition.html

W3C. (2004). W3C Working Draft March 24, 2004, http://www.w3.org/TR/2004/WD-ws-chor-model-20040324/

KEY TERMS

Bioinformatics: The application of the ICT tools to advanced biological problems, like transcriptomics and proteomic, involving huge amounts of data.

Cooperative Information Systems: Independent, federated information systems that can either autonomously execute locally or cooperate for some tasks towards a common organizational goal.

E-Experiment: Scientific experiment executed on an ICT distributed environment centered on cooperative tools and methods.

E-Science: Modality of performing experiments in silico in a cooperative way by resorting to information and communication technology (ICT).

Drug Discovery: Forecasting of the properties of a candidate new drug on the basis of a computed combination of the known properties of its main constituents.

Grid Computing: Distributed computation over a grid of nodes dynamically allocated to the process in execution.

Interoperability: Possibility of performing computation in a distributed heterogeneous environment without altering the technological and specification structure at each involved node.

Web Services: Software paradigm enabling peer-to-peer computation in distributed environments based on the concept of "service" as an autonomous piece of code published in the network.

ENDNOTES

[1] Clients are implemented by standalone Java applications that make use of existing libraries (J2EE application client container) in charge of the low-level data preparation and communication (HTTP, SOAP, WSDL).
[2] http://www.mygrid.org.uk
[3] http://www.mathworks.com
[4] http://ws.apache.org/axis

Chapter VII
Evaluating the Potential of Free and Open Source Software in the Developing World

Victor van Reijswoud
Uganda Martyrs University, Uganda

Emmanuel Mulo
Uganda Martyrs University, Uganda

ABSTRACT

Development organizations and international nongovernmental organizations (NGOs) have been emphasizing the high potential of free and open source software (FOSS) for the less developed countries (LDCs). Cost reduction, less vendor dependency, and increased potential for local capacity development have been their main arguments. In spite of its advantages, FOSS is not widely adopted on the African continent. In this chapter the experiences of one of the largest FOSS migrations in Africa is evaluated. The purpose of the evaluation is to make an on-the-ground assessment of the claims about the development potential of FOSS and draw up a research agenda for a FOSS community concerned with the LDCs.

INTRODUCTION

Over the past years the issue of free and open source software (FOSS)[1] for development in LDCs is receiving more and more attention. Where in the beginning the benefits of FOSS for less developed countries (LDCs) was only stressed by small groups of idealists like Richard Stallman (Williams, 2002), now it is moving into the hands of the large international organizations like the World Bank (Dravis, 2003) and the United Nations. In the *E-Commerce and Development Report* that was released at the end of 2003, it was stated that FOSS is expected to dramatically affect the evolving information and communication technology (ICT) landscape for LDCs. UNCTAD believes that FOSS is here to stay and LDCs should benefit from this trend and start to recognize the importance of FOSS for their ICT policies (UNCTAD, 2003).

Leading organizations in the software and ICT consulting industry have embraced FOSS at a rapid speed. IBM is now the major champion of FOSS, and in 2002 IBM announced the receipt of approximately US$1 billion in revenue from the sale of Linux-based software, hardware, and services. Other technology leaders, including Hewlett-Packard, Motorola, Dell, Oracle, Intel, and Sun Microsystems, have also made major commitments to FOSS (UNCTAD, 2003). The major player objecting the FOSS paradigm at the moment is Microsoft Corporation.

For a brief understanding of what FOSS means, we shall adopt David Wheeler's definition stated in the FOSS primer (Wong & Sayo, 2003) as:

FOSS programs are programs whose licenses give users the freedom to run the program for any purpose, to study and modify the program, and to redistribute copies of either the original or modified program (without having to pay royalties to previous developers).

The terms "free" and "open," in this definition, are representative of the two major philosophies in the FOSS world. Free implies a user should have certain freedoms to do as they please with a piece of software. It should be noted that free does not necessarily imply freedom of cost, even though most software available as FOSS is usually accessible without one having to directly pay software or license fees. Open implies that software source code should be available to whoever is interested in viewing, modifying, and redistributing a particular piece of software.

The advantages of FOSS are diverse, but the most often quoted benefit in relation to LDCs is the reduction of purchase and license costs of the software. Software and licenses are paid for in hard currency and put an extra burden on the, often dismal, financial situation of LDCs. Other advantages are; reduction of vendor lock-in, adherence to open standards, increased transparency, minimizing security risks, increasing technical self reliance, and provision of a good starting point for local capacity development (Dravis, 2003).

The last advantage is probably the most important benefit of FOSS. Local capacity is needed to understand the technical foundation of the digital divide and start initiatives to bridge it.

Despite the obvious advantages mentioned, the adoption of FOSS, in LDCs, has been low (Bruggink, 2003; Van Reijswoud, 2003; Van Reijswoud & Topi, 2004). In Africa, no country other than South Africa, has explicitly mentioned FOSS in their ICT policy. On the contrary, governments of several of the richer countries on the continent are considering large deals with proprietary software vendors (see: www.fossfa.net). At present it seems that FOSS is on the agenda of the donor organizations and international NGOs but not on the agenda of the decision makers in LDCs. Although there are a growing number of initiatives to promote FOSS for LDCs in general and Africa in particular like Free and Open Source Software Foundation for Africa (www.fossfa.net) and the East African Center for Open Source Software (www.eacoss.org), there are very few organizations that consider and actually implement FOSS.

In this chapter we evaluate the experiences of an organization in Uganda, East Africa, that has decided to migrate its ICT infrastructure to FOSS. The purpose of the evaluation is to make an on-the-ground assessment of the claims about the development potential of FOSS. We, therefore, start the chapter with an overview of FOSS and the role it can play in the development of LDCs. Against this background we describe the case study, the progress the organization has made and the problems that were encountered. Finally, we will draw some conclusions on the experiences in the case study and set out an agenda for a successful rollout of FOSS in LDCs, especially in Africa.

FOSS FOR DEVELOPMENT: AN OVERVIEW

When we consider the role of FOSS for development, we have to distinguish multiple levels in order to get a good understanding of the impact of the different initiatives. The implementation and the propagation of FOSS can be performed at micro, meso, and macro levels. At the micro level we like to think about individual users and/or developers that opt for FOSS. At the meso level we consider organizations that take actions to integrate FOSS into their software infrastructure. Finally, the macro level where IT policies and actions at a national level are considered. We will start with the macro level.

FOSS from a Macro Perspective

Governments provide a huge potential for FOSS, not only as a site for implementation of the software, but, more importantly, as propagators of the philosophy behind the FOSS movement.

Over the past 5 years, a growing number of countries are starting to consider FOSS as a serious alternative (APC, 2003). Brazil has been one of the countries that has actively pursued initiatives along this line. It was in Brazil that the first law, in the world, regarding the use of FOSS, was passed in March 2000. Brazil is one of the countries where policies regarding adoption of FOSS have been successful, notably in the states of Rio Grande do Sul and Pernambuco. Also, the Brazilian Navy has been using FOSS since 2002 (see http://www.pernambuco.com/tecnologia/arquivo/softlivre1.html).

In Africa, the South African government is in the forefront. In September 2002, a policy framework document was developed by the Open Source Software Work Group of the South African Government Information Officers' Council (GITOC) (see details about FOSS in South Africa—www.oss.gov.za). The GITOC policy document (GITOC, 2002) recommends that government "explicitly" support the adoption of open source software (OSS) as part of its e-government strategy after a comprehensive study of the advantages and pitfalls of FOSS for government requirements.

Next to adopting FOSS software, GITOC also recommends that the government promotes the further development of FOSS in South Africa. This can be through the involvement of South Africa's SME industry that has the potential to play a role in the production and implementation of FOSS as well as setting up of user training facilities. Some success factors need to be considered in order to ensure that this potential is tapped:

- **OSS implementations should produce value:** Value can either be economic value, for example, reduction of costs and saving of foreign currency; or social value, for example, a wider access to information and computer training.
- **Adequate capacity to implement, use, and maintain:** There is a need for trained people to support and use the FOSS solution. Training of users and developers should be a high priority.
- **Policy support for a FOSS strategy:** Support for FOSS needs to expand to all key players at a governmental level, departmental level, IT professionals and computer users in general.

The FOSS approach represents a powerful opportunity for South Africa companies and government to bridge the technological gap at an acceptable cost. With these success factors driving FOSS initiatives, the development impact can quickly become evident. The South African government's Department of Communication has already begun the move to FOSS by adopting Linux as their operating system. The government plans to save 3 billion Rand a year (approximately €383 million), increase spending on software developed locally, and increase programming skills inside the country. South Africa reports that its small-scale introductions have already saved the country 10 million Rand (approximately €1.27 million).

Other countries are following. Worldwide, similar moves are being discussed in Taiwan, China, Peru, the UK, France, and Germany.[2]

FOSS from a Meso Perspective

The International Institute for Communication and Development (IICD), a Dutch NGO promoting the use of ICTs in LDCs, investigated the use of FOSS in organizations in three countries in Africa: Uganda, Tanzania, and Burkina Faso (Bruggink, 2003). The objective of the research was to find out how, where, and why organizations from all kind of sectors use FOSS, the problems they encounter, and possible opportunities for development. The findings of the research show that the use of FOSS is not yet very widespread. FOSS is mostly found at the server side of Internet service providers (ISPs) and is sometimes used by government and educational institutions. This means that FOSS (operating systems, mainly Linux and derivatives, Web servers, e-mail servers, and file servers) is not visible to the day to day computer users. Large and hierarchical organizations that have migrated completely from proprietary software to FOSS (server side and user side) have not been found. Most of the organizations that are using FOSS are relatively small organizations. When the three countries are compared, it is concluded that Tanzanian organizations show the most initiative, while, in Burkina Faso, organizations do not show interest in moving away from proprietary software.

The research of the IICD highlighted several reasons why organizations do not take up the challenge of FOSS. In the first place, there are some false perceptions. Many organizations take FOSS and Linux to be synonymous and consider it suitable only for the ICT specialist. Secondly, there is limited access to FOSS due to the fact that it (FOSS) is mostly distributed through the Internet and yet the countries under consideration have scarce and/or low bandwidth Internet connections. Software companies (FOSS based

companies included) see little market potential in Africa (outside South Africa) and so the availability of software is low. This is also reflected in the amount of resellers of FOSS. Finally, there is little expertise available to provide training and support for FOSS and consultancy in migration processes.

With steps taken to increase the awareness of FOSS along with more documentation of case studies where FOSS has been successfully implemented in organizations, adoption at this level could greatly increase. In the next section we will elaborate on an example of an implementation of FOSS in Uganda.

FOSS from a Micro Perspective

Most FOSS initiatives start out as small scale projects of individuals or small organizations. A growing number of individuals throughout the African continent are becoming aware of the potential of FOSS from a strategic point of view. Together with relevant advantages from an economic and technical point of view, FOSS represents an excellent opportunity for changing the position of African countries within the information society.

At user level, and for many individuals, the challenges of FOSS provide new opportunities for development, both at personal and community levels. Now that most countries in Africa are connected to the Internet, individual FOSS initiatives, which rely on it, are finally thriving. An initiative with good potential that tries to bring together the scattered FOSS society is the Free and Open Source Foundation for Africa (FOSSFA—www.fossfa.net). The initiative started as the offspring of an ICT policy and civil society workshop in Addis Ababa, Ethiopia, in February 2003. During the workshop the participants agreed that FOSS is paramount to Africa's progress in the ICT arena. The mission of FOSSFA is to promote the use and implementation of FOSS in Africa. Herewith it began to work on a coordinated approach to unite interested individuals and to support FOSS development, distribution, and integration. FOSSFA envisions a future in which governments and the private sector embrace FOSS and enlist local experts in adapting and developing appropriate tools, applications, and infrastructure for an African technology renaissance. A South-to-South cooperation, is predicted, in which students from Ghana to Egypt, and Kenya to Namibia develop programs that are then adopted by software gurus in Nigeria, South Africa, and Uganda.

On a similar line a number of Internet mailing lists and user groups are emerging, that focus on bringing together FOSS developers and users in Africa. At the moment there are active groups working in Burkina Faso, Ghana, Kenya, South Africa, Tanzania, Uganda, Zambia, and Zanzibar. Internet portals that aim at being a starting point for knowledge on FOSS in Africa are emerging as well.

At the commercial level, an initiative has been launched by DireqLearn (www.direqlearn. org). DireqLearn promotes FOSS as an alternative for the education sector. By adopting FOSS, the company can offer new solutions to the educational sector at low costs. Finally, even if only to a limited extent, some African FOSS development projects have been launched. Most of these are situated in South Africa, for reasons connected to the presence of adequate infrastructure. Outside South Africa, a project that is worthy of mention is RULE (Run Up to-date Linux Everywhere—www.rule-project.org). The aim of this project is to enable the running of modern free software on old computer hardware. In order to achieve the goal, the developers are modifying a standard Red Hat Linux distribution, trying to allow the greatest real functionality with the smallest consumption of CPU and RAM resources. The modified distribution is mainly intended for schools and other organizations in LDCs. At present, the RULE project provides a FOSS solution with GPL license that is able to transform 5 year old computer models (Pentium 75MHz, 16 MB RAM, 810 MB Hard disk) into useful machines.

The increasing interest in FOSS is also driving the emergence of FOSS-specific organizations. In several African countries, like Nigeria, Ghana, Uganda, and South Africa, specialized software and consulting companies have started up. Meanwhile, young professionals with a background in computing are embracing the FOSS approach and trying to reform the accepted practice of buying pirated proprietary software. At present, the market share of FOSS is still small and it is a struggle for these specialized companies to survive. However, when the benefits become clear and FOSS is implemented on a larger scale, the capacity to implement the systems shall be ready.

IMPLEMENTING FOSS: A CASE STUDY

There are hardly any documented, large scale organizational implementations of FOSS in LDCs. FOSS is mostly implemented in small donor funded projects or relatively simple organizations. See, for example, the projects described in Dravis (2003). The case study presented here describes a relatively large organization, Uganda Martyrs University (UMU), that made the strategic decision to move away from proprietary software to FOSS.

The decision to migrate to FOSS was, primarily, an ideological one rather than financial. Through the use of FOSS, the university hoped to offer an alternative to the use of pirated software and showcase the benefits. The financial benefits were specified in general terms, like "no software license fees in the future" and "perhaps we can buy more hardware if we save on software." Costs of the migration were mainly specified in social terms, like "will we still be able to communicate and share information with the rest of the world?" and "will the new software have the same functionality as the old one?"

The goal of the case study is to evaluate whether the high expectations of the use of FOSS for development translate well in a practical situation.

The case study is based on documentation and interviews with the main stakeholders at the university. Since both researchers are employed at the university and participated in the migration project, their views and experiences are also included. We have tried to avoid being subjective.

Uganda Martyrs University

Uganda Martyrs University is a private, catholic-founded university in the central province of Uganda. The university opened its doors in 1993 after the government, in a bid to improve the quality and the capacity of higher education in Uganda, allowed private universities to exist next to the government-owned universities.

At the time of writing (February, 2005) the university had 2,200 students enrolled in full time and part time programs at diploma and degree levels. The university's main campus is located in Nkozi village, 80 km outside the Ugandan capital city, Kampala. This location (Nkozi) can be characterised as rural. When the university started, there were no telephone connections, no steady water supply and electricity was unreliable. This has changed over the years and now the university is recognized for its good and reliable facilities. The university has a smaller campus in Kampala, where some postgraduate programs are offered on a part time basis, and several outreach offices are available for students who cannot easily travel to the main campus.

The university employed 86 full time academic staff and 117 administrative and support staff. With this size Uganda Martyrs University qualifies as a large organization in the private sector of Uganda.

The case study mainly focuses on the Nkozi campus of the university.

FOSS at Uganda Martyrs University: The Initial Stages

The FOSS project at Uganda Martyrs University had an informal start in 2001 when foreign assistance was offered to set up a mail server at the main campus. Since there was only money in the budget available for hardware and no provision for software, it was decided to equip the server with FOSS. The mail server was configured with Red Hat Linux, Sendmail as the mail transfer agent (MTA), and Neomail as the web based mail client. A Web server, to host the local intranet, was configured with SuSE Linux and Apache Web server software. When the new systems administrator was hired, he was trained to use and maintain the implemented configurations. The new systems administrator picked up interest in FOSS and later extended other parts of the system with available alternatives. In the beginning of 2002, the systems administrator incorporated FOSS for the proxy server (Squid) and the firewall (SuSEFirewall) for Internet access and some other minor applications.

In mid-2002, the project got a new impulse when several guest lecturers from universities and other organizations in Europe started to visit the university for lectures in a newly started Master of Science in Information Systems program. These lecturers encountered installations of pirated software on most computers of the university and raised questions about the institution's ICT policy. The university administration did not have an ICT policy formulated but realized that there was need to take action. This is when the FOSS project started formally.

In the course of the 2002-2003 academic year the ICT Department, the Office of the Vice Chancellor, and the Department of Computer Science and Information Systems (CSIS), outlined a software policy based on FOSS. The policy was designed with two underlying principles in mind:

Table 1. Main proprietary software used and open source software alternatives selected

Task	Proprietary Software	Open Source Alternative
Operating system	Windows 9x, 2000, XP	GNU/Linux
Office productivity suite	Microsoft Office	Open Office
Mail client	Microsoft Outlook Express	Kmail, Mozilla Mail
Internet browser	Internet Explorer	Konqueror, Mozilla
Database	Microsoft Access	MySQL/phpMyAdmin
Programming	Wordpad	Kate
	Borland Builder	Eclipse
Statistical analysis	SPSS	Open Office Calc
Webdesign	Microsoft Front Page	Bluefish / NVU

1. To optimize access to ICT for students and staff within the limited funds available
2. To stop supporting the use of pirated software on university property.

This was derived from the Christian values on which the university is based, that is, one shall not steal (not even software).

FOSS was considered a good alternative to work within these two principles, therefore, in May, 2003, the university senate officially agreed on the FOSS policy and preparations started for a full migration of both the server-side and desktop applications.

Migrating the Desktops

The major challenge for the university was the migration of the desktop applications. Literature review revealed very little reference material and few success stories. Documented experiences with similar migration projects in other LDCs were not available. The university received help from the FOSS Group of the University of Huddersfield, in the United Kingdom, as a response to a message sent to one of the Linux mailing lists. Other than that, the university ICT staff was on their own to plan and execute the most difficult part of the migration project.

At the start of the project, all computers in the university were using Microsoft Windows (98, 2000, and XP), Microsoft Office Suite, and other proprietary software applications. One of the first steps in the project was to identify the main applications and their usage, in order to select FOSS alternatives. It was observed that the university staff and students used very few "exotic" applications. This made the selection of alternatives relatively straightforward. Table 1 shows the alternatives that were selected to replace proprietary software.

Since the operating system would also be migrated, a decision needed to be made on which Linux distribution would be the standard at the university. Several distributions were evaluated and finally the Knoppix distribution was selected. The main reasons for this decision was that Knoppix is a one-disk, bootable distribution that can also be installed easily. The distribution could be handed out to the students and used as a CD-ROM-bootable program on any available computer (regardless of whether another operating system was already installed). Research on the Internet showed that the Knoppix distribution would work well on older machines, of which the university had quite a lot (Pentium II's). Finally, the Knoppix distribution came already bundled with most of the packages that would provide alternatives for the proprietary software being used at the university.

It was decided that the implementation strategy for the migration would be staged. First, all the public access computers (library and computer labs) would be migrated. Once this was completed, the academic staff would be migrated and finally, the administration (financial and student affairs administration units) of the university. This strategy was chosen to avoid endangering the university's critical operations in case of any setbacks.

The first phase was scheduled to take place during the absence of the students (June-August, 2003). This phase would then be evaluated before starting the second. The second phase was scheduled for the long vacation (JuneAugust, 2004). A time frame for the third phase was not determined.

Problems Encountered during the Migration

The project encountered unexpected technical and organizational problems in the first phase that delayed the time frame for the further implementation. The major problems are listed as follows:

- **Installation:** Although several claims were made about the installation of Linux on older machines (Pentium II/Dell), it was not as smooth as these descriptions seemed to suggest. Many of the machines did not have CD-ROM drives or were not able to boot from the CD for the installation. Bootable floppy disks had to be created to solve this but for about 20% of the older computers the installation failed. There were also problems of maintenance at a later stage for the computers without CD-ROM drives.

- **Performance:** Limited disk space and RAM handicapped the performance of the machines. The machines installed with Linux did not perform much better than

similar hardware configurations with Microsoft Windows installed on them. The users, therefore, did not consider this an improvement and, as a result, there was a negative impact on their acceptance of the new software.

- **Usability:** Although it was anticipated that the GUI (KDE 3.2) would not cause problems for the more experienced Windows users, the slight differences became bigger hurdles than expected. The most common problem was that the Knoppix distribution requires users to mount and unmount their floppy disks. Windows does not require this. After losing information due to (un)mounting improperly, the users started to question and resist the user friendliness of the new systems.

- **External factors:** A special problem was caused by the frequent power cuts in Uganda resulting in improper (hard) shutdown of the computers. When this happened, a file system failure was created with the result that the operating system no longer started up properly. In order to boot the computer, a root password was needed and the file systems needed to be checked and/or repaired. This procedure always took a long time since the repair program was checking the entire hard disk. In the newsgroups it was explained that the problem was caused by the default use of the ext2 file system. When the file systems were converted to more robust alternatives, the problem was solved.

- **Lack of alternative software:** There were some cases where there were no available alternatives for the software being used. Computers had to be installed with a dual boot system (two different operating systems installed on the same computer) setting Linux as the default option. The same FOSS applications had to be installed on both operating systems which meant twice the work per computer for the ICT staff.

Students were still working in Microsoft Windows a lot and so in order to discourage them from choosing Linux, Internet (and as a result Web-based e-mail) access was restricted to the Linux operating system.

- **Compatibility:** Finally, differences between the file formats of the office applications (Microsoft Office and Open Office) caused a problem. Students were able to read Microsoft Word documents, however, since the staff and the administration had not yet migrated, the files sent by the students could not be read. Moreover, when the students converted the documents, the format was changed from that originally intended by the students. Also, when attempting to save files into Microsoft Office suite formats, the following worrying message appeared: "Saving in external formats may have caused information loss. Do you still want to close?" The message was confusing to the users.

Evaluation of Phase I

Although solutions to most of the technical problems with the installation of the new FOSS system were found, the evaluation showed that the acceptance of the new systems was not as high as expected.

The mounting and unmounting of floppy disks was a major cause for resistance, especially since forgetting to unmount the disk caused the loss or corruption of files. This problem was overcome by adopting the SuSE Linux 9.1 distribution that had an auto-mount and unmount feature.

Students, especially the freshers (first-year students), responded very positively to the new systems. Although they had a choice between Windows and Linux (dual boot system), observations in the labs showed that most of them decided to use the Linux side. Among students that had already had some experience with computing in Microsoft Windows, the resistance to the new software was extremely high. Some of the

postgraduate students wrote complaint letters to university management about the use of "inferior" software. The resistance to the use of FOSS remained until that class of students graduated. For the incoming students, a compulsory FOSS computer literacy course was introduced based on a manual (Easy Linux Introductory Guide to Computers) developed by the university. This greatly reduced the resistance.

On the technical side, the problem of maintaining computers without CD-ROM drives was solved by adopting SuSE Linux 9.1. It provided the option to perform installation and upgrading of software over a network. One only had to ensure the computers had a network interface card and a connection point. This saved the technical staff having to carry around and/or keep copies of many installation CDs and external CD-ROM units.

Overall, we underestimated the importance of awareness creation of the underlying motives of the university to move to FOSS. The explanation of these reasons needs to be taken extremely seriously to secure commitment of the users. We also underestimated the need to have existing, continuous, and constantly available support to ease the users into the new system. This meant that even with the introduction of the improved system that performed the auto (un)mounting for the users, they already had a negative impression and were still somewhat reluctant to trust the system. The university has embarked on an active promotion of the ideas behind FOSS.

Phase II: The Staff

The second phase, the migration of the staff computers, was planned for the period June-August, 2004 but was delayed due to the problems in the first phase. In order to keep the migration on track it was decided to concentrate on the newly purchased computers. FOSS was installed on all new computers. Since almost all computers that the university was purchasing came pre-installed with Microsoft Windows operating system, a

Table 2. Applications without satisfactory or compatible FOSS alternatives

Task	Proprietary Software	Open Source Alternative
Financial application	Tally	-
Architectural design	Vector Works	-
Wordprocessing	Corel Word Perfect	-

dual boot system was installed with Linux as the default option of the two.

Some of the computers needed to continue to operate on Microsoft Windows because certain applications (see Table 2) were being used that have no satisfactory FOSS alternative yet.

The staff of the ICT department went around the university to install FOSS applications for Microsoft Windows platform on the staff computers. This was needed to support the format of documents that the students were sending to the staff. The staff was also informed that no support would be given to illegal proprietary software. Unfortunately, no member of staff, other than those in the CSIS and ICT departments, allowed their "personal computer" to be migrated to Linux. Only official work computers were migrated.

For the installations that were done on the university property being used by the staff, it was rare to find them using the FOSS alternatives that were provided for them. The few who tried using these alternatives had lots of complaints about the software not being able to perform the kind of tasks that they wanted.

Evaluation of Phase II

The second phase turned out to be even more difficult than the first phase. Although there were relatively few technical problems, the high level of resistance of the staff at the university virtually stalled the project.

The biggest hindrance in the whole project and especially in the second phase, is the acceptance of the new software by the staff. The users of Microsoft Windows find it difficult to switch to the new system. They feel that they are migrating to an inferior system and, as a result, small differences are capitalized upon, for example, the fact that the settings for the page layout are in a different location for Open Office makes them feel that the new package is inferior to the well-known Microsoft Office Suite. Arguments that the location of the page characteristics in Open Office display a more logical user-interface design are not accepted. The migration team concluded that the differences in the user interface were underestimated and too little information was provided on the reasons and consequences of the migration to get full user commitment. When introducing a new software environment—even when the differences are small—several training workshops highlighting the reasons and consequences of the changes should be planned.

The project also underestimated the number of Corel Word Perfect users and the problem of migrating their documents. Open Office can read and display Microsoft Office file formats relatively well, but there is no facility for doing the same with Word Perfect files. The fact that these files could not be displayed makes users hesitant to migrate regardless of the varying number of documents they have available in Word Perfect format. The ICT department is looking at ways to handle this problem. Some considerations at the moment include encouraging the staff to use Corel Word Perfect as a document reader only and to adopt Open Office for creating and editing new documents. The other consideration is to get document converters that can create PDF versions of older documents that the staff may need to keep as archives.

At the moment we observe a growing divide between the staff and the students in terms of the software used. Staff tends to continue to use proprietary software while students move more on the FOSS side.

LESSONS LEARNED: CRITICAL ANALYSIS

The migration at Uganda Martyrs University allowed us to draw some important lessons about a large scale migration to FOSS.

Installation of FOSS on the server-side proved to be a big technical challenge. There was little hands-on guidance and support available to help the system administrators in the university. In a country where the university was the first organization to migrate, there was no possibility to hire local technical experts to assist the staff on-site. Hiring support on the international market was not considered feasible due to financial limitations (the daily fee of international consultants is, in most cases, higher than the monthly salary of the local staff). Online support by the FOSS community proved to be too unreliable and often not applicable for the situation at the university. Therefore, the staff of the ICT department had to rely on their own research and much of the implementation was done through trial-and-error. The speed of the migration was, therefore, slow and demanded a lot of patience from the users.

Whereas Microsoft software applications provide a standard environment for the desktops, FOSS leaves more room for choice. Advantages and disadvantages of the different FOSS desktop applications are not well documented. At the university, this led to changing standards. Where Konqueror was the first choice for web browser, Mozilla was later chosen when it became clear that Konqueror had problems with viewing some commonly visited pages on the intranet and Internet that contained javascripts. We also observe a change from Bluefish to NVU for editing Web pages. These changing standards were confusing for most users. As far as end-users go, therefore, it would be helpful to pick standard well-developed packages taking into consideration the users possible future needs. End-users would want to spend most of their time being productive rather than learning the computer environment. However, there are no guarantees because new FOSS projects start up all the time and a better alternative might be developed.

The introduction and roll-out of the migration project at the university revealed that continuous information to users is needed. Their commitment and support of the project is essential for success. The approach at the university was a top-down approach with a presentation for management and senate, an initiation workshop, a mid-semester workshop for all staff and individual support for all users. This approach was not enough. Although the resistance to the changes seemed to diminish after the workshop and presentations, it proved to come back quickly, and stronger than before. The fact that the migration team was composed of technical personnel, but with strong support from the top management of the university and the vice chancellor as champion did not guarantee complete success.

The migration of the students before the migration of the staff seems to have been disadvantageous. The expectation that the staff would support new software and request for installation of FOSS on their machines turned out to be a miscalculation. Instead, several staff pushed students into using proprietary software formats, for example, when handing in assignments. Documents saved in Open Office format were not accepted. From our experiences it may be a wise option to get staff acceptance and migrate them before any attempts to migrate the students.

CONCLUSION AND RESEARCH AGENDA

In spite of the high expectations of the policy makers about the development potential of FOSS, the reality of implementing FOSS in an LDC is difficult. The route to the implementation of FOSS is one with a lot of hurdles. Some of these hurdles are general and some are specific to LDCs.

At a general level we observe that there is a strong resistance to changing to FOSS applications. Many users start a migration with the idea that they are confronted by an imperative of the "technical people" to use "inferior software." Their judgment is solely based on the experiences that they have with the desktop applications. It is prudent to gain user commitment and understanding before and during the migration phases. On the server-side where the migration is driven by the technical staff, the clear advantages are a strong motivator for the change to FOSS.

On the desktop the portability of files between FOSS and proprietary software is still a problem. Until this issue is solved, desktop migration will remain difficult. It is high time that proprietary software producers are coerced to adhere to internationally certified standards or to completely open up their own standards.

The need for education material for FOSS is high. The material currently available is mostly very technical and not understandable for the general users. Availability of student material, for example on Linux, Open Office, MySQL/php-MyAdmin, GIMP, and Bluefish, as replacements for the proprietary tools, may greatly improve the use of FOSS tools.

In the context of the LDCs the need for appropriate support in implementing FOSS is high. Experiences at Uganda Martyrs University show that the help received from the international mailing list community was insufficient since the questions posted were considered basic and not challenging to members on the list. On the other side, the discussions in the mailing lists were too

difficult and not (yet) applicable to the situation at hand. It seemed difficult to bridge the knowledge gap, and implementers felt isolated in their problems. In order to support the migration in LDCs international organizations like the World Bank or UNCTAD need to consider setting up a support center that deals with the questions of the system administrators and users in these countries.

Another specific problem in the context of the LDCs is the feeling that the access to the "good" tools from the West is denied. A question that was often asked was: "Why are the people in the West not using these (FOSS) programs when you are saying they are so good?" This argument is difficult to counter until there are success stories available from Western organizations. The situation gets even worse when the international organizations that promote the use of FOSS in LDCs only accept files in proprietary software formats (.doc, .xls, .ppt), have webservers that run on proprietary software and Web sites that can only be browsed optimally with Microsoft Internet Explorer.

Finally, pirated software is commonplace in LDCs. Pirated software is readily available at very low prices, and low cost support for installation often accompanies the sales. Many of the new computers that are bought in Uganda, for example, have full installations of pirated software. The computers that have valid licenses cost more than the individual is willing to part with. This applies to both servers as well as desktops. From a purely economic point of view, an individual is more likely to choose the "cheaper" option.

At present the development potential of FOSS for LDCs is still a theoretical potential. At the practical level more research, more support and a changed attitude of the organizations in developed countries is needed. Research should focus on the development of better tools to bridge the compatibility issues. More support is paramount to the success of the acceptance of FOSS in LDCs. Support should focus on practical help with the implementation of FOSS, but also for lecturers who want to use FOSS applications in their courses. More educational material, preferably published under the Open Content license, could act as a catalyst in an environment where the need for textbooks is extremely high. Finally, organizations working with LDCs should set an example by adopting FOSS as a standard in their organization. As long as organizations in the developing world need to communicate with their counterparts in the developed world by proprietary software standards and proprietary tools, the development potential of FOSS will remain a myth and never a real possibility.

REFERENCES

Association for Progressive Communications (APC). (2003). *ICT policy, A beginners handbook.* Johannesburg: Author.

Bruggink, M. (2003). *Open source in Africa: A global reality; take it or leave it?* IICD Research Brief (No UICT01). International Institute for Communication and Development. Retrieved from http://www.iicd.org

Dravis, P. (2003). *Open source software: perspectives for development.* Washington, DC: World Bank infoDev.

Government Information Officers' Council (GI-TOC). (2002). *Using open source software in the South African Government.* A Proposed Strategy Compiled by the Government Information Technology Officer' Council.

UNCTAD. (2003). *E-Commerce and development report 2003.* New York; Geneva: United Nations.

Van Reijswoud, V. (2003). *Open source software: the alternative for Africa* (Working Paper). Uganda Martyrs University, Nkozi.

Van Reijswoud, V., & Topi, C. (2004). Alternative routes in the digital world: Open source software in Africa. In P. Kanyandago & L. Mugumya (Eds.), *Mtafiti Mwafrika (African Researcher)* (pp. 76-94). Nkozi: Uganda Martyrs University Press.

Williams, S. (2002). *Free as in freedom: Richard Stallman's crusade for free software.* Sebastopol, CA: O'Reilly.

Wong, K., & Sayo, P. (2003). Free/open source software: A general introduction. *United Nations Development Program—Asia Pacific Development Information Programme (UNDP-APDIP).* New Delhi: Elsevier.

KEY TERMS

Africa: Africa is the world's second-largest and second-most populous continent, after Asia.

Desktop: In graphical computing, a desktop environment (DE, sometimes desktop manager) offers a graphical user interface (GUI) to the computer. The name is derived from the desktop metaphor used by most of these interfaces, as opposed to the earlier, textual command line interfaces (CLI). A DE typically provides icons, windows, toolbars, folders, wallpapers, and abilities like drag and drop.

Case Study: A case study is a particular method of qualitative research. Rather than using large samples and following a rigid protocol to examine a limited number of variables, case study methods involve an in-depth, longitudinal examination of a single instance or event: a case. They provide a systematic way of looking at events, collecting data, analyzing information, and reporting the results. As a result the researcher may gain a sharpened understanding of why the instance happened as it did, and what might become important to look at more extensively in future research.

Free and Open Source Software (FOSS): Free software is the term introduced by Richard

Stallman in 1983 for software which the user can use for any purpose, study the source code of, adapt to their needs, and redistribute—modified or unmodified.

Less Developed Countries (LDC): A developing country is a country with a relatively low standard of living, undeveloped industrial base, and moderate to low Human Development Index (HDI). The term has tended to edge out earlier ones, including the Cold War-defined "Third World," which has come to have negative connotations associated with it.

Productivity Software: Consumer software that enhances the productivity of the computer user. Examples are word processor, spreadsheet, software development environments and personal database software.

Software Migration: The managed process where a situation is changed into another situation. Migrating software means that the installed software is replaced by a newer or changed version with similar or extended functionality.

Uganda: Uganda, officially the Republic of Uganda, is a country in East Africa, bordered in the east by Kenya, in the north by Sudan, by the Democratic Republic of Congo in the west, Rwanda in the southwest and Tanzania in the south. The southern part of the country includes a substantial portion of Lake Victoria, within which it shares borders with Kenya and Tanzania. Uganda takes its name from the Buganda kingdom, which encompasses a portion of the south of the country, including the capital Kampala.

ENDNOTES

[1] The authors are well aware of the paradigmatic differences between free software and open source software. However, it is often difficult to clearly distinguish these differences. We, therefore, prefer to use the term free and open source software (FOSS) to capture both paradigms.

[2] Bundesrechnungshof fordert Einsatz von Open Source, 25.02.2002, http://www.heise.de/newsticker/data/anw-25.02.02-004

Chapter VIII
Open Source Software:
A Developing Country View

Jennifer Papin-Ramcharan
The University of the West Indies – St. Augustine Campus, Trinidad and Tobago

Frank Soodeen
The University of the West Indies – St. Augustine Campus, Trinidad and Tobago

ABSTRACT

This chapter presents issues that relate to developing countries' use of open source software (OSS) and the experience of these countries with OSS. Here the terms open source software (OSS), free/libre and open source software (FLOSS) and free software (FS) are used interchangeably. It describes the benefits of FLOSS including its superior quality and stability. Challenges to FLOSS use particularly for developing countries are described. It indicates that despite the greater benefits to developing countries of technology transfer of software development skills and the fostering of information and communication technology (ICT) innovation, the initial cost of acquiring FLOSS has been the key motivation for many developing countries adopting FLOSS solutions. It illustrates this by looking at the experience of a university in a developing country, The University of the West Indies, St. Augustine Campus in Trinidad and Tobago. Strategies for developing countries to benefit "fully" from FLOSS are presented including the implementation of formal organized programmes to educate and build awareness of FLOSS. The authors hope that by understanding some of the developing country issues that relate to OSS, solutions can be found. These countries could then fully benefit from OSS use, resulting in an increase in size of the global FLOSS development community that could potentially improve the quality of FLOSS and indeed all software.

INTRODUCTION

Open source software (OSS) is understood by many to mean software or computer programs where the source code is distributed and can be modified without payment of any fee by other programmers. The term OSS first came into use in 1998 and is attributed to Eric Raymond (Feller & Fitzgerald, 2002). The Open Source Initiative (OSI) has been formed to promote the use of OSS in the commercial world (www.opensource.org/).

The terminology related to software that is released with its source code and is modifiable and distributable without payment and then developed by a group of users or community can be confusing. For example the literal meaning of *open source* implies access to the source code, without necessarily implying permission to modify and distribute. Also, the Free Software Foundation (FSF) (founded in 1985) which predates the OSI refers to these programs not as OSS but as *free software* (www.fsf.org/). The term *free software* was created by Richard Stallman where *free* refers to the freedoms to use, modify, and distribute the programs and does not have anything to do with the cost of acquiring the software. Therefore, *free* software does not necessarily mean *zero cost,* and *open source* does not just mean access to the source code.

The differences between free software and OSS have been well documented (Fuggetta, 2003). Stallman (2002) gives details about these differences at www.gnu.org/philosophy/free-software-for-freedom.html. What really then defines software as OSS? Generally, OSS is a software product distributed by license, which conforms to the Open Source Definition[1]. The best known of these licenses are the GNU General Public License (GPL) and the Berkeley Software Distribution (BSD) license. Unlike traditional commercial or proprietary software (e.g., Microsoft Word, Windows XP, or Internet Explorer), these licenses permit OSS to be freely used, modified, and redistributed. The source code for these programs must also be freely accessible.

The term *free/libre open source software* (FLOSS) is used to refer to both free and open source software and was first coined in 2002 by Rihab Ghosh in a study undertaken for the University of Maastricht (Ghosh, Glott, Kreiger, & Robles, 2002). Libre here is the French word for liberty making clear the *free* as in freedom and not free as in "no cost." It is also common to use the term FOSS (free/open source software) for such programs.

Many countries do not have the luxury of debating the philosophical differences between the OSS or free software movement and so are content to use the all encompassing term of FLOSS (see e.g., www.floscaribbean.org/). For the purposes of this discussion the terms Free Software, Open Source Software and FLOSS are used interchangeably[2].

The development of FLOSS has often been contrasted to that of proprietary software. FLOSS has primarily been developed by individuals who volunteer their time to work on FLOSS projects. Normally, modified versions of the source code are posted on the Internet and are available for free to anyone who wants to use or modify it further. Thus a community of developers is created all working on modifications, bug fixing, and customizations of the initial code. An extensive explanation and analysis of OSS development can be found in Feller and Fitzgerald (2002).

The number of open source software projects can be gleaned by visiting http://sourceforge.net/index.php where there are over a hundred thousand such projects registered to date. Thus, FLOSS is not a fad or fringe phenomenon. It is important to note that FLOSS has penetrated major markets in countries worldwide. Indeed, some open source products like Linux and Apache are market leaders globally, and major ICT companies like IBM, Sun, and Oracle have adopted the open source model (Bruggink, 2003). In some countries, governments have even made the decision to support the use of FLOSS (Brod, 2003; Evans & Reddy, 2003).

Because of its free cost and its freedoms, FLOSS should be an obvious choice for widespread use in developing countries. In fact, these countries should be virgin territory for FLOSS deployment. What do the developing countries themselves say and what has been their experience? This chapter presents the point of view and experience of developing countries with FLOSS.

Table 1. Telecommunications infrastructure for Internet access; comparison of selected countries for 2002 (Source: United Nations Statistics Division-Millennium Indicators [ITU estimates] from http://unstats. un.org/unsd/mi/mi_series_list.asp rounded to the nearest whole number)

Countries	Telephone Lines and Cellular Subscribers/100 Population	Personal Computers/100	Internet Users/100
Trinidad and Tobago	53	8	11
United Kingdom	143	41	42
United States	114	66	55
Singapore	126	62	50
Sweden	163	62	57
Venezuela	37	6	5
Brazil	42	8	8
Chile	66	12	27
China	33	3	6
Guyana	19	3	14
India	5	1	2
Nigeria	2	1	0

BACKGROUND

Developing Countries

As is well-known, the term *developing country* applies to most African, Latin American, Caribbean, and Asian countries, as well as some countries in the Middle East and Eastern Europe. The definition of a developing country is generally based on that country's annual per capita income. Indeed, developing countries are most often defined following the World Bank classification (World Bank, 2006). According to the World Bank, the developing country's annual per capita Gross National Income (GNI) can range from:

- US$875 or less (low income)
- US$ 876-3,465 (middle income)
- US$ 3,466-10,725 (upper middle income)

Thus developing countries are not as homogeneous a group as some may think. Yet there are some common problems in all developing countries. For example, many such countries have unreliable electricity supplies (Ringel, 2004). Additionally, ownership of computers and access to the Internet is low when compared to developed countries (Table 1).

In simple terms, how useful is FLOSS without hardware or electricity or trained and skilled personnel? The vision of developing countries being able to leapfrog from the use of proprietary software into using FLOSS and benefiting from all its "freedoms" must be tempered with these realities (Steinmueller, 2001). This chapter presents many of these realities as they relate to FLOSS in developing countries. If the problems with FLOSS in developing countries could be solved, then such countries could fully participate in FLOSS development. This would increase the size of the global FLOSS development community thereby creating the potential for an increase in the quality of FLOSS and software in general.

Benefits of FLOSS

The benefits of FLOSS are well-documented in the literature particularly by Raymond (2002). These benefits include:

- Free or small cost of acquisition; future upgrades are free
- Flexibility of its license vs. restrictive licenses of proprietary software; the General Public License (GPL) used to license most open source software is much more flexible than the End-User License Agreement (EULA) of proprietary counterparts, giving more freedom to users to customize and to install on as many computers as needed without incurring added costs
- Superior quality and stability of FLOSS; because the source code is open to full and extensive peer review, open source software is known for its superior quality and stability
- Effectiveness as a teaching tool vs. closed proprietary software; users of FLOSS learn team work; importance of intellectual property protection and ethical use of software in addition to programming skills (Rajani, Rekola, & Mielonen, 2003)
- Potential as a solution to the software crisis; the "software crisis" refers to "software taking too long to develop, costing too much, and not working very well when delivered" (Feller & Fitzgerald, 2000, p. 58)
- Reduces the dependence of public administration and international governments in particular on specific software providers (Fuggetta, 2003); according to Nolle (2004), internationally, where Microsoft is viewed with more alarm than it is in the United States, FLOSS is seen as a defense against U.S. and Microsoft domination
- Stimulates innovation; FLOSS encourages the mastering of the technology of software by enabling the development and expression of creativity in the modification of the software by its users
- Improves commercial software
- Develops and enables applications that leverage local knowledge; because it can be freely modified, FLOSS is easier to translate, or localize (Bruggink, 2003)
- Fosters the creation of local software industry and entrepreneurs; the potential exists for the creation of local companies and small businesses supplying services associated with FLOSS in training, support, customization, and maintenance (Ghosh, 2003; Rajani, et al., 2003)

FLOSS Challenges

There are those who question most of the stated benefits of FLOSS particularly its claim to be innovative (Boulanger, 2005; Evans & Reddy, 2003; Fuggetta, 2003). Those on the side of proprietary software suggest that FLOSS is less secure, not as high in quality, stable, or dependable as its advocates insist. The very model of development of FLOSS that results in its best qualities can also lead to concerns about lack of support (Lai, 2006), security, and possible intellectual property violations by incorporating FLOSS into proprietary software (Kramer, 2006).

Compatibility concerns are also common. For example, although most FLOSS runs on both Microsoft Windows and Mac OSX, some run only on the Linux operating system. FLOSS may not come with as complete documentation and ready support as proprietary alternatives. Fees may have to be paid for substantial technical support. It should also be noted that there are fewer trained people available to provide technical support since most ICT training programmes prepare students to work with the most commonly used proprietary software packages, such as those from Microsoft (Bruggink, 2003). Additionally, FLOSS may require more learning and training time as well as skill to deploy and maintain. Large scale migration from proprietary software installations to FLOSS can be problematic, particularly if there is a lack of practical experience and support and ready information on migration issues (Bruggink, 2003; Van Reijswoud & Mulo, 2005).

Cost as the main driver for the adoption of FLOSS in developing countries cannot be ignored. Ghosh (2003) demonstrates this vividly by comparing license fees for proprietary software with the income per capita of selected countries. He concludes that in developing countries, "even after software price discounts, the price tag for proprietary software is enormous in purchasing power terms." This is further supported by the Free and Open Source Software Foundation for Africa (FOSSFA) (as cited in May, 2006) who report that countries in sub-Saharan Africa each year pay around US$24 billion to (mainly U.S.-based) software companies for the rights to use proprietary software.

Thus FLOSS provides an opportunity for developing country institutions to find cost effective solutions in many areas that could include electronic governance to online health and learning. But there is an even greater benefit of FLOSS to these countries. Following the old adage that it is better to teach a man to fish than to give him fish, there is some appreciation that OSS can be even more beneficial to developing countries because it can be a vehicle for technology transfer of software development skills, thus building local IT capacity and stimulating innovation (Camara & Fonseca, 2006; Ghosh, 2003). Yet, for many end-users and even institutions in these countries, the choice is not between FLOSS and proprietary software but between FLOSS and cheap pirated software. When faced with this choice there is very little incentive to consider FLOSS (Heavens, 2006).

Furthermore, limited Internet access and bandwidth may not allow regular interacting with FLOSS online communities for updates, documentation and help with problems (Heavens, 2006). In addition, jobs in the IT industry in these countries are often confined to large companies that place a high premium on skills in traditional proprietary software (e.g., Microsoft Certification and experience). Also for those uninformed about FLOSS in developing countries, there is much skepticism about its use since "free" is often equated with poor quality and expensive software with high quality and reliability. This is confirmed by Gregg Zachary (as cited in Fitzgerald & Agerfalk, 2005) in his personal communication about unsuccessful attempts to introduce FLOSS projects in Ghana.

Are these difficulties peculiar to some developing countries? As a contribution to the FLOSS debate it may be useful to present the experience of a major university in a developing country.

EXPERIENCE IN THE WEST INDIES

The University of the West Indies (UWI)

The University of the West Indies (UWI) was first established in 1948, as a college with a special relationship with the University of London to serve the British territories in the Caribbean area. There are three UWI Campuses, in three different West Indian islands: Mona in Jamaica, Cave Hill in Barbados and St. Augustine in Trinidad and Tobago.

FLOSS at the University of the West Indies – St. Augustine Campus

The St. Augustine campus of the UWI is located in the middle income developing country of Trinidad and Tobago. Rampersad (2003) gives a succinct description of FLOSS in Trinidad and Tobago and reports that "Proprietary software is used most in Trinidad and Tobago, and as such, Microsoft and its many applications have a strong grip on the IT market." The University of the West Indies just like other employers of IT personnel in Trinidad and Tobago places high value on proprietary software certification (e.g., MCSE). Additionally, agreements have been made with computer manufacturers like Dell for the supply of computers campus wide and these are naturally shipped with proprietary software.

It is therefore not surprising that, like many similar developing country institutions, the UWI, St. Augustine campus has no formal institutional policy for the use or deployment of FLOSS. Individual IT personnel and other staff members at UWI who become aware of FLOSS solutions have tried using these in their various departments or units. The main motivation for this has been the cost of FLOSS versus proprietary software particularly when licensing per-seat costs are considered in deploying software in large computer labs. The FLOSS software used so far at the university is shown in Table 2.

Were the other vaulted outcomes of FLOSS use in developing countries experienced at the UWI? Modification of source code, customization, and so forth, implies that there exists a certain level of programming skills locally. In Trinidad and Tobago, practical computer programming skills are in very short supply and so FLOSS is sometimes seen as just a cheap alternative to the high cost of proprietary software, nothing more.

Also, as is the case with most developing countries, UWI has a small IT staff fully engaged at any time on a multiplicity of projects. There is often no time to invest in modifying source code. A good example of how limited resources can affect the progress of FLOSS projects in particular is UWI, St. Augustine's Institutional Repository Project which is based on the open source DSpace software (www.dspace.org). The initial impetus for the implementation of an institutional repository at the UWI, St. Augustine campus was a need to expose the unique Caribbean resources housed in the West Indian collection of the library to the world via digitization.

DSpace was acquired in 2004 and was installed first on a test server at the UWI Main Library in early 2005. Yet the installation is still "ongoing" since it involves a steep learning curve for the staff charged with the technical implementation. Knowledge and skills in Linux, Apache, Tomcat, and Java programming required for a successful DSpace repository deployment are not readily available. Thus, progress on implementation of the repository has been slow (Papin-Ramcharan & Dawe, 2006). Like most developing countries which do not have in place a well developed IT infrastructure and highly skilled IT personnel, it has been found that the true total cost of ownership (TCO) of DSpace as a FLOSS institutional repository solution has been high.

FUTURE TRENDS

It seems clear that the initial cost of acquiring FLOSS has been the key motivation for many developing countries adopting FLOSS solutions. It is also clear that there are greater benefits that can be derived from FLOSS in terms of encouraging the development of local IT skills, the creation of jobs locally to support FLOSS, and the eradication of piracy of proprietary software. Independence from being hostage to a single proprietary vendor is also beneficial to such countries.

The benefits to a country and its citizens from FLOSS adoption can possibly be viewed along a spectrum. Some countries which are relatively

Table 2. FLOSS used at UWI St. Augustine

FLOSS	Type
Linux	Operating System
Open Office/Star Office	Productivity Software
PHP, PERL	Middleware
MySQL	Database
Moodle	Courseware
DSpace	Institutional Repository
Apache	Web Server

new to FLOSS will take time to fully exploit its potential, whereas those that are farther along will work on higher value FLOSS activities like customization. Eventually, developing country users could move from being just consumers of FLOSS to being equal participants in the global community by becoming initiators and creators of FLOSS projects (i.e., FLOSS developers). Further along the spectrum, local jobs and small businesses could be created to sell FLOSS support and maintenance services.

It also seems likely that for developing countries and others, there probably will never be a FLOSS-only or proprietary-only market. The future will be about choice, where both FLOSS and proprietary software will co-exist and decisions to acquire software will not be based on philosophy alone but follow the standard criteria used to select any software package.

CONCLUSION

The literature while emphasizing that FLOSS is obviously a cost effective solution for developing countries also extols its higher benefits. These include: its technology transfer potential, the creation of jobs, fostering of innovation and creativity, the reduction in piracy of proprietary software, the independence achieved from being hostage to a single proprietary vendor, and the ability to localize software products to local languages and conditions. These outcomes will not be achieved for most developing countries unless there are enhanced supporting mechanisms to foster FLOSS use. These can emanate from international agencies like those of the UN and World Bank whose interest lie (for example) in the sustainable development of developing countries. The mechanisms could include:

- Formal organized programmes to educate and build awareness of FLOSS in developing countries; this should not just be targeted to

IT personnel but to common users, governments, and other decision makers
- International agencies working presently to upgrade ICT skills and infrastructure in developing countries should work closely with the FLOSS "movers and shakers" to ensure that training is provided in these countries on commonly used FLOSS with emphasis on programming skills.
- Sponsoring agencies that support nongovernmental organizations (NGO) or other community organizations should require that FLOSS be considered for use in their operations and projects.
- Procurement agencies of governments and other bodies should be educated about FLOSS so that it can be seen as a viable alternative when procurement decisions are made.
- Examination and other education bodies must be encouraged in an organized and targeted manner to change the computer studies and science programmes in these countries from being mostly Microsoft-centric to include the study and use of FLOSS.

REFERENCES

Boulanger, A. (2005). Open-source versus proprietary software: Is one more reliable and secure than the other? *IBM Systems Journal, 44*(2), 239-248.

Brod, C. (2003). *Free software in Latin America: Version 1.2.* Retrieved August 18, 2006, from http://www.brod.com.br/file_brod//heisinki.pdf

Bruggink, M. (2003). *Open source software: Take it or leave it?* International Institute for Communication and Development (IICD) Report. Retrieved July 6, 2006, from http://www.ftpiicd.org/files/research/reports/report16.pdf

Camara, G., & Fonseca, F. (2006). Information policies and open source software in developing countries. *Journal of the American Society for Information Science and Technology (JASIST)* (pre-print version). Retrieved August 26, 2006, from http://www.dpi.inpe.br/gilberto/papers/ca-mara_fonseca_jasist.pdf

Evans, D. S., & Reddy, B. J. (2003). Government preferences for promoting open-source software: A solution in search of a problem. *Michigan Tele-communications and Technology Law Review,* 9(2). Retrieved August 22, 2006, from http://www.mttlr.org/volnine/evans.pdf

Feller, J., & Fitzgerald, B. (2000). A framework analysis of the open source development paradigm. In *Proceedings of the 21ˢᵗ ACM International Conference on Information Systems*, Brisbane, Queensland, Australia (pp. 58-69). Atlanta, GA: Association for Information Systems .

Feller, J., & Fitzgerald, B. (2002).*Understanding open source software development.* Reading, PA: Addison Wesley.

Fitzgerald, B., & Agerfalk, P. J. (2005, January 3-6). The mysteries of open source software: Black and white and red all over? In R. H. Sprague (Ed.), *Proceedings of the 38ᵗʰ Hawaii International Conference on System Sciences*, Big Island, Hawaii [CD-ROM]. Los Alamitos, CA: IEEE Computer Society Press.

Fuggetta, A. (2003). Open source software—An evaluation. *The Journal of Systems and Software, 66,* 77-90.

Ghosh, R. A. (2003, December). Licence fees and GDP per capita: The case for open source in developing countries. *First Monday, 8*(12). Retrieved August 20, 2006, from http://firstmonday.org/issues/issue8_12/ghosh/index.html

Ghosh, R. A., Glott, R., Kreiger, B., & Robles, G. (2002). *Free/libre and open source software study: FLOSS final report.* International Institute of Infonomics, University of Maastricht. Retrieved August 16, 2006, from http://www.flossproject.org/report/

Heavens, A. (2006, July 10). Ubuntu in Ethiopia: Is free such a good deal? [Blog post]. *Meskel square.* Retrieved August 13, 2006, from http://www.meskelsquare.com/archives/2006/07/ubuntu_in_ethiopia_is_free_such_a_good_deal.html

Kramer, L. (2006, April). The dark side of open source. *Wall Street & Technology*, 43-44.

Lai, E. (2006). Lack of support slowing spread of open-source applications. *Computerworld, 40*(8), 20.

May, C. (2006). The FLOSS alternative: TRIPs, non-proprietary software and development. *Knowledge, Technology & Policy, 18*(4), 142-163.

Nolle, T. (2004). Time to take open source seriously. *Network Magazine, 19*(4), 82-83.

Papin-Ramcharan, J., & Dawe, R. A. (2006). The other side of the coin for open access publishing—A developing country view. *Libri, 56*(1), 16-27.

Rajani, N., Rekola, J., & Mielonen, T. (2003). *Free as in education: Significance of the free/libre and open source software for developing countries: Version 1.0.* Retrieved August 6, 2006, from http://www.itu.int/wsis/docs/background/themes/access/free_as_in_education_niranjan.pdf

Rampersad, T. (2003). Free- and open-source software in Trinidad and Tobago. *Linux Journal.* Retrieved August 22, 2006, from http://www.linuxjournal.com/article/6619

Raymond, E. S. (2002). The cathedral and the bazaar. In *The cathedral and the bazaar,* 23. Retrieved July 18, 2006, from http://catb.org/~esr/writings/cathedral-bazaar/cathedral-bazaar/

Ringel, M. (2004). The interlinkage of energy and poverty: evidence from India. *International Journal of Global Energy Issues, 21*(12), 2746.

Stallman, R. (2002). *Free software, free society: Selected essays of Richard M. Stallman* (J. Gay, Ed.). Boston: Free Software Foundation.

Steinmueller, W. E. (2001). ICTs and the possibilities for leapfrogging by developing countries. *International Labour Review, 140*(2),193-210.

Van Reijswoud, V., & Mulo, E. (2005, March 14-15). *Free and open source software for development myth or reality? Case study of a university in Uganda.* Paper presented at a seminar on Policy Options and Models For Bridging Digital Divides Freedom, Sharing and Sustainability in the Global Network Society, University of Tampere, Finland. Retrieved August 22, 2006, from http://www.globaldevelopment.org/papers/Artikel%20OSS-UMUv2%5B1%5D.1.pdf

World Bank. (2006). *Data and statistics: Country classification.* Retrieved August 24, 2006, from http://web.worldbank.org/WBSITE/EXTERNAL/DATASTATISTICS/0,,contentMDK:20420458~ menuPK:64133156~pagePK:64133150~piPK:64133175~theSitePK:239419,00.html

KEY TERMS

Developing Countries: Developing countries are those that have an annual per capita income (Gross National Income [GNI]) between US$875 and US$10,725.

Free/Libre Open Source Software (FLOSS): Used to refer to both free and open source software making no distinction between them.

Free Software (FS): Computer programs that are not necessarily free of charge but give access to the source code and permit users the freedom to freely use, copy, modify, and redistribute.

Open Source Software (OSS): Software that meets the terms of the Open Source Definition (www.opensource.org/docs/definition.php). To be open source, the software must be distributed under a license that guarantees users the right to read, redistribute, modify, and use freely.

Proprietary Software (PS): Software that is normally owned by a company that typically restricts access to the source code to protect the company's intellectual property. The software is distributed as the "compiled" source code or executable code (the binary form of the program). Its use, redistribution, or modification is prohibited or severely restricted (e.g., Microsoft Word, Norton Antivirus).

Source Code: The list of instructions that make up a computer program written in a high level programming language (like C, Java or PHP) that humans can read, understand and modify.

Total Cost of Ownership (TCO): The full cost of deploying, maintaining and using a system (or software) over the course of its lifespan.

ENDNOTES

[1] Open Source Definition, Version 1.9. Retrieved July 15, 2006, from http://www.opensource.org/docs/definition.php
[2] It is important that FLOSS is not confused with terms like freeware and shareware. These terms are usually used to describe software which is available at no cost, but its source code usually is closed. Internet Explorer is one example of freeware that is proprietary.

Chapter IX
The Social and Economical Impact of OSS in Developing Countries

Alfreda Dudley-Sponaugle
Towson University, USA

Sungchul Hong
Towson University, USA

Yuanqiong Wang
Towson University, USA

ABSTRACT

Computing practices in developing countries can be complex. At the same time, open source software (OSS) impacts developing countries in various ways. This chapter examines the social and economic impacts of OSS on three such nations: China, South Korea, and India. In so doing, the chapter discusses and analyzes benefits as well as downsides of the social, political, and financial impacts on these developing countries. Topics covered in this chapter are piracy, software licensing, software initiatives, social and political components involved in OSS implementation, and software compatibility issues.

INTRODUCTION

Some countries, particularly the economically challenged, are still behind regarding hardware and software technologies. This chapter looks at the social and economic impacts of OSS on three technologically developing countries: China, South Korea, and India. The focus of the chapter is on how OSS is changing the social and economical structures in each of these countries. This chapter discusses and analyzes benefits as well as downsides of the social, political, and financial impacts on these developing countries. Topics covered in this chapter are piracy, software licensing, software initiatives, social and political components involved in OSS implementation, and software compatibility issues.

BACKGROUND

OSS in Developing Countries

Open source software impacts developing countries in various ways. Some impacts are positive for example, cost savings, flexibility of software, obtaining negotiation power against big software companies, fighting piracy, building its own software industry, and even increase national security by less dependence on a few foreign companies. The negative impacts would be maintaining the software quality and providing updating or service when the software environment is changed.

Many international governments are increasingly supportive of the use of OSS. "Open source software is often touted to be ideal for accelerating the growth of low-income countries' IT sectors, with the expectation that it will increase their propensity to innovate" (Kshetri, 2004, p. 75). In countries like China, Japan, South Korea, and India, there is political incentive toward the use of OSS. To insure commitment in the use of OSS, these governments have enacted policies and laws. In June 2002, the European Union's position on this issue was that governments (or public administrations) are not promoting OSS over proprietary software, but are optimizing investments in sharing developed software (Drakos, Di Maio, & Simpson, 2003). Whereas each government has its own political motivation toward the adoption of OSS, the decision must be carefully examined. Governments, specifically in developing countries, that are quick to implement OSS over commercial software for development must take into consideration whether the choice will bring about the required end results.

The popularity of OSS is driving vendors to meet the high demands, especially from the developing countries. For example, Sun Microsystems' executives have suggested that they are considering making their entire software stack open source over time (Galli, 2005). However, future changes in the way OSS is distributed (i.e., different types of licensing fees for support and maintenance, compatibility issues with other software and hardware technologies, licensing issues in software development) will bring about major changes in structure and costs.

Most open-source companies have long offered their software free and built business around value-added services and support. A much smaller number have been selling open-source software with premium-level add-on components for years; that model is not new. But the number of companies falling into the latter category appears to be increasing, which could eventually change the underlying structure of the open-source community, as we know it. (Preimesberger, 2005, p. 1)

With possible changes in the marketing and applications of open source software, the need for reassessment of policies will be eminent for developing countries in order to stay in the competitive technological global market.

MAIN FOCUS OF THE CHAPTER

Open Source Software and China

IT Status in China

China has presented a constant GDP growth of 8% over the past decade. China's GDP was more than $1.7 trillion in 2005. In addition, China's information technology industry has increased. Information technology (IT) in China has been moving forward as planed in their tenth Five-Year Plan (2001-2005) for economic development. The plan states "Information technology should be used extensively in all circles of society and the use of computers and Internet should be wide spread" (http://news.xinhuanet.com/zhengfu/2001-10/30/content_82961.htm). The tenth Five-Year Plan earmarks 1.7 trillion yuan (about $200 billion) for spending on information and communications technologies (ICT). According to a report published by IDC (2005), China's spending on IT exceeded $30 billion in 2005, will reach $35.08 billion in 2006 (a 13.7% increase over 2005), and is expected to be over $51 billion by 2009.

The Ministry of Information Industry of the People's Republic of China (www.mii.gov.cn/art/2005/12/18/art_942_1970.html) reported that sales revenue has surpassed 2063.15 billion yuan (about $250.7 billion) during the first nine months of 2005, a 21.3% increase compared to previous years. Meanwhile, the China Internet Information Center (CNNIC) has reported an increase in China's Internet population and broadband users with over 100 million new users by the end of 2005. This Internet population is the second largest in the world. Among Internet users in China, the number of broadband users has reached 64.3 million (www.cnnic.net.cn/html/Dir/2006/01/17/3508.htm).

China's domestic firms have been working hard to take over the personal computer market, which was once dominated by foreign firms. Currently, Legend, a Chinese computer company, has taken control of over 30% of the market share.

In China's software market, Microsoft, IBM, Oracle, Sybase, and China-based UFSoft and Kingsoft have all been big players, in which Microsoft has been dominating the market for quite a while. The domestic Chinese software market was $800 million in 2002 and was predicted to grow 25% a year. In 2005, the software market and IT services experienced a growth rate of 19% and 22% respectively. IDC (2005) predicted that China's software market and IT services opportunity will reach $7 billion and $10.6 billion respectively.

OSS Adoption in China

As part of a 5 year economic development plan, China has identified software as a strategic sector for development. One of the projects involves promotion of Linux applications.

CCIDNET.com, a popular Chinese Web site for technology news, reported that by year 2000, there were already over 2 million Chinese computer users taking up Linux (about 10% of the number using Windows). Recent statistics released by the Chinese government's Beijing Software Industry Productivity Center (BSTC) have shown Linux sales in China growing at more than 40% a year—increasing from $6.3 million in 2002 to $38.7 million by 2007 (GridToday, 2004). Most of this growth will come from the server environment. As more and more OSS is being adopted by Chinese businesses and government agencies, the exploration of the adoption of OSS in China shows various benefits why company and government agencies are willing to adopt OSS as well as problems associated with its adoption.

Benefits of Using OSS in China

Social and Political

Piracy

The problems of software piracy have plagued China for years. As part of their commitment in joining the World Trade Organization (WTO), China has promised to protect intellectual property. However, converting all pirated software into licensed software still presents a big challenge for the government. Because of its low cost, freedom to access, and flexibility to modify, open source software presents major opportunities to China's social and economic growth and development. As result, IDC (2006) reported "China, with one of the fastest growing IT markets in the world, dropped four points between 2004 and 2005" while worldwide piracy rate remains stable. This has been "the second year in a row where there has been a decrease in the PC software piracy rate in China. This is particularly significant, considering the vast PC growth taking place in the Chinese IT market" (IDC, 2006).

Building Its Own Software Industry

Moreover, OSS offers the opportunity for developing countries like China to get their software industry off the ground. By promoting OSS, China has developed its own operating system, database system, middleware, office automation software, and embedded systems. In addition, China has started exporting software to other countries. China has exported over 7,000 sets of the Chinese Linux product to 8 countries during its tenth five-year period.

Improving National Security

Microsoft signed a contract with the Chinese government to allow controlled access to source code and technical information (following similar agreements with India, Russia, NATO, and UK governments) in 2003 (Sina, 2003). However, the Chinese government may still be concerned with Microsoft's true intention to "get Chinese customers addicted to Microsoft software and then figure out how to charge them for it later" (Dedrick & Kraemer, 2001). Like other countries that emphasize the preference of open source software, such as Linux, China believes the adoption of OSS will reduce the dependency on Microsoft while keeping control of their systems.

Financial

Cost Savings

As previously discussed, proprietary software companies, especially Microsoft, have dominated China's software market. However, only small portions of these products are actually legally purchased. Microsoft has been known to request high licensing fees and force their users to upgrade or risk losing support. With over 59,000 PCs sold in year the 2003 alone, a tremendous amount of money will be needed to upgrade the system to a later version. This obviously poses a big threat to business in China.

Unlike the huge cost involved in buying and maintaining a proprietary system, OSS has built an image of "free" software—free to install while getting the flexibility to customize. Although there are costs associated with the software support and maintenance of OSS, it is still expected to have much lower costs than proprietary software. For example, Linux has been expected to cost up to 70% less than Windows (Einhorn & Greene, 2004). Not only China, other developing countries in the region have also presented their preference for OSS, such as Linux.

Flexibility

Unlike proprietary software, OSS such as Linux, developed with a GNU General Public License (GPL), usually allows people to copy, study, modify, and redistribute software. Migrating to this kind of system gives companies the opportunity to look inside and make changes to fit their special requirements. With the Chinese governments encouragement, Red Flag, a Chinese version of Linux, has been created by the Institute of Software at the Chinese Academy of Science in 1999. Since then, Red Flag has been gaining ground in the server operating environment (SOE), PC and handheld computers.

According to the China Ministry of Information Industry (MII), almost 70% of all software purchases in 2004 were of open source based products. Linux was adopted on about 45,000 systems in provincial government institutions.

For example, the China Ministry of Railways has deployed Turbolinux operating systems in 14 railway bureaus, 230 railway stations and more than 440 package processing stations, to encourage standardization for package delivery operation and management. This initiative was the first large-scale Linux implementation by the ministry (Trombly, 2005).

One of the branches under the China Construction Bank (CCB) has adopted Linux system on over 3,600 computers. The Industrial and Commercial

Bank of China (ICBC), China's largest bank, also announced its plan to switch to Linux for all of its front-end banking operations over a 3 year period. The reason behind this decision was the better performance and vendor support (Lemon & Nystedt, 2005). IDC reported that Linux server shipments in China rose to $9.3 million last year, up 20% from 2003. By 2008, it is expected that the Chinese Linux server and client-end operating system market will reach $41.9 million.

Obtaining Negotiation Power

The availability of OSS brings another benefit for government and business institutions: increased negotiation power. By adopting OSS in a small scale, users can reduce their dependency on a single vendor. When customers present the possibility of adopting OSS, their negotiation position increases in a vendor-driven market.

Drawbacks of Using OSS in China

Social and Political

Security and Stability

Although the government has been encouraging the adoption of OSS in China, compared to the business model Microsoft adopted, most end users are not familiar with OSS. People are suspicious about the stability and security of the OSS.

Financial

Maintenance Cost

Although one of the biggest perceived benefits of OSS is the cost savings, lack of qualified personnel to handle the system development and maintenance has contributed to the higher cost in system maintenance. Based on the discussion with five companies that tracked their total cost, Forrester research (2004) reported that Linux

could be between 5% and 20% more expensive than Windows unless the company is migrating from UNIX to Linux or is deploying Linux from the beginning.

Support and Application

Lack of Available Device

Another problem associated with OSS adoption is the availability of device support and application.

What happens quite often is that a vendor provides a Linux solution to a company, but the printer the company is using is not supported on Linux. Also many companies have already developed Web sites that are not following W3C standards or are tailored to (Microsoft's) Internet Explorer. If companies use Firefox, they cannot read these Web sites properly. (Marson, 2006)

This lack of support is also illustrated by the weak support of the Chinese interface. A survey conducted by Beijing Software Industrial Park shows that Chinese interface in Linux still does not work well. For example, it is difficult to input Chinese on Linux although it can display Chinese characters; only few Chinese fonts were supported; it cannot even display the whole Goggle China homepage correctly on a Linux supported computer (Han, 2006).

Summary of OSS and China

With the benefit of governmental support, OSS, represented by Linux, has been the basis of much of China's information technology growth. Hence, OSS has pushed Chinese software industry to move forward with its own technological innovations. Moreover, the introduction of OSS has created a more competitive environment in the software industry, which helps reduce domination by a single vendor, increases cost savings, and opportunities in training software personnel.

Open Source Software and South Korea

IT Status in South Korea

Open source software is considered an alternative to proprietary software, for example, Microsoft in South Korea. According to a news report in 2003, the South Korean government encourages the use of OSS (ZDnet, 2003). Specifically, the South Korean government is very interested in using OSS in various government branches and government-supported systems. According to South Korea's OSS recommendation, OSS will have higher priority in government supported software projects (www.partner.microsoft. co.kr/Partner_Portal/newsclipping/Uploaded_ Files/050120.htm).

One of the government branches that support OSS is South Korea's IT Industry Promotion Agency (KIPA). Among teams in KIPA, the OSS support team is leading the way in South Korea's technological advancements. The practices of the OSS support team in KIPA are acquiring OSS and supporting the software industry by creating the OSS markets such as Linux-based systems.

Stimulated by the South Korean government's recommendation, many software companies in South Korea, for example, IBM, HP, SUN, and Samsung are preparing open source applications. Companies like Samsung SDI, KT, Postdata, and SK C&C show great interest in the OSS market.

"The South Korean government has announced that by 2007 it plans to replace proprietary software with open-source alternatives on a substantial number of its PCs and servers" (ZDnet, 2003). The authors assert that thousands of computers in ministries, government-based organizations, and universities in South Korea will replace Microsoft's Windows operating system and Office products with open source alternatives. Kim and Song (2003) further state that if change is successful then the South Korean government may save $300 million a year. Despite skepticism, from Microsoft, the South Korean government's main impetus will be to promote competition in the software market. South Korea is not alone in this endeavor. The countries of Japan, China, and South Korea met in the Cambodian capital of Phnom Penh to sign an agreement to jointly research and develop non-Windows, open source operating systems (ZDnet, 2003).

OSS Adoption in South Korea

The most dominant OSS is Linux. Linux sales have increased sharply in recent years. The number of Linux server sales in 2002 is 2,216 and it increased to 4,878 in 2003 as shown in Table 1. The total OS sales in 2003 are 1,703,904,000,000 Won, which is around US$1,725,000,000. The Linux sales figure is 38,368,000,000 Won, which is around US$39,000,000.

Moreover, the market share of Linux (shipment base) has increased from 12.1% in 2003 to 18.5% in 2004 with the expectation to reach 21.2% in 2007 (IDC, 2006; KIPA, 2003).

Another adoption example is the PDA market. The picture for the use of OSS in the Korean PDA market does not look promising. Only one company, Gmate sells PDAs which uses Linupy as its operating system. Linupy is a Linux-based operating system. Gmate sold 4,520 units of PDAs in 2003 and its Korean market share is 2.1% (IDC, 2006; KIPA, 2003). The market share demonstrates that OSS in the Korean PDA market is low. However, unlike other companies that use licensed operating systems, Gmate does pay licensing fees and its operating system can be modified and expanded freely. Because of the flexibility and low costs associated with this open source operating system, it can be an advantage to other companies (KIPA, 2003).

Table 1. Server OS Sales in Korea, 2001-2003 (Source: IDC, KIPA)

OS	Data	Year			Total
		2001	**2002**	**2003**	
Linux	Unit	2,235	2,216	4,878	9,329
	Revenue	29,619	28,406	38,368	96,393
Others	Unit	3,478	2,260	1,968	7,701
	Revenue	397,263	324,005	271,588	992,856
Unix	Unit	12,796	12,034	15,861	40,691
	Revenue	1,179,445	1,231,500	1,101,065	3,512,010
Windows	Unit	27,477	20,479	28,915	76,868
	Revenue	377,424	273,502	292,883	943,810
Unit Total		45,982	36,986	51,622	134,589
Revenue Total		1,983,751	1,857,413	1,703,904	5,545,068

Revenue Unit: 1,000,000 (Won)

Note: 1 $ = 987.80 Won (Jan. 13, 2006)

Benefits of Using OSS in South Korea

Social and Political

One of the South Korean government's priorities is to obtain advanced software technologies for their software industry market. The South Korean government recognizes that there is a technology gap between developing countries and advanced countries. However, the government perceives OSS as a vehicle to minimizing this technological gap for their country. Additionally, OSS is a good resource to create markets for other information technology developments.

Financial

The biggest benefit of OSS is saving money. This fact is well demonstrated in the survey results shown below. Companies in South Korea do not need to pay the royalty to proprietary software companies. In addition to the price, upgradeability, stability, and availability of special software are also important issues.

Although the previous section indicates that Linux growth and sales have increased, businesses in South Korea have been slow to adopt Linux. In previous years, the acceptance of Linux among South Korean companies was pretty low. Kim and Song (2003) conducted a survey to measure the use of Linux in South Korea. There were 124 South Korean IT professionals respondenting to this survey.

The results of the survey indicated the following:

- Eighty-Five percent of surveyed companies do not use Linux.
- Only 3% of companies use Linux all the time.

The major reasons for this low acceptance of desktop Linux were:

- Not familiar (54.9%)
- Difficult to use (17.6%)
- Limited number of applications (13.7%)

The major reasons why companies select desktop Linux were:

- Price (33.3%),
- Stability (26.7%)
- Educational purposes (26.7%)

The major reasons for selecting Linux servers were:

- Safety (72.2%)
- Stability (61.1%)

Drawbacks of Using OSS in South Korea

Social and Political

The biggest factor for resisting the use of OSS in South Korea is low confidence. Customers do not trust the quality of the software especially interoperability and security. Moreover, the difficulties of software installation, lack of office productivity, type of software, lack of various tools, data compatibility, and unfamiliarity of UNIX commands are considered weak points (Kim, Yea, & Kim, 2003).

Financial

Even though OSS does not require royalty for its usage, maintaining and upgrading require monetary investment. However, countries like South Korea can use domestic manpower to solve this problem by establishing government-backed technical institutions devoted to OSS support and training.

Summary of OSS in South Korea

South Korea's OSS market is in the early stages and its market share is small. However, various data shows that the gains in the market share will increase. In the future, OSS will gain its market share rapidly because of governmental incentives and low costs of OSS. The South Korean government will continue to promote OSS because of its price and chance of acquiring developing software technology. In addition, the South Korean government hopes that an increase in OSS competition will alleviate Microsoft's domination in the software market.

Open Source Software and India

IT Status in India

India is one of the largest democratic governments and the most impoverished country in the world. The 2001 World Development Report indicated that "the average GNP per capita in India was only US$450 per year, 45% of adults were illiterate, and about one out of twelve children die before the age of five" (as cited in Warschauer, 2004). The status of India's poor economy has remained constant.

Interestingly, India is also becoming a country known for information technology. "India has one of the largest and most developed information technology industries in the world. This industry has created a tiny group of multimillionaires and a small middle class of network and software engineers, computer programmers, and computer-assisted design specialists" (Warschauer, 2004, p. 23). Table 2 shows the size and growth rates for India's ICT markets:

India has shaped the model for global marketing of information and communication technology for companies in Europe and the United States. India has become one of the United States' largest outsourcing countries for information and communications technology services. To indicate this trend, the following are examples of India's dominance in the outsourcing markets:

- *India already accounts for the largest number of IBMers outside of the U.S. (it recently surpassed Japan).*

Table 2. Indian domestic enterprise ICT market size and growth (excluding the offshore IT outsourcing and business process outsourcing markets) (Source: Iyengar, 2005)

$ In billions	2003	2004	2005
Total	*$16.73*	*$19.61*	*$22.88*
By Segment:			
Hardware	2.40	2.75	3.34
Software	0.40	0.44	0.52
Telecommunications	12.22	14.46	16.70
IT Services	1.71	1.96	2.32

- *In 2004, Big Blue acquired India's Daksh e-Services, whose 6,000 employees operate call centers for companies like Amazon.com and Citicorp.*
- *Goldman Sachs calculates that by the end of next year, IBM Services' headcount in India will top 52,000. That would be more than one-fourth of all its services personnel and about one-sixth of IBMers worldwide. It would put IBM in India on a par with Wipro, the largest local software company, and make it bigger than Infosys and Tata Consultancy Services.* (Kirkpatrick, 2005, p, 129)

Only China has surpassed India in economic and technological areas. However, India surpasses China in commercial software development. This gap is substantial. "India's software exports exceeded $12 billion in 2003, compared to China $2 billion" (Kshetri, 2004, p. 86). Globalization is an aspect in which choosing the type of software is a critical decision for businesses and governments. Globalization can affect decisions at every level of software development (Kshetri, 2005).

OSS Adoption in India

Open source software use has proliferated throughout the India IT culture. OSS is highly supported by the government and businesses in India. An example of adoption of OSS in India is the following: OSS groups are distributing free copies of desktop productivity software with the assistance of the Indian government. The software package contains an open source version of e-mail, word processing applications, and optical character recognition that can be run on Linux or Windows. By developing localized versions of these products for several regions of the country, this distribution would be easy, fast, and most importantly, free. There are some proprietary issues associated with the distribution of this software. However, open source advocates believe that this creates an "opportunity to proliferate free software" (Bucken, 2005).

Benefits of Using OSS in India

Social and Political

Piracy

The Indian government looks to OSS to assist in solving concerns such as software piracy and digital divide issues. The perception is that software piracy is practiced in these countries primarily due to high costs associated with commercial or proprietary software. It is the Indian government's assessment that the increased use of OSS will decrease piracy. It is believed that because OSS is free it has no limitations and/or contractual restrictions. This is not the case because OSS, like commercial software, has solid copyright protections (i.e., Open Source Initiative, GNU GPL, etc.). Piracy is the lack of respect for intellectual property. Piracy practices with OSS are evident in a different manner. Just like piracy is practiced with commercial software by illegal copying and distribution, combining open source code within proprietary source code and distributing it as new code is pirating. This practice is contrary and illegal to OSS and commercial software copyright licenses. In this regards, OSS is not the solution to combat piracy. The change in piracy practices would be more beneficial by education and enforcement of intellectual property laws. "Without a fundamental appreciation of the importance of intellectual property to a nation's economic growth, the mere promotion and adoption of open source solutions may not, in and of itself, lower piracy levels in a particular country, nor necessarily create an environment that its conducive to the growth of a domestic software industry" (Business Software Alliance, 2005, p. 18).

Digital Divide

India employs OSS as a strategy to combat the digital divide by providing access to information technology. Many developing countries are adopting policies to make low cost technology available to low-income populations. OSS is seen as way to limit costs and increase productivity to the main populace. However, the digital divide problem is multi-layered and cannot be solved without looking at all variables involved. Open source can be one approach to the solution of the digital divide, but it cannot be the only approach to this problem.

In most poor developing countries, there are no infrastructures in place to support information and communication technology. Substantial portions of India's population live in rural areas. To address this problem, India's government has implemented Simputer (simple, inexpensive, mobile computer) to carry out its initiatives to provide low-cost information technology. The Simputer project was introduced at an international conference on information technology in Bangalore, October, 1998 (Warschauer, 2004). Simputer uses convergent and collaborative technologies, which are adaptable to India's rural infrastructure and fit the needs of the lower class. The cost of the technology is still too high for the poor and lower classes; however, the technology is created so that several users can share it. India has used

OSS to create a citizen access project called the Simputer personal digital assistant www.simputer.org/simputer/faq (cited in Drakos, Di Maio, & Simpson, 2003).

Education

Another important need for developing countries to close the digital divide gap is to increase information literary skills. OSS has the flexibility to be adapted in educational projects and public assistance projects.

OSS companies are assisting in providing software and training to poor counties. As an incentive to equip classrooms in Goa, a colony on the west coast of India, Red Hat provided software and training in GNU/Linux. The India Linux User's Group supports the project.

Financial

India has adopted OSS as an alternative to using commercial software products. OSS allows India to compete in the software industry, which has been subjugated by proprietary software products. The flexibility of open source gives India the freedom to participate in the software development market, as well as the services industry.

The financial aspects in using proprietary or OSS are of the utmost importance to the Indian government. In regards to OSS use in developing regions, such as India, the frequent premise is that it is free. However, there are costs associated with some types of OSS, such as Linux. Embedded in these costs are the support and maintenance fees of this type of open source software. The expectation of these costs is still lower than the cost of proprietary software. Countries in developing regions (i.e., China, South Korea, and India) have "publicly stated a preference for the lower costs and higher security" that Linux provides (Wahl, 2004, p. 15). The common practices in India with the use of proprietary software are "also free due to piracy and the legitimacy of software [which] becomes an issue only when the funding source requires it" (Brewer, Demmer, Du, Ho, Kam, Nedevschi, et. al, 2005, p. 36). Even though the costs associated with licensing fees are only a part of the total price, it seems like a significant factor in countries where labor and support costs are lower.

Drawbacks of Using OSS in India

There are still economic and social inequalities within India. While India's information technology growth has been beneficial to some segments of the country, it has had little economical and social impact on the country's overall poor and lower class populations.

Outsourcing information technology, especially software development, to countries like India has proven to be financially beneficial to businesses and governments. However, outsourcing software development can be a two-edged sword. India has lax or non-existent laws and policies regarding software piracy and privacy. This can be potentially dangerous to entities that deal with the processing of critical information. Trying to enforce laws to rectify these problems are still problematic.

Summary of OSS in India

The utilization of OSS will continue to be a prominent factor in the growth of India's economy and society. India has invested in the OSS industry. OSS gives India the ability to compete with and benefit from wealthy countries. OSS provides India, as well as other developing counties, with options in software use and development. It is the authors' positions that India's government and private businesses' estimation of OSS is that the benefits outweigh any associated negatives.

CONCLUSION

From a financial point of view, the benefits of the perceived cost savings and flexibility associated with the adoption of OSS are common to countries discussed in this chapter. These benefits have been the very reasons for governments in developing countries to encourage the adoption of OSS. Some developing countries have also used OSS as a kind of negotiation tool to get better technology deals in the global community. However, the possible increase of costs associated with considerable maintenance costs, lack of qualified personnel as well as the shortage of supporting applications and devices, and language support have presented problems for OSS adoption.

The adoption of the OSS in developing countries discussed in this chapter has also presented some social and political benefits and drawbacks. Because of the nature of these developing countries, they are facing a bigger technology gap as compared to more technology-developed countries. The OSS adoption has been regarded as one of the possible ways for these countries to train their own personnel and to build their own IT industry.

Developing countries are using the adoption of OSS to combat the proliferation of software piracy. Although the adoption of OSS alone cannot eliminate piracy, it certainly has contributed to the decrease in the number of piracy cases in these developing countries (e.g., China).

Overall, cost savings in OSS initialization, personnel training, the promising future of developing a software industry, and fighting piracy issues are impetus of OSS adoption in developing countries. However, before successfully adopting OSS, consideration should be taken on the issues of training and obtaining qualified personnel, seeking more applications, device and language support to break the barrier of adopting OSS.

REFERENCES

2005 Information Industry Economic Report. (2005). Retrieved December 10, 2005, from http://www.mii.gov.cn/art/2005/12/18/art_942_1970.html

Brewer, E., Demmer, M., Du, B., Ho, M., Kam, M., Nedevschi, S., et al. (2005, June). The case for technology in developing regions. *IEEE Computer Society*, 25-36.

Buckin, M. (2005, November 28). Paris Government plots next open-source move. *Computerworld.* Retrieved from http://itreports.computerworld.com/action/article.do?command=viewArticleBasic&taxonomyName=&articled=106527&taxonomyId=015&intsrc-kc_li_story

Business Software Alliance. (2005). *Open source and commercial software: An in-depth analysis of the issues.* Washington, DC. Retrieved from http://www.bsa.org/usa/report

CCIDNET.com. (2005). Retrieved September 14, 2005, from http://www.ccidnet.com

Dedrick, J., & Kraemer, K. (2001). *China IT Report: 2001.* Center for Research on Information Technology and Organizations. Globalization of I.T. (Paper 252). Retrieved January 3, 2006, from http://repositories.cdlib.org/crito/globalization/252

Drakos, N., Di Maio, A., & Simpson, R. (2003, April 24). *Open-source software running for public office.* (Gartner Research ID. No. AV-19-5251).

Einhorn, B., & Greene, J. (2004, January 19). Asia is falling in love with Linux; as more IT managers and state agencies ditch Windows, Microsoft is scrambling. *Business Week, 3866,* 42. Retrieved September 14, 2005, from http://proquest.umi.com/pqweb?index=O+did=526134651+srchmode=l+sid=l+fmt.3+vinst=prod+vtype=PQD+R

QT=309+VName=PQD+TS=1174921439+clientID=41150

Galli, P. (2005, November 30). Sun gives away Java Enterprise system, other software. *eWeek.* Retrieved July 10, 2006, from http://www.eweek.com/article2/0,1895,1894747,00.asp

Giera, J. (2004, April 12). *The costs and risks of open source: Debunking the myths.* Forrester Research Best Practices Excerpt. Retrieved January 3, 2006, from http://www.forrester.com/Research/Document/Excerpt/0,7211,34146,00.html

GridToday. (2004, March 29). Beijing software testing center joins OSDL. *Daily News and Information for the Global Grid Community, 3*(13). Retrieved January 3, 2006, from http://www.gridtoday.com/04/0329/102947.html

Han, Z. (2006). *Market war between Linux and Windows.* Retrieved June 20, 2006, from http://www.bsw.net.cn/data/news/f8eWb64F0Lp/index.htm (Chinese Version).

IDC. (2006, May 23). *Study finds PC software piracy declining in emerging markets while worldwide piracy rate remains stable.* IDC Press Release. Retrieved June 3, 2006, from http://www.idc.com/getdoc.jsp?containerId=prUS20178406

Iyengar, P. (2005, March 28). *State of the information and communication technology industry in India.* (Gartner Research ID. No. G00126192).

Kirkpatrick, D. (2005). IBM shares its secrets. *Fortune, 152*(5), 128-136.

Kim, S. G., Yea, H. Y., & Kim, J. (2003, August). Survey on usage, and obstacles to introduction of open source software. *KIPA Report.* Retrieved from http://kipa.re.kr/eng%20site/publication/report.asp

Kim, I. W., & Song, Y. (2003, December). A study on the structure of TCO for the open source software. *Internal Report for KIPA.* Retrieved from http://kipa.re.kr/eng%20site/publication/report. asp

KIPA. (2003). *PDA market trend.* Retrieved from http://kipa.re.kr/eng%20site/publication/report. asp

Kshetri, N. (2004, Winter). Economics of Linux adoption in developing countries. *IEEE Software,* 74-80.

Lemon, S., & Nystedt, D. (2005, July 18). Global Linux: Asia. *Computerworld.* Retrieved September 22, 2005, from http://www.computerworld. com/printthis/2005/0,4814,103185,00.html

Marson, I. (2006). The business of Linux in China. *ZDNet UK.* Retrieved May 10, 2006, from http://www.zdnetasia.com/insight/software/0,39044822,39351644,00.htm

Partner Portal. (n.d.). Retrieved January 30, 2006, partner.microsoft.co.kr/Partner_Portal/newsclipping/Uploaded_Files/050120.htm

Preimesberger, C. (2005, September 23). Will profit motives fragment open-source community? *eWeek.*

Prothero, E., Farish, R., & Yang, D. (2005). ICT market outlook in four emerging markets: Brazil, Russia, India, and China (IDC #LA3617). Retrieved January 3, 2006, from http://www.idc. com/getdoc.jsp?containerID=LA3617

Sina news report. (2003). Retrieved from http:// tech.sina.com.cn/s/n/2003-02-28/1420168775. shtml

Trombly, M. (2005). Chinese companies pick Linux to boost their own skills. *CIO Central.* Retrieved December 10, 2005, from http://www. cioinsight.com/print_article2/0,1217,a=161108,00. asp

Wahl, A. (2004). The Linux gambit. *Canadian Business, 77*(7), 15.

Warschauer, M. (2004). *Technology and social inclusion: Rethinking the digital divide.* Cambridge, MA: TMIT Press.

ZDNet News. (2003, October 1, 8:05am PT). *Korea launches a switch to open source.* Retrieved September 15, 2005, from http://news.zdnet. com/2100-3513_22-5084811.html

KEY TERMS

CCIDNET: An online service provider, China's IT portal.

Convergent Technologies: The combination of several industries, (i.e., communications, entertainment, and mass media) to exchange data in a computerized format.

Collaborative Technologies: Combination of hardware and communications technologies that allow linkage among thousands of people and businesses to form or dissolve anytime, anywhere.

GDP: Gross domestic product.

GNU: Free licensing software initiative (i.e., operating systems software, OSS, etc.).

ICT: Information and communication technology.

IDC: International Data Corporation.

Linupy: Linux-based OSS.

Section II
Development Models and Methods for Open Source Software Production

Chapter X
Dependencies, Networks, and Priorities in an Open Source Project

Juha Järvensivu
Tampere University of Technology, Finland

Nina Helander
Tampere University of Technology, Finland

Tommi Mikkonen
Tampere University of Technology, Finland

ABSTRACT

Dependencies between modern software projects are common. Jointly, such dependencies form a project network, where changes in one project cause changes to the others belonging to the same project network. This chapter discusses the issues of dependencies, distances, and priorities in open source project networks, from the standpoint of both technological and social networks. Thus, a multidisciplinary approach to the phenomenon of open source software (OSS) development is offered. There is a strong empirical focus maintained, since the aim of the chapter is to analyze OSS network characteristics through an in-depth, qualitative case study of one specific open source community: the Open Source Eclipse plug-in project Laika. In our analysis, we will introduce both internal and external networks associated with Laika, together with a discussion of how tightly they are intertwined. We will analyze both the internal and the external networks through the elements of mutuality, interdependence, distance, priorities, different power relations, and investments made in the relationships—elements chosen on the basis of analysis of the network studies literature.

INTRODUCTION

Dependencies between modern software projects are commonplace. Jointly, such dependencies form a network, where changes in one project, or part thereof, cause changes in others. In using formal contracts applicable in the traditional industrial setting, these dependencies are defined by legalities and customer/subcontractor relations, which can be easily managed. However, in an open source project, dependencies are based not on some explicitly defined formalization but instead on how the different developer communities view and use each other and themselves. Furthermore, the issue of project priorities requires similar consideration.

In this chapter, we discuss dependencies, networks, and priorities in OSS development. As an example community we use Laika, an Open Source Eclipse plugin project that eases code creation for the Maemo platform. We discuss both external networks, consisting of communities that relate to (or are related to by) Laika, and internal networks that include the developers of the system. The contribution of the chapter and its underlying research question lie in establishing a connection between established network theory and practices in OSS development, on the one hand, and in discussing the organization, evolution, and values leading to the priority selection established in the Laika community, on the other. More precisely, we address the rationale of establishing a mode of cooperation between different developer communities as well as internal networking within a single community where several organizations are involved. This supplies a context in which to study the approach taken to work allocation, which will also be addressed. This chapter is inspired by the background in which two of the authors were directly associated with Laika, with the other having the role of an external observer.

The rest of the chapter is structured as follows. Next we introduce a related theory of networks that we use as a guide in analyzing the properties of Laika. We then discuss Laika and its internal and external networks, and provide a discussion of the goals of the chapter. Finally, we discusses future avenues of research, and offer final remarks.

BACKGROUND

Network Approach as Theoretical Framework

Networks are a contemporary topic that has been studied from several different perspectives and under various scientific disciplines. The term "network" can refer to, for example, an information network in the form of interconnection of layers of computer systems (Macdonald & Williams, 1992; Meleis, 1996); a social network in the form of a social structure among actors, mostly individuals or organizations (Baker, 1994; Barnes, 1972; Hill & Dunbar, 2002; Scott, 2000; Wasserman & Faust, 1994); or a business network in the form of a set of exchange relationships between organizations (Achrol & Kotler, 1999; Easton, 1992; Håkansson & Snehota, 1995; Möller & Halinen, 1999; Möller, Rajala, & Svahn, 2002). In this chapter, we use the term network theory to refer to the so-called network approach introduced by a group of scholars basing their work on theories of social exchange coupled with more economically oriented industrial insights (Möller & Wilson, 1995). The network approach discussed in this chapter aims at providing conceptual tools for analyzing both structural and process characteristics of networks formed among different open source projects and within a single specific open source project, Laika.

Early developers of the network approach, Håkansson and Snehota (1989) point out that the network approach takes into consideration the relationships among various actors. All of the actors, their activities, and their resources are bonded, linked, and tied up together, and in this way they build up a wide network. A basic assumption with the network approach involves the relationship as a fundamental unit, from which proceeds understanding of the network as a sort of cluster of relationships.

Easton (1992) illustrates the basic elements of the network approach from four different angles: networks as relationships, positions, structures, and processes. These basic elements are useful tools for analysis of network dependencies. Here, these elements are considered in the context of open source projects.

Relationships are characterized by four basic elements: mutuality, interdependence, power relations, and investments made in the relationship (Easton, 1992). Mutuality, interdependence, and power relations may vary a great deal from one open source project to the next. Dependencies between two projects can be two-way, leading toward mutuality and usually more balanced power relations between the projects. However, one-way dependencies are also commonplace (i.e., an open source project is dependent on another open source project but not vice versa). This usually leads to unbalanced power relations between the two projects since only one of the parties of the dyad is dependent on the other. Additionally, such asymmetrical power can be present even within a single open source project. In fact, it is more common for there to be, at the heart of the project, a few central developers with more power in the community than the peripheral developers have. These powerful actors can then influence the future direction of the system developed, work allocation, and equally important decisions made within the project.

Another important issue in considering networks as relationships is the nature of the effects of the relationships on the functionality of the whole network; the effects of a relationship can be both positive and negative. Additionally, both primary and secondary functions can be found in relationships. Primary functions refer to the relationship's effects on the members of the dyad, whereas secondary functions refer to the effects that the relationship has on other actors in the network (Anderson, Håkansson, & Johanson, 1994). The latter can be seen in the open source environment in the form of, for example, conflict between two central actors in a project creating difficulties for the functionality of the whole community.

Networks as structures are concretized through the interdependencies of the actors. If there are no interdependencies between actors, neither will there be any network structure. The greater the interdependence of the actors, the clearer the structure of the network. Thus, there can be tight and loose networks. Tight networks are characterized by a great number of bonds between the actors, along with well-defined roles and functions for actors. Loose networks, on the other hand, manifest the opposite characteristics (Easton, 1992). Also, the structures of projects within the open source environment can vary rather a lot in their level of tightness or looseness, as is discussed by Eric Raymond (1999).

Analysis of networks as positions mainly involves examination of the network from the viewpoint of a single actor. Within one open source project, the position analysis is performed mainly at the level of individuals. But when we leverage the analysis from one project to several, the level of analysis changes to that of entire communities; that is, we analyze the positions of different open source projects against the background of each other. The level used in network analysis is an interesting issue that has been discussed a great deal by network researchers in general, also outside the open source context (see Tikkanen, 1998; Möller et al., 2002). In our study, we differentiate between two levels of network analysis, examination within the context of a single open source project and consideration involving several open source projects.

Consideration of networks as processes mirrors the nature of the networks themselves: Networks are stable but not static. Due to the interrelationships among actors in the network, evolutionary changes are more characteristic of networks than radical changes are (Easton, 1992). Thus, from a network perspective, all changes take place gradually.

This also means that stability and development are intimately linked to each other in the network; in certain areas, development is based on stability, while in others stability rests on development (Håkansson & Johanson, 1992). In addition to the other network analysis tools we have discussed, the issue of stable and radical change is going to be addressed in the empirical analysis of the Laika project and its network.

MAIN FOCUS OF THE CHAPTER

Laika: Eclipse Plugin Project

Laika is an open source development project aimed at the creation of an integrated development environment for developing applications for embedded Linux devices that run on the Maemo platform (Laika, 2006). The main idea of the project is to integrate the work of several open source projects in a single software tool. The communities related to Laika and their roles are listed in.

Maemo, which acts as a software platform for the Nokia 770 Internet Tablet, is composed of popular OSS components that are widely deployed in today's leading desktop Linux distributions (Maemo, 2006). It consists of a precompiled Linux kernel, platform libraries, and Hildon user interface framework. The Hildon UI framework is based on GTK+, and the whole platform is binary compatible with GTK+ binaries (GTK, 2005). In the Maemo environment, applications can be compiled by using cross-compilation techniques. The basic idea of cross-compilation is to use some processor ("host") to compile software for some other processor ("target") that uses a different architecture. Maemo applications can be compiled using Scratchbox, a cross-compilation toolkit designed to make embedded Linux application development easier (Scratchbox, 2005).

The Eclipse platform was chosen for the base platform of Laika because of its flexible plugin architecture, even if other alternatives like Anjuta (2005) were available. The Eclipse platform is a vendor-neutral open development platform that provides tools for managing workspaces and building, debugging, and launching applications for building integrated development environments (IDEs) (Eclipse, 2005). In general terms, the Eclipse platform is a framework and toolkit that provides a foundation for running third-party development tools. Eclipse supports plugin architecture, which means that all development tools are implemented as pluggable components. The basic mechanism of extensibility in Eclipse is adding new plugins, which can add new processing elements to existing plugins. The Laika

Table 1. Communities related to Laika

Community	Role	Description
CDT	The Laika IDE uses CDT source code in its implementation	C/C++ development environment for the Eclipse platform
Eclipse	Laika is implemented as an Eclipse plug-in	Vendor-neutral open development platform
Gazpacho	Laika supports Gazpacho for visual design of user interfaces	Graphical user interface builder for the GTK+ toolkit
Maemo	Maemo provides the application framework and platform libraries used in the Laika IDE	Development platform to create applications for the Nokia 770
PyDev	Laika Python integration is based on the PyDev project	Python development environment for the Eclipse platform
Scratchbox	Laika is utilizing Scratchbox to cross-compile software for embedded Linux devices	A cross-compilation toolkit used by the Maemo platform

plugin is based on the C/C++ development tools (CDT) plugin, which is a subproject of the Eclipse community (CDT, 2005).

External Network

As already discussed, the Laika project is dependent on many other open source communities and projects. Together, Laika, Maemo, Scratchbox, Eclipse, and CDT form a network in which changes in one project create changes in others. For example, the CDT project is dependent on the changes in the Eclipse IDE, and Maemo and Scratchbox are closely related in that sense. Changes in the projects are especially important from Laika's point of view. Laika acts as a *glue* between Eclipse, Scratchbox, and Maemo and thus is even more sensitive to the changes in each. Moreover, with Laika acting as the glue, it would not seem practical to expect changes in Laika to lead to rapid changes in other tools that can also be used independently; this implies a one-way dependency. Figure 1 illustrates the network formed by communities related to Laika.

Laika, Maemo, and Scratchbox form the core of the external network. However, communication between actors that involves admitting that communities are dependent on each other has proceeded without any formal agreements. The basic principle for this cooperation is voluntary participation, and communication has been handled via

mailing lists, interactive relay chat (IRC) channels, discussion boards, and e-mail, for the most part. This type of collaboration is suitable only if actors see that the partners' actions yield some benefit for them, too. If a partner's achievements are deemed useless, it is not worth participating in the partnership. In this case, all core actors have the same goal: making embedded Linux application development faster and easier.

Close cooperation may also cause changes in priorities. At the same time, projects utilize more and more of each other's features, and connections between actors are becoming more and more complex. Therefore, an actor's *own* project is not always the project with the highest priority. Sometimes it is more important to give support to another, related project than to continue to develop one's own project. One such example was seen when Maemo released a new version of its software development kit (SDK). The Laika project was interrupted for a couple of weeks while the whole team tested the new SDK. At the same time, the team was able to get familiar with the new version of Maemo and thus was able to quickly prepare a new version of Laika that supported this latest version of Maemo. In addition, the Maemo team offered help when testing a new version of Laika a few weeks after the new Maemo was released. As a result, both Laika and Maemo were released sooner than could have been possible with the traditional approach.

Although several actors helped the Laika project in many ways, there is no such thing as a free lunch. A great deal of extra work has been invested to ensure compatibility among applications. This matter is particularly challenging in the case of open source projects. Typically, new versions are released more often and without an exact release time available in advance. For example, Eclipse issued five release versions between June, 2004 and December, 2005. If we assume that all actors release a new version, on average, two times per year, as an effect of that a new version of Laika has to be released almost

Figure 1. External network originally formed by communities

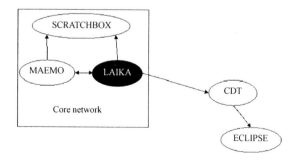

every month. Given that the number of actors is very likely to increase in the near future, this could cause unexpected problems.

Internal Network

The Laika community was established in April, 2005 by the Tampere University of Technology and a participating company sharing the same goals. In the beginning, the community was composed of three developers and coordinated by the university. Two developers were working for the university, and one was sponsored by the company. All of the developers were working on university premises. After some advertisement, other companies expressed an interest in taking part in the project. Companies were willing to add some new features to Laika and thus make it more appropriate for their use.

The first release of Laika was published in August, 2005. This first version offered very basic features needed in every project, such as building and debugging applications. At the same time, one company, with a focus on OSS development, tested the beta version of Laika and announced that it wished to take part in the development of the plugin. It was willing to add localization support and a Gazpacho tool (for graphical user interface design) integration to Laika. From Laika's point of view, the new partner was welcome; consequently, the roadmap for the second release was rewritten. The deadline for the second phase was agreed on at the end of the year. In addition to the new resources for Laika's development, the new partner brought some new knowledge of embedded Linux application development.

The network comprising the original Laika team and the new partner can be considered a tight network, where both actors are strongly dependent on each other but on the other hand do their job in a very independent way. In other words, every time new code is written or existing source code modified, there is a risk of losing compatibility with features created by other actors. However,

the development work of an actor is not dependent on what features other actors have already implemented. The most important problem in this situation for Laika lay in deciding how to keep the compatibility level and the quality of program code as high as possible. To ensure compatibility, it was decided to do integration once a week. In practice, all code was merged every Friday, which ensured that all possible incompatibilities were found quickly.

Until the point described above, all cooperation had been carried out without any legal agreements, and no money moved between the parties. However, when the second phase of Laika was to be finished, another company contacted the community and offered money for implementation of new features such as Python scripting language support for the plugin. The third version of Laika contains features paid for by a sponsor but also some other "nice to have" features and fixes to bugs reported by users. For the development team, accepting the monetary reward resulted in an approach wherein the paid features were committed to first and therefore their priority was increased over that of voluntary ones developed by the same team. At the time of writing, the second version of Laika is soon to be published, and the project team is researching how to add Python support to the plugin. Also, some course material on Laika will be produced in the near future.

Changes in the internal network and new requirements have caused some extensions to the structure of the external network, too. For example, Python support is based on an open source plugin called PyDev. New dependencies extend the external network, and the whole network is going to be more complicated than before. Figure 2 illustrates the network formed for Laika-related communities' future. However, the core of Laika remains unchanged, and it would probably survive even if the new extensions were outdated, since they play an ancillary role only.

Figure 2. Future network formed by communities

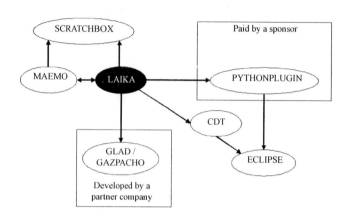

DISCUSSION

We have introduced general network theory as an analytical framework for explaining how open source communities work and organize themselves. As a practical example we used the community developing Laika, an integrated development environment for Maemo, a Linux platform for mobile devices used in, for example, the Nokia 770 Internet Tablet.

The lessons learned from the experimentation of the community are many. To begin with, it seems obvious that network theory is a key to understanding how communities work. In fact, sometimes communities can share responsibilities and create tightly coupled entities that aim at development toward a common goal. In our example, the development of an open source mobile device platform benefits from the work of all software developers involved. We can consider this kind of establishment a *supercommunity*, or a community of communities that share schedules, goals, and interests. From this perspective, Laika can be seen as a member of the Maemo supercommunity. In network theory terms, it seems in the case of Laika that networks of single communities will broaden into macro networks that have some rather loose network structures but also some very tight ones.

Another interesting discovery is that it is the communities that set their priorities themselves to best benefit the network to which they belong. In the case of the Maemo supercommunity, various communities have sometimes adopted supporting roles to benefit some key community. In exchange, these communities have then received mutual assistance in some other phase of development. This mutuality element has been part of the foci of the network theory literature, and, through the research on OSS communities and networks, we can add new insights to the theoretical debate on networks. In Table 2, a summary of the application of network theory to the Laika context in the form of network elements is presented.

FUTURE TRENDS

We believe there is much work that we can carry out in the field described in this chapter. We have provided an outline for future activities concerning Laika, the community maintaining it, and research into the progress of Laika's development.

Concerning Laika, our best prediction is that it will become more and more entangled in the network of Maemo development. Furthermore, while one could assume that actions should be

Table 2. Network elements in Laika

Network Element	Laika-External Network	Laika-Internal Network
Relationship mutuality and interdependence	Mutual relationship and high interdependency between Maemo and Laika; one-way dependency between Laika and the other projects (e.g., Eclipse)	Mutually oriented, close relationships; highly interdependent
Relationship investments	Shared goals as drivers of fruitful cooperation—for example, sometimes priority has been given to the work of another project instead of one's own	Mostly non-monetary (voluntary SW development), though some actors have made monetary investments; no legal commitments
Network position and power relations	Laika: critical position as *glue* between other projects but has no power in the other projects	Equal actors
Network structure	Mostly loose networks	Tight network consisting of individuals and organizations
Network processes	Evolution—radical when the supercommunity experiences major changes, static otherwise	Constant change and rapid evolution

taken to extend the scope of the community to other mobile and embedded Linux environments, we believe that Laika is directly associated with Maemo and that no support is being considered for alternative environments, even if they could benefit from Scratchbox development support. Therefore, assuming that more and more Maemo-based devices are placed on the market, we expect other developers to join Laika, either directly or via plugin technologies that can be integrated into Laika. In a financially oriented environment, such a commitment to a single seminal platform could be considered strategically unwise, which clearly separates community-oriented development from traditional frameworks. At the same time, however, it is conceivable for some development platform other than Eclipse to be supported as well, since this would not alter the mission of the community.

In terms of network theory, we plan to continue monitoring the evolution of the Laika community, as well as the actors participating in the development work. We also wish to study, in the long term, how companies can participate in the development, as well as to observe how funding issues affect the community, potentially leading to the establishment of a company that can take responsibility for some aspects of the community's

work, such as helping developers who use the tool. Then, it would be interesting to observe whether the introduction of financial responsibilities changes the manner in which development is organized and how priorities are chosen.

Another direction for further research arises from the network theory perspective. To begin with, we wish to study networks of other communities as well. This will give us a better understanding of how communities are born and evolve, which in turn enables the creation of long-lived open source communities fostering growth at other levels. Furthermore, the relationship of communities and companies building on the community contributions is considered an important subject for future study.

CONCLUSION

In this chapter, we have addressed networking associated with OSS development practices. We feel that network theory fits well as the foundation in explaining the way open source projects function and cooperate. This theory can be applied both at the level of communities and at that of individual contributors. As a practical example we considered Laika, a small community with a well-defined mission of supporting some more major communities and associated software development. The chapter explained how the different networks involved with Laika have evolved and how various external stimuli have affected the community. We have considered the internal and external networks of Laika in terms of relationship mutuality and interdependency, relationship investments, network position and power relations, network structures, and network processes—which are the main elements of network theory.

Toward the end of the chapter, we also outlined some directions for future research that we hope to perform to improve understanding of how open source communities work. In our future work, we wish to further develop understanding of the managerial implications of open source involvement on the basis of the lessons learned from the Laika case as well as to pay special attention to what kind of theoretical contribution the open source phenomenon can bring to the industrial network literature.

REFERENCES

Achrol, R. S., & Kotler, P. (1999). Marketing in the network economy. *Journal of Marketing, 63* (Special Issue), 146-163.

Anderson, J. C., Håkansson, H., & Johanson, J. (1994). Dyadic business relationships within a business network context. *Journal of Marketing, 58*(4), 1-15.

Anjuta. (2005). *Anjuta DevStudio.* Retrieved June 15, 2006, from http://anjuta.sourceforge.net/

Baker, W. (1994). *Networking smart: How to build relationships for personal and organizational success.* New York: McGraw Hill.

Barnes, J. (1972). Social networks In *Addison-Wesley Module in Anthropology* (Vol. 26, pp. 1-29). Boston: Addison-Wesley.

CDT. (2005). *C/C++ development tool.* Retrieved February 6, 2006, from http://www.eclipse.org/cdt/

Easton, G. (1992). Industrial networks: A review. In B. Axelsson & G. Easton (Eds.), *Industrial networks: A new view of reality* (pp. 3-34). London: Routledge.

Eclipse. (2005). *Eclipse.* Retrieved February 6, 2006, from http://www.eclipse.org/

GTK. (2005). *GTK+ toolkit.* Retrieved June 31, 2006, from http://www.gtk.org/

Hill, R., & Dunbar, R. (2002). Social network size in humans. *Human Nature, 14*(1), 53-72.

Håkansson, H., & Johanson, J. (1992). A model of industrial networks. In B. Axelsson & G. Easton (Eds.), *Industrial networks: A new view of reality* (pp. 28-34). London: Routledge.

Håkansson, H., & Snehota, I. (1989). No business is an island: The network concept of business strategy. *Scandinavian Journal of Management, 4*(3), 187-200.

Håkansson, H., & Snehota, I. (1995). *Developing relationships in business networks.* London: Routledge.

Laika. (2006). *Laika—Scratchbox Eclipse-plugin project.* Retrieved February 6, 2006, from http://www.cs.tut.fi/~laika/

Macdonald, S., & Williams, C. (1992). The informal information network in an age of advanced telecommunications. *Human Systems Management, 11*(2), 77-87.

Maemo. (2006). *Maemo.org.* Retrieved February 6, 2006, from http://www.maemo.org/

Meleis, H. (1996). Toward the information network. *Computer, 29*(10), 59-67.

Möller, K., & Halinen, A. (1999). Business relationships and networks: Managerial challenge of network era. *Industrial Marketing Management, 28*, 413-427.

Möller, K., Rajala, A., & Svahn, S. (2002). Strategic business nets—Their types and management. *Journal of Business Research, 58*(9), 1274-1284.

Möller, K., & Wilson, D. T. (Eds.). (1995). *Business marketing: An interaction and network perspective*. Kluwer Academic Publishers.

Raymond, E. (1999). *The bazaar and the cathedral*. Sebastopol, CA: O'Reilly.

Scott, J. (2000). *Social network analysis: A handbook* (2nd ed.). Newberry Park, CA: Sage.

Scratchbox. (2005). *Scratchbox—Cross-compilation toolkit project*. Retrieved February 6, 2006, from http://www.scratchbox.org/

Tikkanen, H. (1998). The network approach in analyzing international marketing and purchasing operations: A case study of a European SME's focal net 1992–95. *Journal of Business & Industrial Marketing, 13*(2), 109-131.

Wasserman, S., & Faust, K. (1994). *Social network analysis: Methods and applications*. Cambridge, UK: Cambridge University Press.

KEY TERMS

External Observer: Member of the research team; this member generally does not participate in the process being studied, but rather assumes the role of an "objective" outsider who is unfamiliar with the nuances of a given process.

Framework: A perspective or context for viewing, observing, and understanding a particular situation or set of events.

Network Studies: Academic review of how connected communities of individuals work together to achieve certain objectives.

Open Source Community: Group of individuals who (often voluntarily) work together to develop, test, or modify open source software products.

Open Source Project: Undertaking generally involving the development of a piece of open source software.

Properties: Attribute or characteristics of a software package; such attributes often relate to performing a particular function and the degree of success users experience when using that software to perform that function.

System: Software or network of software packages used to perform a variety of tasks.

Chapter XI
Patchwork Prototyping with Open Source Software

M. Cameron Jones
University of Illinois at Urbana-Champaign, USA

Ingbert R. Floyd
University of Illinois at Urbana-Champaign, USA

Michael B. Twidale
University of Illinois at Urbana-Champaign, USA

ABSTRACT

This chapter explores the concept of patchwork prototyping: the combining of open source software applications to rapidly create a rudimentary but fully functional prototype that can be used and hence evaluated in real-life situations. The use of a working prototype enables the capture of more realistic and informed requirements than traditional methods that rely on users trying to imagine how they might use the envisaged system in their work, and even more problematic, how that system in use may change how they work. Experiences with the use of the method in the development of two different collaborative applications are described. Patchwork prototyping is compared and contrasted with other prototyping methods including paper prototyping and the use of commercial off-the-shelf software.

INTRODUCTION

The potential for innovation with open source software (OSS) is unlimited. Like any entity in the world, OSS will inevitably be affected by its context in the world. As it migrates from one context to another, it will be appropriated by different users in different ways, possibly in ways in which the original stakeholders never expected. Thus, innovation is not only present during design and development, but also during use (Thomke & von Hippel, 2002). In this chapter, we explore an emerging innovation through use: a rapid prototyping-based approach to requirements gathering using OSS. We call this approach *patchwork prototyping* because it involves patching together open source applications as a means of creating high-fidelity prototypes. Patchwork prototyping combines the speed and low cost of paper prototypes, the breadth of horizontal prototypes, and the depth and high functionality of vertical, high-fidelity prototypes. Such a prototype is necessarily crude as it is composed of stand-alone applications stitched together with visible seams. However, it is still extremely useful in eliciting requirements in ill-defined design contexts because of the robust and feature-rich nature of the component OSS applications.

One such design context is the development of systems for collaborative interaction, like "cybercollaboratories." The authors have been involved in several such research projects, developing cyberinfrastructure to support various communities, including communities of learners, educators, humanists, scientists, and engineers. Designing and developing such systems, however, is a significant challenge; as Finholt (2002) noted, collaboratory development must overcome the "enormous difficulties of supporting complex group work in virtual settings" (p. 93). Despite many past attempts to build collaborative environments for scientists (see Finholt for a list of collaboratory projects), little seems to have been learned about their effective design, and such environments are notorious for their failure (Grudin, 1988; Star & Ruhleder, 1996). Thus, the focus of this chapter is on a method of effective design through a form of rapid, iterative prototyping and evaluation.

Patchwork prototyping was developed from our experiences working on cybercollaboratory projects. It is an emergent practice we found being independently redeveloped in several projects; thus, we see it as an effective ad hoc behavior worthy of study, documentation, and formalization. Patchwork prototyping is fundamentally a user-driven process. In all of the cases where we saw it emerge, the projects were driven by user groups and communities eager to harness computational power to enhance their current activities or enable future activities. Additionally, the developers of the prototypes had no pretence of knowing what the users might need a priori. As

a result, patchwork prototyping's success hinges on three critical components:

1. Rapid iteration of high-fidelity prototypes
2. Incorporation of the prototypes by the end users into their daily work activities
3. Extensive collection of feedback facilitated by an insider to the user community

In this chapter, we focus on how the method worked from the developers' point of view. It is from this perspective that the advantages of using OSS are most striking. However, one should bear in mind that the method is not just a software development method, but also a sociotechnical systems (Trist, 1981) development method: The social structures, workflows, and culture of the groups will be coevolving in concert with the software prototype.

REQUIREMENTS GATHERING IN COLLABORATIVE SOFTWARE DESIGN

Software engineering methods attempt to make software development resemble other engineering and manufacturing processes by making the process more predictable and consistent. However, software cannot always be engineered, especially Web-based applications (Pressman et al., 1998). Even when application development follows the practices of software engineering, it is possible to produce applications that fail to be used or adopted (Grudin, 1988; Star & Ruhleder, 1996). A major source of these problems is undetected failure in the initial step in building the system: the requirements-gathering phase. This is the most difficult and important process in the entire engineering life cycle (Brooks, 1995).

In designing systems to support collaborative interaction, developers are faced with several complex challenges. First, the community of users for which the cyberinfrastructure is being developed may not yet exist and cannot be observed for one to see how the users interact. In fact, there is often a technological deterministic expectation that the computational infrastructure being created will cause a community to come into existence. Even in the case where there is a community to study, many of the activities expected to occur as part of the collaboration are not currently being practiced because the tools to support the activities do not yet exist. As a result, developers gain little understanding about how the users will be interacting with each other or what they will be accomplishing, aside from some general expectations that are often unrealistic.

Gathering requirements in such an environment is a highly equivocal task. While uncertainty is characterized by a lack of information, which can be remedied by researching an answer, collecting data, or asking an expert, equivocal tasks are those in which "an information stimulus may have several interpretations. New data may be confusing, and may even increase uncertainty" (Daft & Lengel, 1986, p. 554). Requirements gathering is one such situation in which the developers cannot articulate what information is missing, let alone how to set about obtaining it. The only resolution in equivocal situations is for the developers to "enact a solution. [Developers] reduce equivocality by defining or creating an answer rather than by learning the answer from the collection of additional data" (Daft & Lengel, p. 554). As Daft and Macintosh (1981) demonstrate, tasks with high equivocality are unanalyzable (or rather, have low analyzability; Lim & Benbasat, 2000), which means that people involved in the task have difficulty determining such things as alternative courses of action, costs, benefits, and outcomes.

RAPID PROTOTYPING

Rapid prototyping is a method for requirements gathering that has been designed both to improve communication between developers and users, and to help developers figure out the usefulness or consequences of particular designs before having built the entire system. The goal of rapid prototyping is to create a series of iterative mock-ups to explore the design space, facilitate creativity, and get feedback regarding the value of design ideas before spending significant time and money implementing a fully functional system (Nielsen, 1993). There are several dimensions to prototypes. One dimension is the range from low-fidelity to high-fidelity prototypes (see Table 1; Rudd, Stern, & Isensee, 1996). Low-fidelity prototypes have the advantages of being fast and cheap to develop and iterate. However, they are only able to garner a narrow range of insights. Perhaps the most popular low-fidelity prototyping technique is paper prototyping (Rettig, 1994). Paper prototypes are very fast and very cheap to produce. They can also generate a lot of information about how a system should be designed, what features would be helpful, and how those features should be presented to the users. However, paper prototypes do not allow developers to observe any real-world uses of the system, or understand complex interactions between various components and between the user and the system. Also, they do not help developers understand the details of the code needed to realize the system being prototyped.

High-fidelity prototypes, on the other hand, can simulate real functionality. They are usually computer programs themselves that are developed in rapid development environments (Visual Basic, Smalltalk, etc.) or with prototyping tool kits (CASE, I-CASE, etc). In either case, these prototypes, while allowing programmers to observe more complex interactions with users and to gain understanding about the underlying implementation of the system, are comparatively slow and expensive to produce and iterate (Rudd et al., 1996). These costs can be offset somewhat by incorporating these prototypes into the development of the final system itself as advocated by RAD (rapid application development; Martin,

Table 1. Advantages and disadvantages of low- and high-fidelity prototyping (Source: Rudd et al., 1996, p. 80)

	Advantages	Disadvantages
Low-Fidelity Prototypes	• Lower development cost • Can create many alternatives quickly • Evaluate multiple design concepts • Useful communication device • Address screen layout issues • Useful for identifying market requirements • Proof of concept	• Limited error checking • Poor detailed specification to code to • Facilitator driven • Limited utility after requirements established • Limited usefulness for usability tests • Navigational and flow limitations • Weak at uncovering functionality- and integration-related issues
High-Fidelity Prototypes	• Complete functionality • Fully interactive • User driven • Clearly defines navigational scheme • Use for exploration and tests • Look and feel of final product • Serves as a living specification • Marketing and sales tool	• More expensive to develop • Time consuming to create • Inefficient for proof of concept designs • Not effective for requirements gathering

1991). However, critics of RAD methods are quick to point out the limited scalability of software built using source code from prototypes (Beynon-Davies, Carne, Mackay, & Tudhope, 1999). Typically low-fidelity and high-fidelity prototypes are used in succession, with developers increasing the fidelity of the prototypes as they develop the specifications. Due to their high cost, high-fidelity prototypes may only be built for a select number of designs generated by low-fidelity prototyping, which precludes the generation of a series of disposable high-fidelity proofs of concepts to test out alternative design ideas.

Another dimension to be considered in the prototyping discussion is scope. Software can be viewed as consisting of a number of layers, from the user interface to the base layer, which interacts with the underlying operating system or platform. Horizontal prototypes encompass a wide scope, spanning the breadth of a system but only within a particular layer (usually the user interface). Users can get a sense of the range of the system's available functions; however, the functionality is extremely limited. This can help both the user and the programmer understand the breadth of the system without plumbing its

depths. Vertical prototypes, on the other hand, take a narrow slice of the system's functionality and explore it in depth through all layers. This allows users to interact with a particular piece of the system, and gives the programmer a detailed understanding of the subtle issues involved in its implementation (Floyd, 1984; Nielsen, 1993).

The high equivocality present when designing collaborative systems makes it difficult to apply rapid prototyping techniques effectively. Because users may not be able to articulate what they want or need, it helps to be able to collaboratively interact with high-fidelity systems in order to test them in real-world situations and see what requirements emerge. Without such an experience, it is unlikely that any feedback the developers get from the users, either through direct communication or observation, will be useful. Thus, low-fidelity prototypes are limited in their power to elicit requirements as the users have difficulty imagining how the system the prototypes represent will work, what it could do for them, or how they might use it. Also, since the majority of tasks involved in collaboration are quite complex and require multiple kinds of functionality to complete, the users need to be

Figure 1. Horizontal and vertical prototypes (Source: Nielsen, 1993, p. 94)

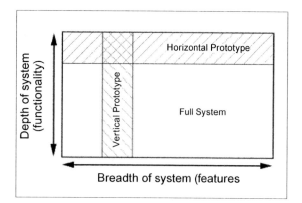

able to interact with the system as a whole and with considerable depth of implementation, thus requiring a prototype that is both horizontal and vertical.

The economics of developing high-fidelity prototypes that are both horizontal and vertical in scope, however, are problematic. Even if the developers were to build a series of high-fidelity, vertical prototypes, they would end up having built the equivalent of an entire system from scratch just to have a functionally sufficient prototype. Not only would it be expensive and time consuming, but the functionality and robustness would be minimal at best. Also, it is likely that the work would need to be discarded and replaced with something new since it is unlikely that the design would be correct on the first, second, or even third try. Thus, the typical methods of prototyping are not sufficient, either because developing all the code would be too expensive, or the prototypes that are developed do not have high enough fidelity.

The proliferation of production-scale OSS systems has created a vast field of growing, reliable, usable, and feature-rich programs, a large number of which support aspects of Web-based collaboration. These programs can be easily stitched together because the code is open and modifiable. Furthermore, they can be treated

as disposable since one application can easily be discarded and replaced with another. This presents an opportunity for developers to rapidly build and evaluate a high-fidelity prototype of a collaborative environment comprising a patchwork of multiple open source applications. Such a prototype spans the breadth of a horizontal prototype and the depth of a vertical prototype within a single system.

ORIGINS AND EXAMPLES OF PATCHWORK PROTOTYPING

Patchwork prototyping is a rapid prototyping approach to requirements gathering that was emergent from practice rather than designed a priori. We have been involved with several groups that were developing cyberinfrastructure to support collaboration, and in each group we observed ad hoc prototyping and development strategies that were remarkably similar and that developed entirely independent of each other. Upon making these observations, we realized that there was a core process at work in each of these projects that could be abstracted out and described as a general approach to requirements gathering for developing cyberinfrastructure. Because patchwork prototyping evolved from practice, however, we believe that it will be much easier to understand our formal description of the approach after we describe some of the relevant details of our experiences. In this section, we describe two projects with which we were involved and the relevant dynamics of each project; in the following section, we describe the patchwork prototyping approach more abstractly.

Project Alpha: Building a Cybercollaboratory for Environmental Engineers

Project Alpha (a pseudonym used to preserve anonymity) was devoted to building a cybercollaboratory for environmental engineers. At the beginning, the project was intended to be a requirements-gathering project, and the goal was to build a functional prototype of the cyberinfrastructure that would be presented to the granting agency as part of a larger proposal. The effort was a success and now, more than a year after the project began, the prototype is being converted into a production-scale system. The cybercollaboratory prototypes were largely designed and built over a period of six months by a team of two developers, with significant contribution to the design by a team of around 12 to 13 other researchers (these researchers, plus the two developers, we call the design team), and some minor programming contributions by undergraduates employed by the project. By the end of the prototyping phase, there was a community of users that included 60 to 70 active users out of approximately 200 registered users, 10 of which comprised a core group of vocal users who provided significant feedback on the design.

The Project Alpha prototype was constructed on the Liferay portal server framework. In addition to using existing portlets, the developers also wrapped other OSS applications in portlet interfaces, enabling their rapid integration into the prototype. A number of different OSS applications were used, including the Heritrix Web crawler, the Lucene search engine, and the MediaWiki wiki system. Other applications were similarly integrated but were not necessarily publicly available OSS. Some were in-house applications developed by other projects for which the developers had source code. These applications were used to prototype data mining and knowledge management functionality in the cybercollaboratory.

The general process by which these tools were incorporated was very ad hoc. The development team might decide on prototyping a particular function, or the programmers might get some idea for a "cool" feature and would set about integrating the feature into the system. This approach had several unexpected benefits. First,

minimal time was spent building portlets so that when a version of the prototype was presented to the design team, minimal effort was lost when particular features or portlets were rejected as being unsuitable. Second, it allowed the design team to choose between several different portlets that had essentially the same function but different interfaces (i.e., were optimized for different types of use). Third, it allowed the developers to easily switch features off when the interface for a portlet was too complex, or turn them back on if they were requested by either the design team or the active users. Fourth, the development community and the associated forums, mailing lists, and Web sites surrounding the OSS applications that were integrated into the prototype served as excellent technical support (Lakhani & von Hippel, 2002).

The fact that the prototype was fully functional was critical to its success in eliciting requirements. By using the prototypes over a period of 6 months, the users were able to incorporate them into their day-to-day work practices. This allowed them to evaluate the utility of the tool in various contexts of actual use. Without functionality, the developers feel that it would have been impossible to effectively gather requirements. However, it was also vital that the users communicate their experiences to the developers, both formally and informally. To this end, the developers conducted several surveys of the users, asking them about the prototype and features they found useful. The developers also used the prototype itself to solicit feedback. On the front page of the prototype was a poll asking users to vote for the features they liked the most. Additionally, on every page of the prototype was a feedback form that allowed users to send quick notes about the system as they experienced it. The users also communicated with the developers via informal means such as e-mail and face-to-face meetings. However, the most important method of obtaining feedback was that one of the PIs in the project acted as an intermediary, actively soliciting feedback from users as an insider to the community of environmental engineers. The PI position allowed the individual to receive more feedback of higher quality and honesty than the developers would have been able to collect on their own.

To illustrate the process in more detail, we describe how one particular piece of OSS was integrated with the cybercollaboratory. The developers wanted to allow users to be able to collaboratively edit documents in the system. The Liferay suite had a wiki system available that the programmers enabled; however, users found that tool to be too difficult to use, partly because of the unintuitive markup syntax of the particular wiki used, and partly because they had no tasks that clearly lent themselves to the use of such a tool. Later during the prototyping phase, some members of the design team wanted to demonstrate the usefulness of scenarios and personas in facilitating requirements gathering, and from prior experience suggested the use of a wiki. In response to this request and the prior difficulties in using the bundled tool, the developers installed MediaWiki on the server and added a link from the cybercollaboratory's menu next to the existing wiki tool pointing to the MediaWiki installation. No time was spent trying to integrate the Liferay and MediaWiki systems; each application had separate interfaces and user accounts.

One benefit of using the MediaWiki system was that it allows people to use the system without logging in, thereby mitigating the need to integrate authentication mechanisms. Users found the MediaWiki system easier to learn and use, and began using it exclusively over the in-built Liferay wiki. The developers then decided to embed the MediaWiki interface in the rest of the cybercollaboratory and wrote a simple portlet that generates an HTML (hypertext markup language) IFRAME to wrap the MediaWiki interface. Each step of integrating the MediaWiki installation took only minimal effort on the part of the developers (sometimes literally only a matter of minutes) and generated insights about the role and design of a collaborative editing tool in the cybercollaboratory. Among the design insights gained by the developers is that the tool should be easy to use with a simple syntax for editing. Also, the tool should support alternate views of the data, offering a unified view of all documents either uploaded to the site's document repository or created and edited on the wiki. The users were able to see how this tool could benefit their jobs, and that shaped the requirements of the tool. As a result of this process, the project is currently implementing a new collaborative editing component. This component will have features like integrated authentication, group- and project-based access control, and integration with other features (e.g., project views and wiki linking). Additionally, the new collaborative writing component will deprecate redundant and confusing features like in-wiki file uploads.

Project Beta: Building Collaborative Tools to Support Inquiry-Based Learning

Project Beta is an ongoing research project aimed at designing and building Web-based tools to support processes of inquiry as described by John Dewey (Bishop et al., 2004). Initiated in 1997, the project has embraced a long-term perspective on the design process and produced a series of prototypes that support inquiry-based teaching and learning. In 2003 the project began exploring the development of tools to support collaborative inquiry within groups and communities. The current prototype is the third major revision of the collaborative cyberinfrastructure, with countless minor revisions on going. Throughout the project's life span, several generations of programmers have joined and left the development team. For a 30-month stretch, the majority of programming was sustained by a single graduate-student programmer. Between four and eight other researchers filled out the design team.

The prototypes are available for anyone to use, and the source code is also distributed under a Creative Commons license. To date, the prototypes have been used to support a large number of communities of users ranging from water-quality engineers to volunteers in a Puerto Rican community library in Chicago, from researchers studying the honeybee genome to undergraduates in the social sciences. There are numerous other groups using the system for any number of purposes. Given this scenario, it is practically impossible to design for the user community or any intended use.

The prototypes were developed in the PHP programming language on an open source platform consisting of Apache, MySQL, and RedHat Linux. In contrast to Project Alpha where the developers initially did very little programming and primarily used readily available tools, the developers of Project Beta spent considerable effort building an infrastructure from scratch, in part because the developers were initially unaware of relevant OSS. However, as the project progressed, several open source tools were incorporated into the prototypes including the JavaScript-based rich-text editors FCK Editor and TinyMCE, the phpBB bulletin board system, and MediaWiki.

To demonstrate the process in more detail, we describe how one particular piece of OSS was integrated with the prototypes. In the earliest version of the cyberinfrastructure, users expressed an interest in having a bulletin board system. The developers selected the phpBB system and manually installed copies of phpBB for each community that wanted a bulletin board; the bulletin board was simply hyperlinked from the community's home page. In the next iteration of the prototype, the phpBB system was modified to be more integrated with the rest of the prototype. Users could now install a bulletin board themselves, without involving the developers, by clicking a button on the interface. Furthermore, the authentication and account management of the bulletin board was integrated with the rest of

the prototype, eliminating the need for users to log in twice. However, the full features of phpBB were more than the users needed. They primarily made use of the basic post and reply functions and the threaded-conversation structure. Users indicated that the overall organization of the board system into topics, threads, and posts made sense to them. In the most recent major revision of the prototype, the phpBB system was replaced by a simpler, more integrated homemade bulletin board prototype that supported these basic features. Had the development progressed in the opposite order (i.e., building the simple prototype first, then adding features), it is possible that developers could have wasted valuable time and energy prototyping features that would only be discarded later for lack of use.

GENERALIZED APPROACH TO PATCHWORK PROTOTYPING

Based on the experiences described above, we have outlined a general approach to building patchwork prototypes using OSS. While our experience has been primarily with Web-based tools, and this process has been defined with such tools in mind, it is likely that a similar approach could be taken with prototyping any kind of software. Like other prototyping methods, this is designed to be iterated, with the knowledge and experience gained from one step feeding into the next. The approach entails the following five stages:

1. Make an educated guess about what the target system might look like.
2. Select tools that support some aspect of the desired functionality.
3. Integrate the tools into a rough composite
4. Deploy the prototype and solicit feedback from users.
5. Reflect on the experience of building the prototype and the feedback given by users, and repeat.

For the most part, these steps are relatively straightforward. Making the first educated guess about what the target system might look like can be the hardest step in this process because it requires the design team to synthesize their collective knowledge and understanding of the problem into a coherent design. In this first iteration of the process, it is often helpful to use paper prototypes and scenarios, but their function is primarily to serve as communications devices and brainstorming aids. The high equivocality of the situation almost guarantees, however, that whatever design they produce will be insufficient. This is not a failure. It is an expected part of the process, and the design will be improved on subsequent iterations. The important thing is to have a starting point that can be made concrete, and not to spend too much time brainstorming ideas. It is essential not to become bogged down in controversies about how the software "ought" to look, but rather to put together a prototype and test it out with users in their everyday environments and let the users figure out what works, what does not, and what is missing.

Selection and Integration of Tools: The Benefits of Using Open Source Software

There are several important considerations to keep in mind when selecting the tools. On first glance, patchwork prototyping as a method does not require OSS; the same general process could theoretically be followed by using software that provides APIs, or by creating prototypes through adapting methodologies for creating production-scale software systems such as COTS (commercial off-the-shelf) integration (Boehm & Abts, 1999). However, using OSS confers several important advantages; in fact, we believe that patchwork prototyping is only now emerging as a design practice because of the recent availability of a significant number of mature, production-scale OSS systems.

Without access to source code, developers are limited in how well they can patch together different modules, the features they can enable or disable, their ability to visually integrate the module with the rest of the system, and their ability to understand the underlying complexity of the code needed to construct such systems on a production scale. High-profile OSS is often of high quality, which means that difficult design decisions have already been made. Given that it is built from the collective experiences of many programmers, less effective designs have already been tried and discarded. In fact, by using and delving into human-readable (compared to that generated by CASE tools, e.g.), open source code, the developers can get a grounded understanding of how particular features can be implemented, which can enable them to better estimate development time and costs.

The Web-based nature of patchwork prototypes affords several ways of integrating the selected software into the prototype, ranging from shallow to deep. Shallow integration consists of either wrapping the tools in an HTML frame to provide a consistent navigation menu between the tools, or customizing the HTML interfaces of the tools themselves to add hyperlinks. Most open source Web applications use HTML templates, cascading style sheets, and other interface customization features, which make adding or removing hyperlinks and changing the look and feel very easy. The advantage of shallow integration is the ease and speed with which the developer is able to cobble together a prototype. A significant drawback to shallow integration is that each application remains independent.

Deeper integration usually requires writing some code or modifying existing source code. This may include using components or modules written for the extension mechanisms designed into the application or other modifications made to the application's source code. If the developers cannot find precisely what they are looking for, they can fashion the code they need by copying and modifying similar extension code, or, in the worst case, the developers will need to write new code to facilitate the integration. However, the amount of code needed is very little in comparison to the amount of code that would have been required of the developers building a prototype from scratch.

For any prototyping effort to be worthwhile, the costs of creating the prototypes must be minimal. OSS systems tend to be fully implemented, stand-alone applications with many features and capabilities that provide a wealth of options to play with when prototyping to elicit requirements. The minimal effort required to add features allows the programmers to treat the features as disposable: Because little effort was needed to implement them, little effort is wasted when they are switched off or discarded. That most OSS are free is also important, both for budgetary reasons and because the developers can avoid complicated licensing negotiations. Additionally, most OSS have very active development communities behind them with members who are often eager to answer the developer's questions in considerable depth, and do so for free, unlike the expensive technical support that is available for commercial products. All of this facilitates the requirements-gathering process because iterations of the prototype can be rapidly created with high functionality at minimal cost, and with minimal effort and emotional investment by the developers.

Deployment, Reflection, and Iteration

During the deployment of the prototype, future users integrate the cyberinfrastructure into their work practices for an extended period of time and explore what they can do with it collaboratively. The collection of feedback on user experiences allows requirements gathering that is not purely need based, but also opportunity and creativity based. By seeing a high-fidelity prototype of the entire system, users can develop new ideas of how to utilize features that go beyond their intended use, and conceptualize new ways of accomplishing their work. In addition, users will become aware of gaps in functionality that need to be filled, and can explain them in a manner that is more concrete and accessible to the developers.

When reflecting on the collected feedback, however, the design team must realize that the prototype does not simply elicit technical requirements; it elicits requirements for the collaborative sociotechnical system as a whole. The existence of the prototype creates a technological infrastructure that influences the negotiation of the social practices being developed by the users via the activities the infrastructure affords and constrains (Kling, 2000). The design team must be aware of how various features affect the development of social practice, and must make explicit the type of interactions that are required but are not currently realized. By allowing the users to interact with the prototypes for extended periods, collecting feedback on their experiences, and paying attention to the social consequences of the cyberinfrastructure, a richer understanding of the sociotechnical system as a whole can emerge. Thus, reflection is a process of attending to the consequences of the design for the broader sociotechnical system, and integrating those consequences into a holistic understanding of how the system is evolving.

Iteration is essential to the rapid prototyping approach. First, iteration allows for the exploration of more features and alternatives. This can uncover overlooked aspects of the system that might be of use. This can also reinforce the importance or necessity of particular features or requirements. Furthermore, iteration provides the users with a constant flow of new design possibilities, which prevents them from becoming overly attached to any single design, giving them the freedom to criticize particular instances of the prototype. Ultimately, it is impossible to reach complete understanding of the system given its evolving

Table 2. Comparison of patchwork prototyping with other methods

Paper Prototyping	Patchwork Prototyping	COTS/API Prototyping
Speed		
Can iterate a prototype multiple times in an afternoon	Can iterate a prototype in less than a week	Can take weeks or months to iterate a prototype
Monetary Costs		
Cost of office supplies	Free, or minimal cost of licenses if in business setting	Purchasing and licensing software can be expensive
Availability of Materials		
Usually already lying around	Large number of high-quality OSS available for free download	Not all commercial systems have APIs
Functionality		
Nonfunctional	High	High
Accessibility		
Anyone can prototype systems using paper, including nontechnical end users	Requires skilled programmers to create patchwork prototypes	Requires skilled programmers to integrate commercial software
Interface		
Not polished, but can provide a consistent and/or innovative interface concept for consideration	Not renowned for excellent usability; assembled components may be inconsistent	Individual elements may be high quality and familiar; assembled components may be inconsistent
Flexibility		
High: can do anything with paper	High: can modify source to create any desired functionality	Low: restricted to what the API allows, which may be limited
Disposability		
High: little investment of time, money, emotions	High: little investment of time, money, emotions	Low: significant effort and money can result in high emotional investment
User Attachment		
Low: users can see it is rough and nonfunctional	Med. to High: upon using it, can get attached to the system unless iterated rapidly	High: cannot be iterated fast enough to avoid attachment

nature. However, by iterating the prototyping process, the design space may narrow, identifying a set of key requirements. At this point the design is not complete, but work on a flexible production-scale system can begin, and further exploration of the design space can be continued within that system.

STRENGTHS AND LIMITATIONS

Patchwork prototyping addresses two major problems that designers face when building new sociotechnical systems. First, it allows the design team to get feedback on the prototype's use in real-world situations. Users interact with the system in their daily activities, which focuses their feedback around task-related problems. In Project Alpha, when members of the design team started using the prototype, the feedback changed from general praise or criticism of the appearance of the interface to more detailed explanations of how particular functionality aided or inhibited task performance. Second, it reduces the equivocality of the design space. By creating a functional prototype, discussions change from being highly suppositional to being about concrete actions, or concrete functionality.

Integration into the real-world context is markedly different from other prototyping and requirements-capture methods. Paper prototypes are typically given to users in a laboratory setting (Nielsen, 1993), thus all the tasks are artificial. While this can give developers important design insights, the drawback is that prototypes can end up optimized for artificial tasks and not for real-world use. More expensive methods such as participatory design (Ehn & Kyng, 1991) and ethnography (Crabtree, Nichols, O'Brien, Rouncefield, & Twidale, 2000) try to incorporate real-world use into the design process, the former by bringing users into the design team, the latter by observing users in their natural work environment. However, when the technology that these methods were used to design is introduced, it inevitably changes the practices and social structures present in the work environment, often in a way that cannot be predicted. Patchwork prototyping overcomes these limitations by being cheap and by providing real-time feedback on both users' problems with the software and the effects the software is having on the broader work context.

The advantages of patchwork prototyping can be seen when comparing it to other prototyping techniques. In Table 2 we compare it to paper prototyping and to prototyping using COTS software. The advantages of patchwork prototyping are that it has many of the benefits of paper prototyping, including low cost and ready availability of materials, yet provides the

high functionality of COTS/API prototyping; the effort needed to create the prototypes and the length of the iteration cycles lies somewhere in between. Thus, while we see the method as being yet another tool for developers and designers to have in their toolbox, in many ways, it combines the best of both worlds.

The patchwork prototyping approach is not without limitations, however. Despite our hope that the visibility of the seams between the applications would be interpreted by the users as an indication that the prototype is a work in progress, our experiences seem to indicate that the users still view it as a finished product due to the fact that it has real functionality. It is possible that such interpretations can be overcome through social means by emphasizing the fact that the system is a prototype to all users who are encouraged to test it. However, since none of the projects we participated in did this, we have no idea whether or not that would be sufficient. One thing that is clear, however, is that visual coherence between applications greatly facilitates the ease of use and positive perceptions of the system as a whole. In fact, in Project Alpha, it was realized that users need different views of the component modules and features depending on the context in which they access the applications, and in some of those views the distinctions between modules must be totally erased.

Patchwork prototyping requires highly skilled programmers to be implemented effectively. Programmers must have significant experience within the development environment in which the OSS applications are coded; otherwise, they will spend too much time reading code and learning the environment, and the speed of implementation will not be as fast. Also, OSS can have security vulnerabilities that can compromise the server on which they are hosted. Project Beta ran into this problem when multiple installations of phpBB succumbed to an Internet worm, bringing down the prototype for several days. Third, patchwork prototyping requires a long-term commitment by users, and a motivated facilitator who is able to convince the users to adopt the prototype and incorporate it into their work practices. The facilitator must collect feedback about the users' experiences. Without willing users and the collection of feedback, the prototyping process will likely fail.

FUTURE TRENDS

The use of patchwork prototyping is still in its infancy. The relative ease with which patchwork prototypes can be constructed means that the method itself affords appropriation into new contexts of use. For example, one of the biggest costs to organizations is buying software systems such as enterprise management systems. Patchwork prototyping offers a cheap and effective method for exploring a design space and evaluating features. Consequently, through prototyping, managers can be more informed when shopping for software vendors and can more effectively evaluate how effective a particular vendor's solution will be for their company (Boehm & Abts, 1999).

Because users have to integrate the prototype into their daily work practices, transitioning from the patchwork prototype to the production-scale system can be highly disruptive. One method of avoiding this is having a gradual transition from the prototype to the production-scale system by replacing prototype modules with production-scale modules. To do this, however, the prototypes must be built on a robust, extensible, modular framework because the latter component is not easily replaced. If this model is used, the system development process need never end. Prototypes of new features can constantly be introduced as new modules, and, as they mature, be transitioned into production-scale systems. As more developers and organizations support open source development, the number and availability of OSS applications will increase. As more modules are written for particular open source, component-

based systems, the costs of doing patchwork prototyping will further decrease, as will the threshold for programming ability—perhaps to the point where users could prototype systems for themselves that embody specifications for software programmers to implement.

CONCLUSION

Patchwork prototyping is a rapid prototyping approach to requirements gathering that shares the advantages of speed and low cost with paper prototypes, breadth of scope with horizontal prototypes, and depth and high functionality with vertical, high-fidelity prototypes. This makes it particularly useful for requirements gathering in highly equivocal situations such as designing cyberinfrastructure where there is no existing practice to support because it allows future users to integrate the cyberinfrastructure into their work practices for an extended period of time and explore what they can do with it collaboratively. It has the benefit of allowing the design team to monitor the sociotechnical effects of the prototype as it is happening, and gives users the ability to provide detailed, concrete, task-relevant feedback.

Patchwork prototyping is an excellent example of how OSS can foster innovation. The affordances of open-source code and a devoted development team create opportunities to utilize OSS in ways that go beyond the functionality of any particular application's design. The cases presented here merely scratch the surface of a new paradigm of OSS use. Further research is needed to understand the specific features of technologies that afford such innovative integration.

REFERENCES

Beynon-Davies, P., Carne, C., Mackay, H., & Tudhope, D. (1999). Rapid application development

(RAD): An empirical review. *European Journal of Information Systems, 8*(3), 211-223.

Bishop, A. P., Bruce, B. C., Lunsford, K. J., Jones, M. C., Nazarova, M., Linderman, D., et al. (2004). Supporting community inquiry with digital resources. *Journal of Digital Information, 5*(3). Retrieved from http://joko.tanu.edu/Articles/v05/i03/Bishop

Boehm, B. W., & Abts, C. (1999). COTS integration: Plug and pray? *IEEE Computer, 32*(1), 135-138.

Brooks, F. P. (1995). *The mythical man-mouth: Essays on software engineering* (Anniversary ed.). Boston: Addison-Wesley.

Crabtree, A., Nichols, D. M., O'Brien, J., Rouncefield, M., & Twidale, M. B. (2000). Ethnomethodologically-informed ethnography and information systems design. *JASIS, 51*(7), 666-682.

Daft, R. L., & Lengel, R. H. (1986). Organizational information requirements, media richness and structural design. *Management Science, 32*(5), 554-571.

Daft, R. L., & Macintosh, N. B. (1981). A tentative exploration into the amount and equivocality of information processing in organizational work units. *Administrative Sciences Quarterly, 26*(2), 207-224.

Ehn, P., & Kyng, M. (1991). Cardboard computers: Mocking-it-up or hands-on the future. In J. Greenbaum & M. Kyng (Eds.), *Design at work* (pp. 169-196). Hillsdale, NJ: Laurence Erlbaum Associates.

Finholt, T. A. (2002). Collaboratories. *Annual Review of Information Science and Technology, 36*(1), 73-107.

Floyd, C. (1984). A systematic look at prototyping. In R. Budde, K. Kuhlenkamp, L. Mathiassen, & H. Zullighoven (Eds.), *Approaches to prototyping* (pp. 1-18). Berlin, Germany: Springer-Verlag.

Grudin, J. (1988). Why CSCW applications fail: Problems in the design and evaluation of organizational interfaces. In *CSCW 88: Proceedings of the Conference on Computer-Supported Cooperative Work* (pp. 85-93).

Kling, R. (2000). Learning about information technologies and social change: The contribution of social informatics. *The Information Society, 16*, 217-232.

Lakhani, K. R., & von Hippel, E. (2002). How open source software works: "Free" user-to-user assistance. *Research Policy, 1451*, 1-21.

Lim, K. H., & Benbasat, I. (2000). The effect of multimedia on perceived equivocality and perceived usefulness of information systems. *MIS Quarterly, 24*(3), 449-471.

Martin, J. (1991). *Rapid application development.* New York: Macmillan Publishing Co.

Nielsen, J. (1993). *Usability engineering.* San Diego, CA: Morgan Kaufman.

Pressman, R. S., Lewis, T., Adida, B., Ullman, E., DeMarco, T., Gilb, T., et al. (1998). Can Internet-based applications be engineered? *IEEE Software, 15*(5), 104-110.

Rettig, M. (1994). Prototyping for tiny fingers. *Communications of the ACM, 37*(4), 21-27.

Rudd, J., Stern, K., & Isensee, S. (1996). Low vs. high-fidelity prototyping debate. *Interactions, 3*(1), 76-85.

Star, S. L., & Ruhleder, K. (1996). Steps toward an ecology of infrastructure: Design and access for large information spaces. *Information Systems Research, 7*(1), 111-134.

Thomke, S., & von Hippel, E. (2002). Customers as innovators: New ways to create value. *Harvard Business Review, 80*(4), 74-81.

Trist, E. L. (1981). The sociotechnical perspective: The evolution of sociotechnical systems as a conceptual framework and as an action research program. In A. H. van de Ven & W. F. Joyce (Eds.), *Perspectives on organization design and behavior* (pp. 19-75). New York: John Wiley & Sons.

KEY TERMS

COTS Integration: The process by which most businesses integrate commercial off-the-shelf software systems in order to create a computing environment to support their business activities.

Equivocality: The name for a lack of knowledge that cannot be mitigated simply by doing research or gathering more information. In an equivocal situation, decisions often need to be made, definitions created, and procedures negotiated by various (often competing) stakeholders.

Paper Prototyping: A rapid prototyping method for creating low-fidelity prototypes using pencils, paper, sticky notes, and other low-tech materials that can be quickly iterated in order to explore a design space. It is often used in interface design.

Patchwork Prototyping: A rapid prototyping method for creating high-fidelity prototypes out of open source software that can be integrated by users into their everyday activities. This gives us-ers something concrete to play with and facilitates a collaborative process of sociotechnical systems development. It is ideal for highly equivocal design situations.

Rapid Prototyping: Rapid prototyping is a method that involves creating a series of prototypes in rapid, iterative cycles. Normally, a prototype is created quickly, presented to users in order to obtain feedback on the design, and then a new prototype is created that incorporates that feedback. This cycle is continued until a fairly stable, satisfactory design emerges, which informs the design of a production-scale system.

Sociotechnical System: Refers to the concept that one cannot understand how a technology will be used in a particular environment without understanding the social aspects of the environment, and that one cannot understand the social aspects of the environment without understanding how the technology being used shapes and constrains social interaction. Thus, one can only understand what is going on in an environment by looking at it through a holistic lens of analysis.

Uncertainty: The name for a lack of knowledge that can be addressed by obtaining more information, such as by researching an answer, looking it up in reference materials, or collecting data.

Chapter XII
An Agile Perspective on Open Source Software Engineering

Sofiane Sahraoui
American University of Sharjah, UAE

Noor Al-Nahas
American University of Sharjah, UAE

Rania Suleiman
American University of Sharjah, UAE

ABSTRACT

Open source software (OSS) development has been a trend parallel to that of agile software development, which is the highly iterative development model following conventional software engineering principles. Striking similarities exist between the two development processes as they seem to follow the same generic phases of software development. Both modes of development have less emphasis on planning and design and a more prominent role for implementation during the software engineering process. This chapter expounds on this connection by adopting an agile perspective on OSS development to emphasize the similarities and dissimilarities between the two models. An attempt is first made to show how OSS development fits into the generic agile development framework. Then, the chapter demonstrates how the development process of Mozilla and Apache as two of the most famous OSS projects can be recast within this framework. The similarity discussed and illustrated between agile and OSS development modes is rather limited to the mechanics of the development processes and do not include the philosophies and motivations behind development.

INTRODUCTION

As conventional software development methodologies struggle to produce software within budget limits and set deadlines, and that fully satisfies user requirements, alternative development models are being considered as potentially more effective. One such model comes under the general umbrella of agile software development, which prescribes a highly iterative and adaptive development process that adapts not only to the changing software requirements and operating environments, but also to the "collective experience and skills" of people working in the development teams (Turk, France, & Rumpe, 2005). Proponents of agile methods advocate the superiority of their model in delivering quality software, produced at an economic cost within a fast development period and meeting evolving customer requirements.

A parallel trend to agile software development has been that of open source software (OSS) development, which looks a priori as a random and chaotic process harnessing the abundance of programmers on the Internet to produce software that is deemed of very high quality. However, upon a closer look at both processes, agile and open source, striking similarities exist in terms of the development process itself. Indeed some research has already pointed out that OSS development, although driven by different motivations and economic considerations than agile methods, follows the same generic phases of agile methodologies (Warsta & Abrahamsson, 2003). In this chapter, we expound on this connection by adopting an agile perspective on OSS development. This is not to confuse the two paradigms, which remain distinct, but to emphasize the similarities and dissimilarities between the two approaches to software engineering.

In the first part of the chapter, we attempt to retrofit OSS development within a generic agile software development framework. In the second part, we demonstrate through the example of two landmark open source projects, Mozilla and Apache, how OSS development processes can be recast within the generic agile development model.

BACKGROUND

An Agile Perspective on OSS Development

Agile development implies developing simple designs and starting the coding process immediately. Frequent stops are made to assess the coding process and gather any new set of features or capabilities from clients in view of incorporating them into the software through iterations rather than following a single formal requirements document (Lindquist, 2005). Some of the most prominent agile software development methods are extreme programming (XP), Scrum, feature-driven development (FDD), and adaptive systems development (ASD; Ambler, 2002). Through plotting these agile software development methods into a generic framework for software development (see Table 1), we identified four common phases to all agile processes, which we termed the generic agile development model (see Figure 1). These phases are outlined as follows:

1. **Problem exploration:** Includes overall planning, requirements determination, and scheduling
2. **Iterative development:** Repeated cycles of simple design, coding, testing, a small release, and refining requirements
3. **Version control:** At the end of one iteration or a few concurrent or consecutive iterations, changes are committed to the final program and documented, probably delivering a working version to the customer (possibly installed for use until development ceases).
4. **Final release:** When changes can no longer be introduced to the requirements or operating conditions

Open Source Development from an Agile Perspective

In general, the fundamental difference between open source and conventional software development is that the extremely emphasized and revisited steps of planning, analysis, and design in software engineering are not part of the general open source life cycle; the "initial project founder" is the one who conducts these steps in a brief and oversimplified manner (O'Gara, 2002).

Table 1. Synthesis of agile methodologies' development phases

Agile Method	Conventional Software Development Phases					
	1. Planning	*2. Requirements Specification*	*3. Design*	*4. Coding*	*5. Testing*	*6. Maintenance*
XP						
	The first activity takes place. "Story cards" are developed to convey the features required by customers. The architectural design is determined.					
		Development iterations include implementing story cards. Each iteration is planned. Tests run after each iteration constitute detailed design.				
			Once the system is released, maintenance iterations proceed to incorporate postponed features, and the design is updated. Once all functionalities are incorporated, documentation is done, and the project is closed.			
Scrum						
	It entails determining and prioritizing software requirements in a backlog list and estimating an implementation schedule. High-level design is done.					
		The development phase includes iterative increments called "sprints."				
				The postgame phase includes integration, system testing, and documentation.		
FDD						
	An overall system model and object models are developed.					
	A feature list is built and planning is done by feature.					
			The project is designed by feature and built by feature, and then is delivered to the end user.			
ASD						
	Project initiation includes setting the mission, schedules, and requirements.					
		Collaboration is carried out concerning concurrent component development.				
						Quality assurance is carried out, and then the product is released.

OSS development consists of seven visible phases: problem discovery, volunteer finding, solution identification, code development and testing, code change review, code commit and documentation, and release management (Sharma, Sugurmaran, & Rajagopalan, 2002). Problem exploration in agile development corresponds to open source problem discovery, volunteer finding, and solution identification combined. Agile iterative development corresponds to code development and testing, and code change review in OSS development. This is where the two processes fully meet. Version control in agile methods corresponds to code commit and documentation in open source, and finally the final release stage in agile development corresponds to release management in open source. The mapping between the two models will be illustrated later through two prominent open source projects.

A question that has been raised many times is whether OSS development could be considered under the umbrella of agile development (Goldman

Figure 1. The generic agile development model

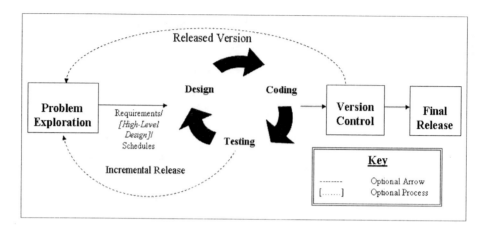

& Gabriel, 2005; Knoernschild, 2006; Rothfuss, 2003; Warsta & Abrahamsson, 2003). Goldman and Gabriel in the most extensive treatment of the matter draw a parallel between the principles underlying the two development approaches and conclude that they are strikingly similar (see Appendix):

Both the agile and open-source methodologies embrace a number of principles and values, which share the ideas of trying to build software suited especially to a class of users, interacting with those users during the design and implementation phases, blending design and implementation, working in groups, respecting technical excellence, doing the job with motivated people, and generally engaging in continuous redesign.

Simmons and Dillon (2003) also argue that the OSS development process is a special type of agile method. This chapter adopts a similar perspective by retrofitting the mechanics of OSS development within those of agile development. There is no attempt, however, to foray into the OSS and agile-development debate and venture either way. Rather, we will adopt an empirical approach to illustrate how landmark OSS projects could be construed as agile development projects. The following cases are an illustration thereof.

MAIN FOCUS OF THE CHAPTER

Case Illustration through Mozilla and Apache

The generic agile development model is applied to two open source projects, namely Apache and Mozilla. Data about the two projects were available on the projects' Web sites and in various other sources on the Web and elsewhere. We attempt hereby to plot the specifics of the two projects against the four-stage process for agile software development, which we outlined earlier. We first give a brief overview of each project, and then illustrate how the development process in each case could be recast as an agile process.

The Mozilla Project

The Mozilla project is an OSS project started in February 1998 after Netscape released most of its code base for the Netscape Communicator browser suite to the open source community; it was done under a license that requires the development of an application called Mozilla, which is coordinated by the then newly created Mozilla Organization.

Since 1998, the Mozilla Organization has succeeded in producing an Internet suite that is not only of high quality, but is also much better than Communicator in both features and stability (*Wikipedia*, 2004c). The Mozilla Organization (http://www.mozilla.org) is responsible for managing, planning, and supporting the development and maintenance environment for the Mozilla project. Each member of the organization undertakes a different task, such as Web site maintenance, documentation, architecture design, and release management. Meanwhile, a different set of people, the Mozilla community, performs coding, testing, and debugging (http://www.mozilla.org).

The Apache Project

For the last 10 years, Apache has been the most popular Web server software on the Internet, running the majority of Web servers (*Wikipedia*, 2004a). The goal of the Apache project was to provide an open source Web server that is adaptable to existing and emerging operating systems in order to be usable by anybody on the Internet. The work on the new server started in 1994 by a group of eight people that came to be known as the Apache Group. During the first release of Apache in 1995, members of the Apache Group formed the Apache Software Foundation (ASF) to provide support for the Apache HTTP (hypertext transfer protocol) server (*Wikipedia*, 2004b).

The Apache project, like any other open source project, depends on collaboration within communities of practice. Each project is managed by a self-selected team of technical experts who are active contributors to the project (*Wikipedia*, 2004b). Major decisions regarding the software developed are taken based on consensus from an inner circle of members.

In the following sections, we attempt to retrofit the development processes of Mozilla and Apache into the generic agile development model outlined earlier.

Problem Exploration in Mozilla and Apache

As mentioned earlier, in agile software development, the problem exploration phase includes overall planning, requirements determination, and scheduling. In open source development, planning, though not formal or comprehensive, takes the form of a central authority setting the overall direction of the project in the form of a to-do list. No resource allocation takes place as contributors are free to either choose the tasks they wish to perform or choose from the to-do list (Stark, 2001). Requirements usually evolve rapidly over time (Norin & Stockel, 1999; Scacchi, 2002; Venugopalan, 2002). Users and developers discuss and negotiate the requirements on mailing lists, newsgroups, or Web sites. If users and developers do not agree on specific requirements, each contributor develops his or her own code resulting in different versions of the software (Raymond, 1999). No fixed design is maintained as the software is more flexible and adaptable to any changes done in requirements (Venugopalan).

Mozilla

On Mozilla.org, the general direction of the Mozilla project is set by defining the objectives of the initiative and any changes to the scope of the project with appropriate justifications. However, no description exists to process the data and control, functions, performance criteria, interface layout, or reliability issues. The Mozilla plan, simply put, is

a quarterly milestone plan that emphasizes regular delivery of stable new-feature releases, ideally with risky changes pushed into an "alpha" minor milestone, followed by stabilization during a "beta" period, then a shorter freeze during which only "stop-ship" bugs are found and fixed. (http://www.mozilla.org/roadmap.html)

The general architecture of the software is also described as a to-do list provided along with a cautionary statement that the "detailed plan of attack should be developed in the newsgroups and via Bugzilla" (Mozilla Organization, 2004). As for scheduling, the Mozilla project only provides the expected dates of releases.

Regarding requirements engineering, high-level requirements are laid down by Mozilla.org management and are generally "few and very abstract" (Reis & Fortes, 2002). This hasty initial requirements specification in Mozilla logically leads to major decisions on including functionalities and implementing changes being discussed piecemeal by the community and major developers, also known as module owners. This ownership implies responsibility for the quality of the overall module as well as for decisions on what gets integrated into the Mozilla release. Mozilla developers maintain some documentation on user-interface requirements and specifications, but most of the other requirements evolve incrementally without any limitations (Reis & Fortes).

Discussion lists for public newsgroups are used to refine requirements. The discussion, in general, acquires a technical flavor, and eventually results in either dropping the requirement or filing a bug for contributors to resolve (a bug in Mozilla does not necessarily refer to an error; it is any change that is incorporated to the browser source code). Yet, those requirements might never be implemented even if they are consensually agreed upon.

Last, design specifications in the Mozilla project for any new contributor take the form of narrative descriptions of what the module is supposed to do, the functionalities the software is supposed to have, as well as previous code developed by other programmers. Similar to the requirements specification phase, most design considerations are taken care of as they occur. Hence, bugs are filed to change the design of a certain component, and the negotiation process takes place over discussion groups (Reis & Fortes,

2002). After changes to design have been introduced and implemented, the only location where contributors can view the evolution of design is in the bug list, which contains a forum for discussion. They can search the bug files by keyword to view all relevant discussions (*Wikipedia*, 2004b).

Apache

Problem exploration for Apache is the responsibility of the Apache Software Foundation (http://www.apache.org/). There is no evidence that the foundation sets any comprehensive plan as with Mozilla. It simply hosts the project, offers support, and provides a "framework for limiting the legal exposure of individual volunteers while they work on behalf of one of the ASF projects" (ASF, 2004). Members set the overall direction of the foundation and leave ad hoc planning to the project initiators or central figures, who acquire their importance based on the complexity and significance of the portions of Apache that they handle. In other words, when a new user wants to join the ASF to work on a favorite project, only the project description is provided with a list of features for developers to start coding right away. Scoping for Apache is a matter of stating the objective of the project.

Regarding requirements engineering, each developer is responsible for identifying an existing problem or a new required functionality and deciding which volunteer will work on a related solution. However, not all contributors will have equal rights. There is a difference between who implements new requirements and who can create a new requirement specification. However, in order to contribute to the specifications of the Apache HTTP project in the first place, the person should be a member of the dev@httpd.apache.org mailing list, which signals that he or she is always up to date on the latest Apache developments. The added specification takes the form of a patch that should be clearly described and attached to the message (ASF, 2004).

Developers, who have a "committed" status, meaning they have committed good code to the server, share in the decision of what specifications get implemented (ASF, 2004). Developers may use a "minimal quorum" (or select group) voting system for resolving any conflicts that might occur (Mockus, Fielding, & Herbsleb, 2000). However, similar to Mozilla, these specifications are not compiled anywhere, and neither are they required to be fixed by a certain date (ASF, 2004).

On the ASF Web site, each major project is divided into subprojects, each of which is supported with basic descriptions of the pseudo code expected for each module (http://www.apache. org). This information is generally found under documentation of previous releases. However, no particular location for current design description or specifications is indicated. For instance, the "Designing Components" location would mention general descriptions of the required functionality or design objectives as opposed to making available design modules. The design of the Apache server software is managed through task lists available on the project's Web site.

Iterative Development in Mozilla and Apache

Iterative development in the generic agile development model entails repeated, and sometimes concurrent, cycles of simple design, coding, and testing, eventually leading to a small release. This is followed by reviews with customers to refine requirements and provide an action plan for the next cycle.

The open source life cycle is argued to be located in the implementation phase (i.e., coding, testing, and maintenance) of the conventional software development life cycle (SDLC). The high level of iteration ensures that planning, specification, and design are performed extensively within the framework of coding, relying on the overabundance of development resources and most notably programmers who develop high-

quality code. This phase constitutes the primary reason why developers join open source projects in the first place (Raymond, 1999). Project "gods," the highly regarded developers in charge of the project under consideration, place no restrictions or specific instructions as to how coding should be implemented (Stark, 2001).

The process also scores very high on external testing (Stark, 2001; Venugopalan, 2002). This is true because it is not developers who conduct exhaustive testing; it is rather huge dispersed communities of good programmers who do it (Stark). This mode of testing is generally described as the best system testing in the industry (Raymond, 1999; Stark), and as Raymond puts it, "given enough eyeballs, all bugs are shallow." Once a module is coded and tested, it is uploaded onto the mailing list or the joint development tool for review by other developers and acceptance into the core distribution. If refinements are required, the module enters another development cycle before it is accepted and integrated.

Mozilla

As is typical of agile development, Mozilla's development is a series of changes that are integrated in successive software releases following intensive and objective peer review (Feller & Fitzgerald, 2002). Before new code is integrated into the final release, it is subjected to a thorough review. Tools such as Bugzilla, Tinderbox, and Bonsai are used for reviewing code in the Mozilla project and as a way to avoid delays and keep incorrect code from entering the repository (Mockus et al., 2000). Moreover, "super reviews" made by a senior engineer who is familiar with the code are included later to provide important design evaluations (Raymond, 1999). Hence, when reviewing the code changes, developers provide either a description of the changes required to improve the quality of the former code (known as "bug comments"), questions about an unclear section, or recommendations on different aspects of the patch.

Hence, most attention is directed toward the quality of coding by enforcing heavy peer review through developers' fear of submitting substandard code to the "supremely talented" open source code gods (Feller & Fitzgerald, 2002). We will not, however, explore all the factors that contribute to the higher quality and the dynamics of coding in the Mozilla project, because it is beyond the scope of this chapter. We should note, though, that in Mozilla, it is in the implementation phase that we actually see the highest level of documentation and iteration (Reis & Fortes, 2002). This has been pointed out as a feature of most agile methods.

Apache

In the implementation (coding and testing) phase of the Apache server development, developers are given the mere freedom to apply changes to the code directly without the need to review the changes before submission. This contributes to an extraordinarily fast development combined with a high sense of autonomy on the part of developers. Yet, if these changes are not acceptable, they will be detected during peer review and eliminated while still being recognized in the documentation of the actual software release (ASF, 2004). Meanwhile, major developers who opt to handle the most critical parts of the server source code tend to have an implicit code ownership on the components they contribute. Consequently, other developers respect the opinions of code owners, creating some sort of a central authority for consultation and guided discussion.

The contribution process takes the form of developers making changes to a local copy of the source code and testing it on their private servers. After changes are done, the developer either saves the changes directly if he or she has a committal status, or posts a patch on the Apache CVS (Concurrent Version System) mailing list for review and acceptance (ASF, 2004). Finally, since all core developers are responsible for checking the Apache CVS mailing list, they generally make sure

that the patch posted is appropriate and changes are valid. Additionally, since the mailing list can be accessed by many people who are not from the core development community, additional useful feedback is provided when reviewing the changes and before the software is formally released.

Version Control in Mozilla and Apache

Since both agile and OSS development involve many developers working simultaneously on various modules in iterations over a relatively long period of time, it is important that changes, resulting from each iteration, be tracked and merged into the overall outcome and documented. Thus, version control is an essential phase for both processes. Version control as a term refers to the ongoing process of

keeping track of the changing states of files over time, and merging contributions of multiple developers by storing a history of changes made over time by different people ... so that it is possible to roll back the changes and see what the [system] looked like [at any previous point in time] before changes were applied. (Nagel, 2005, p. 4)

This process is normally supported by a version control system such as CVS, Subversion (SVN), and Arc (Nagel).

In agile development, the term version control refers to the phase that follows iterative development in order to commit and integrate the changes made during iterations to the overall software in the form of a released version for customer review. This release is either accepted by the customer as finalized and complete, hence the initiation of the final release phase, or requires further refinement, hence triggering the problem exploration phase and the subsequent iterative cycles. In the latter case, the released version might be installed and maintained for the customer to use by the time enhancements are completed.

In OSS development, whenever a module has been developed in a single iteration or multiple concurrent iterations, changes are submitted for review by project leaders and developers who in turn act as users or customers. The result is either acceptance of the module for integration into the core distribution, or a request for refinements, which triggers a series of further development efforts or iterations. Many versions of the software may be developed before an acceptable final release is uploaded on the official project Web site. While there seems to be a difference in version control for OSS as the software is not released to the final customer while the control system is put in place, in reality, many of the volunteer developers are the final users of the software themselves. Both agile and OSS development rely on quick cycles of version releases contrary to conventional software development where quality assurance is mostly performed before the software is released.

Mozilla

Mozilla developers use CVS for the version control of the Mozilla project. There are three classifications of releases produced in CVS. First are the alpha releases, which are testing patches that may have many missing features or numerous bugs. The project maintainer or the "sheriff" works on each of these test versions along with the testing community to provide early feedback on the major changes that the developer should fix before the final release. The second classification is the beta releases. All features are added to the patch in order to give a general idea of what the final release will look like. It is mainly provided for early adopters and testing audiences to provide some feedback on the most important needed changes. Last, before announcing the final release, the sheriff presents release candidates that are duplicates of alphas or betas days or weeks before the actual release. Developers viewing these release candidates may then identify any

new bugs to be fixed. If there are no changes to be made, the release candidate is set as the final release (Mozilla Project, 2006).

Apache

The Apache project utilizes the Internet-accessible version control system called Subversion. Developers have read access to the code repositories from which they can download the most up-to-date version of the server, work on it, and submit their work for review by Apache committers, who have write access to the SVN repository (ASF, 2005b).

After a number of iterations and concurrent development efforts from project initiation, Apache committers, with the aid of Subversion, are able to put together a working version; however, it is unusable due to the presence of serious problems, and is labeled an alpha version. Alpha versions are intended for further development work after review by committers and major programmers. Further refinements are specified to guide future iterations to ultimately arrive at a better version labeled beta, which compiles and fulfills basic functionalities. Once the version is reviewed and refinements are specified, coding and testing cycles resume until a final version is arrived at and designated as ready for general availability (GA), in other words, available for use by the public. It might take many alpha versions before a beta version can be reached, and many beta versions before a GA version is approved and uploaded on the project's Web site for production use (ASF, 2005a). This pattern of controlling the versions resulting from iterative development follows the generic agile development model.

FUTURE TRENDS

Final Release in Mozilla and Apache

With regard to final release, in agile development, the final release phase involves closing the project and ceasing all development efforts; it is when all requirements have been included, reliability and performance have been accepted by the customer, and no changes in the operating environment can be anticipated. Documentation is completed, and any future requirements or changes the customer wants would be done as a maintenance service under a separate agreement (Warsta, Abrahamsson, Salo, & Ronkainen, 2002). This is more in line with conventional software development than with OSS development, where projects never reach a final release in the strict sense of the word, at least from a theoretical point of view. Production versions of the software (e.g., final releases of Mozilla and GA for Apache) are always open to updates and enhancements as long as there is an interest in the developers' community to work on them, and areas for enhancement exist in the first place. Therefore, iterations resulting in patches or bug fixes may still occur in the final release phase and can be thought of as ongoing maintenance to the post-beta versions of the software. However, one would normally expect that once the program matures and reaches good stability, developers might as well direct their efforts into other more challenging projects. That would amount in some way to the final release of agile methods.

CONCLUSION

OSS development has led many to reconsider their old beliefs and conceptions about software, a critical aspect of which is its development process. While proprietary software seems to cling onto its paradigm of intellectual property rights, through agile methods, it seems to be edging closer to open source as both development processes seem to converge toward less emphasis on planning and design and more on a prominent role for implementation during the software engineering process. It should be emphasized, however, that the similarities between the two and the consideration

of OSS development as yet another agile method is limited to the mechanics of the two processes. Indeed, the philosophies driving both processes remain quite different as agile methods are generally used for proprietary software development. Moreover, it is also unlikely that agile methods achieve the agility of OSS development. Indeed, the abundance of programmers and reviewers of program code in open source projects cannot be offset by the flexibility of a proprietary development process, no matter how agile it is.

With agile methods becoming even more agile as a result of the increasing fluidity of proprietary development and the integration of open source components into proprietary software (exp. MacOS) on one hand, and the increasing pressure being put on OSS development to recognize the individual property rights of contributors, hence reducing the pool of participant programmers, on the other hand, it is likely that the gap in agility between both processes will be further bridged, irremediably bringing OSS development within the fold of agile methods as could be inferred from earlier developments in this chapter. In the meantime, both approaches can come closer by standardizing common tools and developing similar quality assurance.

REFERENCES

Ambler, S. (2002). Agile development best dealt with in small groups. *Computing Canada, 28*(9), 9.

Apache Software Foundation (ASF). (2004). *Frequently asked questions.* Retrieved from http://www.apache.org/foundation/faq.html#how

Apache Software Foundation (ASF). (2005a). *Release notes.* Retrieved from http://httpd.apache.org/dev/release.html

Apache Software Foundation (ASF). (2005b). *Source code repository.* Retrieved from http://www.apache.org/dev/version-control.html

Feller, J., & Fitzgerland, B. (2002). *Understanding open source software development.* London: Addison Wesley.

Goldman, R., & Gabriel, R. (2005). *Innovation happens elsewhere: Open source as business strategy.* San Francisco: Morgan Kaufmann Publishers.

Knoernschild, K. (2006). Open source tools for the agile developer. *Agile Journal.* Retrieved from http://www.agilejournal.com/component/option,com_magazine/func,show_article/id,36/

Lindquist, C. (2005). Required: Fixing the requirements mess. *CIO, 19*(4), 1.

Mockus, A., Fielding, R., & Herbsleb, J. (2000). *A case study of open source software development: The Apache server.* Retrieved from http://opensource.mit.edu/papers/mockusapache.pdf

Mozilla Organization. (2004). *Mozilla development roadmap.* Retrieved from http://www.mozilla.org/roadmap/roadmap-02-Apr-2003.html

Mozilla Project. (2006). *Firefox 2 roadmap.* Retrieved from http://www.mozilla.org/projects/firefox/roadmap.html

Nagel, W. (2005). *Subversion version control: Using the subversion version control in development projects.* Upper Saddle River, NJ: Pearson Education, Inc.

Norin, L., & Stockel, F. (1999). *Open source software development methodology.* Retrieved from http://www.ludd.luth.se/users/no/os_meth.pdf

O'Gara, M. (2002). The OSS development life cycle: Part 4 of 4. *Linuxworld.* Retrieved from http://www.linuxworld.com/story/34356.htm

Raymond, E. S. (1999). *The cathedral and the bazaar: Musings on Linux and open source by an accidental revolutionary* (Rev. ed.). Sebastopol, CA: O'Reilly Associates, Inc.

Reis, C., & Fortes, R. (2002). *An overview of the software engineering process and tools in the Mozilla project.* Retrieved from http://www.async.com.br/~kiko/papers/mozse.pdf

Rothfuss, G. (2003). *Open source software engineering: Beyond agile?* Retrieved from http://greg.abstrakt.ch/docs/wyona_oss_development.pdf/

Scacchi, W. (2002). Understanding the requirements for developing open source software systems. *IEE Proceedings—Software, 149*(1), 24-39.

Sharma, S., Sugurmaran, V., & Rajagopalan, B. (2002). A framework for creating hybrid-open source software communities. *Information Systems Journal, 12*(1), 7-25.

Simmons, G. L., & Dillon, T. (2003). Open source development and agile methods. In *Proceedings of the Conference on Software Engineering and Applications.*

Stark, J. (2001). *Peer reviews in open-source software development.* Retrieved from http://ecommerce.cit.gu.edu.au/cit/docs/theses/JStark_Dissertation_OSS.pdf

Turk, D., France, R., & Rumpe, B. (2005). Assumptions underlying agile software-development processes. *Journal of Database Management, 16*(4), 62-84.

Venugopalan, V. (2002). *FAQ about open source projects.* Retrieved from http://www.magic-cauldron.com/opensource/faq-on-osp.html

Warsta, J., & Abrahamsson, P. (2003). Is open source software development essentially an agile method? In *Proceedings of the Third Workshop on Open Source Software Engineering*, Portland, OR (pp. 143-147).

Warsta, J., Abrahamsson, P., Salo, O., & Ronkainen, J. (2002). Agile software development methods: Review and analysis. *Espoo 2002*, 18-81.

Wikipedia. (2004a). *The Apache HTTP server.* Retrieved from http://www.factbook.org/wikipedia/en/a/ap/apache_http_server.html

Wikipedia. (2004b). *The Apache Software Foundation.* Retrieved from http://www.factbook.org/wikipedia/en/a/ap/apache_software_foundation.html

Wikipedia. (2004c). *Reference library. Encyclopedia: Mozilla.* Retrieved from http://www.campusprogram.com/reference/en/wikipedia/m/mo/mozilla.html

KEY TERMS

Agile Software Development: A software development methodology characterized by continuous adaptation to changes in software requirements, operating environments, and the growing skill set of developers throughout the project.

Apache HTTP Server: An open source Web server for almost all platforms. It is the most widely used Web server on the Internet. Apache is developed and maintained by the Apache Software Foundation.

Final Release: The fourth (last) phase in agile development when changes can no longer be introduced to the requirements or operating conditions.

Iterative Development: The second phase in agile development consisting of repeated cycles of simple design, coding, testing, a small release, and requirements refinement.

Mozilla: Generally refers to the open source software project founded to create Netscape's next-generation Internet suite. The name also refers to the foundation responsible for overseeing development efforts in this project. The term is often used to refer to all Mozilla-based browsers.

Open Source Software (OSS): Software whose source code is freely available on the Internet. Users can download the software and use it. Unlike proprietary software, users can see the software's source code, modify it, and redistribute it under an open source license, acknowledging their specific contribution to the original.

Problem Exploration: The first phase in the agile development model that includes overall planning, requirements determination, and scheduling.

Version Control: The third phase in agile development wherein at the end of one iteration or a few concurrent or consecutive iterations, changes are committed to the final program and documented, delivering a working version to the customer (possibly installed for use until development ceases).

APPENDIX

Principles of Agile Methods and Open-Source Development (Adapted from Goldman & Gabriel, 2005):

- "Our highest priority is to satisfy the customer through early and continuous delivery of valuable software."
  - o Open source does not address the customer, but in general, open source projects include nightly builds and frequent named releases, mostly for the purpose of in situ testing.
- "Welcome changing requirements, even late in development. Agile processes harness change for the customer's competitive advantage."
  - o Open source projects resist major changes as time goes on, but there is always the possibility of forking a project if such changes strike enough developers as worthwhile.

- "Deliver working software frequently, from a couple of weeks to a couple of months, with a preference to the shorter time scale."
  - Open source usually delivers working code every night, and an open source motto is "release early, release often."
- "Business people and developers must work together daily throughout the project."
  - Open source projects do not have a concept of a businessperson with whom the team must work, but users who participate in the project serve the same role.
- "Build projects around motivated individuals. Give them the environment and support they need, and trust them to get the job done."
  - All open source projects involve this, almost by definition. If there is no motivation to work on a project, a developer will not. That is, open source projects are purely voluntary, which means that motivation is guaranteed. Open source projects use a set of agreed-on tools for version control, compilation, debugging, bug and issue tracking, and discussion.
- "The most efficient and effective method of conveying information to and within a development team is face-to-face conversation."
  - Open source differs most from agile methodologies here. Open source projects value written communication over face-to-face communication. On the other hand, open source projects can be widely distributed and do not require collocation.
- "Working software is the primary measure of progress."
  - This is in perfect agreement with open source.
- "Agile processes promote sustainable development. The sponsors, developers, and users should be able to maintain a constant pace indefinitely."
  - Although this uses vocabulary that open source developers would not use, the spirit of the principle is embraced by open source.
- "Continuous attention to technical excellence and good design enhances agility."
  - Open source is predicated on technical excellence and good design.
- "Simplicity—the art of maximizing the amount of work not done—is essential."
  - Open source developers would agree that simplicity is essential, but they also do not have to worry quite as much about scarcity as agile project developers do. There are rarely contractually committed people on open source projects—certainly not the purely voluntary ones—so the amount of work to be done depends on the beliefs of the individual developers.
- "The best architectures, requirements, and designs emerge from self-organizing teams."
  - Open source developers would probably not state things this way, but the nature of open source projects depends on this being true.
- "At regular intervals, the team reflects on how to become more effective, and then tunes and adjusts its behavior accordingly."
  - This is probably not done much in open source projects, although as open source projects mature, they tend to develop a richer set of governance mechanisms. For example, Apache started with a very simple governance structure similar to that of Linux and now there is the Apache Software Foundation with management, directors, and officers. This represents a sort of reflection, and almost all community projects evolve their mechanisms over time.

Chapter XIII
A Model for the Successful Migration to Desktop OSS

Daniel Brink
University of Cape Town, South Africa

Llewellyn Roos
University of Cape Town, South Africa

Jean-Paul Van Belle
University of Cape Town, South Africa

James Weller
University of Cape Town, South Africa

ABSTRACT

Although open source software (OSS) has been widely implemented in the server environment, it is still not as widely adopted on the desktop. This chapter presents a migration model for moving from an existing proprietary desktop platform (such as MS Office on an MS Windows environment) to an open source desktop such as OpenOffice on Linux using the Gnome graphical desktop. The model was inspired by an analysis of the critical success factors in three detailed case studies of South African OSS-on-the-desktop migrations. It provides a high-level plan for migration and is illustrated with an example. This chapter thus provides a practical guide to assist professionals or decision makers with the migration of all or some of their desktops from a proprietary platform to an OSS environment.

INTRODUCTION

The growing market share of open source software (OSS) can be attributed to the rising prices of Microsoft products, the increased availability of OSS, the increased quality and effectiveness of desktop OSS software, and the drive for open standards in organisations and governments (Wheeler, 2005). However, though OSS has been widely accepted as a viable alternative to proprietary software (PS) in the network server market for some time, desktop usage of OSS still remains fairly limited (Prentice & Gammage, 2005). Unlike many server OSS installations where the organisational impacts are relatively minor due to their isolation in the server room, moving to an OSS desktop generally requires an organisation-wide migration involving a large number of users. Correspondingly, there has been an increased interest and awareness in guidelines to assist with the migration from proprietary desktop platforms to OSS. (Bruggink, 2003; Government Information Technology Officers Council [GITOC], 2003).

This need for migration guidelines was the inspiration for our research. This chapter thus proposes a practical model to assist with the migration to desktop OSS. The model is based on an in-depth analysis of the critical success factors (CSFs) in three migration case studies in South Africa. However, the model that emerged from this research should prove useful in other contexts, specifically so—but not only, it is hoped—in other developing-country contexts.

For clarity, the term desktop OSS (or OSS on the desktop) will be used to refer to those OSS applications that are utilised by everyday users to perform daily work tasks. This must be contrasted to server-side OSS, which comprises those OSS applications that traditionally reside on a server as opposed to a client (or workstation) and are used primarily by technical staff such as systems administrators to fulfill back-office functions such as e-mail routing and Web hosting. Typical desktop OSS applications include productivity software (e.g., OpenOffice), e-mail clients (e.g., Mozilla Thunderbird), Internet browsers (e.g., Mozilla Firefox), and a variety of other utilities. Although many PC (personal computer) users use one or several OSS applications, the proposed model deals with situations where fairly significant desktop OSS migrations are implemented, that is, those that include at least an OSS operating system (Linux) as well as at least a full productivity software suite.

BACKGROUND

For many organisations, the decision to migrate to OSS from a proprietary platform is a strategic one (Wiggins, 2002). Potential advantages associated with the use of OSS are summarized by Gardiner, Healey, Johnston, and Prestedge (2003), but include lower cost or free licenses, lower total cost of ownership (TCO), access to source code, reliability and stability, support by a broad development community, scalability, and security. The authors also list the following potential disadvantages: lack of vendor support, difficult installation, lack of integration, hardware compatibility problems, security, insufficient technical skills, user resistance, and warranty or liability issues.

Migration requires analysis of the expected return on investment (ROI) in terms of the current and expected TCO and the associated migration costs (Fiering, Silver, Simpson, & Smith, 2003). One of the bigger costs of migrating to a business OSS desktop, such as Novell Linux Desktop, is that proprietary business applications have to be rewritten to run on Linux.

Migration does not have to be an all-or-none decision. For some users, Linux desktops are more appropriate, while for others, there are too many proprietary, non-Linux-compatible applications in use for a migration to make sense. Companies must decide which user groups to migrate and may have to provide support for both the proprietary and OSS products simultaneously (Goode, 2004).

An illustrative ROI analysis by Gartner (Prentice & Gammage, 2005) shows that migration costs are significant when compared to savings. It is possible to reach a breakeven in 1.3 years in the best-case scenario of migrating users from Microsoft Windows 95 to locked Linux desktops, while payback for knowledge workers may still be unattainable in many circumstances. Structured-task users are more likely to take to a locked desktop without impacting their productivity and use significantly fewer applications. These calculations led Gartner to claim that OSS (Linux) on the desktop for data-entry (or structured-task) workers has reached the plateau of productivity on the Gartner hype cycle, while Linux on the desktop for mainstream business users is only reaching the peak of inflated expectations. Gartner predicts that business use of Linux for the knowledge worker will only mature in the next 3 to 5 years (Prentice & Gammage, 2005).

Some Typical Obstacles in Migrating to Desktop OSS

Bruggink (2003) highlights some of the typical obstacles that need to be overcome when migrating to OSS on the desktop. Most appear to be even more pronounced in developing countries:

- There is little published guidance available as to how to go about migrating from proprietary software to OSS.
- Desktop OSS is not widely used in many countries. This leads to a huge problem in finding qualified staff to support and maintain the desktop OSS.
- There are few resellers of desktop OSS, especially in some developing countries. Although most, if not all, desktop OSS can be downloaded from the Internet, it is not always an option as reliable, fast Internet connections are not always available in developing countries, and bandwidth tends to be very expensive.
- Few OSS certification programmes exist for computer and network support professionals, which leads to the lack of technical support available. However, this situation is currently improving.
- Many countries have a very risk-averse corporate culture, which is slowing down the OSS migration process.
- There is also a widespread perception that Linux is the only OSS product and that it

is not very user friendly, requiring in-depth technical skill to operate it.

OSS Migration Guidelines

A number of researchers have proposed methodologies or guidelines to implement desktop OSS. Lachniet (2004) published a framework for migration to OSS on the desktop. The Lachniet framework focuses on the prework that needs to be done before migrating to desktop OSS in a corporate environment. Another framework is suggested by Wild Open Source Inc. (2004) consultants. Their methodology consists of three phases: the planning phase, design phase, and implementation phase. Finally, NetProject (2003) proposes an OSS migration methodology that divides the migration into the following five exercises: the data gathering and project definition phase, justification for the migration and estimation of migration costs, the piloting phase, complete rollout, and implementation monitoring against the project plan. These are discussed in more detail below.

Lachniet Framework for Migrating to OSS on the Desktop

Lachniet (2004) published a fairly detailed framework or set of guidelines for migration to OSS on the desktop. The focus is on the prework that needs to be done before migrating to desktop OSS in a corporate environment. The framework divides the premigration tasks into three sections: administrative tasks, application development tasks, and information technology tasks:

- Administrative tasks are primarily focused on creating support from top management levels for the migration to desktop OSS:
  - Develop a high-level policy of support by upper management; upper management supporting the migration to OSS will serve as added incentive.

o Implement purchasing requirements, and consider OSS when making new software purchases.

o Implement hiring requirements, favouring employees with Linux skills.

o Develop a Linux team to continue the analysis and implementation of OSS.

o Hire an expert open source project manager.

o Establish a budget.

- Application development tasks aim to ensure that future OSS migrations are made possible as most difficulties experienced during such a migration are related to internally developed software:

  o Identify a portable development platform for future development, and identify and standardise on a development platform and language that is portable to multiple architectures, such as Java.

  o Cease and desist all nonportable development.

  o Obtain training for application developers.

  o Identify a migration strategy for previously developed applications.

- Information technology tasks ensure that the back-end servers and network services are in place before implementation of the desktop OSS commences. In addition, the information technology tasks ensure compatibility and functionality through testing programmes:

  o Identify and migrate back-end applications.

  o Obtain training for IT staff.

  o Pilot open source software with willing user communities.

The Lachniet (2004) framework for the migration to OSS in a corporate environment is very comprehensive in the tasks that need to be done before migration to OSS in general, but it does not specify the tasks to be performed or the sequence of events that needs to take place to effect the actual migration.

Wild Open Source Migration Methodology

The following methodology is for the migration to OSS in general, as used by Wild Open Source Inc. (2004). Its methodology consists of three phases: the planning phase, design phase, and implementation phase (see Figure 1):

1. **Planning phase:** Here the client's mission, strategy, and objectives for the migration to OSS are identified. This is followed by a detailed assessment of the client's functional and architectural requirements. The phase is completed by the generation of a high-level solution.

2. **Design phase:** This involves the creation of a detailed systems design and engineering specification. All hardware and software needed for the migration is also specified at this time.

3. **Implementation phase:** Before implementing the OSS, a detailed implementation strat-

Figure 1. Wild Open Source Inc. methodology (2004)

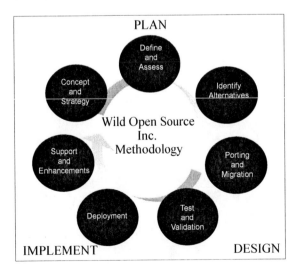

egy and plan is formulated. Following the implementation, the results are documented and a postimplementation audit is done. The purpose of the audit is to identify work done, lessons learned during the migration, and what remains to be finished. After the implementation is completed, the users are trained in the operation of the OSS.

When examining the methodology, two clear problems become apparent. First, the users are not involved in the migration process and only receive training at the very end of the migration. Second, the methodology does not specify how the migration should be performed, merely stating that it should be planned and documented.

NetProject IDA OSS Migration Methodology

The NetProject (2003) IDA OSS migration methodology is primarily focused on the migration prework, dividing the migration into the following five exercises:

- Create a description of the existing software and hardware in use, the required functionality, and the implementation plan in the data gathering and project definition phase.
- Define the justification for the migration and estimate migration costs.
- Test the implementation and project plan in the piloting phase.
- Roll out the OSS to all users and servers.
- Monitor the actual results against the implementation and project plan.

This OSS migration methodology is fairly high level and focuses on the technical tasks. It does not detail management or user acceptance and training issues.

THE PROPOSED MODEL FOR DESKTOP OSS MIGRATION

The authors analysed three case studies of desktop OSS in South Africa: one in the educational sector, one in a governmental sector, and one in a commercial organisation. A number of common themes were identified (Van Belle, Brink, Roos, & Weller, 2006). The following lists some of the more salient themes and compares them with the general literature on the benefits and pitfalls of OSS.

Consistent with the literature, the main driver for deciding to migrate to desktop OSS was the promise of financial benefits, such as decreased license costs and the ability to redistribute funds that would have been spent on software licenses to other areas. No evidence of any of the migrations being motivated by political or social responsibility factors could be found.

Benefits encountered as a result of the migration were primarily described as financial. There was also mention of intangible benefits, such as the freedom from vendor lock-in and the ability to customise the software should one wish to do so. Other supposed benefits identified in the literature, such as improved security, did not appear to be important to the organisations studied.

Consistent with the literature findings, the potential savings on hardware costs, due to the ability of desktop OSS to run on older hardware, were only identified in the education-sector case study. In fact, hardware upgrades were required in the government-sector case study in order to run the desktop OSS efficiently. Similarly, user resistance was fairly low in the education sector.

The problems of user resistance and legacy applications preventing total migration were identified in all of the case studies (although resolving these problems was reasonably successful in the Novell case). This adds to the findings of the literature review, which did not specifically identify the commonality of these problems across all three sectors studied. The problems of high support costs and availability of support were common across all three case studies, which is again consistent with the literature.

A new set of problems related to training, specifically the general perception of nonusefulness of training and the lack of a hands-on, practical approach to training, was identified. This was not covered in the literature and is therefore an important element of the proposed model.

Based on these common themes, critical success factors could be identified for migration to desktop OSS (Table 1; listed in the typical sequence that a migration to desktop OSS would follow).

It is clear that some of these are not addressed in the migration guidelines and frameworks identified earlier. Thus, the following empirically inspired model was created to assist organisations in successfully migrating to desktop OSS. The model is given in the form of a diagram. A detailed explanation of its components and an example of its usage will be given next.

The BRW migration model is focused on the implementation of desktop OSS where desktop PS has been the standard, but the model should not be viewed as cast in stone; it can easily be adapted to any type (environment and situation) of migration to desktop OSS. As mentioned, the tasks that make up the BRW migration model are primarily based on the critical success factors for migration to desktop OSS identified in the

previous section, but the model also incorporates elements from the migration frameworks found in the literature. A detailed description of each task follows.

Obtain Organisational Commitment to the Migration

For any project that will result in a significant change, it is vital to have support and commitment from the top management levels. This commitment from the top management level to a change forces the lower levels of the organisation to conform to the decision and is eased through the use of a project champion. The role of the project champion is to promote the change and ensure it is completed. By nature, most humans are resistant to change, and if there is no commitment to the change from the top management levels, the lower level employees and users will know that they can get away with not migrating to the desktop OSS and will most likely opt to stay with the PS that they know and are used to.

During this initial migration task, it is also important to acquire the necessary resources needed for the migration. These should include a financial budget and staff with project-management and technical OSS skills. It is also vital to document the way forward through the creation a project plan that includes the reasoning for the migration to desktop OSS, the goals of the project, a project timeline, and estimated costs.

Table 1. Critical success factors for desktop OSS migration

1.	Obtain top-level management support for the migration project.
2.	Practice change management, specifically the following.
	1.1. Create user awareness.
	1.2. Conduct proper user training.
	1.3. Communicate constantly with target users.
2.	Conduct a detailed analysis of application, business, and technical requirements.
3.	Ensure that the systems architecture hardware and software are prepared and tested before users start using the system.
4.	Conduct a pilot project and/or perform a phased migration.
5.	Provide ongoing user support and training.

Figure 2. OSS-on–the-desktop BRW migration model

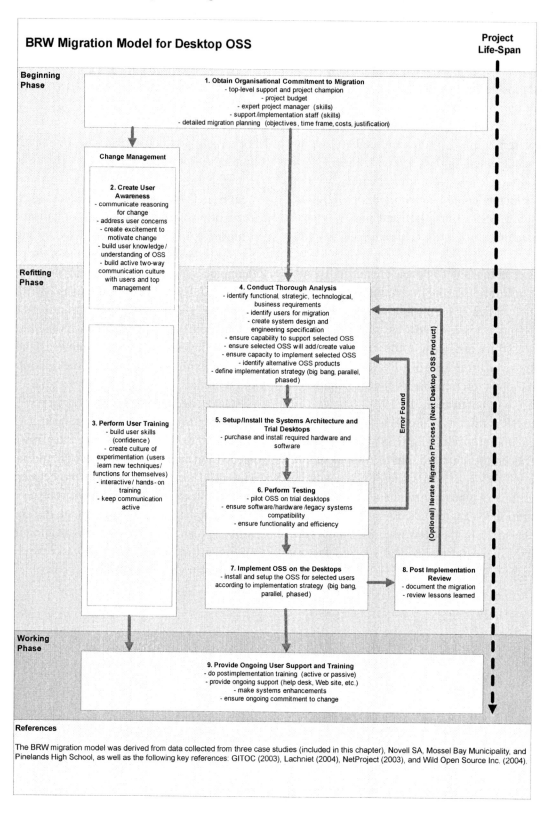

160

Create User Awareness

In order to create user buy-in for the migration and reduce resistance, users need to be included in the whole migration process. This is done through plenty of two-way communication, making the users understand why the migration is necessary, what they can expect to happen during the migration, and what will change. It is very likely that the users will have concerns regarding the migration; it is important that these concerns be addressed as early in the migration as possible and that excitement about the migration is created in order expedite the migration process.

Perform User Training

Adequate training is vital to ensure that the users are capable of efficiently using the new desktop OSS products. More importantly, the training, by building the users' computer skill levels, also increases their confidence. The best way for users to learn how to use the new software is through experimentation and helping each other, as formal training is often found to be tedious, boring, and compulsory, thus not very effective.

Conduct Thorough Analysis

By far the biggest task of the entire migration, analysis is the key to success. The role of the analysis task is to ensure that the OSS is implemented only where it can promote efficiency and meet the needs of the users as well as the organisation. It is not possible to state the subtasks that need to be performed during analysis as the tasks will vary according to the situation. There are, however, some general tasks that should always be performed. These are as follows:

- Perform an audit of all existing software and hardware in use.
- Identify the functional requirements.

- Identify the technical requirements.
- Identify the organisation's business strategy.
- Identify users who could be migrated to OSS.
- Identify possible OSS products and select the best option according to the above requirements.
- Create a system design and engineering specification.
- Define the best suitable implementation strategy (big bang, parallel, or phased) depending on the size and scope of the migration.

Setup and Install the Systems Architecture and Trial Desktops

Before a desktop migration to OSS can be undertaken, the back end of the systems architecture, such as file servers, needs to be set up, installed, and updated in preparation for the change to desktop OSS on the user side. This is important as network problems during the actual desktop implementation could cause extensive delays. Often, new hardware and/or software will also have to be purchased and installed (server side, back end) at this stage.

Perform Testing

Proper testing is crucial to ensure that the selected desktop OSS meets the users' functional needs. Testing should be done through piloting the selected product(s) on a suitable number of desktops. This pilot testing could be done in a laboratory environment, but it is preferable to pilot the selected software in a real-world scenario by having the end users attempt to use it to perform their day-to-day jobs. Any hardware, software, or legacy system incompatibilities; lack of functionality; or inefficiencies will become apparent during testing and should be resolved before the desktop OSS is deployed to all of the final end users.

Implement OSS on the Desktop

Implementing the OSS on the desktops of the final users should only be done after all possible problems with the use of the OSS have been resolved. The implementation should be done in accordance with the implementation strategy as defined during the analysis tasks.

Postimplementation Review

The postimplementation review is not actually part of the physical migration, but assists the organisation in learning from the implementation, preventing the same problems from occurring in future desktop OSS implementations. The review is somewhat tedious and often ignored, but can yield great rewards in the future through the building of organisational knowledge and wisdom. This is of particular importance if only one functional area or regional branch of a larger organisation has migrated. The review is performed by documenting the entire migration process and should include how the change management was done (including training), what analysis was done and what the findings where, what changes were made to the systems architecture, the problems that were identified during testing, and most importantly, how these problems were resolved.

Provide Ongoing User Support and Training

It is vital that support and training are continued after the implementation of the OSS on the desktops of the earmarked users is completed. This is because it is often the case that not all of the problems related to the use of the desktop OSS are identified during testing. Also, at this stage in the migration, the users are expected to use the OSS to do their jobs and will find it frustrating if they can no longer perform a certain function. The main purpose of the ongoing support and training is to prevent a relapse back to the original software and maintain commitment to the change.

APPLYING THE MODEL TO A CASE STUDY

The following looks at how the model would have applied in the context of one of the three case studies on which the migration model was based (Van Belle et al., 2006). The case study concerns a relatively small municipality along the south coast of South Africa. This municipality was underlicensed in terms of its Microsoft Windows 98 licenses (150 against 60 legal licenses) with a similar situation for their Microsoft Office productivity software. The Business Software Alliance (BSA) threatened legal action unless the situation was regularised. The following looks at the different actions that were taken to effect a partial migration and, from an ex post perspective, how they would have fit into the model.

Obtain Organisational Commitment to Migration

In response to the letters from the BSA and the threat of possible legal issues, the municipality's IT manager arranged a meeting with the Mossel Bay management team, consisting of all the heads of the various municipality departments, in which they discussed the current licensing situation.

As a solution to the licensing issues and limited funds available, the IT manager presented a cost comparison of buying the needed Microsoft product licenses vs. migrating selected users to Linux and related OSS products. RPC Data (an external IT-support contractor) advised the IT manager that a complete migration to Linux is not always feasible due to the application limitations, user resistance, and training budget constraints. Therefore, a mixed IT environment of Novell Linux, related OSS, Microsoft products, and other proprietary products would be more suitable to Mossel Bay Municipality's needs.

The management team provisionally gave their approval for the partial migration to OSS as the regional government is supporting migration to OSS and Mossel Bay Municipality wants to align their IT environment to this strategy. It was found that the majority of the PCs in the financial department could be migrated to Linux as the users use primarily network-based financial systems, e-mail, and spreadsheets. Thus, the number of Microsoft product licenses required in order to meet software licensing requirements could be reduced. In total, 55 computers were identified for migration to OSS.

Given the identification of the PCs to be migrated and the potential reduction in licensing costs, the management team presented their proposal to migrate to OSS to the Mossel Bay Council, who gave permission for the migration to proceed.

Create User Awareness

In this case, a special introduction session was included in the initial training session to make the users aware of the plans to migrate to OSS and explain what they could expect and the reasoning behind the migration: the reduction of license costs and the legalisation of software products in use. During these sessions, the end users expressed several concerns about migrating to Linux, fearing that which is new and unfamiliar. The personnel from Creative Minds handled all resistance to the migration through the dissemination of information, informing the staff members of the actions taken by the BSA and the need to reduce license costs. The users were also reassured that they would still have the same functionality and would still be able to perform their jobs.

Initiate Training while Migration is Still in Early Stages

The Mossel Bay Municipality contracted Creative Minds, a local computer training firm, to give training sessions to the Mossel Bay end-user staff before migration. The training was done in the form of a seminar, which all staff members who would be migrated to OSS were required to attend. Creative Minds, through the seminars, introduced the staff to Linux, using an overhead projector linked to a PC running SUSE Linux 8.2. The seminar covered the SUSE Linux functionality the staff members would be required to use and how it differed form the other proprietary products that they had been using in the past. This seminar also included an introduction session on the rationale and reasons behind the decision to move to OSS.

Conduct Thorough Analysis

This step was not fully executed in the case study. Exploratory but fairly high-level analysis of system and user requirements was done to motivate the proposal in Step 1. A concrete result from this was the identified need to replace an aging Novell server (they did not have Novell licenses for all 150 end-user PCs, either). The new Linux server's primary role would be a central file server, sharing files between Novell SUSE Linux and Microsoft-based PCs.

However, the implications of moving the desktops was not fully investigated and last-minute hardware upgrades for a number of workstations were required to cope with the graphical interface requirements. Also, it was found that users experienced incompatible file formats and/or documents appearing different in different computer environments (OSS vs. Microsoft). This problem was specifically experienced with Corel WordPerfect documents and Quattro-Pro spreadsheets, as initially the SUSE Linux PCs had K-Word and K-Spread installed for word processing and spreadsheets. The KOffice suite was not capable of opening the aforementioned file formats at the time, nor could KOffice save files in Microsoft Word or Excel file formats, as required by the other departments. In order to solve this problem, Star-Office was installed on all the SUSE Linux 8.2 PCs, but this proved to be too system-resource intensive. A more thorough analysis of the software in use before (i.e., user requirements) might have prevented these mishaps.

Setup Systems Architecture and Trial Desktops

Before handing over the desktop-OSS-based PCs to all staff, four to five staff members were selected to be the first trial recipients. Mossel Bay again contracted Creative Minds, who supplied a Linux expert to assist the initial users. The Linux expert's responsibilities included assisting the trial staff in performing their daily jobs using the OSS and noting where problems occurred. After several critical errors in the setup and the installed software on the OSS-based PCs were observed, the PCs were temporarily removed and rebuilt to have all the settings and software needed for the users on the desktop.

Testing

During testing it was discovered that the initial configuration of Star-Office on SUSE Linux 8.2 ran very slowly on the old PCs still in use at Mossel Bay Municipality. Star-Office was eventually replaced with OpenOffice.org, which has proven to meet Mossel Bay Municipality's file formatting requirements and is running smoothly on the low-specification PCs. In addition, the original SUSE Linux 8.2 was replaced with Novell SUSE Linux 9.2 in order to solve some of the finer issues identified by the IT staff and end users.

Implement OSS on Desktops

The total physical migration only took about 2 weeks, and during this time, all training was completed. The expert from Creative Minds (outside contractor) assisted the individual users, and RPC Data (outside contractor) performed the installation of the Novell SUSE Linux PCs and handled technical problems. The overall cutover phase took about 4 to 5 weeks while users grew accustomed to the new Novell SUSE Linux 9.2. Most of the issues that users experienced surrounding the use of the new OSS were systematically resolved.

Perform a Postimplementation Review

The total duration of the migration to OSS was about 3 months. The IT manager classifies the migration as a success, but admits that it "did not solve all the problems; at this point in time [the municipality is] still underlicensed."

No formal postimplementation review was held by the municipality. However, different users were polled about their experience as part of this research project and there was a wide difference in opinion between various users on the success of the project.

The clerk of the debt collections department experienced some initial problems with the first Linux implementation, which were solved with the second Linux implementation (Novell SUSE Linux 9.2). The clerk commented on the fact that she did not receive as much training as she would like to have had before the migration, but due to the intuitive nature of SUSE Linux 9.2, she was capable of using it productively within a few days, learning for herself. She was initially hesitant about migrating to Linux, but had no choice in the matter. Her reason for resisting the migration was primarily due to the fact that the desktop OSS was new and unfamiliar. She would recommend that others migrate to Linux, but that they should make sure that they would still be able to perform their jobs efficiently before doing so. On the other hand, the senior data operator, whose primary job involves capturing financial data into the Promon Financial system (networked based), is very much against the move to OSS. She still insists that Linux cannot do everything Microsoft Windows could. She complained that she is no longer capable of changing dates in the payment-detail files she works with and has to go and use another Microsoft Windows machine to perform some of her job requirements. In addition, she is experiencing problems opening Microsoft Word documents with OpenOffice.org. She is experiencing a variety of usability issues with the OSS and attributes it to lack of training; she only received a 2-hour introduction session to Linux and had a Linux expert who showed her how to use Linux. She was, however, one of the staff members who did not attend the additional training sessions.

Provide Ongoing User Support and Training

It appears that most of the support is done by approaching the IT manager directly. This is certainly possible in a small organisation. However, a number of problems cannot be solved easily since they are built into the software. Configuration problems are normally addressed fairly quickly. The balance of the issues generally relates to training, which is often outsourced.

The IT support and maintenance staff raised the argument that Novell SUSE is doing exactly what the OSS community accuse Microsoft of doing, namely, adding "bloatware" to their products. All the extra software (bloatware) is good, but most of it is unnecessary and simply slows the PC down by using up the system resources, such as RAM (random-access memory). A strong grievance they have with Novell is the problem of having to pay for support, which may offset the cost of the Microsoft licenses. They also claimed that the support offered by Novell is very limited and not

meeting their needs. In addition, the default Linux distribution does not have WINE (a Microsoft Windows emulator) installed, which they need to run legacy applications on Linux PCs. On the plus side, Mossel Bay looked at other Linux distributions, such as Ubuntu Linux, but they found that these Linux distributions were not at the same level of maturity as SUSE Linux.

FUTURE TRENDS

The model proposed above is applicable to small-scale migration. Its post hoc illustration to one of the initial case studies shows its real-world, albeit high-level, applicability. However, the model has not yet been validated in larger organisations or outside South Africa. It is anticipated that such application of the model will result in further refinements.

It is also envisaged that the model will have to be specified in much more detail to cater for a more structured and specific approach, especially where migration affects larger organisations and requires longer term and more elaborate plans.

CONCLUSION

This chapter looked at a number of publicised models or frameworks that aim to assist organisations with migrating proprietary desktop platforms to desktop OSS environments. However, research by the authors revealed that a number of critical success factors for desktop OSS migration were not at all addressed in these frameworks. Thus, an alternative yet practical model for a proposed full or partial desktop OSS migration was proposed and illustrated by means of a practical example. In particular, the need for project champions, an adequate system architecture, and pilot testing, but especially the involvement and training of users throughout the migration process, are critical elements of the proposed model.

It is hoped that a number of practitioners will adopt (and possibly adapt or extend) the model. This would provide very useful validation information.

REFERENCES

Bruggink, M. (2003). *Open source software: Take it or leave it? The status of open source software in Africa* (Research Rep. No. 16). International Institute for Communication and Development (IICD). Retrieved April 16, 2005, from http://www.ftpiicd.org/files/research/reports/report16.doc

Fiering, L., Silver, M., Simpson, R., & Smith, D. (2003). *Linux on the desktop: The whole story* (Gartner Research ID AV-20-6574). Retrieved April 15, 2005, from http://www.gartner.com/DisplayDocument?id=406459

Gardiner, J., Healey, P., Johnston, K., & Prestedge, A. (2003). *The state of open source software (OSS) in South Africa.* Cape Town, South Africa: University of Cape Town, Department of Information Systems.

Goode, S. (2004). Something for nothing: Management rejection of open source software in Australia's top firms. *Information & Management, 42*, 669-681.

Government Information Technology Officers Council (GITOC). (2003). *Using open source software in the South African government: A proposed strategy compiled by the Government Information Technology Officers Council, 16 January 2003, version 3.3.* Retrieved from http://www.oss.gov.za/OSS_Strategy_v3.pdf

Lachniet, M. (2004). *Desktop Linux feasibility study overview.* Retrieved from http://lachniet.com.desktoplinux

NetProject. (2003). *IDA OSS migration guidelines, version 1.0.* Retrieved August 29, 2005, from http://www.netproject.com/docs/migoss/v1.0/methodology.html

Prentice, S., & Gammage, B. (2005). *Enterprise Linux: Will adolescence yield to maturity?* Paper presented at Gartner Symposium/ITxpo 2005.

Rockart, J. F. (1979). Chief executives define their own data needs. *Harvard Business Review, 57*, 81-93.

Van Belle, J. P., Brink, D., Roos, L., & Weller, J. (2006, June). Critical success factors for migrating to OSS-on-the-desktop: Common themes across three South African case studies. In *Proceedings of the Second International Conference in Open Source Software*, Como, Italy.

Wheeler, D. (2005). *How to evaluate open source software/free software (OSS/FS) programs.* Retrieved April 16, 2005, from http://www.dwheeler.com/oss_fs_eval.html

Wiggins, A. (2002). *Open source on the business desktop: A real world analysis.* Retrieved April 15, 2005, from http://desktoplinux.com/articles/AT9664091996.html

Wild Open Source Inc. (2004). *Steps to take when considering a Linux migration.* Retrieved August 25, 2005, from http://www.wildopensource.com/technology_center/steps_linux_migration.php

KEY TERMS

Business Software Alliance (BSA): A trade group representing the interests of the largest software companies operating internationally. One of their main aims appears to be the combating of software piracy through educational campaigns, software legalisation processes, and legal action. Its funding comes from members and settlements from successful legal actions.

Critical Success Factor (CSF): This term was coined in 1979 by Rockart, who defined critical success factors as "the limited number of areas in which results, if they are satisfactory, will ensure successful competitive performance for the organization" (Rockart, 1979, p. 85). Generally, it is used in an organisational context to refer to those factors that need to be in place for a project to succeed, that is, for the project to achieve its stated objective or goal. Factors can relate to business processes, key resources, products, or any other dependency.

Desktop OSS (OSS on the Desktop): This is comprised of those OSS applications that are utilised by everyday users to perform daily work tasks. This stands in contrast to server-side OSS, which are those OSS applications that traditionally reside on a server as opposed to a client (or workstation) and are used primarily by technical staff such as systems administrators to fulfill back-office functions such as e-mail routing and Web hosting. Typical desktop OSS applications include productivity software (e.g., OpenOffice), e-mail clients (e.g., Mozilla Thunderbird), Internet browsers (e.g., Mozilla Firefox), and a variety of other utilities. Although many PC users use one or several OSS applications, generally only fairly significant desktop OSS implementations are considered, that is, those that include at least an OSS operating system (Linux) as well as at least a full-productivity software suite.

Total Cost of Ownership (TCO): A financial measure (in monetary terms) that aims to capture the sum of all the costs relating to a business (usually IT related) investment over its entire lifetime. For an information system, this includes costs such as hardware, software, training, maintenance, upgrades, and management. It is typically used to make potential buyers aware of longer term financial implications when using the initial purchase price as the main criterion when deciding between two or more alternatives.

Chapter XIV
The Social Order of Open Source Software Production

Jochen Gläser
Australian National University, Australia

ABSTRACT

This chapter contributes to the sociological understanding of open source software (OSS) production by identifying the social mechanism that creates social order in OSS communities. OSS communities are identified as production communities whose mode of production employs autonomous decentralized decision making on contributions and autonomous production of contributions while maintaining the necessary order by adjustment to the common subject matter of work. Thus, OSS communities belong to the same type of collective production system as scientific communities. Both consist of members who not only work on a common product, but are also aware of this collective work and adjust their actions accordingly. Membership is based on the self-perception of working with the community's subject matter (software or respectively scientific knowledge). The major differences between the two are due to the different subject matters of work. Production communities are compared to the previously known collective production systems, namely, markets, organizations, and networks. They have a competitive advantage in the production under complete uncertainty, that is, when neither the nature of a problem, nor the way in which it can be solved, nor the skills required for its solution are known in advance.

INTRODUCTION

This chapter contributes to the sociological understanding of open source software (OSS) production by identifying the social mechanism that creates social order in OSS communities. The concept of social order is used here in its most basic sense as describing a situation in which actors have adjusted their actions to each other. This order is indeed very high in OSS communities, who produce large and highly complex software from many independent contributions. It is even astonishingly high when we take into account how few of the most common tools for creating order—rules, commands, and negotiations—are used. Therefore, most analysts agree that OSS is produced in a distinct "new" mode that is qualitatively different from the "corporate way" of software production.

However, none of the four strands of literature on OSS production has produced a consistent explanation of the way in which this amazing order is achieved. The *participant-observer* literature has proposed metaphors that emphasize the decentralized, democratic, open, and communal nature of OSS, notably the "cooking pot market" (Ghosh, 1998) and the "bazaar" (Raymond, 1999). These metaphors, while suggestive, are not grounded in social theory. *Economics* is still fascinated by the voluntary contributions to a public good, and has consequently focused on motivations to contribute to OSS (Dalle & Jullien, 2003; Lerner & Tirole, 2002; von Hippel & von Krogh, 2003). However, neither these investigations nor the generalized questions about transaction costs (Demil & Lecocq, 2003) or about the allocation of efforts to modules (Dalle, David, Ghosh, & Steinmueller, 2004) capture the specific ways in which an ill-defined group of people manages to produce a complex good. These ways have been looked at primarily in the context of *management and software engineering analyses*, which produced interesting case studies of the coordination of individual OSS projects such as Linux, Apache, Perl, Sendmail, Mozilla, and others (Holck & Jørgensen, 2005; Iannacci, 2003; Jørgensen, 2001; Koch & Schneider, 2002; Lanzara & Morner, 2003; Mockus, Fielding, & Herbsleb, 2002). Some analysts tried to compare OSS communities to "traditional organizations" (Sharma, Sugumeran, & Rajagopalan, 2002) or to catch the specific mode of OSS production with generalized concepts such as "virtual organization" (Gallivan, 2001) or "distributed collective practice" (Gasser & Ripoche, 2003). However, these concepts are similar to the metaphors in the observer-participant literature in that they are ad hoc generalizations that are not embedded in theories of social order or of collective production. Finally, sociological analyses have contributed the idea of a *gift economy* (Bergquist & Ljungberg, 2001; Zeitlyn, 2003), various concepts of community (Edwards, 2001), social movements (Hess, 2005; Holtgrewe & Werle, 2001), the hacker culture (Lin, in this volume), and applications of actor-network theory (Tuomi, 2001). These sociological accounts focus on the specificity of social relations in OSS communities and more or less entirely disregard the specific mode of production employed by these communities.

Missing from the numerous case studies on OSS production is a description of the social mechanisms that create order by enabling the adjustment of actions. Following Mayntz (2004, p. 241), we define a social mechanism as *a sequence of causally linked events that occur repeatedly in reality if certain conditions are given and link specified initial conditions to a specific outcome* (for a similar definition, see Hedström, 2005, p. 11). Heroically simplifying, we can think of social mechanisms as *subroutines of the social* that are activated under certain conditions and produce specific results. Only by describing the social mechanism at work can we explain how a specific outcome is produced under certain conditions (Hedström). In order to explain how a well-ordered collective production of OSS is achieved under conditions of shifting membership, incomplete information, and autonomous decision making by contributors, we need to find the social mechanisms that create order under these conditions.

Theoretical analyses of this kind are still rare. Only Benkler's (2002) proposal to regard OSS production as an instance of *commons-based peer production* comes close to describing a social mechanism of OSS production. According to Benkler, commons-based peer production "relies on decentralized information gathering and exchange to reduce the uncertainty of participants," and "depends on very large aggregations of individuals independently scouring their information environment in search of opportunities to be creative in small or large increments. These individuals then self-identify for tasks and perform them for a variety of motivational reasons" (pp. 375-376).

The focus on information processes contributes an important insight in the process of OSS production. However, Benkler (2002) applies an extremely diffuse notion of production, which makes him subsume every personal communication, every electronic mailing list, and every online computer game to his model of commons-based peer production. Consequently, he is not able to describe the specific way in which the individual contributions are integrated into a common product. The distinctiveness of OSS production gets lost in his very general model.

Thus, while some important elements of the social mechanism that leads to OSS have been identified, we still do not have a consistent theoretical model of OSS production. Sociology can provide such a model if it supplements its analyses of social relations and cultures of OSS communities with an analysis of the social order of collective production. In this chapter, I provide a description of the distinct mechanism of collective production underlying OSS and compare it to the known mechanisms of markets, organizations, and networks.

Analyzing the social order of a collective production system requires delineating the producing collective and establishing how the actions of members are adjusted to each other. I will do this for OSS production by establishing how tasks for members of the producing collective emerge, how requirements of later integration affect the conduct of tasks, and how the integration of individual contributions into a common product is achieved. Thus, I will aim not at providing a complete description of OSS production, but rather at identifying the select social phenomena that enable the adjustment of actions in a dispersed collective whose members are only incompletely informed about each other but nevertheless manage to jointly produce a complex good. The "thin description" of OSS production that is provided in the following section enables the identification of features that make it a distinct type of collective production.[1] The description is based on published studies of OSS production from which I extracted the important elements of the social mechanism. I then use this description for two comparisons. First, I compare OSS production to its archetype, namely, the mode of production

employed by scientific communities, and introduce the concept *production community* for this mode of production. In the subsequent section, I compare production communities to markets, organizations, and networks, and tentatively discuss the specific efficiency of each mode. As a conclusion, I argue that OSS production is both a role model for much creative work of the future and a promising subject matter for studying production communities.

BACKGROUND

How Do Open Source Software Communities Produce?

The Emergence of Tasks

A crucial problem of all production processes is the definition of tasks for members of the producing collective. Each of these tasks must describe *utilizable contributions*, that is, contributions that are useful additions to the common product, can later be integrated into this product, and can be produced by the specific member of the collective. How is this achieved in OSS production?

One of the characteristic features of OSS production is that individual producers define their own tasks. Nobody is forced to produce a specific contribution. Instead, an individual producer perceives that something needs to be done, believes he or she is able to do it, and defines it as a task to do personally. Thus, individuals decide for themselves to produce the contributions they think are necessary, and offer them to the community. Necessary tasks are also publicly announced in a variety of ways. However, no mechanism exists that could force community members to solve one of these tasks (Bonaccorsi & Rossi, 2003; Mockus et al., 2002; Sharma et al., 2002). The creation of tasks for individual producers is essentially a decentralized, local activity by autonomous contributors. This has been expressed by characterizing OSS production as a "distributed collective practice" (Gasser & Ripoche, 2003).

The decentralized autonomous decision making about tasks makes it likely that tasks match the abilities of producers. However, the autonomously defined contributions must also fit together and be integrated in a common product, which requires a mutual adjustment of individual actions. Since direct mutual adjustment between producers requires extensive information about participants, their current actions, and their plans, it is obviously impossible in OSS communities. OSS communities solve this problem by *mediated adjustment*, that is, by all producers adjusting to the common subject matter of work, which they observe and from which they derive their tasks. The common subject matter of work is a complex body of knowledge; at its core we find the current version of the source code, which often has a modular design in order to enable independent parallel work by many developers (Lanzara & Morner, 2003; Lerner & Tirole, 2002). This core is surrounded by a variety of corollary information that refers to the source code, such as bug reports or, more generally, "pending work tasks" (Holck & Jørgensen, 2005, pp. 7-8), documentation, other documents describing the software,[2] and discussions in mailing lists. Members of the producing collective draw on this knowledge when they formulate tasks for themselves. Based on these observations and their tests of the program, they perceive that additional code, a change in the existing code, or corollary work (e.g., on documentation) is needed for the software to function as planned.

Thus, the shared subject matter of work mediates the adjustment of producers' actions by providing them with a common point of reference. This effect is implicitly described in numerous case studies, and has been explicitly discussed for the code (de Souza, Froehlich, & Dourish, 2005; Lanzara & Morner, 2003), and for the explicit and implicit descriptions of software requirements in

Web-based descriptions of the software (Scacchi, 2002). Since the subject matter of work orders the decentralized autonomous definition of tasks, OSS projects need to start with a significant initial submission of code, as has been observed by Raymond (1999) and by Lerner and Tirole (2002, p. 220): "The initial leader must also assemble a critical mass of code to which the programming community can react. Enough work must be done to show that the project is doable and has merit." West and O'Mahony (2005) consider the amount of structured code initially available as a major advantage of the "spinout model," where previously internally developed software is released under an open source software license, inviting the community to join the project.

Conduct of Work

While the production of code is the core task of OSS production, it is accompanied by a variety of other activities that are necessary for the production to succeed. Existing code must be tested and evaluated, which leads to information about bugs and suggestions for fixing them (new or changed code). Even the mere report of problems with the software is an important contribution. A documentation of the software must be produced, and users must be advised.

The conduct of these tasks is ordered in basically the same way as their emergence. Members of the producing collective are guided by the subject matter of work, that is, by the existing code and corollary information. The structure of the code that is used in the conduct of work poses highly specific requirements (de Souza et al., 2005; Mockus et al., 2002). Apart from these requirements inherent to the code, standards and general rules of good programming govern the conduct of work. Core developers of OSS projects set up guidelines, which are sometimes formalized (put in writing) and distributed (Bonaccorsi & Rossi, 2003). The fit of contributions is ensured by standard protocols, standardized interfaces, guidelines for problem reports, and so forth (Bonaccorsi & Rossi; Iannacci, 2003; Jørgensen, 2001; Lanzara & Morner, 2003).

Integration of Contributions

The decentralized task definition leads to a significant redundancy. Many producers carry out the same pending task simultaneously, and thus submit reports of or solutions to the same problem (Bonaccorsi & Rossi, 2003; Holck & Jørgensen, 2005). It is important to notice that this is a redundancy only insofar as many offered contributions are solutions to the same problem. The solutions themselves differ from each other, and the best solution can be chosen from what is offered.

The integration of contributions is subject to an explicit decision process. Proposed code is published by submitting it to a mailing list, testing and evaluation by peers (other developers of code), and thereafter having it submitted to the software by one of the maintainers. The testing continues after submission because the members of the community who download the code and use it locally are testing it at the same time.

The quality of contributions to the common product is thus maintained by two different mechanisms. The first mechanism is a peer review procedure that is explicitly directed at establishing the quality of a proposed contribution. In the peer review process, a few authorized peers (some of the core developers) analyze the code, judge it explicitly, and decide on its integration into the software. The second mechanism is an implicit one, namely, quality control by use in a potentially infinite variety of different settings. It is possible because of the partial overlap of and close contact between the producer and user collectives. The common product is downloaded, used, and tested at every stage of its development by a potentially large audience. Since the local conditions (hardware and software environments) vary, the software is submitted to tests under a variety of conditions. The people who use the code,

encounter problems, and report these problems contribute to the production process.

Whenever the maintainers are of the opinion that a sufficiently advanced, comprehensive, and error-free code has been produced, they decide to release the code by assigning a version number to it and treating it as a product. These official releases are addressed to a wider audience of users and thus create a distinct user community that might not take part in any production activities.

Membership, Roles, and Social Structure

Students of OSS communities usually deal implicitly with the question of membership by applying a specific empirical method for identifying members. Depending on the method applied, a member is someone who has been identified as such by having an ID at source forge, being subscribed to a mailing list, having posted a message or a bug report, or having contributed code. For a theoretical model of OSS production, however, we need a theoretical answer to the membership question. What social phenomenon establishes membership in an OSS community? The obvious answer to that question is that a member is someone who participates in the collective production. However, this answer merely transforms the question. What constitutes participation in the production process?

While a variety of actions can be considered as participation in OSS production, none of them provides the opportunity of a clear-cut delineation that is both theoretically and empirically satisfying. Thus, while contributing code obviously constitutes participation in the production process, much of the offered code is not used, and the corollary work (testing, documentation, and other contributions) must be taken into account. Furthermore, some users report bugs but do not fix them, test the software but do not find bugs, or discuss the software but do not offer code. Some of these activities are invisible to both the OSS community and the sociological observer.

From a theoretical point of view, the weakest form of participation is the conscious test of the software in its current version, which can be either an official release or a version in between releases. A conscious test means using the software while being aware of the existence of a community *out there* to whom problems can be communicated and from which advice could be received. While this description does not apply to all users of OSS, it includes the large number of them who never turn into active contributors of any sort because they never encounter problems with the software. In the communal production of OSS, successful tests of the software are rarely reported, the only systematic exception being the classification of reported bugs as invalid or "works for me" (Holck & Jørgensen, 2005, p. 11). Conducting tests, however, is a contribution to the communal production regardless of their outcome.

Thus, membership in OSS production communities is constituted by perception-based self-selection. One is a member of an OSS community if one perceives oneself as contributing, and the least possible contribution is a test of the software. This concept of membership is significantly wider than those that have been applied so far. It is consistent with empirical investigations of membership and social structure. Figure 1 provides an overview of the composition of an OSS community and the various roles in the production process. The major formalized distinction between work roles is that between the people who have permission to change the code and to release versions of the software (maintainers), and all other members of the community who may propose code but are not able to integrate it into the current version of the software themselves. OSS communities are basically meritocratic because membership to the core group of an OSS community depends on programming abilities as demonstrated in submitted code. However, the current activities of members need not coincide with the formalized

roles. Crowston, Wei, Li, and Howison (2006, p. 6) found that "the formal list of developers is not an accurate representation of contribution to the teams, at least as regards interactions around bug fixing" (see also de Souza et al., 2005, for transitions between positions in OSS communities).

The distribution of activities in OSS communities is highly skewed, with a very small proportion of members making most of the contributions. For example, a case study on the Apache project found that 3,060 people reported problems, 249 submitted new code, and 182 changed existing code (Mockus et al., 2002). This pattern repeats itself with regard to the lines of code submitted, number of problems reported, number of messages posted, and so forth (Crowston et al., 2006; Ghosh & Prakash, 2000; Koch & Schneider, 2002; Lerner & Tirole, 2002; von Krogh, Spaeth, & Lakhani, 2003). Furthermore, advanced projects appear to be surrounded by a larger user community whose members observe the production process, download and use the software, but become visible (turn into an active user) only if they choose to take part in discussions or report problems.

The perception-based membership is responsible for the fuzzy and fluid boundaries of OSS communities. In larger communities, no member knows all other members. One can easily become a member or fade out of a community. Simultaneous memberships in more than one OSS community are frequent (Robles, Scheider, Tretkowski, & Weber, 2001). Another important feature of OSS communities that is linked to the perception-based membership is the decoupling of subsistence and contributions. Even in the significant number of cases where producers are paid for participating in OSS production, payment is not received as a reward for a specific contribution (a problem report, code, etc.), but for the time that is devoted to the project. The payment is exogenous to the community, which creates its own nonmaterial reward (reputation). OSS communities are therefore based on a principal decoupling of offered and accepted contributions

Figure 1. Work roles and contributions in OSS communities (Source: Combination of information from Crowston et al., 2006, p. 1; Gacek & Arief, 2004, p. 36)

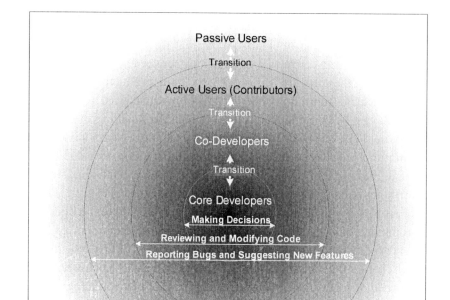

on the one hand, and the producers' subsistence on the other hand. This is in perfect agreement with the great variety of motives to contribute to OSS that has been empirically observed (Dalle & Jullien, 2003; Lerner & Tirole, 2002; von Hippel & von Krogh, 2003). The functioning of production communities is not affected by the specific motives of its members as long as enough people choose to contribute.

MAIN FOCUS OF THE CHAPTER

Production Communities

OSS is produced by collectivities whose members are incompletely informed about each other,

decide autonomously about what task they are going to solve in which way, and propose contributions that are rarely ever used. Social order in the sense of adjusted individual actions emerges primarily because people adjust their actions to the community's common subject matter of work, that is, to the shared code and to the knowledge about that code. I have proposed the concept of *production community* for these identity-based producing collectivities, and *communal production* for the mode of production that employs autonomous decentralized decision making on contributions and the autonomous production of contributions while maintaining the necessary order by adjustment to the common subject matter of work (Gläser, 2001, 2006). The concept of a production community challenges traditional community theory, which assigns communities features such as multiplex relationships, shared values, frequent face-to-face interactions, and emotional bonds (Brint, 2001). However, recent studies on interest communities suggest that it is useful to make theoretical room for communities that are based on quite specific identities and do not necessarily feature shared lives and emotional bonds (Gläser, 2001, 2006). The numerous relatively small collectivities that produce OSS without much communication in open forums (Healy & Schussman, 2003) are groups rather than communities. Whether an OSS project is conducted by a group appears to be due to chance or the attractiveness of the proposed software. The differences between community and group OSS projects have not yet been explored.

OSS development indeed appears to be a unique mode of production. However, it is not. It has been repeatedly mentioned in the literature that OSS communities are similar to scientific communities (Benkler, 2002; Bezroukov, 1999; Dalle et al., 2004; Dalle & Jullien, 2003; Seiferth, 1999). However, these comparisons have been metaphorical rather than analytical because they lack suitable theoretical frameworks and exact descriptions of both modes of production. The following comparison of OSS production communities and scientific communities is based on a larger study that identified scientific communities as production communities (Gläser, 2006).

Both OSS and scientific communities produce new knowledge and consist of members who not only work on a common product, but are also aware of this collective work and adjust their actions accordingly. Membership is based on the self-perception of working with the community's subject matter (software, respectively scientific knowledge). That is why both communities also have in common the fuzziness of their boundaries and incomplete information about members. Another similarity is the decoupling of contributions from contributors' subsistence. Research has been professionalized for a long time. However,

the organizations that employ scientists and pay them salaries are not the social contexts in which tasks are defined and contributions are used. The production of scientific knowledge takes place in the scientific communities that do not provide material rewards for offered or accepted contributions. Thus, payments to scientists are exogenous to the production community, as are payments for OSS developers.

In both OSS production and research, task creation is a decentralized and autonomous process. Both communities are similar in that the elite may (and indeed do) define tasks of special importance or urgency, but has no means to force other members to work on these tasks. The adjustment of individual, dispersed, local activities is achieved by individuals' reference to the joint subject matter of work. In order to become such a reference point for an open and unknown audience, the common subject matter of work must be available to all potential members of a production community. Since membership is unknown, the only way to guarantee availability to all is publication, which is realized by scientific books and journals (and increasingly via the Internet) in scientific communities, and by Internet publication in OSS communities.[3] The common subject matter also guides the conduct of work, which is additionally informed by standards and rules of conduct. In scientific communities, standards and rules for experimentation have functions similar to the rules of conduct for OSS production.

Both OSS production and the production of scientific knowledge apply the same mechanism of quality control, namely, peer review as a first, explicit preliminary check, and a subsequent implicit and possibly infinite series of checks in a variety of settings.

The important differences between scientific and OSS communities can be summarized by stating that structures and processes that remain implicit and informal in science are explicit and partly formalized in OSS communities. Thus, while both scientific and OSS communities have been described as meritocracies, the elite of a scientific community do not have a formal status and no formal right to decide on the current *true* state of a community's knowledge. While scientific elites exert significant influence on the opinions of their communities, they cannot make binding decisions or establish binding rules of conduct. In OSS production, there are official versions of the common product, and only formally recognized elite members have both the right and the opportunity to change these versions.

A similar difference exists between the ways in which contributions are integrated into the common product, which are more explicit in OSS communities. Peer review of contributions is ubiquitous but relatively unimportant for the integration of new contributions into the knowledge of scientific communities. The main way of integrating new contributions is implicit integration by using each others' results. When members of a scientific community use a contribution in their further production of knowledge, they implicitly accept it and integrate it into the community's common body of knowledge, which is the basis of further knowledge production. In the case of OSS production, members of the elite must decide explicitly about the integration of contributions.

Thus, we observe an explicitly defined elite who makes decisions by adhering to explicit and partly formalized procedures and creates formalized rules and standards in OSS communities, while the analogous processes in scientific communities remain more informal and implicit. This difference is due to the fact that OSS is not only knowledge, but also a technical product. While scientists can proceed regardless of gaps, inconsistencies, and contradictions in their shared body of knowledge, a computer program's functioning would be endangered. Garzarelli and Galoppini (2003) have argued that in OSS production, hierarchy is necessary to manage the interaction of modules in large OSS projects. The analogous process in science—the interaction of knowledge

from different fields—is left to the self-organization of research. Moreover, the OSS production is intended to lead to a standardized mass product at some stage, namely, when a new version of the software is released. The overlap of the production and use of OSS, albeit considerable, is not complete, and an unproblematic output must be created for users who are not able to change the product. Therefore, consistency and reliability are much more important in OSS production than in science, where the only further use is by the producers themselves in the very same activity of knowledge production.

The Competitive Advantage of Production Communities

If the production community is a distinct type of collective production system, it must be both comparable to and qualitatively different from the three known systems of collective production. Table 1 applies the questions answered for production communities to the other three types.[4]

It becomes apparent that the collective production systems can be grouped according to a distinction introduced by Hayek (1945, 1991), who observed that social order can be either made or spontaneous. Organizations and networks are *made orders* because they both employ dedicated actions (coordination) to create order. The characteristic forms of coordination in organizations and networks are hierarchical decisions and negotiations, respectively. Markets and production communities both rely on decentralized task definition and ex post selection of contributions. They are *spontaneous orders* and rely on "parametric adjustment" (Lindblom, 1965) to a situation rather than coordination. However, production communities differ from markets in their use of concrete information about the production system's subject matter of work rather than extremely reduced abstract information as provided by market prices. Another significant difference that is not shown in Table 1 is that markets are characterized by feedback between contributions and producers' subsistence, feedback that does not exist in production communities.

While this comparison does indeed prove that production communities can be introduced as a fourth distinct collective production system, it does not answer an important question. Which production system is the most efficient under which conditions? We have not yet a taxonomy that links types of production tasks to types of collective production systems. Therefore, only a few exploratory comments on this topic are possible.

One of production communities' striking features is their apparent waste of effort. In both OSS and scientific communities, many contributions are offered that will never be used. The decentralized task definition is liable to both misperceptions, that is, the production of contributions no one is interested in, and solutions of roughly the same problem by a multitude of contributors.

However, this apparent waste of effort and resources has major advantages for a collective production system that operates under *complete uncertainty*, which is the case for both OSS production communities and scientific communities. Complete uncertainty means that at any stage of the production process, the following are not clear:

- What exactly is the problem that needs solving (how should it be formulated)?

- Is there a solution to the problem at the current stage of knowledge?
- How could the problem be solved?
- What knowledge can be regarded as valid and reliable and should therefore be used for solving the problem?
- Who can solve the problem?

This is the case in science where each researcher formulates a problem in a particular way based on prior research and the current work environment, and in OSS production where the software informalisms are subject to idiosyncratic interpretations, and where the hardware and software environments of an OSS are constantly changing. Under these conditions, the apparent *waste of effort* is actually a very effective and probably even efficient use of resources. First, when nobody can say for sure what the problem is, decentralized task creation appears to be feasible because members of the concept of *production community* make as many independent attempts to formulate and solve problems as possible. While many (and sometimes most) of the attempts are bound to fail or to become redundant, the decentralized approach provides the highest likelihood that the problem is solved as quickly as possible.

The second advantage of production communities is that under the conditions of complete uncertainty, tasks are assigned to producers by "self-identification" (Benkler, 2002, pp. 414-415).

Table 1. Comparison of collective production systems

	Organization	Network	Market	Community
Membership constituted by	Formal rules	Negotiation	Exchange offer	Perception
Tasks created by	Ex ante division of labor		Decentralized autonomous decisions	
	Hierarchical	Negotiated		
Actions adjusted primarily by	Coordination by		Parametric adjustment to	
	Hierarchical decisions	Negotiations	Price	Subject matter of work
Integration of contributions based on	Preproduction decisions		Postproduction exchange	Postproduction peer review and use

The problem of who should do what, which would be impossible to solve by centralized decision making, is decided by every individual producer autonomously. The decentralized decision making on tasks guarantees the best possible fit of tasks and producers. Naturally, misperceptions and therefore a misallocation of tasks are still possible.

The third advantage of production communities that can outweigh the waste of effort is their mechanism of quality control. While it is often the case that peer review is seen as the major mechanism of quality control and an advantage of production communities, the quality control by use is much more important because it is more thorough in the long run. In production communities, contributions (code, respectively knowledge claims) are used in the further production process. Since production occurs in local settings, which vary from each other, each producer's work environment constitutes a specific test site for the community's common subject matter of work. The numerous tests of the common product that are permanently performed at these sites significantly enhance the quality of the product. In the case of scientific knowledge, this aspect of quality can be described without reference to truth as robustness, that is, as stability across different environments (Star, 1993). In the case of software, the major concern is the quality of the code. While there is still limited empirical evidence to back the claim that OSS is superior to proprietary software in terms of bug detection and fixing, one comparative study indicates that this advantage might indeed exist (Mockus et al., 2002).

FUTURE TRENDS AND CONCLUSION

OSS communities are production communities that apply a distinct mode of production that has so far been neglected by the literature. Production communities rely on decentralized task definition

that is ordered by the common subject matter of work, which is observed by all members. Decisions about the integration of contributions are made ex post by peer review and subsequent use. The use of contributions in a variety of local work environments is also the major mechanism of quality control.

The analysis has also revealed gaps in our knowledge about systems of collective production in general and about production communities in particular. Since OSS and scientific communities belong to the same type of collective production system, ideas and research questions could be exchanged. For example, strategies used in science studies for analysing knowledge structures and knowledge flows could be adopted for the investigation of OSS. In science, relationships between contributions are reflected in citations, which in turn enable the study of structures and flows of knowledge. Citation analyses have demonstrated that all scientific knowledge is interconnected and can be thought of as one body of knowledge, which is internally structured (Small & Griffith, 1974) and has fractal characteristics (Van Raan, 1990). The pendant to citation in science appears to be coupling, which reflects connections between modules. The analysis of coupling with techniques such as cross-referencing tools (Yu, Schach, Chen, & Offutt, 2004), design structure matrices (MacCormack, Rusnak, & Baldwin, 2004), and call-graph analysis (de Souza et al., 2005) should provide opportunities to study structures in OSS in a way similar to citation analysis.

Studies of production communities could benefit from the fact that OSS communities operate more explicitly than scientific communities. The explicit negotiations, decisions, and rules that characterize the production processes of OSS communities make them a suitable research object for studying the organizing functions of a production community's shared subject matter of work. For example, the social mechanisms that are at work in the branching out of knowledge production into different directions (*forking*) and in the *dying* of a line of knowledge production (of an OSS project) can be assumed to be at least similar to those occurring in scientific communities, where they are less easy to identify and to observe because of the implicitness of knowledge structures and cognitive developments.

The introduction of the *production community* also challenges the theory of collective production systems. While the superiority of the communal mode of production for some types of software development tasks is felt by many observers, this point must also be made theoretically in the comparison of collective production systems. We need a generalized approach that relates types of production problems to types of collective production systems and enables a comparison of their advantages and disadvantages. Thus, in order to get more out of the many case studies on OSS production, we need more abstract and more comparative theory.

REFERENCES

Benkler, Y. (2002). Coase's penguin, or, Linux and the nature of the firm. *Yale Law Journal, 112*, 369-446.

Bergquist, M., & Ljungberg, J. (2001). The power of gifts: Organizing social relationships in open source communities. *Information Systems Journal, 11*, 305-320.

Bezroukov, N. (1999). Open source software development as a special type of academic research (critique of vulgar Raymondism). *First Monday, 4*. Retrieved from http://www.firstmonday.dk/issues/issue4_10/bezroukov/index.html

Bonaccorsi, A., & Rossi, C. (2003). Why open source software can succeed. *Research Policy, 32*, 1243-1258.

Brint, S. (2001). Gemeinschaft revisited: A critique and reconstruction of the community concept. *Sociological Theory, 19*, 1-23.

Crowston, K., Wei, K., Li, Q., & Howison, J. (2006). *Core and periphery in free/libre and open source software team communications.* Paper presented at the 39th Annual Hawaii International Conference on System Sciences, Waikoloa, HI.

Dalle, J.-M., David, P. A., Ghosh, R. A., & Steinmueller, W. E. (2004). *Advancing economic research on the free and open source software mode of production* (SIEPR Discussion Paper 04-03). Stanford: Stanford Institute for Economic Policy Research.

Dalle, J.-M., & Jullien, N. (2003). "Libre" software: Turning fads into institutions? *Research Policy, 32*, 1-11.

Demil, B., & Lecocq, X. (2003). *Neither market nor hierarchy or network: The emerging bazaar governance.* Retrieved from http://opensource.mit.edu/papers/demillecocq.pdf

De Souza, C., Froehlich, J., & Dourish, P. (2005). Seeking the source: Software source code as a social and technical artifact. In *Proceedings of the ACM International Conference on Supporting Group Work (GROUP 2005)* (pp. 197-206).

Edwards, K. (2001). *Epistemic communities, situated learning and open source software development.* Retrieved from http://opensource.mit.edu/papers/kasperedwards-ec.pdf

Gacek, C., & Arief, B. (2004). The many meanings of open source. *IEEE Software, 21*, 34-40.

Gallivan, M. J. (2001). Striking the balance between trust and control in a virtual organization: A content analysis of open source software case studies. *Information Systems Journal, 11*, 277-304.

Garzarelli, G., & Galoppini, R. (2003). *Capability coordination in modular organization: Voluntary FS/OSS production and the case of Debian GNU/Linux. In Economics Working Paper Archive at WUST, Industrial Organization.* Retrieved February 23, 2006, from http://opensource.mit.edu/papers/garzarelligaloppini.pdf

Gasser, L., & Ripoche, G. (2003). *Distributed collective practices and free/open-source software problem management: Perspectives and methods.* Paper presented at the 2003 Conference on Cooperation, Innovation & Technologies (CITE2003). Retrieved from http://www.isrl.uiuc.edu/~gasser/papers/cite-gasser.pdf

Ghosh, R. A. (1998). Cooking pot markets: An economic model for the trade in free goods and services on the Internet. *First Monday, 3.* Retrieved from http://www.firstmonday.org/issues/issue3_3/ghosh/index.html

Ghosh, R. A., & Prakash, V. V. (2000). *The Orbiten free software survey.* Retrieved from http://www.orbiten.org/ofss/01.html

Gläser, J. (2001, December). *"Producing communities" as a theoretical challenge.* Paper presented at the TASA 2001 Conference, Sydney, Australia.

Gläser, J. (2006). *Wissenschaftlich produktionsgemeinschaften: Die soziale ordnung der forschung.* Frankfurt am Main, Germany: Campus, im Erscheinen.

Hayek, F. A. (1945). The use of knowledge in society. *The American Economic Review, 35,* 519-530.

Hayek, F. A. (1991). Spontaneous ("grown") order and organized ("made") order. In G. Thompson, J. Frances, R. Levacic, & J. Mitchell (Eds.), *Markets, hierarchies and networks: The coordination of social life* (pp. 293-301). London: SAGE Publications Ltd.

Healy, K., & Schussman, A. (2003). *The ecology of open-source software development.* Retrieved from http://opensource.mit.edu/papers/healy-schussman.pdf

Hedström, P. (2005). *Dissecting the social: On the principles of analytical sociology.* Cambridge, UK: Cambridge University Press.

Hess, D. J. (2005). Technology- and product-oriented movements: Approximating social movement studies and science and technology studies. *Science, Technology, and Human Values, 30,* 515-535.

Holck, J., & Jørgensen, N. (2005). Do not check in on red: Control meets anarchy in two open source projects. In S. Koch (Ed.), *Free/open source software development* (pp. 1-25). Hershey, PA: Idea Group Publishing.

Hollingsworth, J. R., & Boyer, R. (1997). Coordination of economic actors and social systems of production. In J. R. Hollingsworth & R. Boyer (Eds.), *Contemporary capitalism: The embeddedness of institutions* (pp. 1-47). Cambridge, UK: Cambridge University Press.

Holtgrewe, U., & Werle, R. (2001). De-commodifying software? Open source software between business strategy and social movement. *Science Studies, 14,* 43-65.

Iannacci, F. (2003). *The Linux management model.* Retrieved from http://opensource.mit.edu/papers/iannacci2.pdf

Jørgensen, N. (2001). Putting it all in the trunk: Incremental software development in the FreeBSD open source project. *Information Systems Journal, 11,* 321-336.

Knorr-Cetina, K., & Merz, M. (1997). Floundering or frolicking: How does ethnography fare in theoretical physics? (And what sort of ethnography?) A reply to Gale and Pinnick. *Social Studies of Science, 27,* 123-131.

Koch, S., & Schneider, G. (2002). Effort, cooperation and co-ordination in an open source software project: GNOME. *Information Systems Journal, 12,* 27-42.

Lanzara, G. F., & Morner, M. (2003). *The knowledge ecology of open-source software projects.* Paper presented at the 19th EGOS Colloquium, Copenhagen, Denmark.

Lerner, J., & Tirole, J. (2002). Some simple economics of open source. *The Journal of Industrial Economics, 50,* 197-234.

Lindblom, C. E. (1965). *The intelligence of democracy: Decision making through mutual adjustment.* New York: The Free Press.

MacCormack, A., Rusnak, J., & Baldwin, C. (2004). *Exploring the structure of complex software designs: An empirical study of open source and proprietary code.* Unpublished manuscript.

Mayntz, R. (1993). Policy-netzwerke und die logik von verhandlungssystemen. In A. Héritier (Ed.), *Policy-analyse: Kritik und neuorientierung* (pp. 39-56). Opladen: Westdeutscher Verlag.

Mayntz, R. (2004). Mechanisms in the analysis of social macro-phenomena. *Philosophy of the Social Sciences, 34,* 237-259.

Merz, M., & Knorr-Cetina, K. (1997). Deconstruction in a "thinking" science: Theoretical physicists at work. *Social Studies of Science, 27,* 73-111.

Mockus, A., Fielding, R. T., & Herbsleb, J. D. (2002). Two case studies of open source software development: Apache and Mozilla. *ACM Transactions on Software Engineering and Methodology, 11,* 309-346.

Powell, W. W. (1990). Neither market nor hierarchy: Network forms of organization. *Research in Organizational Behavior, 12,* 295-336.

Raymond, E. (1999). The cathedral and the bazaar. *Knowledge, Technology & Policy, 12,* 23-49.

Robles, G., Scheider, H., Tretkowski, I., & Weber, N. (2001). *Who is doing it? A research on Libre software developers.* Retrieved from http://widi.berlios.de/paper/study.html

Scacchi, W. (2002). Understanding the requirements for developing open source software systems. *IEE Proceedings: Software, 149,* 24-39.

Scharpf, F. W. (1997). *Games real actors play: Actor-centered institutionalism in policy research.* Boulder, CO: Westview Press.

Scott, W. R. (1992). *Organizations: Rational, natural, and open systems.* Englewood Cliffs, NJ: Prentice-Hall.

Seiferth, C. J. (1999). Open source and these United States. *Knowledge, Technology & Policy, 12,* 50-79.

Sharma, S., Sugumeran, V., & Rajagopalan, B. (2002). A framework for creating hybrid-open source software communities. *Information Systems Journal, 12,* 7-25.

Simon, H. A. (1991). Organizations and markets. *Journal of Economic Perspectives, 5,* 25-44.

Small, H., & Griffith, B. C. (1974). The structure of scientific literatures I: Identifying and graphing specialities. *Science Studies, 4,* 17-40.

Star, S. L. (1993). Cooperation without consensus in scientific problem solving: Dynamics of closure in open systems. S. Easterbrook (Ed.), *CSCW: Cooperation or conflict?* (pp. 93-106). London: Springer.

Tuomi, I. (2001). Internet, innovation, and open source: Actors in the network. *First Monday, 6.* Retrieved from http://firstmonday.org/issues/issue6_1/tuomi/index.html

Van Raan, A. F. J. (1990). Fractal dimension of co-citations. *Nature, 347,* 626.

Von Hippel, E., & von Krogh, G. (2003). Open source software and the "private-collective" innovation model: Issues for organization science. *Organization Science, 14,* 209-223.

Von Krogh, G., Spaeth, S., & Lakhani, K. R. (2003). Community, joining, and specialization in open source software innovation: A case study. *Research Policy, 32,* 1217-1241.

Von Krogh, G., & von Hippel, E. (2003). Editorial: Special issue on open source software development. *Research Policy, 32*, 1149-1157.

West, J., & O'Mahony, S. (2005). Contrasting community building in sponsored and community founded open source projects. In *Proceedings of the 38th Annual Hawaii International Conference on System Sciences*, Waikoloa, HI. Retrieved from http://opensource.mit.edu/papers/westomahony.pdf

Yu, L., Schach, S. R., Chen, K., & Offutt, J. (2004). Categorization of common coupling and its application to the maintainability of the Linux kernel. *IEEE Transactions on Software Engineering, 30*, 694-706.

Zeitlyn, D. (2003). Gift economies in the development of open source software: Anthropological reflections. *Research Policy, 32*, 1287-1291.

KEY TERMS

Community: A group of actors who share a collective identity that is based on the perception of having something in common, and who adjust some of their actions because of this identity.

Peer Review: Process in which people engaged in the same kind of work judge the quality of one's work, comment on it, and make decisions on these judgements.

Production Community: A community whose members jointly produce a good by autonomously deciding about their contributions by adjusting their decisions to the common subject matter of work.

Social Mechanism: A sequence of causally linked events that occur repeatedly in reality if certain conditions are given and link specified initial conditions to a specific outcome (Mayntz, 2004, p. 241).

Social Order: A state of a group of actors that is characterized by a mutual adjustment of actions.

ENDNOTES

[1] The concept of "thin description" has been developed by Merz and Knorr-Cetina (Knorr-Cetina & Merz, 1997; Merz & Knorr-Cetina, 1997) in their analysis of the work of theoretical physicists.

[2] Each open source software is accompanied by a variety of Web-based descriptions that contain requirements for the software (the so-called "software informalisms," Scacchi, 2002; see also Gasser & Ripoche, 2003).

[3] While the openness of source may be an ideology that is rooted in the Hacker movement (Holtgrewe & Werle, 2001, pp. 52-53; von Krogh & von Hippel, 2003, pp. 1150-1151), it is also a logistic prerequisite for the communal production of software.

[4] As is the case with OSS production, the description of the other systems of collective production needs to be synthesized from a dispersed literature (Gläser, 2006). Key publications that support the interpretation applied here are Hayek (1945) for markets, Simon (1991) and Scott (1992) for organisations, Powell (1990) and Mayntz (1993) for networks, and Hollingsworth and Boyer (1997) and Scharpf (1997) for the comparative approach.

Section III
Evaluating Open Source Software Products and Uses

Chapter XV
Open Source Software:
Strengths and Weaknesses

Zippy Erlich
The Open University of Israel, Israel

Reuven Aviv
The Open University of Israel, Israel

ABSTRACT

The philosophy underlying open source software (OSS) is enabling programmers to freely access the software source by distributing the software source code, thus allowing them to use the software for any purpose, to adapt and modify it, and redistribute the original or the modified source for further use, modification, and redistribution. The modifications, which include fixing bugs and improving the source, evolve the software. This evolutionary process can produce better software than the traditional proprietary software, in which the source is open only to a very few programmers and is closed to everybody else who blindly use it but cannot change or modify it. The idea of open source software arose about 20 years ago and in recent years is breaking out into the educational, commercial, and governmental world. It offers many opportunities when implemented appropriately. The chapter will present a detailed definition of open source software, its philosophy, its operating principles and rules, and its strengths and weaknesses in comparison to proprietary software. A better understanding of the philosophy underlying open source software will motivate programmers to utilize the opportunities it offers and implement it appropriately.

INTRODUCTION

Open source software (OSS) has attracted substantial attention in recent years and continues to grow and evolve. The philosophy underlying OSS is to allow users free access to, and use of, software source code, which can then be adapted, modified, and redistributed in its original or modified form for further use, modification, and redistribution. OSS is a revolutionary software development methodology (Eunice, 1998) that involves developers in many locations throughout the world who share code in order to develop and refine programs. They fix bugs, adapt and improve the program, and then redistribute the software, which thus evolves. Advocates of OSS are quick to point to the superiority of this approach to software development. Some well-established software development companies, however, view OSS as a threat (AlMarzouq, Zheng, Rong, & Grover, 2005).

Both the quality and scope of OSS are growing at an increasing rate. There are already free alternatives to many of the basic software tools, utilities, and applications, for example, the free Linux operating system (Linux Online, 2006), the Apache Web server (Apache Software Foundation, 2006; Mockus, Fielding, & Herbsleb, 2000), and the Sendmail mail server (Sendmail Consortium, 2006). With the constant improvement of OSS packages, there are research projects, even complex ones, that entirely rely on OSS (Zaritski, 2003). This opens new research and educational opportunities for installations and organizations with low software budgets.

Incremental development and the continuity of projects over long periods of time are distinctive features of OSS development. The software development processes of large OSS projects are diverse in their form and practice. Some OSS begins with releasing a minimal functional code that is distributed for further additions, modification, and improvement by other developers, as well as by its original authors, based on feedback from other developers and users. However, open source projects do not usually start from scratch (Lerner & Tirole, 2001). The most successful OSS projects, like Linux and Apache, are largely based on software provided by academic and research institutions. In recent years, more and more OSS has been derived from original software provided by for-profit companies.

A large potential-user community is not enough to make an OSS project successful. It requires dedicated developers. In Raymond's (1998) words, "The best OSS projects are those that scratch the itch of those who know how to code." For example, the very successful Linux project attracted developers who had a direct interest in improving an operating system for their own use. Similarly, webmaster developers contributed to the development of the Apache Web server project.

Despite the characterization of the OSS approach as ad hoc and chaotic, OSS projects appear, in many cases, to be highly organized, with tool support that focuses on enhancing human collaboration, creativity, skill, and learning (Lawrie & Gacek, 2002). The good initial structural design of an OSS project is the key to its success. A well-modularized design allows contributors to carve off chunks on which they can work. In addition, the adoption of utility tools and the use of already existing OSS components are necessary if an OSS project is to succeed.

The growing interest of commercial organizations in developing and exploiting OSS has led to an increased research focus on the business-model aspects of the OSS phenomenon. There are a number of business models for OSS, all of which assume the absence of traditional software licensing fees (Hecker, 2000). The economics of OSS projects is different from that of proprietary projects (Lerner & Tirole, 2002). Models of effort and cost estimation in the development of projects involving OSS are needed (Asundi, 2005).

In the past, most OSS applications were not sufficiently user friendly and intuitive, and only very knowledgeable users could adapt the software to their needs. Although the use of OSS is growing, OSS is still mainly used by technically sophisticated users, and the majority of average computer users use standard commercial proprietary software (Lerner & Tirole, 2002). The characteristics of open source development influence OSS usability (Behlendorf, 1999; Nichols, Thomson, & Yeates, 2001; Raymond, 1999), which is often regarded as one of the reasons for its limited use. In recent years, the open source community has shown increased awareness of usability issues (Frishberg, Dirks, Benson, Nickell, & Smith, 2002). Existing human-computer interface (HCI) techniques and usability improvement methods appropriate for community-based software development on the Internet can be used to leverage distributed networked communities to address issues of usability (Nichols & Twidale, 2003). Some OSS applications, such as the Mozilla Web browser (Mozilla, 2006; Reis & Fortes, 2002) and OpenOffice (OpenOffice, 2006), have made important advances in usability and have become available for both Windows and Linux users.

As the stability and security of open source products increase, more organizations seem to be adopting OSS at a faster rate. There are many open source community resources and services online. When implemented appropriately, OSS offers extensive opportunities for government, private-sector, and educational institutions. OSS appears to be playing a significant role in the acquisition and development plans of the U.S. Department of Defense and of industry (Hissam, Weinstock, Plakosh, & Asundi, 2001).

For many organizations, integrating the revolutionary OSS developmental process into traditional software development methods may have a profound effect on existing software development and management methodologies and activities.

The remainder of this chapter will review the history of OSS and define some key terms and concepts. It will discuss the incentives to engage in OSS and its strengths and weakness. Finally, it will review some OSS business models.

BACKGROUND

Software source code that is open has been around in academic and research institute settings from the earliest days of computing. Feller and Fitzgerald (2002) provide a detailed historical background to open source since the 1940s. The source code of programs developed in universities, mainly as learning and research tools, was freely passed around. Many of the key aspects of computer operating systems were developed as open source during the 1960s and 1970s in academic settings, such as Berkeley and MIT, as well as in research institutes, such as Bell Labs and Xerox's Palo Alto Research Center, at a time when sharing source code was widespread (Lerner & Tirole, 2002).

The free software (FS) movement began in the 1980s in academic and research institutes. The Free Software Foundation (FSF) was established by Richard Stallman of the MIT Artificial Intelligence Laboratory in 1984. The basic idea underlying the foundation was to facilitate the development and free dissemination of software. It is important to note that free in this case relates not to price, but to freedom of use. The developers of the Linux operating system bought in to the FS concept. Linux, initiated by Linus Tovalds in 1991, was the first tangible achievement of the FS movement (Stallman, 1999). This successful operating system has, through the collaboration of the global FS community, grown into the second most widely used server operating system.

The OSS movement evolved from the FSF during the 1990s. OSS has more flexible licensing criteria than the FSF. The widespread use of the Internet led to acceleration in open source activities. Numerous open source projects emerged, and interaction between commercial companies and the open source community became commonplace. Unlike the FS community, the OSS movement does not view itself as a solution for proprietary software, but rather as an alternative to it (Asiri, 2003). This has led to the acceptance of selective open sourcing, in which companies may elect to make specific components of the source code, rather than the entire code, publicly available, an approach which appeals to the business community. This allows companies to package available OSS products with other applications and extensions, and sell these to customers. Profit can also be made on the exclusive support provided with the retail packages, which may include manuals, software utilities, and support help lines.

For example, Red Hat Software (Red Hat, 2006), the leading provider of the Linux-based operating system, founded in 1995, based its business model on providing Linux software for free and selling extras such as support, documentation, and utilities, making it easy for users to install and use the software.

Definitions: FS and OSS

According to the FSF, FS involves users' freedom to run, copy, distribute, study, change, and improve software. The FSF defined four kinds of freedom for software users (Free Software Foundation, 2006).

1. The freedom to run the program for any purpose
2. The freedom to study how the program works, and adapt it to one's needs; access to the source code is a precondition for this
3. The freedom to redistribute copies so one can help a neighbor
4. The freedom to improve the program and release improvements to the public so that the whole community benefits; access to the source code is a precondition for this

The FSF recommends the GNU (a Unix-compatible operating system developed by the FSF) General Public License (GPL; Free Software Foundation, 1991) to prevent the GNU operating system software from being turned into proprietary software. This involves the use of "copyleft," which Stallman (1999) defines as follows:

The central idea of copyleft is that we give everyone permission to run the program, copy the program, modify the program, and distribute modified versions—but not permission to add restrictions of their own. Thus, the crucial freedoms that define "free software" are guaranteed to everyone who has a copy; they become inalienable rights.

The GPL permits the redistribution and reuse of source code for unfettered use and access as long as any modifications are also available in the source code and subject to the same license.

The term OSS was adopted in large part because of the ambiguous nature of the term FS (Johnson, 2001). On the most basic level, OSS simply means software for which the source code is open and available (Hissam et al., 2001), and that anyone can freely redistribute, analyze, and modify while complying with certain criteria (AlMarzouq et al., 2005). However, OSS does not just mean access to source code. For a program to be OSS, a set of distribution terms must apply.

A comprehensive Open Source Definition (OSD) was published by the Open Source Initiative (OSI). The OSD differentiates itself from FS by allowing the use of licenses that do not necessarily provide all the freedoms granted by the GPL. According to the updated version of the OSD (1.9), the distribution terms of OSS must comply with all 10 of the following criteria (Open Source Initiative, 2005).

1. **Free redistribution:** The license shall not restrict any party from selling or giving away the software as a component of an aggregate software distribution containing programs from several different sources. The license shall not require a royalty or other fee for such sale.
2. **Source code:** The program must include source code, and must allow distribution in source code as well as compiled form. Where some form of a product is not distributed with source code, there must be a well-publicized means of obtaining the source code for no more than a reasonable reproduction cost—preferably, downloading via the Internet without charge. The source code must be the preferred form in which a programmer would modify the program. Deliberately obfuscated source code is not allowed. Intermediate forms such as the

output of a preprocessor or translator are not allowed.

3. **Derived works:** The license must allow modifications and derived works, and must allow them to be distributed under the same terms as the license of the original software.

4. **Integrity of the author's source code:** The license may restrict source code from being distributed in modified form only if the license allows the distribution of patch files with the source code for the purpose of modifying the program at build time. The license must explicitly permit distribution of software built from modified source code. The license may require derived works to carry a different name or version number from the original software.

5. **No discrimination against persons or groups:** The license must not discriminate against any person or group of persons.

6. **No discrimination against fields of endeavor:** The license must not restrict anyone from making use of the program in a specific field of endeavor. For example, it may not restrict the program from being used in a business, or from being used for genetic research.

7. **Distribution of license:** The rights attached to the program must apply to all to whom the program is redistributed without the need for execution of an additional license by those parties.

8. **License must not be specific to a product:** The rights attached to the program must not depend on the program's being part of a particular software distribution. If the program is extracted from that distribution and used or distributed within the terms of the program's license, all parties to whom the program is redistributed should have the same rights as those that are granted in conjunction with the original software distribution.

9. **License must not restrict other software:** The license must not place restrictions on other software that is distributed along with the licensed software. For example, the license must not insist that all other programs distributed on the same medium must be open source software.

10. **License must be technology neutral:** No provision of the license may be predicated on any individual technology or style of interface.

Although there are some differences in the definitions of OSS and FS, the terms are often used interchangeably. Neither OSS nor FS pertains to the source code and its quality, but rather to the rights that a software license must grant. Various licensing agreements have been developed to formalize distribution terms (Hecker, 2000). Open source licenses define the privileges and restrictions a licensor must follow in order to use, modify, or redistribute the open source software. OSS includes software with source code in the public domain and software distributed under an open source license. Examples of open source licenses include the Apache license, Berkeley Source Distribution (BSD) License, GNU GPL, GNU Lesser General Public License (LGPL), MIT License, Eclipse Public License (EPL), Mozilla Public License (MPL), and Netscape Public License (NPL).

Table 1 provides a comparison of several common licensing practices described in Perens (1999).

The OSI has established a legal certification for OSS, called the OSI certification mark (Open Source Initiative, 2006b). Software that is distributed under an OSI-approved license can be labeled "OSI Certified."

Incentives to Engage in OSS

A growing body of literature addresses the motives for participation in OSS projects. Lerner and Tirole

Table 1. Comparison of licensing practices (Source: Perens, 1999)

License	Can be mixed with non-free software	Modifications can be made private and not returned	Can be relicensed by anyone	Contains special privileges for the original copyright holder over others' modifications
GPL				
LGPL	X			
BSD	X	X		
NPL	X	X		X
MPL	X	X		
Public Domain	X	X	X	

(2001) describe incentives for programmers and software vendors to engage in such projects:

Programmers' Incentives

- Programmers are motivated by a desire for peer recognition. Open source programmers' contributions are publicly recognized. By participating in an OSS project, programmers signal their professional abilities to the public.
- Programmers feel a duty to contribute to a community that has provided a useful piece of code.
- Some programmers are motivated by pure altruism.
- Some sophisticated OSS programmers enjoy fixing bugs, working on challenging problems, and enhancing programs.
- OSS is attractive to computer science students who wish to enter the market as programmers in higher positions.

Software Vendors' Incentives

- Vendors make money on OSS complementary services such as documentation, installation software, and utilities.
- By allowing their programmers to get involved in OSS projects, vendors keep abreast of open source developments, which allows them to better know the competition.
- Vendors benefit from efficient use of global knowledge. Many companies can collaborate on a product that none of them could achieve alone.

MAIN FOCUS OF THE CHAPTER

OSS Strengths and Weaknesses

OSS has a number of strengths and weaknesses compared to traditional proprietary software.

Strengths

The strengths of OSS can be classified into five main categories: freedom of use; evolution of software; time, cost, and effort; quality of software; and advantages to companies and programmers.

Freedom of Use

- It allows free access to the software source code for use, modification, and redistribution in its original or modified form for further use, modification, and redistribution.

189

- OSS users have fundamental control and flexibility advantages by being able to modify and maintain their own software to their liking (Wheeler, 2005).
- OSS allows independence from a sole source company or vendor. It provides users with the flexibility and freedom to change between different software packages, platforms, and vendors, while secret proprietary standards lock users into using software from only one vendor and leave them at the mercy of the vendor at a later stage (Wong & Sayo, 2004).
- It eliminates support and other problems if a software vendor goes out of business.
- It prevents a situation in which certain companies dominate the computer industry.
- Users can get free upgrade versions of the software, switch software versions, and fix and improve software (Perens, 1999).

Evolution of Software

- OSS contributes to software evolution due to the parallel process of many developers being simultaneously involved rather than a single software team in a commercial proprietary software company (Feller & Fitzgerald, 2002).
- It enables programmers all over the world to fix bugs.
- It evolves continuously over time as opposed to proprietary software whose development takes place in a series of discrete releases under the control of the authors.
- OSS represents a viable source of components for reuse and to build systems.

Time, Cost, and Effort

- It involves a joint effort by contributors from countries all over the world, collaborating via the Internet.

- There is a lower cost of software development in comparison to proprietary software.
- Open source initiatives allow software to be developed far more quickly and permits bugs to be identified sooner.
- The OSS approach is not subject to the same level of negative external process constraints of time and budget that can often undermine the development of dependable systems within an organizational setting (Lawrie & Gacek, 2002).
- OSS reduces the cost of using the software as the licensing is not limited compared to the limited licensing of proprietary software. The licensing cost, if any, is low, and most OSS distributions can be obtained at no charge. On a licensing cost basis, OSS applications are almost always much cheaper than proprietary software (Wong & Sayo, 2004). Open source products can save not-for-profit organizations, such as universities and libraries, a lot of money.
- It reduces development time, cost, and effort by reusing and building on existing open source code.
- It reduces maintenance and enhancement costs by sharing maintenance and enhancements among potential users of the same software application.

Quality of Software

- OSS reduces the number of bugs and enhances software quality by using the feedback of many users around the world and other qualified developers who examine the source code and fix the bugs.
- OSS is under constant peer review by developers around the world. Linus' law states the following: "Given enough eyeballs, all bugs are shallow" (Raymond, 1998).
- Programmers, knowing in advance that others will see the code they write, will be

more likely to write the best code they can possibly write (Raymond, 1998).

- Security vulnerabilities are more quickly solved when found in OSS than in proprietary software (Reinke & Saiedian, 2003).
- OSS represents an alternative approach to distributed software development able to offer useful information about common problems as well as possible solutions (Johnson, 2001).

**Advantages to Companies
and Programmers**

- There is an efficient use of global knowledge.
- Programmers learn from existing source code how to solve similar problems.
- Students, especially computer science students, can gain excellent programming experience and make contributions to open source software by becoming involved in open source projects (Zaritski, 2003).
- OSS allows groups of companies to collaborate in solving the same problem.
- Companies gain leverage from developers who contribute free improvements to their software.
- Companies using OSS benefit from its very rapid development, often by several collaborating companies, much of it contributed by individuals who simply need an improvement to serve their own needs (Perens, 1999).

Weaknesses

OSS weaknesses are mainly related to management, quality, and security.

Management

- Given the difficulty in managing resources in closed source proprietary software projects,

planning and delivering projects based on an open source community can be a much bigger challenge (Asundi, 2005). The separation between distributed developers creates difficulties in coordination and collaboration (Belanger & Collins, 1998; Carmel & Agarwal, 2001).

- Some OSS projects are developed without concern for the process of accepting or rejecting changes to the software.
- Resource allocation and budgeting are more complex than in proprietary software projects.
- There is higher fluidity in the membership of the development team. OSS developers are not bound to projects by employment relationships and therefore may come and go more often (Stewart, Darcy, & Daniel, 2005).
- Existing effort and cost models for proprietary projects are inadequate for OSS projects, and there is a need to develop new models.
- Commercial proprietary projects generate income and thus enable companies to hire high-quality and motivated programmers. This is not the case in open source projects.

Quality and Security

- OSS programmers are not always enthusiastic about providing and writing documentation, therefore some OSS have inadequate documentation, far below commercial standards.
- Some OSS applications are not sufficiently intuitive and user friendly, and are thus accessible only to very knowledgeable users.
- It appears that there is sometimes a race among many current OSS projects, which often results in rapid releases with the software consequently containing many bugs (Hissam et al., 2001).

- The OSS movement has made the life of cyberterrorists somewhat easier. Since the source code is open and available, cyberterrorists can learn about vulnerabilities in both OSS and proprietary closed source software (CSS) products. The knowledge that some components of CSS are descendants of similar OSS components, or share the same root code base or the same architecture, design, or specification provides clues as to what attacks could be possible against such software (Hissam et al., 2001).
- There is less variety of applications as compared to proprietary applications.

OSS Business Models

The open source model has a lot to offer the business world. For a company considering adopting an open source strategy, open source needs to be evaluated from a business point of view. It requires being clear on the advantages and disadvantages of open source relative to the traditional proprietary model. There are a number of business models for OSS, all of which assume the absence of traditional software licensing fees. As published by the Open Source Initiative (2006a), there are at least four known business models based on OSS.

1. **Support sellers:** In this model, the software product is effectively given away, but distribution, branding, and after-sales service are sold. This is the model followed by, for example, Red Hat (2006).
2. **Loss leader:** The open source is given away as a loss leader and market positioner for closed software. This is the model followed by Netscape.
3. **Widget frosting:** In this model, a hardware company (for which software is a necessary adjunct but strictly a cost rather than profit center) goes open source in order to get better drivers and cheaper interface tools. Silicon

Graphics, for example, supports and ships Samba (2006).

4. **Accessorizing:** This involves selling accessories such as books, compatible hardware, and complete systems with open source software preinstalled. It is easy to trivialize this (open source T-shirts, coffee mugs, Linux penguin dolls), but at least the books and hardware underlie some clear successes: O'Reilly Associates, SSC, and VA Research are among the companies using this model.

So far, the exemplars of commercial success have been service sellers or loss leaders. Nevertheless, there is good reason to believe that the clearest near-term gains in open source will be in widget frosting. For widget makers (such as semiconductor or peripheral-card manufacturers), interface software is not even potentially a revenue source. Therefore, the downside of moving to open source is minimal. (Hecker, 2000, proposes more models potentially usable by companies creating or leveraging OSS products.)

CONCLUSION

OSS is an alternative method of development that makes efficient use of global knowledge. It has captured the attention of academics, software practitioners, and the entire software community. Some OSS products have proven to be as reliable and secure as similar commercial products, and are a viable source of components from which to build OSS and CSS systems. Unfortunately, through OSS products, cyberterrorists also gain additional information about these components and discover vulnerabilities in products based on them.

There are a number of business models for OSS. Software development companies are beginning to support OSS-style development. They tend to try to profit through providing additional value to OSS products, such as value-added software, professional documentation, packaging, and support.

Both the quality and scope of OSS are growing at an increasing rate and there are already free alternatives to many of the fundamental software tools, utilities, and applications that are able to compete with traditional proprietary software. However, there is still controversy about whether OSS is faster, better, and cheaper than proprietary software. Adopters of OSS should not enter the realm blindly and should know its benefits and pitfalls. Further empirical and theoretical research is needed on developing and managing OSS projects. Identifying and explicitly modeling OSS development processes in forms that can be shared, modified, and redistributed appears to be an important topic for future investigation (Jensen & Scacchi, 2005). The open development process can provide a suitable environment for investigation of software development processes.

LIST OF ACRONYMS

BSD: Berkeley Source Distribution

CSS: Closed source software

FS: Free software

FSF: Free Software Foundation

GNU: GNU Not Unix (recursive acronym)

GPL: General Public License

LGPL: Lesser General Public License

MPL: Mozilla Public License

NPL: Netscape Public License

OSD: Open Source Definition

OSI: Open Source Initiative

OSS: Open source software

REFERENCES

AlMarzouq, M., Zheng, L., Rong, G., & Grover, V. (2005). Open source: Concepts, benefits, and challenges. *Communications of the Association for Information Systems (CAIS), 16*, 756-784.

Apache Software Foundation. (2006). *Apache HTTP server project.* Retrieved January 8, 2006, from http://httpd.apache.org/

Asiri, S. (2003). Open source software. *ACM SIGCAS Computers and Society, 33*(1), 2.

Asundi, J. (2005). The need for effort estimation models for open source software projects. In *Proceedings of the Fifth Workshop on Open Source Software Engineering (5-WOSSE)*, 1-3.

Behlendorf, B. (1999). Open source as a business strategy. In M. Stone, S. Ockman, & C. DiBona (Eds.), *Open sources: Voices from the open source revolution* (pp. 149-170). Sebastopol, CA: O'Reilly & Associates.

Belanger, F., & Collins, R. W. (1998). Distributed work arrangements: A research framework. *The Information Society, 14*(2), 137-152.

Carmel, E., & Agarwal, R. (2001). Tactical approaches for alleviating distance in global software development. *IEEE Software, 18*(2), 22-29.

DiBona, C., Ockman, S., & Stone, M. (Eds.). (1999). *Open sources: Voices from the open source revolution.* Sebastapol, CA: O'Reilly and Associates.

Eunice, J. (1998). *Beyond the cathedral, beyond the bazaar.* Retrieved January 10, 2006, from http://www.illuminata.com/public/all/catalog.cgi/cathedral

Feller, J., & Fitzgerald, B. (2002). *Understanding open source software development.* London: Addison Wesley.

Free Software Foundation. (1991). *GNU general public license, version 2.* Retrieved January 8,

2006, from http://www.gnu.org/licenses/gpl.html

Free Software Foundation. (2006). *Definition of free software.* Retrieved January 8, 2006, from http://www.fsf.org

Frishberg, N., Dirks, A. M., Benson, C., Nickell, S., & Smith, S. (2002). Getting to know you: Open source development meets usability. In *Extended Abstracts of the Conference on Human Factors in Computer Systems (CHI 2002)* (pp. 932-933).

Hecker, F. (2000). *Setting up shop: The business of open source software.* Retrieved May 31, 2006, from http://www.hecker.org/writings/setting-up-shop.html

Hissam, S., Weinstock, C. B., Plakosh, D., & Asundi, J. (2001). *Perspectives on open source software* (Tech. Rep. No. CMU/SEI-2001-TR-019). Retrieved January 10, 2006, from http://www.sei.cmu.edu/publications/documents/01.reports/01tr019.html

Jensen, C., & Scacchi, W. (2005, May 27). Experiences in discovering, modeling, and reenacting open source software development processes. In M. Li, B. W. Boehm, & L. J. Osterweil (Eds.), *Unifying the software process spectrum, ISPW 2005,* Beijing, China (LNCS Vol. 3840, pp. 449-462). Berlin, Germany: Springer-Verlag.

Johnson, K. (2001). *Open source software development.* Retrieved January 8, 2006, from http://chinese-school.netfirms.com/computer-article-open source.html

Lawrie, T., & Gacek, C. (2002). Issues of dependability in open source software development. *Software Engineering Notes (SIGSOFT), 27*(3), 34-37.

Lerner, J., & Tirole, J. (2001). The open source movement: Key research questions. *European Economic Review, 45*(4-6), 819-826.

Lerner, J., & Tirole, J. (2002). Some simple economics of open source. *Journal of Industrial Economics, 46*(2), 125-156.

Linux Online. (2006). *Linux.* Retrieved January 8, 2006, from http://www.linux.org/

Mockus, A., Fielding, R. T., & Herbsleb, J. (2000). A case study of open source software development: The Apache server. In *Proceedings of the 22nd International Conference on Software Engineering* (pp. 263-272).

Mozilla. (2006). *Mozilla.* Retrieved January 8, 2006, from http://www.mozilla.org/

Nichols, D. M., Thomson, K., & Yeates, S. A. (2001). Usability and open source software development. In *Proceedings of the Symposium on Computer Human Interaction* (pp. 49-54).

Nichols, D. M., & Twidale, M. B. (2003). The usability of open source software. *First Monday, 8*(1). Retrieved from http://firstmonday.org/issues/issue8_1/nichols/index.html

OpenOffice. (2006). *OpenOffice.* Retrieved January 10, 2006, from http://www.openoffice.org/

Open Source Initiative. (2005). *The open source definition.* Retrieved January 8, 2006, from http://www.opensource.org/docs/definition.php

Open Source Initiative. (2006a). *Open source case for business.* Retrieved May 31, 2006, from http://www.opensource.org/advocacy/case_for_business.php

Open Source Initiative. (2006b). *OSI certification mark and program.* Retrieved January 8, 2006, from http://www.opensource.org/docs/certification_mark.php

Perens, B. (1999). The open source definition. In C. DiBona, S. Ockman, & M. Stone (Eds.), *Open sources: Voices from the open source revolution* (1st ed., pp. 171-188). Sebastopol, CA: O'Reilly and Associates.

Raymond, E. S. (1998). *The cathedral and the bazaar.* Retrieved January 8, 2006, from http://www.catb.org/~esr/writings/cathedral-bazaar/cathedral-bazaar/

Raymond, E. S. (1999). The revenge of the hackers. In M. Stone, S. Ockman, & C. DiBona (Eds.), *Open sources: Voices from the open source revolution* (pp. 207-219). Sebastopol, CA: O'Reilly & Associates.

Red Hat. (2006). *Red Hat.* Retrieved May 31, 2006, from http://www.redhat.com/

Reinke, J., & Saiedian, H. (2003). The availability of source code in relation to timely response to security vulnerabilities. *Computers & Security, 22*(8), 707-724.

Reis, C. R., & Fortes, R. P. d. M. (2002). An overview of the software engineering process and tools in the Mozilla project. In C. Gacek & B. Arief (Eds.), *Proceedings of the Open Source Software Development Workshop* (pp. 155-175).

Samba. (2006). *Samba.* Retrieved May 31, 2006, from http://www.sgi.com/products/software/samba/

Sendmail Consortium. (2006). *Sendmail^{TM}.* Retrieved January 10, 2006, from http://www.sendmail.org/

Stallman, R. (1999). The GNU operating system and the free software movement. In C. DiBona, S. Ockman, & M. Stone (Eds.), *Open sources: Voices from the open source revolution* (pp. 53-70). Sebastopol, CA: O'Reilly & Associates.

Stewart, K. J., Darcy, D. P., & Daniel, S. L. (2005). Observations on patterns of development in open source software projects. In *Proceedings of the Fifth Workshop on Open Source Software Engineering (5-WOSSE)* (pp. 1-5).

Wheeler, D. A. (2005). *Why open source software/free software (OSS/FS, FLOSS, or FOSS)? Look at the numbers!* Retrieved January 10, 2006, from http://www.dwheeler.com/oss_fs_why.html

Wong, K., & Sayo, P. (2004). *Free/open source software: A general introduction.* UNDP, Asia-Pacific Development Information Programme. Retrieved January 10, 2006, from http://www.iosn.net/downloads/foss_primer_print_covers.pdf

Zaritski, R. M. (2003). Using open source software for scientific simulations, data visualization, and publishing. *Journal of Computing Sciences in Colleges, 19*(2), 218-222.

KEY TERMS

Closed Source Software (CSS): Non-OSS for which the source code is not available and not open. It is closed to modification and distribution by licenses that explicitly forbid it. The term CSS is typically used to contrast OSS with proprietary software.

Copyleft: Permission for everyone to run, copy, and modify the program, and to distribute modified versions, but no permission to add restrictions of one's own.

Free Software (FS): Free relates to liberty and not to price. It is similar to OSS but differs in the scope of the license. FS does not accept selective open sourcing in which companies may elect to make publicly available specific components of the source code instead of the entire code.

General Public License (GPL): License that permits the redistribution and reuse of source code for unfettered use and access as long as any modifications are also available in the source code and subject to the same license.

Open Source Software (OSS): Software for which the source code is open and available. Its licenses give users the freedom to access and use the source code for any purpose, to adapt and modify it, and to redistribute the original or the modified source code for further use, modification, and redistribution.

Proprietary Software (PS): Software produced and owned by individuals or companies, usually with no provision to users to access to the source code, and licensed to users under restricted licenses in which the software cannot be redistributed to other users. Some proprietary

software comes with source code—users are free to use and modify the software, but are restricted by licenses to redistribute modifications or simply share the software.

Source Code: The original human-readable version of a program, written in a particular programming language. In order to run the program, the source code is compiled into object code, a machine-readable binary form.

Chapter XVI
Open Source
Software Evaluation

Karin van den Berg
FreelancePHP, The Netherlands

ABSTRACT

If a person or corporation decides to use open source software for a certain purpose, nowadays the choice in software is large and still growing. In order to choose the right software package for the intended purpose, one will need to have insight and evaluate the software package choices. This chapter provides an insight into open source software and its development to those who wish to evaluate it. Using existing literature on open source software evaluation, a list of nine evaluation criteria is derived including community, security, license, and documentation. In the second section, these criteria and their relevance for open source software evaluation are explained. Finally, the future of open source software evaluation is discussed.

INTRODUCTION

The open source software market is growing. Corporations large and small are investing in open source software. With this growth comes a need to evaluate this software. Enterprises need something substantial to base their decisions on when selecting a product. More and more literature is being written on the subject, and more will be written in the near future.

This chapter gives an overview of the available open source evaluation models and articles, which is compounded in a list of unique characteristics of open source. These characteristics can be used when evaluating this type of software. For

a more in-depth review of this literature and the characteristics, as well as a case study using this information, see van den Berg (2005).

OPEN SOURCE SOFTWARE EVALUATION LITERATURE

The name already tells us something. Open source software is open—not only free to use but free to change. Developers are encouraged to participate in the software's community. Because of this unique process, the openness of it all, there is far more information available on an open source software package and its development process.

This information can be used to get a well-rounded impression of the software. In this chapter we will see how this can be done.

Though the concept of open source (or free software) is hardly new, the software has only in recent years reached the general commercial and private user. The concept of open source evaluation is therefore still rather new. There are a few articles and models on the subject, however, which we will introduce here and discuss more thoroughly in the next section.

Open Source Maturity Models

Two maturity models have been developed specifically for open source software.

The first is the Capgemini Expert Letter open source maturity model (Duijnhouwer & Widdows, 2003). The model "allows you to determine if or which open source product is suitable using just seven clear steps." Duijnhouwer and Widdows first explain the usefulness of a maturity model, then discuss open source product indicators and use these in the model. The model steps start with product research and rough selection, then uses the product indicators to score the product and determine the importance of the indicators, combining these to make scorecards. Finally it ends with evaluation.

Second, there is the Navica open source maturity model, which is used in the book *Succeeding with Open Source* (Golden, 2005). This model uses six product elements in three phases: assessing element maturity, assigning weight factors, and calculating the product maturity score.

Open Source Software Evaluation Articles

Aside from the two models, a number of articles on open source software evaluation have been written.

Crowston et al. (2003) and Crowston, Annabi, Howison, and Masango (2004) have published articles in the process of researching open source software success factors. In these articles, they attempt to determine which factors contribute to the success of open source software packages.

Wheeler's (n.d.) *How to Evaluate Open Source/Free Software (OSS/FS) Programs* defines a number of criteria to use in the evaluation of open source software, as well as a description of the recommended process of evaluation. Wheeler continues to update this online article to include relevant new information.

Another article defining evaluation criteria for open source software is *Ten Rules for Evaluating Open Source Software* (Donham, 2004). This is a point-of-view paper from Collaborative Consulting, providing 10 guidelines for evaluating open source software.

Finally, Nijdam (2003), in a Dutch article entitled "Vijf Adviezen voor Selectie van OSS-Componenten" ("Five Recommendations for Selection of OSS Components"), gives recommendations based on his own experience with selecting an open source system.

Literature Summary

Table 1 summarizes the criteria derived from the literature mentioned in the previous two sections and how they are discussed.

EVALUATING OPEN SOURCE SOFTWARE

The open source software market is in some ways very different from the traditional software market. One of the differences is that there is an abundance of information available concerning the software and its development process that is in most cases not available for traditional software.

The evaluation of traditional software is usually focused on the functionality and license cost of the software. In the open source world, the evaluation includes information from a number of other resources, giving a well-rounded picture

Table 1.

Criterion	Duijnhouwer and Widdows (2003)	Golden (2005)	Crowston et al. (2004)	Wheeler, (2005)	Donham (2004)	Nijdam (2003)
Community	Y	Y	Team size and activity level	In support	-	Active groups
Release Activity	-	Activity level	Activity level	Maintenance	-	Active groups
Longevity	Age	Y	-	Y	Maturity	Version
License	Y	In risk	-	Y	Y	Y
Support	Y	Y	-	Y	Y	-
Documentation	In ease of deployment	Y	-	In support	Y	-
Security	Y	In risk	-	Y	Y	-
Functionality	Features in time	Y	-	Y	Y	Y
Integration	Y	Y	-	In functionality	In infrastructure	-

of the software, its development, and its future prospects.

Using the existing evaluation models and articles discussed in the previous section, an overview is given here of the characteristics of open source software relevant to software evaluation and the information available on an open source software project concerning these characteristics.

Community

According to Golden (2005, p. 21), "One of the most important aspects of open source is the community."

The user community for most open source projects is the largest resource available. The community provides developers, user feedback, and ideas, and drives the project team. An active community helps the project move forward. It also shows the level of interest in the project, which can provide a measurement of quality and compliance with user requirements. A well-provided-for community also shows the team's interest in the user, allows the user to participate, and gives voice to the user's wishes and requirements.

The user community of an open source project consists of the people that use the software and participate in some way, from answering user questions to reporting bugs and feature requests. Users in the community sometimes cross the line into the developer community, which is often a line made very thin by encouraging participation and making the developer community accessible to anyone who is interested. In some cases, the user and developer community interact fully in the same discussion areas.

The community of an open source project is very important because it is the community that does most of the testing and provides quality feedback. Instead of using financial resources to put the software through extensive testing and quality assurance (QA), like a proprietary vendor will do, the open source projects have the community as a resource. The more people that are interested in a project, the more likely it is that it will be active and keep going. A large and active community says something about the acceptance of the software. If the software was not good enough to use, there would not be so many people who cared about its development (Duijnhouwer & Widdows, 2003).

The community is mostly visible in terms of the following (Crowston et al., 2004; Duijnhouwer & Widdows, 2003; Golden, 2005; Nijdam, 2003):

- **Posts:** Number of posts per period and number of topics
- **Users:** Number of users and the user-developer ratio in terms of the number of people and number of posts; if only users post, the developers are not as involved as they should be
- **Response time:** If and how soon user questions are answered
- **Quality:** The quality of posts and replies; are questions answered to the point, and are the answers very short or more elaborate? Is there much discussion about changes and feature additions?
- **Friendliness:** How friendly members are toward each other, especially to newcomers, also known as "newbies"; the community should have an open feel to it, encouraging people to participate

The depth of conversations, as mentioned in the fourth item, gives a good impression of how involved the community is with the ongoing development of the project. Much discussion about the software, in a friendly and constructive manner, encourages the developers to enhance the software further. The community activity is also reflected in other areas such as support and documentation.

Release Activity

The activity level of a project consists of the community activity and the development activity. The community was discussed above. The development activity is reflected in two parts:

- The developer's participation in the community

- The development itself—writing or changing the source code

The latter activity is visible mostly in the release activity. All software projects release new versions after a period of time. The number of releases per period and their significance, meaning how large the changes are per release (i.e., are there feature additions or just bug fixes in the release), illustrates the progress made by the developers. This gives a good indication of how seriously the developers are working on the software.

The open source repositories SourceForge[1] and FreshMeat[2], where project members can

share files with the public, provide information that could be useful to evaluate the release activity (Wheeler, n.d.).

An open source project often has different types of releases:

- **Stable releases:** These are the most important type for the end user. They are the versions of software that are deemed suitable for production use with minimal risk of failure.
- **Development versions:** These can have different forms, such as beta, daily builds, or CVS (Concurrent Version System) versions, each more up to date with the latest changes. These versions are usually said to be used "at your own risk" and are not meant for production use because there is a higher possibility of errors. A project that releases new versions of software usually publishes release notes along with the download that list all the changes made in the software since the previous release. Other than the release notes, the project might also have a road map, which usually shows what goals the developers have, how much of these goals are completed, and when the deadline or estimated delivery date is for each goal. Checking how the developers keep up with this road map shows something about how

well the development team can keep to a schedule.

Though a project might stabilise over time as it is completed, no project should be completely static. It is important that it is maintained and will remain maintained in the future (Wheeler, n.d.).

The project's change log can give the following information (Chavan, 2005):

- **The number of releases made per period of time:** Most projects will make several releases in a year, sometimes once or twice a month. A year is usually a good period in which to count the releases.
- **The significance of each release:** The change log or release notes explain what has changed in the release. These descriptions are sometimes very elaborate, where every little detail is described, and sometimes very short, where just large changes are listed. A good distinction to make is whether the release only contains bug fixes or also contains enhancements to features or completely new features. One thing to keep in mind here is that fewer, more significant releases is in most cases better than a large number of less significant releases leading to the same amount of change over time since the users will have to upgrade to new versions each time a release is made, which is not very user friendly. There should be a good balance between the number of releases and the releases' significance. If the project is listed on SourceForge and/or FreshMeat, some of the release activity information is available there.

Longevity

The longevity of a product is a measure of how long it has been around. It says something about a project's stability and chance of survival. A project that is just starting is usually still full of bugs (Golden, 2005). The older a project, the less likely the developers will suddenly stop (Duijn-houwer & Widdows, 2003). However, age is not always a guarantee of survival. First of all, very old software may be stuck on old technologies and methods, from which the only escape is to completely start over. Some software has already successfully gone through such a cycle, which is a good sign in terms of maturity. One thing that needs to be taken into account when products are not very young is whether or not there is still an active community around it.

The age and activity level of a project are often related. Young projects often have a higher activity level than older ones because once a project has stabilised and is satisfactory to most users, the discussions are less frequent and releases are smaller, containing mostly bug and security fixes. This does not mean that the activity should ever be slim to none. As mentioned before, no project is ever static (Wheeler, n.d.). There is always something that still needs to be done.

Longevity is checked using the following criteria (Golden, 2005; Nijdam, 2003):

- **Age of the product:** The date of the first release
- **Version number:** A 0.x number usually means the developers do not think the software is complete or ready for production use at this time.

If the project is very old, it is worthwhile to check if it has gone through a cycle of redesign, or if it is currently having problems with new technology.

Keep in mind that the version number does not always tell the whole story. Some projects might go from 1.0 to 2.0 with the same amount of change that another project has to go from 1.0 to 1.1. The fast progression of the version number might be used to create a false sense of progress. Other software products are still in a 0.x version even after a long time and after they are proved suitable for production use (Nijdam, 2003).

License

The licenses in the open source world reflect something of the culture. The most important term in this context is "copyleft," introduced by Richard Stallman, which means that the copyright is used to ensure free software and free derivative works based on the software (Weber, 2004). In essence, a copyleft license obligates anyone who redistributes software under that license in any way or form to also keep the code and any derivative code under the license, thus making any derivatives open source as well.

The most well-known example of a copyleft license is the GNU GPL (General Public License; Weber, 2004). This is also one of the most used licenses. On SourceForge, a large open source public repository where over 62,000 projects reside, almost 70%[3] of projects use the GNU GPL as their license. There are some large and well-known products that do not use SourceForge, and some of these have their own license, such as Apache, PHP, and Mozilla (Open Source Initiative [OSI], 2005).

Because copyleft in the GNU GPL is very strong, an additional version was made called the LGPL (library GPL, also known as lesser GPL), which is less restrictive in its copyleft statements, allowing libraries to be used in other applications without the need to distribute the source code (Weber).

A non-copyleft license that is much heard of is the BSD (Berkeley source distribution) license. It has been the subject of much controversy and has had different versions because of that. Components that are licensed under the BSD are used in several commercial software applications, among which are Microsoft products and Mac OS X (Wikipedia, 2005a). The license of the software in use can have unwanted consequences depending on the goal of the use. If the user plans

to alter and redistribute the software in some way but does not want to distribute the source code, a copyleft license is not suitable. In most cases, however, the user will probably just want to use the software, perhaps alter it to the environment somewhat, but not sell it. In that case, the license itself should at least be OSI approved and preferably well known. The license should fit with the intended software use.

As just mentioned, the license should preferably be an OSI-approved license. If it uses one of the public licenses, the better known the license, the more can be found on its use and potential issues (Wheeler, n.d.).

Support

There are two types of support for a software product:

- **Usage support:** The answering of questions on the installation and use of the software
- **Failure support or maintenance:** The solving of problems in the software

Often, the two get mixed at some level because users do not always know the right way to use the product. Their support request will start as a problem report and later becomes part of usage support (Golden, 2005).

The way support is handled is a measure of how seriously the developers work on the software (Duijnhouwer & Widdows, 2003). One way to check this is to see if there is a separate bug tracker[4] for the software and how actively it is being used by both the developers and the users. When the developers use it but hardly any users seem to participate, the users may not be pointed in the right direction to report problems. Aside from community support, larger or more popular projects may have paid support options. The software is free to use, but the user has the option to get professional support for a fee, either on a service-agreement basis where a subscription fee is paid for a certain period of time, or a per-incident fee for each time the user calls on support. The project leaders themselves may offer something like this, which is the case for the very popular open source database server MySQL (2005).

There are companies that offer specialised support for certain open source software. This is called third-party support. For example, at the Mozilla support Web page, it can be seen that DecisionOne offers paid support for Mozilla's popular Web browser FireFox, the e-mail client Thunderbird, and the Mozilla Suite (Mozilla, 2005). The fact that paid support exists for an open source product, especially third-party support, is a sign of maturity and a sign the product is taken seriously.

Support for open source software is in most cases handled by the community. The community's support areas are invaluable resources for solving problems (Golden, 2005). Mature products often have paid support options as well if more help or the security of a support contract is required.

Community Support

The usage support is usually found in the community. Things to look for include the following (Golden, 2005):

- Does the program have a separate forum or group for asking installation- and usage-related questions?
- How active is this forum?
- Are developers participating?
- Are questions answered adequately?
- Is there adequate documentation (see the documentation section)?

Responses to questions should be to the point and the responders friendly and helpful. In the process of evaluating software, the evaluator will probably be able to post a question. Try to keep to the etiquette, where the most important rule is to search for a possible answer on the forum before posting a question and to given enough relevant information for others to reproduce the problem (Golden, 2005; Wheeler, n.d.).

The way the community is organised influences the community support's effectiveness. A large project should have multiple areas for each part of the project, but the areas should not be spread to thin. That way, the developers that are responsible for a certain part of the project are able to focus on the relevant area without getting overwhelmed with a large amount of other questions. If the areas are too specialised and little activity takes place in each, not enough people will show interest and questions are more likely to remain unanswered.

Failure support within the project is often handled by a bug tracker by which problems are reported and tracked. Statistical studies have shown that in successful projects, the number of developers that fix bugs in open source software is usually much higher than the number of developers creating new code (Mockus, Rielding, & Herbsleb, 2000).

Paid Support

Paid support might be available from the project team itself (Golden, 2005). There may have been people who have given their opinion about the quality of this support.

One of the strong signs of the maturity of open source software is the availability of third-party support: companies that offer commercial support services for open source products (Duijnhouwer & Widdows, 2003). Some companies offer service contracts, others offer only phone support on a per-incident basis. Check for paid support options whether they will be used or not (Duijnhouwer & Widdows). How the situation may be during actual use of the software is not always clear and it can give a better impression of the maturity of the software.

Documentation

There are two main types of documentation (Erenkratz & Taylor, 2003):

- User documentation
- Developer documentation

User documentation contains all documents that describe how to use the system. For certain applications, there can be different levels in the user documentation, corresponding with different user levels and rights. For example, many applications that have an administrator role have a separate piece of documentation for administrators. Additionally, there can be various user-contributed tutorials and how-tos, be it on the project's Web site or elsewhere. The available documentation should be adequate for your needs. The more complex the software, the more you may need to rely on the user documentation.

The other main type of documentation, which plays a much larger role in open source software than in proprietary applications, is developer documentation. A voluntary decentralised distribution of labour could not work without it (Weber, 2004). The developer documentation concerns separate documents on how to add or change the code, as well as documentation within the source code by way of comments. The comments usually explain what a section of code does, how to use and change it, and why it works like it does. Though this type of documentation may exist for proprietary software, it is usually not public.

If it is possible that you may want to change or add to the source code, this documentation is very valuable. A programmer or at least someone with some experience in programming will be better able to evaluate whether this documentation is set up well, especially by the comments in the source code. It is a good idea to let someone with experience take a look at this documentation (n.d., 2005).

A third type of documentation that is often available for larger server-based applications is maintainer documentation, which includes the install and upgrade instructions. These need to be clear, with the required infrastructure and the steps for installing the software properly explained. This documentation is needed to set up the application. For this type, again, the complexity of the application and its deployment determines the level of documentation that is needed. Documentation is often lagging behind the status of the application since it is often written only after functionality is created, especially user documentation (Scacchi, 2002). It is a good idea to check how often the documentation is updated, and how much the documentation is behind compared to the current status of the software itself.

The documentation for larger projects is often handled by a documentation team. A discussion area may exist about the documentation, giving an indication of the activity level of that team.

Security

Security in software, especially when discussing open source software, has two sides to it. There are people who believe security by obscurity is better, meaning that the inner workings of the software are hidden by keeping it closed source, something that open source obviously does not do. The advocates of security by obscurity see the openness of open source software as a security hazard. Others argue that the openness of open source actually makes it safer because vulnerabilities in the code are found sooner. Open source software gives both attackers and defenders great power over system security (Cowan, 2003; Hoepman & Jacobs, 2005).

Security depends strongly on how much attention the developers give to it. The quality of the code has much to do with it, and that goes for both proprietary and open source software. If the code of proprietary software is not secure, the vulnerabilities may still be found. There are plenty of examples where this occurs, such as the Microsoft Windows operating system (OS). The vulnerabilities are often found by hackers who try to break the software, sometimes by blunt force or simple trial and error. In this case, a vulnerability might get exploited before the vendor knows about it. The attack is the first clue in that case. The open source software's vulnerabilities, however, could be found by one of the developers or users just by reviewing the code; he or she can report the problem so it can be fixed (Payne, 2002). It is important that the developers take the security of their software seriously and respond swiftly to any reported vulnerabilities.

There are various security advisories to check for bugs in all types of software that make it vulnerable to attacks. A couple of well-known advisories are http://www.securityfocus.com and http://www.secunia.com. Keep in mind that more popular software will have a higher chance of having vulnerability reports, so the mere lack of reports is no proof of its security. On the project's Web site, it can be seen, for instance in the release notes, how serious the project is about security.

Functionality

Though functionality comparison is not specific to open source software evaluation and is properly covered in most traditional software evaluation models, there are some points to take into consideration. Open source software often uses the method described by the phrase "release early and often" (Raymond, 1998). This method enables faster error correction (Weber, 2004) by keeping the software up to date as much as possible. It also encourages people to contribute because they see the result of their work in the next release much sooner (Raymond). However, this often means that the software is incomplete during the first releases, at least more so than is customary with proprietary software. Where vendors of proprietary software will offer full functionality descriptions for their

software, open source projects might not have the complete information on the Web site (Golden, 2005). Just like with documentation, the information on the Web site might be lagging behind the actual functionality. Other means of checking the current functionality set might be needed. Fortunately, open source software that is freely available gives the added option of installing the software to enable the full testing of the functionality, an option that is mostly not available with proprietary software, for which at most only limited versions, in terms of functionality or time, are given freely for trying it out.

One problem with open source projects is that the documentation is not always up to date with the latest software. Look beyond the feature list on the Web site to find out what features the software has. Two options are to query the developers and ask the user community (Golden, 2005). Eventually the software itself should be investigated. If it is a Web-based application, an online demo might be available, though installing it on a test environment could be useful because it also gives insight on how well the software installs.

A list of functional requirements for the goal of the software can be used to check if the needed functionality is available. If such a list is not given, there may be one available from technology analyst organisations (Golden, 2005). It is wise to make a distinction in the list between features that are absolutely necessary, where the absence would lead to elimination, and those that would be a plus, which results in a higher score. If there is something missing, there is always the option to build it or have it built.

When comparing functionality, those features that are part of the functional requirements should take priority, but additional features may prove useful later. The features used or requested by the users in the future are not really predictable. While evaluating the software, features may be found in some of the candidates that are very useful for the goal. These can be added to the functional requirements.

Part of the functionality is localisation. The languages to which the interface and documentation are translated are a sign of the global interest taken in the software.

Integration

Duijnhouwer and Widdows (2003) mention three integration criteria. These are most important for software that is being used in collaboration with other software, and for people who are planning on adapting the software to their use, such as adding functionality or customising certain aspects so that it fits better in the organisation's environment. The three criteria are discussed in the next three subsections.

Modularity

Modularity of software means that the software or part of the software is broken into separate pieces, each with its own function. This type of structure has the following advantages:

- Modular software is easier to manage (Garzarelli, 2002; Mockus, Fielding, & Herbsleb, 2002).
- With a base structure that handles the modules well, people can easily add customised functionality without touching the core software.
- Modular software enables the selection of the needed functionality, leaving out those that are not necessary for the intended use. This way, the software can be customised without the need for a programmer.
- Modular software can be used in commercial applications. By making software modular, not everything needs to be given away as open source. It is can be used to give away only parts of software as open source while the add-on modules are sold as proprietary software (Duijnhouwer & Widdows, 2003). This is also called the razor model, as in

giving away the razor for free and charging for the blade (Golden, 2005).

Evidence of a modular structure can often be found in several places, such as the source code, the developer documentation, or the download section, where modules might be available for download separate from the core software.

Standards

In the software market, more and more open standards emerge to make cooperation between software easier (Golden, 2005). If the software vendors use these standards in their software, it makes it easier to communicate between different software packages, and to switch between software packages. In some industries, standards are far more important than in others. For some software, there may not even be an applicable standard.

The use of current and open standards in open source software is a sign of the software's maturity (Duijnhouwer & Widdows, 2003). The feature list of the software usually lists what standards are used and with which the software complies.

Collaboration with Other Products

Closely connected to standards is the collaboration with other products. As mentioned before, not every software type has applicable standards, and sometimes the formal standards are not used as much as other formats. Examples of such formats are the Microsoft Word document format, and Adobe's PDF (portable document format). The office suite OpenOffice.org (2005) has built-in compatibility for both formats.

Software Requirements

Most software is written for a specific OS, for example, Microsoft Windows or Linux (Wheeler, n.d.). Certain types of software also rely on other software, such as a Web server or a database. The requirements of the software will state which software and which versions of that software are compatible. If these requirements are very specific, it could lead to problems if they are incompatible with the organisation's current environment.

THE FUTURE OF OPEN SOURCE SOFTWARE EVALUATION

Open Source Software Evaluation Literature

More is being written on open source software evaluation at the time of writing. For example, another model called the business readiness rating (OpenBRR, 2005), aimed at open source software, was released recently. The research of Crowston and others is still ongoing, so there will be more results in the near future to include in the open source software evaluation process. Given how recent the rest of the literature discussed in this chapter is, it is likely that more will be published on the subject in the next few years.

The Future of Open Source Software

Open source software is being used increasingly by corporations worldwide. There is now some literature available to help with the evaluation of open source software, and the number of articles and models is increasing. With this growth in the field comes more attention from companies, especially on the enterprise level, which will cause more demand for solid evaluation models. Because open source software and the process around it provide much more information than traditional software, there is certainly a need for such models.

This literature will help justify and solidify the position of open source software evaluation in a corporate setting, giving more incentive to use open source software. Most likely, more companies will be investing time and money in its development, like we are seeing today in examples such as Oracle investing in PHP and incorporating this open source Web development language in its products (Oracle, 2005), and Novell's acquisition of SUSE Linux (Novell, 2003). The open source software evaluation literature can help IT managers in adopting open source.

CONCLUSION

The field of open source software evaluation is growing, and with that growth more attention is gained from the large enterprises. With this attention comes more demand for evaluation models that can be performed for these corporations, which will give more growth to the open source software market as well. In this chapter, an overview is given of the current literature and the criteria derived from that literature that can be used in open source software evaluation. For each of the criteria—community, release activity, longevity, license, support, documentation, security, and functionality—this chapter explains why it is important in the market and what to do to evaluate it. This information can be used on its own or in conjunction with more traditional evaluation models and additional information referenced here by companies and individuals that wish to evaluate and select an open source software package. It helps to give insight into the open source software sector.

REFERENCES

Chavan, A. (2005). Seven criteria for evaluating open source content management systems. *Linux Journal*. Retrieved August 9, 2005, from http://www.linuxjournal.com/node/8301/

Cowan, C. (2003). Software security for open source systems. *Security & Privacy Magazine, 1*(1), 38-45.

Crowston, K., Annabi, H., & Howison, J. (2003). Defining open source software project success. In *Twenty-Fourth International Conference on Information Systems, International Conference on Software Engineering (ICIS 2003)* (pp. 29-33). Retrieved March 30, 2005, from http://opensource. mit.edu/papers/crowstonannabihowison.pdf

Crowston, K., Annabi, H., Howison, J., & Masango, C. (2004). Towards a portfolio of FLOSS project success measures. *Collaboration, Conflict and Control: The Fourth Workshop on Open Source Software Engineering, International Conference on Software Engineering (ICSE 2004),* 29-33. Retrieved March 30, 2005, from http://opensource.ucc.ie/icse2004/Workshop_on_ OSS_Engineering_2004.pdf

Donham, P. (2004). Ten rules for evaluating open source software. *Collaborative Consulting.* Retrieved August 8, 2005, from http://www.col-laborative.ws/leadership.php?subsection=27

Duijnhouwer, F., & Widdows, C. (2003). *Capgemini open source maturity model.* Retrieved February 12, 2006, from http://www.seriouslyo-pen.org/nuke/html/modules/Downloads/osmm/ GB_Expert_Letter_Open_Source_Maturity_ Model_1.5.3.pdf

Erenkratz, J. R., & Taylor, R. N. (2003). *Supporting distributed and decentralized projects: Drawing lessons from the open source community* (Tech. Rep.). Institute for Software Research. Retrieved August 9, 2005, from http://www.erenkrantz. com/Geeks/Research/Publications/Open-Source-Process-OSIC.pdf

Garzarelli, G. (2002, June 6-8). *The pure convergence of knowledge and rights in economic organization: The case of open source software development.* Paper presented at the DRUID Summer Conference 2002 on Industrial dynam-ics of the new and old economy—Who embraces whom?, Copenhagen.

Golden, G. (2005). *Succeeding with open source.* Boston: Addison-Wesley Pearson Education.

Hoepman, J., & Jacobs, B. (2005). *Software security through open source* (Tech. Rep.). Institute for Computing and Information Sciences, Radboud University Nijmegen. Retrieved August 9, 2005, from http://www.cs.ru.nl/~jhh/publications/oss-acm.pdf

Mockus, A., Fielding, R. T., & Herbsleb, J. (2000). A case study of open source software development: The Apache Server. In *Proceedings of the 22nd International Conference on Software Engineering (ICSE 2000).* Retrieved on March 30, 2005, from http://opensource.mit.edu/papers/ mockusapache.pdf

Mockus, A., Fielding, R. T., & Herbsleb, J. (2002). Two case studies of open source software development: Apache and Mozilla. *ACM Transactions on Software Engineering and Methodology, 11*(3), 309-346.

Mozilla. (2005). *Mozilla.org support.* Retrieved February 16, 2005, from http://www.mozilla. org/support/

MySQL. (2005). *MySQL support Web site.* Retrieved February 16, 2005, from http://www. mysql.com/support/premier.html

Nijdam, M. (2003). Vijf adviezen voor selectie van oss-compontenten. *Informatie: Maandblad voor Informatieverwerking, 45*(7), 28-30.

Novell. (2003). *Novell announces agreement to acquire leading enterprise Linux technology company SUSE LINUX.* Retrieved August 8, 2005, from http://www.novell.com/news/press/ archive/2003/11/pr03069.html

OpenBRR. (2005). *Business readiness rating for open source: A proposed open standard to facilitate assessment and adoption of open source software (RFC1).* Retrieved August 10, 2005, from http://www.openbrr.org/docs/BRR_white-paper_2005RFC1.pdf

OpenOffice.org. (2005). *OpenOffice.org writer product information.* Retrieved August 10, 2005, from http://www.openoffice.org/product/writer.html

Open Source Initiative (OSI). (2005). *Open Source Initiative: Open source licenses.* Retrieved August 9, 2005, from http://opensource.org/licenses/

Oracle. (2005). *Oracle and Zend partner on development and deployment foundation for PHP-based applications.* Retrieved February 12, 2006, from http://www.oracle.com/corporate/press/2005_may/05.16.05_oracle_zend_partner_finalsite.html

Payne, C. (2002). On the security of open source software. *Information Systems Journal, 12*(1), 61-78.

Raymond, E. S. (1998). The cathedral and the bazaar. *First Monday, 3*(3). Retrieved March 30, 2005, from http://www.firstmonday.org/issues/issue3_3/raymond/

Scacchi, W. (2002). Understanding the requirements for developing open source software systems. In *IEEE Proceedings: Software, 149*, 24-29. Retrieved March 30, 2005, from http://www1.ics.uci.edu/wscacchi/Papers/New/Understanding-OS-Requirements.pdf

Van den Berg, K. (2005). *Finding open options: An open source software evaluation model with a case study on course management system.* Unpublished master's thesis, Tilburg University, Tilburg, The Netherlands. Retrieved August 30, 2005, from http://www.karinvandenberg.nl/Thesis.pdf

Weber, S. (2004). *The success of open source.* Cambridge, MA: Harvard University Press.

Wheeler, W. (n.d.). *How to evaluate open source/free software (OSS/FS) programs.* Retrieved February 17, 2005, from http://www.dwheeler.com/oss_fs_eval.html

KEY TERMS

Community: A group of people with shared interests that interact. In case of open source software, the community is the group of developers and users that come together, mostly on a Web site, to discuss, debug, and develop the software.

Documentation: The documents that are associated with a piece of software. There is usually user documentation, in the form of help files, tutorials, and manuals, and there can be developer documentation, such as programming guidelines and documents explaining the structure and workings of the software (source code). In some cases there is administrator documentation, which explains how to install and configure the software. The latter is more important for large pieces of software, where one installation will be used by many users, such as Web applications.

License: An agreement that is attached to the use of a product. In case of software, the software license agreement defines the terms under which you are allowed to use the software. For open source software, there are a number of common licenses, not bound to a specific piece of software, that can be used for almost any type of open source software. These licenses are well known so users and developers usually know the conditions of these licenses.

Maturity Model: Not to be confused with the capability maturity model (CMM), a maturity model as discussed in this chapter is a model that can be used to assess the maturity of a software package, evaluating the software using several criteria.

Software Longevity: The life expectancy of software, measured by various factors among which is its age.

Software Release Activity: The number and significance of releases that are made for a certain software package. A release can be a minor change such as a bug fix, or a major change such as added functionality.

Software Security: How well a piece of software is built in terms of vulnerabilities and defense against them. Any software will have some type of security hole in it that allows a person, often with hostile intentions, to break into the software and use it for purposes that are unwanted. It is necessary for developers to minimize these holes and fix them if they are discovered. In case of open source software, because the source is public, the users may help in discovery by examining the source code. This, however, also means that a person with hostile intentions can also find these holes by examining the source code. Thus, it is always important to keep a close eye on security.

ENDNOTES

[1]   http://www.sourceforge.net

[2]   http://www.freshmeat.net

[3]   Established using the SourceForge Software Map on April 20, 2005, at http://sourceforge.net/softwaremap/trove_list.php?form_cat=13

[4]   A bug tracker is an application, often Web based, through which the users can report problems with the software, the developers can assign the bug to someone who will handle it, and the status of the bug can be maintained. Bugzilla is one such package that is often used for this purpose.

Chapter XVII
Open Source Web Portals

Vanessa P. Braganholo
DCC/UFRJ, Brazil

Bernardo Miranda
COPPE/UFRJ, Brazil

Marta Mattoso
COPPE/UFRJ, Brazil

ABSTRACT

Open source software is required to be widely available to the user community. To help developers fulfill this requirement, Web portals provide a way to make open source projects public so that the user community has access to their source code, can contribute to their development, and can interact with the developer team. However, choosing a Web portal is not an easy task. There are several options available, each of them offering a set of tools and features to its users. The goal of this chapter is to analyze a set of existing Web portals (SourceForge.net, Apache, Tigris, ObjectWeb, and Savannah) in the hopes that this will help users to choose a hosting site for their projects.

INTRODUCTION

One of the main sustaining pillars of the open source (Perens, 1997) philosophy is that software must be widely available to the user community. In order to mature, open source projects need collaboration from the user community, and this is hard to achieve just by publishing a project on a developer's personal home page. An efficient way of reaching these requirements of availability and collaboration is by hosting the software on an open source Web portal. There are several portals that address these requirements, offering free hosting to open source projects.

Besides giving access to a project's source code, these portals also offer tools to help the development of the projects they host. Among such tools, we can cite task management tools, issue trackers, forums, mailing lists, tools to support feature requests, and version control servers.

The different portals offer different advantages to the projects they host. It is difficult for a developer who is not used to contributing to open source projects to choose the one that best fits his or her needs. This is because there are many portal features that are only visible to those who actively contribute to an open source project. Additionally, a portal may have particular requirements that the developer must be aware of. For example, some portals require that the project be under the protection of a specific open source license. The goal of this chapter is to help such users in choosing a portal to host their projects. We analyze five Web portals and compare them in terms of the services they offer. The analyzed portals are as follows:

- SourceForge.Net (Open Source Technology Group, 2005),
- Apache (Apache Software Foundation, 1999)
- Tigris (Tigris, 2005)
- ObjectWeb (Object Web Consortium, 2005)
- Savannah (Free Software Foundation, 2000b)

They were chosen for several reasons. First, they host projects for free. Second, they are general in the sense that they host general free or open source software (Savannah hosts even nonsoftware projects). Third, they have been online for enough time for one to assume that they probably will not disappear and leave users helpless.

It is important to emphasize that this kind of analysis is new in literature. To the best of our knowledge, there is no work in the literature that provides similar analysis (DiBona, Stone, & Cooper, 2005).

It is also important to state that some of the portals may be focused on free software (Free Software Foundation, 1996) while others focus on open source software (Perens, 1997). Although their way of looking at the world is different (Stallman, 2002), the philosophies are similar. In this chapter, we do not intend to make any distinction between them. Thus, we use the term FOSS (free and open source software) as synonymous of free software and open source software.

The subsequent section describes briefly the way most Web portals work. Then we discuss the methodology of our study and the features of each portal. Next we discuss future trends and conclude with a tabular comparison of the Web portals.

BACKGROUND: HOSTING SOFTWARE ON WEB PORTALS

In this section, we describe how portals work in essence, hoping this will give readers a better understanding of our proposal in this chapter.

Web portals dedicated to hosting software projects are basically Web pages that offer a set of functionalities to its users. Usually, the entrance page explains the purpose of the portal and provides links to documentation, instructions to users who want to host a project, a news section, and links to the hosted projects. Such links are usually presented within categories. Figure 1 shows a cut of the main page of the SourceForge.Net portal. Notice the news section and the links to software categories (at the bottom of the figure). Such categories link to hosted projects classified under them.

Each hosted project has its own page within the portal with a URL (uniform resource locator) similar to http://www.portal.org/project, where *portal* is the portal name, and *project* is the project name. It is through this page that the portal provides tools and services to developers. Also, such pages play the role of advertising the project. Users will find projects they may be interested in through such pages.

The main page of a project within any portal has basic information about the project, news, and a list of links to source-code downloads and mailing lists, among other features. As we will discuss later on, it is a choice of the project's administrator what will appear on the main page.

Figure 1. A cut of the main page of SourceForge.Net

Another important aspect of hosted project pages is that they have public and private areas. Public areas can be seen by any person. The public area is basically the project's main page. This way, any user may read about a specific project and download its source code. The private area of a project is exclusively for the project's contributors. The list of people that may contribute to it is maintained by the project administrators. In order to contribute to a project or even to create one, you must have a user account with the portal. All of the portals we analyzed allow you to create a user account for free. Once logged in, you are in your private area. In this area, you can create a new project, or see all the projects you contribute to.

If you are not part of a project, then your private area may include several links, including one to create a new project. Figure 2 shows a private area at SourceForge.Net. Notice the "Register a new project" link on the left-hand side.

If you have access to the private area of a given project, you can solve bugs, write documentation, check in modifications to the source code, and so forth. In other words, the private area gives you ways to contribute to software projects.

Project administrators may include you as a developer, as a documentation writer, as a project administrator, and so forth. Each role you assume grants you access to certain parts of the project's private area. The project administrator has full power in choosing what you may or may not edit or develop in his or her project.

An important point that needs to be made here is on how a user can contact the project administrator and ask to be included in the project. This is usually made through the help-wanted section. Administrators can explicitly ask for the help of other developers through a specific section on the project's Web page. Interested developers (users) respond to such requests by clicking on the offer he or she is interested in. This opens a Web form

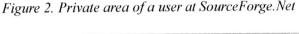

Figure 2. Private area of a user at SourceForge.Net

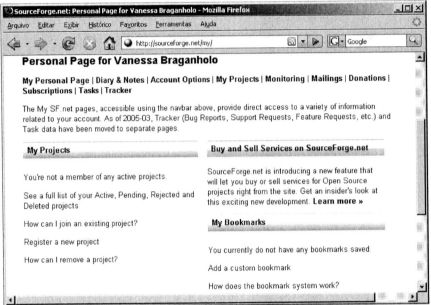

that sends an e-mail do the project's administrator. We will return to this point later on.

METHODOLOGY

We have studied the documentation of each of the five Web portals trying to answer 13 questions:

1. What are the requirements for the registration of a new project on the portal?
2. Does the portal offer version control systems?
3. Does it offer forums?
4. Does it offer mailing lists?
5. Does the portal supply a Web page for the project?
6. Does it offer issue tracking?
7. Does it have tools to support the documentation of the project?
8. Does the portal preserve the intellectual property of the project's owner?
9. Does it require the developers to provide support even after the project is finished?
10. Does it have tools to support task management?
11. Does it provide automatic backups of the repositories in the version control system?
12. Does it allow the developer to customize the public area (remove unwanted items from the public view)?
13. Does it have any license restrictions?

To answer these questions, we used two approaches. The first approach was the analysis of the requirements of a new project. Were there any categories the project should fit in? Were there any license restrictions? We analyzed the submission process of a new project using fictitious data in order to know what the requirements were for registering a new project. We first created a user in each of the portals and followed the submission process until the last step. We did not actually submit the project since we did not have a real project to submit at that time. We could not have used a fake project because, in all of the portals, submitted projects first go through an evaluation step where it is approved or rejected by the Web portal managers. Only approved projects are hosted. Our fake project would probably be rejected.

The second approach was the analysis of the features of the Web portal. Since most of the features are on private areas, we mainly used the Web portal documentation to find out their features. We collected all the answers we could find about all the portals we were evaluating, and contacted the Web portal administrators, asking for confirmation. This way, we would be sure we were not making any misjudgment. Most portals (SourceForge.Net, Savannah, and ObjectWeb) replied to us promptly with feedback. In the next section, we present our evaluation in detail.

MAIN FOCUS OF THE CHAPTER: WEB PORTALS

In this section, we describe each of the five portals in detail. First, however, we focus on the features we found in all of the portals we analyzed:

- **Registered users:** Only registered users may submit projects to be hosted on the Web portals.
- **Projects must be approved:** One of the common features among the portals is the requirement of submitted projects to be approved by the portal. No project is hosted without being approved. This is done to avoid fake projects (spam), but mainly to avoid projects that are not FOSS.
- **Formal submission procedure:** Due to the necessity of approval, the portals require a series of information about the project that is being submitted. The amount and type of required information may vary from a simple description to a detailed document analyzing

similar projects, planned features, components that will be used, and so forth.

- **Distribution license:** During project submission, the portals require the definition of the license under which the project results are to be distributed. In all cases, there is a list of licenses you may choose. It is usually possible to choose one of the standard licenses approved by the Open Source Initiative (OSI, 2006) such as GPL (General Public License), LGPL (Lesser General Public License), BSD (Berkeley Source Distribution), MIT, and MPL (Mozilla Public License), among others. It is also possible to specify a new license, but this usually increases the evaluation time of the submitted project. A license needs to be studied by Web portal administrators to check if it violates the FOSS definition. For example, the Savannah administrators check for compatibility with GPL Version 2 (Savannah-help-private@gnu.org, 2006). ObjectWeb is an exception. It requires you to choose a given license (LGPL, in this case); another license may be chosen if you explain the reasoning (Object Web Consortium, 1999).
- **Source code publication:** No portal requires that there be source code available at project submission time. The goal of these portals is to support the development of FOSS, so it is understandable that projects start with no source code at all.
- **Software development support:** All of the portals offer tools to support the development of software projects. All of the portals we analyzed offer version control systems, mailing lists, Web pages for the project, bug tracking, and task management. Some of the portals offer additional tools. We will refer to them in the sections that follow.
- **Help from external users:** All of the portals allow you to request help from the user community. This is usually done by opening

a help-wanted section in your project (only project administrators have permission to do so). After that, you may inform about the kind of help you want (interface designer, developer, tester, and support manager, among others) and wait for people holding that skill to contact you. We must warn, however, that user help is usually restricted to those projects that already provide features that may attract new developers' attention. Please do not count on external help to develop your project at the beginning.

Now we are ready to look at each portal in detail.

SourceForge.Net

SourceForge.Net (http://www.sourceforge.net) is the world biggest FOSS development site. It hosts thousands of projects developed by people from several different countries. The main goal of this portal (and also of the other portals) is to offer a centralized place where developers can control and manage the development of FOSS (Open Source Technology Group, 2005).

The philosophy of SourceForge.Net is centered on the FOSS ideas:

- **Facilitate the maintenance of projects:** The user community has the right to use and give support to a FOSS project, even after its activities have ceased.

- **Help to achieve the license requirements:** Some FOSS licenses require the source code to be available for a certain amount of time (usually longer than the development time period). SourceForge.Net keeps the files of finished projects to help developers to accomplish this requirement.

- **Promote reuse:** The rights of use, modification, and redistribution are guaranteed by all FOSS licenses. These rights help to promote the reuse of source code. An old project that

is available at SourceForge.Net may help other developers to avoid reimplementing and testing pieces of software that other people have already implemented.

- **Allow the continuation of orphan projects:** When a project is finished, there are usually users who are interested in continuing its development. SourceForge.Net allows this to happen. Notice, however, that the project owner has to agree with this.

- **Allow project alternatives:** A project fork with alternative features may be created from a preexisting project. Both can be maintained in parallel.

To register a new project at the portal (Open Source Technology Group, 2002), it is necessary to determine the type of the project (i.e., software, documentation, Web site, peer-to-peer software, game, content management system, operational system distribution, precompiled package of existing software, software internationalization). After this, it is necessary to go through a term agreement step, and then provide a description of the project and choose the project name (which cannot be changed later). The registration process is quite simple and fast. Once registered, Source-Forge.Net will take about 2 days to approve or reject the request.

After approval, the project can start taking advantage of the benefits offered by SourceForge. Net: forums, CVS (Concurrent Version System), mailing lists (public or private), the project Web page, documentation (DocManager), task management, automatic backup of the version control repository, a donation system, news and trackers for bugs, support requests, features requests, and patches (Open Source Technology Group, 2001b).

All of these tools are straightforward except for the donation system, which deserves a more detailed explanation. The donation system allows site users and projects to receive donations from other projects or site users (Open Source

Figure 3. Project at SourceForge.Net with public tools

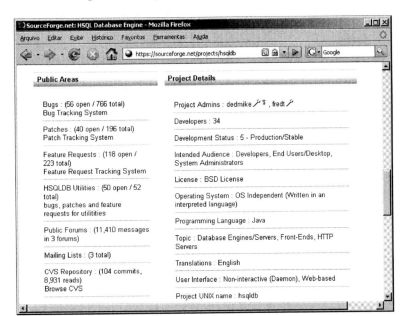

Technology Group, 2001a). The purpose of these donations is to help projects to survive. Specific projects and users may justify why they need donations. Donations are processed by the PayPal (1999) system. Both PayPal and SourceForge.Net charge fees for the donations that go through their system. The PayPal fee may vary from country to country, while SourceForge.Net charges 5% for each donation, with a minimum fee of $1. However, one cannot donate arbitrary quantities. The allowed donation values are $5, $10, $20, $50, $100, and $250 (Open Source Technology Group, 2001a).

Initially, all of these tools are visible to external users (the ones that are not registered as developers in the project). In fact, the default configuration allows even anonymous CVS checkout. However, all of this can be configured by the project administrator. This means that if necessary, some tools can be completely removed from public view. Some developers, for example, prefer to grant anonymous access to the CVS repository together with the first release, but not

before that. Figure 3 shows a project for which tools are visible to external users, and Figure 4 shows a project for which every tool has been hidden from external view (only developers from that project can access the tools).

Apache

Apache Software Foundation (1999) also keeps a portal to host FOSS projects. However, Apache's stance on intellectual property is unique. Projects hosted at Apache must be donated to the Apache Software Foundation. The foundation is then responsible for deciding the project road map (Apache Software Foundation, 2005). We think this is not a disadvantage. It is just a different way of looking at things. By assuming the intellectual property, Apache takes the responsibility for the project. It can legally answer for the project and fight for the project's and the FOSS community's interests. Additionally, the project certainly gains visibility. There are cases where projects became

Figure 4. Project at SourceForge.Net with no public tools

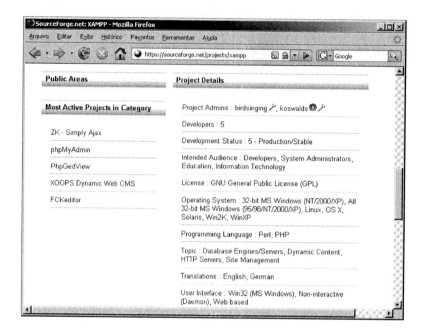

de facto industrial standards, like Apache Web server, Tomcat, and Ant. It is also worth mentioning that the original project owner can still be involved in the development of the project.

Every project hosted by Apache must be submitted through the Apache Incubator Project (Apache Software Foundation, 2002a). The Incubator Project is responsible for informing how the Apache Foundation works, and what paths the project will go through until it is transformed into an official Apache Foundation project (or die before that). Projects currently incubated (together with unsuccessful projects) are listed in Apache Software Foundation (2006).

The registration process of a new project is quite complex. To incubate, the new project must meet the following criteria (Apache Software Foundation, 2002b):

- Be indicated by a member of the Apache Foundation
- Be approved by a sponsor

The sponsor can be one of the following:

- The board of the Apache Software Foundation
- A top-level project (TLP) within the Apache Foundation, where the TLP considers the candidate project to be a suitable subproject
- The Incubator Project management committee

To initiate the hosting request process, it is necessary to submit a proposal that describes the project to the sponsor. There are no fixed items that need to be provided since the Apache Incubator documentation does not specify the level of the project detailing in the proposal or what it must contain.

After being accepted, the Incubator Project management committee is responsible for all decisions regarding the new project. Only after this point does the project receive a CVS account and a Web page under the Incubator Project.

The Apache portal offers a version control system (CVS or Subversion), mailing lists (which

Figure 5. Project incubated at Apache

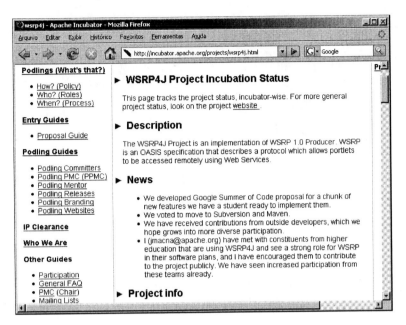

can be exclusively for the project or in conjunction with the Incubator Project), a Web page, documentation (Apache Forrest), bug tracking, and task management.

Figure 5 shows a project incubated at Apache.

Tigris

Tigris (2005) is a FOSS community focused on building tools for collaborative software development that only hosts projects related to that mission. Tigris is hosted at Collabnet (Collabnet Inc., 2006), which is a provider of solutions in software development. Collabnet is currently responsible for hosting OpenOffice and Subversion, two very popular FOSS projects. It is important to notice that hosting a project at Tigris is free while it is not when hosted directly under Collabnet. As Collabnet charges a fee for this service, we do not analyze it here.

Projects hosted at Tigris must fit in one of the following categories:

- **Construction:** Tools for coding, testing, and debugging
- **Deployment:** Tools for software deployment and update
- **Design**
- **Issue tracking**
- **Libraries:** Reusable components
- **Personal use:** Personal projects of Tigris collaborators
- **Processes:** Projects related to software development processes
- **Professional use:** Professional software engineering (courses, certificates, professional practices)
- **Requirements:** Software requirement management tools
- **Software configuration management**
- **Student use:** Student class projects
- **Technical communication**
- **Testing**

The only requirements for the registration of a new project are that it falls into one of the listed

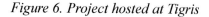

Figure 6. Project hosted at Tigris

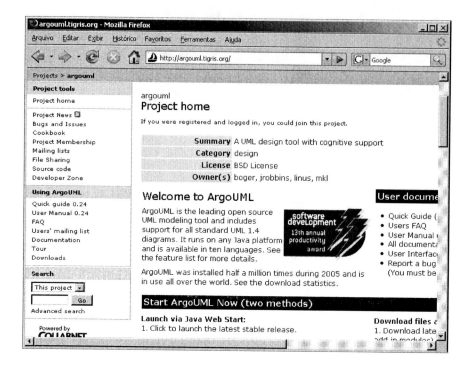

categories and that it is a collaborative software development tool. To register, users must log in and then access the link "start new project."

Tigris offers the following features: mailing lists, task management, bug tracking, a Web page for the project, news, CVS or Subversion, and forums. Figure 6 shows a project hosted at Tigris.

ObjectWeb

ObjectWeb (Object Web Consortium, 2005) is a consortium created in 1999 to promote the development of FOSS. It is maintained by the French National Institute for Research in Computer Science and Control (INRIA) and hosts projects such as Active XML (extensible markup language), C-JDBC, and JoNaS (Java Open Application Server), among others. The consortium is composed of a hierarchy (Cecchet & Hall, 2004):

- The board is comprised of representatives, both individuals and from companies, who are members of the consortium. The board is responsible for the policies, strategies, and direction of the consortium. The executive committee is in charge of the daily operations.

- The College of Architects is comprised of individuals chosen for their expertise and abilities. It is responsible for technically orienting the consortium, leading the development of the ObjectWeb code base, overseeing the evolution and architectural integrity of the code base, and approving new projects.

Projects on ObjectWeb, in the same way as Tigris, must be categorized. The available categories are communications, databases, desktop environments, education, games and entertainment,

Figure 7. A project hosted at ObjectWeb

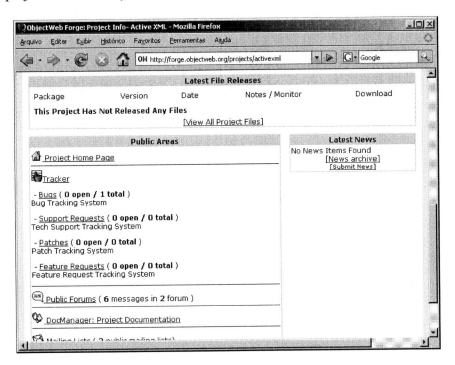

Internet, multimedia, office and business use, other unlisted topics, printing, religion, science and engineering, security, software development, systems, terminals, and text editors.

In order to be hosted at ObjectWeb, the result of the project must be a middleware component that can be reused by a great variety of software platforms and application domains. Besides this, the project members must participate in the discussions of the evolution of the code base of ObjectWeb, participate in the definition of this evolution, and apply the architectural principles and frameworks provided by ObjectWeb to maximize the reuse of the project's source code. The discussions are made through the Web portal mailing list (Object Web Consortium, 2006).

The registration process of new projects in ObjectWeb involves several project descriptions. Detailed information about the project is required, including synergies with the projects already hosted by ObjectWeb, internationalization issues, a description of similar projects, the project team and support, the user community, and the technologies and standards implemented, among others. The list of requirements is much like a formal project submission. Additionally, the LGPL is the recommended license, but a different license may be accepted if you can justify the use of another.

ObjectWeb offers several advantages to the projects it hosts. Among them, we can cite CVS, a Web page, a forum, a mailing list, task management, backup and trackers for bugs, support requests, patches, and feature requests. Figure 7 shows a project hosted at ObjectWeb. In addition, they promote annual events to gather its College of Architects and a demonstration conference that aims at approximating potential users or developers to the projects hosted at ObjectWeb.

Figure 8. Project hosted at Savannah

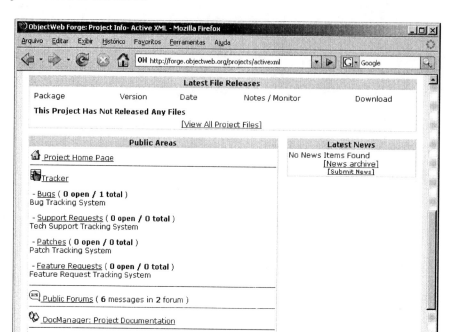

As with SourceForge.Net, ObjectWeb also allows projects to request help from external developers.

Savannah

Different from the other portals we analyzed here, the Savannah portal (Free Software Foundation, 2000b) is focused on free software projects. It hosts projects that fall into one of the following four categories:

- **Software project:** A software project that runs over a free operational system without depending on any non-FOSS; versions to non-free operational systems can be provided as long as there is also a (possibly more complete) version for free systems.
- **Documentation project:** A FOSS documentation project distributed under a free

documentation license (Free Software Foundation, 2000a)
- **Free educational book:** A project to create free educational textbooks distributed under a free documentation license
- **Free Software Foundation/GNU project:** A project approved by the GNU project coordinator, Richard Stallman

Non-GNU projects are hosted at http://savannah.nongnu.org, but the functionalities of both portals are the same.

The registration process of a new project requires a detailed description of the project. If you already have existing source code, you must include a URL to it and a list of libraries used in the source code. This is done to make sure no non-free library is used. However, the existence of source code is not an obligation.

Table 1. Comparisons of the portals

	SourceForge.Net	Apache	Tigris	ObjectWeb	Savannah
Project registration	Depends on approval	Approved by Apache Incubator	Depends on approval	Depends on approval	Depends on approval
Version control	V	V	V	V	V
Customization of tools to avoid external access	V	V	V	V	V
Forum	V	--	V	V	--
Mailing list	V	V	V	V	V
Project Web page	V	V	V	V	V
Issue tracking	V	V	V	V	V
Documentation	V (DocManager)	V (Forrest)	--	V	--
Intellectual property	owner	Apache Foundation	owner	owner	owner
No need to support project after termination	V	--	V	V	?
Task management	V	V	V	V	V
Backup	V	?	?	V	--
Restrictions regarding the project	Categories	Find a sponsor	Collaborative software development tool	Formal submission process	Categories
Restrictions regarding license	OSI-approved license	Apache Software License (ASL)	OSI-approved license	LGPL	GNU GPL compatible

Savannah offers a smaller list of advantages to its users when compared with other portals: CVS, a Web page, a mailing list, bug tracking, support request management, and task management. As with the other portals, it is also possible to hide some of these functionalities from external users. Figure 8 shows the public area of a project hosted at Savannah.

Help from external developers can be achieved by a process similar to the ones at SourceForge. Net and ObjectWeb.

FUTURE TRENDS

Web portals play a major role in the success of free and open source software. Considering the service business around FOSS, we believe that portals will tend to follow ObjectWeb's line of FOSS promotion. We believe portals will increasingly offer more services to users in addition to hosting projects. Such services will probably include dissemination of the FOSS they host and promotion of the approximation of potential users or developers. ObjectWeb nowadays promotes this by organizing architectural meetings with its associates, where people are encouraged to approximate and collaborate. These meetings usually include presentations of newcomer projects so that the community knows what is happening and what the new projects are.

CONCLUSION

In this section, we present a comparison of the analyzed portals. The criteria for this comparison were specified previously. Table 1 summarizes the comparison. A question mark (?) indicates that there was not enough information to evaluate the item.

Regarding support, FOSS development is voluntary. This means that you are not (and should not be) obligated to maintain your code. Some of the portals we have analyzed make this point clear by explicitly saying that you need not offer support for your project, and will not be penalized if or when you discontinue your project. Among these portals are SourceForge.Net, Savannah, and ObjectWeb. The remaining portals do not clearly state this, but they probably follow this criterion since they allow projects to be removed from the portal. The removal is not complete though. All the public information of the project prior to removal remains at the portal (existing file releases, CVS history, forums, etc.).

Intellectual property is another important issue. All of the portals (except for Apache) preserve the intellectual property of the project owner.

Portals that offer the major number of advantages are SourceForge.Net and ObjectWeb. If you pretend to host your project on ObjectWeb, you would have to consider using LGPL. Another issue to be considered in ObjectWeb is the complex registration process. Nevertheless, ObjectWeb has good reputation in academia because of the strong collaboration of INRIA. ObjectWeb requires that a new project have financial supporters in order to guarantee the continuation of the project development. As result, we found at this portal a group of well-known projects, for example, JOnAS (1999), C-JDBC (2002), and eXo Platform (2005).

Regarding automatic backup, Savannah does not have a formal backup policy. However, it does back up the data (including CVS repositories) on a nonregular basis (Savannah-help-private@gnu.org, 2006).

We hope this analysis will be useful for developers who need to choose a Web portal for their projects. We have done this study to find a Web portal to host ParGRES (http://pargres.nacad.ufrj.br/), a free software project supported by FINEP and Itautec (Brazil). After conducting a careful analysis of the hosting options (having also analyzed a Brazilian Web portal), we came to the decision of hosting ParGRES at ObjectWeb. Despite all of the advantages it offers to users, there are mainly two reasons for our decision. First, ObjectWeb opens the possibility of collaboration with a similar project (C-JDBC) that is already hosted there; second, since ParGRES is an academic project, we took the good reputation of ObjectWeb in academia as an important plus to our case. ParGRES has been available at ObjectWeb since November 2005 (http://forge.objectweb.org/projects/pargres/).

ACKNOWLEDGMENT

This research was partially supported by FINEP, Itautec, and CNPq.

REFERENCES

Apache Software Foundation. (1999). *The Apache Software Foundation.* Retrieved January 23, 2006, from http://www.apache.org/

Apache Software Foundation. (2002a). *Apache incubator project.* Retrieved January 28, 2005, from http://incubator.apache.org/

Apache Software Foundation. (2002b). *Incubation policy.* Retrieved January 23, 2006, from http://incubator.apache.org/incubation/Incubation_Policy.html

Apache Software Foundation. (2005). *How the Apache Software Foundation works.* Retrieved January 28, 2005, from http://apache.org/foundation/how-it-works.html#management

Apache Software Foundation. (2006). *Incubated projects.* Retrieved January 23, 2006, from http://incubator.apache.org/projects/

Cecchet, E., & Hall, R. S. (2004). *Objectweb projects life cycle: A practical guide for Object-web projects.* Retrieved February 2, 2005, from https://forge.objectweb.org/register/ObjectWeb-project-lifecycle-v0.3.pdf

C-JDBC. (2002). *Clustered JDBC.* Retrieved June 26, 2006, from http://c-jdbc.objectweb.org/

Collabnet Inc. (2006). *Collabnet: Distributed development on demand.* Retrieved January 23, 2006, from http://www.collab.net/

DiBona, C., Stone, M., & Cooper, D. (2005). *Open sources 2.0: The continuing evolution.* Sebastopol, CA: O'Reilly.

eXo Platform. (2005). *eXo platform.* Retrieved June 26, 2006, from http://c-jdbc.objectweb.org/

Free Software Foundation. (1996). *GNU project.* Retrieved January 5, 2005, from http://www.gnu.org/

Free Software Foundation. (2000a). *Free software and free manuals.* Retrieved February 14, 2006, from http://www.gnu.org/philosophy/free-doc.html

Free Software Foundation. (2000b). *Savannah.* Retrieved February 1, 2005, from http://savannah.gnu.org/

Java Open Application Server (JOnAS). (1999). *Java open application server.* Retrieved June 26, 2006, from http://jonas.objectweb.org/

Object Web Consortium. (1999). *Objectweb forge: Project information.* Retrieved January 17, 2006, from http://forge.objectweb.org/register/projectinfo.php

Object Web Consortium. (2005). *Objectweb open source middleware.* Retrieved January 31, 2005, from http://www.objectweb.org/

Object Web Consortium. (2006). *Projects life cycle.* Retrieved June 29, 2006, from https://wiki.objectweb.org/Wiki.jsp?page=ProjectsLifeCycle

Open Source Initiative. (2006). *The approved licenses.* Retrieved January 17, 2006, from http://opensource.org/licenses/

Open Source Technology Group. (2001a). *Sourceforge.Net: Donation system* (Document d02). Retrieved January 23, 2006, from http://sourceforge.net/docs/D02/en/

Open Source Technology Group. (2001b). *Sourceforge.Net: Service listing* (Document b02). Retrieved January 23, 2006, from http://sourceforge.net/docs/B02/en/

Open Source Technology Group. (2002). *Sourceforge.Net project hosting requirements and the project registration process.* Retrieved January 21, 2006, from http://sourceforge.net/docman/display_doc.php?docid=14027&group_id=1

Open Source Technology Group. (2005). *Sourceforge.Net.* Retrieved January 5, 2005, from http://www.sourceforge.net

PayPal. (1999). *Paypal.* Retrieved January 23, 2006, from http://www.paypal.com/us/

Perens, B. (1997). *The open source definition.* Retrieved January 17, 2006, from http://www.opensource.org/docs/definition.php

Savannah-help-private@gnu.org. (2006). *Comparison of open source Web portals.*

Stallman, R. M. (2002). Why free software is better than open source. In J. Gay (Ed.), *Free software, free society: Selected essays of Richard M. Stallman* (chap. 6, pp. 55-60). Boston: GNU Press.

Tigris. (2005). *Tigris open source software engineering.* Retrieved January 31, 2005, from http://www.tigris.org/

Wikipedia. (2006). Retrieved from http://www.wikipedia.org

KEY TERMS

Forum: A discussion board on the Internet (Wikipedia, 2006).

Intellectual Property: Umbrella term used to refer to the object of a variety of laws, including patent law, copyright law, trademark law, trade-secret law, industrial design law, and potentially others (Wikipedia, 2006).

Issue Tracking: Also known as bug tracking, it is a system designed to manage change requests of a software. It can also be used to manage bugs.

Mailing List: A collection of names and addresses used by an individual or an organization to send material to multiple recipients. The term is often extended to include the people subscribed to such a list, so the group of subscribers is referred to as the mailing list (Wikipedia, 2006).

Task Management: Software capable of managing lists of pending tasks.

Version Control System: A system that tracks and stores changes on files (source code, binary files, and text files, among others). Such systems are able to retrieve old versions of a given artifact as long as such version has been stored some time before in the system.

Web Portal: A site on the World Wide Web that typically provides personalized capabilities to visitors (Wikipedia, 2006).

Chapter XVIII
Curious Exceptions?
Open Source Software and "Open" Technology

Alessandro Nuvolari
Eindhoven University of Technology, The Netherlands

Francesco Rullani
Sant'Anna School of Advanced Studies, Italy

ABSTRACT

The aim of this chapter is to explore the differences and commonalities between open source software and other cases of open technology. The concept of open technology is used here to indicate various models of innovation based on the participation of a wide range of different actors who freely share the innovations they have produced. The chapter begins with a review of the problems connected to the production of public goods and explains why open source software seems to be a "curious exception" for traditional economic reasoning. Then it describes the successful operation of similar models of innovation (open technology) in other technological fields. The third section investigates the literature in relation to three fundamental issues in the current open source research agenda, namely, developers' motivations, performance, and sustainability of the model. Finally, the fourth section provides a final comparison between open source software and the other cases of open technology.

INTRODUCTION

Over the last 10 years, open source software development has increasingly attracted the attention of scholars in the fields of economics, management, and social sciences in general (for sociological contributions, see Himanen, Torvalds, & Castells, 2001; Weber, 2004; see Maurer & Scotchmer, 2006, for an account of the phenomenon from the economist's perspective).

Although the significance of the software industry in modern economic systems can partially explain the increasing number of research contributions in this area, it is clear that the chief reason behind this growing interest is the fact that open source software development seems to represent a form of innovation process that challenges many facets of the current conventional wisdom concerning the generation of innovations in market economies (Lerner & Tirole, 2001).

Traditionally, economists have considered technological knowledge as a public good, that is, a good endowed with two fundamental features: (a) nonrivalry and (b) nonexcludability. Nonrivalry states that when one actor consumes or uses the good, this does not prevent other actors from consuming or using it. Obviously, this does not hold for standard economic goods: If Paul eats the apple, it is clear that Nathan cannot eat the same apple. On the other hand, both Paul and Nathan can breathe the fresh air of the park. Nonexcludability refers to the fact that when technological knowledge is in the public domain, it is no longer possible to prevent other actors from using it. Again, while Paul may force Nathan to pay for the apple, he cannot (legally) prevent Nathan from breathing the fresh air of the park. The traditional economist's viewpoint contends that market economies are characterized by a systematic underprovision of public goods as their production is, due to the two properties described above, not profitable for private firms. In these circumstances, the standard prescription is that governments should intervene, using tax revenues to supply directly the appropriate quantity of public goods. This reasoning is at the heart of the argument that is commonly used in making the case for the public support of scientific research (Nelson, 1959). It is worth noting that, historically, the allocation of public resources for the production of scientific knowledge has been organized around a rather particular institutional arrangement ("open science") capable of producing both incentives to create new knowledge and the public disclosure of scientific finding (Dasgupta & David, 1994).

Public funding, however, is not the only answer. Another solution put forward by the literature is based on the idea of inducing private firms to invest in the production of technological knowledge by means of an *artificial* system of property rights (Arrow, 1962). The most common example, in this respect, is the patent system. A patent assigns temporarily to its inventor the complete control of the new technological knowledge discovered.

The rationale for this institutional device is straightforward: The prospect of the commercial exploitation of this temporary monopoly right will induce private firms to invest resources in inventive activities, that is, in the production of new technological knowledge.

In this context, open source software represents a case of the production of new technological knowledge (high-quality computer programs) carried out by individuals without any direct attempt of "appropriating" the related economic returns. Clearly, all this is at odds with the conventional wisdom summarized above.

Recent research has, however, shown that the innovation process characterizing open source software is not an isolated case. Instead, at least since the industrial revolution, similar types of innovation processes have been adopted in other industries in different periods. Following Foray (2004), we will refer to these episodes as cases of "open technology" in order to stress their similarity with open source software. It is worth warning the reader that in the literature, a variety of other terms and definitions such as "collective invention" or "community based innovation" are frequently used.[1] There is a growing awareness that these cases do not represent just "curious exceptions" to the traditional models of innovation based on public funding or on commercial exploitation by means of exclusive property rights. The aim of this chapter is to provide a compact overview of this literature and to compare these cases of open technology with open source software. Our belief is that this broader perspective can enrich our understanding of open source software.

BACKGROUND

Open Technology: A Neglected Model of Innovation

In a seminal paper, Robert C. Allen (1983) presented a detailed case study of technical change in the iron industry of Cleveland (United Kingdom) during the period of 1850 to 1870. According to Allen, the Cleveland iron industry was characterized by a particular model of innovation, which he labeled collective invention. In the Cleveland district, iron producers freely disclosed to their competitors technical information concerning the construction details and performance of the blast furnaces they had installed. Information was normally shared both through formal (presentations at meetings of engineering societies, publication of design details in engineering journals, etc.) and informal channels (visits to plants, conversations, etc.). Additionally, new technical knowledge was not protected using patents so that competing firms could freely make use of the released information when they had to construct a new blast furnace. The consequence of this process of information sharing was that the blast furnaces of the district increased their performance very rapidly. Allen noted three essential conditions at the basis of the emergence of the collective-invention regime. The first condition refers to the nature of the technology. In the period considered, there was no consolidated understanding of the working of a blast furnace. The best engineers could do when designing a new blast furnace was to come up with some design guidelines on the basis of previous experiences. Obviously, the sharing of information related to the performance of a large number of furnaces allowed engineers to rely on a wider pool of information in their extrapolations, leading to a more rapid rate of technological progress. Second, blast furnaces were designed by independent consulting engineers who were normally employed on a one-off basis. In this context, the most talented engineers had a strong incentive to disseminate the successful design novelties they had introduced in order to enhance their professional reputation and improve their career prospects. Third, iron producers were often also owners of iron mines. As a consequence, improvements in the efficiency of blast furnaces would have led to an enhancement in the value of

the iron deposits of the region. Thus, there was a keen interest in the improvement of the average performance of blast furnaces, as only improvements in the average performance would have influenced the value of iron deposits.

Following Allen's work, other scholars have pointed out the existence of a collective-invention regime in other industries. In a recent study, Nuvolari (2004) has shown that the three conditions of Allen's model of collective invention were also at work in the Cornish community of steam engineers during the first half of the 19th century. This case is particularly interesting because some evidence suggests that the emergence of the collective-invention regime was triggered by a widespread dissatisfaction toward the traditional model of innovation based on patents (in particular, James Watt's patent of 1769).

Other cases of collective invention have been noted in the historical literature, for example, the Western steamboat (Hunter, 1949) and the Lyon silk industry in the 19th century (Foray & Hilaire-Perez, 2000). Scholars have also noted similar knowledge-sharing episodes in several contemporary high-technology districts (most prominently in Silicon Valley; see Saxenian, 1994). However, it is worth noting that in these cases, the very dense knowledge flows between firms may be due to user-producer interactions (Lundvall, 1988) or episodes of know-how trade between engineers (von Hippel, 1987) rather than to the existence of a collective-invention regime in Allen's sense.[2]

A related stream of literature has highlighted the growing importance of user communities as sources of innovation (Franke & Shah, 2003; Shah, 2005). The starting point of these investigations is the observation that in many fields, a sizable share of inventions is due to the users of a specific product and not to its manufacturers (von Hippel, 1988, 2005). One interesting feature of the innovation processes centered on users is that they are often based on very intense knowledge exchanges in the context of specific communities.

Again, within these communities, inventions are normally released in the public domain, and there are no attempts of exploiting them by means of exclusive property rights. Research in this field (see Franke & Shah, 2003, for a detailed study of four user communities in sport equipment) has noted a variety of motivations for the emergence of this type of behavior. First, users belonging to these communities have a keen interest in the performance level of the product. Hence, as in the case of collective invention, the community seems to be characterized by a widespread belief that a mutual cooperative attitude toward inventive activities will enhance the rate of innovation. Second, the social structure of these communities seems to favor the emergence of an ethos prescribing reciprocity and mutual aid.

Apart from the field of sports equipment, in which this type of (user) community-based innovation seems to be prominent, research has identified the existence of this particular model in other industries, such as geographic information systems, astronomic instruments, and early computer and automobile users (see Maurer & Scotchmer, 2006; Shah, 2005).

MAIN FOCUS OF THE CHAPTER

Open Source Software: A Synthesis of Recent Research

One of the main issues to be explored in order to understand the existence and the success of open source software can be stated as follows: Why are developers willing to develop open source software if the typical licenses of this regime[3] prevent them to extract any direct monetary gain from the diffusion of their work? In other words, a study of the open source software phenomenon requires an understanding of developers' motivations.

In order to describe the structure of the landscape of developers' motivations, a first useful distinction has been put forward by Lakhani and

Wolf (2005). In this chapter, the authors, following the work by Deci and Ryan (1985), Amabile (1996), and Lindenberg (2001), classify the motivations driving developers' participation into two main groups: intrinsic and extrinsic motivations. When the development activity is undertaken because it enhances developers' utility directly, providing a gain in terms of fun or creativity fulfillment, or a feeling of identity and belongingness to a group, the underlying incentives are said to be intrinsic because the actions they trigger have an intrinsic value for the agent. On the contrary, when the production of code is undertaken instrumentally to reach other goals, such as increasing wages, enhancing the agent's reputation on the job market, or fulfilling specific needs the existing software cannot satisfy, the motivations behind the action are defined as extrinsic because the increase in the individual utility is not due to action itself, but to its consequences.

Each one of the two regions of the developers' motivational landscape can be further structured to isolate the different mechanisms at work in each field. The FLOSS (free/libre open source software) surveys developed by Ghosh, Krieger, Glott, and Robles (2002; answered by 2,784 developers) and by David, Waterman, and Arora (2003; answered by 1,588 developers) offer a finer grain point of view on the motivational landscape. In both the surveys, the most popular answers to questions related to developers' incentives span from "I thought we should all be free to modify the software we use" to "As a user of free software, I wanted to give something back to the community," "I saw it as a way to become a better programmer," "to participate in a new form of cooperation," "to improve OS/FS [open source/free software] products of other developers," and "to improve my job opportunities." Thus, a series of different intrinsic and extrinsic motivations emerges.

Lakhani and Wolf (2005; see also Lakhani, Wolf, Bates, & DiBona, 2002), using survey data collected from 684 developers working on

287 open source projects, were able, by means of a cluster analysis exercise, to identify a number of archetypical cases of open source software developers. They find four clusters, each one approximately the same size as the others. For the members of the largest cluster, a personal sense of creativity and fun are crucial determinants of their contribution to the open source movement. Two other elements emerge as important in this group: the learning opportunities the community offers them, and the possibility to enhance their career through the diffusion of the code they produce. The population of the second cluster resembles the user communities described in the previous section: Skilled developers with specific needs the existing software cannot fulfill are pushed to create the program answering their needs (i.e., lead users). The third cluster is instead composed of paid developers who receive a wage connected to their production of open source products. Eventually, the fourth cluster gathers together individuals strongly committed to the community, moved by the willingness to reciprocate the received help and code, and having a strong ideological position in favor of the open source movement (e.g., believing that code should be open and participating in order to beat proprietary software).

From the empirical studies just described, some subsets of the two main motivation sets emerge. On the one hand, intrinsic motivation can have a psychological nature when it takes the form of fun or creativity fulfillment (Lakhani & Wolf, 2005; Torvalds & Diamond, 2001), or a social nature when it is a product of the interaction between community members and between them and the whole social body of the community, that is, its culture, its shared rules, its ideology, its debate, and so on. In such a thick social environment, developers are willing to participate because they identify with the community, they belong to the hacker culture and feel the need to follow its rules, they believe in the common enterprise they are

undertaking, or simply because they care about their status or reputation in the community and are sensitive to peers' regard (Bagozzi & Dholakia, 2006; Dalle & David, 2005; Dalle, David, Ghosh, & Wolak, 2004; Hertel, Niedner, & Hermann, 2003; Himanen et al., 2001; Raymond, 1998; Weber, 2000, 2004; Zeitlyn, 2003). On the other hand, extrinsic motivations can be diversified into subcategories such as career concerns (Lerner & Tirole, 2002), when developers' production of code and diffusion is determined by the willingness to be recognized in the job market as valuable programmers; own use, when the open source community is conceived as a user community à la von Hippel (Hertel et al., 2003; Jeppesen & Frederiksen, 2006; Lakhani & von Hippel, 2003; von Hippel, 2001); and paid contributions (Roberts, Hann, & Slaughter, 2006), when developers are employees of firms active in the open source software environment.

A further element emerged from the cluster analysis by Lakhani and Wolf (2005): learning (von Hippel & von Krogh, 2003). Developers are often driven by the desire to improve their skills and perceive the community as a social environment where they can get help in solving problems by studying collectively new solutions and finding new challenges. Learning can be considered both an intrinsic and an extrinsic incentive, and it cannot be placed easily in one of the subsets defined above. It certainly has an individual and psychological nature, but since it develops alongside the agents' interaction, its nature is much broader. Once the open source community is conceived as a "community of practice" or an "epistemic community" (Cohendet, Creplet, & Dupouët, 2001; Lin, 2004), where the body of knowledge of the whole community interacts and coevolves with each individual's knowledge, learning can be clearly identified as a social process. The same blurred result can be found when conceiving learning as an extrinsic incentive: It can be an instrument for most of the goals typical of the extrinsic motivations described above. Thus, it should be considered as a third group, intersecting all the previous sets. Figure 1 shows the structure of the motivations set as drawn from the quoted literature.

The description of the community proposed above is mainly focused on developers as individuals. However, other subjects are active in the open source environment: firms. Even in an open environment as open source, it is possible for firms to generate profits. The most famous example is given by firms assembling and distribution a ready-to-install version of the GNU/Linux operating system, like Red Hat or Novell. The literature has highlighted several ways by which firms can create value from their participation in the open source movement, but has also shown the instrumental use of their adherence to the community norms and ideology (Rossi & Bonaccorsi, 2005). In other words, as long as incentives are concerned, firms have a much narrower set of incentives, being motivated, as expected, by profit maximization. However, even if the participation of the firms in the open source community is only

instrumental, they play an increasingly important role in the open source scene. As we will see in the following sections, they can be fundamental sources of code or related services the community is not willing to produce.

So far, we have given a brief account of the motivations sustaining developers' production of open source software. However, even if developers can decide to dedicate a high amount of effort and time to the production of code, this does not mean that open source represents a successful model of innovation. Thus, our next step is to focus on the performance of open source software as an innovation model.

The first thing to be noticed is that the distribution of open source projects in terms of the main performance indicators—the number of developers and forum or mailing-list discussions (Krishnamurthy, 2002), downloads (Healy & Schussman, 2003), and CVS (Concurrent Version System) commits and file releases (Giuri, Ploner, Rullani, & Torrisi, 2005)—is extremely skewed. Most of the projects remain small individual enterprises without a serious impact on the

Figure 1. Structure of developers' motivational landscape

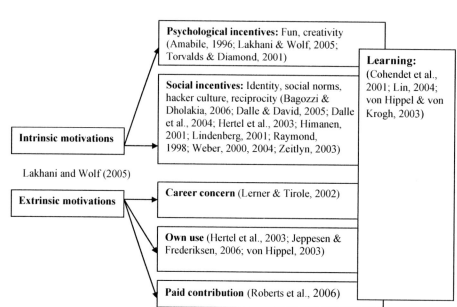

landscape of the software industry. However, as argued by David and Rullani (2006), open source software should be regarded as a "dissipative" system, burning more resources than those used to produce the actual results. This characteristic is typical of self-organized social structures, where individuals choose on a voluntary basis how much effort to devote to what task and when. In order for the whole system to produce an outcome, several combinations of the available resources have to be worked out before the valuable ones can be selected by the environment.

Thus, on the one hand, the disproportion between the inactive projects and the successful ones characterizes the open source model as dissipative rather than an unsuccessful model of innovation. On the other hand, the same argument calls for a definition of the drivers of open source projects performance in order to be able to reduce the gap between the mobilized resources and those that are actually used.

A first result along this line of inquiry states that projects adopting a restrictive license like the GPL (General Public License) tend to have lower performance (Comino, Manenti, & Parisi, 2005; Fershtman & Gandal, 2004; see also Lerner & Tirole, 2005). This result could be due to a detrimental impact of the excessive openness of the GPL projects, which may be unable, for example, to attract firms and sponsored developers. A hybrid model of innovation, where the adopted license scheme is able to create a synergy between the community and other economic actors, should be then considered as a valuable configuration (Bonaccorsi, Giannangeli, & Rossi, 2006). A second result is that the division of labor has a significant positive impact on project performance. However, the variety and the level of members' skill sets (Giuri et al., 2005) and the costs connected to the coordination of many developers (Comino et al., 2005) have to be taken into account in order to avoid a net negative effect. Modularity at the level of the code structure has been analyzed by Baldwin and Clark (2006), who find that a modular

architecture is able to attract more voluntary effort and reduce free riding. Applying an ecological perspective, Grewal, Lilien, and Mallapragada (2006) look at the developers' membership in different projects to draw a network of relationships between projects and developers. They show that projects with a central position in the network are more likely to exhibit high technical performance, but the network is not so crucial in determining the commercial success (i.e., number of downloads) of the produced software.

Having established what moves developers and what the drivers of the open source software innovative performance are, a last question regards the possibility to sustain such a structure over time. The contributions moving in this direction are scarce, and there is need for further research. A first contribution has been given by Gambardella and Hall (in press). The authors show that a coordination device is needed to assure the stability of collaboration. The adoption of the GPL can be thought of as such a mechanism, preventing any developer joining the project after the founder to adopt opportunistic behavior. This argument points out an interesting trade-off between performance and sustainability: Less restrictive licenses can induce higher performance, but can undermine the sustainability of the community. A second point has been made by David and Rullani (2006), showing that developers undertaking their activity on the SourceForge.net (http://sourceforge.net/) platform during the period of 2001 to 2002 exhibit a robust, nontransient tendency to participate in existing projects and also to create new projects. Sustainability, then, can be conceived at least as a valuable working hypothesis.

Open Technology and Open Source Software: A Comparison

Various researchers have noted the existence of important parallels between the model of open technology discussed previously and the findings emerging from ongoing studies of the open source communities (see, among others, Foray, 2004; Nuvolari, 2005). In a nutshell, these are the main points of interest:

a. Both collective-invention regimes and the open source movement seem to emerge thanks to a perceived dissatisfaction toward the innovative performance delivered by traditional regimes based on exclusive property rights.

b. Case studies of collective invention and user communities seem generally characterized by remarkable performances in terms of rates of innovation. The same remarkable innovative performance has characterized some open source software projects, at least since the 1990s, when GNU/Linux was born.

c. However, only a restricted number of open source software projects are successful. Similarly, only few innovations coming from the users are really valuable, as well as only few contributions added to the common pool of collective inventions are really improving the performance of the sector. Thus, the models share the dissipative property described for the open source model of innovation.

d. Both collective invention and a number of open source software projects are characterized by high levels of complexity and uncertainty. In these conditions, a model of innovation based on knowledge sharing, cooperation, and continuous feedback permits the timely identification of the most promising lines of development.

e. Cases of collective invention, user-based innovation models, and open source software are forms of innovation processes involving heterogeneous sets of actors (in particular, engineers, lead users, and developers with different skills and talents, and firms) organized into communities.

f. Collective invention, open source software, and user communities rely on a complex set of motivational drivers, spanning from economic incentives, dissatisfaction toward tight intellectual property-rights regimes, psychological and social motives, and so on. Even if the open source software and the other examples of open innovation seem to rely on different compositions of the aforementioned motivational factors, it might well be that this plurality of motives represents one of the fundamental ingredients for sustaining both open source and open-technology regimes.

CONCLUSION

The Core of the Difference and the Challenges to Sustainability

The main difference between the three regimes of innovation can be found in the relationship between the communities of innovative agents and the involved firms. In a collective-invention regime, firms strategically suspend appropriation of the produced knowledge in order to deal with technological problems that an individual firm could not handle. In this sense, firms are the fundamental actors of collective-invention regimes. Accordingly, these regimes usually disappear when the collective effort to overcome the radical uncertainty in the technological space is not necessary anymore (i.e., when a specific technological trajectory or paradigm emerges; Dosi, 1982), and each firm is willing to return to the proprietary regime that will assure higher individual profits. On the contrary, the nexus between manufacturers and users is much tighter in user communities. Users innovate around the products of the firms, which in turn try to sustain users' involvement. Sometimes, these communities are originated directly by the firms, and other times they emerge spontaneously through users' interaction. In the open source software case, the leading role is instead played by users and developers, and firms are mainly active in those spaces that the community does not or cannot reach. Firms have to adapt to the rules of the community and do not directly control on the product (Dahlander & Magnusson, 2005; Shah, 2006). Thus, the basic difference between the three models of innovation is in the balance between the roles of firms and of users and developers.

These considerations shed new light on the relative sustainability of these regimes. Collective inventions can exist only as long as firms do not profit enough from a traditional proprietary regime; this happens mostly in conditions of radical technological uncertainty (emerging phases of a novel technological paradigm). Instead, user communities and open source software seem to be characterized by different sustainability conditions (Osterloh & Rota, 2005). The sustainability of the former depends directly on the ability of communities and firms to involve individual users and keep their participation at a certain level; in the case of the latter, several factors, still to be fully identified, can induce a decay of the phenomenon or strengthen its sustainability. The foregoing discussion on the trade-offs between open source sustainability, performance, and level of openness (as defined by the license) clearly bears out this point.

REFERENCES

Allen, R. C. (1983). Collective invention. *Journal of Economic Behaviour and Organization, 4*, 1-24.

Amabile, T. M. (1996). *Creativity in context.* Boulder, CO: Westview Press.

Arrow, K. (1962). Economic welfare and the allocation of resources for invention. In R. Nelson (Ed.), *The rate and direction of inventive activity: Economic and social factors.* Princeton, NJ: Princeton University Press.

Bagozzi, R. P., & Dholakia, U. M. (2006). Open source software user communities: A study of participation in Linux user groups. *Management Science, 52*(7), 1099-1115.

Baldwin, C. Y., & Clark, K. B. (2006). The architecture of participation: Does code architecture mitigate free riding in the open source development model? *Management Science, 52*(7), 1116-1127.

Bonaccorsi, A., Giannangeli, S., & Rossi, C. (2006). Entry strategies under competing standards: Hybrid business models in the open source software industry. *Management Science, 52*(7), 1085-1098.

Chesbrough, H., Vanhaverbake, W., & West, J. (Eds.). (2006). *Open innovation: Researching a new paradigm.* Oxford, UK: Oxford University Press.

Cohendet, P., Creplet, F., & Dupouët, O. (2001, June). *Communities of practice and epistemic communities: A renewed approach of organizational learning within the firm.* Paper presented at the Workshop on Economics and Heterogeneous Interacting Agents, Marseille, France.

Comino, S., Manenti, F. M., & Parisi, M. L. (2005). *From planning to mature: On the determinants of open source take off* (Discussion Paper 2005-17). Trento, Italy: Trento University.

Dahlander, L., & Magnusson, M. (2005). Relationships between open source software companies and communities: Observations from Nordic firms. *Research Policy, 34*(4), 481-493.

Dalle, J.-M., & David, P. A. (2005). The allocation of software development resources in "open source" production mode. In J. Feller et al. (Eds.), *Making sense of the bazaar: Perspectives on open source and free software.* Cambridge, MA: MIT Press.

Dalle, J.-M., David, P. A., Ghosh, R. A., & Wolak, F. A. (2004, June). *Free & open source software developers and "the economy of regard": Participation and code-signing in the modules of the Linux kernel.* Paper presented at the Oxford Workshop on Libre Source, Oxford, UK.

Dasgupta, P., & David, P. A. (1994). Towards a new economics of science. *Research Policy, 23,* 487-521.

David, P. A., & Rullani, F. (2006, June). *Open source software development dynamics: Project joining and new project generation on Source-Forge.* Paper presented at the Meetings of the International Schumpeter Society (ISS), Sophia-Antipolis, France.

David, P. A., Waterman, A., & Arora, S. (2003). *The free/libre/open source software survey for 2003* (Preliminary draft). Unpublished manuscript.

Deci, E. L, & Ryan, R. M. (1985). *Intrinsic motivation and self-determination in human behavior.* New York: Plenum Press.

Dosi, G. (1982). Technological paradigms and technological trajectories: A suggested interpretation of the determinants and directions of technical change. *Research Policy, 11*(3), 147-162.

Fershtman, C., & Gandal, N. (2004). *The determinants of output per contributor in open source projects: An empirical examination* (Discussion Paper 4329). London: CEPR.

Foray, D. (2004). *The economics of knowledge.* Cambridge, MA: MIT Press.

Foray, D., & Hilaire-Perez, L. (2000, May). *The economics of open technology: Collective invention and individual claims in the "fabrique lyonnaise" during the old regime.* Paper presented

at the conference in honour of Paul A. David, Turin, Italy.

Franke, N., & Shah, S. (2003). How communities support inventive activities: An exploration of assistance and sharing among end-users. *Research Policy, 32,* 157-178.

Gambardella, A., & Hall, B. (in press). Proprietary vs public domain licensing of software and research products. *Research Policy.*

Ghosh, R. A. (1998). Cooking pot markets: An economic model for the trade in free goods and services on the Internet. *First Monday, 3*(3). Retrieved March 22, 2007, from http://www.firstmonday.org/issues/issue3_3/ghosh

Ghosh, R. A., Krieger, B., Glott, R., & Robles, G. (2002). *Free/libre and open source software. Part IV: Survey of developers.* International Institute of Infonomics, Berlecom Research GmbH.

Giuri, P., Ploner, M., Rullani, F., & Torrisi, S. (2005, September). *Skills, division of labor and performance in collective inventions: Evidence from the open source software.* Paper presented at the EARIE Conference, Porto, Portugal.

Grewal, R., Lilien, G. L., & Mallapragada, G. (2006). Location, location, location: How network embeddedness affects project success in open source systems. *Management Science, 52*(7), 1043-1056.

Healy, K., & Schussman, A. (2003). *The ecology of open source software development* (Working Paper). AZ: Department of Sociology, University of Arizona.

Hertel, G., Niedner, S., & Hermann, S. (2003). Motivation of software developers in open source projects: An Internet-based survey of contributors to the Linux kernel. *Research Policy, 32*(7), 1159-1177.

Himanen, P., Torvalds, L., & Castells, M. (2001). *The hacker ethic and the spirit of the information age.* London: Secker & Warburg.

Hunter, L. (1949). *Steamboats on the Western rivers: An economic and technological history.* Cambridge, MA: Harvard University Press.

Jeppesen, L. B., & Frederiksen, L. (2006). Why firm-established user communities work for innovation? The personal attributes of innovative users in the case of computer-controlled music instruments. *Organization Science, 17*(1), 45-64.

Krishnamurthy, S. (2002). Cave or community? An empirical examination of 100 mature open source projects. *First Monday, 7*(6). Retrieved March 22, 2007, from http://www.firstmonday.dk/issues/issue7_6/krishnamurthy

Lakhani, K. R., & von Hippel, E. (2003). How open source software works: "Free" developer-to-developer assistance. *Research Policy, 32.*

Lakhani, K. R., & Wolf. (2005). Why hackers do what they do: Understanding motivations and effort in free/open source software projects. In J. Feller, B. Fitzgerald, S. Hissam, & K. R. Lakhani (Eds.), *Perspectives on free and open source software.* Cambridge, MA: MIT Press.

Lakhani, K. R., Wolf, R. G., Bates, J., & DiBona, C. (2002). *The Boston Consulting Group hacker survey* (Release 0.73). Author.

Lerner, J., & Tirole, J. (2001). The open source movement: Key-research question. *European Economic Review, 45,* 819-826.

Lerner, J., & Tirole, J. (2002). Some simple economics of open source. *The Journal of Industrial Economics, L*(2), 197-234.

Lerner, J., & Tirole, J. (2005). The scope of open source licensing. *Journal of Law, Economics, and Organization, 21,* 20-56.

Lin, Y. (2004). Contextualising knowledge-making in Linux user groups. *First Monday, 9*(11).

Retrieved March 22, 2007, from http://www.firstmonday.org/issues/issue9_11/lin/index.html

Lindenberg, S. (2001). Intrinsic motivation in a new light. *Kyklos, 54*(2/3), 317-342.

Lundvall, B.-Å. (1988). Innovation as an interactive process: From user-producer interaction to the national system of innovation. In G. Dosi, C. Freeman, R. Nelson, G. Silverberg, & L. Soete (Eds.), *Technical change and economic theory.* London: Pinter.

Maurer, S. M., & Scotchmer, S. (2006). *Open source software: The new intellectual property paradigm* (NBER Working Paper 12148).

Nelson, R. (1959). The simple economics of basic scientific research. *Journal of Political Economy, 67,* 297-306.

Nuvolari, A. (2004). Collective invention during the British industrial revolution: The case of the Cornish pumping engine. *Cambridge Journal of Economics, 28,* 347-363.

Nuvolari, A. (2005). Open source software development: Some historical perspectives. *First Monday, 10*(10). Retrieved March 22, 2007, from http://www.firstmonday.org/issues/issue10_10/nuvolari/index.html

Osterloh, M., & Rota, S. G. (2005). *Open source software development—Just another case of collective invention?* (CREMA Working Paper 2005-08).

Raymond, E. (1998). Homesteading the Noosphere. *First Monday, 3*(10). Retrieved March 22, 2007, from http://www.firstmonday.org/issues/issue3_10/raymond/index.html

Roberts, J. A., Hann, I., & Slaughter, S. A. (2006). Understanding the motivations, participation, and performance of open source software developers: A longitudinal study of the Apache projects. *Management Science, 52*(7), 984-999.

Rossi, C., & Bonaccorsi, A. (2005). Intrinsic vs. extrinsic incentives in profit-oriented firms supplying open source products and services. *First Monday, 10*(5). Retrieved March 22, 2007, from http://www.firstmonday.org/issues/issue10_5/rossi

Saxenian, A. (1994). *Regional advantage: Culture and competition in Silicon Valley and in Route 128.* Cambridge, MA: Harvard University Press.

Shah, S. (2005). Open beyond software. In C. DiBona, D. Cooper, & M. Stone (Eds.), *Open sources 2.0: The continuing evolution.* Sebastopol, CA: O'Reilly.

Shah, S. (2006). Motivation, governance, and the viability of hybrid forms in open source software development. *Management Science, 52*(7), 1000-1014.

Torvalds, L., & Diamond, D. (2001). *Just for fun: The story of an accidental revolutionary.* New York: Texere.

Von Hippel, E. (1987). Cooperation between rivals: Informal "know how" trading. *Research Policy, 16*, 291-302.

Von Hippel, E. (1988). *The sources of innovation.* Oxford, UK: Oxford University Press.

Von Hippel, E. (2001). Innovation by user communities: Learning from open source software. *MIT Sloan Management Review, 42*(4), 82-86.

Von Hippel, E. (2005). *Democratizing innovation.* Cambridge, MA: MIT Press.

Von Hippel, E., & von Krogh, G. (2003). Open source software and the "private-collective" innovation model: Issues for organization science. *Organization Science, 14*(2), 209-223.

Weber, S. (2000). *The political economy of open source software* (BRIE Working Paper 140).

Weber, S. (2004). *The success of open source.* Cambridge, MA: Harvard University Press.

Zeitlyn, D. (2003). Gift economies in the development of open source software: Anthropological reflections. *Research Policy, 32*, 1287-1291.

KEY TERMS

Collective Invention: An innovation model in which private firms engaged in the production or use of a specific good freely share one another's inventions and other pertinent technical information.

Dissipation: We call dissipation an innovation model that mobilizes (or "burns") more resources than those actually used to produce the outcome. Dissipation is typical of self-organizing and explorative organizations.

Intrinsic/Extrinsic Motivations: When an activity is undertaken because it enhances agents' utility directly, the underlying incentives are intrinsic because the actions they trigger have an intrinsic value for the agent. On the contrary, when an action is undertaken instrumentally to reach other goals, the motivations behind the action are defined as extrinsic because the increase of the individual's utility is not due to action itself, but to its consequences.

Sustainability: We call sustainable an innovation model that re-creates over time the premises for its own reproduction, that is, if it is endowed with a mechanism able to re-create incentives for the participants to continually invest in innovation. In this sense, the patent system as well as the public-funded research system can be conceived as sustainable.

User Community: An innovation model where a community of users of a particular product are the main source of innovation and where innovations are normally freely shared within the community.

ENDNOTES

[1] Another term that is becoming increasingly popular in the management literature is "open innovation" (see Chesbrough, Vanhaverbeke, & West, 2006). The concept of open innovation refers to the fact that firms are increasingly making use of external sources of knowledge in their innovation processes. Clearly, this is somewhat related to the phenomenon of open technology sketched above as firms, in order to gain access to these external sources, are frequently required to adopt a more relaxed attitude toward the appropriation of their inventions. In this chapter, we will not deal with the literature on open innovation.

[2] In know-how trading, information is typically exchanged by engineers belonging to competing firms on a bilateral basis. In collective-invention regimes, all the competing firms have free access to the potentially proprietary know-how.

[3] The possibility for subsequent developers to change the open regime established by the initial choice of an open source license depends on the terms of each specific license. We refer the reader to Lerner and Tirole (2005) for a futher discussion on the different types of licenses.

Chapter XIX
Reducing Transaction Costs with GLW Infrastructure

Marcus Vinicius Brandão Soares
NECSO Research Group – Federal University of Rio de Janeiro, Brazil

ABSTRACT

This chapter introduces the hybrid GLW information infrastructure as an alternative to proprietary-only information infrastructures with lower costs. The author argues that the use of FLOSS servers in a client-server infrastructure reduces the transaction costs relative to the data processing and the contract management that organizations have to support, preserving the investment already made with the installed base of clients in comparison to the use of proprietary managed servers. Transaction costs of two real-world proprietary infrastructures, Netware 5.0 and Windows NT 4.0, and of GLW, all with Windows 98 clients, are described and compared to give elements for the reader to analyze and decide.

INTRODUCTION

Firms, or more generally, organizations, develop and become larger over time, using more and more computers. Information systems of an organization turn into an information infrastructure, and the growth of the number of computers leads to a growth of software use (operating systems and their applications, e.g.), resulting in the growth of the number of software use and access licenses.

When all of the software used by the organization is proprietary, this growth leads to a greater supervision of users to regulate lawful access to software for the owners of software intellectual property rights since these rights are regulated by contracts, in this case, license agreements. This results in some costs associated with contracting—transaction costs—that are not usually taken into account by administrators and managers. They are used to paying much more attention to the costs of software licenses. However, what happens if FLOSS[1] is used?

This chapter aims to show a hybrid[2] information infrastructure named GLW[3] as a lower cost alternative to proprietary information infrastructures. GLW reduces the transaction costs of the organization in two ways: (a) by eliminating the access control mechanisms that are embedded in proprietary software, which reduces transaction costs in terms of computational costs, and (b) by reducing the number of managed contracts by half in comparison with some other proprietary information infrastructures.

BACKGROUND

What is an Information Infrastructure?

Once upon a time, computers existed as stand-alone devices. There was no (or very little) communication between computers within an organization. All work was truly local, based on local needs and standards. As organizations grew, personnel and computers multiplied. Methodologies were developed to all individuals within organizations to communicate to reduce the duplication of data and work, including models such as structured systems analysis (Gane & Sarson, 1983), modern structured analysis (Yourdan, 1990), structured systems design (Page-Jones, 1988), and, most recently, RUP (rational unified process) and UML (unified modeling language; Booch, Rumbaugh, & Jacobson, 1999).

All these methodologies (and many others) have been used for at least 20 years to model work and data to a certain size, time, and place. In the words of Hanseth (2002), "in short: IS methodologies aim at developing a closed system by a closed project organization for a closed customer organization within a closed time frame."

Planning information systems by scratching, designing, specifying, implementing, and "big-banging" becomes harder because it is not possible to change the installed base of hardware and software immediately. The solutions are (a) to improve the installed base by adding new functionalities, or (b) to extend the installed base by adding new elements to it (Hanseth, 2002).

An information system evolves into an information infrastructure, which is defined as "a shared, evolving, open, standardized, and heterogeneous installed base" (Hanseth, 2002). Its extension or improvement depends strongly

on the existing information infrastructure, that is, the installed base. One can notice that the working definitions of information infrastructure and installed base depend on each other. Hanseth does not ignore information systems; they are treated in another way: as local phenomena.

Hanseth (2002) splits information infrastructures into two levels: application infrastructures and support infrastructures, with the first at the top of the second. The support infrastructure is split into two categories: transport and service infrastructures. These levels are depicted in Figure 1.

For example, Web browsers are part of the application infrastructure, TCP/IP (transmission-control protocol/Internet protocol) is part of the transport infrastructure, and DNS[4] is part of the service infrastructure.

Gateways are elements that "link different infrastructures which provide the same kind of service based on different protocols/standards. The infrastructures that are linked this way are called neighboring infrastructures" (Hanseth, 2002) and are used to escape from certain situations where an organization is locked into a certain software or hardware platform (Shapiro & Varian, 1999). This situation is depicted in Figure 2. One example of a gateway is Samba, software that connects Microsoft Windows and GNU/Linux infrastructures. This will be discussed later in the chapter.

This chapter will focus on infrastructures with up to 250[5] users and in which the installed base depends strongly on proprietary software, contracts, license agreements, and associated costs.

Modes of Interaction in Infrastructures

We can describe two main models of computational interaction found in computer science literature: peer to peer (collaborative, with de-

Figure 1. Information infrastructures (Source: Hanseth, 2002)

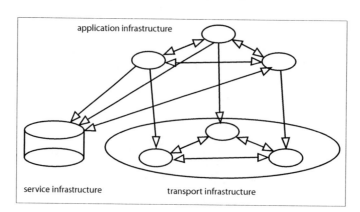

Figure 2. Ecologies of infrastructures (Source: Hanseth, 2002)

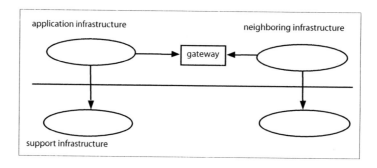

centralized control) and client-server (service oriented, with centralized control).

In a peer-to-peer infrastructure, all components of this infrastructure work together cooperatively so that there is no centralized processing unit. All elements work together without hierarchy. The cost of adding one more component to the infrastructure involves the cost of hardware, plus the cost of the software license, plus the cost of support.

In a client-server infrastructure, servers handle requests for computational processing, acting as a locus for specific programs and data, and acting as a gateway to other infrastructures. Clients send requests to the server for specific tasks; servers treat these requests in a centralized manner. The cost of adding one more component to this infra-

structure involves not only the additional cost of hardware plus the cost of the software license, but also the server access license as well as a support professional to make everything work.

This chapter will focus on the client-server approach because it is a hierarchical, firm-like structure, with someone who is responsible for decision making. Firms, as stated by Zylbesztajn (1995, p. 49),

almost always achieve the creation of a control structure, under their inner contracts, that minimizes the demand of support by third party or arbitration for the solution of their problems. The area of labor contracts presents a grand demand to the Courts. Anyway, what is intended is that the existence of a centralized control which permits the accomplishment of productive activity seems as a result of a series of sequential contracts in a lower cost then if each of these contracts was made outside the firm. There is no other reason for the existence and the advantages of the power of decision of the firm (FIAT).[6]

The server, in this case, plays the role of the structure controller, deciding which clients can have access to the infrastructure as well as which requisitions from the clients will be rejected or accepted and executed.

The infrastructure of an imaginary organization called ACME will serve as the installed base. ACME has a single server and 247 clients; the server offers file sharing, shared access to printers, and access to the Internet. The hypothesis for server and client software will be exposed further in this chapter.

Transactions and Transaction Costs

Williamson (1985) explains that a transaction occurs "when a good or service is transferred across a technologically separable interface. One stage of processing or assembly activity terminates and another begins" (p. 1). We rarely observe the boundaries of a transaction. We think of a transaction as an isolated operation and we do not think about the details that make a transaction possible and sustainable.

Coase (1960) exposes the fact that

in order to carry out a transaction, it is necessary to discover who it is that one wishes to deal with, to inform people that one wishes to deal and on what terms, to conduct negotiations leading up to a bargain to draw up the contract, to undertake the inspection needed to make sure that the terms of the contract are being observed and so on. These operations are often extremely costly, sufficiently costly at any rate to prevent many transactions that would be carried out in which the price system worked without costs. (p. 7)

These costs are called transaction costs. Prochnik (2003) classifies transaction costs:

as pre- and post-costs to a transaction. Pre-costs include searching and contractual costs while the post-costs include management and legal costs. Searching costs include those costs for locating and evaluating another party. ... Contractual costs include those costs for negotiating and writing a contract. ... Management costs include those for supervising the terms of the contract. ... If these terms are not met, then there are legal costs ...[7] (p. 12)

We may view most transactions as involving tangible goods. However, computationally, we can also look at transaction costs for information goods (Shapiro & Varian, 1999). For example, Demil and Lecocq (2003) note that "downloading Linux constitutes a transaction" (p. 10). Hence, any communication mediated by an infrastructure constitutes a transaction.

Kenneth Arrows defined transaction costs as the "costs of running the economic system" (cited in Williamson, 1985, p. 8). So, transaction costs of infrastructures are the costs of running the infrastructure. This chapter will address the transaction costs of running the transport infrastructure between the client side and server side, which are the communication transaction costs and the management costs of the contracts that regulate these transactions.

MAIN FOCUS OF THE CHAPTER

Communication: The Connection Transaction

Torvalds and Diamond (2001) explain that:

in your brain, each component is very simple, but the interactions between all of these components generate a very complex system. It is akin to the problem that notes that a given set is bigger than its parts. If you take a problem, divide it in half to solve it, you will ignore the need to communicate the solutions in order to solve the whole problem.[8] (p. 126)

There are two communication transactions in the infrastructure of ACME: the connection transaction, which begins with the user's authentication (log-in) and is maintained until the user leaves the infrastructure (log-out), and the printing transaction, which starts when any connected user sends data to a printer. For the purposes of this chapter, only the connection transaction will be studied.

There are four possibilities of software specification in a client-server approach: free or proprietary for the client-side software, and free or proprietary for the server-side software. This work compares client-side and server-side proprietary software (which is called proprietary infrastructure) with client-side proprietary software and server-side FLOSS (which is called hybrid infrastructure). In both cases, the client-side software remains proprietary.

The main reasons for this choice are (a) the most divulged operating system for client-side software (or desktop software) is Microsoft Windows, which is a proprietary software, making this operating system's installed base considerable, and (b) according to the definitions of professor Hanseth (2002), the existent installed base serves as the basis for the next installed base. In the case of this chapter, Windows 98 will be the client-side software.

Since the client-side installed base is chosen, another choice must be made for the server side of the proprietary-only infrastructure. Novell Netware 5.0 server software and Microsoft Windows NT 4.0 server software were chosen because (a) once more, their existent installed base is considerable, (b) there is considerable material for research and study on their end-user license agreements, and (c) most of the material mentioned before can be found on the Internet, which reduces the transaction costs of search for the researcher (the material can be found with a search of the sites of the enterprises or by using Google, http://www.google.com, or AltaVista, http://www.altavista.com, e.g.[9]). These two infrastructures will be compared to GLW.

Connection Transaction in Proprietary Infrastructures: Novell Netware 5.0

The Netware 5.0 server works with one or more Netware clients and runs on a dedicated computer. A Netware client operates on each computer running Windows 98. Netware literally only talks to Netware, and the Netware server does not work stand-alone. Netware is a network operating system. L. F. Soares, Colcher, and Souza (1995) explain that:

when networks appeared, computers, that worked in an isolated manner before, already had their local operating systems—LOS. Hence, a basic starting point of the software introduced was the possibility of offering new services as to disturbing the local environment as little as possible, mainly the interface that this environment offered to the users. In this context the network operating systems (NOS) broke up as an extension of the local operating systems, complementing them with the set of basic functions, and of general use, necessary to the operation of the stations, to make the use of shared resources seem transparent to the user. ... Among the functions of the network operating system one can accentuate the management of the access to the communication system and, consequently, to the use of hardware and software of the remote stations ...[10] (p. 423)

When a user enters his or her user name and password, the Windows 98 interface is already modified by Netware (Starlin, 1999). It is important to notice that the system that receives the request is Windows 98, not the Novell Netware client. The Windows interface sends the request to the Novell Netware client, which in turn sends a request to the Novell Netware server. A positive response from the server means essentially that the user has access to a given infrastructure. This process of user identification for accessing the infrastructure is referred to by the term authentication (Tanenbaum, 1992).

For authentication to be approved, Novell (1998b) explains that a connection to the server will happen only if the number of active concurrent connections to it is lower than the "the number of user licenses that you have lawfully purchased or acquired" (p. 1). The network management software monitors the number of active connections. The license agreement of Novell (1998b) describes this management software as follows:

Connection Management Software means computer programs provided as part of the Software that monitor the number of connections permitted under this License and that are designed to prevent more connections than the number of licensed connections specified by the User Count. (p. 1)

If there is any attempt to connect over the user count, the server rejects it. For example, the 251st request for a network with 250 computers running Windows 98 and Netware clients based on the lawful acquisition of 250 user licenses will be rejected. Netware clients can be installed without limitation, but the number of connections to the server is limited. A disk contains the software that informs the server software of the number of lawfully purchased licenses provided by Novell (Starlin, 1999).

Each time communication between a client and the server takes place, the server management software will be activated to monitor the connection. The CPU[11] of the server is used for the benefit of the user, but also to monitor the connection. Hence, the user's hardware is spending time and effort in terms of transaction costs. Attempting to reduce these costs by deactivating the management

software in any way is expressly forbidden by the licensing agreement of Novell (1998a):

Ownership: No title to or ownership of the Software is transferred to you. Novell, or the licensor through which Novell obtained the rights to distribute the Software, owns and retains all title and ownership of intellectual property rights in the Software; including any adaptations or copies of the Softwares. Only the License is purchased to you. ... License Restrictions: Novell reserves all rights not expressly granted to you. Without limiting generality of the foregoing, You may not modify the Connection Management Software to increase the number of connections supported by the Software; use any device, process or computer program in conjunction of the Host Software that increases, either directly or indirectly, the number of connections of the Host Software; reverse engineer, decompile or disassemble the Software, except and only to the extent it is expressly permitted by applicable law; or rent, timeshare or lease the Software unless expressly authorized by Novell in writing. (pp. 1-2)

Novell provides licenses of Netware 5.0 in multiples of five in the case of enterprises similar to ACME.[12] ACME needs only 247 licenses, but it will have to purchase 250 licenses, and even though the number of licenses is greater than the number of possible active connections, the server will continue monitoring all connections.

Microsoft Windows NT 4.0

Unlike Novell Netware 5.0, Windows NT 4.0 is not a network operating system, so there is no need for clients because the system can work standalone. Clients are only needed when working in a client-server infrastructure. When it occurs, Windows NT 4.0 server receives requests for authentication from Windows clients, in this case, Windows 98 clients. It is clear that the Windows

98 client and the Windows NT 4.0 server work on different computers and that there is a single server working with multiple clients.

The Windows infrastructure requires three kinds of licenses to work lawfully: a license for the Windows NT 4.0 server, a license for the Windows 98 client, and a license for the client to access the server, known as the Client Access License (CAL). There are two means of purchasing licenses for a Microsoft Windows NT 4.0 server: per seat or per server. According to Microsoft (2004a):

With per server licensing, each client access license is assigned to a particular server and allows one connection to that server for the use of that product. With Per Seat licensing, a client access license is assigned to each specific computer that accesses the server.

In our ACME example, the firm has only one server and 247 clients. In this case, Microsoft (2004a) would recommend the use of per server licensing:

The licensing mode you select depends on which applications you will be using. For example, if you use a Windows NT Server mainly for file and print sharing and on multiple servers, you may be better off with the Per Seat option. However, if you use it as a dedicated Remote Access Server computer, you can select the Per Server concurrent connections option. Use the following guidelines for selecting a licensing mode: - If you have only one server, select the Per Server option because you can change later to the Per Seat mode; - If you have multiple servers and the total number of Client Access Licenses across all servers to support the Per Server mode is equal to or greater than the number of computers or workstations, select or convert to the Per Seat option.

A CAL must be assigned to each computer that is running a client and connecting with the server:

Client Access Licenses are separate from the desktop operating system software you use to connect to Microsoft server products. Purchasing Microsoft Windows 95, Windows NT Workstation, or any other desktop operating system (such as Macintosh) that connects to Microsoft server products does not constitute a legal license to connect to those Microsoft server products. In addition to the desktop operating system, Client Access Licenses must also be purchased. (Microsoft, 2004a)

Hence, it is necessary to purchase CALs equivalent to the number of potential connections to the server at the same time; in the ACME case, it would be 247 CALs. But what happens if a 248th computer attempts a connection? A mechanism monitoring the number of concurrent connections comes into action and locks out connections to the server, obliging the administrator to interfere and select which connections will be aborted to maintain the integrity of the infrastructure:

With Per Server licensing, once the specified limit for concurrent connections is reached, the server returns an error to the client's computer and does not allow more computer connections to that server. Connections made by administrators are also considered as part of the total number of concurrent connections. When the limit is reached, though, administrators are still allowed to connect to manage the lockout situation. New users, however, cannot connect again until enough users (including administrators) have disconnected to get below the specified limit. (Microsoft, 2004a)

The license manager is the mechanism that monitors and manages connections to the server (Jennings, 1997). It is clear that each time communication between a client and the server takes place, the license manager is activated to monitor the connection. Again, transaction costs are being absorbed by the user in a variety of ways, with server CPU time and effort for monitoring functions. As might be expected, deactivating the license manager in any way is expressly forbidden by the license:

DESCRIPTION OF OTHER RIGHTS AND LIMI-TATIONS. ... Limitation on Reverse Engineering, Decompilation, and Disassembly. You may not reverse engineer, decompile, or disassemble the SOFTWARE PRODUCT, except and only to the extent that such activity is expressly permitted by applicable law notwithstanding this limitation. (Microsoft, 2004b)

Connection Transaction in a Hybrid Infrastructure: GLW

A hybrid infrastructure, or GLW, performs a transaction using GNU/Linux operating system software at the server side, as well as a piece of software called Samba, which allows the use of Microsoft Windows 98 at the client side. Collier-Brown, Eckstein, and Jay (2003) explain that Samba

is a suite of Unix applications that speak the Server Message Block (SMB) protocol. Microsoft Windows operating systems and the OS/2 operating system use SMB to perform client-server networking for file and printer sharing and associated operations. By supporting this protocol, Samba enables computers running Unix to get in on the action, communicating with the same networking protocol as Microsoft Windows and appearing as another Windows system on the network from the perspective of a Windows client. (p. 3)

Some considerations must be made on Samba, exposed by Collier-Brown et al. (2003):

Samba can help Windows and Unix computers coexist in the same network. However, there are some specific reasons why you might want to set up a Samba server on your network: You don't want to

pay for—or can't afford—a full-fledged Windows server, yet you still need the functionality that one provides; the Client Access Licenses (CALs) that Microsoft requires for each Windows client to access a Windows server are unaffordable. (p. 3)

So, once using Samba, there is no need for CALs.

Comparing Connection Transaction Costs

From now on, connection transaction costs in proprietary and hybrid infrastructures will be described separately and compared. It will be shown that GLW (a) preserves most of the existent installed base, (b) performs tasks similar to the proprietary infrastructure, (c) produces minimal impact on the routines of the firm, and (d) works with lower transaction costs in comparison with the proprietary installed base.

Netware 5.0 and Windows 4.0

In both cases, the server will monitor operations at all times, with incurred connection transactions costs for all parties involved. Law, economics, and technology interfere with each other in such a way that contractual obligations do not allow technological modifications (Law, 2000). The software obliges the customer to fulfill the agreement; using the words of Lessig (1999, p. 3), "code is law." See Figure 3.

It is important to notice the proprietary software is written with the network management software embedded. This piece of software works to avoid opportunistic behavior (Williamson, 1985) by a customer who tries to connect beyond the number of connections permitted.

For the customer who did not read the end-user license agreement (EULA; or, at least, did not pay attention to such technical details), the connection monitoring software may be acting in an opportunistic way, overloading the CPU of the customer with its work; however, it plays the

Figure 3. Proprietary infrastructure

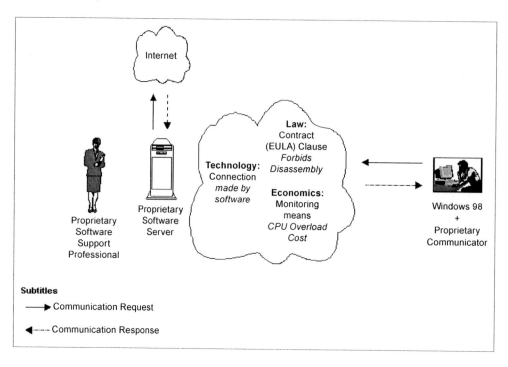

role agreed upon with the customer, who pays the transaction cost of monitoring.

In respect to the management of contracts, Netware 5.0 and Windows NT 4.0 cases have to be analyzed separately.

In the case of Netware 5.0, it would be necessary to license 247 copies of Windows 98 client software, plus one copy of the Netware 5.0 server software, along with 250 licenses for Netware client software. Additionally, there should be a Netware support professional to manage it all, which introduces one more contract for support. So, there are 497 (247 + 250) contracts to be managed by a Novell-trained support professional, plus one more contract to be managed by the firm, which results in 498 contracts.

In the case of Windows NT 4.0, it would be necessary to license 247 copies of Windows 98 client software, plus one copy of the Windows NT 4.0 server software, plus 247 CALs. A Windows support professional is needed to manage it all.

So, there are 495 (247 + 1 + 247) contracts to be managed by a Microsoft-trained support professional, plus one more contract to be managed by the firm, which results in 496 contracts.

The responsibilities of managing the licenses, either for Netware 5.0 client connections or Windows 98 client connections, are shared between the server CPU and the support professional. The server CPU pays the transaction cost of monitoring the connections performed by the embedded software, and the support professional pays, if needed, the search costs to find a license vendor in the market for this proprietary software, as well as the contracting costs with this vendor. In these two cases, there will always exist costs associated with the process of linking a new client to the installed base. These costs are the cost of an additional client license, plus the time and effort of the professional to configure the client, plus the time and effort of the server to monitor the link.

GLW

GLW allows the preservation of all Windows 98 clients of the existent installed base. These clients will interact with the GNU/Linux server as if it were a Windows NT 4.0 operating system software at the server side. Windows 98 clients will experience few alterations in the migration to this hybrid environment. The alterations are concentrated at the administration level. Hence, training costs are minimal.

The reduction of transaction costs is achieved in two ways: through the use of CPU computing capacity and through the management of contracts, in this case, license agreements.

In respect to the CPU computing capacity, Samba is licensed under a GPL (general public license; Free Software Foundation, 1991), which obliges source code to be totally accessible and open to modification. GPL eliminates the problems with proprietary software exposed before, in which there is embedded software using the CPU to monitor connections with unaffordable source code. Since there are no more connection monitors and the source code is affordable to be audited, the transaction costs of connection monitoring can be completely eliminated. Software limitations on the number of connections with Samba become merely an issue of CPU computing capacity.

Thus, it would be necessary to license 247 copies of Windows 98 client software, plus a copy of GNU/Linux. There should be a GNU/Linux support professional to manage it all. Using the initial hypothesis on the Windows 98 client operating system installed base, the GNU/Linux support professional will turn into a GLW support professional and manage the Windows 98 installed base, too. So, there are 248 (247 + 1) contracts to be managed by the GLW support professional, plus one more contract to be managed by the firm, which results in 249 contracts. See Figure 4.

This occurs because the costs of adding clients to the installed base are limited to costs of the hardware needed plus the effort of the GLW support professional to configure the Windows 98 client. No additional connection licenses are

Figure 4. GLW infrastructure

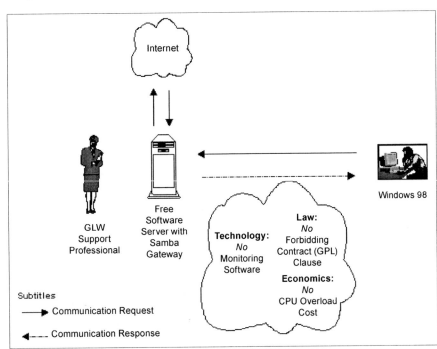

Table 1. Infrastructures and their connection transaction costs

Transaction Costs ► Infrastructures with ▼	Connection Monitoring	CPU Overloading	Number of Client Access Licenses	Total Number of Contracts Managed
Netware 5.0 Server	YES	YES	Implicit - 250	498
Windows NT 4.0 Server	YES	YES	Explicit – 247	496
GNU/Linux Server (GLW)	NO	NO	NO CALs - 0	249

necessary because Samba does not need to monitor the connections, freeing the server to actually work for the firm and not bother with network policing. Hence, GLW supports the growth of the infrastructure without spending unnecessary communication transaction costs.

The responsibility of managing the licenses for GLW becomes a task restricted only to the GLW support professional and, as long as the software is free, this task can be summarized to finding the sites on the Internet from which the software can be downloaded using a search engine, navigating to the site, and downloading the software.

Another point to be observed is that as free projects rarely fork,[13] once the support professional has visited the main site that hosts the project and its source codes,[14] this professional does not have to search for it again. The search costs are lowered and the contracting costs with vendors are eliminated. Additional benefits obtained from using free software are the addition of the maintainability of FLOSS by its community of users, collaborators, and supporters to the overall infrastructure and the benefits of peer-to-peer computing (Benkler, 2003). Table 1 summarizes the connection transaction costs of the three infrastructures presented in this chapter related to ACME.

FUTURE TRENDS AND CONCLUSION

Information system costs are largely based on the acquisition of software and hardware and on the training of personnel to operate the system. Often, transaction costs are not completely considered. As information systems grow into information infrastructures, these transaction costs, which involve interferences among technology, economics, and law, can no longer be ignored.

These transaction costs can vary greatly depending on the kind of software used and its impact on hardware and personnel. The study of transaction costs cannot be restricted to the costs of the management of formal contracts: There must be special attention paid to mechanisms embedded in software, like the ones that were analyzed in this chapter. We have indicated that there are some considerable savings in transaction costs in using a GLW infrastructure where FLOSS is applied, either in the case of software use or in the case of the management of licenses and other contracts.

Future topics of research may be the modification of the transaction costs exposed in this chapter that come with the next versions of the server software; transaction costs in infrastructures with more than two proprietary servers, or even with two or more different proprietary servers; opportunistic behaviors that can come from the emerging GLW support professional and infrastructure bilateral monopoly constituted over time; demanding transaction costs in either proprietary-only or GLW infrastructures in case of software malfunction; and more detailed quantitative descriptions of the impacts of FLOSS on transaction costs.

ACKNOWLEDGMENT

The author would like to thank Professor Ivan da Costa Marques, PhD, who introduced him to the New Institutional Economics and who advised him in his MSc course in COPPE-UFRJ in the computation and systems engineering program.

REFERENCES

Benkler, Y. (2002). Coase's penguin, or, Linux and the nature of the firm. *Yale Law Journal, 3*, 112.

Booch, G., Rumbaugh, J., & Jacobson, I. (1999). *The unified modeling language user guide.* Reading, MA: Addison-Wesley.

Coase, R. H. (1937). The nature of the firm. *Economica, 4*, 386-405.

Coase, R. H. (1960). The problem of social cost. *Journal of Law and Economics, 3*, 1-44.

Collier-Brown, D., Eckstein, R., & Jay, T. (2003). *Using Samba* (2nd ed.). Sebastopol, CA: O'Reilly.

Demil, B., & Lecocq, X. (2003). *Neither market nor hierarchy or network: The emerging bazaar governance.* Retrieved June 12, 2006, from http://opensource.mit.edu/papers/demillecocq.pdf

Diamond, D., & Torvalds, L. (2001). *Só por prazer. Linux: Os bastidores da sua criação* (F. B. Rössler, Trans.). Rio de Janeiro, Brazil: Campus.

Free Software Foundation. (1991). *GNU general public license, version 2 (June).* Retrieved June 12, 2006, from http://www.gnu.org/licenses/gpl.txt

Gane, C., & Sarson, T. (1983). *Análise estruturada de sistemas.* Rio de Janeiro, Brazil: LTC.

Hanseth, O. (2002). *From systems and tools to networks and infrastructures: From design to cultivation. Towards a theory of ICT solutions and its design methodology implications.* Retrieved June 12, 2006, from http://heim.ifi.uio.no/~oleha/Publications/ib_ISR_3rd_resubm2.html

Jennings, R. (1997). *Usando Windows NT Server 4: O guia de referência mais completo* (4th ed.). Rio de Janeiro, Brazil: Campus.

Latour, B. (1994). *Jamais fomos modernos* (C. Irineu da Costa, Trans.). Rio de Janeiro, Brazil.

Law, J. (2000). *Economics as interference.* Retrieved June 12, 2006, from http://www.lancs.ac.uk/fss/sociology/papers/law-economics-as-inteference.pdf

Lessig, L. (1999). *Code and other laws of cyberspace.* New York: Basic Books.

Microsoft. (2006a). *Licensing and license manager.* Retrieved June 12, 2006, from http://www.microsoft.com/resources/documentation/windowsnt/4/server/proddocs/en-us/concept/xcp12.mspx?mfr=true

Microsoft. (2006b). *NT server end user license agreement.* Retrieved June 12, 2006, from http://proprietary.clendons.co.nz/licenses/eula/windowsntserver-eula.htm

Novell. (1998a). *Netware 5.0 Server Communication Version 5 manual.* Provo, Utah: Author.

Novell. (1998b). *Netware 5.0 software license and limited warranty.* Provo, Utah: Author. Retrieved June 12, 2006, from http://www.novell.com/licensing/eula/nw_5.pdf

Page-Jones, M. (1988). *Projeto estruturado de sistemas.* São Paulo, Brazil: McGraw–Hill.

Prochnik, V. (2003). *Economia dos contratos: Princípios da teoria dos custos de transação* (classroom notes). Rio de Janeiro, Brazil: Federal University of Rio de Janeiro, Institute for Economics.

Raymond, E. S. (2003). *The new Hacker's dictionary*. Retrieved June 12, 2006, from http://www.catb.org/~esr/jargon/html/F/fork.html

Shapiro, C., & Varian, H. R. (1999). *Economia da informação: Como os princípios econômicos de aplicam à era da Internet* (6th ed., R. Inojosa, Trans.). Rio de Janeiro, Brazil: Campus.

Soares, L. F., Colcher, S., & Souza, G. (1995). *Redes de computadores: Das LANs, MANs e WANs às redes ATM*. Rio de Janeiro, Brazil: Campus.

Soares, M. V. B. (2001). *Concepção e adoção da metodologia GNU/Linux: O caso WirelessNet*. Paper presented at the First Meeting of Production Engineering of North Rio de Janeiro State University, Rio de Janeiro, Brazil.

Starlin, G. (1999). *Netware 5 completo: Manual do administrador CNE* (2nd ed.). Rio de Janeiro, Brazil: Book Express.

Tanenbaum, A. S. (1992). *Modern operating systems*. Upper Saddle River, NJ: Prentice-Hall.

Williamson, O. E. (1985). *The economic institutions of capitalism*. New York: Free Press.

Zylbesztajn, D. (1995). *Estruturas de governança e coordenação do agribusiness: Uma aplicação da Nova Economia das Instituições*. Tese (Livre-Docência), Universidade de São Paulo. Retrieved June 14, 2006, from http://www.pensa.org.br/pdf/teses/Tese_Livre_Docência.pdf

KEY TERMS

Client Access Licenses (CALs): Licenses required by Microsoft to connect each Microsoft client software to a Microsoft Windows NT 4.0 server software.

Connection Manager: A Novell Netware server embedded software that monitors the concurrent connections from Netware client software to the Netware server software to make sure that there will be no more concurrent connections than those that were lawfully acquired.

Contracts: Agreements signed by two parties in which it is described what they may, can, or have to do and what they may not or cannot do in order to accomplish their objectives. Contracts make laws between parties.

Gateways: Elements that "link different infrastructures which provide the same kind of service based on different protocols/standards" (Hanseth, 2002).

GLW: The initials of the expression GNU/Linux-Windows, meaning a mixing of them.

Infrastructure: As in an information infrastructure, which, in the words of Hanseth (2002), is "a shared, evolving, open, standardized, and heterogeneous installed base."

License Manager: A Microsoft Windows NT 4.0 server embedded software that monitors the concurrent connections from Microsoft client software to the Microsoft Windows NT 4.0 server software to make sure that there will be no more concurrent connections than those that were lawfully acquired.

Rights: Prerogatives that can be exercised by persons under the observation of the laws.

Samba: Software that connects Microsoft Windows and GNU/Linux infrastructures.

Transaction Costs: The costs of accomplishing a transaction. A transaction occurs "when a good or service is transferred across a technologically separable interface. One stage of processing or assembly activity terminates and another begins" (Williamson, 1981, p. 552).

ENDNOTES

[1] FLOSS stands for free/libre open sorce software.

[2]   Latour (1994) explains that a hybrid results from the composition of heterogeneous elements. This work focuses in the pieces of software.

[3]   GLW stands for GNU/Linux-Windows.

[4]   DNS stands for dynamic naming system.

[5]   The number 250 for users was chosen because this is the number of PCs that Microsoft considers a medium size business (from 50 to 250). For more information, look at http://www.microsoft.com/technet/itsolutions/midsizebusiness/default.mspx retrieved 12 2006.

[6]   The Latin word FIAT means, in English, the power of decision.

[7]   It is translation of the following: "[...] Os custos de transação podem classificados em anteriores e posteriores (ex–ante e ex–post) à realização da transação propriamente dita. Os custos anteriores são os custos de busca e de contratação e os posteriores são os de monitoração e de fazer cumprir o contrato. Os custos de busca abrangem o custo de encontrar e avaliar um parceiro. [...] Os custos de contratação incluem negociar e escrever o contrato. [...] Os custos de monitoração são os custos de fiscalizar o contrato, observando seu cumprimento pelo parceiro. [...] Por último, os custos de fazer cumprir o contrato são os custos de implantar uma solução quando o contrato não está sendo seguido. [...] Ao nível da economia nacional, entre os custos de transações, estão todos os gastos com advogados, contadores, bancos, mensuração da qualidade, comércio e seguros. [...]".

[8]   It is translation of the following: "Pense no seu cérebro. Cada peça é simples, porém as interações entre as peças geram um sistema muito mais complexo. É aquele problema que diz que o conjunto é maior do que as partes. Se você pegar um problema, dividi-lo pelo meio e disser que as partes são complicadas pela metade, estará ignorando o fato de que é preciso acrescentar a comunicação entre as duas metades."

[9]   The Internet address (URL) of the Web sites where public material can be found is in the references at the end of the chapter.

[10]   It is translation of the following: "[...] quando surgiram as redes, os computadores, antes funcionando isoladamente, já possu'am seus respectivos sistemas operacionais locais—SOL. Portanto, uma premissa básica do software introduzido para fornecer os novos serviços foi perturbar o menos poss'vel o ambiente local, principalmente a interface que esse ambiente ofereceria a seus usuários. Neste contexto surgiram os sistemas operacionais de redes (SOR) como uma extensão dos sistemas operacionais locais complementando-os com o conjunto de funções básicas, e de uso geral, necessárias à operação das estações, de forma a tornar transparente o uso dos recursos compartilhados [...]. Dentre as funções do sistema operacional de redes destaca-se, assim, o gerenciamento do acesso ao sistema de comunicações e, conseqüentemente, as estações remotas para utilização de recursos de hardware e software remotos. [...]".

[11]   CPU stands for central processing unit.

[12]   Novell provides other licensing agreements over specific periods of time and with other conditions. These sorts of agreements are outside the scope of this chapter. For more details, see http://www.novell.com.

[13]   Raymond (2003) exposes that "In the open-source community, a fork is what occurs when two (or more) versions of a software package's source code are being developed in parallel which once shared a common code base, and these multiple versions of the source code have irreconcilable differences between them. This should not be confused with a development branch, which may later be folded back into the original source code base. Nor should it be confused with what

happens when a new distribution of Linux or some other distribution is created, because that largely assembles pieces than can and will be used in other distributions without conflict. Forking is uncommon; in fact, it is so uncommon that individual instances loom large in hacker folklore. Notable in this class were the Emacs/XEmacs fork, the GCC/EGCS fork (later healed by a merger) and the forks among the FreeBSD, NetBSD, and OpenBSD operating systems.

Soares (2001) explains that this process is avoided by free softwares development com-

munities because there is a loss of time and effort in a competition that does not favor the development of the software.

[14] The expression "main site" was used to distinguish this site from the sites that are known as "mirror sites," that is, sites that contain a copy of the main sites. For example: http://www.kernel.org is the main site from which Linux kernel source code can be downloaded and http://www.br.kernel.org/pub/linux/ is a mirror site from which Linux kernel source code can be downloaded too.

Chapter XX
Issues to Consider when Choosing Open Source Content Management Systems (CMSs)

Beatrice A. Boateng
Ohio University, USA

Kwasi Boateng
University of Arkansas at Little Rock, USA

ABSTRACT

This chapter examines the main issues that have to be considered when selecting an open source content management system. It involves a discussion of literature and the experiences of the authors after installing and testing four widely used open source CMSs (Moodle, Drupal, Xoops, and Mambo) on a stand-alone desktop computer. It takes into consideration Arnold's (2003) and Han's (2004) suggestions for the development of CMSs, and identifies six criteria that need to be considered when selecting an open source CMS for use.

INTRODUCTION

Content management systems (CMSs) have gained prominence and are used for database-driven Web sites for all kinds of electronic communication activities.[1] A content management system is a nonstatic, dynamic, database-driven system that is used for electronic management and the publishing of information and resources in an organized manner. The features of a CMS-run Web site permit Web site administrators and authorized users to log into an electronic system to author or approve posted materials. Similarly, they can facilitate access to archival, confidential, or password-protected materials that are hosted on the Internet.

The emergence of open source software (OSS) applications and the culture of making source code available for all to use is causing a stir in the software development industry and among software users. According to the Open Source Definition (OSD) Web site (http://www.opensource.org/docs/definition.php), an OSS application generally complies with a set of criteria that include the following:

- Source code should be available to users.
- Software is available for free download and distribution.
- Users should be able to modify source code and possibly create new applications.
- Software is available under the "copyleft" licensing agreement (http://www.debian.org/social_contract).

In this chapter, we examine the main issues that have to be considered when selecting an open source content management system. We draw upon the literature and our experiences after the installation and testing of four widely used open source CMSs, namely, Moodle,[2] Drupal, Xoops, and Mambo, on a stand-alone desktop computer. Through our installation, we were able to verify the installation processes, and understand how the back end of the select CMSs work in order to address issues that a potential adopter of open source CMS should consider. We chose Mambo, Xoops, Drupal, and Moodle based on the fact that these CMSs come up often in discussions and the literature on open source CMSs. Also, from our observation, these CMSs have well-organized product software, support, documentation, training, integration, and professional services. These are standards considered necessary for determining the maturity of OSS as determined by Golden (2005). Also, information available on the CMSMatrix Web site, a site for the discussion, rating, and comparison of content management systems, prominently feature Mambo, Xoops, Drupal, and Moodle among the frequently used and efficient systems based on system requirements, security, support, ease of use, performance, management, interoperability, flexibility, built-in application, and commerce (http://cmsmatrix.org). Our examination of information on CMSMatrix on the four CMS candidates indicates that on the average, Drupal and Xoops are rated higher on all

the criteria. Similarly, Mambo, Xoops, Drupal, and Moodle are listed on the site http://opensourcecms.com, a Web site that displays a comprehensive list of open source CMSs, and provides a free administrative platform for real-time testing of various open source systems.

BACKGROUND: WHAT ARE CMSs AND HOW DID THEY EMERGE?

The need for online platforms ideal for the dissemination of information, document delivery, and electronic archiving of materials has necessitated the development of content management systems that support the publishing of materials in different and easily accessible electronic formats. Discussing CMSs as tools for such online organization of materials, Arnold (2003) opined that "when the needs and requirements [for the electronic delivery of materials] are understood, a system to manage the creation, approval, and dissemination of text, images, and even streaming video can make life in today's fluid environment somewhat more orderly" (pp. 36-37). The development of CMSs is relatively new in the software industry (Arnold; Han, 2004). The origins of CMSs lie in (a) the records management field and (b) the need to move content from the desk of the creator to an organization's Web site (Arnold). Performing these two tasks effectively in a 21st century online environment could be daunting. CMSs usually comprise of modules and/or components as add-ons of an application that allow a programmer or an adopter of such software to build an electronic system that could be used to perform various functions electronically. The functionality of a CMS is dependent on the structure of the CMS and the processes that occur within the CMS (Friedlein, 2003). For instance, the CMS could be configured to allow a system administrator and users to manage, compile, and publish resources, as well as to facilitate online interaction among users as illustrated in the process part of Figure 1. Figure

Figure 1. The basic architectural configuration and structural functions of a CMS

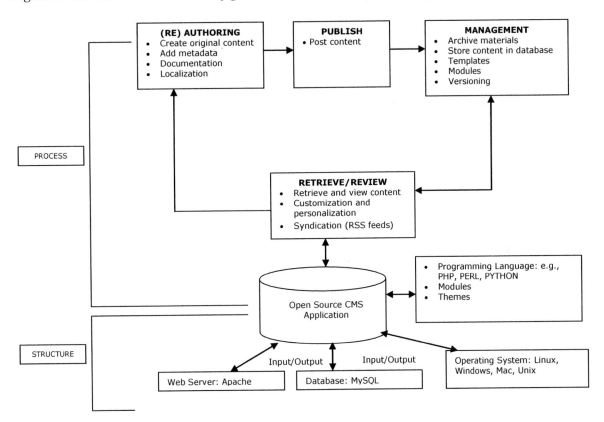

1 captures the basic architectural configuration and the structural functions of a CMS.

To understand how to choose an open source CMS, we explain the process of installing the CMSs we tested, what we observed, and the technical installation approaches that can be used to try a CMS before installing online. A CMS can be set up on a desktop or even a laptop on a trial basis. This could help the installer learn and understand how a particular CMS candidate functions. This approach requires a full download of a CMS candidate for installation. This allows for testing and evaluation before making a final decision to adopt a candidate. Alternatively, such software can be tested by using the online platform created by the Web site http://opensourcecms.com. The platform allows the testing of diverse CMSs created to perform specialized functions.

Should a prospective adopter of an open source CMS decide to do an in-house testing of some CMS candidates on a personal computer (PC), it will require downloading Easy PHP (a program that packages the Apache Web server and MySQL database application together) and a copy of the CMS candidate to be tested, as shown by the structure part of Figure 1. We undertook that task to install the CMS candidates under discussion on our PC as "hand-holding" (Stallman, 2003) for prospective adopters of OS CMSs.

We installed Moodle, Drupal, Xoops, and Mambo and documented the steps for configuration. We downloaded and installed Easy PHP 1.7. EasyPHP runs a Web interface for the MySQL database application called PHPMyAdmin. This feature of EasyPHP renders unnecessary the knowledge of MySQL coding or programming used in deploying a database-driven Web portal. The downloaded CMS software were unzipped in order to gain access to the CMS files for installations using the Web interface of EasyPHP to mount the CMSs onto a computer. At this point, the Web interface of EasyPHP walked us through the installation steps and the creation of the CMS database. After the installation, the rest of the development of these CMSs involved identifying a specific design theme and modules or components needed for customizing the functions and user interface of the CMS. The process of adding on modules to a Drupal, Xoops, and Mambo CMS are similar with minor variations (Boateng & Boateng, in press). In Drupal, Xoops, and Mambo, it requires identifying and downloading zipped modules or component files, and uploading such files for installation and activation on a Web site. On the other hand, Moodle installation files are packaged with all interactive features that can be further developed through coding:

- Xoops module add-ons require unzipping the files and saving them into a specific modules folder.
- Mambo has an inbuilt function that unzips the modules and installs them on the CMS platform.
- The installation of modules in the Drupal CMS environment is done using the Web interface of the PHPMyAdmin.

The processes of installing Drupal, Xoops, Mambo, and Moodle are user friendly. Files could be uploaded onto a site using FTP (file transfer protocol) for configuration. The installation and customization on a PC and online can be done without knowing any programming language. Various open source CMSs have been developed for specific services, for instance, osCommerce and Zen Cart are for e-commerce; bBlog, BLOG: CMS, Simplog, and WordPress are for blogging; Moodle and Atutor are noted for e-learning; and Greenstone and Koha are for digital libraries.

Table 1 offers a list of Web sites where open source CMS downloads are available. Having installed a select number of CMSs, we identified six criteria to be considered when selecting an open source CMS for adoption.

CRITERIA FOR SELECTING AN OSS CMS

Choosing a CMS candidate for use requires an assessment of the resources for such an initiative, what Han (2004) describes as the functional or nonfunctional requirements for developing CMSs. The functional requirements refer the need for a content management system, and the nonfunctional requirements are related to costs and skill sets (p. 357). The cost of managing a content management system could be enormous. Arnold (2003) found that Jupiter Research in June 2002 indicated that some companies spend $25,000 per nontechnical employee per year to manage simple content on a Web site. According to Arnold, "a company with 5-people in the customer support chain would translate to $1.25 million, excluding software license fee" (p. 38). He explained that an April 2002 Jupiter executive survey indicated that "53 percent of companies will have deployed a new documents, content, or media asset management system by the end of 2002." The survey indicated that one tenth or 19% of Web site managers stated "they will be involved in content management consolidation projects" that "unify systems to manage multiple web properties." The case made by Arnold is that "the pay-off from content management can be a savings of 20 percent or more compared to pre-content management system [online resource management]." The license fees for a proprietary content management system could cost between a few hundred dollars per month to about a $1 million for a license (Arnold). Lerner and Tirole (2004) indicated that "the importance of open source software can be illustrated by considering a few

Table 1. OSS CMS resources

Resources		
Try CMSs before you install:	http://www.opensourcecms.com/	
OSS CMS list:	http://www.la-grange.net/cms	
OSS CMS management directory:	http://www.cmsreview.com/OpenSource/directory.html	
Content management comparison tool:	http://www.cmsmatrix.org/	
International association for OSS CMS management:	http://www.oscom.org/	
CMS info:	http://www.cmsinfo.org/index.php	
Different types of CMSs	**Portals**	Drupal: http://drupal.org/ Exponent: http://www.exponentcms.org/ Joomla: http://www.joomla.org/ Mambo: http://www.mamboserver.com/ Xoops: http://www.xoops.org/ Plone: http://plone.org/
	Blogs	bBlog: http://www.bblog.com/ BLOG:CMS: http://blogcms.com/ Simplog: http://www.simplog.org/ WordPress: http://wordpress.org/
	Forums	MyBB: http://www.mybboard.com/ phpBB: http://www.phpbb.com/
	Wiki	DokuWiki: http://wiki.splitbrain.org/wiki:dokuwiki MediaWiki: http://www.mediawiki.org/wiki/MediaWiki
	Image Galleries	Coppermine: http://coppermine-gallery.net/index.php Plogger: http://www.plogger.org/
	Groupware	eGroupWare: http://www.egroupware.org/ NetOffice: http://netoffice.sourceforge.net/ phpGroupWare: http://www.phpgroupware.org/ WebCollab: http://webcollab.sourceforge.net/
	E-Commerce	osCommerce: http://www.oscommerce.com/ Zen Cart: http://www.zen-cart.com
	E-Learning	Moodle: http://moodle.org/ ATutor: http://www.atutor.ca/ Claroline: http://www.claroline.net/ OLAT: http://www.olat.org/public/index.html
	Digital Libraries	Greenstone: http://www.greenstone.org/cgi-bin/library Koha: http://www.koha.org/

examples." Taking into consideration Arnold's and Han's suggestions, and after installing and testing four content management systems on a PC, we identified six criteria that need to be considered when selecting an open source CMS for use. They are interoperability; software licenses; user community; documentation; versatility, stability, and availability of source code; and security.

Interoperability

The interoperability of a content management system is its ability to support content from different hardware and software vendors. According to Lagoze and Van de Sompel (2001), interoperability facilitates the dissemination of content efficiently and the "construction of innovative services" (p. 2). All the CMSs we installed can be deployed in Windows, Mac, Linux, and Unix environments. The nature of most open source CMS configurations on servers makes their interoperability quick to determine at the deployment stage. However, what a prospective adopter needs to know is that specific OS CMSs have specific modules and components that can be appended onto the system. The possibility of installing add-ons or modules is one of the features that indicate the interoperability of such CMS. Xoops, Drupal, Mambo, and Moodle allow developers or users to create their own themes, modules, or components using Web authoring software or programming code that can be read in a Web browser. The ability to adopt different versions of modules, components, or add-ons depends on the compatibility of a specific add-on with the version of the main application engine. We realized that specific Xoops, Drupal, and Mambo versions are compatible with specific modules, components, or add-ons. For example, Mambo modules and components are designed and tested on specific core applications. A module developed and tested for Mambo 4.5.4 may not be compatible with Mambo 4.5.1 as was determined during our testing and evaluation of the CMSs.

Software Licenses

Software licenses are conceived within the notion of ownership and rights to use. OSS licensing issues are fundamentally different from the traditional software licensing rules and legalities. Weber (2004) noted that

the conventional notion of property is, of course, the right to exclude [a person] from using something that belongs to [somebody else] ... property in the open source [movement] is [conceived] fundamentally around the right to distribute, not the right to exclude. (p. 1)

Not only does the OS public license guarantee the right to distribute OSS, it also guarantees access to the code and the right to alter it to meet specific needs of the administrator of such software. This is an unprecedented approach to licensing, one that promotes easy access and software innovation in an unconventional and public manner. A significance of such a licensing approach supports the notions of do-it-yourself and creative principles that drive the current practice of electronic communication. This practice is inherent in the operations of the open source movement, which attempts to reduce the communication process to the level where software creation and availability is increasingly decentralized, giving more people the opportunity to become creators, publishers, and users of electronic content and software.

This egalitarian value of OSS is expressed in the GNU (GNU is not Unix) manifesto, which was declared at the release of the kernel of the GNU operating software, an initiative that set the open source movement in motion. Stallman (2003) declared:

once GNU is written, everyone will be able to obtain good system software free, just like air ... .Complete system sources will be available to everyone. As a result, a user who needs changes in the system will always be free to make them himself, or hire any available programmer or company to make them for him. Users will no longer be at the mercy of one programmer or company which owns the sources and is in sole position to make changes. (p. 547)

Xoops, Drupal, Mambo, and Moodle all fall under the broad GNU license agreements. Discussing the OS approach to software licensing, Lerner and Tirole (2004) found the following:

• Restrictive licenses are more common for applications geared toward end users and system administrators, like desktop tools and games.

- Restrictive licenses are significantly less common for those applications aimed toward software developers.
- Restrictive licenses are also less common for projects operating in commercial environments or that run on proprietary operating systems.
- Projects whose natural language is not English, whose community appeal may be presumed to be much smaller, are more likely to employ restrictive licenses.
- Projects with less restrictive licenses tend to attract more contributors.

Generally, OS copyleft law is more liberating than restrictive. It is the attractive side of such software. OS CMSs offer an accessible and convenient way to run a sophisticated electronic communication system with room for innovation in the delivery of electronic materials. All the CMSs we tested are available for free download and for further development. Most of the add-ons, themes, and the like are available for free download and require acknowledgment of the original developers when used.

User Community

A major concern raised regarding the wisdom in using OSS is related to continuity and longevity in terms of the development of various OSS. The usual contention is whether the community that develops specific OSS will survive the test of time and continue to work on its development as the years go by. There are no guarantees to this problem. However, there are indications that OS-driven projects cannot and will not be wished away overnight. The OS user-developer communities for successful and widely adopted software have stood the test of time. Having said that, there is the likelihood that some existing communities could break up and, in fact, others have disbanded. As a result, it is now more important than ever for users of such software to learn how to master

the programming languages used in developing such software. Based on our online research of the four applications, we found the four CMSs have very wide user and developer communities. There is usually a core development team that works on the core engine, while add-ons and template themes are often developed and shared by numerous volunteers across the world on different Web sites.

To the issue of sustenance with respect to longevity and continuity, Stallman (2003) stated that "even allowing for Murphy to create a few unexpected problems, assembling these [software] components will be a feasible task ... The kernel will require closer communication and will be worked on by a small, tight group" (p. 546). Similarly Weber (2004) noted:

[c]ollaborative open source software projects such as Linux and Apache have demonstrated that a large and complex system of software code can be built, maintained, developed, and extended in a nonproprietary setting in which many work in a highly parallel, relatively unstructured way. (p. 2)

The extensive adoption of OSS is dependent on the awareness of the existence of such software among the public, and the availability of information about how to access and deploy them. Awareness about CMSs and OSS is minimal. Even among societies that have easy access to and high use of the Internet, knowledge about OSS is still minimal. Although the knowledge about OS and OS CMSs is growing, the tendency for it to be known among tech-savvy and highly motivated users of computer software is higher than among average users of software. As OS CMSs and OSS become mainstream, more publicity is needed in educational institutions and among teachers in colleges and in high schools. There is the need to educate teachers about OSS and encourage them to use OSS as alternatives to proprietary software in their schools.

Documentation

Moodle, Drupal, Mambo, and Xoops all have extensive documentation on how to install the applications that can be downloaded online. Documentation on Moodle, Drupal, Mambo, and Xoops are fragmented in comparison to their proprietary alternatives. However, there is undoubtedly a huge catalog of documentation online on all four. Lerner and Tirole (2002) are of the conviction that the nature of incentives available to open source programmers could be a reason for the fragmented nature of documentation and user interfaces in open source software.

The nature of Xoops, Mambo, Drupal, and Moodle documentation suggests an approach to support that says, "Good products need no publicity." There appears to be a suggestion that the economic and technical values of Xoops, Mambo, Drupal, and Moodle and similar OS CMS will make them popular and lead to their extensive adoption as alternatives to proprietary software. Among programmers and many avid users of computer software, this maxim could be true, but among the larger population of software users, Xoops, Mambo, Drupal, and Moodle and similar OSS have yet to make major impact as alternatives to existing proprietary software. What appears to be helping in popularizing Xoops, Mambo, Drupal, and Moodle is the promotion of OSS adoption by some governments around the world. For instance, Brazil has declared an open source country status, and all government establishments are adopting and installing OSS. Also, well-established Web sites like http://source-forge.org, http://cmsmatrix.org, and http://open-sourcecms.com have valuable information about OSS for adopters and prospective adopters of OSS. The Web sites of Xoops, Mambo, Drupal, and Moodle have cataloged developer discussions and documented how to deploy such applications. Robertson (2004) speaks for the need for better documentation for CMS applications. He commented that "most community-based open source

CMS products provide only a small amount of documentation, and other support information...it remains to be seen how this work can be funded within the open source business model" (p. 4).

The documentation on OS CMSs and indeed OSS falls under the user support services. The user support service offered by OSS developer communities are in three forms: on-demand help, forum contributions, and online published materials. These support services suggest that the developer communities and providers of such support services assume that prospective adopters of Xoops, Mambo, Drupal, and Moodle and other OSS need just the bare minimum of information on how to deploy such software in order to install and execute complete configuration.

Versatility, Stability, and Availability of Code

Weber (2004) described software source code as "the list of instructions that make up the 'recipe' for a software package" (p. 4). Robertson (2004) indicated that "having access to all the code of [a] content management system provides an unparalleled degree of flexibility" (p. 2) and invaluable access to resources sold under other license agreements. Access to open source code allows for easier integration of such software with other software. For example, with Web authoring software like NVu (open source), we were able to adapt some portions of Xoops, Mambo, Drupal, and Moodle for our needs. Although we are not programmers, having but a fair understanding of PHP, CSS, and HTML (hypertext markup language) coding, we were able to create our own templates, modify existing templates made available online, and adjust some functionalities on the core engine and add-ons of Xoops, Mambo, Drupal, and Moodle. For people who are not tech savvy, there is a large community of programmers and discussion groups online that people can tap into. Also, there is an emergence of resources and services in the form of books, consultants, and companies that offer suggestions for customization. Although some users and adopters of OS CMS may customize certain functions, the applications we reviewed can be used as is with a theme of the user's or adopter's choice.

The administrator back ends of Xoops, Mambo, Drupal, and Moodle are simple to maneuver. Other functions of the candidates can be activated or deactivated mainly through clicks, especially in the case of Moodle. Xoops, Mambo, and Drupal requires uploading modules and components for installation after the main engine has been installed. This could be a bit tricky since it requires that the site developer ensure the module or component he or she intends to deploy is compatible with the version of the engine that has been installed as was discussed under interoperability. Therefore, prior to selecting a CMS for use, it would be beneficial if an adopter makes a functional analysis on what the CMS-driven site will be used for, and search for a stable and non-beta version of the core engine of the CMS to be adopted. It will also require finding compatible add-ons with the version of the adopted CMS. In order to populate the Xoops, Mambo, Drupal, and Moodle CMSs with content, the back-end facilities provide uploading and publishing functions that allow for the publishing of digital images and documents, specifying terms for meta tags, composing disclaimer notices, updating user and administrator profiles, and specifying user access parameters. The basic administrator maintenance could be done using buttons that facilitate the submission, reviewing, updating, uploading, installing, and uninstalling of modules, and the editing and deleting of content. Depending on the user privileges, users could be granted the same access as the administrator or limited access that allows them to perform specific functions. These back-end features are indicative of the fact that most electronic content managed through CMSs require the creation and packaging of content outside of the CMSs.

Security

The security of content management systems centers on issues of dependability, that is, the inability of unauthorized individuals to gain access to the back-end operations of the application, and the ability to control the content that flows within the system to ensure integrity and credibility. However, accessibility to the source code of OSS is often cited as a weakness to such software (Raymond, 1999; Stallman, 2003). Such assertions are refuted on the grounds that the openness of the system should not have any real significance on software's security (Anderson, 2002; Raymond; Stallman). Xoops, Drupal, Mambo, and Moodle provide unlimited access to back-end operations to system administrators, and also provide log-in access to registered users in order to give them access to materials available on a site, or to upload, download, or post materials into the system. In the same manner, Xoops, Drupal, Mambo, and Moodle allow the site administrator to ensure the credibility of content on the site and in the system by regulating access to content online, and by providing secure access to features of the system. Han (2004) affirms that security issues in a CMS environment "generally [consist] of authentication and authorization. Authentication means the process of a system validating a user logon information, while authorization [involves] the right granted a user to use a system" (p. 356).

Advocates for open source like Payne (2002) contend that open source software is more secure than proprietary software due to the perceived strength of the peer review process. Programmers continuously work together to find bugs and develop patches to make OSS more robust and secure. Also, OS project development Web sites have concurrent versioning system (CVS) programs that track the development and changes of the software. A CVS works by allowing programmers to make available bugs and fixes to the CMS on the CMS documentation site or directly to the user community through other online resources.

Regarding the dependability or reliability of Xoops, Drupal, Mambo, and Moodle, the nature of open source projects is such that at any point of the development of the product, there are several volunteers working on them. Fuggetta (2003) asserts that the fact that open source codes are in the public domain ensures extreme scrutiny that leads to bug fixing and the discovery of code errors. Similarly, Hansen, Köhntopp, and Pfitzman (2002) contend that open source promotes software transparency that facilitates the quick detection of bugs and the development of patches as remedial measures through Internet access. They emphasized that "the availability or disclosure of source code is a necessary condition for security" (p. 467). In the same way, the flexibility of OS products makes it possible for security flaws to be determined by anyone who has access. For instance, Payne (2002) noted that "when the FTP site containing Wietse Venema's TCP wrapper software was broken into and the attackers modified the source code to contain a backdoor, the malicious code was discovered and corrected within a day"[3] (p. 64). Studies conducted by Kuan (2001), Bessen (2002), and Johnson (2004) at least point to the strengths of the OS approach to software development and maintenance. Bessen's claim of good security in open source software is premised on the fact that heterogeneous users have the opportunity to customize the OSS. This was confirmed by Kuan's study on a comparison between three open source software and their proprietary alternatives. She determined that in two of the three applications she studied, the rate of bug fixing was significantly faster in the open source projects than in their proprietary counterparts. Johnson argued that the open source process is more likely to lead to improvement in software development because the process and reward system makes the development communities less susceptible to cover-ups and connivance to conceal software

defects and programming errors. In spite of the fact that open source products have proved to be secure, Lerner and Tirole (2004) cautioned that much research into the superiority of OS in comparison to proprietary software has not yielded any conclusive evidence in favor of either approach to software development and maintenance. Xoops, Mambo, Drupal, and Moodle are doing well as CMSs. Activities on their project Web sites are indicative of their strengths and prospects. These software are under consistent periodic reviews for improvement and new releases. Having installed and tested Xoops, Mambo, Drupal, and Moodle, we recommend that prospective adopters of open source CMS should consider the strengths of CMS candidates based on the following issues:

- **Interoperability:** Can OSS CMSs be used across platforms? What are the database server requirements?
- **Software licenses:** What is the nature of the legal obligation for using the CMS? Is it GNU?
- **User community:** How popularity is its use? Is there a community frequently developing and using it?
- **Documentation:** Are there user guides? What support systems exist for users?
- **Versatility, stability, and availability of code, and security:** How resourceful are the CMSs for users and launchers for online use? Are they multipurpose in terms of usage and adaptation for specific services and online interaction? How robust are the back-end and front-end infrastructure? Is the source code available and what are the implications in terms of long-term support, the ability to add new features and fix bugs, and the ability to understand how the components work? The CMS must be able to limit access to certain content online and/or provide secure access to certain features.

FUTURE TRENDS

OS CMSs, specifically Drupal, gained prominence and was used significantly during the 2004 presidential elections. Howard Dean's presidential campaign team and the Deaniacs used Drupal extensively. Xoops, Mambo, Drupal, and Moodle are being used for interactive Web sites for various activities. They are popular for blogs used by journalists, and for online social interaction services and business activities. The Xoops, Mambo, Drupal, and Moodle Web sites have links to sites that are run on such software. For instance, sites that have adopted Moodle are obtainable at http://moodle.org/sites. They have proved to be functional for managing commercial sites similar to trends started by successful e-commerce sites like E-bay and Amazon. Trends in the development of OSS indicate that OS activities are on the rise. Data available on http://cmsmatrix.org indicate that Xoops, Mambo, Drupal, and Moodle are among the preferred CMSs. The site has statistical information generated from user responses to questions related to system requirements, security, support, ease of use, performance, management, interoperability, flexibility, built-in application, and commerce (including statistical information on Xoops, Mambo, Drupal, and Moodle). It also contains comparative information on various CMSs. Information generated from the site could be helpful in understanding the strengths and weakness of Xoops, Mambo, Drupal, Moodle, and many CMSs.

OS is championing the drive for more access to software technology, more participation in the development of software, and the localization and customization of software and content. For instance, Drupal and Moodle are noted for their multilingual functions that allow almost everyone to access it to create, edit, or publish content in their own languages. This approach to electronic publishing is a complete overhaul of conventional publishing, and allows for the localization of content and the use of culture-specific and sensi-

tive approaches to content creation. Such OS approaches to knowledge creation and dissemination are very radical and are considered subversive. However, they are generating dynamic political and economic responses among individuals and governments. There are countless numbers of online projects that run on OSS CMSs, and governments of countries like Brazil, Venezuela, and Argentina have declared their countries open source and are adopting OSS in government institutions. Arnold Schwarzenegger, governor of California, has expressed interest in OSS and advocates OSS as alternatives to proprietary software, a choice that could help deal with the budgetary woes of his state. Clearly, OSS has huge economic prospects, ones that can be explored only if investment is channeled into OSS development initiatives alongside a drive to increase the use of CMSs like Xoops, Drupal, Mambo, and Moodle. Although, OS may promise cuts in cost in terms of software purchasing, their adoption requires doing what it takes to install and configure such software for use. Not all configurations of OSS need programming experts, and our installation of Xoops, Mambo, Drupal, and Moodle is symbolic of this fact. OSS only seeks to promote self-help and do-it-yourself opportunities for everyone motivated enough to take up an electronic project. The OS movement has unleashed a digital production dragon that promotes alternative licensing rights that render software piracy useless and encourage collaboration. It has declared a new order of social and economic organization, and the creation and use of software that promise to give the proprietary software industry a good run for its money. Weber (2004) explained that the emergence of OSS is part of an almost epochal battle over who will control what is in the midst of a technological revolution.

Stallman's (2003) rebuttals to the objections to GNU goals as listed and explained in "The GNU Manifesto" capture the prospects of open source software in general. He emphasized:

If people would rather pay for GNU [OSS] plus service than get GNU [OSS] free without service, a company to provide just service to people who have obtained GNU [OSS] free ought to be profitable. ... Such services could be provided by companies that sell just hand-holding and repair service. If it is true that users would rather spend money and get a product with service, they will also be willing to buy the service having got the product free ... The sale of teaching, hand-holding and maintenance services could also employ programmers. (pp. 547-550)

Regarding the future of OSS, Lerner and Tirole (2004) refer to a rise of corporate investment into open source projects. (IBM is reported to have spent over $1 billion in 2001 alone on such projects; Microsoft, Adobe, Google, Hewlett-Packard, and many more have expressed interest in and supported open source projects.) Also, there is an upsurge of political interest around the world in OSS by governments including the United States, the European Union, China, Brazil, Mexico, South Africa, Uganda, India, and many more. However, there is the need to improve the documentation and publicity of OSS. The sale of programming and customization services could and should be on the rise. Also, the forking of OS projects give rise to concern about the future of OSS. Forking refers to splits that can and does happen among OSS development groups that lead to the concurrent development of new applications that are derivatives of older ones, for instance, the creation of Joomla, an emerging CMS based on the Mambo core engine. The nature of OS applications is such that the availability of source code makes it possible for the emergence of new applications from existing ones. The outcomes of forking are twofold: It could disrupt the development of the initial application due to the splitting up of core developers, or it could lead to the emergence of better applications.

CONCLUSION

We examined four open source content management systems to determine a set of criteria that an adopter could consider when selecting an open source CMS for use. We noted that, in using open source content management systems, it is important to identify the purpose of the dynamic site before embarking on installing and implementing the application. Having adequate documentation and a supportive user community are highly important. Although the application may be obtained at no cost, deploying it may require dedication and know-how to ensure success. Having knowledge about the security; versatility, stability, and availability of the source code; documentation; user community; software licenses; and interoperability of a CMS candidate are essential for success.

REFERENCES

Anderson, R. (2003). *Security in open source versus closed systems: The dance of Boltzman, Coase and Moore.* Unpublished manuscript.

Arnold, S. E. (2003). Content management's new realities. *Onlinemag.*

Bessen, J. (2002). *Open sources software: Free provision of complex public goods.* Unpublished manuscript.

Boateng, K., & Boateng, B. A. (2006). Open source community portals for e-government. In M. Khosrow-Pour (Ed.), *Encyclopedia of e-commerce, e-government and mobile commerce* (pp. 884-889). Hershey, PA: Idea Group Reference.

Friedlein, A. (2003). *Maintaining and evolving successful commercial Web sites: Managing change, content, customer relationships and site measurement.* San Francisco: Morgan Kaufmann Publishers.

Fuggetta, A. (2003). Open source software: An evaluation. *The Journal of Systems and Software, 66*, 77-90.

Gambardella, A., & Hall, B. H. (2005). *Proprietary vs. public domain licensing of software and research products* (NBER Working Papers Series No. 11120). Cambridge, MA: National Bureau of Economic Research. Retrieved from http://www.nber.org/papers/wl1120

Garfinkel, S. (1999). *Open source: How secure?* Retrieved from http://www.wideopen.com/story/101.html

Golden, B. (2005). *Making open source ready for the enterprise: The open source maturity model.* Retrieved May 10, 2006, from http://www.navicasoft.com/Newsletters/OSMM Whitepaper.pdf

Han, Y. (2004). Digital content management: The search for a content management system. *Library Hi Tech, 22*(4), 355-365.

Hansen, M., Köhntopp, K., & Pfitzman, A. (2002). The open source approach: Opportunities and limitations with respect to security and privacy. *Computers and Security, 21*(5), 461-471.

Johnson, J. P. (2004). *Collaboration, peer review and open source software.* Unpublished manuscript.

Kuan, J. (2001). *Open source software as consumer integration into production.* Unpublished manuscript.

Lagoze, C., & Van de Sompel, H. (2001). The open archives initiative: Building a low-barrier interoperability framework. *First ACM/IEEE-CS Joint Conference on Digital Libraries (JCDL'01),* 54-62.

Lerner, J., & Tirole, J. (2002). Some simple economics of open source. *Journal of Industrial Economics, 52*, 197-234.

Lerner, J., & Tirole, J. (2004). *The economics of technology sharing: Open source and beyond*

(NBER Working Papers Series No. 10956). Cambridge, MA: National Bureau of Economic Research. Retrieved from http://www.nber.org/papers/wl0956

Payne, C. (2002). On the security of open source software. *Information Systems Journal, 12*, 61-78.

Raymond, E. (1999). *The cathedral and the bazaar: Musing on Linux and open source by an accidental revolutionary.* O'Reilly.

Robertson, J. (2004). *Open source content management systems.* Retrieved February 10, 2006, from http://www.steptow.com.au

Schneier, B. (1999). *Crypto-gram.* Retrieved from http://www.counterpane.com/crypto-gram9909.html#OpenSourceandSecurity

Stallman, R. (2003). The GNU manifesto. In N. Wardrip-Furin & N. Montfort (Eds.), *The new media reader.* Cambridge, MA: The MIT Press.

Weber, S. (2004). *The success of open source.* Cambridge, MA: Harvard University Press.

KEY TERMS

Back Door: A code that can be attached to an application or software to enable the bypass of security mechanisms.

Beta Version: An application or software at the testing stage.

Concurrent Versioning System (CVS): A control system used by open source developers to record the history of source files and documents.

Engine: Codes and files that form the heart of an application.

Forking: The emergence of new software from other applications.

GNU: GNU is not UNIX. It primarily stands for ideas for free software.

Modules or Components: A CMS element that is already available within an OS CMS or can be appended to enable specific functionalities.

ENDNOTES

[1]   E-learning, e-library, e-commerce, e-news, e-government

[2]   Moodle is primarily considered a learning management system. It is used mostly for managing online education activities. However, it is multipurpose software that could be used for most online activities. It could be used for managing Web sites in general.

Chapter XXI
Evaluating Open Source Software through Prototyping

Ralf Carbon
Fraunhofer Institute for Experimental Software Engineering (IESE), Germany
Software Engineering: Processes and Measurement Research Group, Germany

Marcus Ciolkowski
Fraunhofer Institute for Experimental Software Engineering (IESE), Germany
Software Engineering: Processes and Measurement Research Group, Germany

Jens Heidrich
Fraunhofer Institute for Experimental Software Engineering (IESE), Germany
Software Engineering: Processes and Measurement Research Group, Germany

Isabel John
Fraunhofer Institute for Experimental Software Engineering (IESE), Germany

Dirk Muthig
Fraunhofer Institute for Experimental Software Engineering (IESE), Germany

ABSTRACT

The increasing number of high quality open source software (OSS) components lets industrial organizations seriously consider integrating them into their software solutions for critical business cases. But thorough considerations have to be undertaken to choose the "right" OSS component for a specific business case. OSS components need to fulfill specific functional and non-functional requirements, must fit into a planned architecture, and must comply with context factors in a specific environment. This chapter introduces a prototyping approach to evaluate OSS components. The prototyping approach provides decision makers with context-specific evaluation results and a prototype for demonstration purposes. The approach can be used by industrial organizations to decide on the feasibility of OSS components in their concrete business cases. We present one of the industrial case studies we conducted in a practical course at the University of Kaiserslautern to demonstrate the application of our approach in practice. This case study shows that even inexperienced developers like students can produce valuable evaluation results for an industrial customer that wants to use open source components.

EVALUATING OPEN SOURCE SOFTWARE THROUGH PROTOTYPING

There is an increasing number of open source software (OSS) projects that release software components which provide almost complete sets of functionality required in particular domains. These components are often also of such high quality that in more and more cases industry is seriously considering to use them as part of their commercial products. In such scenarios, OSS components must certainly compete with any similar component on the market including other OSS projects and commercial solutions.

The model behind OSS is generally more attractive to companies than commercial business models, especially for small and medium-sized companies, due to the free distribution of OSS, the full access to sources and documentation, as well as quick responses and support by the community consisting of developers and other users. The implementation of this OSS model and the quality of the software, however, varies significantly from one OSS project to another. Hence, it is crucial for an organization to systematically investigate the implementation of the OSS model for the particular OSS projects whose software it considers to reuse.

Reusability of any type of software (including OSS, in particular) depends on the quality of the software itself as well as that of its documentation. Code quality is affected, for example, by code comments, structuring, coding style, and so forth. The quality of available documentation is defined by its readability, comprehensibility, or technical quality, and by its suitability for the intended reuse scenarios involving OSS. Besides documentation, the community supporting particular OSS projects is a crucial element, too. Response time and quality of community feedback depend on the overall size of the user group and the skill level of its members. All these aspects should be explicitly evaluated before an

OSS is reused in real projects, let alone in critical projects. Note that all of these aspects may not only vary significantly from one OSS project to another, but also heavily depend on the concrete context and reuse scenarios of the OSS.

This chapter reports on a way to evaluate OSS in a holistic way. That is, OSS components are firstly evaluated like any other potential COTS (commercial off-the-shelf) component; and secondly they are used in a prototype project similar to, but smaller than the intended product developments, including an evaluation of the product in the context of the projected architecture to avoid architectural mismatches, as well as an evaluation of the support provided by the related community. To minimize the costs of such a pre-project evaluation, an evaluation team consisting of a group of graduate computer science students may be deployed. A prototyping approach can also be used to gather more detailed information on the adequacy of COTS components for a specific context. But especially for the selection of OSS components a prototyping approach pays off. The quality of the source code and the development documentation can be evaluated, for instance. This increases trust in the quality of the component. Furthermore, it is even possible to evaluate if the OSS component can be easily adapted by oneself to better fulfill the specific requirements.

The chapter presents experience from several OSS evaluation projects performed during the last few years in the context of a one-semester practical course on software engineering at the University of Kaiserslautern. The systematic and sound evaluation was supported by researchers of the Fraunhofer Institute for Experimental Software Engineering (IESE).

Each evaluation employed a temporary team of students to conduct a feasibility study, that is, realizing a prototypical solution based on OSS to be evaluated as specified by industrial stakeholders. The industrial stakeholder always provided a set of functional and quality requirements and a projected architecture for the envisioned products; optionally, it already referred to an initially selected OSS component potentially suitable for the given task.

The question to be answered eventually by each evaluation project is whether the OSS component(s) under consideration is (or are) usable as a basis for the kind of products the industry stakeholder envisions.

The chapter first provides an overview of evaluation approaches relevant to the evaluation projects presented, then the approach itself is presented and its application is shown exemplarily by means of an evaluation project performed in 2004/2005 in cooperation with an industrial stakeholder. Finally, an overview of open issues and future trends is given.

BACKGROUND

The background and work related to our work described here can be seen in two areas:

- COTS evaluation
- Open source evaluation

As we describe the evaluation of software in this chapter, related contributions can be found in the area of COTS evaluation. Current research does no longer strongly distinguish between evaluation frameworks for COTS and for open source software components (di Giacomo, 2005; Li, Conradi, Slyngstad, Bunse, Torchiano, & Morisio, 2006; Paulson, Succi, & Eberlein, 2004). A range of COTS-based evaluation methods has been proposed, the most widely used ones being off-the-shelf option (OTSO) and procurement-oriented requirements engineering (PORE). The OTSO method (Kontio, 1995) provides different techniques for defining evaluation criteria, comparing the costs and benefits of alternative products, and consolidating the evaluation results for decision-making. OTSO assumes that clear requirements already exist, since it uses a require-

ments specification for interpretation. The PORE method (Ncube & Maiden, 1999) is a template-based approach to support requirements based COTS acquisition. The method uses an iterative process of requirements acquisition and product evaluation. The method proposed by Ochs, Pfahl, Chrobok-Diening, and Nothelfer-Kolb (2001) uses a risk analysis approach for COTS selection and explicitly takes risks into account but has no means for dealing with unclear requirements. A method that focuses on requirements in COTS is the social-technical approach to COTS evaluation (STACE) framework (Kunda & Brooks, 1999). It emphasizes social and organizational issues in the COTS selection process, but has no clear requirements integration process.

Some seminal work focusing on the evaluation and selection of OSS for an industrial context can be found in the literature. First thoughts on the parallel evaluation of COTS and open source can be found in (Sciortino, 2001), where the author identifies strategic decisions when selecting a certain software component. Kawaguchi, Garg, Matsushita, and Inoue (2003) propose a method for the categorization of open source components from their code basis. This can be seen as the first step towards an evaluation framework, but their work aims in a different direction. Wang and Wang (2001) emphasize the importance of requirements when selecting an open source component. They identified several open source software characteristics, but leave the process for evaluation open (Ruffin & Ebert, 2004) also define characteristics specific to open source software, but do not focus on evaluation in their work. A more elaborate approach is the business readiness rating (BRR) (Business Readiness Rating for Open Source, 2006). Here a process is presented to assess open source software for industrial use. The four phase assessment and filtering process results in a final business readiness rating for all selected open source software packages and also takes into account soft factors like quality and community. Different from the approach we propose here, the requirements and the metrics have to be clear from the beginning of the assessment. With our prototyping-based approach it is possible to gain knowledge step by step and to try the open source software package in the context of the intended use. Nevertheless, it is imaginable to combine both approaches and to use our prototyping-based approach during step three of the BRR for data collection and processing.

Most of these approaches mainly focus on functional, hard characteristics that have to be clear from the beginning of the evaluation process. They disregard non-functional and contextual factors like performance of the components and development community aspects and do not consider changing requirements. Those aspects can only be captured and evaluated in a prototyping approach where the open source software is actually used to realize functionality in the context of the user and developer.

MAIN FOCUS OF THE CHAPTER

In this section, we present an approach to evaluate OSS components in an industrial context. Users and developers are often concerned whether OSS candidates provide adequate functionality and enough quality to be integrated into the products they use or sell. We propose an approach where OSS candidates are applied in a practical context by means of prototyping, in other words, a prototype of the system an OSS candidate is supposed to be integrated into is developed. Based on the results, final decision support can be provided. Our approach is goal- and feedback-oriented according to the goal-question-metric (GQM) approach (Basili, Caldiera, & Rombach, 1994a) and the quality improvement paradigm (QIP) (Basili, Caldiera, & Rombach, 1994b). Figure 1 gives an overview of the approach. It takes into account functional and non-functional requirements of the software system to be developed, architectural constraints, context factors, OSS

candidates, and OSS specifics as input products. In the preparation phase, an initial requirements analysis is done, OSS candidates are selected, and evaluation teams are set up.

The prototyping phase follows an iterative approach, in other words, the prototypes are developed in several iterations. At the end of each iteration, an evaluation step is performed. If several OSS candidates are available, one prototype per OSS candidate is developed in parallel. The prototyping phase is followed by the final evaluation, where all evaluation results from the prototyping phase are consolidated and integrated into an evaluation report.

The approach has been applied and validated five times so far by request of different industrial customers, namely Maxess Systemhaus GmbH (Ciolkowski, Heidrich, John, Mahnke, Pryzbilla, & Trottenberg, 2002), BOND Bibliothekssysteme, market maker Software AG (two times), and the city council of Kaiserslautern.

The input for the evaluation process was provided by the customers, the prototyping was done by students, who were typically in their 3rd year of computer science study. Researchers of the University of Kaiserslautern and Fraunhofer IESE managed the evaluation projects.

In the following, our evaluation approach is presented in detail and we describe the input and output products and the evaluation process itself in detail. After that, we demonstrate the feasibility of our approach by means of one of the five evaluation projects mentioned previously.

Input Products

- **Functional requirements:** The functional requirements specify the software system that is supposed to incorporate OSS candidates. The functional requirements of the prototype to be developed can be a subset of them, for example, only including the mandatory requirements. The required functionality is the primary indicator for

Figure 1. Overview of the OSS evaluation approach

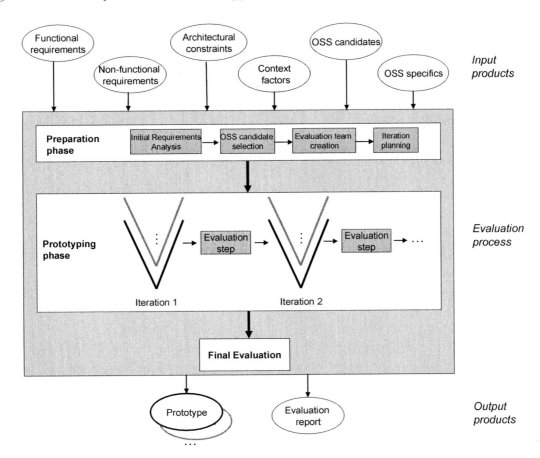

the applicability of an OSS candidate in the customer's context. If the OSS candidate does not cover the functionality that is requested by it, it will be rejected.

- **Non-functional requirements:** The non-functional requirements specification defines the quality required of the software system to be developed. Such quality characteristics are, for instance, reliability, performance, usability, or flexibility. The non-functional requirements already need to be fulfilled by the prototype. In our case, the contribution of the OSS candidates to the quality of the whole system has to be evaluated carefully. If the OSS candidates are not capable of providing the required quality level, they will be rejected.

- **Architectural constraints:** The architecture describes the planned architecture of the software system to be developed. The architecture of a software system provides several views on a software system and describes, for instance, its structure, in other words, how it is composed of logical components, and how these interact with each other. By means of an appropriate architecture, several non-functional requirements can be guaranteed right from the beginning of a project. In the context of the evaluation process of OSS candidates, at least the architectural constraints need to be defined up-front. They provide valuable information on where and how OSS candidates can be plugged into the planned system. The architectural constraints help to refine the requirements for the OSS candidates, for instance, which interfaces an OSS candidate has to provide. Furthermore, it can be determined whether the OSS candidate is critical to the fulfillment of important non-functional requirements in the actual context.

- **Context factors:** Every OSS component is used in a specific context. The context is characterized by a set of context factors. Examples of context factors are the application domain a software system is developed for, the experience of the developers of the software system, or the technologies that are supposed to be used for the realization of the software system. For instance, if the OSS candidates were to be used in a domain with high security requirements for the first time, for instance, this should be taken into account during evaluation. If the developers have no experience with a specific technology used in an OSS candidate, this could even be a reason to reject the OSS candidate at once, because they would not be able to adapt it without spending an unreasonable amount of effort.
- **OSS candidates:** The OSS candidates are of central interest in our evaluation approach. Sometimes the customer, for instance, has already done a preliminary selection of OSS candidates. Such a set of OSS candidates is then input to our evaluation process. If no preliminary selection has been performed, the evaluation process itself has to start with the search for OSS candidates.
- **OSS specific issues:** OSS components are concerned with recurring issues. Usually, components have to be adapted before they can be integrated into a new system. Crucial questions are, therefore, whether an OSS component is documented adequately or whether competent support is provided. Our evaluation process explicitly takes such OSS-specific issues into account.

Output Products

- **Prototype:** The outputs of the prototyping phase are executable prototypes of the system to be developed. These prototypes can be used to demonstrate whether the specified functional and non-functional requirements can be fulfilled by means of the used OSS candidates. The customer, for example, can gain confidence that an OSS candidate is applicable in the respective context.
- **Evaluation report:** The evaluation report gives a detailed summary of the results of the evaluation process. The evaluation report is handed over to the customer and gives comprehensive support for the final decision on the selection of an OSS candidate.

Evaluation Process

The evaluation process is organized in an iterative manner. If more than one consistent set of OSS components has to be evaluated, several evaluation teams are created, who perform the development process in parallel with certain synchronization points. We call a consistent set of OSS components an OSS candidate; that is, this combination of OSS products is a possible candidate for creating the final product the customer is interested in. After each iteration, the prototype, which is based on a consistent set of OSS components, is evaluated with respect to the functional and non-functional requirements as well as OSS-specific issues. Based on this evaluation, the customer has the possibility to adapt his requirements (e.g., stop evaluation of an OSS candidate; perform a deeper evaluation of a certain OSS candidate). The adapted requirements are the starting point for the next iteration. At the end, a final evaluation report is created. You can find an overview of the whole evaluation process in Figure 1. In detail, we distinguish between the following activities:

- **Initial requirements analysis:** During initial requirements analysis, the students identify a list of functional and non-functional requirements by performing structured interviews with the customer. In some cases, the customer provides an initial list of requirements and even architectural constraints that have to be incorporated into a requirements document. It is important to list all requirements that are important for making a decision about using a certain OSS candidate. For the prototypes, it is not necessarily useful to list all requirements a final product should have, which is created out of the positively evaluated OSS candidates later on. Quite on the contrary, it is important to first focus on essential requirements the final product must have. To support this focus, we also prioritize all requirements. The final requirements document, which is the starting point for all evaluation teams, must also contain a specification of all non-functional requirements (preferably in a quantitative form by means of, for instance, the GQM approach) that are important for evaluating the OSS candidates (such as data throughput and latency). These requirements must be assessed after each iteration of the development process later on.

- **OSS candidate selection:** During this activity, a survey of possible OSS candidates is conducted in order to find suitable OSS components based on customer requirements. The students perform a Web-based search for OSS components that match the functional customer requirements and create an overview of components found. In doing this, especially OSS-specific issues are addressed, such as community size and support. This overview is discussed with the customer and a set of components is identified. For practical reasons, the number of candidates is limited by the number of evaluation teams that can be grouped out of the students later on. Usually, two candidates are evaluated and compared against each other. If the customer already has a predefined set of OSS components that should be evaluated, this step can be skipped.

- **Evaluation team creation:** When the basic requirements have been analyzed and the components to be evaluated have been defined, the evaluation teams are created. Each evaluation team consists of about four to seven students. One student assumes the role of the project manager and coordinates all others. Usually, a quality assurance manager is also determined, who is responsible for all verification and validation activities performed as well as for checking that the prototype satisfies the non-functional requirements.

- **Iteration planning:** Based on the actual requirements document, the development iterations are planned accordingly. The requirements to be implemented in certain iterations are determined based upon the assigned priority. Usually, two to three iterations are planned.

- **Iterative prototype development:** When the evaluation teams have been created and iterations have been planned, the actual development work starts. Based upon the requirements document, each evaluation team starts to create a prototype for the corresponding set of requirements to be implemented for the current iteration. This includes application design based upon application constraints by the customer, implementation, and, finally, testing. Usually, each document created is reviewed with respect to its correctness, completeness, and consistency. Therefore, checklist-based inspections are performed. During the development process, the customer is involved in project

275

meetings in a previously defined manner (usually, once a week) in order to keep track of the development and to answer open questions. The project meetings are usually conducted together with all evaluation teams in order to discuss problems which a candidate may have during development. During each iteration, a specific role, the quality assurance (QA) manager, continuously checks all non-functional customer requirements and OSS-specific issues (such as response time of support requests in news groups or email lists). At the end of each development iteration, the QA manager of each evaluation team prepares a small presentation in order to give an overview of the pros and cons of the OSS candidates evaluated. The end of one individual development iteration marks a synchronization point for each evaluation team. Based upon the QA presentation, the project managers and the customer discuss which requirements to implement next and, in extreme cases, whether a certain candidate or components of a candidate should be replaced in future development.

- **Final evaluation:** After the last iteration is done, a final evaluation report is created. This report contains a detailed discussion of all OSS-specific issues in the form of a lessons-learned table for each issue. For each quantitatively expressed non-functional requirement, test results are included. In general, the pros and cons of the OSS candidates are addressed, as is a recommendation on how to overcome the limitations found and which candidate to use. So, the final assessment evaluates the OSS candidates with respect to their applicability for the intended software product, based on joint analyses of all relevant characteristics (comprising functionality and quality) of the resulting prototype, as well as on the experience gained during implementation.

CASE STUDY

In this section, we present in detail one of the five projects where we applied our OSS evaluation approach in practice. We conducted this project during the winter semester 2004/2005 at the University of Kaiserslautern, Germany, as a practical course. Sixteen students conducted an OSS evaluation project according to our approach for the market maker Software AG, a provider of Web-based stock market information systems. The company demanded a test tool for Web applications and wanted to evaluate the feasibility of two OSS components in this context. The focus in this chapter is on the presentation of our approach and not on the discussion of the pros and cons of two concrete OSS components. Thus, we call the two components pointed out by the customer component A and component B.

The students worked full-time on this OSS evaluation project for 8 weeks. The prototyping phase usually consumes most of the time of an evaluation project. In this case, the preparation phase consumed one week, the prototyping phase six weeks, and the final evaluation again one week. The project was managed by two researchers. In the beginning, the customer provided them with a detailed problem statement. A short summary of the information included in the problem statement according to our classification of the input products of our approach can be found in Table 1.

The project started with a preparation phase. In this case, a half-day requirements workshop was conducted together with the students, the customer, and the researchers in order to detail and prioritize the requirements and consolidate a common understanding of all stakeholders. The selection of OSS candidates had already been done by the customer. Two OSS candidates had been selected and thus two prototypes needed to be built according to our approach. The students were split into two teams of 8 students each, who then performed the prototyping concurrently.

One team developed a prototype of the test tool based on component A, the other team developed a prototype based on component B. It was decided to perform two iterations.

In the first iteration, the two teams mainly evaluated functional aspects and the extensibility of the OSS candidates. First, they developed a solution for generalizing/parameterizing test case building blocks and then they evaluated whether their solution can be supported by component A/component B. The main evaluation results after the first iteration were:

1. Both of the OSS candidates provide the expected functionality to trace user interaction and perform load tests, but additional functionality is necessary to satisfy all customer-specific requirements, for instance, regression test functionality.
2. The tool-specific scripting language used by component B for the specification of test plans and the test output needed to be mapped to an XML format as specified by the customer. Component A already uses an XML format for the specification of test plans and outputs, which is a key advantage of component A over component B.
3. Both of the OSS candidates are documented insufficiently from a developer's point of view; for instance, the documentation of the code is incomplete and neither architecture nor design documentation is available. A modification of the existing functionality of the OSS candidates seemed to be too risky. Thus, the OSS candidates would be plugged into the prototype without modification of the existing functionality.
4. The two components differ in their extension mechanisms. Component A explicitly provides several extension points that can be used to add additional functionality. In the case of component B, the only way to extend the functionality seemed to be to add new functionality to additional components.

Table 1. Short summary of the project inputs

Functional requirements	• Traces of the user interaction with the system via a Web browser can be captured.
	• Traces can be divided into semantically cohesive, generic/parameterized building blocks of test suites, for example, user login.
	• Test suites can be derived from the generic/parameterized building blocks.
	• Test suites can be executed:
	  o Load and performance tests
	  o Regression tests
	• Test results are documented.

Non-functional requirements	• **Extensibility:** The OSS components must be easily extendable. • **Performance:** The OSS components must be capable of hitting the test candidate with a large amount of requests per time unit.
Architectural constraints	• Traces of the user interaction must be captured by the OSS component. • The execution of test cases must be performed by the OSS component. • Test candidates must be accessed via http. • The data format of the test building blocks must be XML based. • The test tool must be integrable into the customer's build environment.
Context factors	**Domain:** Web-based stock-market information systems **Programming language:** Java **Experience of student developers:** Low **Experience of the customer's developers:** High
OSS candidates	Component A, component B
OSS specifics	OSS under GNU license may not be used in this case.

This has to be evaluated in more detail in iteration 2.

In the second iteration, the teams detailed their solutions, developing additional customer-specific functionality not supported by the OSS candidates so far, for instance, regression test functionality. Another evaluation of the extensibility of the two OSS candidates was conducted and the performance of the prototypes was investigated. The main results of iteration 2 were:

1. The extensibility of component A was evaluated to be higher than that of component B. As already assumed after the first iteration, it turned out that one crucial problem with component B was the usage of its tool-specific scripting language. The mapping of the test building blocks specified in XML to a test script could be realized much easier for the XML-based format of component A. This led to a significant gap in productivity between the two teams in iteration two. The team using component A was capable of providing much more functionality. In addition, the team using component A also benefited from the predefined extension points of this component.

2. A significant difference in performance could not be observed between the two prototypes. Both OSS candidates provide a clustering functionality for load testing, in other words, the tested Web page can be queried with requests from several clients distributed across several machines. Thus, a high load can be produced efficiently.

In the end, the results of the prototyping phase were consolidated and integrated during a final evaluation. The prototypes and the evaluation results were presented to the customer. Based on this presentation, the customer decided to elaborate a solution based on component A. Because the OSS candidates differ only little in function-ality and performance, the main reason for the decision of the customer was the extensibility of component A, which was enabled by the XML-based format for test plans and the predefined extension points.

This case study demonstrates the application of our approach for evaluating OSS through prototype development. It shows that the application of the OSS candidates leads to a comprehensible decision on choosing one of them for further elaboration.

FUTURE TRENDS

In addition to writing code, evaluation of OSS components will grow in importance and be considered a valuable contribution to the OSS community. With the growing widespread usage of OSS, the evaluation of OSS components will become ever more crucial to fostering the adoption of OSS by society. Therefore, it lies within the interest of the OSS community to provide evaluations of the applicability of their OSS components. Consequently, systematic evaluations of OSS components present a valuable contribution to OSS projects.

In addition, empirical evaluation and its systematic documentation will continue to gain importance in the area of OSS (Virtuelles Software Engineering Kompetenzzentrum, 2006; CeBASE—NSF Centre for Empirically Based Software Engineering, 2006). This includes systematic case studies, as presented in this chapter, as well as controlled experiments (Boehm, Rombach, & Zelkowitz, 2005) to investigate certain aspects in a laboratory setting; for example, a controlled experiment could investigate the importance of documentation for the maintenance of a particular OSS component (Endres & Rombach, 2003). Another aspect of empirical evaluation is the installation of testbeds (Lindvall et al., 2005). Testbeds are environments or platforms that allow carrying out case studies in a more controlled and comparable way. For example, such a platform could be configured to run projects on CMMI level 3-5, while integrating teams of students and practitioners. Such a setting would allow evaluating more detailed aspects of OSS components under more reproducible and more realistic industrial settings than the approach presented in this chapter. Currently, Fraunhofer IESE is setting up such a platform.

CONCLUSION

The approach presented has proven to produce real-life experience, which is significant to industry organizations when deciding on serious usage of OSS components. Especially the short time period required and the deployment of a non-expert team led to well-grounded and thus better decisions. The selected set-up thus provides a great return-on-investment. Hence, we recommend for any organization to perform such a systematic evaluation of OSS through prototyping projects before using OSS in their mission-critical projects.

REFERENCES

Basili, V.R., Caldiera, G., & Rombach, H.D. (1994a). The goal question metric approach. In J. J. Marciniak (Ed.), *Encyclopedia of software engineering* (2nd ed., Vol. 1, pp. 578-583). New York: John Wiley & Sons.

Basili, V. R., Caldiera, G., & Rombach, H. D. (1994b). The experience factory. In J. J. Marciniak (Ed.), *Encyclopedia of software engineering* (2nd ed., Vol. 1, pp. 511-519). New York: John Wiley & Sons.

Boehm, B., Rombach, D., & Zelkowitz, M. (Eds.). (2005). *Foundations of empirical software engineering: The legacy of Victor R. Basili*. Berlin: Springer-Verlag.

Business Readiness Rating for Open Source (n.d.). Retrieved June 29, 2006, from http://www.openbrr.org

CeBASE—NSF Centre for Empirically Based Software Engineering (n.d.). Received June 29, 2006, from http://www.cebase.org

Ciolkowski, M., Heidrich, J., John, I., Mahnke, W., Pryzbilla, F., & Trottenberg, K. (2002). *MShop—Das Open Source Praktikum* (IESE-Report, 035.02/D). Germany: Fraunhofer Institute for Experimental Software Engineering.

Demming, E. (1986). *Out of the crisis*. Cambridge, MA: MIT Press.

Endres, A., & Rombach, D. (2003). *A handbook of software and systems engineering: Empirical observations, laws and theories*. Amsterdam: Addison-Wesley.

di Giacomo, P. (2005). COTS and open source software components: Are they really different on the battlefield? In *Proceedings of the Fourth International Conference on COTS-Based Software Systems*, Bilbao, Spain (pp. 301-310).

Kawaguchi, S., Garg, P. K., Matsushita, M., & Inoue, K. (2003). On automatic categorization of open source software. In *Proceedings of the Third WS on Open Source SE* (pp. 79-83). Portland, OR.

Kontio, J. (1995). A COTS selection method and experiences of its use. In *Proceedings of the 20th Annual Software Engineering Workshop*, Maryland, USA.

Kunda, D., & Brooks, L. (1999). Applying social-technical approach for COTS selection. In *Proceedings of the Fourth UKAIS Conference* (pp. 552-565). UK: University of York.

Li, J., Conradi, R., Slyngstad, O., Bunse, C., Torchiano, M., & Morisio, M. (2006). An empirical study on off-the-shelf component usage in industrial projects. In R. Conradi (Ed.) *Software process improvement: Results and experience in*

Norway (pp. 54-68). Berlin, Germany: Springer LNCS.

Lindvall, M., Rus, I., Shull, F., Zelkowitz, M., Donzelli, P., Memon, A., Basili, V., Costa, P., Hochstein, A., Asgari, S., Ackermann, C., & Pech, B. (2005). An evolutionary testbed for software technology evaluation. *Innovations in Systems and Software Engineering, 1*(1), 3-11.

Ncube, C., & Maiden, N. A. M. (1999). PORE: Procurement-oriented requirements engineering method for the component-based systems engineering development paradigm. In *Proceedings of the International Workshop on Component-Based Software Engineering*, Los Angeles (pp. 1-12).

Ochs, M., Pfahl, D., Chrobok-Diening, G., & Nothhelfer-Kolb, B. (2001). A method for efficient measurement-based COTS assessment and selection-method description and evaluation results. In *Proceedings of the Seventh IEEE International Software Metrics Symposium*, London (pp. 285-290).

Paulson, J. W., Succi, G., & Eberlein, A. (2004). An empirical study of open-source and closed source software products. *IEEE Transactions on Software Engineering, 30*(4), 246-256.

Ruffin, M., & Ebert, C. (2004). Using open source software in product development: A primer. *IEEE Software, 21*(1), 82-86.

Sciortino, M. A. (2001). *COTS, open source, and custom software: Which path to follow COTS, open source, and custom software.* Retrieved June 29, 2006, from http://www.acsu.buffalo.edu/~ms62/papers/softwarepath.htm

Virtuelles Software Engineering Kompetenzzentrum (n.d.). Retrieved June 29, 2006, from http://www.softwarekompetenz.de

Wang, H., & Wang, C. (2001). Open source software adoption: A status report. *IEEE Software, 18*(2), 90-95.

KEY TERMS

Case Study: An observational empirical study, which is done by observation of an on-going project or activity. A case study typically monitors a project or assignment. Case studies are normally aimed at tracking a specific attribute or establishing relationships between different attributes.

Context (Factor): The context is the environment in which an empirical study is run. Context factors are influences in the context (such as the experience of developers) that may have an impact on the phenomenon under observation.

Empirical Evaluation: A study where the research ends are based on evidence and not just theory. This is done to comply with the scientific method that asserts the objective discovery of knowledge based on verifiable facts of evidence. This includes observing a phenomenon under laboratory conditions (e.g., in a controlled experiment) or in the field (e.g., in a case study).

Goal-Question-Metric Paradigm (GQM): The goal-question-metric (GQM) paradigm has been proposed to support the definition of quantifiable goals and the interpretation of collected measurement data. It is a goal-oriented approach to derive metrics from measurement goals to ensure that collected data is usable and serves a purpose.

Iterative Software Development: Denotes a software development process that splits system development into several parts (or iterations). The basic idea behind iterative development is to develop a software system incrementally, allowing the developer to take advantage of what was being learned during the development of earlier, incremental, deliverable versions of the system.

Open Source Software (OSS): Software whose code is developed collaboratively, and is freely available to the public under specific license conditions.

Prototyping: Denotes the process of quickly putting together a working model (a prototype) in order to test various aspects of a design, illustrate ideas or features and gather early user feedback. Prototyping is often treated as an integral part of the system design process, where it is believed to reduce project risk and cost. Its characteristic is that prototypes are typically developed without adhering to software engineering principles, which typically results in products that are not maintainable.

Quality Improvement Paradigm (QIP): The quality improvement paradigm is a general improvement scheme tailored for the software business. It is a goal-driven feedback-oriented improvement paradigm for software engineering based on total quality management principles, and on the plan/do/check/act cycle (Demming, 1986).

Chapter XXII
Motives and Methods for Quantitative FLOSS Research

Megan Conklin
Elon University, USA

ABSTRACT

This chapter explores the motivations and methods for mining (collecting, aggregating, distributing, and analyzing) data about free/libre open source software (FLOSS) projects. It first explores why there is a need for this type of data. Then the chapter outlines the current state-of-the art in collecting and using quantitative data about FLOSS project, focusing especially on the three main types of FLOSS data that have been gathered to date: data from large forges, data from small project sets, and survey data. Finally, the chapter will describe some possible areas for improvement and recommendations for the future of FLOSS data collection.

INTRODUCTION

Numbers, statistics, and quantitative measures underpin most studies of free/libre open source (FLOSS) software development. Studies of FLOSS development usually require the researchers to have answered questions like: How many FLOSS projects are there? How many developers? How many users? Which projects are dead, which are flourishing? What languages are popular for development? How large are development teams, and how are these teams structured?

These questions are fun to answer in the context of FLOSS development because project teams are self-organized, widely-distributed geographically, and use many different programming languages

and software development methodologies. Teams are organized in an ad hoc, decentralized fashion. Projects can be very hard to track, and changes can be difficult to follow. Developers primarily use the Internet for communication, and teams are organized around the idea that anyone can contribute. Since the organization of the teams is done via the Internet and since the source code is open for anyone to view, it may seem as though data about these projects is as open as the projects themselves.

This is in direct contrast to the way proprietary projects are most often structured, and consequently, data about proprietary projects are collected and analyzed in a different way. Empirical software engineering researchers have, in the past, typically used metrics from a single company or a single proprietary project. This data was collected systematically and distributed in a tightly controlled manner, consistent with the proprietary nature of the software being developed. Whereas data analysis about proprietary software practices was primarily a problem of scarcity (getting access and permissions to use the data), collecting and analyzing FLOSS data becomes a problem of abundance and reliability (storage, sharing, aggregation, and filtering of the data).

Thus, this chapter will explore the motivations and methods surrounding the mining of FLOSS data, specifically how and why the collection, aggregation, distribution, and analysis of this data takes place. We will first discuss motives: why does software engineering research rely on metrics at all, and why do we need FLOSS metrics in particular? We will then study methods: what is the current state-of-the-art in FLOSS data mining? Finally, we note some possible future trends, and propose some general recommendations for measuring FLOSS projects quantitatively.

BACKGROUND

Importance of Metrics to Software Engineering

The collection and aggregation of real-world and historical data points are critical to the task of measurement in software engineering. Quantitative and empirical approaches to software engineering require real-world data; for example, the branch of software engineering concerned with estimation will use empirical or historical data to seed the estimate calculation. More generally, the four reasons for measuring software creation processes are commonly listed as a characterization, evaluation, prediction, or improvement on these processes (Park, Goethert, & Florac, 1996). All of these goals require useful data (measurements) in order to be carried out effectively. Interest-

ing measures of the software process can vary depending on the goals of the research, but they could include things like the number of errors in a particular module, the number of developers working in a particular language or development environment, or the length of time spent fixing a particular code defect (Yourdon, 1993). The collection domain of a research project will differ as well; measures can be collected for a group of products, a group of developers, a single software product, a single release of a software project, or even for a single developer.

The empirical software engineering literature is replete with examples of how gathering metrics about projects can lead to important insights. Software engineering metrics can be used to avoid costly disasters, efficiently allocate human and financial capital, and to understand and improve business processes. One famous example of a software error that caused significant financial and property damage was the European Ariane 5 flight 501 disaster of 1996 (Jezequel & Meyer, 1997). The European rocket crashed 40 seconds after liftoff, reportedly due to an error in the way software components were reused within the system. This was a US$500 million software engineering error. In 1975, Fred Brooks made famous another software engineering debacle: the management of the IBM OS/360 project (Brooks, 1975). His conclusions about the inefficiencies in the way programmers were added to the development team became known as *Brooks' Law*, and this remains one of the tenets of software engineering practice to this day. Using metrics about team composition, communication, and productivity, Brooks concluded that work done by a set of programmers will increase linearly as programmers are added to a project, but communication and coordination costs will rise exponentially. Brooks' Law is most often remembered as: "adding manpower to a late project makes it later."

There are hundreds of these examples in the software engineering literature about metrics in proprietary projects, but where are the metrics and measurements for studying FLOSS development practices? We know that FLOSS projects are fundamentally different from proprietary projects in several important ways: they are primarily user-driven as opposed to driven by a hierarchically-organized for-profit corporation (von Hippel, 2001). These user-programmers work in loosely defined teams, rarely meet face-to-face, and coordinate their efforts via electronic media such as mailing lists and message boards (Raymond, 1999). These are all fundamentally different arrangements than the way proprietary software is traditionally developed.

Importance of FLOSS Metrics

Recognizing this unique separation between proprietary and FLOSS software engineering traditions, and building on a strong foundation of measurement in software engineering literature, there are then several compelling reasons to collect, aggregate, and share data about the practice of FLOSS software development. First, the relative novelty of the FLOSS movement means that there is a high degree of unfamiliarity with development processes and practices, even within the larger software engineering domain. Studying FLOSS development practices can be useful in its own right, in order to educate the larger research and practitioner communities about an important new direction in the creation and maintenance of software (Feller, 2001). FLOSS researchers have noticed that many of the practices of FLOSS teams are not well-understood (Scacchi, 2002; von Hippel 2003) or, when they are, they seem to directly oppose traditional wisdom about how to build software (Herbsleb & Grinter, 1999). At the very least, this situation indicates something interesting is afoot, and in the best case will foreshadow an important methodological shift for software development.

Additionally, the lessons learned through studying the organizational characteristics and motivations of FLOSS development teams are applicable to many other fields. Much research has been conducted on the economic (Lerner & Tirole, 2002; Raymond, 1999) and policy aspects of FLOSS development, especially as the reason for various licensing choices (Rosen, 2004) or about their implications for intellectual property (Dibona, Ockman, & Stone, 1999; Kogut & Meitu, 2001; Lerner & Tirole, 2001; Weber, 2004). Additional research has been conducted on the motivations of FLOSS developers (Raymond, 1999; Torvalds, 1998; Ye & Kishida, 2003), which is an interesting question to consider since these developers are working without pay. There are also implications for other types of distributed teams and computer-mediated group work (Crowston, Annabi, Howison, & Masango, 2004a, 2005; Crowston & Howison, 2005; Annabi, Crowston, & Heckman, 2006; Crowston & Scozzi, 2006), such as gaining a better understanding of the role of face-to-face meetings in highly distributed work teams, or analyzing the leadership hierarchies that work best for distributed teams. Studying development team dynamics in the context of social networking continues to be a popular application for FLOSS data also (Howison, Inoue, & Crowston, 2006).

One recent significant development in the practical application of FLOSS software metrics is the use of metrics in software package evaluation frameworks for business, such as the business readiness rating (BRR) (Wasserman, Pal, & Chan, 2006). Evaluation frameworks like the BRR are designed to give businesses unfamiliar with FLOSS products a technical rationale for choosing particular products. For example, the BRR attempts to assign an overall quality rating to various products based on each product's score on various factors that may predict success. This attempt to quantify FLOSS product quality has resulted in a flurry of publications that either support or extend the BRR (Cau, Concas, & Marchesi, 2006; Monga & Trentini, 2006), or which urge caution in assessing quality in this way (German, 2006; Robles & Gonzalez-Barahona, 2006).

With this in mind, many of the questions that have been asked about FLOSS development require quantitative data in order to be answered: What are the programming languages being used for FLOSS development? Are these the same languages being used to create proprietary software? Are bugs fixed faster or slower on FLOSS teams or on a proprietary team? What is the most *common* size of a FLOSS team, and how does this relate to the *ideal* size for a FLOSS team? How, and at what rate, do new developers join a team? How do workers on a FLOSS team divide the work? What computer-mediated discussion activities are being used to manage workflow, and are they effective? As researchers gather the answers to these questions, they can begin to answer even bigger questions: Why does this particular team structure work better? Can we learn anything from FLOSS methods that can be applied to the construction of proprietary software?

MAIN FOCUS OF THE CHAPTER

Methods: The State of the Art in FLOSS Data

Quantitative FLOSS data appears to be highly available, and appears easier to access for research than proprietary data. Researchers who wish to study FLOSS development issues (for example, the adoption rates of various programming languages or the speed of bug-fixing) know that, in theory, they probably *should* have access to this information. The perception is that since the code is free and open to everyone, and because the general attitude of FLOSS developers tends toward openness, therefore the data should be straightforward to find and gather. For researchers, then, studying FLOSS development teams can have advantages over studying proprietary teams; specifically, with FLOSS, it is no longer necessary to find a corporation willing to provide researchers access to their in-house development

databases and source code control systems. How then do researchers go about getting this FLOSS data, and what are some of the problems with these methods? This section outlines the current state-of-the-art in FLOSS data gathering. It is divided into three sections: tools and studies which focus on the traversal of large forges, tools and studies which focus on a single project or a few projects, and survey-based studies.

Studying Large Forges

For a researcher who needs a large sample size of FLOSS projects or developers, the large code repositories, or *forges*, may seem like a good place to collect data. The researcher might know that there are over 100,000 FLOSS projects hosted on Sourceforge[1], a large Web-based project repository and suite of developer tools[2]. Each project hosted on Sourceforge has a general information Web page which holds basic information about the project: its license type, programming language, database environment, date it was registered, number of downloads, list of developers working on the project and their roles and skills, and so forth. As convenient as it may seem to use this forge data, the realities of gathering FLOSS data from a forge can make this a very unappealing exercise (Howison & Crowston, 2004). First, there are the obvious practical issues about how to traverse (or spider) the repositories efficiently without violating spidering rules (robots.txt) or the terms of service (TOS) of the Web sites being spidered, and where to store the massive amount of data that such an exercise generates. But the biggest limitation of spidering data like this is that the data model is always open to being changed by whoever is in control of the repository and there is no way to know of changes in advance. Thus, maintaining control over this free and open data is actually a hugely inefficient process for a researcher.

Nonetheless, numerous research papers have been written using data collected from forges (some early examples are: Crowston, Annabi, Howison, & Masango, 2004b; Ghosh & Prakash, 2000; Krishnamurthy, 2004; Weiss 2005a; Weiss, 2005b; Xu, Gao, Christley, & Madey, 2005), and tools have been developed to assist researchers in spidering these large forges (Conklin, Howison, & Crowston, 2005; Howison, Conklin, & Crowston, 2005). One early example of a forge-based tool is the Orbiten project (Ghosh & Prakash, 2000), undertaken in 1999 to "provide a body of empirical data and analysis to explain and describe this [free and open source] community." This automated source code review was designed to accumulate statistics on open source software development, including number of projects, number of developers, how often code is being changed and by whom, etc. Unfortunately, the Orbiten project is now defunct. This fact introduces a serious problem with relying on published-but-proprietary data sources for research: data can disappear. Though the original article links to a Web site that is supposed to provide both the software and the data, this site is no longer operational. A researcher wishing to duplicate, validate, or extend the methods of Orbiten would be at a loss to do so. Using FLOSS development methodologies, such as the tradition of "passing the baton" (Raymond, 1999), would have reduced the likelihood of this information becoming extinct. A subsequent section discusses additional recommendations for making forge-based data collection work well.

Studying Single Projects

Despite the vast quantities of information available inside FLOSS code forges, much of the FLOSS research to date requires a different approach. In some cases, FLOSS researchers take a similar approach to proprietary software researchers: they analyze some feature of a single software project (or a few related projects), such as the source code or the bug databases, and extrapolates some lesson or advancement which can then be applied to other projects. For example, Koch and Schneider (2000) look at the source code repository and mailing lists for the Gnome project and attempt to extract traditional software engineering metrics (function points) from this data. Mockus, Fielding, and Herbsleb (2000) study the change logs for two projects: Apache and Mozilla. In fact, the Apache Web server continues to be a single project very heavily used by researchers (Weiss, Moroiu, & Zhao, 2006; Annabi et al., 2006). Ye and Kishida (2003) study the social structure of programmers working on the GIMP graphics package. German (2004a) investigates the way the GNOME team has structured itself, and what lessons other project teams can learn based on the GNOME experience. The study by deGroot, Kugler, Adams, and Gouisos (2006) uses KDE as a test case for a general quality metric they call the SQO, or software quality observatory.

The study by Nakakoji, Yamamoto, Nishinaka, Kishida, and Ye (2002) looks at four open source projects all related to the same company. Lerner and Tirole (2002) studied four different open source projects, some of which also appear in other studies (Koch & Schneider, 2000; Mockus et al., 2000). Recent work by den Besten, Dalle, and Galia (2006) studies the artifacts of code maintenance in 10 large projects.

Recognizing that serious inefficiencies occur when every FLOSS research team writes a new tool for analyzing source code or bug reports, several research teams have also developed tools that are designed to be used generically with any given project that uses a particular source code versioning system or bug-tracking system. Examples include CVSAnalY (Robles, Koch, & Gonzalez-Barahona, 2004) and SoftChange (German, 2004b) for analyzing CVS repositories and GlueTheos (Robles, Koch, & Ghosh, 2004) for retrieving and analyzing code artifacts. The biggest benefit to these general-purpose tools is that they can be used to gather metrics on *any* project that uses the underlying source-control or bug-tracking system being studied. This is a great advantage to researchers who may have an idea for a study, but would not be able to obtain the metrics they need to begin the study without spending time to write their own retrieval or analysis system. Indeed, Robles and Gonzalez-Barahona (2006) study the contributor turnover of 21 projects using their own CVSAnalY system.

Studies Based on Survey Data

Finally, it is clear that not every research question requiring quantitative data can be answered using purely electronic artifacts. Research on intrinsic developer motivations, for example, will rely on metrics perhaps better gleaned from personal interviews or surveys. For example, the Lakhani and Wolf study (2003) was based on a survey of 684 developers on 287 FLOSS projects, while the Hars and Ou paper (2001) describes a survey of 81 developers working on an unspecified number of open source projects. Gosain and Stewart (2001) interview project administrators to show how developer ideologies impact team effectiveness. Another survey (Scacchi, 2002) intended to find out how project requirements were set, involved a dozen software projects in four different research areas. Elliot and Scacchi (2004) followed this survey with an in-depth analysis of the social artifacts (IRC logs and email discussions) of the GNUe project. Crowston and Scozzi (2006) surveyd teams with more than seven core developers as part of a study of mental models. Berdou (2006) interviewed paid and volunteer KDE and Gnome contributors.

The largest surveys of this kind to date are the 2700-person survey done by Ghosh, Glott, Krieger, and Robles (2002) and the 1500-person survey done by David, Waterman, and Arora (2003). In both of these surveys, developers answered questions about their motivations for working on FLOSS software, as well as basic demographic information.

FUTURE TRENDS

Impacts of Continued Growth in Size and Complexity of Data Sets

It becomes clear from reading the preceding section that over time, studies of FLOSS projects have enjoyed an upward trajectory in the amount of data surveyed, frequency of the surveys, and depth of the surveys. While the occasional single-project study is still common (Apache Web server is a very popular topic, e.g.), it is increasingly common for research teams to structure studies around a dozen or more projects, and to study these projects from every possible angle: communication artifacts, bug databases, code quality, as well as public metadata. Studies of individual developers are now expected to contain results for hundreds, if not thousands of participants, and surveys can ask questions about every possible aspect of a developer's life.

What are the impacts on the project leaders, developers, and the project infrastructure of this increased research interest? Do project leaders enjoy being studied? Do the project leaders enjoy the benefits of the results of the studies in which their projects are used? Is there any developer backlash against research surveys? There is some vigorous debate in the research community about breaching developer privacy in a large system of aggregated data like FLOSSmole (Robles, 2005). For example, if we aggregate several code repositories and are now able to show in a colorful graph that Suzy Developer is ten times more productive than Bob Coder, does this violate Bob's privacy? If we can show that Suzy's code changes are five times more likely to cause errors than Bob's, does that violate Suzy's privacy? Robles suggests that the next generation of community data repositories should have the ability to hash the unique keys indicating a developer's identity. Do project leaders and developers demand this level of privacy? Do they have other concerns about the research in which they are (often unwitting) participants? These answers will have to be researched, implemented, and documented for our community.

Increased Emphasis on Sharing within the Community

One of the biggest challenges for new FLOSS researchers is to figure out what data is already available and how to find it so that time is not wasted on duplicative efforts. This is especially true in light of the previous discussion about the increases in the amount of data expected and available, as well as the increased frequency of data collection efforts. Multiple research teams have already worked on analyzing the data in large forges, gathering and massaging data from individual project artifacts such as CVS, and collecting data through surveys. How can researchers leverage each other's work to reduce redundancy?

In the interests of actively promoting data sharing among research teams, one group (for which the author is a principal developer) has developed the FLOSSmole project (Conklin, Howison, & Crowston, 2005; Howison, Conklin, & Crowston, 2005). The founding principle of the FLOSSmole project is that its data, code, and database schemas should be made accessible to other researchers (http://ossmole.sf.net). This reduces redundant efforts for the research community as a whole. Since the data about FLOSS teams and projects is public to begin with, it makes sense that the data remain public after being collected from forges or from project repositories. (Presumably, some data taken from surveys or survey results can also be made public, assuming that dissemination of the results is part of the survey protocol.) Researchers or practitioners who wish to use FLOSS data should be able to look at our system and quickly procure the data they need, rather than having to go through the complicated process of gathering their own—oftentimes redundant—data. This

stance reflects the principles behind open source software itself; if a user wants to look at the code or data, she is free to do so. Having the FLOSSmole system open and easily accessible also lowers the barriers to collegial comment and critique. Because our code and data are easily accessible by anyone at any time, and because we use a source code control system and a public mailing list for discussing code and schema changes, this means that we are accountable to the public for what we create. Papers written using our data have a verifiable paper trail on which to rest.

Bridging the Gap between Disparate Data Sources

A second challenge for FLOSS research teams dealing with quantitative data is integrating disparate data sources. For example, we occasionally have access to data from now-defunct projects and from published FLOSS research studies, and we know these are valuable for historical analyses. Can these be integrated into an existing (and active) community database (such as FLOSSmole, or another project)? Even if this donated or historical data were complete, clean, and well-labeled, such as the data scraped from a large forge, integrating it could still be problematic because different repositories store different data elements. Different forges can have projects with the same names; different developers can have the same name across multiple forges; the same developer can go by multiple names in multiple forges. In addition, forges have different terminology for things like developer roles, project topics, and even programming languages.

For example, there has been some effort to coordinate the work of the CVSAnalY (Robles, Koch, & Ghosh, 2004) efforts with FLOSSmole when analyzing Sourceforge projects. Specifically, Sourceforge projects are identified by a value called the project unixname, which is unique among all Sourceforge projects. This unique identifier helps unify these two disparate data sources. The job of joining disparate data sources becomes more complex when there are multiple forges involved, however.

Current data collection and integration efforts have also not begun to address the best way to extract knowledge from published research. Is this possible, ethical, or desirable? What is the best way to express the quantitative knowledge in a domain and integrate multiple sources of this knowledge? How will we create sufficient metadata about each data source so that the results can be used together? Can any of this be done in an automated fashion? What query tools should be used so that the user can fully explore both data sets? These are big questions with no easy answers.

CONCLUSION

This chapter first reviews why quantitative data is useful in software engineering, the outlines some of the reasons why researchers are particularly interested in getting metrics and quantitative data about FLOSS development projects and practices. Next, we point out the three main types of quantitative data available for FLOSS projects: data gleaned from large code forges, data based on quantitative analyses done on single projects or a few similar projects, and data gathered from surveys. Finally, we outline what the next steps should be for creating a truly valuable and transformative community data repository: the data (and the tools used to collect and analyze the data) should be shared, and multiple data sources should be integrated.

REFERENCES

Annabi, H., Crowston, K., & Heckman, R. (2006). From individual contribution to group learning: the early years of Apache Web server. In E. Damiani, B. Fitzgerald, W. Scacchi, M. Scotto, & G.

Scucci (Eds.), *Open source systems* (pp. 77-90). New York: Springer.

Berdou, E. (2006). Insiders and outsiders: paid contributors and the dynamics of cooperation in community led F/OS projects. In E. Damiani, B. Fitzgerald, W. Scacchi, M. Scotto, & G. Scucci (Eds.), *Open source systems* (pp. 201-208). New York: Springer.

Brooks, F. (1975). *The mythical man-month*. Reading, MA: Addison-Wesley.

Cau, A., Concas, G., & Marchesi, M. (2006, June 10). Extending OpenBRR with automated metrics to measure object-oriented open source project success. In A. I. Wasserman (Ed.), *Proceedings of the Workshop on Evaluation Frameworks for Open Source Software (EFOSS) at the Second International Conference on Open Source Systems*, Lake Como, Italy (pp. 6-9).

Conklin, M., Howison, J., & Crowston, K. (2005). Collaboration using OSSmole: A repository of FLOSS data and analyses. In *Proceedings of the Workshop on Mining Software Repositories (MSR 2005) at the 27th International Conference on Software Engineering (ICSE2005)*, St. Louis, Missouri, USA (pp. 1-5).

Crowston, K., Annabi, H., Howison, J., & Masango, C. (2004a). Effective work practices for software engineering: Free/libre/open source software development. In *WISER Workshop on Interdisciplinary Software Engineering Research (SIGSOFT 04)*, Newport Beach, California, USA.

Crowston, K., Annabi, H., Howison, J., & Masango, C. (2004b). Towards a portfolio of FLOSS project success metrics. In *Proceedings of the Open Source Workshop of the 26th International Conference on Software Engineering (ICSE2004)*, Edinburgh, Scotland.

Crowston, K., Annabi, H., Howison, J., & Masango, C. (2005). *Effective work practices for*

FLOSS development: A model and propositions. Paper presented at the Hawaii International Conference on System Science (HICSS 2005), Hawaii, USA.

Crowston, K., & Howison, J. (2005). The social structure of free and open source software development. *First Monday, 10*(2). Retrieved March 20, 2007, from hppt://www.firstmonday.org/issues/issue10_2/crowston

Crowston, K., & Scozzi, B. (2006). The role of mental models in FLOSS development work processes. In E. Damiani, B. Fitzgerald, W. Scacchi, M. Scotto, & G. Scucci (Eds.), *Open source systems* (pp. 91-98). New York: Springer.

David, P. A., Waterman, A., & Arora, S. (2003). *Free/libre/open source software survey for 2003.* Stanford, CA: Stanford Project on the Economics of Open Source Software, Stanford University. Retrieved from http://www.stanford.edu/group/floss-us

de Groot, A., Kugler, S., Adams, P. J., & Gousios, G. (2006). Call for quality: Open source software quality observation. In E. Damiani, B. Fitzgerald, W. Scacchi, M. Scotto, & G. Scucci (Eds.), *Open source systems* (pp. 35-46). New York: Springer.

den Besten, M., Dalle, J-M., Galia, F. (2006). Collaborative maintenance in large open-source projects. In E. Damiani, B. Fitzgerald, W. Scacchi, M. Scotto, & G. Scucci (Eds.), *Open source systems* (pp. 233-241). New York: Springer.

DiBona, C., Ockman, S., & Stone, M. (1999). *Open sources: Voices from the open source revolution.* Sebastopol, CA: O'Reilly & Associates.

Elliott, M. S., & Scacchi, W. (2004). Communicating and mitigating conflict in open source software development projects. *Projects and Profits, 10*(4), 25-41.

Feller, J. (2001). Thoughts on studying open source software communities. In N. L. Russo et al. (Eds.), *Realigning research and practice in information systems development: The social and organizational perspective* (pp. 379-288). Dordrecht, The Netherlands: Kluwer Academic Publishers.

German, D. M. (2004a). Decentralized open source global software development, the GNOME experience. *Journal of Software, Process: Improvement and Practice, 8*(4), 201-215.

German, D. M. (2004b). Mining CVS repositories: The Softchange experience. In *Proceedings of the Workshop on Mining Software Repositories (MSR 2004) at the 26th International Conference on Software Engineering (ICSE 2004),* Edinburgh, Scotland (pp. 17-21).

German, D. M., (2006, June 10). The challengers of automated quantitative analyses of open source software projects. In A. I. Wasserman (Ed.), *Proceedings of the Workshop on Evaluation Frameworks for Open Source Software (EFOSS) at the Second International Conference on Open Source Systems,* Lake Como, Italy (pp. 10-14).

Ghosh, R. A., & Prakash, P. P. (2000). The Orbiten free software survey. *First Monday, 5*(7). Retrieved March 20, 2007, from hppt://www.firstmonday.org/issues/isse5_7/ghosh

Ghosh, R. A., Glott, R., Krieger, B., & Robles, G. (2002). *Free/libre and open source software: Survey and study.* International Institute for Infonomics, University of Maastricht, The Netherlands. Retrieved from http://www.infonomics.nl/FLOSS/report/

Gosain, S., & Stewart, K. (2001). An exploratory study of ideology and trust in open source development groups. In *Proceedings of the International Conference on Information Systems (ICIS)* (pp. 507-512).

Hars, A., & Ou, S. (2001). *Working for free? Motivations of participating in open source projects.* Paper presented at the Thirty-Fourth Annual Hawaii International Conference on System Sciences (HICSS 2001), Hawaii, USA.

Herbsleb, J. D., & Grinter, R. E. (1999). Splitting the organization and integrating the code: Conway's law revisited. In *Proceedings of the International Conference on Software Engineering (ICSE 1999)*, Toronto, Canada (pp. 85-95).

Howison, J., & Crowston, K. (2004). The perils and pitfalls of mining Sourceforge. In *Proceedings of the Workshop on Mining Software Repositories (MSR 2004) at the 26th International Conference on Software Engineering (ICSE 2004)*, Edinburgh, Scotland (pp. 7-11).

Howison, J., Conklin, M., & Crowston, K. (2005). OSSmole: A collaborative repository for FLOSS research data and analyses. In *Proceedings of the First International Conference on Open Source Systems (OSS 2005)*, Genova, Italy (pp. 54-59).

Howison, J., Inoue, K., & Crowston, K. (2006). Social dynamics of free and open source team communications. In E. Damiani, B. Fitzgerald, W. Scacchi, M. Scotto, & G. Scucci (Eds.), *Open source systems* (pp. 319-332). New York: Springer.

Jezequel, J.-M., & Meyer, B. (1997). Design by contract: The lessons of Ariane. *Computer, 30*(1), 129-130.

Koch, S., & Schneider, G. (2000, August 2-6). Results from software engineering research into open source development projects using public data. In H. R. Hansen & W. H. Janko (Eds.), *Diskussionspapiere zum Tät igkeitsfeld Informationsverarbeitung und Informationswirtschaft* (Discussion Paper No. 22). Vienna: Vienna University of Economics and BA. Retrieved March 20, 2007, from http://epub.wu-wien.ac.at/dyn/virlib/wp/eng/mediate/epub-wu--01_c3.pdf?ID=epub-wu-01_c3

Kogut, B., & Meitu, A. (2001). Open-source software development and distributed innovation. *Oxford Review of Economic Policy, 17*(2), 248-264.

Krishnamurthy, S. (2004). Cave or community? An empirical examination of 100 mature open source projects. *First Monday, 7*(6). Retrieved March 20, 2007, from http://www.firstmonday.org/issues/issue7_6/krishnamurthy

Lakhani, K., & Wolf, R. G. (2003). *Why hackers do what they do: Understanding motivation effort in free/open source software projects* (Working Paper 4425-03). Cambridge, MA: Sloan School of Management, Massachusetts Institute of Technology.

Lerner, J., & Tirole, J. (2001). The open source movementL Key research questions. *European Economic Review, 45*, 819-826.

Lerner, J., & Tirole, J. (2002). Some simple economics of open source. *Journal of Industrial Economics, 50*(2), 197234.

Mockus, A., Fielding, R. T., & Herbsleb, J. (2000). A case study of open source software development: The Apache server. In *Proceedings of the 22nd International Conference on Software Engineering (ICSE 2000)*, Los Angeles, California (pp. 263-272).

Monga, M., & Tentini, A. (2006, June 10). Weighing the value of changeability in open source software. In A. I. Wasserman (Ed.), *Proceedings of the Workshop on Evaluation Frameworks for Open Source Software (EFOSS) at the Second International Conference on Open Source Systems*, Lake Como, Italy (pp. 15-18).

Nakakoji, K., Yamamoto, Y., Nishinaka, Y., Kishida, K., & Ye, Y. (2002). Evolution patterns of open-source software systems and communities. In *Proceedings of the International Workshop on Software Evolution (IWPSE 2002)*, Orlando, Florida (pp. 76-85).

Park, R. E., Goethert, W. B., & Florac, W. A. (1996, August) *Goal Driven Software Measurement—A Guidebook* (CMU/SEI-96-BH-002). Pittsburgh, PA: Software Engineering Institute, Carnegie Mellon University.

Raymond, E. (1999). *The Cathedral and the Bazaar.* O'Reilly and Associates: Sebastopol, California.

Robles, G. (2005). Developer identification methods for integrated data from various sources. In *Proceedings of the International Workshop on Mining Software Repositories (MSR2005) at the 27th International Conference on Software Engineering (ICSE2005)*, St. Louis, Missouri (pp. 1-5).

Robles, G., & Gonzalez-Barahona, J. (2006). Contributor turnover in libre software projects. In E. Damiani, B. Fitzgerald, W. Scacchi, M. Scotto, & G. Scucci (Eds.), *Open source systems* (pp. 273-286). New York: Springer.

Robles, G., Koch, S., & Gonzalez-Barahona, J. M. (2004). Remote analysis and measurement of libre software systems by means of the CVSAnalY tool. In *Proceedings of the Second ICSE Workshop on Remote Analysis and Measurement of Software Systems (RAMSS'04) at the 26th International Conference on Software Engineering (ICSE2004)*, Edinburgh, Scotland (pp. 51-55).

Robles, G., Koch, S., & Ghosh, R. A. (2004). GlueTheos: Automating the Retrieval and Analysis of Data from Publicly Available Repositories. In *Proceedings of the Mining Software Repositories Workshop (MSR'04) at the 26th International Conference on Software Engineering (ICSE2004)*, Edinburgh, Scotland (pp. 28-31).

Rosen, L. (2004). *Open source licensing: Software freedom and intellectual property law.* Englewood Cliffs, NJ: Prentice Hall.

Scacchi, W. (2002). Understanding the requirements for developing Open Source Software systems. *IEE Proceedings on Software, 149*(1), 24-39.

Torvalds, L. (1998). FM interview with Linus Torvalds: What motivates free software developers? *First Monday, 3*(3). Retrieved http://firstmonday.org/issues/issue3_3/torvalds

von Hippel, E. (2001, Summer). Innovation by user communities: Learning from open-source software. *Sloan Management Review, 42*(4), 82-86.

von Hippel, E. (2003). Exploring the open source software phenomenon: Issues for organization science. *Organization Science, 14*(2), 209-223.

Wasserman, A. I., Pal, M., & Chan, C. (2006, June 10). The business readiness rating: A framework for evaluating open source. In A. I. Wasserman (Ed.). *Proceedings of the Workshop on Evaluation Frameworks for Open Source Software (EFOSS) at the Second International Conference on Open Source Systems*, Lake Como, Italy (pp. 1-5).

Weber, S. (2004). *The success of open source.* Cambridge, MA: Harvard University Press.

Weiss, D. (2005a). Quantitative analysis of open source projects on Sourceforge. In *Proceedings of the First International Conference on Open Source Systems (OSS 2005)*, Genova, Italy (pp. 140-147).

Weiss, D. (2005b). Measuring success of open source projects using Web search engines. In *Proceedings of the First International Conference on Open Source Systems (OSS 2005)*, Genova, Italy (pp. 93-99).

Weiss, M., Moroiu, G., & Zhao, P. (2006). Evolution of open source communities. In E. Damiani, B. Fitzgerald, W. Scacchi, M. Scotto, & G. Scucci (Eds.), *Open source systems* (pp. 21-34). New York: Springer.

Xu, J., Gao, Y., Christley, S., & Madey, G. (2005). A topological analysis of the open source software development community, In *Proceedings of the 38th Hawaii International Conference on Systems Science (HICSS 2005)*, Hawaii, USA.

Ye, Y., & Kishida, K. (2003). Toward an understanding of the motivation of open source software developers. In *Proceedings of the 25th International Conference on Software Engineering (ICSE 2003)*, Portland, Oregon.

Yourdon, E. (1993). *Decline and fall of the American programmer.* Englewood Cliffs, NJ: Prentice-Hall.

KEY TERMS

Data Analysis: Reviewing collected information to identify trends or patterns.

Data Mining: Collecting information in order to use that collected information for a specific purpose.

Development Practices: Systems for creating a software product.

Free Software (FS): Software that others are open to use, copy, or modify.

Open Source Software (OSS): Software designed in such a way that users can access/review the underlying operating code that allows that software to perform certain processes.

Quantitative Methods: Research based on the collection of numeric data.

Software Engineering: Creating/developing software products.

ENDNOTES

[1] Sourceforge (http://sf.net) describes itself as the "world's largest software development Web site." It is a centralized repository for thousands of open source projects. The site includes source code control features (CVS), community building features (forums and mailing lists), and facilities for bug tracking, feature requests, and downloading packages of the software projects hosted on the site.

[2] Some examples of other repositories include Tigris for software engineering tools (http://tigris.org), CPAN for programs written in the perl language (http://cpan.org), RubyForge for projects written in the Ruby language (http://rubyforge.net), Freshmeat for popular open source projects (http://freshmeat.net), and Savannah for free software projects (http://savannah.gnu.org).

Chapter XXIII
A Generalized Comparison of Open Source and Commercial Database Management Systems

Theodoros Evdoridis
University of the Aegean, Greece

Theodoros Tzouramanis
University of the Aegean, Greece

ABSTRACT

This chapter attempts to bring to light the field of one of the less popular branches of the open source software family, which is the open source database management systems branch. In view of the objective, the background of these systems will first be briefly described followed by presentation of a fair generic database model. Subsequently and in order to present these systems under all their possible features, the main system representatives of both open source and commercial origins will be compared in relation to this model, and evaluated appropriately. By adopting such an approach, the chapter's initial concern is to ensure that the nature of database management systems in general can be apprehended. The overall orientation leads to an understanding that the gap between open and closed source database management systems has been significantly narrowed, thus demystifying the respective commercial products.

INTRODUCTION

The issue of data storage, organization, protection, and distribution has grown in importance over the years. This is justified by the fact that data, in increasing quantities and of multiple origins, serving possibly different operational divisions, were required to be processed by companies and organizations in order to be viable and, if that

was achieved, to flourish appropriately (Loney & Bryla, 2005).

This chapter will initially examine the field of database software, while pinpointing and briefly examining the most important representatives of both open source and commercial origins. Subsequently, a generalized structure of the database model will be deployed and the most significant database system software will be evaluated according to the model's component specifications. The chapter will conclude by presenting the results of the comparison along with our views on the future of open source database software.

BACKGROUND

The open source vs. closed source (alternatively called proprietary development) debate has been a topic of continuous quarrel between experts affiliated to either of the two camps.

The notion of making money through traditional methods, such as the selling of individual copies is incompatible with the open source philosophy. Some proprietary source advocates perceive open source software as damaging to the market of commercial software. However, this complaint is countered by a large number of alternative funding streams such as (Wikipedia. org, 2006a):

- Giving away the software for free and, in return, charging for installation and support as in many Linux distributions
- Making the software available as open source so that people will be more likely to purchase a related product or service you do sell (e.g., OpenOffice.org vs StarOffice)
- **Cost avoidance/cost sharing:** Many developers need a product, so it makes sense to share development costs (this is the genesis of the X-Window System and the Apache Web server).

Moreover, advocates of closed source argue that since no one is responsible for open source software, there is no incentive and no guarantee that a software product will be developed or that a bug in such a product will be fixed. At the same time, and in all circumstances, there is no specific entity either of individual or organizational status to take responsibility for such negligence.

However, studies about security in open source software vs. closed source software (Winslow, 2004) claim that not only each significant commercial product has its counterpart in the open source arsenal but also that open source software usually provides less time for flaw discovery and, consequently, for a relative patch or fix.

Besides, open source advocates argue that since the source code of closed source software is not available, there is no way to know what security vulnerabilities or bugs may exist.

The database system software twig of the open source software family has been highly criticized especially during the last 10 years. This is due to the fact that the early versions of such products included relatively few standard relational database management system (RDBMS) features. This has led some database experts, such as Chris Date (Wikipedia.org, 2006b), a database technology specialist, who was involved in the technical planning of DB2, to criticize one of the major representatives of the field, MySQL, as falling short of being a RDBMS. Open source RDBMSs advocates reply (BusinessWeek.com, 2006) that their products serve their purposes for the users, who are willing to accept some limitations (which are fewer with every major revision) in exchange for speed, simplicity, and rapid development. Developers and end-users alike have been using more and more open source database management systems (DBMSs). Such experimentation has laid the groundwork for open source DBMSs to follow in the footsteps of Apache and Linux, two open source code products that have already penetrated the enterprise wall. Nonetheless, analysts Scott Lundstrom, Laura Carrillo and David O'Brien are of the opinion that open source DBMSs are not going to get the boost from IBM and Oracle that Linux and Apache did (Informationweek.com, 2004) due to the apparent competitive adversity of the former with the database commercial products published by these two companies.

Another group of experts (Wikipedia.org, 2006b) claims that another, perhaps simpler, explanation for open source DBMSs popularity is that it they are often included as a default component in low-end commercial Web hosting plans along side with PHP or Perl.

MAIN THRUST OF THE CHAPTER

The Competitors

DB2

DB2 is IBM's family of information management software products. Most often, though, when people say DB2, they are referring to IBM's flagship RDBMS, DB2 Universal Database (DB2 UDB). The software is available on many hardware and operating system platforms, ranging from mainframes and servers to workstations and even small hand-held devices. It runs on a variety of IBM and non-IBM operating systems. Besides the core database engine, the DB2 family consists of several other products that provide supplementary support to the DBMS's functionality such as administration and replication tools, distributed data access, online analytical processing (OLAP) and many others. The origin of DB2 can be traced back to the System R project at the IBM's Almaden Research Centre. The first official release took place in 1984 and was designed to operate on IBM's mainframe platform (Silberschatz, Korth, & Sundarsham, 2002).

DB2 is available in several editions, in other words, licensing arrangements. By opting for a reduced-feature edition, IBM allows customers to avoid paying for DBMS features which they do not need. Sample editions include the Workgroup, Workgroup Unlimited, and Enterprise Server Edition. A high-end edition is called DB2 UDB Data Warehouse Enterprise Edition, or DWE for short. This edition includes several business intelligence features such as data mining, OLAP, and in line-analysis.

On January 30, 2006, (IBM.com, 2006) IBM released a "community" edition of DB2 called DB2 Universal Database Express-C. This was an expected response to the recently announced free versions of Oracle 10g and Microsoft SQL Server. Express-C has no limit on number of users or database size. It's deployable on machines with up to two processors and up to 4GB of memory.

DB2 can be administered from either the command-line or a graphical user interface (GUI).

Oracle

Oracle Corporation founded in 1977 produces and markets the Oracle RDBMS, which many database applications use extensively on many popular computing platforms.

Larry Ellison, Bob Miner, and Ed Oates—of Software Developer Laboratories (SDL)—developed the original Oracle DBMS software. They called their product Oracle after the code name of a CIA-funded project they had worked on while previously employed by Ampex Company. Their product was the first to reach the market, and, since then, has held a leading position in the relational database market (Silberschatz, Korth, & Sundarsham, 2002).

In 2003, the Oracle Corporation released Oracle Database 10g. The g stands for grid, emphasizing a marketing thrust of presenting 10g as "grid computer ready."

As of June 2005, the Oracle Corporation has been supporting a wide array of operating systems including Windows and the majority of Unix-based operating systems.

The Database distribution includes many built-in tools, including a Java-based utility (Figure 1) and a Web-based tool serving the same purpose.

In addition, the company sells a set of added value add-on products (Loney & Bryla 2005) that expand the DBMS capabilities, providing specialized tools such as query and analysis tools,

Figure 1. The Oracle Java-based administration console

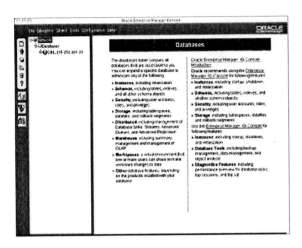

data mining and sophisticated security tools, and so forth.

The development of applications utilizing the Oracle RDBMS commonly takes place in Java, which is inherently supported by the database. Oracle Corporation has started a drive toward wizard-driven environments with a view to enabling non-programmers to produce simple data-driven applications. Oracle, as of January 2006, offers Database 10g Express Edition (Oracle Database XE) an entry-level, small-footprint database-based on the Oracle Database 10g Release 2 code base that is free to develop, deploy, and distribute; and is fast to download; and simple to administer. Furthermore, Oracle's flagship the Enterprise edition is also a free download, but its use is, as with express edition, restricted to development and prototyping purposes.

Commercial usage must be accompanied with an appropriate license from the Corporation. However, Oracle database software is considered to be one of the most expensive. As of January 2006, the list price for the Enterprise Edition is $40,000 per processor. Additional features and maintenance costs may add to the price substantially. As computers running Oracle often have

eight or more processors, the software price can be in the hundreds of thousands of dollars. The total cost of ownership is much more, as Oracle databases usually require highly trained administrators to operate.

SQL Server

Microsoft SQL Server is a RDBMS produced by Microsoft. It is commonly used by businesses for small- to medium-sized databases, and—in the past five years—some large enterprise databases.

The code base for Microsoft SQL Server (prior to version 7.0) originated in Sybase SQL Server, and was Microsoft's entry to the enterprise-level database market, competing against Oracle and IBM. About the time Windows NT operating system was coming out, Sybase and Microsoft parted ways and pursued their own design and marketing schemes. Several revisions have been done independently since, with improvements for the SQL Server. The SQL Server 7.0 was the first true GUI-based DBMS server (Spenik &d Sledge 2002).

The Microsoft SQL Server product is not just a DBMS, it also contains (as part of the product) an enterprise ETL tool (Integration Services), Reporting Server, OLAP and messaging technologies specifically Service Broker.

Microsoft released the SQL Server Express product (Microsoft.com 2006), which included all the core functionality of the SQL Server, but places restrictions on the scale of databases. It will only utilize a single CPU, 1 GB of RAM, and imposes a maximum size of 4 GB per database. SQL Express also does not include enterprise features such as Analysis Services, Data Transformation Services, and Notification Services.

Microsoft's primary competition includes Oracle and DB2. The SQL Server, as of January 2006, has been ranked third in revenue share among these *big three* DBMSs' vendors. A significant drawback of the SQL Server is that it runs only on the Windows Operating System.

Firebird

Firebird (sometimes called FirebirdSQL) is a RDBMS offering many ANSI SQL-99 and SQL-2003 features. It runs on Linux, Windows, and a variety of Unix platforms. Firebird was programmed and is maintained by Firebird Foundation (formerly known as FirebirdSQL Foundation). It was forked from the open sources of InterBase from Borland. Firebird's first release took place back in 1984 and, as of January 2006, the product has evolved to being a very mature DBMS requiring minimal administration, providing advanced features, and compliant database engine that implements most of the SQL-2003 standard (The Inquirer. net, 2005, Firebirdsql.org, 2006a). Firebird is expandable, utilizing specialized modules that are licensed under the Initial Public Developers License (IDPL). The original modules released by Inprise are licensed under the Interbase Public License. Both licences are modified versions of the Mozilla Public License.

In April 2003, Mozilla decided to rename their Web browser from Phoenix to Firebird. This decision caused concern within the Firebird DBMS project because of the assumption that a DBMS and Web browser using the Firebird name would confuse users. The dispute continued until the Mozilla developers, on February of 2004, renamed their product as Firefox thus clearing up confusion (Wikipedia, 2006b).

MySQL

MySQL is considered the most popular open source RDBMS with an estimated six million installations (BusinessWeek.com, 2006). Its first release took place unofficially in 1995. Swedish company MySQL AB is responsible for MySQL making their product available as free software under the GPL License. At the same time they

also dually license it under traditional proprietary licensing arrangements for cases where the intended use is incompatible with the GPL (MySQL.com, 2005). A license of this type might for example be suitable for companies that do not want to release the source code of their MySQL-based application.

The company MySQL AB also develops and maintains the system, selling support and service contacts as well as proprietary licensed copies of MySQL, and employing people all over the world who collaborate via the Internet. Among its strong points are its speed, ease of installation, and as of January 2006 MySQL's version 5 included for the first time many new enterprise level features. MySQL is also highly popular for Web applications and acts as the DBMS component of the LAMP platform (Linux/Apache-MySQL-PHP/Perl/Python). Its popularity as a Web application is closely tied to the popularity of PHP, which is often combined with MySQL and nicknamed the Dynamic Duo.

To administer MySQL one can use the included command-line tool and free downloadable separate GUI administration tools. One of them, MysqlAdministrator, is depicted in Figure 2.

MySQL works on many different platforms, including Windows, Linux, and UNIX based op-

Figure 2. The MySQL Administration Console

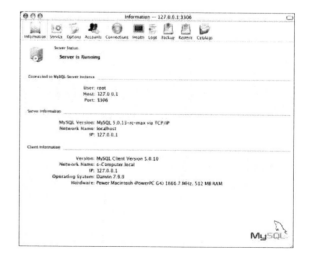

Table 1. Generic Information regarding DBMSs

RDBMS	Maintainer	Supported Platforms	First release	Licence	URL
DB2	IBM	Windows, Linux, Unix	1982	Proprietary	http://www-306.ibm.com/software/data/db2/
Firebird	Firebird Corporation	Windows, Linux, Mac OS X, Unix BSD	2000	IDP	http://www.firebirdsql.org/
MySQL	MYSQL AB	Windows, Linux, Mac OS X, Unix BSD	1996	GPL or Proprietary	http://www.mysql.com/
Oracle	Oracle Corporation	Windows, Linux, Mac OS X, Unix	1977	Proprietary	http://www.oracle.com/technology/software /products/database/oracle10g/index.html
PostgreSQL	PostgreSQL Global Development Group	Windows, Linux, Mac OS X, Unix BSD	1989	BSD	http://www.postgresql.org/
SQL Server	Microsoft	Windows	1989	Proprietary	http://www.microsoft.com/sql/default.asp

erating systems. MySQL features have attracted a set of distinguished customers including Yahoo!, CNET networks, Amazon, Cox Communications, and others. These firms have adopted MySQL as a reliable solution to support some of their internal operations.

PostgreSQL

PostgreSQL is a free object-relational database management system (ORDBMS) released under flexible BSD License. It offers an alternative to other open source database systems as well as to commercial systems. Similar to other open source projects such as Apache and Linux, PostgreSQL is not controlled by any single company, but relies on a community of global developers and companies to develop it.

PostgreSQL is based on POSTGRES Version 4.2 1, developed at the University of California in the Berkeley Computer Science Department. POSTGRES pioneered many concepts, such as functions, inheritance, and other object-oriented features that only became available in some commercial database systems much later. PostgreSQL is an open source descendant of this original Berkeley code. It supports a large part of the SQL standard and offers many advanced features. Furthermore

PostgreSQL supports a number of add-on modules and packages such as geographic objects, full text search, replication packages and XML/XSLT support that greatly enhance the products' capabilities (PostgreSQL.com, 2005a).

Moreover, PostgreSQL has provided the base for the development of EnterpriseDB (EDB). The latter is a most promising enterprise-class RDBMS compatible with Oracle—and costing as a base product only a minor fraction, varying from 10% to 20%, of the price of a commercial system.

On the down side, the product suffers from an image problem (The Inquirer.net 2005). This is on account of the fact that PostgreSQL remains a project and there is no company accountable for offering respective services and support. Moreover, even though it is regarded by many as the most advanced open source DBMS, and despite commercial support by many smaller companies, it has a relatively small base of installations.

A Fair DBMS Model

As seen above, all the competitors have been on track for years and this justifies the popularity and recognition that these DBMSs enjoy. Some useful information regarding these systems is summarized in Table 1.

Over the years, vendors kept improving their software by adding new features and increasing performance and stability (Fermi National Accelerator Laboratory, 2005). Unfortunately, this furthermore obscures the situation, as no database software can prove to be better than the others. Taking into account marketing and software promotion, the situation becomes even more complicated as vendors attempt to prove the dominance of a product. In an attempt to resolve the issue, a five-component DBMS comparison model was conceived and used as a protractor in order to produce fair, accurate and valuable results, setting open source against commercial in the scientific field of database software. The model's architecture was influenced by all time classic DBMS standards (Johnson, 1997) as well as by the requirements (BusinessWeek.com, 2006) of low to high-populated organizations from database software.

The first component includes the fundamental features that modern database system software should provide. Among these are elementary data type support, SQL standard compliance data constraint, index, and transaction protocols support.

The second component is made up of advanced DBMS features such as special data types, stored procedures, triggers, cursors, sequences, user-defined data types, OLAP and inherent support for object oriented, spatial, and XML databases.

The third component is related to database administration robustness and optimization. Evaluation on this component is based on provision of the appropriate access control, backup, and data migration mechanisms as well as replication support and recovery capabilities of the software products.

The fourth component consists of customizability criteria like scalability, reliability, and database performance according to data set size.

The fifth component features DBMS support and acceptance. Software training, operation,

administration and maintenance manuals, as well as programming interfaces, external libraries and product popularity around the world are considered to belong to this evaluation component.

Following are comparisons and evaluations, mapping every DBMS model's components to respective tiers. The results of this appraisal are presented in the final part of the section.

The Comparison

Tier 1

All DBMSs perform, with respect to these particular component standards, within very high levels (Devx.com, 2005). They fully support the latest, as of January 2006, SQL—2003 Standard, and their transactions comply with the ACID protocol. MySQL could be taken as an exception, as both transactions and, as a result, ACID, along with referential integrity constraints, are supported on Tables utilizing the INODB storage engine and not on the other available ones like MYISAM (PostgreSQL.org, 2005b). Additionally, MySQL, PostgreSQL and Firebird support the 2-phase commit protocol to achieve concurrency control while commercial systems offer more options. Furthermore, commercial DBMSs alongside with PostgreSQL and MySQL support save points during transactions. Finally, with respect to indexes Oracle is known for the amount of tweaking it allows for databases, especially when it comes to indexing. Other systems support single column, multi-column, unique, full text and primary key indexes.

The results from the comparison at the first tier are summarized in Table 2.

Tier 2

All commercial systems support advanced data types like large objects, which have become increasingly popular over the years. Proprietary

Table 2. Tier 1 comparison results

RDBMS	SQL Standard Compliance	ACID Compliance	Constraint Support	Transaction and Lock Support	Indexes
DB2	VERY HIGH	YES	YES	VERY HIGH	VERY HIGH
Firebird	VERY HIGH	YES	YES	HIGH	HIGH
MySQL	VERY HIGH	YES/NO	YES/NO	HIGH	HIGH
Oracle	VERY HIGH	YES	YES	VERY HIGH	VERY HIGH
PostgreSQL	VERY HIGH	YES	YES	HIGH	HIGH
SQL Server	VERY HIGH	YES	YES	VERY HIGH	VERY HIGH

Table 3. Tier 2 comparison results

RDBMS	Advanced Data types	Advanced Features	OpenGIS Support	XML Support	OLAP Support	Object-Oriented Features
DB2	VERY HIGH	VERY HIGH	YES	YES	YES	YES
Firebird	HIGH	HIGH	NO	NO	NO	NO
MySQL	HIGH	VERY HIGH	YES	NO	NO	NO
Oracle	VERY HIGH	VERY HIGH	YES	YES	YES	YES
PostgreSQL	VERY HIGH	VERY HIGH	YES	YES	NO	YES
SQL Server	VERY HIGH	VERY HIGH	NO	YES	YES	YES

DBMSs and PostgreSQL have network-aware data types that recognize Ipv4 and Ipv6 data types. Moreover, MySQL and PostgreSQL also both support the storing of geographic features, data types and operations of the Open Geodata Interchange Standard (OpenGIS). All systems support enterprise level features such as triggers, views, stored procedures, cursors while PostgreSQL and Commercial systems additionally support inheritance, sequences, and user-defined data types as well as. Additionally all systems use a procedural extension to the SQL query language to allow developers to implement routines that transfer some application logic to the database. Examples of using these routines are stored procedures that are written in a respective database procedural language. Among them, Oracle Database's choice, named PL/SQL although considered most difficult

to use, is also thought of as the most powerful one. Firebird Database, using the Compiere module (Firebirdsql.org, 2006b) is capable of executing natively Oracle PL/SQL code, while MySQL and PostgreSQL use their own versions of procedural language in their DBMSs.

On the other hand, MySQL alone in the open source camp supports the advanced feature of data partitioning within a DBMS. All open source DBMSs, save PostgreSQL, fall short when it comes to XML support. This consistutes an issue that will certainly be addressed in future releases of these systems. Finally, all open source systems lack OLAP support to perform high demanding large enterprise business intelligence operations. On the commercial base, IBM, Microsoft and Oracle supply their products with in-house OLAP modules that expand the capabilities of their software to serve organizations that require such services.

The results from the comparison at this tier are summarized in Table 3.

Tier 3

This specific tier shows some of the features that should be addressed at the open source DBMSs in order for it to become more competitive. With respect to security, open source DBMSs support access control mechanisms data encryption, views, roles and other security methods that can undoubtedly constitute a reliable backbone for any organization. On the other hand they lack the sophisticated security mechanisms offered by commercial products such as Oracle's added value add on "Oracle Advanced security" (Oracle.com, 2006) which offers more options and supports some industry standard authentication methods such as PKI. The SQL Server, on the other hand, even though it often uses Windows authentication and is subject to OS-based vulnerabilities that can compromise its operation, has received a C2 certificate from the U.S. government's National Security Agency that recommend it for use in government projects. When it comes to backup, open source DBMSs come with appropriate scripts to facilitate a simple text dump of database data and its schema like Firebird's NBackup module. At the same time all products provide methods for doing a hot-database backup or, in other words, backing up the database without shutting it down.

However, they still lack the array of options during a backup procedure that commercial systems offer, allowing the generation of automatic selective and customisable backups.

On the contrary, open source DBMSs prove to offer high data migration capabilities, allowing data hosted in their system to be formatted appropriately for usage in another database. Commercial systems support data migration, often via commercial third party tools. Another major feature of enterprise-level DBMSs is support for replication. Both MySQL and PostgreSQL have support (Devx.com, 2005) for single-master, multi-slave replication scenarios. Commercial systems offer more replication methods, although these methods are not considered of outmost necessity for the majority of users and organizations. Finally, with respect to recovery in MySQL, only InnoDB tables have automatic crash recovery of a running database in background, without setting any locks or using replication methods. PostgreSQL uses a system called Write Ahead Logging to provide database consistency checking and point-in-time recovery (PiTR) that allows recovery either to the point of failure or to some other in the past (PostgreSQL.com, 2005a). Firebird uses third party tools (FreeDownloadsCenter, 2006) that can be used for automatically diagnosing and repairing corrupted data due to failures during normal operation. Commercial DBMSs provide automated and manual recovery capabilities that allow the database to return to any chosen state,

Table 4. Tier 3 comparison results

RDBMS	Security Features	Backup	Data Migration	Replication	Recovery
DB2	VERY HIGH	VERY HIGH	HIGH	VERY HIGH	VERY HIGH
Firebird	HIGH	HIGH	HIGH	MEDIUM	HIGH
MySQL	HIGH	HIGH	VERY HIGH	HIGH	HIGH
Oracle	VERY HIGH	VERY HIGH	HIGH	VERY HIGH	VERY HIGH
PostgreSQL	HIGH	HIGH	VERY HIGH	HIGH	VERY HIGH
SQL Server	VERY HIGH	VERY HIGH	HIGH	VERY HIGH	VERY HIGH

according to specific log files, like the REDO LOGS in Oracle Database. The results from the comparison at tier 3 are summarized in Table 4.

Tier 4

The fourth component consists of quality criteria such as scalability, reliability, and database performance.

DB2, Oracle and SQL Server can scale to terabytes of data storage fairly easily supporting millions of users. This is achieved (Microsoft. com, 2006) by supporting scale up on symmetric multiprocessor (SMP) systems, allowing users to add processors, memory, disks and networking to build a large single node, as well as scale out on multinode clusters. Thus, it makes possible for a huge database to be partitioned into a cluster of servers, each server storing part of the whole database, and each doing a portion of the work, while the database remains accessible as a single entity. Various sources (IBM.com, 2006), (Oracle.com, 2006) give an edge on one commercial system over the other, but these systems are considered by many to belong to the same high quality class (Wikipedia.org, 2006b).

MySQL using the MySQL Cluster option (MySQL.com, 2006) and PostgreSQL are known to run very fast, managing up to more than 500 transactions per second when dealing with databases hosting gigabytes of data and can perform adequately enough even when the size of the databases exceeds that threshold.

Firebird, as of January 2006, offers some baseline multiprocessor support although it uses a standard process-based architecture. This decreases significantly its performance when the hosted data become of terabyte magnitude.

With respect to reliability, MySQL, because of a large installed base and as a result of the knowledge and experience surrounding it, is perceived to be a highly reliable system. Looking in the same direction, PostgreSQL, although less popular, has proved to be a very dependable system, a fact that

can be credited to the rich set of features and the maturity of this software product (BusinessWeek. com, 2006). Firebird on the other hand, although with smallest installed base, has demonstrated a remarkable stability and consistency.

Commercial systems then again, are accompanied by industry standard verification certificates that ensure the product's reliability and quality of service. An example of this is the Common Criteria Certification awarded to SQL Server 2005 (Microsoft.com, 2006).

The TPC (Transaction Processing Council) is an independent organization that specifies the typical transactions and some general rules these transactions should satisfy. The TPC produces benchmarks that measure transaction processing and database performance in terms of how many transactions a given system and database can perform per unit of time, for example, transactions per second or transactions per minute. As of June 2006, Oracle is the fastest commercial DBMS around, outperforming DB2 and SQL Server (Transaction Processing Performance Council, 2006) and maintaining the place that it had the previous years (Eweek.com, 2002; Burlseon Consulting, 2003; Promoteware.com, 2004). Open source DBMSs did not participate in this comparison: according to the party who benchmarked (Promoteware.com, 2004) this was because of their limitations when dealing with large data sets. However, at the level of data sizes of small to medium enterprises, several gigabytes, it has been shown that the open source DBMSs perform equivalently to proprietary ones. Among open source DBMSs, MySQL is believed to be the fastest (Eweek.com, 2002). The results from the comparison at tier 4 are summarized in Table 5.

Tier 5

On the whole, commercial products enjoy high support from their respective owners varying from initial training to real-time diagnostic and monitoring capabilities that serve optimization ends.

Table 5. Tier 4 comparison results

RDBMS	Scalability	Reliability	Performance
DB2	VERY HIGH	VERY HIGH	VERY HIGH
Firebird	HIGH	HIGH	HIGH
MySQL	VERY HIGH	VERY HIGH	VERY HIGH
Oracle	VERY HIGH	VERY HIGH	VERY HIGH
PostgreSQL	MEDIUM	VERY HIGH	HIGH
SQL Server	VERY HIGH	VERY HIGH	VERY HIGH

Additionally many third-party affiliate consultants can be easily located all around the world.

The issue of support faces mitigated point of views for open source software in the enterprise. Many do not realize that support is available for many open source products—beyond Web sites and mailing lists. MySQL AB provides support for MySQL, and several companies and PostgreSQL Inc. provide support for PostgreSQL. These offers include support levels that rival commercial DBMSs, many providing 365x24 support.

Training is an important issue in commercial DBMSs. IBM, Microsoft and Oracle set up courses and issue the relevant exams for approval and qualifications to administer the database (Dbazine.com, 2005). MySQL AB provides training in cities around the world and, in some cases, provides in-house education. PostgreSQL training is also available from third parties.

Administration is an additional issue, where open source DBMSs shine. The use of smart graphical administration tools facilitates the management of the database. These tools can either be applications that run natively on the operating system or Web-based tools. Many of these tools are modelled closely on tools available to commercial DBMSs with the appropriate modifications. Out of the latter, Oracle is believed to run on the most complex administration, requiring significant knowledge on the part of the administrator of the system's internal structure.

With respect to external library and API support, all systems enjoy the privilege of having implementations of all major programming interfaces such as ODBC, JDBC, C and C++ libraries and others (PostgreSQL.org, 2005b). This allows developers to select their programming language and database of choice when creating applications that utilize a database server.

In conclusion, the cost of acquiring a license to use database software should not be omitted. PostgreSQL and Firebird can offer their services for free even though third party commercial modules may change that. MySQL AB dual licenses their DBMS, while the commercial version of MySQL consists of a small fraction of the costs of even the cheapest commercial DBMS.

In the commercial camp, DB2 is the most expensive product (Microsoft.com, 2006), when considering base product, maintainability, and additional enterprise level capabilities reaching in July 2006 a total of $329.00. Oracle is also an expensive product as the enterprise edition version bundled with the enterprise level add-ons sells at approximately $266.00. On the other hand Microsoft offers SQL Server accompanied by their respective business intelligence support at a significantly lower price of $25.00. The results from the comparison at tier 5 are summarized in Table 6.

Table 6. Tier 5 comparison results

RDBMS	Training	Administration	Technical Support	Interfaces	Cost
DB2	HIGH	HIGH	VERY HIGH	VERY HIGH	VERY HIGH
Firebird	MEDIUM	VERY HIGH	LOW	VERY HIGH	-
MySQL	HIGH	HIGH	HIGH	VERY HIGH	-/MEDIUM
Oracle	VERY HIGH	HIGH	VERY HIGH	VERY HIGH	VERY HIGH
PostgreSQL	LOW	HIGH	MEDIUM	VERY HIGH	-
SQL Server	HIGH	HIGH	VERY HIGH	VERY HIGH	HIGH

Assumptions

Open source DBMSs have evolved to a considerable degree. The gap between these systems and their proprietary rivals has been narrowed but not totally closed. Currently, the leading open source database engines, considering all possible aspects, are still inferior in terms of performance and features to DB2, Oracle and SQL Server. However, their capabilities may certainly offer enough to meet the needs of most small and medium sized companies or even large ones, serving supplementary purposes. A living example of this practice is encountered at NASA that uses MySQL to store information about public contracts, and the American Chemical Society that uses PostgreSQL to store specific documents. It is important to note that most users and companies do not require some of the state of the art advanced features, and scalability options found exclusively on commercial DBMSs. Moreover, as with all open source software, Firebird, MySQL, PostgreSQL and other open source DBMSs are free, easy to try out and have lots of freely available online documentation to help each individual to learn how to use them. While these DBMSs may not be optimal for every possible project, they could prove to be acceptable and satisfying to others.

FUTURE TRENDS

It is strongly believed that the open source movement will transform the software business in the next five to ten years, according to top industry executives speaking at the AO 2005 Innovation Summit at Stanford University (Wheeler, 2005). A group of analysts claims that the reasons for such an adoption are not entirely of an ideological nature. Stability, performance and security will be other drivers of open source software, according to BusinessWeek.com 2006. Sun Microsystems President Jonathan Schwartz claimed that the software industry must adopt open standards for it to thrive. "Open standards mean more than open source" (CNET NEWS.COM, 2005).

This stream will inevitably influence the scientific sector of database software. As DBMSs built from open source code are gaining in capabilities with every new release and enjoying rapid adoption by various users of new technology, it is almost certain that open source DBMSs will eventually level with commercial ones with respect to all possible aspects. As a result, many companies will adopt these systems instead of commercial ones allowing them to save money and reduce their operational costs while forming at the same time a current that will definitely threaten commercial DBMS vendors. When

Oracle Corporation announced its acquisition of Innobase (The Inquirer.net, 2005) it gave notice that MySQL's license to use the InnoDB storage mechanism would be renegotiated when it comes up for renewal next year. Some in the industry see this purchase as a way for Oracle to align MySQL AB towards their politic and influence their future direction. Furthermore, Sun Microsystems announced that they will add support for the PostgreSQL Database and that it would add it to the Solaris operating system.

What is sure is that the open source boat that carries along the database system software as one of its open source passengers is sailing fast, towards its growing recognition and adoption. Commercial firms that once neglected the presence of open source projects are now on the move to approach and somehow contain open source initiatives, either by embracing them or trying to tame them. Many field experts believe (BusinessWeek.com, 2006) that open source databases software has a bright future, not as standalone products but as fundamental blocks in commercial database software, that also includes proprietary elements.

CONCLUSION

Many could question the interest surrounding open source DBMSs. And this is due to the fact that, in many ways, the open source label is attached to initiatives such as Linux and Apache. Unfortunately for many commercial firms and fortunately for the rest, pen source is much more than these two representatives.

After many years of hard work and little attention, these open source DBMSs are starting to have a noticeable impact on the largest DBMS companies. Long criticized for not having advanced enterprise features, reliability and customer support, open source DBMS kept on becoming more and more competitive with the release of each new version. Taking into consideration,

this criticism, these products strived to improve and include the so-far lacking features, while maintaining their strong aspects. As a result, it is only a matter of time before open source DBMSs could stand against their proprietary software counterparts as equals and even perform better in some sectors. This has alarmed many commercial organizations that, in one way or another, laid their hands on these open source products. Even though the results cannot be absolutely foreseen, its can be asserted without any doubt that open source DBMSs will scale up from the status of attracting intellectual curiosity that led them in 2003 to become widespread. Either as standalone products or as subsystems of commercial DBMSs, open source DBMSs will continue to support the IT community for the years to come, as they have always done.

REFERENCES

Burlseon Consulting. (2003). *Oracle vs. SQL Server*. Retrieved from http://www.dba-oracle.com/oracle_tips_oracle_v_sql_server.htm#jambu

BusinessWeek.com. (2006). *Taking on the database giants*. Retrieved from http://www.businessweek.com/technology/content/feb2006/tc20060206_918648.htm

CNETNEWS.COM. (2005). *Tech VIPs say future belongs to open source*. Retrieved from http://news.com.com/Tech+VIPs+say+future+belongs+to+open+source/2100-7344_3-5798964.html

Dbazine.com. (2005). *DBA certifications compared: Oracle vs. DB2 vs. SQL Server*. Retrieved from http://www.dbazine.com/ofinterest/oi-articles/fosdick2

Devx.com. (2005). *PostgreSQL vs. MySQL vs. commercial databases: It's all about what you need*. Retrieved from http://www.devx.com/db-zone/Article/20743/1954?pf=true

Europa.eu.int. (2005). *A big step forward.* Retrieved from http://europa.eu.int/idabc/en/document/5220/469

Eweek.com. (2002). *Server databases clash.* Retrieved from http://www.eweek.com/article2/0,4149,293,00.asp

Fermi National Accelerator Laboratory. (2005). *Comparison of Oracle, MySQL and PostgreSQL DBMS.* Retrieved from http://www-css.fnal.gov/dsg/external/freeware/mysql-vs-pgsql.html

Firebirdsql.org. (2006a). *Firebird—Relational database for the new millennium.* Retrieved from http://www.firebirdsql.org/

Firebirdsql.org. (2006b). *Firebird user documentation.* Retrieved from http://www.firebirdsql.org/manual/index.html

FreeDownloadsCenter. (2006). *Free InterBase downloads.* Retrieved from http://www.freedownloadscenter.com/Search/interbase.html

IBM.com. (2006). *IBM software—DB2 product family.* Retrieved from http://www-306.ibm.com/software/data/db2/

Informationweek.com. (2004). *Popularity growing for open source databases.* Retrieved from http://www.informationweek.com/story/showArticle.jhtml?articleID=18312009

Johnson J. (1997). *Database: Models, languages, design.* Oxford, UK: Oxford University Press.

Loney K., & Bryla B. (2005). *Oracle Database 10g DBA Handbook.* Oracle Press.

Microsoft.com. (2006). *Microsoft SQL server home.* Retrieved from http://www.microsoft.com/sql/default.mspx

MySQL.com. (2005). *MySQL manual.* Retrieved from http://dev.mysql.com/doc/mysql/en/index.html

Oracle.com. (2006). *Oracle database security.* Retrieved from http://www.oracle.com/technology/deploy/security/db_security/index.html

PostgreSQL.com. (2005a). *PostgreSQL manual.* Retrieved from http://www.postgresql.org/docs/8.0/interactive/index.html

PostgreSQL.org. (2005b). *Open source database software comparison.* Retrieved from http://jdbc.postgresql.org/

Promoteware.com. (2004). *SQL server comparison chart (SQL vs MySQL vs Oracle).* Retrieved from http://www.promoteware.com/Module/Article/ArticleView.aspx?id=23

Silberschatz A., Korth H. F., & Sundarsham S. (2002). *Database System Concepts* (4th ed.). McGraw Hill.

Spenik M., & Sledge O. (2002), *Microsoft SQL Server 2000 DBA Survival Guide* (2nd ed.). Sams Press.

The Inquirer.net. (2005). *Open source databases rounded up.* Retrieved from http://www.theinquirer.net/?article=28201

Transaction Processing Performance Council. (2006). Retrieved from http://www.tpc.org/

Wheeler, D. (2005). *How to evaluate open source software/free software (OSS/FS) programs.* Retrieved from http://www.dwheeler.com/oss_fs_eval.html

Wikipedia.org. (2006a). *Open source software.* Retrieved from http://en.wikipedia.org/wiki/Open_source_software

Wikipedia.org. (2006b). *Comparison of relational database management systems.* Retrieved from http://en.wikipedia.org/wiki/Comparison_of_relational_database_management_systems

Winslow, M. (2004). *The practical manager's guide to open source.* Lulu Press.

KEY TERMS

Atomicity, Consistency, Isolation, and Durability (ACID): Considered to be the key transaction processing features/properties of a database system. Without them, the integrity of the database cannot be guaranteed.

Database: An organized collection of data (records) that is stored in a computer in a systematic way, so that computer software might consult it to answer questions. The database model in most common use today is the relational model.

Grid Computing: A computing model that provides the ability to perform higher throughput computing by taking advantage of many networked computers to model a virtual computer architecture that is able to distribute process execution across a parallel infrastructure. Grids use the resources of many separate computers connected by a network to solve large-scale computation problems.

GNU General Public License (GPL): It is the most popular free software license originally written by Richard Stallman for the GNU project. The GPL grants the recipients of computer software the following rights:

- Freedom to run the program, for any purpose
- Freedom to study how the program works, and modify it. (Access to the source code is a precondition for this)
- Freedom to redistribute copies
- Freedom to improve the program, and release the improvements to the public (access to the source code is a precondition for this)

Graphical User Interface (GUI): It refers to computer software that offers direct manipulation of graphical images and widgets in addition to text.

Object-Relational Database Management System (ORDBMS): It is a database management system that allows developers to integrate the database with their own custom data types and methods.

Online Analytical Processing (OLAP): It is an approach to quickly provide the answer to complex analytical queries. It is part of the broader business intelligence category that also includes data mining. The typical applications of OLAP are in business reporting for sales, marketing, management reporting business performance management (BPM), budgeting and forecasting, financial reporting, and similar areas.

Open Source Software (OSS): Computer software available with its source code under an open source license to study, change and improve its design. The open source philosophy further defines a boundary on the usage, modification, and redistribution of open source software. Software licenses grant rights to users, which would otherwise be prohibited by copyright. These include rights on usage, modification, and redistribution. Several open source software licenses have qualified within the boundary of the Open Source Definition.

Relational Database Management System (RDBMS): It is a database management system that is based on the relational model as introduced by Edgar F. Codd. The model represents all information in the form of multiple related tables, every one consisting of rows and columns.

Structured Query Language (SQL): It is the most popular computer language used to create, modify and retrieve data from relational database management systems. The language has evolved beyond its original purpose to support object-relational database management systems. It is an ANSI/ISO standard.

Chapter XXIV
Evaluation of a Migration to Open Source Software

Bruno Rossi
Free University of Bozen-Bolzano, Italy

Barbara Russo
Free University of Bozen-Bolzano, Italy

Giancarlo Succi
Free University of Bozen-Bolzano, Italy

ABSTRACT

The chapter discusses the adoption and assimilation process of open source software as a new form of information technology. Specifically, the case reports a general positive attitude towards the widely used technology, the OpenOffice.org suite for office automation. Nevertheless, it shows the difficulties of the first early adopters to lead the innovation process and push other users. Different usage patterns, interoperability issues, and, in general, the reduction in personal productivity typical of the early phases of adoption are also remarked. The aim of this chapter is to give the reader an overview of the adoption process by means of the analysis of quantitative and qualitative data gathered during real world experimentation, and to shed some light on how empirical data can corroborate or challenge the existing literature about open source software and technology adoption.

INTRODUCTION

Open source software (OSS) and open data standards (ODS) have emerged in recent years as a viable alternative to proprietary solutions. There are many cases in which the adoption of OSS has proved advantageous for companies deciding to adopt it in replacement or in conjunction with closed source software (CSS). Unfortunately, at our knowledge, these studies often report only about server-side migrations or give very little empirical evidence of the benefits of the new solution. Among case studies that report successful transitions on the desktop side we can surely mention as pioneers the Extremadura, Munich, and Vienna case studies (Marson, 2005; Lande-

Table 1. Large deployments of OSS inside public administrations

Region	Clients to migrate	Side	Distribution
Extremadura	80000	Desktop/Servers	gnuLinex
Munich	14000	Desktop	Debian
Vienna	7500	Desktop	Wienux (Debian/KDE)
Largo, FL	900	Desktop/Servers	Linux KDE 2.1.1

shauptstadt München, 2003; Stadt Wien, 2004). All these cases have in common the intention of a large migration inside a single public administration (PA). Furthermore, the migration to OSS in all these cases has been already performed or is in the process of being deployed. We summarise the most famous deployments in Table 1, three are European, while one is U.S.-based.

One of the most remarkable deployments of OSS on the desktop side is surely the one of the Extremadura region in Spain, recently installing 80,000 Linux systems, 66,000 for the educational system and 14,000 for administrative workstations. The local administration even created their Linux distribution called gnuLinex[1]. According to their IT department, the savings have been of the order of €18M (Marson, 2005). Another case of success is the one of the city of Largo, FL (USA) where the migration has involved 900 clients; the savings have been estimated in $300,000-$400,000 (Miller, 2002). The migration of the city of Munich and the one of the city of Vienna are currently underway (Landeshauptstadt München, 2003; Stadt Wien, 2004). As the delay of the Munich migration seems to demonstrate, a transition to OSS is not a process to underestimate. There are also cases where the proprietary solution has been considered more convenient, like the city of Nürnberg, where according to their own migration study, the transition from Windows 2000/Office 2000 to Windows XP/Office XP was considered as €4.5M cheaper than the transition to Linux/OpenOffice.org (Stadt Nürnberg, 2004).

Another case of interest that emerged recently is the decision of the state of Massachusetts to abandon closed data standards (CDS) in favour of ODS, in particular to adopt the open document format for office automation documents exchange activities starting from January 2007 (Massachusetts State, 2005). According to the Organization for the Advancement of Structured Information Standards (OASIS) the purpose of the format is "to create an open, XML-based file format specification for office applications" (OASIS, 2005). Following this case and the increasingly requests coming from the European Commission to reduce e-government barriers, Microsoft decided to open the formats supported by its office automation suite in the upcoming months (Palmer, 2005).

The goal of this chapter is to provide an insight on two different experimental migrations to OSS inside European PAs. In particular, we don't consider a full migration, but the introduction of OSS in the office automation field. Throughout a constant monitoring of the software employed, we derive some indications on software usage that can be useful to provide more information on the migration process and the adoption of OSS.

In the next sections, we will provide first an overview of the existing literature about technology adoption and then start reviewing the experimentation details providing background information about the two Public Administrations involved. The last part will be devoted to the discussion of the results.

Technology Adoption and Assimilation

Before entering the discussion about the experimentation and the migration performed, an overview of the existing literature about technology adoption and assimilation will be useful. This will also provide a framework in which the results of the experimentation will be inserted.

Technology adoption, diffusion and acceptance research bases its foundation on the early work of Everitt Rogers, in the book titled *Diffusion of Innovations*. Rogers (1995) interest lies in studying the diffusion process that characterises technology adoption. In his seminal work, technology adopters are categorised according to the phase in which they make the adoption decision. The main distinction is among innovators, early adopters, early majority, late majority, and laggards. In particular, the author models the diffusion as an S-shaped curve characterised by an initial adoption speed and a later growth rate. The claim is that different technologies will lead to different adoption patterns.

Interesting for our study are various factors that affect the level of technology adoption inside organisations, like the *organisational age* (Chatterjee, Grewal, & Sambamurthy, 2002), *organisational size* (Fichman, & Kemerer, 1997), *industry type* (Chatterjee, Grewal, & Sambamurthy, 2002; Fichman, & Kemerer, 1997), and *sophistication of the IT infrastructure* (Armstrong, 1999; Chau & Tam, 1997). To some extent, the evidence seems to report that organisations that are younger, larger and belong to certain industry types are more willing to invest and adopt new technology. The existence of a sophisticated IT infrastructure will also lead to an easier adoption path.

Furthermore, Fichman and Kemerer (1999) report two critical factors that influence the technology assimilation process: knowledge barriers and increasing returns. The first effect relates to the effort necessary to acquire the necessary knowledge and skills to properly adopt a certain technology. This effect leads to what are known as knowledge barriers (Attewell, 1992; Fichman & Kemerer, 1999). Being a new and still somewhat unexplored field, we think that OSS is subject heavily to knowledge acquisition barriers that can in some way hinder its adoption.

As a second macro-level phenomenon, the adoption of certain technologies is subject not only to supply-side benefits due to economies of scale (Shapiro & Varian, 1999) but also to a demand-side effect called increasing returns effect (Arthur, 1989). The effect leads to an increase of utility in adoption for each successive adopter, based on the number of previous adopters. Arthur (1989) goes further in this analysis, claiming that "[e]conomy, over time, can become locked-in by 'random' historical events to a technological path that is not necessarily efficient, not possible to predict from usual knowledge of supply and demand functions, and not easy to change by standard tax or subsidy policies" (p. 2). In this sense, it may not be possible to easily switch from a certain technology once a certain critical level of adoption has been reached.

Open source software and software in general is one of the goods that are particularly sensible to economies of scale, increasing returns and knowledge barriers. To understand fully the adoption process, all these effects have to be considered.

BACKGROUND INFORMATION

Experimentation on the migration to OSS in the office automation field has been performed in two different European public administrations (PAs). We will discuss briefly the background details of the two public administrations involved and for simplification purposes, we will refer to the first public administration as PA1 and the second as PA2.

PA1 is a large public administration, counting globally over 5,000 employees. The budget allocated for the ICT (information and communication technology) services is high, but the experience with OSS is still limited. The reason for the interest in a possible migration to OpenOffice.org and in general to other OS applications is threefold:

- Spare the money spent yearly to cover the license costs.
- Reduce the effort needed to handle the licenses.
- Provide a benefit to the local economy, by means of the adoption of OSS.

PA2 is composed by a large number of municipalities spread across the territory of its region. Nearly all the municipalities in the consortium are small and count on the average 50 desktop machines. The maintenance is performed remotely by the central IT (information technology) department. In this case, the budget available for ICT services in such small municipalities is low, but a great experience in OSS has been built in recent years, mostly based on server-side solutions. The objectives of a possible migration to OSS are the following:

- Reduce the costs of ICT services in the long term.
- Ensure the accessibility of generated documents also in the future, not relying on proprietary data standards.

To summarise the characteristics of the the PAs that took part to the experimentation, both share a similar organisational size, while differences exist in prior OSS experience and budget allocated for ICT services.

FOCUS OF CHAPTER

Experiment Design

The experimentation performed has involved the market leader Microsoft Office[2] and OpenOffice.org[3], an OSS suite offering ODS support. The decision to use these applications has been done in accordance with the relevance of office automation inside PAs (Drakos, Di Maio, & Simpson, 2003) and the guidelines given by IDABC (Interoperable Delivery of European eGovernment Services to public Administrations, Businesses and Citizens) for a gradual transition to OSS. One of the main suggestions is to "introduce applications in a familiar environment" (IDBAC Report, 2003, p. 23). The introduction of OpenOffice.org is seen as a necessary step for a successive complete migration to OSS.

To monitor the behaviour of users with both solutions, we adopted the PRO Metrics (PROM) software as a mean to collect and analyse software metrics and personal software process data (Sillitti, Janes, Succi, & Vernazza, 2003), software that permits to collect metrics on software usage in a non-invasive manner. It allows the collection of the measures of time spent on documents, name of the document and other useful information about the general software usage. To protect the privacy of the users several measures were taken in accordance with the local union representatives:

- Data collected has been encrypted by means of the strong AES algorithm (Pfleeger & Pfleeger, 2002).
- Usernames were randomly generated.
- Data of single users were not given to single PAs, the analysis presented has been only given in aggregated form and with the aim to provide an evaluation of the migration.

In Table 2, a comparison of both experimentations is performed.

The number of users involved in the experimentation has been equivalent, both PAs decided to install OpenOffice.org in order to evaluate the possible future migration. The suite has been installed on a large number of workstations in both PAs; however our study has been performed on a smaller subset of users. The total events that are reported in table refer to the smallest unit that the

Table 2. Comparison of both experimentations

	PA1	PA2
Users experimenting	1486	1475
Total OOo installations	~4000	~2000
Total MSO installations	~4000	~2000
Days	30	40
Total events	1518150	1435553
Maturity	Starting Ooo introduction	Already using Ooo
ID generation	Per user	Per machine

data collection software details; a single event refers to the application's window release of focus.

This number details the amount of data that have been collected during the experimentation. The maturity row refers to the situation in which the experimentation has been performed; PA2 was already in a more advanced state of technology adoption, offering the open solution for several months prior to the experimentation. As a last annotation, the details of username generation have been slightly different between the two installations, in PA2 the usernames have been generated on a per machine basis, different users working on a single workstation are mapped as a single entity. This will result in higher documents per day or time per day per single username compared to PA1, where the usernames map directly to a single user. Nevertheless, the results have not been influenced by this approach, since the common practice in PA2 is to have a single workstation per user.

The experimentation protocol followed the same schema in both experimentations:

- Installation of OpenOffice.org; the version of the suite installed is OpenOffice.org 1.1.3 in both PAs; various versions of the Microsoft Office suite were available on the target systems
- Installation of the PROM agent to monitor the software adoption level

- Training on the OpenOffice.org suite, mostly performed to show how to perform the same task in the new office automation environment
- A questionnaire on the attitude towards Open Source Software submitted to users
- Support provided to users by means of forums and hot-lines

Methodology and Limitations

The methodology applied is mainly empirical; the analysis is based on quantitative data collected through a non-invasive software agent and on qualitative data collected by means of questionnaires. A full controlled experiment could not be performed, as it would not have been possible to control all exogenous factors that could affect the final results (Campbell & Stanley, 1990). For a controlled experiment, but on a more limited number of users during a migration to OSS (see Rossi, Scotto, Sillitti, & Succi, 2005). A comparison of the functionalities of Sun StarOffice Writer and Microsoft Office Word[4] can be found in (Everitt & Lederer, 2001). Also in this case the comparison is on a limited number of users, focusing on the functionalities offered by both solutions and how users could perform the same task. Researchers found that "[w]hile overall ratings for both products were comparable, participants were more comfortable and satisfied with Microsoft Word and found it easier to use than StarOffice Writer" (Everitt & Lederer, 2001, p. 2).

In the following sections we will perform first a comparison of the initial attitude of users towards OSS, the comparison of the two solutions by means of the quantitative data collected and in the end evaluate possible interoperability issues that can raise in case of a full migration.

Initial Attitudes Toward OSS

The experimentation has been supported by qualitative data coming from one questionnaire submitted to users; the aim of the questionnaire was to evaluate the attitude of the users towards OSS, as it can have a great impact on the successive acceptance of OSS. The questionnaire has been submitted in electronic format. We report here the results that may be interesting to evaluate the attitude of users before entering the experimentation. Data in this section refers to 282 users of PA1.

The first two questions related to the knowledge of OSS, in particular the familiarity with the concept and the general users' perception. The answers are represented in Figure 1.

Surprisingly, more than 60% of the users that filled the questionnaire depict themselves either as very familiar or fairly familiar with the concept. One of the reasons can be that users with more attitude towards OSS were the ones that filled the questionnaire earlier. The second question about the perception of OSS leads to a group of users neutral or positive towards the new concept; after the experimentation it is possible that users acquire a more sharp view on the subject; in this sense, we should expect at that point, neutral users to represent the minority.

The third question in Figure 2 further investigates the knowledge of users in the field, we asked whether users know OS products and whether they can name at least one.

Not surprisingly, the majority of users report no application. The most known products are OpenOffice.org and the Linux operating system.

In showing the results of the remaining questions, we divided the users in two categories, users

Figure 1. PA1—Results of Question A

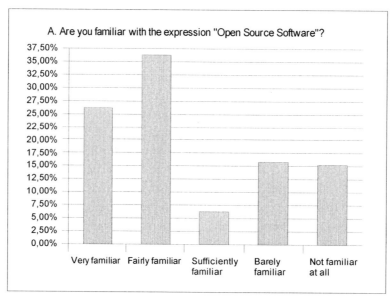

continued on the following page

Figure 1. continued

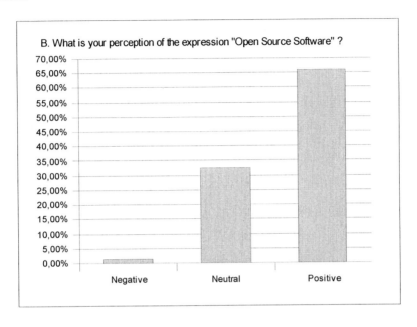

Figure 2. PA1—Results of Question C and Question D

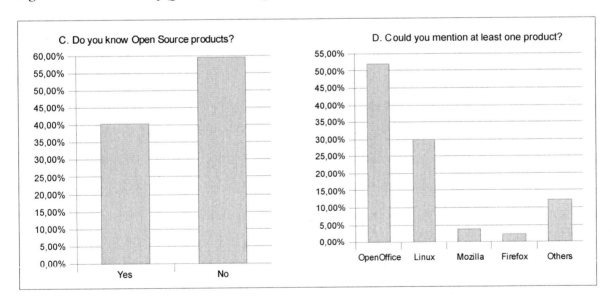

that had already an opinion on OSS and users that don't know the phenomenon. In this sense the first category consisted of all users considering themselves either as familiar or very familiar with OSS (see Figure 1, Question A) and naming at least one application (see Figure 2, Question C). For the reader's convenience, in the upcoming tables we named these groups OSS and non-OSS users. In Figure 3, the experimenters are questioned about the purchasing criteria of software in general, without particular reference to OSS.

In this case, users already with knowledge of OSS seem to be more aware of the customisation requirements of software inside PAs. The next two questions are related to a full migration to OSS. In Figure 4 users are posed in a situation of a generalized introduction of OSS and its effects on the organizational aspects of the PA.

The results of Figure 4 are comparable across both groups: the majority of users consider the introduction of OSS as a chance of reorganization of the IT department of the PA. Furthermore, 15% of users consider the introduction as non important in terms of organizational impact. The last question in Figure 5 is very similar to the previous one, but this time is related to the impact of the migration on the single user.

The results also in this case do not report a large difference between the two groups, more than half of the users are convinced that the substitution will have a negative impact on the workload in the short period, the advantages will be evident only in the long period. Users of the OSS group seem more conscious about the effort that a migration causes.

Overall the results of the questionnaires report users in general positive towards OSS. It would be interesting as an additional study to evaluate the impact of the experimentation on the users' attitude, to see how the perception of users changes after the influence of a full migration.

Comparison of the Solutions

Both softwares for office automation have been running during the whole experimentation, users were free to choose the solution more appropriate for the task to perform. A limitation on this decision was given by the large number of files

Figure 3. PA1—Results of Question E

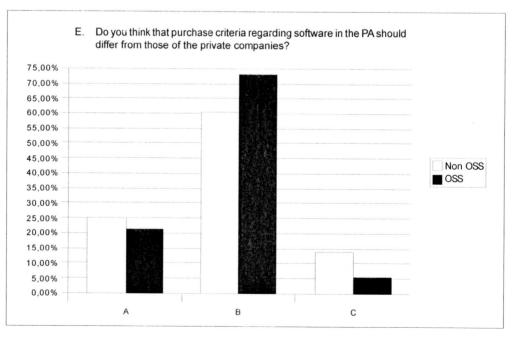

Note: Possible answers are (a) No, they should follow the same criteria; (b) Yes, they should take into account the peculiar needs of the PA; (c) I don't know.

Figure 4. PA1—Results of Question K

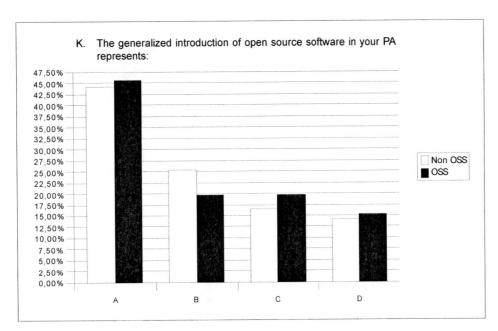

Note: *Possible answers are (a) A chance for the reorganization of the IT structure; (b)A chance for the redefinition of the organizational structure in a wide perspective; (c) A further load of work for the single units; (d) The introduction will not be important.*

Figure 5. PA1—Results of Question L

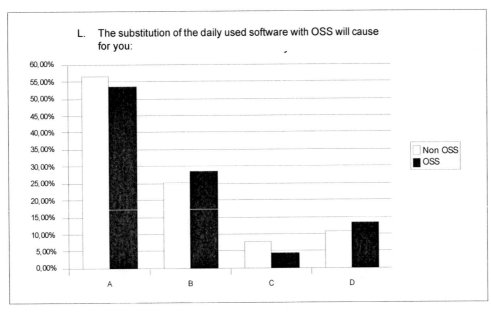

Note: *Possible answers: (a) More work in the short run, but advantages in the long run; (b) More work in the short run and no advantages in the long run; (c) Less work; (d) More work.*

already available in the original data format. More of these interoperability details will be analyzed in the apposite section.

From the analysis performed, we observed that the average time spent with the new solution tends to be minimal in PA1, where the software has been introduced with the experimentation. In PA2 instead, where the solution has been installed for several months, daily average minutes per day tend to be above 50 minutes per user.

The events generated have been aggregated in two different kinds of measures, the average number of documents worked per day by each user during the whole period and the average time in minutes spent on the documents. In Figures 6 and 7 we see the mapping of each user for PA1 and workstation for PA2 in this space. In each figure on the left the mapping is for Microsoft Office, while on the right the mapping is for OpenOffice. org. Each point represents a user.

In PA1, 90% of all users lie in the space between 20 documents per day and 200 minutes per day spent using Microsoft Office. In PA2 90% of all clients lie between 24 documents and 240 minutes.

Figure 6. PA1—Distribution of users across average documents (x-axis) and average time (y-axis)

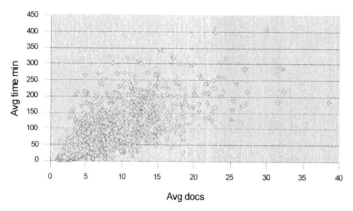

(a) Microsoft Office documents handling

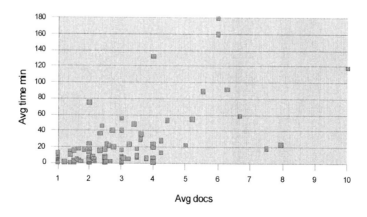

(b) OpenOffice.org documents

Figure 7. PA2—Distribution of users across average documents (x-axis) and average time (y-axis)

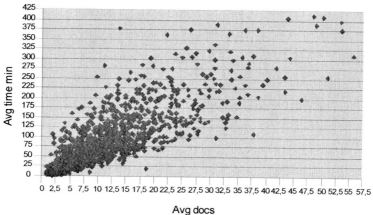

(a) Microsoft Office documents handling

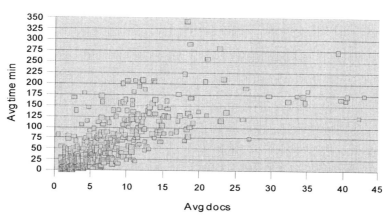

(b) OpenOffice.org documents

Regarding OpenOffice.org, 90% of PA1's users lie between four documents and less than 60 minutes of usage. In PA2, as we should expect, the usage is stabilized at higher levels, with the limit of eight documents and 135 minutes that encompasses 90% of the users.

From the temporal evolution of the software usage, we note that the software usage is constant in both PAs during the whole experimentation. This is due to the short time frame we are analysing. At this stage of the experimentation, the

difference in usage of OpenOffice.org between the two PAs is clearly evident.

Furthermore, the distribution of users across time and documents gives a better idea of the grouping of users. As a further step, the application of clustering techniques in order to group users according to further variables and in accordance to their attitude, might also shed some lights on the usage pattern of the different applications (Duda, Hart, & Stork, 2001).

Figure 8. PA1—Functions per office automation software

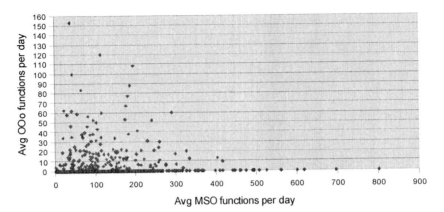

Functionalities

A further study on the functionalities[5] has been performed in both PAs, the goal was to gather information on how users evaluate the office automation application's features. Indeed, one of the major critics to OSS on the desktop-side, is its supposed lack of usability compared to CSS (Nichols & Twidale, 2003). The aim of this section is to evaluate the difference in functionalities usage between the two applications and whether from this distinction we can derive some indications about the usability.

In Figure 8, a first representation of the situation in PA1 is plotted. Users are mapped according to the average Microsoft Office functions per day (x-axis) and average OpenOffice.org functions per day (y-axis).

From the distribution of users in Figure 8, we can notice that users tend to use daily more functions in Microsoft Office. To further investigate this issue, we then compared both situations normalizing the functions used per time unit. To perform this operation, we set-up the following metric:

$$\frac{1}{n} \cdot \sum [\frac{\sum f}{t}]$$

where f is a single function, t is the time spent on documents and n is the total number of users. By using the time we can compare the results among the two solutions. As a result of this formula, we get the distribution results shown in Table 3.

The number of normalised functionalities used is in general lower with Microsoft Office than with OpenOffice.org; an explanation can be the fact that users are more acquainted to shortcuts in order to perform certain operations. On the other side newcomers to OpenOffice.org have yet to acquire the necessary confidence in the functionalities offered.

These considerations cannot alone denote a possible usability problem of OpenOffice.org. However, they can indicate the difference in usage of the new technology introduced, a difference

Table 3. PA1—Results of functions calling between Microsoft Office and OpenOffice.org

	MSO	OOo
Min	0,03	0.08
Max	5,45	5,45
Mean	0,41	1,71
Std. Dev	0,45	1,26

Figure 9. PA—Representation of the Microsoft Office documents opened by using OpenOffice.org

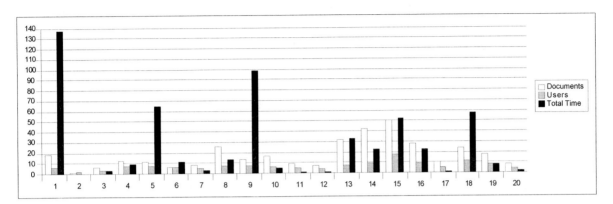

Note: For each day the figure reports number of documents (in white), users adopting this feature at least once in that day (grey) and the total time for the day spent in minutes on the documents after the opening (black). Extensions considered are .doc and .rtf (Microsoft Word), .ppt (Microsoft Powerpoint) and .xls (Microsoft Excel).

that will obviously reflect on users' productivity during the early phases of a migration.

FUTURE TRENDS

Interoperability Considerations

One of the strategies that a software vendor entering a market can exploit to emerge in a situation where users are in a situation of lock-in, is to provide higher compatibility with the standards already offered on the market. This strategy has its drawback in the fact that some performance of the application has to be sacrificed in favour of the compatibility, entering a mechanism of trade-off (Shapiro & Varian, 1999).

In this sense, OpenOffice.org offers compatibility also with the closed data standards of the Microsoft Office suite. It is interesting in the study proposed to see how users adopted this compatibility feature. To gain a measure of this interoperability issue, we computed as a first step the number of Microsoft Office proprietary formats documents opened by means of

OpenOffice.org. Further data, as the time spent with the documents opened with this method and the number of users adopting the feature also add detail to this analysis.

In Figure 9, data are reported for each day: the number of foreign documents opened per day (in white), the total time in minutes spent on the documents opened (in black) and the number of users adopting this solution at least once per day (in grey). The Microsoft Office formats considered are the ones handled by Microsoft Word, Excel and Powerpoint, namely files with doc, rtf, xls, and ppt extensions.

The results of this kind of analysis are not encouraging; probably one of the reasons is that users are not aware of this possibility. A minimal number of users is adopting this feature in his everyday work, to be precise only 10.90% of OpenOffice.org users (17 out of 156) with 6.76% of the global time spent in OpenOffice.org (nearly nine hours out of 138 hours). This last aspect seems to justify also that users tend to open documents of the foreign format for viewing purposes only; editing is seen as dangerous due to the different application used.

Figure 10. PA2—Representation of the Microsoft Office documents opened by using OpenOffice.org

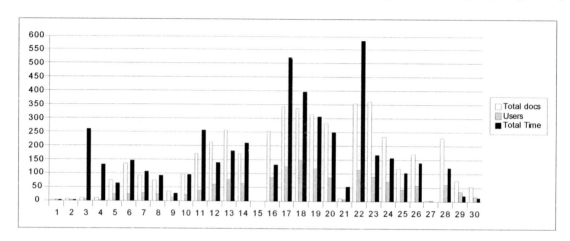

Note: For each day the figure reports number of documents (in white), users adopting this feature at least once in that day (in grey) and the total time for the day spent in minutes on the documents after the opening (in black). Extensions considered are .doc and .rtf (Microsoft Word), .ppt (Microsoft Powerpoint) and .xls (Microsoft Excel.)

Figure 11. PA1—Representation of the Microsoft Access documents handled by users

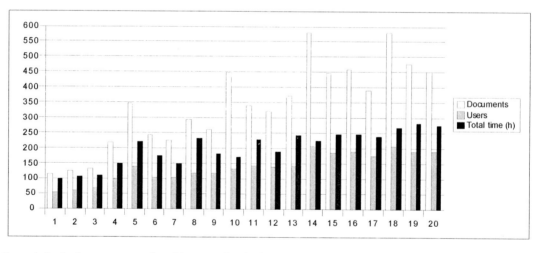

Note: For each day the figure reports number of documents (in white), users opening such a document at least once in that day (in grey) and the total time for the day spent in hours on the documents after the opening (in black).

The same analysis is represented in Figure 10 also for the second PA. In this case we see that users more trained and adopting OpenOffice.org for a longer time have a clearer idea of the functionalities offered.

The same considerations of the previous group report that 57.46% of OpenOffice.org users (447 out of 778) used this feature, but only 2.26% of the time spent in OpenOffice.org (nearly 78 hour over 3.594 hours). In this case the result confirms that users are more aware of the interoperability features.

Another important interoperability issue in the migration in the office automation field is due to the different applications available in both suites. While Microsoft Office offers a small personal database application called Access, OpenOffice.org in the version available to experimenters doesn't offer a comparable alternative[6] In Figure 11 the use of Microsoft Access is reported, with number of documents (in white), number of users using the application in that particular day (in grey) and total time spent by all users in hours (in black). Time has been reported in hours to facilitate the reading.

What can be seen is that the software is still very used; nearly 30% of all Microsoft Office users used at least once the application during the experimentation period. If we consider only users employing it for a period greater than five days, the percentage drops to 15%. In this kind of analysis we cannot perform a comparison with PA2 as the software for data collection installed was not configured to collect this kind of information.

The results of this section report that a more focused training on the interoperability features offered by the OpenOffice.org suite can lead to a broader diffusion of the suite. It is still to understand the reasons of the lack of confidence in editing documents in the other application source format.

CONCLUSION

The results of both experimentations show that open source software (OSS) can represent a viable alternative to closed source software (CSS) even on the desktop side. The analysis was focused on four different levels of technology adoption, the level of the users' attitude towards OSS, level of adoption and usage of both solutions during the period, functionalities adopted and the interoperability issues. Where possible, all levels have

been considered for both PAs that participated to the experimentation:

- The attitude was in general positive; users had a positive attitude before starting the experimentation. However, we should expect a change of the attitude at the end of the experimentation. Neutral users will probably join the groups of enthusiastic or sceptics about OSS.

- The adoption and usage of both solutions has seen the predominance of the market-dominant Microsoft Office, although in the experimentation where OpenOffice.org was already introduced users started to use it in everyday work. This is due also to network effects in IT markets that have been exploited by the early adopter PA of our study (Katz & Shapiro, 1985). Implementing a strategy of documents exchange in the new format is a key decision to widen the diffusion of the new application. The results obtained show that the migration path will be more difficult in absence of a proper strategy of documents exchange.

- The analysis of the functionalities used has shown that there are different patterns between the groups of the two suites. The group of Microsoft Office users has more confidence in the software, performing their task mainly through shortcuts. Such a confidence is not present in OpenOffice.org users. The results cannot be used to evaluate the usability of OSS, however they do report the reduction in productivity that is typical of the early phases of software migration.

- The analysis on interoperability shows that there are still interoperability issues, mainly in the form of personal databases creation. Furthermore, users don't seem to evaluate positively the compatibility with the foreign format offered by OpenOffice.org. The strategy to increase the diffusion of the software by providing a greater level of compatibility with the existing data standards doesn't seem to provide the results expected.

Overall, the data collected have granted the possibility to evaluate the adoption levels of OSS inside two different PAs. In the cases reported, the initial levels of adoption are low and interoperability issues exist that can potentially hinder OSS adoption.

NOTE

This work has been partially supported by COSPA (Consortium for Open Source Software in the Public Administration), EU IST FP6 project nr. 2002-2164

REFERENCES

Armstrong, C. P., & Sambamurthy, V. (1999). Information technology assimilation in firms: The influence of senior leadership and IT infrastructure. *Information Systems Research, 10*(4), 304-327.

Arthur, W. B. (1989). Competing technologies, increasing returns, and lock-in by historical events. *Economic Journal, 99*, 116-131.

Attewell, P. (1992). Technology diffusion and organizational learning. *The Case of Business Computing. Organization Science, 3*(1), 1-19.

Campbell, D. T., & Stanley, T. D. (1990). *Experimental and quasi-experimental design*. Boston: Houghton Mifflin Company.

Chatterjee, D., Grewal, R., & Sambamurthy, V. (2002). Shaping up for e-commerce: Institutional enablers of the organizational assimilation of web technologies. *MIS Quarterly, 26*(2), 65-89.

Chau, P., & Tam, K. (1997). Factors affecting the adoption of open systems: An exploratory study. *MIS Quarterly, 21*(1), 1-24.

Drakos, N., Di Maio, A., & Simpson, R. (2003). *Open source software running for public office* (Gartner Research Report AV-19-5251). Retrieved December 2005, from www4.gartner.com/resources/114500/114562/114562.pdf

Duda, R. O., Hart, P. E., & Stork, D. G. (2001). *Pattern classification*. New York: John Wiley & Sons.

Everitt, K., & Lederer, S. (2001). *A usability comparison of Sun StarOffice Writer 5.2 vs. Microsoft Word 2000*. Retrieved November 2, 2005, from http://www.sims.berkeley.edu/courses/is271/f01/projects/WordStar/

Fichman, R. G., & Kemerer, C. F. (1997). The assimilation of software process innovations: An organizational learning perspective. *Management Science, 43*(10), 1345-1363.

Fichman, R. G., & Kemerer, C. F. (1999). The illusory diffusion of innovation: An examination of assimilation gaps. *Information Systems Research, 10*(3), 255-275.

IDABC. (2003). *The IDA open source migration guidelines*. Retrieved February 15, 2006, from http://ec.europa.eu/idabc/servlets/Doc?id=1983

Katz, M. L., & Shapiro, C. (1985). Network externalities, competition, and compatibility. *The American Economic Review, 75*(3), 424-440.

Landeshauptstadt München. (2003). *Clientstudie der Landeshauptstadt München*. Retrieved February 2, 2006, from http://www.muenchen.de/aktuell/clientstudie_kurz.pdf

Marson, I. (2005). *Linux brings hope to Spain's poorest region*. Retrieved January 10, 2006, from ZDNetUK Web site: http://insight.zdnet.co.uk/software/linuxunix/0,39020472,39197928,00.htm

Massachussets State. (2005). *Enterprise technical reference model.* Retrieved on February 2, 2006, from http://www.mass.gov/portal/site/massgovportal/menuitem.769ad13bebd831c/14db4a11030468a0c?pageID=itdsubtopic&L=4&L0=Home&L1=Policies%2c+Standards+%26+Legal&L2=Enterprise+Architecture&L3=Enterprise+Technical+Reference+Model+-+Version+3.5&sid=Aitd

Miller, R. (2002). *Largo loves Linux more than ever.* Retrieved February 2, 2006, from Newsforge Web site: http://www.newsforge.com/print.pl?sid=02/12/04/2346215

Nichols, M., & Twidale, M. B. (2003). The usability of open source software. *First Monday, 8*(1), Retrieved July 6, 2006, from http://firstmonday.org/issues/issue8_1/nichols/

Palmer, M. (2005). *Microsoft to give Office access to rivals.* Retrieved February 2, 2006, from Financial Times Online Web site: http://news.ft.com/cms/s/e9f5c0f8-5ab7-11da-8628-0000779e2340.html

OASIS—Organization for the Advancement of Structured Information Standards. (2005). *OASIS Open Document Format for Office Applications (OpenDocument).* Retrieved December 5, 2005, from http://www.oasis-open.org/home/index.php

Pfleeger, C. P., & Pfleeger, S. L. (2002). *Security in computing* (3rd ed.). Upper Saddle River, NJ: Prentice Hall.

Rogers, E. (1995). *Diffusion of innovations.* New York: The Free Press.

Rossi, B., Scotto M., Sillitti, A., & Succi, G. (2005). Criteria for the non invasive transition to OpenOffice. In *Proceedings of OSS2005,* Genova, Italy.

Shapiro, C., & Varian H. R. (1999). *Information rules: A strategic guide to the network economy.* Cambridge, MA: Harvard Business School Press.

Sillitti, A., Janes, A., Succi, G., & Vernazza, T. (2003, September 1-6). Collecting, integrating and analyzing software metrics and personal software process data. In *Proceedings of EUROMICRO 2003,* Belek-Antalya, Turkey.

Stadt Nürnberg. (2004). *Strategische Ausrichtung im Hinblick auf Systemunabhängigkeit und Open Source Software.* Retrieved February 2, 2006, from http://online-service.nuernberg.de/eris/agendaItem.do?id=49681

Stadt Wien. (2004). *Open Source Software am Arbeitsplatz im Magistrat Wien.* Retrieved February 15, 2006, from http://www.wien.gv.at/ma14/pdf/oss-studie-deutsch-langfassung.pdf.

KEY TERMS

Assimilation: Passive adoption of a new practice or behaviour, generally resulting from participating in activities where such behaviour is used or is expected.

Data Standard: Denotes a standard to store data in information science. The most important classification is between open/closed data standards according to the publishing of the specification, although the exact classification is still controversial.

Deployment: Use of an item on a relatively large scale.

Lock-In: In economics, denotes a situation in which a consumer cannot change his buying decision without incurring in high switching costs. For example, a user may be bound to a certain software provider for the services offered, by switching to another provider he may incur in high switching costs to change his system infrastructure.

Network Effect: In economics, denotes a demand-side effect, by which the utility given to a certain good increases with the number of successive users adopting it. Information goods are a typical example of good that manifest this behaviour.

Migration: Transitioning from one particular software package to another.

Office Automation: The set of software necessary to provide the necessary integration between the information system and the standard office activities. The minimal set of instruments includes a word-processor, a spreadsheet, software for presentations, and a small personal database application.

ENDNOTES

1   GnuLinex, http://www.linex.org/

2   Microsoft Office, http://www.microsoft.com/office/editions/prodinfo/default.mspx

3   OpenOffice.org, http://www.openoffice.org

4   The study is dated November-December 2001 and refers in particular to the comparison between Sun StarOffice Writer 5.2 and Microsoft Word 2000.

5   As functionalities we intend the opening of one window inside an application, as for example—to remain in the office automation field—the paragraph options or the Save As screen. At the time of both experimentations we could not collect more fine-grained data, like the invocation of keyboard shortcuts that gives us the exact correspondence with the functions used in a program.

6   Starting from version 2.0 OpenOffice.org offers also the Base component to provide simple database functionalities.

Section IV
Laws and Licensing Practices Affecting Open Source Software Uses

Chapter XXV
Legal and Economic Justification for Software Protection

Bruno de Vuyst
Vrije Universiteit Brussel, Belgium

Alea Fairchild
Vrije Universiteit Brussel, Belgium

ABSTRACT

This chapter discusses legal and economic rationale in regards to open source software protection. Software programs are, under TRIPS[1], protected by copyright (reference is made to the Berne Convention[2]). The issue with this protection is that, due to the dichotomy idea/expression that is typical for copyright protection, reverse engineering of software is not excluded, and copyright is hence found to be an insufficient protection. Hence, in the U.S., software makers have increasingly turned to patent protection. In Europe, there is an exclusion of computer programs in Article 52 (2) c) EPC (EPO, 1973), but this exclusion is increasingly narrowed and some call for abandoning the exclusion altogether. A proposal by the European Commission, made in 2002, called for a directive to allow national patent authorities to patent software in a broader way, so as to ensure further against reverse engineering; this proposal, however, was shelved in 2005 over active opposition within and outside the European parliament. In summary, open source software does not fit in any proprietary model; rather, it creates a freedom to operate. Ultimately, there is a need to rethink approaches to property law so as to allow for viable software packaging in both models.

INTRODUCTION

Copyright Protection of Software

A software program is foremost a sequence of orders and mathematical algorithms emerging from the mind of the innovator, hence creating a link with copyright law as a prime source of intellectual property protection.

According to Article 10 TRIPS, computer programs, whether in source or object code, shall be protected as literary works under the Berne

Convention provided that they are (1) original and (2) tangible. In light of Article 9 TRIPS, which states that copyright protection shall extend to expressions, but not to ideas, procedures, methods of operation or mathematical concepts as such, copyright protects the actual code of the computer program itself, and the way the instructions have been drawn up, but not the underlying idea thereof (Overdijk, 1999).

Hence, an author can protect his original work against unauthorized copying. Consequently, an independent creation from another person would not automatically be seen as a copyright infringement (Kirsch, 2000a; Leijnse, 2003). With respect to software programs this could have as consequence that a person disassembles and decompiles an existing software program to determine the underlying idea and uses this idea to build his own program (reverse engineering). As he only uses the idea, which is not copyright-able, no infringement will result.

BACKGROUND

Patent Law Protection of Software

Software is a novel form in the technology world, and may make a claim to patent protection from that angle. The conditions to be met to enjoy patent protection are more stringent than those to enjoy copyright protection. In Europe[3], for example, an invention will enjoy protection from patent law provided that the invention (1) is new (i.e., never been produced before), (2) is based on inventor activity (i.e., not have been before part of prior art), and (3) makes a technical contribution (i.e., contribute to the state of the art). In the U.S., the patent requirements to be met are (1) novelty, (2) non-obviousness, and (3) the innovations must fall within the statutory class of patentable inventions.

Pursuant to patent law, a patent holder can invoke the protection of his patent to exclude others from making, using or selling the patented invention. As opposed to copyright protection, the inventor's patent is protected regardless whether the software code of the patented program was copied or not.

The Evolution of the Legal Protection of Software

Prior to the 1980s, U.S. courts unanimously held that software was not patentable and that its only protection could be found in copyright. Indeed, the U.S. Supreme Court ruled in two landmark decisions, Gottschalk vs. Benson (1972) and Parker vs. Flook (1978), that software was similar to mathematics and laws of nature (both excluded from being patented) and, therefore, was unpatentable.

In Diamond vs. Diehr (1981), however, the court reversed course, deciding that an invention was not necessarily unpatentable simply because it utilized software. Since this decision, U.S. courts as well as the US Patent Office gradually broadened the scope of protection available for software-related inventions (Kirsch, 2000). The situation evolved to the current status in which it is expected to obtain a patent for software-related inventions. Since the State Street Bank and Trust Co. vs. Signature Financial Group Inc. (1996) case even mathematical algorithms and business methods have been found to be patentable (see also the Amazon One-click case IPXL Holding, plc vs. Amazon.com, Inc., 2005; Bakels, 2003). As from this decision, the U.S. focus, for patentability, is "utility based," which is defined as "the essential characteristics of the subject matter" and the key to patentability is the production of a "useful, concrete and tangible result" (Hart, Holmes, & Reid, 1999). The evolution resulted in a rush of patent applications for software-related inventions and business methodologies.

Contrary to the U.S., Europe has been unwilling to grant patents for ideas, business processes and software programs. The most important reasons are their (in-) direct exclusion from patent protection, as stated in Article 52 (2)(c) European Patent Convention (EPC)[4]. Nevertheless, the European Patent Office (EPO) also reversed course. Its view on patentability of software programs and, more particularly, the interpretation of the "as such" limitation as described below, has been under revision, especially driven by the context of computer programs (the so-called computer-implemented inventions).

Following three landmark cases, Vicom/Computer Related Invention (1987), Koch & Sterzel/X-ray apparatus (1988), and SOHEI/General purpose management system (1995), the European Patent Office concluded:

a claim directed to a technical process is carried out under the control of a program (whether implemented in hardware or software) cannot be regarded as relating to a computer program as such *within the meaning of Article 52 EPC,* (emphasis added)

and

an invention must be assessed as a whole. If it makes use of both technical and non-technical means, the use of non-technical means does not detract from the technical character of the overall teaching.

Notwithstanding this enlargement in European patent law, patens have, contrary to the U.S., never been granted for software programs "as such," the main reason being that in Europe an invention has to be technical in nature. This requirement of technicality is not explicitly stated in the EPC, but can be deduced from Article 52 (2) EPC. Indeed, this provision contains a list of subject matters that are not patentable "as such" (among them programs for computers). The list is not meant to be exclusive, as it only gives examples of materials that are non-technical and abstract

in nature and, thus, cannot be patented (Sarvas & Soininen, 2002).

In the U.S. on the other hand, a patentable invention must simply be within the technological arts. No specific technical contribution is required. The mere fact that an invention uses a computer or software makes it become part of the technological arts if it also provides a "useful, concrete and tangible result" (Hart et al., 1999; Meijboom, 2002).

In Europe, a number of software developers desire patent protection to be enlarged in such a way that software programs become eligible. One of the arguments of supporters of the patentability of software is that patent law provides inventors with an exclusive right to a new technology in return for publication of the technology. Thus, patent law rewards innovators for the investment and encourages continued investment of time and money. Opponents of patent protection argue that such protection is not needed, indeed appropriate in an industry such as software development, in which innovations occur rapidly, can be made without a substantial capital investment and tend to be creative combinations of previously-known techniques (Pilsch, 2005).

The opponents of software patents also indicate practical problems in administering the patent system, as software is voluminous and incremental. Indeed, an invention can only enjoy patent protection provided that it is not part of the prior art. To verify whether this condition is met or not, it is required to know the prior art. However, knowledge about software is widespread and unbundled (very often either tacit or embedded) and may thus be insufficiently explicit for the patent system to work well. In other words, there is too much software, not enough information about it, and what there is, is hard to find (Kahin, 2003). As transaction costs are high, a patent system will favor those with enough resources to verify whether their software can be patented and, afterwards, to search for and deal with possible infringers.

Next to these financial impediments, there are some theoretical issues that concern the installing of a system of software patents. These have to do, first, with the basic, global instrument for intellectual property protection, in other words, TRIPS, and second with the specific legislation in Europe and the U.S.

Although according to Article 10 TRIPS, computer programs are protected by copyright, it is the intention of TRIPS not to exclude from patentability any inventions, whether products or processes, in all fields of technology, provided that they are new, involve an inventive step, and are capable of industrial application (Article 27 TRIPS) (Janssens, 1998). Consequently, TRIPS states, implicitly, that computer programs may also be the subjects of patent protection.

From what is stated previously, it is clear that the U.S. legislation allows patentability of software. In Europe, however, Article 52 (2) EPC remains an obstacle for such a protection, however regretted by even EPO. Indeed, in its decision of February 4, 1999, the Board of Appeals of EPO (hereafter the "Board") stated[5]:

The fact that Article 10 is the only provision in TRIPS which expressly mentions programs for computers and that copyright is the means of protection provided for by said provisions, does not give rise to any conflict between Articles 10 and 27 TRIPS. Copyright and protection by patents constitute two different means of legal protection, which may, however, also cover the same subject matter (e.g., programs for computers), since each of them serves its own purpose. (...) The Board has taken due notice of the developments in the U.S. (and Japanese) patent offices, but wishes to emphasize, that the situation under these two legal systems differs greatly form that under the EPC in that it is only the EPC which contains an exclusion as the one in Article 52 (2) and (3). Nevertheless, these developments represent a useful indication of modern trends. In the Board's opinion they may contribute to the further highly desirable (worldwide) harmonization of patent law.

This decision makes it clear that if software "as such" must be protected on the basis of patents, the exclusion under Article 52 (2)c EPC shall have to be deleted. Which brings one to the question of whether one should want this to happen and whether one perceives its consequences favorably.

Supporters of software patents would like to win a first battle in the race for software patentability by endorsing the proposal for a directive on the protection by patents of computer-implemented inventions currently being discussed within the European Union. They are aware that approving this directive will not immediately result in patentability of software "as such," however, it will form "a new development that may contribute to the further highly desirable (world-wide) harmonization of patent law," which can end up in a deletion of the exclusion now stated in Article 52 EPC. Obviously, opponents will do anything to avoid this evolution, however oblique, from taking place.

Currently the latter have the wining hand: a proposal to allow patents for computer-implemented inventions was rejected on July 6, 2005 by the European Parliament, and any new proposal will take time to develop, if ever again (Perens, 2005)[6].

MAIN FOCUS OF THE CHAPTER

Economic Justifications for Software Protection as Part of Intellectual Property (IP) Protection?

If the European Union may want to strike a balance, it must on the one hand take into account societal needs, and on the other hand the reward of the inventor.

The theoretical foundations of intellectual property rights are debatable, to say the least. Classical philosophy has attempted to explain why intellectual property rights exist, but neither Hegel—the Germanic, idealistic school—nor Locke—the English, empirical school—has been able to provide a coherent, suitable philosophical basis for intellectual property rights—not for lack of trying by themselves or their more recent adherents (Radin, 1982; Schnably, 1993; Hettinger, 1989; Gordon, 1993). The explanation may ultimately occur, not out of law, which is in any event but a mechanic's framework, and not out philosophical theory, but out of the theory of economic pragmatism. Indeed, it may be argued that intellectual property rights, among them software protection, are what they are because they are based on, and fundamentally about, incentives to create and invest.

The United States Supreme Court summed it up in Marer vs. Stein (1954):

"The copyright law, like the patent statutes, makes reward to the owner a secondary consideration." United States v. Paramount Pictures, 334 U.S. 131, 158. However, it is "intended definitely to grant valuable, enforceable rights to authors, publishers, etc., without burden-some requirements; 'to afford greater encouragement to the production of literary [or artistic] works of lasting benefit to the world.'" Washingtonian Co. v. Pearson, 306 U.S. 30, 36.

The economic philosophy behind the clause empowering Congress to grant patents and copyrights is the conviction that encouragement of individual effort by personal gain is the best way to advance public welfare through the talents of authors and inventors in "Science and useful Arts." Sacrificial days devoted to such creative activities deserve rewards commensurate with the services rendered. (emphasis added)

Hence, it appears that the utilitarian, economic incentive perspective is the key driver in the granting of intellectual property laws. But are these incentives really necessary to ensure and sustain creation and invention? In the 1970s, professor (and later justice) Stephen Breyer argued that lead time advantages and the threat of retaliation reduced the cost advantages of copiers, hence obviating if not eliminating the need for copyright protection of books (Breyer, 1970, 1972; Tyerman, 1971). Advances in technology may not have strengthened Breyer's argument. George Priest argued that economic analysis (in his case of patent law) is "one of the least productive lines of inquiry in all of economic thought" because of the lack of adequate empirical bases for the assessment of theoretical models of innovation (Priest, 1986). Still, this view does not undo the fact that the pragmatic utilitarian/economic incentive perspective may remain if not the only, then at least the most useful underpinning for IP rights. These rights inescapably clash with a libertarian view that "information wants to be free" (Barlow, 1994) while those arguing against such freedom cry insist that creation and incentive will be hampered by the diminishing of intellectual property rights. Extremism has polarized views on both sides of the argument, while in the end balanced IP laws may be what are being sought (Lessig, 2004).

Enter Open Source Software

The open source software approach differs radically from the IP protection approach stated above in that it, in the words of Richard Stallman, flips it over to serve the opposite of its usual purpose: instead of a means of "privatizing" software—as Stallman puts it—through IP (copyright or patent) protection, it becomes a means of keeping software free (Stallman, 1999).

In effect, it is meant to create, at least initially, what in patent terms would be deemed a freedom to operate. In other words, open source creates an at least initial space that is open for users. Whether this is a public domain space or another, similar form may be debated. If it is a public domain space, such an approach does not necessarily keeps open all that it touches (Friedman & Kreft, 2000). Open source software as such does not, as it does with a General Public License (GPL), have this "viral" effect.

If open source may mean the creation of a space to all users, the effect of the different licenses granted to users limits the grant of freedom to operate. There may be free software under a GPL. There may also be Open Source Initiative (OSI) licenses, which require nine elements to qualify for approval as an OSI certified license:

1. Free redistribution (no royalties or fees)
2. Access for any party to a source code
3. The license must allow modifications and derived works.
4. The license may restrict a source code, however, from being distributed in modified form but only if certain conditions are fulfilled.
5. There may be no discrimination against groups or persons.
6. Or against fields of endeavour
7. It is prohibited to require any additional licenses from users to whom a program is redistributed.
8. The license must not be specific to a product.
9. The license must not restrict other software.

Next to OSI licenses, there are others that may be copying features of such licenses, but differ, for example, as to treatment of derivative work[7]. In effect, there appear to be over 50 different open source licenses, and no clear guide (Gormulkiewicz, 1999, 2002, 2004). This is a first challenge—one that has not yet been overcome by any centralized system or standardisation.

This first challenge leads to a second one, as the licensing of open source software poses a number of legal challenges that are not necessary resolved at present. First, there is the issue of validity—a classical one that is also known in the proprietary would and goes back to the use of shrink wrap and click wrap agreements to use—assuming acceptance of the user when she/he opens the software. If this constitutes sufficient an agreement remains a question (Trompenaars, 2000).

Coupled with validity is the issue of enforceability, which is more pregnant in a open source model, because the end user may not have, or even be aware, of any license agreements unless he downloads the source code and starts using the software.

Open source software moral rights—rights related to the inventor's personality—include in patent law the right to attribution (also known as a paternity right) and in copyright include in addition at least the right to resist deformation and defamation (de Vuyst & Steuts, 2005; Metger & Jaeger, 2001). Moral rights being inalienable (i.e., non transferable) they may never be put to a user. If a user were to apply a software package for a use that the author/inventor did not like, the latter could, particularly on the European continent, where the notion of moral rights remains strongest, enforce an injunction for such use (e.g., in violent games or pornographic displays).

A third issue that is particular to an open source software approach, from a legal viewpoint, is that of representations and warranties. In a proprietary, particularly a patented would, it is inherent that the invention has been reduced to practice—that is works—before patent publication. No such warranty can necessarily be given in an open source model. In effect, the GPL does not state explicitly that the GPL code can be run—paradoxically, it does state that one may modify and redistribute. As liability is inherent in a proprietary atmosphere, it is not so inherent in an open source model (Kennedy, 2001) where limitations of liability and disclaimers of warranties may be more rightfully expected—but many limit attractiveness to users.

However, open source licensing on the basis of the principle of "no liability" is paramount to the success of open source software development. The fact that a developer of open source code has the ability to distribute work accompanied by little or no warranties effectively shifts the risk from the licensor of the code to the recipient of the code.

This is important:

Valid reasons underlie this risk-shifting strategy ... . Individual hackers are unwilling to assume the risk of a multi-million dollar class action law suit as the consequences of pursuing their passion for hacking code. "Low Risk" also means low barriers to entry; anyone can contribute code to the process, not just those that can afford insurance or lawyers ... . (Gormulkiewicz, 2002)

If open source software developers were not able to disclaim liability on their code, it might substantially increase development costs on account of legal risks and greatly discourage open source development.

It is however questionable whether disclaimers and limitations of liability work in all jurisdictions. Choice of jurisdiction in business open source software licenses is therefore essential, but remains problematic in a consumer atmosphere and certainly in cross-border licensing.

Last but not least, the issue of derivative work is, in all cases but in the GPL, an issue. The GPL's "viral effect," in which any modification to the source code must also be under the GPL is unique—and hard to enforce: who can find out but a disgruntled customer faced with a violation? The explanations given by the Free Software Foundation[8] refer to the judiciary for an answer to the question as to what constitutes a contribution of two parts into another program. But the courts may not be the best, certainly not

the most efficient means of interpreting open source model licenses (Costello, 2002). In effect, a single German case giving effect to the GPL may not yet be an end to the relative quiet for GPL infringers (Shankland 2004). The lack of clarity in GPL licensing (not to speak of other forms, e.g., BSD or LGPL licensing) makes for uncertainty in derivative work's proprietary or non-proprietary status.

In a proprietary model, the answer, under patent law, is forthright: any derivative work is an infringement of the patent owner's rights. In the open source model, the question is the reverse: can derivative work, including interoperable work as decompiled from the source code, become proprietary in nature? In other words, can reverse engineering lead to proprietary software as it is based only on the idea encompassed in open source software?

If one reverses this risk, one should acknowledge that open source software may have the ability to expose software developers to risk if they use open source licenses improperly. This may be a disincentive to use open source software as a base platform for future development.

FUTURE TRENDS

Discussion: Open Source and the Balance of IP Rights

Free distribution and an open source code propel the open source software movement. It is a fact that software patents are expensive to prosecute and take time to publish, hampering development effort—even being bypassed by events in a quickly developing world.

The benefits of open source software also provide it its soft legal underbelly. The plethora of open source licenses, the lack of clarity and a dependence on court interpretation are a disincentive to users, but more importantly, to developers.

The solution may be in a more formal description—a restatement or standardisation, if one wants—of open source software terms of use and licenses. But this appears not to occur yet or in the near future, although efforts to state best practices are in the make (Kennedy, 2005).

More fundamentally to the discussion of open source versus proprietary rights, a restatement of the debate as one on excessive rent seeking and a consequent imbalance of rights may point to a way to set the stage for a meaningful discussion which may lead to long-term policy framing, likely at the level of a WTO's TRIPS—a new one, with a more globally accepted balance of interest. If one addresses, in this mindset, software protection and IP rights in general, in terms of its value as an economic good to society, one is bound to find a balanced view. As Judge Posner put it:

granting property rights in intellectual property increases the incentive to create such property, but the downside is that those rights can interfere with the creation of subsequent intellectual property (because of the tracing problem and because the principal input into most intellectual property rights is previously created intellectual property). Property rights can limit the distribution of intellectual property and can draw excessive resources into the creation of intellectual property, and away from other socially valuable activities, by the phenomenon of rent seeking.

Striking the right balance, which is to say determining the optimal scope of intellectual property rights, requires a comparison of these benefits and costs—and really, it seems to me, nothing more:

The problems are not conceptual; the concepts are straightforward. The problems are entirely empirical. They are problems of measurement. In addition, we do not know how much intellectual property is in fact socially useful, and therefore we do not know how extensive a set of intellec-tual property rights we should create. For all we know, too many resources are being sucked into the creation of new biotechnology, computer software, films, pharmaceuticals, and business methods because the rights of these different forms of intellectual property have been too broadly defined. (Posner, 2002)

The socio-economic measurements, in empirical studies to be undertaken, set a daunting task in terms of methodology as well as in terms of execution. But they point to the way forward: a need to measure the impact of IP rights to determine the optimally needed scope vis-à-vis society.

For software protection, a know-nothing attitude that denies all rights to inventors will be a disincentive to valorise. As became clear to the first author during his tenure at the Common Fund for Commodities, what is put in the public domain is most often left there, as it is difficult to invest the time and energy to shape a competitive advantage from an invention known to and ready to use by all.

That appears to be the case at present: open source would not exist if it was not out of dissatisfaction with an excessively proprietary business model that ignores societal needs. The balance is indeed one between rights and societal needs. Unbalancing in one way or another risks, the wrong investment, or disinvestments, in software development or any other form of IP.

It is open source software that pushes towards this balance by its very existence on alternative to a proprietary business model.

Indeed, it may in three ways contribute to this balancing act already: first, though the so-called Open Patent Review it may involve the citizen in making sure patents represent progress over prior art. This is being accomplished by an alerting system that activates from the USPTO Web site (front of the AppFT database). It will enhance a public view, and review, of applications and assists in preserving safeguards against flooding strategies by would-be patent holders.

Second, open source software Prior Art, an initiative by IBM, Novell, Red Hat and Source-Force aims at developing a system that stores source code in an electronically searchable format, exposing open source software—millions of lines of publicly available computer source code—as prior art to examiners and the public, so as to assist in ensuring that patents are issued only for actual software inventions and not for appropriations—expropriations, if one wants—of existing open source software (Noveck, 2004).

Finally, a patent quality index, in other words, a numeric index in respect of the quality of patents and patent applications, as a resource for the patent system[9], may be of interest if it proves an objective and data-driven tool. Patent rights, if put to the test, might show their true worth—and there may be a readiness to re-evaluate European and U.S. patent law—if there is a tool which assists citizens and examiners in finding the necessary balance between property rights and innovative freedom to operate, through the application of economic tools, such as a rent-seeking methodology.

CONCLUSION

While it may not be a panacea, open source software is clearly a sufficient counterweight to excessive IP creation, which in the words of professor Jeremy Phillips, does to the public domain—and to innovation—what men do to the Amazon forest (Phillips, 1996).

REFERENCES

Bakels, R. B. (2003). Van software tot erger: Op zoek naar de grenzen van het octrooirecht. *IER, August*(4), 214.

Barlow, J. P. (1994). The economy of ideas. *2.03 Wired*, 84.

Breyer, S. (1970). The uneasy case for copyright: A study in copyright of books, photocopies and computer programs. *Harvard Law Review*, 281.

Breyer, S. (1972). Copyright: A rejoinder. *UCLA Law Review*, 75.

Costello, S. (2002). *Settlement nears in open source GPL suit.* Retrieved January, 20 2006, from http://www.networkworld.com/news/2002/0305settleGPL.html

Diamond vs. Diehr, 450 US 175. (1981).

De Vuyst, B., & Steuts, L. (2005). De notie morele rechten in een internationale, vergelijkende en transactionele context. *Intellectuele Rechten—Droits Intellectuels*, 8.

EPO. (1973). *Convention on the grant of European patents (European Patent Convention) of 5 October 1973.* Retrieved January, 20 2006, from http://www.european-patent-office.org/legal/EPC/e/ma1.html

Fink, M. (2003). *The business of economics of limited open source.* Upper Saddle River, NJ: Prentice Hall.

Friedman, D., & Kreft, B. M. (2000). *Open source software: Background, licensing and practical implications.* Retrieved May 10, 2000, from http://www.daviddfriedman.com/Academic/Course_Pages/21st_century_issues/legal_issues_21_2000_pprs_web/Kreft_Open_Source.html

Gomulkiewicz, R. W. (1999). How copyleft uses licence rights to succeed in the open source software revolution and the implications for Article 2B. *Houston Law Review*, 179.

Gormulkiewicz, R. W. (2002). De-bugging open source software licensing. *University of Pittsburg Law Review*, 75.

Gormulkiewicz, R. W. (2004). *Entrepreneur open source software hackers: MYSQL and its dual licensing.* Retrieved from http://www.law.washington.ed/faculty/gomulikiewicz/publicaters/entopensourcesl.pdf

Gordon, W. J. (1993). A property right in self-expression: Equality and individualism in the natural law of intellectual property. *Yale Law Journal*, 1533.

Gottschalk vs. Benson, 409 US 63 (1972).

Hart, R., Holmes, P., & Reid, J. (1999). *The economic impact of patentability of computer programs (Report to the European Commission).* Retrieved February 18, 2004, from http://europa.eu.int/comm/internal_market/en/indprop/comp/study.pdf (pp. 20-23 of full report)

Hettinger, E. C. (1989). Justifying intellectual property. *Phil. & Publ. Aff.*

IPXL Holding, plc vs. Amazon.com, Inc., US Fed. App. Ct. 05-1009, -1487, November 21, 2005.

Janssens, M. C. (1998). Bescherming van computerprogramma's: (lang) niet alleen maar auteursrecht. *T.B.H.*, 421-422.

Kahin, B. (2003). Information process patents in the U.S. and Europe: Policy avoidance and policy divergence. *First Monday, 8*(3). Retrieved March 6, 2004, from http://www.firstmonday.org/issues/issue8_3/kahin/

Kennedy, D. M. (2001). *A primer on open source licensing legal issues: Copyright, CopyLeft, copyfuture.* Retrieved from http://www.denniskennedy.com/opensourcedmk.pdf

Kennedy, D. (2005). *Best legal practices for open source software.* Retrieved November 20, 2005, from http://www.llrx.com/features/opensource.htm

Kirsch, G. J. (2000a). *Software protection: Patents versus copyrights.* Retrieved January 20, 2006, from http://www.gigalaw.com/articles/2000/kirsch-2000-03.html

Kirsch, G. J. (2000b). *The software and e-commerce patent revolution.* Retrieved February 18, 2004, from http://www.gigalaw.com/articles/2000/kirsch-2000-01.html

Koch & Sterzel/X-ray apparatus T 0026/86, OJ EPO 1988, 19.

Leijnse, B. (2003, January 16). Een patente oplossing voor uw patentprobleem. Softwarepatenten.be. *Trends.* Retrieved February 27, 2004, from http://www.softwarepatenten.be/pers/trends_20030116.html

Lessig, L. (2004). Be wary of IP extremists. *Computerworld.* Retrieved March 26, 2004, from http://www.computerworld.com.au/index.php?id=43841790&fp=16&fpid=0

Marer vs. Stein 347 U.S. 201 (1954).

Meijboom, A. P. (2002). Bang voor software-octrooien. *Computerrecht, 2002*(2), 66.

Metger, T., & Jaeger, A. (2001). Open source software and German copyright law. *Int'l Journal of Industrial Property and Copyright Law, 32*(1).

Noveck, B. (2004). Unchat: Democratic solutions for a wired world. In P.M. Shane (Ed.), *The prospects of political renewal through the Internet.* Oxford, UK: Routledge.

Overdijk, T. F. W. (1999). Europees Octrooibureau verruimt mogelijkheden voor octrooiering van computersoftware. *Computerrecht, 1999*(3), 158-159.

Parker vs. Flook, 437 US 584 (1978).

Perens, B. (2005, January 31). The open-source patent comundrum. *News.Com.*

Phillips, J. (1996). The diminishing domain. *European Intellectual Property Review*, 429.

Pilch, H. (2005). *Quotations on software patents. Logical Patent Web site.* Retrieved February 4, 2006, from http://swpat.ffii.org/vreji/quotes/index.en.html

Posner, R. A. (2002). The law & economics of intellectual property. *Daedalus, 5*, 12.

Priest, G. (1986). What economists can tell lawyers about intellectual property. *Res. Law & Econ.,* 19.

Radin, M. J. (1982). Property and personhood. *Stanford Law Review,* 957.

Sarvas, R., & Soininen, A. (2002, October 15-16). *Differences in European and U.S. patent regulation affecting wireless standardization.* Paper presented at the International Technology and Strategy Forum Workshop on Wireless Strategy in the Enterprise: An International Research Perspective, Berkeley, CA. Retrieved on March 9, 2004, from http://www.hiit.fi/de/core/PatentsWirelessStandardization.pdf

Schnably, S. J. (1993). Property and pragmatism: A critique of Radin's theory of property and personhood. *Stanford Law Review,* 347.

Shankland, S. (2004, April 22). GPL gains clout in German legal case. *News.com.*

Sohei/General Purpose Management System T 0796/92, OJ EPO 1995, 525.

Stallman, R. (1999). The GNU operating system and the free software movement. In C. Dibona et al. (Eds.), *Open sources: Voices from the open source revolution.* O'Reilly. Retrieved from http://www.oreilly.com/catalog/opensources/book/stallman.html

State Street Bank and Trust Co. vs. Signature Financial Group Inc., 927 F. Supp. 502, 38 USPQ2d 1530 (D. Mass. 1996).

Trompenaars, W. M. B. (2000). Legal support for online contracts. In B. Hugenholtz (Ed.), *Copyright and electronic commerce. Legal aspects of electronic copyright management.* Amsterdam, The Netherlands: Kluwer.

Tyerman, L. (1971). The economic for copyright protection for published books: A reply to Professor Breyer. *UCLA L. Revs.,* 1100.

Vicom / Computer Related Invention, T O208/84, OJ EPO 1987, 1.

WIPO. (1971). *The Berne Convention for the protection of literary and artistic works.* Paris.

WTO. (1994). *The Agreement on Trade-Related Aspects of International Property, Annex 1C of the Marrakech Agreement of April 15, 1994 establishing the World Trade Organization ("WTO").* Retrieved January 20, 2006, from http://www.wto.org/english/docs_e/legal_e/04-wto.pdf

KEY TERMS

Copyright: A set of exclusive rights regulating the use of a particular expression of intellectual property.

EPC: Convention on the grant of European patents (European Patent Convention) of October 5, 1973.

EPO: European Patent Office (München) established by the EPC.

Patent: A grant made by a government that confers upon the creator of an invention the sole right to make, use, and sell that invention for a set period of time, through letters patent which protect an invention by such a grant.

TRIPS: The Agreement on Trade-Related Aspects of International Property, Annex 1C of the Marrakech Agreement of April 15, 1994 establishing the World Trade Organization ("WTO").

USPTO: United States Patent and Trademark Office.

WTO: World Trade Organization, established on April 15, 1994.

ENDNOTES

[1]   The Agreement on Trade-Related Aspects of International Property, Annex 1C of the Marrakech Agreement of April 15, 1994 establishing the World Trade Organization ("WTO").

[2]   The Berne Convention for the Protection of Literary and Artistic Works (1971).

[3]   By Europe, we mean the European national patent and the system pursuant to the European Patent Convention so that patents can be applied centrally for all contracting states of the European Patent Office (EPO).

[4]   Article 52 (2) (c) EPC states that programs for computers shall not be regarded as inventions within the meaning of Article 52 (1) EPC and are, therefore, excluded from patentability. Article 52 (3) EPC establishes, however, an important limitation to the scope of this exclusion. According to this provision, the exclusion applies only to the extent to which a European patent application or a European patent relates to programs for computers "as such."

[5]   Technical Chamber of the Board of Appeals of EPO, February 4, 1999, *Computerrecht* 1999/6, 306-310 with Note of D.J.B. BOSS-CHER, 310-312.

[6]   See also Free Software Foundation, Software patents in Europe at http://www.fsfeurope.org/projects/swput/swput.en.htm.

[7]   See An overview of "Open Source" Software License, Report of the Software Licensing Committee of the American Bar Association's Intellectual Property Section, at http://www.abanet.org/intelprop/opensource.html.

[8]   Free Software Foundation, FAQ on the GNU GPL, at http://www.fsf.org/licenses/GPL-faq.html.

[9]   See the work of Prof. Polk at www.law.upenn.edu/blogs/polk/pqi/documents/2006_1_presentation.pdf

Chapter XXVI
OSS Adoption in the Legal Services Community

Ray Agostinelli
Kaivo Software, Inc., USA

ABSTRACT

This chapter provides an anecdotal case study of the adoption of open source software by government-funded nonprofit organizations in the legal services community. It focuses on the Open Source Template, a Web site system that provides information to the public on civil legal matters, and collaborative tools for legal aid providers and pro bono attorneys. The successful aspects of the adoption within this community are traced to the funders' emphasis on developing re-usable, non-proprietary technology tools, the strong communitarian ethic which nonprofits share with the open source community, and the presence of an active support network to broadly leverage intellectual capital. It is hoped that this chapter will assist those considering the adoption of open source software by identifying the specific factors that have contributed to the success within the legal services arena and the real-world benefits and challenges experienced by the members of that community.

INTRODUCTION

An instructive case study in OSS adoption is afforded by the experience of a number of government-funded nonprofit organizations in the legal services community. Since 2001, over 20 such organizations have established community Web site systems built entirely on OSS technologies. These Web sites are designed with a twofold purpose: to serve as a portal where individuals who cannot otherwise afford legal representation can find information to help with civil legal problems and questions, and to facilitate collaboration be-tween providers of legal assistance through the use of online tools.

The subjects of this case study have some particular characteristics which have shaped their involvement with the technology and suggest where their experiences are apt to be shared. Those characteristics include: a small number of thought leaders motivated to pursue OSS alternatives to existing proprietary solutions; a broader community receptive to the advice and guidance of those leaders; a technology stack sufficiently mature to minimize the need for technical support while also sufficiently open to allow for flexible use and customization; continuing support, financial, and consultative, from government funders committed to spreading the use of the technology and integrating it into wider initiatives; and the establishment of a robust and active community of users around the technology committed to principles of sharing, collaboration, and self-reliance.

The objectives of this chapter are to examine the successes and failures associated with OSS adoption by government-funded nonprofits by exploring:

- The philosophical appeal of OSS to the nonprofit legal services community
- The practical application of the OSS development methodology within the community, including re-use and sharing of code
- The cost benefits realized by this community
- The challenges in adopting OSS, including the importance of a supportive infrastructure and quality technical training
- Vendor relations, as they bear upon issues of control and independence

This case study is meant to provide insight into the real-world use of OSS within the nonprofit community and, by anecdote if not by rigorous scientific analysis, to draw out some important themes and implications of OSS adoption for those considering a similar path.

BACKGROUND

The Legal Services Corporation (LSC) is a private, nonprofit corporation established by Congress to provide civil legal assistance in areas such as family law, housing, and consumer issues to those who otherwise would be unable to afford it. Since 2000, as part of its Technology Initiative Program (TIG), the LSC has awarded grants to nonprofit legal services organizations nationwide to subsidize innovative uses of technology to improve the delivery of legal services to their client population. A central focus of the TIG program has involved the development of statewide Web site portals where clients can obtain legal information and where legal aid and pro bono attorneys throughout the state can collaborate and share resources.

Recognizing that statewide organizations across the nation have common needs that can be served by common tools, the LSC early on decided to support technology solutions that could be shared and reused by multiple groups, thereby minimizing costs for development, training, and support. In evaluating the merits of grant applications, replicability and reusability have always been important criteria.

In 2000, the LSC awarded two grants that subsidized the development of Web portal systems for Pine Tree Legal Assistance in Portland, Maine and Ohio State Legal Services Association in Columbus, Ohio. The vendor for those projects was selected via an open RFP process with preference given to proposals that included the use of OSS tools. The resulting sites—www.helpmelaw.org and www.oslsa.org, respectively—incorporate a wide array of informational resources and collaborative tools, including a full-text searchable document library, office locator, calendaring system, jobs database, and interest group areas with discussion forums. Both sites were built exclusively on OSS technologies, including Zope (a leading OSS content management system), MySQL, Apache, Python, and Linux.

In 2001, the components that comprised these portal systems were re-packaged as a template that could be easily customized for use by other legal aid organizations. The Open Source Template (OST), as it came to be known, was subsequently endorsed by the LSC as one of two systems that could be used by recipients of future grants subsidizing statewide Website projects in other states. (Hereafter in this chapter, the term OST shall be used to refer to the combination of Web-based tools that comprise the template system.)

As of February 2006, the OST had been adopted by over 20 states. Each state organization employs a Website administrator (typically a part-time position) to coordinate content on the site. Depending upon the technical abilities within a given organization, technical changes and support are either performed in-house or outsourced to the vendor who developed the original system, who remains active within the community.

Traffic to the Web sites varies fairly dramatically by state, ranging in 2005 from just over 20,000 unique visits for the least popular sites to 895,000 unique visits for the single most popular (the Pine Tree Legal Assistance site in Maine) which has experienced steady increases in traffic in each of its first five years, including a 20% increase from 2004 to 2005.

MAIN FOCUS OF THE CHAPTER

Philosophical Appeal

While flexibility and cost savings—two traditional benefits of OSS—are important to their successful adoption in the legal aid arena, it is interesting to note that its original appeal to members of this community were as much *philosophical* in nature as purely material or financial. The non-proprietary ownership model, the emphasis on collaborative development, and the benefits of re-use and re-purposing are as conspicuous in the legal aid arena as they are central to the OSS development model.

Hugh Calkins, the director of research and development at Pine Tree Legal Assistance, is a respected leader on technology issues among legal aid providers. An early champion of OSS who argued forcefully for its adoption both within his own organization as well as in the broader community, Mr. Calkins points to the communitarian ethic of the OSS development model as the single most important factor in determining its applicability for his uses. "I think of legal aid attorneys as open source lawyers. The ethic of open source software fits perfectly with the way we try to work in the legal aid community. We are not proprietary about our work and we try to build on what other people are doing around the country."

In the wide world of open source software development, some projects seem to accumulate robust communities of contributors while others languish and eventually disappear. The power of Mr. Calkins' insight is that the universe of legal aid providers across the nation *already formed* a robust community dedicated to sharing intellectual property and working toward common goals—the network did not need to be grown from scratch—and so its members were therefore well positioned to reap the benefits of OSS principles and methodologies.

Practical Application

From its inception, the OST was designed so that innovations and advances made by one organization could be shared with many others. Practically, this is enabled by a combination of collaborative technologies and dedicated support personnel.

A portal Web site dedicated to this community includes an area where programming code can be posted and downloaded. This portal predominantly includes code that enhances or extends the template's core functionality—a feature that concatenates existing search routines into a global search, for example; code to generate new reports of resources in the document library; code to alter the user authentication system so that self-registering users must enter a private identifier before gaining access to the site's restricted content. (The portal site does not contain *major* new component functionality, which is typically beyond the ability of the individual Web site administrators to integrate without outside assistance.) The community also makes use of a listserv on which questions are raised and opinions are voiced.

For any technical problem or challenge associated with their use of the system, community members are encouraged to first consult the shared code library to see if their problem has been solved by others before them, then to solicit assistance from the listserv, and finally, if those avenues do not prove fruitful, to obtain support from the primary vendor of the OST or another outside technical resource.

In addition to the online collaborative tools, an important part of the supporting infrastructure for this community is a dedicated resource with a national focus—a "circuit rider"—who combines domain knowledge of the legal aid arena with in-depth understanding of the OST platform. Her job is to assist individual organizations in enriching the content on their Web sites and to expand the sites' usage via stakeholder committees and marketing efforts. While her primary purpose is not to support the technical aspects of the sites, she plays an important coordinating role in making members aware of what is happening elsewhere in the community and to help ensure that time is not wasted solving problems that have been solved elsewhere.

One interesting corollary of the application of OSS development methodology in this community has been the increased ability for disparate organizations to share *content* in addition to technology. The common platform has allowed informational resources to be cross-posted easily on related, though independent, sites. For example, in Maine, a sub-section of their Website dealt with Medicare Part D, the prescription drug plan that went into effect in January, 2006. That content was immediately made available to other sites, including Legal Services for the Elderly, Maine Equal Justice Partners, and Vermont Law Help, where local administrators were able to quickly post that content on their own sites, modifying those portions that were state-specific and

retaining the federally focused content that was generally applicable to all states.

Cost Benefits

Earlier, we cited the communitarian ethic as a primary incentive for adopting OSS in the legal aid community. As is typical in the world of nonprofits generally, reduced cost has also been an important factor.

Joyce Raby, a program analyst at the LSC responsible for overseeing many of the statewide Web site grants, cites decreased costs for both initial development and subsequent enhancements as critical factors in awarding OST grants: "New Web sites get created very quickly as there is a basic foundation already in place which is open and accessible to everyone. No one has to start from scratch. In addition, enhancements to the Web sites are much less costly. As we have improved and refined how the Web sites would operate, each improvement only has to be created once and is immediately available to all other users. The incremental cost for replication around the country is negligible compared to what it would have been to do custom development for many individual standalone sites."

Speaking to the issue of maintenance, Ms. Raby claims, "Support for the upkeep and administration of the Web sites is also less costly. We funded a single circuit rider to provide ongoing project management assistance to the Web site implementation teams around the country. Had each state organization created custom Web site solutions, we might have been forced to fund multiple types of assistance to ensure our programs got the help they needed."

While it is difficult to quantify the cost impacts directly ascribable to the use of OSS, some general parameters are worth noting:

- The two original sites which served as the foundation of the OST were developed at a combined cost of approximately $175,000. (Since the vendor for those projects were selected via an open RFP/Proposal process, it is fair to assume this cost accurately reflects market rates for those development projects.)

- The resulting template system was thereafter made available to other organizations at a cost of $10,000 per site. (This cost was not a license fee but rather applied to the consulting services associated with the initial setup and configuration of the site; the code itself was available free of charge.)

The cost differential is a result, at least partly, of the fact that (1) the original systems were explicitly developed to be as re-usable and as replicable as possible, and (2) the source code for those systems was not held proprietarily by either the vendor, the client organizations or the LSC, but rather was explicitly made available for adoption, modification, and expansion by the other organizations with similar needs.

There are no license fees associated with the OST. On-going costs associated with the system scale according to the technical support needs of a given organization. These typically range from approximately $1,000 per year (for hosting services only) to approximately $10,000 (for hosting plus significant technical support).

Some additional cost impacts of the OST are suggested in the following comparison. The alternative Web template system endorsed by the LSC for funding under TIG grants is a system built on proprietary technology platforms and offered according to an application service provider (ASP) model. The initial start-up cost for this system, as with the OST, is $10,000 per site, which includes a use license for the first year. On-going subscriber fees are $5,000-$15,000 per year, scaling according to several factors including the size of the organization.

This is not to say that the OST is by definition dramatically less expensive than an alternative

ASP model. It is not. The OST model is, however, more flexible, allowing individual organizations to pay only for what they need. The flexibility is a direct outgrowth of the non-proprietary nature of the underlying code.

Challenges

Adoption of the OST has not been without its challenges, many of which are common to the broader world of open source software. The supporting infrastructure described above (the shared code portal; the listserv; the circuit rider) which are critical to the platform's success, has taken several years to develop and mature. In the early years of the platform's existence, the user community was hierarchical and shallow, a few true experts at the top guiding a much wider range of less-skilled users at the base. Only in time has a true peer-to-peer network developed, one in which a listserv posting might realistically receive a reply from all quarters, and where innovations are occurring on multiple fronts, not exclusively in the laboratories of the early adopters.

Another obstacle has been a direct by-product of the open nature of the code. The relatively greater flexibility of the OST (than, e.g., the alternative ASP platform) requires greater technical expertise within the user community. With access to every level of the programming code, the platform can be customized to any individual organization's needs. However, making customizations does require some level of technical knowledge, whether that is the CMS's scripting language, database/SQL, Python, or Linux shell. Roughly speaking, about one half of the client community has technical expertise sufficient to perform minor customizations without recourse to outside vendors, and only about one quarter can make major changes on their own.

Understanding that, training the OST users on the CMS scripting language has been a high priority, and trainings have been conducted at least once a year since 2002. The effectiveness of the training has been limited by two factors: firstly, many of the Web site administrators for the OST sites do not possess the requisite technical backgrounds to become extremely productive working with the system's source code, and secondly (and perhaps more representative of the nonprofit world in general) a high degree of staff turnover has resulted in the need to re-train new administrators. The result is that the OST's flexibility—a key reason that OSS was embraced—is not always exploited to its fullest.

A final challenge, also common to the broader world of OSS, is the perpetual danger of forking. The community consists of over 20 sites, all sharing a common code base, each of which can be customized in significant ways to fit the local needs. Enhancements made by one organization, or by the OST vendor, always rely to some extent on a certain baseline set of code in order to function properly. If that baseline code has been altered, then implementing features developed elsewhere becomes prohibitively difficult, and the site will have effectively orphaned itself from the community and lost the benefits of code re-use.

For the most part, forking has been kept to a minimum within this community, largely via constant consultation with the vendor, but also because most organizations are either not inclined, or do not possess the technical capability, to veer dramatically from the mainstream. Still, the importance of being attentive to the dangers of forking and managing the development process carefully cannot be underestimated, particularly as the technical expertise level of the community increases.

Vendor Relations

There is no doubt that the success of the OST has relied to date on a very close relationship between the vendor and the client community. The importance of this relationship has been recognized by the LSC, and their model for funding the OST sites has from the start included follow-on grants that individual state organizations can use for continued maintenance of the Web sites, some of which can be applied to technical support activities.

Since ownership of source code and vendor independence are important aspects of OSS, it is worthwhile to consider the OST community's relation to the OST vendor. Do the community members really have control? Are they really independent? The answers are: yes and almost.

From a *control* standpoint, the members really are in full command of their site's underlying technology. Every aspect of the application stack is open and available to them. Individual sites can be changed according to local needs, priorities, and schedules. For example, OST administrators at Legal Action of Wisconsin developed significant enhancements to the security model on their own initiative, according to client confidentiality requirements specific to their state. This can be contrasted with the ASP model where any change to the system, because it affects all users, must be vetted by the community at large and implemented according to the vendor's development schedule.

OST clients are *independent*, too, to the extent that they are not contractually bound to the original vendor beyond the initial cost for implementation. For subsequent assistance, if they so choose, they can return to the original vendor or draw from the worldwide community of developers proficient in the underlying OSS technologies.

Practically, many of the client groups do continue to rely on outside assistance—predominantly the OST vendor—because their needs for some level of technical support still outstrip the user community's capacity to provide it. It is fair to say that the "umbilical cord" connecting clients to vendor is weakening, although it is not yet fully cut. And while a full severance is not necessarily the ultimate goal, the ability for these organizations to maintain their sites without regular recourse to outside consultation does speak to the sustainability of the platform.

FUTURE TRENDS

The sustainability of the sites was in fact studied in 2005 by outside consultants tasked to examine how the Web sites could be financially supported in the future in the absence of LSC funding. One of their conclusions (Melton, 2005, p. 20):

Support networks remain critically important to maintaining the momentum to sustainability. The opportunities for sharing strategies, best practices, and lessons learned that the ... networks provide are invaluable to the ongoing maintenance and growth of the state Web sites.

The network is key. Without the network—the surrounding community of like-minded institutional players with common goals invested in the success of the technology—the long-term prospects of the system are very much in doubt. In this way, the OST experience reinforces a truism of the OSS world, that absent a critical mass of interest and traction, projects fall by the wayside, grow outdated and are eventually abandoned. Openness alone is no guarantee of success.

Technically, the growth path of the OST is planned to mirror that of the underlying CMS platform, Zope. In 2003, the OST was upgraded to work under the then-current major version release of Zope which allowed the creation of more powerful search and reporting functions in the document library. In the future, other enhancements and extensions to the platform that emerge from the Zope community (which as an open source project in its own right has thousands of participants worldwide) including personalization and workflow features, will be integrated into the template and will be made available only for the cost of integration, with no associated license or upgrade fees.

The move toward OSS in the legal services community is expanding in other ways as well, independent of the OST platform. LSTech.org, a major support site for technical advice and collabo-ration, has fully standardized on OSS technologies (including the Zope CMS, Sympa mail list and MediaWiki collaboration platforms). In addition, LSTech staff members regularly manage an Open Source CyberCafe at industry conferences and trade shows where attendees can browse the Web and check their email using donated PC's running Linux and Mozilla Firefox.

CONCLUSION

To the extent that the legal aid arena is representative of the broader world of government-funded nonprofit organizations, the following general conclusions can be drawn:

- The communitarian ethic of some sectors of the nonprofit world has a direct analog in the shared, non-proprietary character of OSS.

- Common philosophy makes nonprofits particularly well suited to realize the benefits of reuse and sharing central to the OSS development model since they are already institutionally committed to an open exchange of intellectual property with like-minded organizations.

- Given that the Web is a marriage of technology and content, a common technical infrastructure also facilitates sharing of information.

- Successful adoption of OSS depends on the existence of an active support network, which may take some time to mature. For government-funded nonprofits, a commitment to supporting the community during that maturation phase is critical.

- Of the benefits typically associated with OSS, the *flexibility* that comes with control over the technology is the most prominent. Operating off a common code base that they are free to customize, individual organizations can reap the benefits of others' work while not being tied to an identical platform.

- From a funder's perspective, real cost savings result from an initial strategic decision to subsidize replicable systems such as templates with re-usable components. From a client organization's perspective, additional cost savings may result from the greater granularity in scaling individual needs to technical support models as opposed to the more fixed cost models typically associated with proprietary licensing structures.

- The benefits of the OSS model are most highly realized when there is *some level* of technical expertise available to individual organizations. This deserves especially close consideration in an arena like legal services where full-time technical support staffs are not the norm.

REFERENCES

Melton, L., Snider, O., & Zorza, R. (2005). *Ensuring the long term viability of the statewide Web site component of the access to justice system*. Unpublished Final Report of the Web site Sustainability Project, prepared for the National Association of IOLTA Programs under a grant from the Legal Services Corporation.

KEY TERMS

Adoption: The process of accepting and using particular software as a standard within an organization.

Community: A group of individuals who come together due to a common interest or a shared focus on a particular item or idea.

Open Source Template (OST): A combination of Web site tools built exclusively on open source technologies, funded by the Legal Services Corporation, that allows clients to obtain assistance on civil legal matters and enables collaboration between legal aid providers within a state.

Legal Services Corporation (LSC): A private, nonprofit corporation established by Congress to provide civil legal assistance in areas such as family law, housing, and consumer issues to those who otherwise would be unable to afford it.

Nonprofit: An organization that does not include the generation of a profit as a core part of its overall organizational focus or strategy.

Technology Initiative Grant (TIG) Program: A program within the LSC which directs funds specifically to technology projects. The program subsidized the development of the Open Source Template.

Zope: A leading open source content management system, written in the python programming language. It serves as the underlying platform of the Open Source Template.

Chapter XXVII
The Road of Computer Code Featuring the Political Economy of Copyleft and Legal Analysis of the General Public License

Robert Cunningham
Sourthern Cross University, Australia

ABSTRACT

This chapter examines the development of open source computer software with specific reference to the political economy of copyleft and the legalities associated with the General Public License (GPL). It will be seen that within the context of computer software development the notion of copyleft provides an important contrast to more traditional uses of copyright. This contrast symbolizes political, economic, and social struggles which are contextualized within this chapter. As the GPL is an important legal embodiment of copyleft, its legalities are preliminarily explored so as to determine its future potential. While there is some scope to further refine the legal strength of the GPL, it will be seen that it remains a strong and subversive legal instrument which will continue to underlie open source initiatives in the years to come.

INTRODUCTION

This chapter has two distinct objectives. Firstly to survey the political economic foundation of copyleft as it applies to open source computer software, and secondly, to provide some preliminary legal analysis in relation to the General Public License (GPL) which legally embodies copyleft principles. The political economic dimension of the chapter embraces a philosophical approach on the basis that "philosophy offers nuance where there was none" (Lehman, 1999, p. 239). In relation to the GPL legal analysis, it should be noted by way of disclaimer that the commentary constitutes legal analysis, not legal advice.

The chapter begins its philosophical exploration by giving a brief overview of copyright as it applies to the language of computer software, specifically source code. This is followed by a discussion that contrasts closed source and open source software development. It will be seen that this contrast is grounded in a political, economic, and social struggle which, almost classically, fits into the right/left political divide. This divide is made explicit by elucidating the anarchist tendencies of open source software development via a discussion of the "tragedy of the anticommons" theory. The political divide is also drawn upon by way of juxtaposing the traditional notion of copyright with the open source notion of copyleft.

Copyleft is an innovative concept derived by the open source movement which draws upon traditional copyrights in an unconventional manner so as to maximize information flow. It is enshrined within Provision 2(b) of the GPL and remains an important legal mechanism of the open source movement. Those readers that are primarily interested in the legal analysis, as opposed to the political economic dimensions of copyleft, are encouraged to turn directly to the second half of this chapter. It will be seen in this part, and thereafter, that the GPL raises a plethora of interesting legal issues specifically arising from its duality as a license *and* as a contract. While the wholesale enforceability of the GPL escapes the parameters of this chapter, the license/contract duality of the license, which has led to it commonly being referred to as a contractual license, is touched upon in order to uncover potential latent legal issues. It will be seen that implicit within the GPL discussion is the apparent self-enforcing nature of the license which makes it at once a strong and subversive legal instrument.

BACKGROUND

Copyright and Computer Software

Copyright is one of the important pillars of the international intellectual property right (IPR) regime. Its practical effect is broad in relation to both its application and its subject matter. As St. Laurent (2004) explains, application of copyright is automatically inferred in a broad manner to the point that a drawing of a flower on a café napkin is copyrighted simultaneously with its creation and, generally speaking, becomes the sole property of its creator. The drawing of the flower cannot be displayed, copied or otherwise commercially exploited by any person other than the creator for the life of the copyright. Under copyright law, no person other than the creator can create "derivative works," which are works that depend upon or develop from the original, copyrighted work. In many cases there is no need for registration of the right as it automatically attaches to every novel expression of an idea, whether through text, sounds or imagery for the period of the life of the copyright. In countries such as the USA and Australia this is the life of the creator plus 70 years.

Over and beyond the broad *application* of copyright, *subject matter* is also far-reaching touching upon a wide-range of endeavours; from the mundane such as timetables and betting coupons; to the truly artistic such as films, literature and music; to technology-based products such as television broadcasts and computer software (Caenegem, 2001). It is the latter form that is the subject of this chapter. Although computer *software* has escaped a statutory definition in many Anglo-American countries, Justice Gibbs in the Australian High Court did define the notion of computer *program* broadly in *Computer Edge Pty Ltd. Vs. Apple Computer Inc.* (1986) 161 CLR 171 at 178-179 as:

*a set of **instructions** designed to cause a computer to perform a particular function or to produce a particular result.* (emphasis added)

The "instructions" manifest in what is called computer code which generally exists in three formats: flowcharts, source code, and object code (Carstens, 1994). Programmers initially draft a new program in flowchart form which symbolizes the idea of the program. Drawing upon the flowchart, the programmer then writes the source code in a high-level programming language, such as BASIC, C++, or Java, which corresponds with the spoken English language (Mc John, 2000; Reger, 2004). These high-level languages primarily use descriptive words, formulas, and mathematical equations which enable the developer to tell the computer what to do (Velasco, 1994; Nadan, 2002). Webbink (2003) observes that it is source code that "links computers and humans" and its legal protection is found in the form of a literary text under copyright law at the time which it is written. Once the source code is complete, a compiler translates the written source code into "executable" code, otherwise known as object code, which is a low-level computer language that is generally unintelligible as it consists primarily of binary ones and zeros read by the computer to run the program (Wilson, 1993; see also *Apple Computer Inc. vs. Franklin Computer Corp.*, 714 F. 2d 1240. 1243. 219 U.S.P.Q. (BNA) 113 (3d Cir. 1983) at 116).

The fact that computer software, or more specifically source code, has been deemed to be a literary work under copyright laws suggests that the creation of source code is indeed a creative act. While computer software can also be protected via the industrial intellectual property instrument of patents, this dimension of protection lies outside the scope of this chapter (N.B. readers interested in the patenting of software issue can turn to literature such as Fitzgerald and Fitzgerald, 2004, and Lemley, 2003).

The intellectual property ownership or otherwise of creative output concerning source code is an issue which will obviously effect the creative development of computer software, and is an issue which lies at the heart of this chapter. This ownership issue is of particular significance within the different software development models, and it is to this subject that the chapter now turns.

MAIN FOCUS OF THE CHAPTER

From the outset it is important to recognize that this chapter, in the name of brevity, has adopted the position of merging the free software movement with the open source movement, since the focus is the juxtaposition of the closed and open source software development models. This position is adopted in full recognition that it is a simplified perspective as there are critical nuances between the free software movement and the open source movement. While these nuances are beyond the scope of this chapter, they remain important and are dealt with elsewhere (see, e.g., Nadan, 2002, pp. 353-363). In using the term *open source*, this chapter is therefore implicitly referring to both the free software and open source movements collectively. While this approach is somewhat simplistic is does allow for the discussion of an important juxtaposition between closed source and open source software development which is an issue to which the chapter now turns.

Closed Source vs. Open Source (2006)

Moglen (1999a) suggests "there is a myth, like most myths partially founded on reality, that computer programmers are all libertarians." According to this myth right-wing libertarians support closed proprietary models of software development, are avid capitalists who play the stock-market and scorn unions, civil rights laws and taxes; the left-wing libertarians support open source software development, detest the market and all government, and hate Bill Gates because he's rich (Moglen, 1999a, 1999b). While this myth is a useful starting point for discussion, the analysis of Vaidhyanathan (2004) provides greater political and philosophical depth and insight. Vaidhyanathan (2004) suggests that at the heart of the tussle between close source and open source software programming is the clashing ideologies of anarchy and oligarchy. As Vaidhyanathan (2004, p. xii) somewhat playfully puts it:

One side invents a device, method, algorithm or law that moves our information ecosystem toward increased freedom of distribution and the other subsequently deploys a method to force information back into its toothpaste tube.

Drawing upon Figure 1 under Vaidhyanathan's (2004) analysis, the close source software programming tends towards corporate capitalism or (economic) fascism whereas the open source software programming tends towards liberal socialism or anarchism.

This perspective reinforces a generalized notion that the clash between closed and open forms of software development is real. The clash is based on a political, social and economic struggle, and is "between those that wish to commodify and exploit creative output and those that wish to be able to access and freely distribute information in an act of social discourse" (Fitzgerald & Fitzgerald, 2004, p. 446). The struggle has manifested in many ways, one of which is the development by the open source community of a "gift culture" which has been built to counteract

and mitigate against "worrisome concentrations of corporate power in the software industry [by disdaining] those who seek to financially profit from the community's shared body of knowledge" (Fitzgerald & Bassett, 2003b, p. 16). The chapter now turns to a juxtaposition of these two different modes of software development.

Closed Source Software Development

Closed source software development methods involve proprietary interests employing a group of programmers to create, test, and debug code. The programmers are generally subject to a non-disclosure agreement, and copyright is claimed over the resulting code (Suzor, Fitzgerald, & Bassett, 2004). Under this method of software development, software is marketed as a copyright license and defined as "any product we make available for license for a fee" (Microsoft Open License Agreement v 6.0, 2001, para. 1). Such proprietary licenses are typically sold under a volume license product key (VPK) which gives the consumer the right to install, copy, access, use or display the product for the number of copies authorized. The consumer is held liable for any unauthorized use of this key.

Figure 1. (Source: Adapted from Stilwell, 2000)

The closed source software model is exemplified by Microsoft software, and is the model used by most software companies today. Economically, Bobko (2000) indicates that this model operates on two assumptions. Firstly, that selling the product will compensate the company for the developer's labour and time, and secondly, that the market price of the software will be proportionate to its economic value. Accordingly, under these assumptions the corporation will theoretically recover its costs of production (Johnston & Grogan, 1994). Under the closed source model it is argued that distributing the program exclusively in object code reduces the risk of exposing the source code, which would allow computer engineers to see the embodiment of the original programmer's skill, effort, creativity, and innovation. The publication of source code would also allow a computer programmer to take and reuse the innovative or labour-intensive aspects of a particular program and use this innovation in that programmer's own competing program (Nadan, 2002). This process of "reverse-engineering" is considered to be a threat to closed source software developers, as they argue that it would diminish the competitive advantage of their original program (Carstens, 1994). This threat is echoed in the following direct testimony by Bill Gates in *State of New York vs. Microsoft Corporation* (see Suzor, Fitzgerald, & Bassett, 2002, p. 1):

a ... competitor who is free to review Microsoft's source code ... will see the architecture, data structures, algorithms and other key aspects of the relevant Microsoft product. That will make it much easier to copy Microsoft's innovations, which is why commercial software vendors generally do not provide source code to rivals.

Open Source Software Development

Frustrated with the monopolization of creativity and innovation, the computer scientific community evolved an open source software move-ment which subscribes to the principles of free modification and distribution of source code (Gomulkiewicz, 1999). Under the open source software development model, the typical arrangement is for a community of developers' to engage with source code so as to create extensions and enhancements and improvements, which are in turn fed back freely into the community so as to be further enhanced, extended and improved. With proprietary software, enhancements and the fixing of bugs are entirely dependent upon the schedule and employees of a single corporation such as Microsoft. On the contrary, under the open source model a plethora of software development communities, connected via the Internet around the world, are available to freely and willingly provide enhancements and bug fixes, and "if you are not satisfied with their pace or performance, you can simply do it yourself" (Nadan, 2002, pp. 352-353). Open source projects are generally facilitated via the original software developer or a small group of interested programmers who typically act as a de facto project manager, by controlling what new code will be incorporated into the evolving software program, as well as ensuring that any new enhancements, extensions, or improvements are suitably well written to be integrated into the official code base. The project manager "may act as an official arbiter of versions, and periodically release official improved versions of the original source code to incorporate other programmers' modifications" (Natoli, 1999, p. 2).

One of the leading figures of this movement, Richard Stallman, argues that common proprietary software "keeps users helpless and divided [since] the inner workings are secret" (Stallman, 2001a). The open source movement therefore is based on the notion that the public should have "the freedom to study, change and redistribute the software" it uses or obtains, since "these freedoms permit citizens to help themselves and help each other, and thus participate in a community" (Stallman, 2001a). Stallman builds upon this point in the following metaphor (Lessig, 2001, p. 50):

So imagine what it would be like if recipes were packaged inside black boxes. You couldn't see what ingredients they're using, let alone change them, and imagine if you made a copy for a friend, they would call you a pirate and try to put you in prison for years. That world would create tremendous outrage from all the people who are used to sharing recipes. But that is exactly what the world of proprietary software is like. A world in which common decency towards other people is prohibited or prevented.

The open source software movement is perhaps best understood in terms of the methods employed to distribute software code. The open source movement vehemently argues that easy access to source code can facilitate efficient detection of bugs and security problems and enhance the positive evolution of a product (Fitzgerald & Fitzgerald, 2004). Presumably on the basis that the "proof is in the pudding", the open source movement spawned a number of free software initiatives in the 1980s as a direct reaction to AT&T's propertisation of the Unix operating system. One such freely available initiative was the BSD Unix system which was a largely modified version of Unix launched by the University of California at Berkeley.

Stallman, in his role as one of the open source visionaries, also reacted to the propertisation of Unix by founding the Free Software Foundation and the GNU project (www.fsf.org). The GNU project, which stood for "GNU's Not Unix" by way of an ironical recursive acronym, was guided by the principle of freely distributed source code and collective creation (Moglen, 1999a). Anyone could freely modify and redistribute such software, or sell it, provided they did not attempt to reduce the rights of others to whom they passed it on. In this manner open source software has become a self-organizing project, in which no innovation can be lost through the proprietary exercises of rights. It has been through this self-organising nature of open source that important software developments have evolved. One such

development is the fore-mentioned GNU/Linux operating system which was completed when Finnish student, Linus Torvalds, added the Linux kernel to the evolving GNU project (Nadan, 2002). While Linux was initially considered little more than a student joke, it presently makes up a significant share of the operating systems market, and is seen as a significant competitor to Microsoft being successful enough to be used in many commercial and government environments (Dusollier, 2003). Indeed, in recognition of the stability of open source software programs, the Australian Capital Territory (ACT) Legislative Assembly, on December 10, 2003, passed the *Government Procurement (Principles) Guideline Amendment Act* 2003, which ensures the government considers the use of open source software when procuring computer software.

Interestingly, the Linux open source operating system "was created, and is continuously updated, by a global network of software developers who contribute their labor for free" (*U.S. vs. Microsoft Corp.,* 84 F.Supp.2d 9, 23 [D.D.C. 1999]). As the Open Source Initiative's Web site states (http://www.opensource.org/index.html):

The basic idea behind open source is very simple: When programmers can read, redistribute, and modify the source code for a piece of software, the software evolves. People improve it, people adapt it, people fix bugs. And this can happen at a speed that, if one is used to the slow pace of conventional software development, seems astonishing.

This mode of development raises interesting questions concerning why people would voluntarily give their time to develop software in such a manner. Perhaps the impetus for such activity is comparable to Mozart's desire to make music, and Monet's yearning to paint pictures. That is, the desire to create, to contribute and to perhaps leave a mark on the world. Raymond (1998) has labeled the phenomena as it applies to computer software development as "egoboo," which he describes as the enhancement to self-esteem and reputation that results from successful participation in the group. Whatever the reason, the development of the Linux operating system is living proof that computer software can evolve through such collaboration and there are more recent examples which prove the same. For instance, within hours after the Netscape browser's source code was released as open source in 1998, a group of Australian programmers had created additional code to enable secure Internet transactions, and within one month the open source community had completed a new version of the browser (Cella & Kelly, 1999).

A significant aspect of the self-organising nature of the free software movement has been its anarchistic tendencies, as well as its licensing mechanism which have evolved as a proactive attempt to diminish the harmful consequences of the tragedy of the anti-commons. We will now turn to a discussion of these aspects of the open source movement.

Anarchy and the Anticommons

As a political philosophy anarchism is especially suited to the open source network society, which has evolved via the Internet, as it represents organization through disorganization, or in other words, order through chaos. Such anarchist principles have been drawn upon in open source software literature as exemplified in the celebrated article of *The Cathedral and the Bazaar* by Raymond (1998) where he quoted Russian anarchist, Kropotkin, when referring to the mode of communal software distribution:

Having been brought up in a serf-owner's family, I entered active life, like all young men of my time, with a great deal of confidence in the necessity of commanding, ordering, scolding, punishing and the like. But when, at an early stage, I had to manage serious enterprises ... I began to appreciate the difference between acting on the principle of command and discipline and acting on the

principle of common understanding. The former works admirably in a military parade, but it is worth nothing where real life is concerned, and the aim can be achieved only through the severe effort of many converging wills.

As Vaidhyanathan (2004) indicates, anarchism is arguably the most misunderstood political philosophy of the 21st century, perhaps because the big political and philosophical skirmishes have been among the forces that oppose anarchy such as capitalism, (state) socialism, and fascism. It is derived from the Greek word *anarchos*, which means "without authority", and while it is commonly associated with bloody violence and rage, it should be understood that anarchists generally believe deeply in an ideology of love (Vaidhyanathan, 2004). As a fact or condition, anarchism is perhaps the original political philosophy of *Homo sapiens* as the world has witnessed many stateless societies by way of groups of people who have lived without a dominant *institutionalized* authority. The cyber world can perhaps be considered as an extension of this stateless condition, and in this regard anarchism is embedded within the open source movement via the principles of voluntary association, mutual aid, cooperation, consensus, collaboration, and anti-possessive individualism. It is the latter principle that has inspired the open source movement, in an ironical twist, to turn the liberal notion of property on its head by subversively drawing upon the notion of (intellectual) property to communalise information.

Heller's (1998) "tragedy of the anticommons" theory is insightful when seeking to appreciate the informational commune aspect of the open source software movement. Heller's theory states that where "too many owners hold rights of exclusion, the resource is prone to under use." This newly emerging discourse can be juxtaposed against Hardin's (1968) original "tragedy of the commons" discourse which states that when too many people have a privilege to use a resource and no one user has a legal right to exclude any other user the end result is over consumption and the depletion of the resource. An important differentiator between Hardin's "tragedy of the commons" and Heller's "tragedy of the anticommons" is the "right to exclude." As Aoki (1998) explains, in the commons situation, part of the problem is that *no one* has the right to exclude, thereby giving rise to over-utilisation and depletion. By contrast, with the anticommons situation, *too many parties* independently possess the right to exclude, which gives rise to under-utilization amounting to the "tragedy of the anticommons."

A corollary of the *anticommons* quandary is that "rational individuals, acting separately, collectively waste a resource by underconsuming it compared with the social optimum" (Heller, 1998, p. 677). One of the examples Heller uses to demonstrate the anticommons phenomena is the post-1989 Moscow storefronts that remain empty, while at the same time flimsy metal kiosks proliferate. In the context of IPR's the corollary of the anticommons is that the demands of many, paradoxically, go unmet. In this way, the tragedy of the anticommons reminds us of the limits inherent in this propertising of information (Wagner, 2003). The late U.S. president Jefferson understood this constraint when he surmised that "inventions then cannot, in nature, be a subject of property" (Washington, 1855, p. 181). Thus, despite the maximalist impulses of the international IPR regime, the attempts to propertise information have not been *entirely* successful because "information really does want to be free" (Wagner, 2003, p. 1003).

Relevantly, the issue of propertising of information and its effect on creativity and innovation lies at the heart of the struggle between closed and open source software development models. On the one hand, the closed source developers argue that an information commons can expand even as proprietary information is increased, since "whereas on Blackacre every square yard that is propertised diminishes the total left in the commons, in the information commons, no such zero-sum game exists" (Wagner, 2003, p. 1002). On the other side of the coin, the open source developers believe that to the extent that information is both costless and nondiscriminatory the costs of further creation will be reduced (Landes & Posner, 1989). In this context, the open source movement argues that information is nonrival in that its use by one person does not deny others from using it (Landes & Posner, 1989; Lemley, 1997). This creates a clear conception within the open source movement that the maximization of the informational commons fuels the fire of human progress since "creation begets more creation [and] invention leads to further invention" (Wagner, 2003, pp. 1001-1002).

The open source arguments concerning the maximization of the informational commons have the effect of undermining the original utilitarian foundation of IPR's. In this way, the anticommons analysis expands the current debate over the appropriate scope of IPR's to consider not just the level of protection, but also the manner in which those rights are designed and held (Elkin-Koren, 1998). This perspective places the focus on the effect of the organization of rights with respect to efficient use of information. Excitingly, in the name of subversive anarchism the open source software movement has redefined the public aspect of IPR's by ensuring that information and work disseminated can be drawn upon in future projects so as to benefit all of human-kind. One critical legal mechanism underlying this subversion is open source licensing, and in particular the General Public License (GPL) which covers a majority of open source projects. It is to the GPL that the chapter now turns.

The General Public License

While a number of legal mechanisms have evolved to accompany the various open source initiatives which have spawned over the last few decades, the GNU General Public License (GPL) and its derivative, the Lesser General Public License

(LGPL), have remained the most popular, covering approximately 65% of all open source initiatives (James, 2003). According to James (2003), other popular open source licenses include the MIT Open Source License (9.4%), the Berkeley Software Distribution License (7.5%), and the Mozilla Public License (6.8%). This chapter seeks to focus on the GNU GPL as it remains the primary legal framework for the distribution of open source software.

The GPL is a unique licensing instrument that governs downstream activity of licensed work (i.e., open source software) creating a strategic mechanism that ensures that information remains "free" as in speech (as opposed to "free" as in beer) (Fitzgerald & Fitzgerald, 2004). It was fashioned by the Free Software Foundation who realised that IPR's could be utilised in a manner which secured open access to knowledge, rather than the simple motif of profiteering. The GPL operates by conditionally granting the user the right to use, reproduce, distribute, and modify the software. Under the GPL, users must consent to supply the source code to anyone they provide the object code, and each copy of the program must include a valid notice of copyright and a warranty exemption (GNU General Public License, 1991, provisions 3a-c). The license applies automatically to each new copy of the software as well as to each derivative work or other variation of the software. A user who modifies software developed and distributed under a GPL cannot impose restrictions other than those tolerated by the original license. This aspect of the GPL disallows software written and distributed under the license from being subsequently appropriated by proprietary interests (Dusollier, 2003). In this manner IPR's, specifically copyright law, is used to create a "copyleft" effect as opposed to a copyright effect by ensuring that code remains accessible (i.e., free and open) for all to use in the development and innovation of software (Fitzgerald & Fitzgerald, 2004). As Stallman (2004) states, "Proprietary software developers use copyright to take away

the users' freedom; we use copyright to guarantee their freedom. That's why we reverse the name, changing copyright into copyleft."

The GPL and Copyleft

The principle of copyleft is enshrined in the Preamble of the GPL and is primarily enacted via Section 2(b) of this license which states:

2. You may modify your copy or copies of the Program or any portion of it, thus forming a work based on the Program, and copy and distribute such modifications or work ... provided that you also:

(b) ... cause any work that you distribute or publish, that in whole or in part contains or is derived from the Program or any part thereof, to be licensed as a whole at no charge to all third parties under the terms of this License.

In this manner the copyleft effect is created because anyone who develops software based on GPL'd code must give the public free use, modification, and distribution of the derived work (Horne, 2001). In the words of Stallman (2004):

To copyleft a program, we first state that it is copyrighted; then we add distribution terms, which are a legal instrument that gives everyone the rights to use, modify, and redistribute the program's code or any program derived from it but only if the distribution terms are unchanged. Thus, the code and the freedoms become legally inseparable.

Copyleft licenses therefore provide that a user may distribute the open source code and any modifications to it, provided the user does so under the same open source license which the user received it. In this way, as the code and modifications to it pass from person to person or entity to entity, they stay open source. This is in deep contrast to the closed distribution model where source code is not released and can only be obtained through the complex tasks of reverse-engineering or de-compilation (Gomulkiewicz, 1999).

In essence, the main objective of the copyleft provision is therefore to prevent anyone from ap-propriating the open source code by, for example, distributing it under a proprietary, non-GPL license. To take the successful Linux operating system discussed above, for example, without copyleft a proprietary interest could obtain the open source code of Linux, make some modifi-cations, and then license the modified operating system under a proprietary model, profiteering from the sale of this new operating system, but not revealing the source code. The effect of this situ-ation is that the proprietary interest has been able to disproportionately benefit from a product they have contributed very little to. Such conduct would undermine the open source software because the free labour that contributes to the evolution of this software would be unwilling to share code enhancements if someone could take the code private and not share their enhancements with the rest of the "community" (Nadan, 2002).

The open source movement argues that the requirement that any *derivative works* of GPL code also be covered by the GPL is reasonable, because if it was not for the GPL, the user would in fact have no rights to create the derivative works in the first place. That is, "the condition on the abandonment of the restriction of the [GPL] is the surrender of the rights granted by the license" (St. Laurent, 2004, p. 152). This self-enforcing nature of the GPL is indoctrinated in Provision 5 of the license which states that:

5. You are not required to accept this License, since you have not signed it. However, nothing else grants you permission to modify or distribute the Program or its derivative works. These actions are prohibited by law if you do not accept this License. Therefore, by modifying or distributing the Program (or any work based on the Program),

you indicate your acceptance of this License to do so, and all its terms and conditions for copying, distributing or modifying the Program or works based on it.

In this way, the GPL "actually has the strength to say no to people who would be parasites on our community" since the potential licensee is faced with a choice: either refuse the GPL, which means they are barred from distributing or modifying the work (except to the limited extent permit-ted by fair use), or accept it, and use the work as permitted by the GPL (Stallman, 2001b; St. Laurent, 2004, p. 152). Importantly, the GPL is only triggered if a user attempts to distribute the software or a derivative work made from GPL'd code. Since no one can ever redistribute without a license, it can be safely presumed that anyone redistributing GPL'd software intended to accept the license (Moglen, 2001). This is especially the case because provisions 1 and 2 of the GPL requires that each copy of GPL'd software include the license text, so as to ensure that everyone is fully informed of the license conditions.

FUTURE TRENDS

Thus, implicit within the GPL is a self-enforc-ing aspect which applies to the license itself, as well as *notice* of the license. This is, of course, provided the GPL is in fact enforceable. While the resolution of the general enforceability of the GPL is beyond the scope of this chapter and there-fore the subject of further research, one relevant and important consideration that can be touched upon is whether the GPL is a license, a contract, or some combination of both (i.e., a contractual license) as this will presumably have some bear-ing on GPL enforceability. This chapter puts forth the view that the future trend of the GPL from a legal perspective will be the culmination of the law of license *and* contract.

The GPL: License or Contract?

The legal classification of the GPL as contract and/or license is important because contract law is subject to the vagaries of various national approaches whereas a copyright license enables products to come under intellectual property laws that have been harmonised by international treaties such as the *Berne Convention, WIPO Copyright Treaty* (1996) and *TRIPS*. The nuance of the contract/license distinction was drawn upon in *Sun Microsystems Inc vs. Microsoft Corp* 188 F. 3d 1115 (9th Circ, 1999) where the court stated that (p. 1121):

Generally a copyright owner who grants a non-exclusive license to use her copyrighted material waives her right to sue the licensee for copyright infringement and can sue only for breach of contract: Graham v James 144 F. 3d 229, 236 (2d. Cir. 1998). If however, a license is limited in scope and the licensee acts outside the scope, the licensor can bring an action for copyright infringement: S.O.S. Inc v Payday Inc 886 F. 2d. 1081, 1087 (9th Cir. 1989).

As the GPL has had limited confrontation with the courts up until this point, it cannot be stated with any degree of certainty how the GPL will actually be classified. Open source commentators such as Eben Moglen, chief legal adviser for the Free Software Foundation, suggest the GPL is not a contract, but a license. As Moglen (2001) states:

Licenses are not contracts: the work's user is obliged to remain within the bounds of the license not because she voluntarily promised, but because she doesn't have any right to act at all except as the license permits.

Moglen (2002) also made the following declaration in *Progress Software Corp vs. MySQL AB* 2002 U.S. Dist. LEXIS 5757:

The GPL is a very simple form of copyright license, as compared to other current standards in the software industry, because it involves no contractual obligations. Most software licenses begin with the exclusive rights conveyed to authors under copyright law, and then allow others access to the copyrighted work only under additional contractual conditions. The GPL, on the other hand, actually subtracts from the author's usual exclusive rights under copyright law, through the granting of unilateral permissions.

While the arguments concerning the GPL as a license are compelling, this chapter argues that the GPL represents a bundle of license *and* contractual obligations. The license occurs, as the discussion above indicates, because the GPL is permitting a user to do something that they would otherwise not be able to do. By way of analogy, if an owner of real property permits a visitor on to their land for a specific purpose, the visitor has obtained a license as they are doing something that they could not have done other than with the owner's permission (Fitzgerald & Bassett, 2003b). This requires no counter obligation from the visitor and the arrangement therefore remains a unilateral permission not a contract.

The contractual aspect of the GPL arises, however, because of the positive obligations placed upon the user by the GPL. As Madison (2003) suggests, this is compatible with the understanding that many conventional lawyers' have of the software license as simply a contract that stipulates the obligations of the licensor and the licensee. The rationale underlying this view is that software licensing relies on a legitimate but purely positive legal framework, "drawn wholesale from the domain of promissory obligation wrapped around a core of property rights" (Madison, 2003, p. 295). Contractual considerations arise because the GPL does not just give rise to permission but also to positive obligations. For example, the GPL necessitates agreement of the "no warranty" provisions, and it also requires that the source code be published and that the GPL attach itself to this source code (GNU GPL, 1991, Provisions 1, 2, 11, and 12). Seemingly, these obligations are positive obligations in the sense that they require the user to take positive action, as opposed to permission for things that could otherwise not be done but for the GPL. Such arguments strongly suggest that positive obligations, particularly the limitations of warranty, are matters of contract and cannot be enforced except in contract law.

One final consideration—whether or not the GPL is considered to be a license, a contract, or a contractual license—is the issue of notice. This is because the lack of notice of the GPL on behalf of the user could potentially increase the risk that no contractual or license agreement has been formed (see Ticketmaster Corp. vs. Tickets.Com, Inc. *54 U.S.P.Q.2d [BNA] 1344 [C.D. Cal. 2000];* Specht vs. Netscape Communications Corp. 00 Civs. 4871 [AKS] 2001 WL 755396 [S.D.N.Y. July 5, 2001]; and GNU GPL Version 2 [June 1991] post terms and conditions for relevant law concerning computer software licensing assent). To minimise this risk—which is according to this chapter, incidentally, the only significant legal risk of the GPL—a code download site could be set up so that the user is forced to click "I accept" to a clickwrap form of the GPL before download-ing. Commercial entities rely on clickwraps and shrinkwraps every day, and employing the same device for the GPL would be a simple, inexpensive and prudent approach (Nadan, 2002).

Notice requirements aside, however, it is worth reiterating that when it comes to the GPL some diligence is required on behalf of the *licensee* since if the licensee honestly believed that there was "no license" applicable to the program, they should have made no use of the program other than the very limited uses permitted by copyright law. It is in this manner that the subversive and self-enforcing strength of the GPL is yet again reinforced.

CONCLUSION

This chapter has surveyed the fascinating subject of copyright as it applies to open source software development. It was seen that source code as computer language has become the subject of an important economic, political, and social struggle that is ultimately concerned with the ownership of creative output in the context of computer software development. The contrast between closed and open source software development models made explicit the nature of this political polarization. It was seen that the experimental nature of the open source movement creates a tendency towards theoretical complexity, and for this reason, the chapter focused on the academic foundation of the open source movement, rather than the closed source model. In this regard, the anarchist tendencies of the open source movement were confirmed, as were the arguments supporting the maximization of the information commons. The chapter also verified that the General Public License (GPL) has become an important legal mechanism to enact the principles and objectives of the open source movement. While the ultimate question concerning the enforceability of the GPL was beyond the scope of this chapter, the contract/license duality of the GPL did provide fuel for a fruitful, if not brief, discussion concern-ing the nature of the GPL as a legal mechanism. One important dimension of this discussion was the GPL as a culmination of licensing *and* contractual arrangements in its capacity as both a permit and as a facilitator of the copyleft no-tion. This perspective provides a useful insight into the complex nature of the GPL, while at the same time highlighting the simple self-enforcing nature of the license. In this way, the GPL was shown to be at once a strong and subversive legal instrument that will continue to underlie the open source movement in the future.

REFERENCES

Aoki, K. (1998). Neo-colonialism, anticommons property and biopiracy in the (not-so-brave) new world order of international intellectual property protection. *Indiana Journal of Global Legal Studies, 6*(11).

Apple Computer Inc. vs. Franklin Computer Corp., 714 F. 2d 1240. 1243. 219 U.S.P.Q. (BNA) 113 (3d Cir. 1983).

Bobko, P. (2000). Open source software and the demise of copyright. *Rutgers Computer & Technology Law Journal, 27*(51).

Caenegem, W. (2001). *Intellectual property.* Sydney, Australia: Butterworth Tutorial Series.

Carstens, D. (1994). Legal protection of computer software: Patents, copyrights and trade secrets. *Journal of Contemporary Legal Issues, 20*(13).

Cella, C., & Kelly, E. (1999). Considerations for companies developing software under the open source model. *Cyber Law 9, 4*(6).

Computer Edge Pty Ltd. vs. Apple Computer Inc. (1986) 161 C.L.R. 171.

Dusollier, S. (2003). Open source and copyleft: Authorship reconsidered. *Columbia Journal of Law and the Arts, 26*(281).

Elkin-Koren, N. (1998). Copyrights in cyberspace: Rights without laws? *Chicago Kent Law Review, 73*(1155).

Fitzgerald, B., & Bassett, G. (2003a). *Leal issues relating to free and open source software: Essays in Technology Policy and Law* (Vol. 1, QUT). Retrieved January 11, 2006, from http://www.law.qut.edu.au/files/open_source_book.pdf

Fitzgerald, B., & Bassett, G. (2003b). Legal issues relating to free and open source software. In B. Fitzgerald & G. Bassett (Eds.), *Leal issues relating to free and open source software: Essays in Technology Policy and Law* (Vol. 1, QUT, chap. 2). Retrieved January 11, 2006, from http://www.law.qut.edu.au/files/open_source_book.pdf

Fitzgerald, B. & Fitzgerald, A. (2004). *Intellectual property in principle.* Sydney, Australia: Lawbook Co.

Gomulkiewicz, R. W. (1999). How copyleft uses license rights to succeed in the open source software revolution and the implications for article 2B. *Houston Law Review, 36*(179).

GNU General Public License, Version 2. (1991, June). Free Software Foundation. Retrieved June 6, 2005, from http://www.fsf.org/licenses/gpl.txt

Hardin, G. (1968). The tragedy of the commons. *Science, 162*(1243).

Heller, M. (1998). The tragedy of the anticommons: Property in the transition from Marx to markets. *Harvard Law Review, 11*(621).

Horne, N. (2001). Open source software licensing: Using copyright law to encourage free use. *Georgia State University Law Review, 17*(863).

James, P. (2003). Open source software: An Australian perspective. In B. Fitzgerald & G. Bassett (Eds.), *Leal issues relating to free and open source software: Essays in Technology Policy and Law* (Vol. 1, QUT, chap. 4). Retrieved January 11, 2006, from http://www.law.qut.edu.au/files/open_source_book.pdf

Johnston, R., & Grogan, A. (1994). Trade secret protection for mass distributed software. *The Computer Law*, November.

Landes, W., & Posner, R. (1989). An economic analysis of copyright law. *Journal of Legal Studies, 18*(325).

Lehman, G. (1999). Disclosing new worlds: A role for social and environmental accounting and auditing. *Accounting, Organizations and Society, 24*(227).

Lemley, M. (1997). Romantic authorship and the rhetoric of property. *Texas Law Review, 75*(873).

Lemley, M. (2003). *Software and Internet law* (2nd ed.). New York: Aspen Publishers.

Lessig, L. (2001). *The future of ideas: The fate of the commons in a connected world.* New York: Random House.

Madison, M. (2003). Reconstructing the software license. *Loyola University of Chicago Law Journal, 35*(275).

Mc John, S. (2000). The paradoxes of free software. *George Mason Law Review, 9*(25).

Microsoft Open License Agreement vs. 6.0 (2001, October 1).

Moglen, E. (1999a). *Anarchism triumphant: Free software and the death of copyright.* Prepared for delivery at the Buchmann International Conference on Law, Technology and Information, at Tel Aviv University, May 1999. Retrieved January 11, 2006, from http://emoglen.law.columbia. edu/my_pubs/anarchism.html

Moglen, E. (1999b). *So much for savages: Navojo 1, Government 0 in final moments of play.* Retrieved January 11, 2006, from http://emoglen.law. columbia.edu/my_pubs/yu-encrypt.html

Moglen, E. (2001). *Free software matters: Enforcing the GPL I.* Retrieved June 6, 2005, from http://emoglen.law.columbia.edu/publications/lu-12.html

Moglen, E. (2002). *Declaration of Eben Moglen in support of defendant's motion for a preliminary injunction on its counterclaims in Progress Software Corp v MySQL AB 2002 U.S. Dist. LEXIS 5757.* Retrieved May 8, 2002, from http://www. fsf.org/press/mysql-affidavit.html

Nadan, C. (2002). Open source licensing: Virus or virtue? *Texas Intellectual Property Law Journal, 10*(349).

Natoli, A. (1999). Open source software, *Multimedia & Web Strategist, 5*(2).

Raymond, E. (1998). *The cathedral and the bazaar.* Retrieved January 11, 2006, from www. firstmonday.org/issues/issue3_3/raymond/

Reger, C. (2004). Let's swap copyright for code: The computer software disclosure dichotomy. *Loyola Entertainment Law Journal, 24*(215).

Specht vs. Netscape Communications Corp. 00 Civ. 4871 (AKS) 2001 WL 755396 (S.D.N.Y. July 5, 2001).

St. Laurent, A. (2004). *Understanding open source & free software licensing.* San Francisco: O'Reilly.

Stallman, R. (2001a). *The GNU GPL and the American way.* Retrieved January 11, 2006, from http://www.gnu.org/philosophy/gpl-american-way.html

Stallman, R. (2001b, May). *Free Software: Freedom and Cooperation.* Speech presented at NYU. Retrieved June 6, 2005, from http://gnu.archive. hk/events/rms-nyu-2001-summary.txt

Stallman, R. (2004). *The GNU operating system, licenses: What is copyleft?* Retrieved January 11, 2006, from http://www.gnu.org/licenses/licenses. html#WhatIsCopyleft

Stilwell, F. (2000). *Changing direction: A new political economic direction for Australia.* Sydney: Pluto Press.

Suzor, N., Fitzgerald, B., & Bassett, G. (2004, January). Free Software and Government Proceedings of Linux and Open Source in Government Conference. In *The Challenges Conference Proceedings,* Adelaide University, Adelaide, South Australia, Australia.

Ticketmaster Corp. vs. Tickets.Com, Inc. 54 U.S.P.Q.2d (BNA) 1344 (C.D. Cal. 2000).

U.S. vs. Microsoft Corp., 84 F.Supp.2d 9, 23 (D.D.C. 1999).

Vaidhyanathan, S. (2004). *The anarchist in the library.* New York: Basic Books.

Velasco, J. (1994). The copyrightability of nonliteral elements of computer programs. *Columbian Law Review, 94*(242).

Wagner, P. (2003). Information wants to be free: Intellectual property and the mythologies of control. *Columbian Law Review, 103*(995).

Washington, H. (Ed.). (1855). *The writings of Thomas Jefferson.* New York: Thorne & Co.

Webbink, M. (2003). Licensing and open source. In B. Fitzgerald & G. Bassett (Eds.), *Legal issues relating to free and open source software: Essays in Technology Policy and Law* (Vol. 1, QUT, chapter 1). Retrieved from http://www.law.qut.edu.au/files/open_source_book.pdf

Wilson, C., Jr. (1993). Software piracy litigation. *Florida Bar Journal, 67*(1 29).

KEY TERMS

Anarchism: Absence of government.

Code: A set of instructions designed to cause a computer to perform a particular function or to produce a particular result.

Copyleft: A type of intellectual property license which uses copyright law to remove restrictions on the distribution of copies and modified versions of a work for others and which also requires the same freedoms be preserved in modified versions.

General Public License (GPL): A unique intellectual property licensing system which governs downstream activity of licensed work by conditionally granting the user the right to use, reproduce, and distribute.

GNU: An ironical recursive acronym which stands for "GNU's not Unix."

Open Source: Refers to practices in production and development that promote access to the end product's sources.

Tragedy of the Anticommons: Where too many owners hold rights of exclusion, the resource is prone to under use.

Tragedy of the Commons: When too many people have a privilege to use a resource and no one user has a legal right to exclude any other user the result is over consumption and depletion of the resource.

Chapter XXVIII
The Evolution of Free Software

Mathias Klang
University of Goteborg, Sweden

ABSTRACT

The more we rely upon software to mediate the many facets of our lives the more important the ability to control and adapt that software to our needs becomes. The Free Software Foundation stands at the forefront for this effort to ensure user empowerment. The main tool of the foundation is the General Public License that has been a fundamental document in software development since its conception in 1989. At present the Free Software Foundation is in the process of launching a new version of their license and the process is similar to the development of an existing social contract—the delicate problem is meeting the new challenges that have appeared since the earlier version while maintaining the spirit of the original.

INTRODUCTION

Most legal documents exist in relative obscurity. Despite their legal effect and control over our lives they receive scant attention and are rarely recognised as unique documents outside the narrow group who are responsible for drafting and interpreting them. On occasion certain documents rise above this obscurity and achieve an iconic status where their actual content is overshadowed by their symbolic value. Arguably the clearest such example is the American Constitution. Its position and fame go beyond its content, it is arguably more important as a symbol than a legal document. The iconic value of this declaration is enhanced by the value society attributes to the ideology they believe to reside within the formulations.

The focus of this chapter is the iconic software license—the GNU General Public License (GPL). Stated objectively the GPL is a widely used free software[1] license, originally written by Richard Stallman for the GNU project. The latest version of the license, version 2, was released in 1991. While this is an accurate statement it fails to capture the importance and status of the document. In a recent statement by the drafters Stallman and Moglen (2005) the GPL was described as fulfilling four important roles: (1) the GPL is a worldwide copyright license, (2) the GPL is the code of conduct for free software distributors, (3) the GPL is the constitution of the free software movement, and (4) the GPL is the literary work of Richard M. Stallman.

This list better captures the iconic status of the GPL and indicates the list of stakeholders that have an interest in the way in which the license develops. At present the development of the GPL is a central issue in the world of software development. The reason for the increased interest is because the organisation in control of the license, the Free Software Foundation (FSF), is presently coordinating the move from version 2 to version 3. Their stated goal is to ensure that the spirit of the license is maintained while the content is updated to better reflect the social-technical developments that have taken place since version 2 was released in 1991.

This chapter will describe the background and spirit of the GPL and also point to its importance. The chapter will then explain some specific socio-technical developments that challenge the effectiveness of the existing license and a description of the process of moving from version 2 to version 3, which is intended to meet these challenges. The goal of this chapter is to arrive at an understanding of the importance of the GPL and to observe how it develops as a regulatory instrument to meet new challenges while maintaining its ability to offer the freedoms the license entails.

BACKGROUND

The Spirit of GPL

Writing about the importance of software is difficult without resorting to what seems to be empty hyperbole. It is important to point out that software is rapidly becoming one of the most fundamental building blocks of human interaction and activity. There remains a common misconception that software is a complex component, which in some sense "lives" within computer hardware. By confining software to the inner workings of the traditional computer most non-technical software users are unaware of the extent to which software permeates their lives.

Moglen (1999) talks of computers being under our social skin but this seems to imply that there are computers everywhere. To most people the computer is still a very specific artefact that only affects their lives in specific, controllable situations. Talking less about the computers and more about software may help bring about an understanding of the omnipresence of software. Also like most other things that surround us this software belongs to someone. The software that fills our homes and our lives is, in almost all cases, the property of someone else and therefore we are dependent upon the property of others for our everyday lives to a much greater extent that we may previously have imagined.

It was in part to counteract this that Richard Stallman wrote the original announcement for the GNU project in 1983. He wrote, "Starting this Thanksgiving I am going to write a complete Unix-compatible software system called GNU (for Gnu's Not Unix), and give it away free to everyone who can use it." In 1985 Stallman launched the Free Software Foundation (FSF), an organisation whose goals it is to promote the computer users' right to use, study, copy, modify, and redistribute computer programs.

The spirit of the GPL is commonly condensed into what has become known as the four freedoms. From the point of view of the FSF software licenses that offer these four freedoms to the user is free software. Software that does not meet all four of these freedom criteria is proprietary software. These freedoms are the freedom: to run the program, for any purpose (called freedom 0), to study how the program works, and adapt it to your needs (called freedom 1), to redistribute copies so you can help your neighbour (called freedom 2) and to improve the program, and release your improvements to the public (called freedom 3). This list has become the mantra of the free software movement are known collectively as the four freedoms.

Despite the relatively clear description offered by the four freedoms and the GPL the term free software has been the subject of some controversy. The fundamental freedom referred to is the freedom from constraints—it is the language of rights rather than the economic term in relation to cost. This understanding of free builds on the concept of software as a fundamental building block in the information society. Therefore to be able to maintain their fundamental freedoms users need to be able to maintain full control over the infrastructure. This egalitarian principle demands that software remains beyond the control of those who would limit its usage.

Before continuing it is important to provide a brief clarification on the concept of freedom the GPL refers to. This is important since the term has earlier caused some discussion (Klang, 2005). The English language recognises two separate, but related, understandings of the term free. The term can both refer to the absence of cost and to the absence to restrictions on liberty. Generally speaking this duality of meaning should not cause problems. The GPL is concerned with freedom in other words, the absence of limitations on liberty. In an attempt to clarify this Stallman (2002) presented a most original analogy in the discourse on freedom by recommending that the term free in GPL should be understood in terms of free speech rather than in terms of free beer.

Among the critics of the term free software are the creators of the term open source as an alternative term. Their argument for this alternative term was that the ambiguity with the term free will reduces the acceptance of free software in business (Weber, 2004; Williams, 2002). In addition to this argument there is a philosophical critique to the freedom granted by free software and the GPL. This argument is built upon the fact that the content of the license creates limitations to what the users may do with their work if this is licensed under the GPL (Klang, 2005). The most widely publicised limitations to the freedoms of the user is the requirement that the user who modify and then distribute free software must provide the same freedoms to other users that they themselves had received (Klang, 2005).

The latter critique of freedom is not a system flaw. Freedom without limitations will not ensure the development and protection of free software. Only freedom within limitations similar to those provided by the GPL can ensure that freedom is maintained for users. Moglen (1999) maintains that this clause ensures users always have the best available software. While critics claim that this means that widespread commercial development cannot take place, nor will commercial companies dare to use any part of GPL software in their products for fear that small parts of GPL software may contaminate the whole. The latter critique has led the GPL to be seen as largely anti-commercial.

MAIN FOCUS OF THE CHAPTER

Inside the License

Is the GPL a license? On the face of it this may seem an odd question. Despite this, a great deal of effort has been spent on this question. However, this question need not concern us here (for more on this question see Metzger & Jaeger, 2001; O'Sullivan, 2002, 2004; Rosen, 2004; St. Laurent 2004). Suffice to say that the basis of the GPL is in copyright law (GPL preamble 5th paragraph & §0). If there is no contract then what remains is copyright. Almost all uses of copyright protected material without the authors' permission constitute copyright violation. Therefore, the GPL can then be seen as a unilateral statement from the programmer not to sue for copyright violation as long as the terms are followed.

The GPL preamble captures the spirit of the license. It begins: "The licenses for most software are designed to take away your freedom to share and change it. By contrast, the GNU General Public License is intended to guarantee your freedom to share and change free software—to make sure the software is free for all its users ..." The purpose of the preamble in the GPL is to provide a general introduction to the license. Generally speaking the legal status of a preamble is questionable. Since preamble texts are not actually part of the license but more a general introduction any court is free to ignore such texts. However, the courts also have the option to utilise the preamble if they are attempt to clarify any ambiguity in interpretation of the license. Therefore while the preamble may fall outside the actual contract between the parties it is not without value since it may be used to clarify the content of the agreement between the parties.

The preamble places the license in a political context. The license positions itself as being the opposite of "most software" (GPL preamble 1st paragraph) since it does not aim to limit people's freedom of use of the software. It is important to realise that despite its tone of equality for all, the group this license is aimed at is not the public at large but a comparatively small group of programmers (O'Sullivan, 2004). The concept of freedom is a naturally ambiguous (Klang, 2005) and in order to clarify the concept the preamble explains that the term free software entails (GPL preamble 2nd paragraph) "freedom, not price" and that this freedom involves "... the freedom to distribute copies of free software (and charge for this service if you wish), that you receive source code or can get it if you want it, that you can change the software or use pieces of it in new free programs; and that you know you can do these things" (GPL preamble 2nd paragraph).

Copying, Modification, and Distribution

Making verbatim copies of software and distributing them (GPL 1§) is covered under this license is permitted under condition that a copy of the GPL accompanies the copies and that the information about copyright and warranties remains intact. Modification of the software is permitted, as is the distribution of modified copies (GPL 2§). However these activities are subject to certain conditions found in GPL 2§a,b & c.

Files, which have been changed, must display the information of who made the change and when the change was made (GPL 2§a). Changes made cannot affect the licensing form. In other words the original permissions granted by the GPL must be maintained, at no additional cost, even after changes have been made (GPL 2§b). If the program reads commands interactively when run it must be made in such a way as to display the copyright notice, warranty disclaimer and information on how to view the license. There is a general exception to this rule. If the licensee wishes to incorporate software protected by the GPL and distribute the derivative under different terms this can be done by obtaining permission from the original copyright holder. If the copyright holder is the FSF the licensee should contact them for permission (GPL 10§).

All copying, modification distribution and sub-licensing of software is permitted only as long as it falls within the scope of the GPL. Any attempts to act in a manner which is not expressly provided in the license is void (GPL 4§) as a result of this the permissions granted to the licensee are revoked and copyright law is enforceable (GPL 4§). Those who have obtained rights through sub-licenses will not automatically loose their permissions unless they to act in a manner which is not expressly permitted by the GPL (GPL 4§).

Copyleft

This is probably the most controversial, and at times misunderstood, concept of the GPL. The term refers to the system by which the GPL intends to create a software commons. From this commons all programmers are free to take code and use it as they wish. The condition for this use is that if the resulting creation has used code from the commons it too must pass into the commons. The fact that the addition of a small piece of GPL licensed code forces the whole software produced to be released under the same terms as the GPL has been referred to as the viral nature (the term is sometimes attributed to Radin, 2000 but was in use in mailing lists earlier than this date) of the GPL. The term has a largely negative association and most pro-copyleft writers prefer the less inflammatory (no pun intended) term vaccinated. The term viral refers to the fact that GPL software in a sense infects any software to which it is added. The term vaccination is so called since the effect of the GPL is to protect software against being appropriated and made into a form less free to its users.

Changes made cannot affect the licensing form. In other words the original permissions granted by the GPL must be maintained even after changes have been made. This is true even if only a small part of the new program contains code from the original. The content of GPL 2§a is further developed (GPL 2§2nd paragraph) which explains that identifiable sections of a program may be distributed as independent programs under other licenses. However if they are distributed as part of the original program or cannot be seen as independent then they must be distributed under the GPL. Therefore the same code can be distributed under different licenses, under the GPL distributed as a section and an alternative license if it is distributed as an independent program. By doing so the GPL attempts to point out that it does not make any claim to works written entirely by individuals but its main interest is to ensure that work released under the permissions granted by the GPL are not limited in any way (GPL 2§3rd paragraph).

While taking code covered by the GPL and adding it to non-GPL code will require the whole package to be distributed under the GPL. However it is important here to point out that storing GPL and non-GPL code on the same storage medium will not require the whole content of the storage medium to be licensed under the GPL.

The principle of copyleft should not be considered legally controversial. The combination of copyrighted works creates derivative works as a result. The creation of a derivative work requires the permission of the copyright holder. Permission from the copyright holder may be granted under certain conditions and the condition referred to as CopyLeft (GPL §2b) is valid condition.

Socio-Technical Developments

Version 2 of the GPL was released in 1991. The social and technical changes have impacted on the license and its ability to provide the four freedoms to the users who rely upon them. Naturally, too many developments relevant to the GPL have taken place to be able to include them all. In an effort to present an idea of what has been happening this section will present three developments of vital impact to the GPL, namely, the legal activity surrounding the GPL, the issue of software patents and finally the effect of TiVo-ization.

GPL in Court

The GPL is fighting an uphill battle. Aficionados and critics discuss the validity of the license on Web pages, academics mimic this in journals, lawyers speculate, and the whole community waits. An important fact, which works in favour of the GPL, is the fact that it has never *lost* in court. This is important because the longer the license can survive unchallenged the stronger it becomes as it works its way towards becoming a de-facto established trade practice.

The defence of the GPL is not limited to the courts. The first line of defence is the legal work being carried out by organisations such as the FSF. The general council of the FSF receives information of GPL violation "dozens of times a year" (Moglen, 2001). Since there is a strong sense of community within the FOSS movement most of these situations are usually rectified voluntarily by the party violating the GPL. This can be interpreted as misunderstandings being cleared up. The approach of the FSF has been to build upon this community and take a non-confrontational approach to violations.

Similar work is being carried out by the GPL-violations.org project. Since 2004 the organisation has dealt with over 30 GPL violations, which have resulted in out-of-court settlement agreements. The project has been the first to test the GPL in court. The case (District Court of Munich I, Judgement of 19/05/2004 (21 0 6123/04) involved a company that distributed GPL licensed software without providing the license. The court found that this was a clear violation of the GPL (Höppner, 2004).

Software Patents

Software patents are viewed as a threat to free software if they are used to limit the freedoms discussed above. To prevent distributors (or re-distributors) obtaining patents based upon free software (in whole or in part) and limiting the scope of freedom provided originally this paragraph of the preamble is intended to emphasise the fact that this is against the terms of the license. Software patents are permissible if the patent is "… licensed for everyone's free use or not licensed at all."

The GPL §7 expands the view presented in the preamble and states that even if the licensee is not released from the obligations of the license even if the licensee is forced by any conditions (not only patents) to contradict the terms of the license. "If you cannot distribute so as to satisfy simultaneously your obligations under this license and any other pertinent obligations, then as a consequence you may not distribute the Program at all." The solution offered by anyone pressured into acting contrary to the license is to stop using the software altogether.

Article §7 of the GPL was intended to prevent certain detrimental effects caused by software patents. Version 2 was therefore aware of the threats posed by software patents. Despite this awareness this version is not sufficient to prevent the numerous ways in which software patents can be implemented to limit the efficiency of free software. It is the goal of the next version to adopt a more comprehensive approach to combating the ills of software patents.

TiVo-ization

Digital products such as mobile telephones, DVD players and televisions all rely on an operating system (OS) that manages the hardware and software resources. The OS performs basic tasks, such as controlling and allocating memory, prioritizing the processing of instructions, controlling input and output devices, facilitating networking, and managing files. Both for reasons of cost and adaptability many of the OS used in consumer electronics are based upon free software.

One such application is the TiVo, a digital product that can automatically find and digitally record selected television programs. The user selects what is to be recorded and the TiVo locates and records the program automatically. From a programming point of view the TiVo is a device based upon a free software base with a small layer of proprietary software. The device has also given rise to a technological process called *TiVo-ization* (Turner, 2006).

The issue that FSF has with the process known as TiVo-ization is neither that TiVo makes proprietary software nor their proprietary software runs on a Free Software operating system (Taylor, 2006). Both these practices are common occurrences and while the FSF would prefer users to use free software exclusively this is not a violation of the letter or the spirit of the GPL. The OS on the TiVo is a modified GNU/Linux operating system. It is in compliance with the GPL. TiVo has released the source code for these modifications and therefore users are able to modify the code and the operation of their product. To this extent the TiVo is GPL compliant.

The issue FSF has with TiVo-ization is the practice of implementing proprietary software in such a manner as to control the ability of the user to practice the freedoms granted by free software.

Free software on the TiVo is covered by the GPL and therefore users can modify it to suit their needs. The same is not so of the proprietary software. By using digital signatures the proprietary code will only interact with code that originates from the TiVo programmers. Therefore when a user attempts to modify the free software (in accordance with the license) the user discovers that the product will not function with these modifications since any modifications also invalidate the digital signature. Such systems of interaction between Free and proprietary software invalidate freedom 1: the freedom to study how the program works, and adapt it to your needs.

The threat of TiVo-ization is that it will effectively prevent the user from putting the four freedoms into practice. Zuck (2006) attempts to downplay this threat by arguing that TiVo-ization is simply "… the merging of free and proprietary software into a single system." However, in practice this is an error bordering on misinformation. TiVo-ization is the building together of proprietary and free software systems with the goal of circumventing the spirit of the GPL. It is a method of systems building that acts in such a manner as to follow the letter of the license while making a mockery of the purpose.

The new version of the GPL (version 3) will prevent the compliance with the letter of the freedoms without the compliance to the purpose and spirit of the GPL. Those developers who want to be able to limit the users' freedom through TiVo-ization will in future not be able to build upon a base of free software.

Renewing the Social Contract

One of the strongest features of the GPL, aside from its clear ideology, is its stability. The license has been used on tens of thousands of programs over the last decade. Its stature has grown to become more than a copyright—the GPL has, for better or worse, achieved the iconic features of a constitution. Despite its clarity, stability,

and widespread acceptance some lawyers insist on attempting to obfuscate and complicate the ideology with law and outlandish claims such as "[the GPL] suffers from drafting errors and too many revisions" (Guadamuz, 2004). Such a comment seems uninitiated considering that the license was launched in 1989, went to version 2 in 1991 and is in the process of moving to version 3 in 2006. Most licenses tend to be changed on a more regular basis than this.

The widespread popularity and iconic status of the GPL create a different type of problem for the FSF. The issue is one of updating the basis of a social contract to encompass the needs of all the stakeholders without loosing sight of the original ideology and clarity. In addition to this the drafters of the GPLv3 are anxious to receive the feedback and comments from the community that they serve. To enable this, a transparent system that promoted discussion was required.

The formal system can best be seen in the overview of the process, which begins with the initial release and presentation of the draft of the GPLv3 with additional documentation such as the overview of the review system and the explanatory documents. In addition to the more formal structure the information needs to be communicated out to the users and to ensure an equality of information transfers was established. The latter was accomplished primarily through the use of the Internet as a distribution method of all texts and additional audio and video material.

The process was formally commenced with the release of the first discussion draft of version 3 of the GPL (including additional explanatory material) at the first International Public Conference in January 2006, at the Massachusetts Institute of Technology (MIT). The two day event at MIT was recorded and the audio video material was also made available online.

To ensure that comments on the GPL are collected and dealt with discussion committees have been formed. The members of the committees were chosen to represent diverse users groups such as "... large and small enterprises, both public and private; vendors, commercial and noncommercial redistributors; development projects that use the GPL as a license for their programs; development projects that use other free software licenses, but are invested in the contents of the GPL; and unaffiliated individual developers and people who use software" (GPL Process Definition). The role of these committees is to organise and analyse the received comments and propose solutions.

The FSF invited the initial members of the discussion committees but granted the committees the power to invite further members and to autonomously organise their work process. The committees work to encourage commentary on the license from the sectors they represent. Once the comments have been collected, organised and analysed the committee is responsible for presenting its results of the deliberations to the FSF.

Aside from this organisational method of soliciting and analysing comments from a wider public the FSF have created an online method of allowing anyone to comment directly on the license draft. This is done by creating commenting system, which allows the user to read the draft text of the GPLv3 online and if the user wishes to provide a comment on the text, the users can mark the section of their choice and add a comment to the section directly online.

Once a user has commented on a section of text, that section becomes highlighted. If no one has commented on the text the background colour is white. After a comment the background is light yellow. The colour of the background becomes progressively darker for each comment added. This colour system allows users to see at a glance which sections of the draft are the most commented.

By holding the cursor over highlighted text the user is informed how many comments have been made on that section. By clicking on highlighted text the comments that have been made appear and can be read. The latter feature has the added benefit of reducing the amount of duplicated

comments since the commentator can see the commentary of others.

FUTURE TRENDS

Seen as a social phenomenon, the GPL is much more than a license—it is a philosophy. Yet despite its iconic nature it is important not to overlook the importance of the dry legal text that makes up the GPL. The purpose of this text is to ensure that the spirit of the license is maintained and protected. The spirit of the GPL rests in the four freedoms intended to ensure that free software remains free even to future users.

The FSF position on freedom has developed the concept of free software and ensured the development of software which allows the user to control fundamental elements of the necessary infrastructure. This control is carried out through the specific terms of the license in particular terms that ensure that the user can adapt and distribute the software. These terms have been developed in a specific period of time and reflected an understanding of the technology at the time of development.

Since the release of version 2 of the GPL, 15 years have passed. These years have also entailed developments in technology and society that affect the way in which the license works to protect the four freedoms. As this work has shown some developments strengthen the position of the license while others undermine it. Therefore, as with all regulatory documents, the GPL needs to be updated to be better suited to the social and technical reality of the day.

Basing regulations on licenses can bring the concern that licenses may be changed rapidly and without rooting the changes with the community. This has been seen here with the comment that the license suffers from too many revisions (Guadamuz, 2004). These types of statements attempt to create uncertainty were there is none. This article has shown that the GPL has not been revised often.

In addition to this the process to draft and adopt version 3 is transparent, based on participation and supported by the user community.

CONCLUSION

Updating regulatory instruments always requires caution. This is even more so in the case of a document such as the GPL. The iconic nature of the GPL makes the development of the license a sensitive affair. This chapter has shown the way in which the FSF is working to ensure that the needs of all the stakeholders are met and that the fundamental freedoms provided by the license are not lost either intentionally or inadvertently. The FSF have chosen to be transparent and open in their process. They are inviting comments from all comers. At the same time they are not going so far as to relinquish control over the drafting process altogether. This open controlled approach is in line with the FSF attitude to controlled freedom that is implemented to ensure that the freedoms granted are not lost through the desire of individuals to maximise their position by profiteering from free software without contributing to the community.

REFERENCES

Guadamuz, A. (2004). Viral contracts or unenforceable documents? Contractual validity of Copyleft licenses. *European Intellectual Property Review, 26*(8), 331-339.

Höppner, J. (2004). The GPL prevails: An analysis of the first-ever court decision on the validity and effectivity of the GPL. *Script-ed, 1*(4), 662-667.

Klang, M. (2005). Free software & open source: The freedom debate and its consequences. *First Monday, 10*(3).

Metzger, A., & Jaeger, T. (2001). Open source software and German copyright law. *International Review of Industrial Property and Copyright Law, 32*(1), 52-74.

Molgen, E. (1999). Anarchism triumphant: Free software and teh death of copyright. *First Monday, 4*(8).

Moglen, E. (2001, September). Enforcing the GPL. *LinuxUser*, 66.

O'Sullivan, M. (2002). Making copyright ambidextrous: An expose of CopyLeft. *The Journal of Information, Law and Technology, 3*(2002).

O'Sullivan, M. (2004). The pluralistic, evolutionary, quasi legal role of the GNU general public licence in free/libre/open source software (FLOSS). *European Intellectual Property Review, 26*(8), 340-348.

Radin, M. J. (2000). Humans, computers, and binding commitment. *Indiana Law Journal, 75*(1125), 1125-1161.

Rosen, L. (2004). *Open source licensing: Software freedom and intellectual property law.* Upper Saddle River, NJ: Prentice Hall.

St.Laurent, A. M. (2004). *Understanding open source and free software licensing.* Sebastopol, CA: O'Reilly Associates.

Stallman, R. (2002). The free software definition. In J. Gay (Ed.), *Free software, free society: Selected essays of Richard M. Stallman.* Boston: Free Software Foundation.

Stallman, R., & Moglen, E. (2005). *GPL version 3: Background to adoption.* Boston: Free Software Foundation.

Turner, D. (2006). What is TiVo-ization? *Welcome to the GPLv3 Update, #5.* Mailing List, Free Software Foundation.

Weber, S. (2004). *The success of open source.* Cambridge, MA: Harvard University Press.

Williams, S. (2002). *Free as in freedom: Richard Stallman's crusade for free software*. Sebastopol, CA: O'Reilly & Associates.

Zuck, J. (2006, March 9). Perspective: GPL 3.0: A bonfire of the vanities? *CNet News.* CNET Networks.

KEY TERMS

Copyleft: Copyleft is a general method for making a program or other work free, and requiring all modified and extended versions of the program to be free as well. Copyleft says that anyone who redistributes the software, with or without changes, must pass along the freedom to further copy and change it. Copyleft guarantees that every user has freedom.

Free Software (FS): A term denoting software which fulfills the four freedoms, a set of standards set by the Free Software Foundation. See more http://www.gnu.org/philosophy/free-sw.html.

Free Software Foundation (FSF): An organisation, established in 1985, dedicated to promoting computer users' rights to use, study, copy, modify, and redistribute computer programs. The FSF promotes the development and use of free software, particularly the GNU operating system, used widely in its GNU/Linux variant.

General Public License (GPL): The fundamental software license of the free software movement. It guarantees that the four freedoms are awarded to the users.

GNU: GNU is a recursive acronym for "GNU's Not UNIX." The GNU Project was launched in 1984 to develop a complete UNIX-like operating system which is free software: the GNU system. Variants of the GNU operating system, which use the kernel called Linux, are now widely used; though these systems are often referred to as "Linux," they are more accurately called GNU/Linux systems.

Linux: Linux is a free Unix-type operating system. Developed under the GNU General Public License, the source code for Linux is freely available to everyone.

Software Licenses: A software license is a license that grants permission to do things with software. The license can be used to grant permissions to do things which are not granted by copyright. The license can also be used to deny users the right to do things to software to a much larger degree than those granted by copyright.

ENDNOTE

[1] The word free in free software in this chapter refers to freedom not cost. Any software that grants the user the four freedoms is free software. The four freedoms are (1) the freedom to run the program for any purpose, (2) the freedom to study how the program works and adapt it to your needs, (3) The freedom to redistribute copies so you can help your neighbor, and (4) the freedom to improve the program and release your improvements to the public so that the whole community benefits. Any software that does not grant the users any of these freedoms is proprietary software.

Chapter XXIX
Free Access to Law and Open Source Software

Daniel Poulin
Université de Montréal, Canada

Andrew Mowbray
University of Technology, Sydney, Australia

Pierre-Paul Lemyre
Université de Montréal, Canada

ABSTRACT

Law consists of legislation, judicial decisions, and interpretative material. Public legal information means legal information produced by public bodies that have a duty to produce law and make it public. Such information includes the law itself (so-called primary materials) as well as various secondary (interpretative) public sources such as reports on preparatory work and law reform and resulting from boards of inquiry and available scholarly writing. The free access to law movement is a set of international projects that share a common vision to promote and facilitate open access to public legal information. The objectives of this chapter are to outline the free access to law movement, to set out the philosophies and principles behind this, and to discuss the role that open source software has played both in terms of its use and development.

INTRODUCTION

The *free access to law movement* is a set of international projects that share a common vision to promote and facilitate open access to public legal information. There are direct synergies between the notion of "freeing the law" by providing an alternative to commercial systems and the ideals that underpin *open source software*. In addition, open source software has been an essential foundation for the work that has been done and new open source code has been developed.

The objectives of this chapter are to outline the *free access to law movement*, to set out the philosophies and principles behind this, and to discuss the role that open source software has played both in terms of its use and development. It concludes with an assessment of what has been achieved and of the similarities between the free access to law and open source software movements.

BACKGROUND

Law consists of legislation, judicial decisions and interpretative material. *Public legal information* means legal information produced by public bodies that have a duty to produce law and make it public. This includes the law itself (so-called *primary materials*) as well as various *secondary* (interpretative) public sources such as reports on preparatory work and law reform and resulting from boards of inquiry and available scholarly writing. It also includes legal documents created as a result of public funding.

Lawyers have been interested in the electronic publication of legal materials and associated information retrieval systems for a very long time. The earliest reported experiment is generally said to have been done by John Horty at the University of Pittsburgh in the late 1950s (Bing, 2004). The first major commercial system appeared in 1973 with the launch of *Lexis* (now *LexisNexis*). This was based on an earlier system developed by the Ohio Bar (OBAR) which had been established in 1969. OBAR was acquired by Mead Data Central and redesigned to become Lexis. LexisNexis is now one of the largest commercial text databases in the world. It is currently owned by the Reid publishing group. Lexis was followed by *Westlaw* in 1976. Westlaw is now owned by Thomson Publishing and is the major business competitor to Lexis. Several other commercial and government based systems also appeared about this time, but were largely ultimately unsuccessful such as the now defunct European system *EUROLEX* and the Australian system *SCALE* (Greenleaf, Mowbray, & Lewis, 1988).

In the 1980s and 1990s, Lexis and WestLaw expanded the scope of their services to include international collections and in their original jurisdiction (the United States) established a near duopoly (McKnight, 1997; Arewa, 2006). Attempts were made in various other places such as Australia and Canada to create either government or government sanctioned commercial monopolies (Greenleaf et al., 1988).

The resulting environment was, and to some extent still is one that is characterised by limited access to basic legal materials. Whilst the commercial systems provide a very sophisticated set of services they are for the most part targeted at the legal profession, they require significant training in order to use them. The services are very expensive and generally are not available for casual use. Non-lawyers seldom access the commercial systems and even lawyers can often not afford to use them.

Why is Free Access to Legal Information Important?

At the most fundamental level, access to public legal information supports the rule of law. Citizens are governed by laws and so have a need and right for effective access to these laws. Businesses also generally operate in a regulated environment and have similar needs. Effective access to basic legal information is essential both from a social perspective and also to facilitate the proper operation of business and commerce.

Apart from being able to access domestic laws, there is also increasingly a need to access law from other jurisdictions. Business operates on an international basis. Corporations need to be aware of international regulatory requirements and countries need to make their legal systems transparent to encourage international investment and trade. Particularly in the case of developing countries, there is a major need for access to international laws to assist with law reform and development (Poulin, 2004).

The Free Access to Law Movement

The *free access to law movement* has grown out of a set of projects that have attempted to address these issues and to provide alternatives to the commercial legal publishers' systems. Most of these projects are called *legal information institutes* (or *LII*s for short).

The earliest initiatives were in the United States and Canada. In 1992, Tom Bruce and Peter Martin established the Cornell *Legal Information Institute* (Bruce, 2000). This service was initially based on *Gopher* and provided free access to decisions of the United States Supreme Court and the United States code. It moved to the Web in 1994. In Canada, Daniel Poulin and his team at *LexUM* started publishing the full text of decisions of the Canadian Supreme Court in 1993 (Poulin, 1995).

Both systems helped to identify a strong demand for free public access to primary legal materials. In Australia, Graham Greenleaf and Andrew Mowbray founded *AustLII* (the *Australasian Legal Information Institute*) in 1995 (Greenleaf, Mowbray, King, & van Dijk, 1995). By the end of the year, AustLII was publishing some 16 databases including the decisions of most of the major Australian federal courts as well as federal and state legislation and by 1998 became the first LII to achieve national coverage. It now includes over 200 databases covering virtually all courts and tribunals in the country.

Other systems adopting a similar approach followed. These included the *British and Irish Legal Information Institute* (*BAILII*) in 1999, the *Pacific Islands Legal Information Institute* (*PACLII*) and the *Canadian Legal Information Institute* in 2000, and the *Hong Kong Legal Information Institute* (*HKLII*) in 2003. Various *meta-systems* were also built that drew upon the information contained in the other LIIs (*WorldLII, Droit francophone* and *CommonLII*).

The *free access to law movement* was proclaimed at the annual *Law via the Internet* confer-ence in Montreal in 2002. The current terms of the *Montreal Declaration* (as amended in Sydney, November 29, 2005 and Paris November 5, 2004) are (in part):

Legal information institutes of the world, meeting in Montreal, declare that:

- *Public legal information from all countries and international institutions is part of the common heritage of humanity maximising access to this information promotes justice and the rule of law;*
- *Public legal information is digital common property and should be accessible to all on a non-profit basis and free of charge;*
- *Independent non-profit organisations have the right to publish public legal information and the government bodies that create or control that information should provide access to it so that it can be published.*

...

Legal information institutes:

- *Publish via the internet public legal information originating from more than one public body;*
- *Provide free, full and anonymous public access to that information;*
- *Do not impede others from publishing public legal information; and*
- *Support the objectives set out in this Declaration.*

...

Each LII is responsible for publishing legal materials for a particular country or geographical region. AustLII, for example, publishes materials for Australasia (i.e., Australia and New Zealand). Apart from providing access to the full-text decisions of all major courts (such as the High Court, Federal Court, and State Supreme Courts), as has been said, AustLII also publishes decisions of nearly all Australian tribunals. Access to consolidated (and in some cases, *point in time*) legislation and regulations from all nine jurisdictions is also available. Other content includes: Law Reform Commission reports from most States; access to most Australian law journals; and a database of all bilateral and multi-party treaties.

Like most of the other LIIs, AustLII uses automated processes to add rich hypertext markup to its materials. In all, the system currently includes around 40 million internal hypertext links. Free text searching is available over the entire system or selected databases. AustLII is the major source of legal information in Australia and accounts for 25-30% of all legally related traffic in the country.

At the time of writing, the various LIIs together publish around 663 databases containing legal materials from 86 countries as well as 21 international collections. The total number of individual documents exceeds 3 million. Total usage is estimated to be in the vicinity of 3.5 million direct hits (or page accesses) per day.

The content of these databases consists mainly of *primary materials*—that is, court decisions, legislation and treaties, but increasingly secondary materials such as law journals, law reform commission reports, and the like are being added.

The LIIs have changed the way that law is made available to the public. Whereas in the past, there was exclusive reliance upon commercial publishers as conduits for the dissemination of this information, primary legal information now flows directly from courts and governments to consumers. The LIIs freely offer a level of value adding that establishes a new baseline for commercial publishers. Examples of this value adding include hypertext markup and search capabilities. The citator created by LexUM for CanLII (Reflex) provides a further example (Poulin, Paré, & Mokanov, 2005).

Each LII concentrates on making available domestic laws, but beyond these local endeavors all LIIs collaborate to expand the freely accessible law space internationally. This collaboration takes many forms. First of all, they all participate in promoting and supporting free access to the law by lobbying data providers such as courts, govern-

ments, and other bodies. They also provide, within their means, technical assistance and advice and training to other organizations. They hold annual conferences in order to exchange information and share knowledge. These conferences are public and all those interested can register to take part. Since many LIIs are based in universities, a significant part of those conferences is set aside for academic exchange of research results.

This cooperative spirit can be easily illustrated by the collaboration between the University of the South Pacific and AustLII to establish PacLII. Robynne Blake had worked for a number of years to build a substantial collection of South Pacific legal materials. AustLII assisted by provision of technical know-how and their software. In 2006, after many years of progress PacLII obtained a large grant from New Zealand Aid to expand its reach towards making the laws of the various states of the area freely accessible on the Internet. PacLII is now (in terms of the number of staff) one of the largest of the LIIs.

Similarly, LexUM collaborated with many interested parties in Burkina Faso to establish Juriburkina. Today, Juriburkina is operated from Ouagadougou by the local bar association and with the support of the higher courts, government general secretariat and a local Internet startup called ZCP informatique. A similar approach is being followed in Senegal and the project has reached implementation stage.

MAIN FOCUS OF THE CHAPTER

Development of Interest in Open Source Software

The LII promoters and developers were not always early adopters of open source software. Although most of the LIIs were Unix based, the significance of open source software only started to become more evident towards the end of the 1990s. Today, not only is most of the software used by LIIs open source, but the LIIs have themselves started to offer elements of their own production software under open source licences.

Many reasons may be put forward to explain the initial caution. First of all, in the early 1990s, open source was not as developed and mature as it is today. At the time, the LIIs rightly set "making the law accessible for free" as their principal agenda item. To achieve this, the most effective software, proprietary or otherwise was deployed. The reluctance towards more generally embracing open source by the LIIs, was partly based on the lack of maturity of the available open source software and partly attributable to the dominant prevailing prejudice towards conventional corporate approaches.

There was a major reappraisal of the initial attitude towards open source software from around 1998. At the time, for example, the operating system of choice for the LexUM's servers was Solaris from Sun Microsystems (this was also in use at AustLII and the Cornell LII). However, LexUM's programmers were mostly undergraduates and some of them had Linux installed on their home computers. These programmers were aware of the value of open source and argued strongly for the adoption of GNU/Linux. In the course of this campaign, they had even installed for demonstration purposes another open source flagship of the time, the already well respected—Apache Web server to replace the Netscape Enterprise server that was then in use. But despite the apparent functioning of Apache, LexUM, was reluctant to abandon the safety of using a proprietary solution for what appeared to be a more risky free alternative.

Then as today, LexUM was working with the Supreme Court of Canada (SCC) to make its decisions available for free in a timely manner. A long-awaited SCC judgment was expected on August 20, 1998 when the court's decision on the legality of a unilateral secession of Quebec from Canada was to be published. The morning the decision became available, the LexUM Netscape Enterprise based server went down at the moment the decision became available. The server was unable to cope with the rise in demand. After over an hour of rebooting the server, LexUM's student programmers brought up the Apache based sever. The move saved the day, and Apache kept running without failure for many weeks. From then on, LexUM used Apache as its Web server. In the following years, LexUM switched all of its servers to Linux and Apache.

The other LIIs had similar experiences. Most either had already or were soon to adopt Apache. Many moved to Linux and to generally adopt open source software as the basis of their production systems.

Current Use of Open Source Software

Although the commitment to open source has never been a religious one, most of the LIIs are nevertheless strongly reliant upon open source software. Although this is partly a matter of simple economics, this is not of itself sufficient to drive the adoption of open source as even free bad software is still obviously a poor choice. The open source orientation leads to a twofold benefit: savings in licence costs, but more importantly it led to the provision of reliable tools and powerful products to achieve the vision of freely accessible law. The current approaches used by the LIIs closely match open source trends. Open source developers develop many tools targeted for the Web that closely meet the needs of LIIs.

As has been said above, most of the LIIs use GNU/Linux and Apache. In addition some commonly used open source programs include database and indexing programs such as PostgreSQL, Open LDAP, Apache Lucene, and Nutch; programming languages and tools that include: perl, python, gcc, Eclipse, mod_perl, and Mason; and various other tools such as FastCGI and Mason.

Proprietary software is still used but only where a suitable open source solution cannot be identified. For example, most LIIs still rely upon proprietary software for a significant part of basic document preparation and conversion (such as Microsoft Word) and for some aspects of network security (for example, AustLII uses Check Point and Tripwire).

FUTURE TRENDS

Development of Open Source Software by LIIs

Prior to the World Wide Web, the publishing of databases of legal information was essentially the work of commercial publishers who used specialised software that had often been developed in-house. The Web brought with it a number of generic publishing tools such as conversion tools, search tools and Web servers. However, tools to support more specialised legal publishing needs remained rare. This led a number of the LIIs to develop the tools they needed.

One of the first of these was *Sino* (short for "size is no object"). Sino is a high performance free text search engine. It was originally written in 1995 and has been mainly used to provide production level search facilities for most of the Legal Information Institutes that form part of the free access to law movement. Sino went to a major rewrite in 2006 that makes it even faster and adds new functionality. Sino from its initial release has always been a very fast search engine and its indexing and searching time have been kept at the level of the fastest proprietary products.

Sino is designed to be easy to interface with via a simple C/Perl API as well as a ready written interactive interface for testing or for actual use on Unix sockets. The tool is relatively small and easy to understand at about 12K lines of ANSI/POSIX.1 compliant C code. Sino concordances (indexes) are portable across platforms with different architectures. Sino has been in use on a number of major Web sites answering many millions of requests for the past 10 years and so is robust and reliable.

Sino is a tool aimed at improving the access to the law. It was at the heart of AUSTLII from the very beginning and has been subsequently adopted by BAILII, PacLII and HKLII. LexUM used it for CanLII for many years. From 1995 until 2006, Sino and its source code were made available for free to anybody wanting to publish the law openly and for free. With its last rewrite, Sino became open source and it is now licensed under the GNU General Public Licence (GPL).

LexUM has also developed a number of pieces of open source software. *LexEDO* is a legal publishing platform aimed at providing a ready-made and easy to use solution for small-scale publication projects particularly in the developing world. LexEDO provides a means to manage legislation, caselaw, and legal periodicals as simple databases, to automatically convert documents to PDF and HTML and to generate a Website accordingly. All of these tasks can be accomplished by lawyers or law students acting as editors through Web-based management interfaces.

LexEDO has been distributed to such organisations as the Bar of Burkina Faso, the government general secretariat of Burkina Faso and the Bar of Senegal. In the context of these projects, the availability of the source code was critical for capacity building purposes. In Burkina Faso for instance, LexEDO has been maintained locally for a period of over two years by a private host called ZCP Informatique. To some extent, the fact that LexEDO source code is available allows ZCP to develop local solutions to local problems without requiring LexUM's assistance. It also provides them with the means to control the evolution of their project, or even to replicate it elsewhere thus spreading free access to law. As is the case for Sino, LexEDO is distributed under the GPL.

LexUM has also developed a program called NOME to assist with the anonymisation of judicial decisions. In many jurisdictions some or all judgments must not contain the names of parties or accused. For instance, anonymisation of judicial decisions involving young offenders is mandatory in Canada. To efficiently achieve this result, LexUM worked with the Computer Science Department at the University of Montreal (Plamondon, Lapalme, & Pelletier, 2004). The result was a small program which is capable of guessing and initialising proper names in Word documents. NOME is now distributed for free with its source code.

In respect to software developed in LIIs, Sino is certainly the most mature. Sino, LexEDO, and NOME are distributed under the GPL. Various other software tools have been developed and are distributed by the LIIs to various partner organisations. As other tools become of more general application, they will become candidates to become new open source offerings.

CONCLUSION

The use of open source software by the LIIs reflects the fact that both movements are well aligned and in many senses similar. The most evident of these similarities can be listed as follows:

Avoiding Monopolistic Control over the Information

Legal information, similarly to source code, *wants to be free* (Williams, 2002). Both the free access to law and the open source software movements were conceived in reaction to the seizure of information by entities (state or commercial) not willing to share it freely with others.

Promote the Reuse of Information by Third Parties

As is the case for source code, legal information is useful only if it can be reused for various purposes. Users need the possibility to save legal documents in different formats, to send them to colleagues and to present them in courts. Some users might even need the right to reuse documents in a commercial context (for the publication of a paper based law report, e.g.).

Promote the Development of Standards

As for software development, the dissemination of legal information is improved by the adoption of standards by the players involved. These standards can take the form of uniform citation mechanisms, drafting practices or workflow models. Historically, LIIs are at the center of such initiatives.

Need to Share Tools

Organizations involved in free access to law all face the same difficulties. They constitute a community tied together by the need to edit and convert large volume of legal documents, to publish them on the Web and to provide information retrieval tools to their users. Similarly to every open source software community, LIIs have incentives to share their efforts in the achievement of common goals.

Proponents do not Derive Revenue from Selling Information as a Product

The source of revenue of LIIs and open source software developers is the same. It flows not from the information they publish but from the expertise they developed doing so.

Considering all these similarities, the use of open source software can easily be seen as a complementary strategy to strengthen free access to law. It allows the LIIs to achieve near complete transparency by opening-up not only the legal information, but also their publication process. By doing so, the LIIs achieve several goals at once: they guarantee (to a certain degree) the integrity of their data; they facilitate interactions with the other players in the field; and finally, they help foster the emergence of additional free access to law projects.

For people or organisations that would like to pursue free access to law projects in their own country or region, the required software is now available. There are many high quality resources available from the open source community that can be used to establish Web services. The major distributions of Linux (and other open source operating systems) and the Apache Web server are of world-class quality. There are a number of suitable search engines available. The Web and the availability of open source software means that it is now relatively straight forward to disseminate information.

For the more specialized requirements involved in publishing the law such as the conversion of data, hypertext markup, metadata extraction, and the like, the LIIs are able to make a contribution. As a result, it is increasingly the case that for those who wish to make the law more accessible, there are available tools.

REFERENCES

Arewa, O. (2006). *Open access in a closed universe, Lexis, Westlaw and the law school.* Case Legal Studies Research Paper No. 06-03. Retrieved from http://ssrn.com/abstract=888321

Bing, J. (Ed.). (1994). *Handbook of legal information retrieval.* Amsterdam: North Holland.

Bruce, T. (2000). Public legal information: Focus and future. *Journal of Information, Law and Technology,* (1).

Greenleaf, G., Mowbray, A., King, G., & van Dijk, P. (1995). Public access to law via Internet: The Australasian Legal Information Institute. *Journal of Law & Information Science, 6*(1).

Greenleaf, G., Mowbray, A., & Lewis, D. (1988). *Australasian computerised legal information handbook.* Sydney: Butterworths.

McKnight, J. (1997). Wexis versus the Net. *Illinois Bar Journal, 85*(4).

Plamondon, L., Lapalme, G., & Pelletier, F. (2004). Anonymisation de décisions de justice. *TALN Conference Proceedings,* Fès. Retrieved from http://www.lpl.univ-aix.fr/jep-taln04/proceed/actes/taln2004-Fez/Plamondon-etal.pdf

Poulin, D. (1995). Legal resources for Canadian lawyers on the Internet. *CSALT Review—Canadian Society for the Advancement of Legal Technology, 9*(1).

Poulin, D. (2004). CanLII: How law societies and academia can make free access to the law a reality. *Journal of Information, Law and Technology,* (1).

Poulin, D, Paré, E., & Mokanov, I. (2005, November 17-19). Reflex: Bridging open access with a legacy legal information system. In *Proceedings of the 7th Law via the Internet International Conference,* Port Vila, Vanuatu.

Williams, S. (2002). *Free as in freedom: Richard Stallman's crusade for free software.* Sebastopol, CA: O'Reilly & Associates.

KEY TERMS

Jurisdiction: The geopolitical region in which the laws of a certain governing body are recognized as legitimate and can be enforced.

Law: A body of knowledge consisting of legislation, judicial decisions, and interpretative material.

License: Permission needed to use or modify materials in a way that is recognized as legitimate by the owner of such materials and by an overall community familiar that recognized a similar understanding of legitimate use.

Primary Materials: Court decisions, legislation, and treaties.

Public Access: Making materials available for all members of the general public to read and review.

Public Legal Information: Legal information produced by public bodies that have a duty to produce law and make it public.

Secondary Materials: Public sources, such as reports on preparatory work, that report on and often interpret legal developments.

Chapter XXX
Examining Open Source Software Licenses through the Creative Commons Licensing Model

Kwei-Jay Lin
University of California, USA

Yi-Hsuan Lin
Creative Commons Taiwan Project, Taiwan

Tung-Mei Ko
OSSF Project, Taiwan

ABSTRACT

In this chapter, the authors present a novel perspective by using the Creative Commons (CC) licensing model to compare 10 commonly used OSS licenses. The authors also propose a license compatibility table to show that whether it is possible to combine OSS with CC-licensed open content in a creative work. By using the CC licensing concept to interpret OSS licenses, the authors hope that users can get a deeper understanding on the ideas and issues behind many of the OSS licenses. In addition, the authors hope that by means of this table, users can make a better decision on the license selection while combining open source with CC-licensed works.

INTRODUCTION

With the rapid growth of the open source software (OSS) community in the past decade, many users now are convinced that OSS is a practical and attractive alternative to proprietary software. Since almost all OSS licenses allow worldwide, royalty-free usage and encourage users to copy, modify, and enhance original codes, OSS has attracted many users and programmers. Some other benefits include significantly lower development and deployment cost, and software quality improvement due to open inspections and discussions.

To meet the needs of various authors and users, different software licenses have been defined. The diversity and complexity of these licenses, on the other hand, create confusions for many potential OSS authors and users. It has been a constant community effort through articles, reviews, and books to discuss and to elaborate on the subtle differences among these licenses.

For non-software publications, such as Web sites, graphics, music, film, photography, literature, courseware, and so on, that normally fall under the current copyright law, some authors may want to open up part of their rights to the public with a spirit similar to those of OSS licenses. To allow for such possibilities, Creative Commons (CC) was founded in 2001 to define the licenses beyond the traditional "all rights reserved" copyright definition. CC licenses, motivated in part by the GNU General Public License (GPL) of the Free Software Foundation (FSF), provide a similar function to OSS licenses for non-software creative works.

Both OSS and CC licensing models are about promoting the ideas of free access. Therefore, it is not a rare case to combine open software released under OSS licenses with CC-licensed creative material. Nevertheless, there are differences between these two models. For users who combine these two types of materials to create a new resulting work, some questions are of deep concern. For example, whether a specific OSS license is compatible with CC licenses? Which license should the resulting work apply to? Unfortunately, so far there is hardly any study discussing these issues in depth.

As participants of the open source movement in Taiwan, we have witnessed the flourishing innovation and creativity of OSS activities in Taiwan. However, the license selection issue has continued to be an obstacle for many potential local contributors. Part of the charters of the Open Foundry project in Taiwan (called OSSF, http://www.openfoundry.org) is to help people easily capture a basic understanding of the licenses that govern OSS, related documentations and open content.

In this chapter, we present a novel perspective by using the CC licensing model to compare 10 commonly-used OSS licenses. Specifically, we have defined a license compatibility table that shows whether it is possible to combine OSS with CC-licensed open content in a creative work. The idea of comparing the two types of licenses is partly inspired by Rosen (2004). In Chapter 10 (pp. 244-251) of his book, Rosen takes four commonly used OSS licenses as examples and discusses the compatibility of these licenses. Similarly, our study may help people understand if they can re-license a resulting work under a specific CC license. The reason for our study on the compatibility table is from the observation that many new OSS contributors are primarily interested in getting their software known and accepted by the community, and circulated as widely as possible. They do not want to interfere with licensees' use of the software nor constraining the licensing of derivative works. Their goal is to create works that people may share and enjoy, much like open content. Therefore, by using the CC licensing concept (such as *attribution* and *share alike*) to interpret OSS licenses, people may get a deeper understanding on the ideas and issues behind many of the OSS licenses, and make a better decision on the license selection.

The rest of the chapter is organized as follows. The following section reviews the basic elements of OSS licenses and CC licenses. Subsequently, the comparison of the two licenses classes is presented. Next, we discuss two new license concepts, then the chapter is concluded in the last section.

BACKGROUND: FROM GPL TO ATTRIBUTION

OSS Licenses and FDL

There are many types of OSS licenses. According to the statistics from FSF and the Open Source Initiative (OSI), there are over 60 OSS licenses. In general, these licenses have three common characteristics (Free Software Foundation, 2005a; The Open Source Initiative, 2006a):

1.   No royalties
2.   No geographical restrictions on distribution
3.   No specific licensees

Among them, we have chosen 10 more commonly-used OSS licenses (including GNU General Public License, GNU Library/Lesser General Public License, BSD license, MIT license, Apache Software License 1.1, zlib/libpng License, Artistic License, Common Public License, Qt Public License, and Mozilla Public License) plus the GNU Free Documentation License (FDL) for discussion in this paper. These licenses, excluding the FDL, have all been approved by the OSI and conform to the Open Source Definition (OSD) (The Open Source Initiative, 2006b).

The most well-known OSS license is GPL, which was drafted by Richard M. Stallman, the founder of the FSF and the Project GNU. The GPL is developed on the basis of the *copyleft* mechanism. According to the *copyleft* mechanism, a licensee has to adopt the same license as that of the licensor for his (or her) program. Using the *copyleft* mechanism, source code can always remain open and royalty free. The GNU Library/Lesser General Public License 2.1 (LGPL) is the other OSS license implementing the *copyleft* mechanism (Free Software Foundation, 2005b). The LGPL is designed specifically for library code, and is less strict than the GPL.

On the other hand, the *copyleft* mechanism does not limit any right arising from *fair use*. Thus, when an author uses GPL-licensed or LGPL-licensed codes as examples in a book or as references, he (or she) may not have to apply the GPL or LGPL to the book as long as the application falls within the scope of *fair use*. Under this circumstance, the author can choose a license at his (or her) will, for example, any traditional proprietary copyright license or any CC license for the book. Similarly, in accordance with the *fair*

use doctrine, when the author attaches a whole copy of the GPL-licensed or LGPL-licensed codes with the book while published or distributed, the license adoption of the book will not be restricted to the GPL or LGPL.

Compared with the GPL, the BSD license, another popular OSS license, does not impose any restriction on the licensee in terms of future license selection. In other words, the licensee is allowed to use any license (even make it proprietary) for his (or her) program and is also allowed to collect royalties. The Apache Software License 1.1 (Apache 1.1), zlib/libpng License (zlib/libpng), and MIT License have similar characteristics as that of the BSD license.

The other four licenses discussed in this chapter are the Mozill Public License 1.1 (MPL), Common Public License 1.0 (CPL), Qt Public License 1.0 (QPL), and the Artistic License (Artistic). Basically, the MPL, CPL, and QPL are all designed for the commercial use of OSS, thus their regulations about licensees' rights and obligations are very similar. The MPL employs a partial *copyleft* mechanism in that the licensee can only use the MPL for his (or her) program in principle (Mozilla.org, 2006). However, the licensee is allowed to adopt another license for certain parts of the program. The CPL adopts the *copyleft* mechanism and is the first license to regulate commercial distribution of OSS with separate terms. Artistic has its own legal logic, which is different from the other nine OSS licenses.

Same as the GPL, the FDL is drafted by Stallman and also adopts the *copyleft* mechanism. However, the FDL is normally used for textbooks or teaching materials written for some equipment or software (Free Software Foundation, 2005b). Wikipedia, a famous online encyclopedia, adopted the FDL for its text content, is a noted example (Wikipedia, 2006).

The Inception of Creative Commons

A group of professionals from various fields, including intellectual property and cyberlaw experts

James Boyle, Michael Carroll, and Lawrence Lessig, and MIT science professor Hal Abelson, founded Creative Commons in 2001 (Lessig, 2005). CC advocates the "some rights reserved" concept in contrast to the default "all rights reserved" in current copyright laws. CC also takes ideas in part from the FSF and produces a series of copyright licenses to help creators declare to the world what freedom they want their works to carry. These freedoms are composed by four elements: *Attribution* (denoted as "by" or ⒝), *noncommercial* ("nc" or ⓢ), *no derivatives* ("nd" or ⊜) and *share alike* ("sa" or ⓞ). When CC licenses v. 1 were first released in December 2002, these four elements were all optional. Later on, CC found that 98% of the adopters have chosen "attribution" as a requisite; thus CC sets "attribution" as a default in v. 2, and offers six licenses (Lessig, 2005).

CC Licenses

CC licenses are designed to bridge creators and users in that users have no need to ask for creators' prior permission to use the works as long as they follow the rules the creators set. For example, if a work is released under the "by-nc" CC license (i.e., attribution and noncommercial), a user can freely make use of the work under the condition that the user uses this work for noncommercial purposes only and must always credit the original creator. The six CC licenses are defined as follows:

1. **Attribution:** It means that a user can freely use the work, provided that he (or she) credits the creator.
2. **Attribution-share alike:** It means that a user can freely use the work, provided that he (or she) credits the creator and also licenses any derivative under the same license as that of the original work.
3. **Attribution-no derivatives:** It means that a user can only make use of verbatim copies of the work and have to credit the creator.

4. **Attribution-noncommercial:** It means that a user can only use the work for noncommercial purpose, and have to credit the creator.
5. **Attribution-noncommercial-no derivatives:** It means that a user can only make use of verbatim copies of the work, for noncommercial purposes only, and have to credit the creator.
6. **Attribution-noncommercial-sharealike:** It means that a user can only use the work for noncommercial purposes, credit the creator, and license any derivative under the same license as that of the original work.

In addition to the above six licenses, CC also offers other licenses for more specialized situations. For example, sampling licenses allow people to use a part of some creative works and mix with some original or other parts to create a new work. One can use founders copyright to free works from copyright completely, after it has been created for 14 or 28 years. In general, CC provides the vehicle that "does not mean giving up your copyright. It means offering some of your rights to any member of the public but only on certain conditions" (Creative Commons, n.d.).

Reviews of Issues on OSS and CC Licenses

Since the OSS licensing model appeared in the 1990s, it has started a lot of discussion, for example: What is the free/open source software movement? How does it run? How does it work with the current legal system? (Hill, 1999). Many have questioned about the enforceability of OSS licenses (Nadan, 2002; Ravicher, 2000). Later, because of the lack of precedents regarding OSS licenses' enforceability, Gomulkiewicz (2002) proposes to create an open source license organization (OSLO) to solve issues relating to OSS licenses. Gomulkiewicz thinks that the OSLO could play a role in calling programmers and lawyers together to built up useful licensing practices of OSS and further solve related licensing problems.

Among these various OSS licenses, the GPL receives the most attention. Stoltz (2005) discusses the scope of derivative works under current US copyright laws and how the extent of derivative works affects the GPL. Besides, along with the rising number of successful open source commercial cases, the issues about OSS license policies started to get much attention. For example, Satchwell (2005) provides users a basic understanding of OSS; how to choose a suitable OSS license and how to establish an appropriate OSS policy.

With the increase of OSS licenses, there are more and more articles discussing issues of OSS licenses. However, among these articles, only a few have addressed the OSS license compatibility. Perens (1999) and Maher (2000) point out that OSS license compatibility is a noteworthy issue, but neither offers any concrete solution to the compatibility problem. Rosen (2004) provides some discussion on the issue of OSS license compatibility but does not come up with any concrete solution. Therefore, sensing the need of a simple and clear explanation for various OSS licenses, we use a relatively intuitive licensing model, CC licenses, to examine OSS licenses.

Since CC licenses' release in December 2002, these licenses spread quickly and dramatically. There have been more than 50 million Web pages linked to CC licenses as of August 2005 (Katz, 2006). However, with the rising popularity of CC licenses, many skeptical views have appeared. Some challenge the compatibilities between certain CC licenses or between different free-content contracts (e.g., Elkin-Koren, 2005; Katz, 2006); while others question that a variety of CC licenses will cause confusion or increase the information cost (e.g., Elkin-Koren, 2005; Katz, 2006). Moreover, license translation and legal adaptation may undermine the success of CC licensing (Valimaki, 2005).

Katz (2006) questions that a variety of CC licenses would puzzle users in that users may run into difficulties in determining which CC license is the most suitable for them. Elkin-Koren (2005) challenges the consistency of CC's strategy over license selection. He argues that CC attempts to reduce external information cost by its license choosing platform, but the variety of CC licenses would on the contrary impose extra informational burden on authors. He uses musical works as an example: In addition to six CC core licenses composed of four elements (i.e., attribution, noncommercial, no derivatives, and share alike), there are the other three sampling licenses (i.e., sampling, sampling plus, and noncommercial sampling plus). To find out which license is the most suitable one for musical works would unwittingly increase information costs. Elkin-Koren (2005) further states that the lack of standardization in CC licenses would increase the cost of ascertaining the rights and obligation related to any specific work.

In addition, Katz (2006) argues about the incompatibilities of CC licenses. He concludes that the viral effect of the "share alike" element will result in the incompatibilities problem between different share alike licenses and would further restrict the distribution of derivative works.

Some of the above mentioned literatures discuss the compatibility among different CC licenses, but none offers a systematic analysis on the problem of CC license incompatibility or combining OSS with CC-licensed content. Therefore, we attempt to illustrate what license a user may choose when he (or she) combines OSS with CC-licensed content by a clearly defined table to be discussed next.

MAIN FOCUS OF THE CHAPTER

Examining Compatibility between OSS and CC Licenses

Our study in this chapter is to define a license compatibility table that shows whether the combination of OSS and CC-licensed open content in a creative work may be properly licensed. The table (Table 1) may help people understand if they can continue to re-license a derived work under a specific CC license.

In Table 1, we use "by," "nd," "nc," and "sa" to denote CC's four elements "attribution," "no derivatives," "noncommercial," and "share alike" respectively. The mark "o" indicates that when a derivative work incorporates two or more works under licenses listed in a specific column and a specific row, it can be re-licensed under the CC license shown in the column. The mark "X," on the other hand, shows that a derivative work, incorporating two or more works under licenses listed in a specific column and a specific row, cannot be re-licensed under the license shown in the column. For example, if one combines a program A released under the GPL, with an open content B issued under the CC attribution license, and produces a new work C. One may not re-license C under the CC attribution license because the GPL requires that GPL-applied program or its derivative work must always be governed by the GPL.[1] Thus, A and C must be licensed similarly, and C work will not be able to release under any CC license.

In the following sections, we discuss the table entries in details.

No Derivatives

The 10 OSS licenses chosen to compare with CC licenses in this chapter are all approved by the OSI. An OSI-certified license must conform to the OSD (The Open Source Initiative, 2006b). Under criterion 3 of the OSD, the license must permit making modifications and derived works (The Open Source Initiative, 2006c). Therefore, these 10 OSS licenses allow modification to the original program. Thus, any of the six CC licenses which contains "No Derivative" element (i.e., CC attribution-no derivatives license, CC attribution-no derivatives-noncommercial license) is not compatible with any of the 10 OSS licenses, and "X" is shown in all cells of the "by-nd" and "by-nd-nc" columns in the table.

Noncommercial

Criterion 1 of the OSD states that an OSS license should not "restrict any party from selling or giving away the software as a component of an aggregate software distribution containing pro-

Table 1. License compatibility table

	by	by-nd	by-nd-nc	by-nc	by-nc-sa	by-sa	FDL
GPL	X	X	X	X	X	X	X
LGPL	X	X	X	X	X	X	X
MPL	X	X	X	X	X	X	X
QPL	O	X	X	X	X	O	O
CPL	O	X	X	X	X	O	O
Artistic	O	X	X	X	X	O	O
Apache	O	X	X	X	X	O	O
Zlib/ libpng	O	X	X	X	X	O	O
BSD	O	X	X	X	X	O	O
MIT	O	X	X	X	X	O	O

grams from several different sources" (The Open Source Initiative, 2006c). It thus implies that any OSD-compliant license should not restrict any use of commercial purposes. This results in the conflict between 10 OSS licenses and any CC license with "noncommercial" element (i.e., CC attribution-noncommercial licenses, CC attribution-noncommercial-share alike licenses). An "X" is shown in all cells of "by-nc" and "by-nc-sa" in the table.

Copyleft

The *copyleft* mechanism provides that anyone will be granted the rights to use, copy, modify, or distribute a program or its derivative works on the condition that when redistributing a program, with or without change, all rights he (or she) gained must be passed on to subsequent users (Free Software Foundation, 2005b). The GPL, LGPL, FDL, and CPL are terms to implement the *copyleft* mechanism. The implemented result will be the original work and its derivative works must be redistributed under the same license as the original work.

Although not originated from the FSF, the MPL partially employs a *copyleft* mechanism. MPL requires that modifications to MPL-licensed program must be governed by MPL (Mozilla.org, 2006). Because of this viral nature of *copyleft*, the GPL, LGP, MPL, and CPL is not compatible with any of the CC licenses, and thus "X" is shown in all cells of the top four rows.

Author Credit

The Apache, zlib/libpng, BSD, and MIT explicitly indicate that the authors of original work must be credited (Lin, Ko, Chuang, & Lin, 2006). QPL, CPL, and Artistic have copyright notices related regulations, and do not exclude the authors' names of the original work from the copyright notices. Yet, CPL implements copyleft mechanism. Therefore, GPL, LGPL, MPL, in addition to CPL

are not compatible with CC attribution licenses. A work incorporating a program licensed under the QPL, Artistic, Apache, zlib/libpng, BSD, or MIT with other works issued under CC attribution license could be re-licensed under CC attribution license. Excluding the GPL, LGPL, CPL, and MPL, "○" is shown in the other cells of the "by" column in the table.

Share Alike

The compatibility between the 10 OSS licenses and the CC attribution-share alike license is discussed next. The GPL, LGPL, CPL, and MPL implement the *copyleft* mechanism. But the other six OSS licenses do not explicitly adopt it and do not have the viral effect on the resulting derivative work. Thus, when a work incorporates a program licensed under the QPL, Artistic, Apache, zlib/libpng, BSD, or MIT with the other work issued under the CC attribution-share alike license, this newly created work may be re-licensed under the CC attribution-share alike license. Except the top four cells, the mark "○" is shown in the other cells of the "by-sa" column in the table.

FDL

Finally, we examine the compatibility between the 10 OSS licenses and the FDL.

Even though the GPL, LGPL and FDL are all developed by the FSF, because of *copyleft* mechanism's viral effect, when a work incorporates a GPL-licensed or LGPL-licensed program with other FDL-released work, the resulting work may not be re-licensed under the FDL. The same result applies to CPL because CPL implements copyleft as well. The MPL partially employs the *copyleft* mechanism; thus, a created derivative work incorporating a MPL-licensed program with the other FDL-released work may not be re-licensed under the FDL, either.

In principle, the FDL enables the same freedom as the CC attribution-share alike license (Creative

Commons, 2005). Due to the *copyleft* mechanism, when a work incorporates a program licensed under the GPL, LGPL, CPL, or MPL with the other FDL-released work, this resulting derivative work may not be re-licensed under the FDL. Except the GPL, LGPL, CPL, and MPL, a work incorporating a program governed by the QPL, CPL, Artistic, Apache, zlib/libpng, BSD, or MIT with the other work issued under the FDL, this new created work could be re-licensed under the FDL. Except the top four cells, "○" is shown in the other cells of the "FDL" column in the table.

Using License Compatibility Table for License Selection

From Table 1, we could identify the following license selection strategies. If a user would like a creative work which combines OSS with CC-licensed open content to be re-licensed under the CC license or the FDL, he (or she) should avoid using OSS licensed under GPL, LGPL, or MPL. In other words, if a creative work is combining OSS licensed under the GPL, LGPL, CPL, or MPL with CC-licensed open content, this work is not possible to be re-licensed under any CC license.

If a user would like a creative work which combines OSS with CC-licensed open content to be re-licensed under some CC licenses, he (or she) should choose OSS licensed under the QPL, Artistic, Apache, zlib/libpng, BSD, or MIT. However, not all CC licenses are compatible with the QPL, Artistic, Apache, zlib/libpng, BSD, or MIT; only CC by and by-sa licenses are compatible with these six licenses. In other words, a creative work combing an OSS license under QPL, Artistic, Apache, zlib/libpng, BSD, or MIT with CC by or by-sa licensed open content, the resulting work could be re-licensed under CC license identical to the original open content. Similar to CC by or by-sa license, FDL is not compatible with the GPL, LGPL, CPL, or MPL, but is compatible with QPL, Artistic, Apache, zlib/libpng, BSD, or MIT.

Therefore, if a user would like a creative work which combines OSS with FDL-licensed open content to be re-licensed under the FDL, he (or she) may use OSS licensed under any of these six OSS licenses.

From the above discussions, we can make two simple conclusions. Firstly, OSD-compliant OSS licenses should not restrict any use of commercial purposes[2], and OSD-compliant OSS licenses must allow modifications and derived works.[3] Therefore, CC licenses containing "noncommercial" or "no derivatives" element are not compatible with 10 OSS licenses discussed in this chapter.

The second conclusion is that the *copyleft*'s viral effect requires the original work and its derivative works to be redistributed under the same license as the original work. Thus, the GPL, LGPL, CPL, and FDL, which adopt the *copyleft* mechanism completely, plus MPL, which partially adopts the *copyleft* mechanism, are not compatible with any CC license.

For authors that are not combining OSS with open content, the above discussion may provide some useful insights as well. The *copyleft* mechanism is a strong license requirement that may prevent others from producing derivative works mixed with even the "share alike" element. It is probably better to select other licenses if such a requirement may present a problem in the future.

FUTURE TRENDS

Open Access Publishing

CC licenses are inspired from the GPL. In addition to OSS and CC licensing models, other models have been developed in different fields sharing similar notions with that of OSS and CC. Open access publishing is one of them.

Typically, publishers charge readers a subscription fee, and sometimes also charge authors a page fee. Open access publishing, on the contrary, allows the author to retain his (or her) article's copyright; at the same time, authors or their sponsors, not the users, pay the publishers.

Although the open access publishing model will make authors bear more cost than traditional publishing models, the charged fees are possible to be transferred to the authors' sponsor institutes or even be waived (Suber, 2004). Moreover, open access publishing will increase the possibility that the authors' articles are searched, and help to build the authors' prestige (Harmel, 2005). Recent studies also show the same result: online articles are more frequently cited (Lawrence, 2001) and more often used than offline articles (Lawrence, 2001; Walker, 2004). Therefore, more and more leading publishers, such as the Public Library of Science (PLoS) and BioMed Central have joined the open access movement.

Science Commons, a newly launched project of Creative Commons, was founded to support the sharing of scientific research, such as the field of biotechnology, medicine, and even law[4], with the same "some rights reserved" spirit Creative Commons holds.

Studies have found that the open access publishing model is practicable (Gonzalez, 2005; Odlyzko, 1998), and there are already publishers gain profits from it (Walker, 2004). It is foreseeable that with the increasing subscription fees of academic journals the open access publishing model will continue to gain more support, especially on academic content.

New OSS Elements

Compared with numerous OSS licenses, the CC licensing model built on the basis of four elements is relatively simple. However, although these four elements are less complex and easier to understand, they are not broad enough to cover all major considerations of OSS licenses. Here we include two new concepts, "no endorsement" and "modification record" that should be considered by OSS users when selecting a license.

No Endorsement

One of the OSS's common characteristics is that anyone is free to create derivative works (The Open Source Initiative, 2006c). Because of this, the quality of derivatives is hard to control. When the quality of a derivative is not as good as the original program, but the name of the original developer or the copyright holder is still shown on the derivative, new users may not have enough acknowledgement of this and relies on the name of the original developer or the copyright holder to evaluate the derivatives. Under the circumstance, it may harm the reputation of the original developer or the copyright holder. Sometimes, the developer of the derivatives may intentionally show the name of the original developer or the copyright holder on the derivative work to endorse or promote his (or her) own works.

Therefore, to prevent OSS adopters from using the authorship to implicitly or explicitly show the support, association of the initial developers or to promote their derivative work, and, even more, to prevent the derivatives from being wrongly trusted, the original program's developer should choose a license which contains "no endorsement" or disclaimer clause[5], such as BSD and Artistic. The Creative Archive License developed by BBC adopts the main ideas of CC licenses but injects such a new element into the license.[6]

Modification Record

We also notice that many OSS licenses have regulations regarding the modification records.[7] Take the 10 OSS licenses we analyze in this chapter for example. Only the QPL does not require that a modification record must be made. Instead, the QPL forbids users to directly make modifications to the original works and requires that all modifications be in a form that is separate from the original works (e.g., patches).[8] All of the other nine licenses have modification record related regulation. These records are very helpful for the convenience of follow-up software modifications. Moreover, they are beneficial to maintain the original works' integrity.[9]

CONCLUSION

OSS licenses have triggered a lot of discussions in the past few years because of their complicacy. The OSI even appeals to reduce the number of approved OSI-licenses to allow programmers and users to understand OSS licenses more easily. In contrast with OSS licenses, CC licenses provide a cleaner licensing model. In this chapter we investigate the compatibility between the six CC licenses and 10 commonly-used OSS licenses including the FDL. OSS authors may use the table to identify which CC license he (or she) can use for his (or her) work that combines OSS with CC-licensed work.

However, CC's four simple elements do not capture all major issues of OSS. We thus raise two new issues, "no endorsement" and "modification record", to address some main concerns by OSS. We believe that by employing CC's four elements, plus our proposed two new elements, OSS community, including both authors and users, will be able to get a more complete picture of OSS licenses.

REFERENCES

Creative Commons. (n.d.). *Choosing a license.* Retrieved March 30, 2006, from http://creativecommons.org/about/licenses/

Creative Commons. (2005). *Discussion draft: Proposed license amendment to avoid content ghettos in the commons.* Retrieved March 30, 2006, from http://creativecommons.org/Weblog/entry/5701

Elkin-Koren, N. (2005). What contracts cannot do: The limits of private ordering in facilitating a creative commons. *Fordham Law Review, 74,* 375-422.

Free Software Foundation. (2005a). *FSF: Licenses.* Retrieved March 30, 2006, from http://www.fsf.org/licensing/licenses/index_html

Free Software Foundation. (2005b). *What is copyleft? GNU Project: Free Software Foundation (FSF).* Retrieved March 30, 2006, from http://www.gnu.org/copyleft/copyleft.html

Gomulkiewicz, R. W. (2002). De-bugging open source software licensing. *University of Pittsburg Law Review, 64,* 75-103.

Gonzalez, A. G. (2005). The digital divide: It's the content, stupid: Part 2. *Computer and Telecommunications Law Review, 11*(4), 113-118.

Harmel, L. A. (2005). The business and legal obstacles to the open access publishing movement for science, technical, and medical journals. *Loyola Consumer Law Review, 17,* 555-570.

Hill, T. (1999). Fragmenting the copyleft movement: The public will not prevail. *Utah Law Review, 1999,* 797-822.

Katz, Z. (2006). Pitfalls of open licensing: An analysis of Creative Commons licensing. *IDEA: The Intellectual Property Law Review, 46,* 391-413.

Lawrence, S. (2001). *Free online availability substantially increases a paper's impact.* Retrieved July 10, 2006, from http://www.nature.com/nature/debates/e-access/Articles/lawrence.html

Lessig, L. (2005). *CC in review: Lawrence Lessig on supporting the commons.* Retrieved March 30, 2006, from http://creativecommons.org/Weblog/entry/5661

Lin, Y. H., Ko, T. M., Chuang, T. R., & Lin, K. J. (2006). Open source licenses and the Creative Commons framework: License selection and comparison. *Journal of Information Science and Engineering, 22,* 1-17.

Maher, M. (2000). Open source software: The success of an alternative intellectual property incentive paradigm. *Fordham Intellectual Property, Media and Entertainment Law Journal, 10,* 619-695.

Mozilla.org. (2006). *MPL FAQ.* Retrieved March 30, 2006, from http://www.mozilla.org/MPL/mpl-faq.html

Nadan, C. H. (2002). Open source licensing: Virus or virtue? *Texas Intellectual Property Law Journal, 10,* 349-377.

Odlyzko, A. (1998). The economics of electronic journals. In R. Ekman & R. Quandt (Eds.), *Technology and scholarly communication.* University of California Press.

Perens, B. (1999). The open source definition. In C. DiBona, S. Ockman, & M. Stone (Eds.), *Open sources: Voices from the open source revolution* (pp. 171-188). Sebastopol, CA: O'Reilly & Associates.

Ravicher, D. B. (2000). Facilitating collaborative software development: The enforceability of mass-market public software licenses. *Virginia Journal of Law & Technology, 115,* 1522-1687.

Rosen, L. (2004). *Open source licensing: Software freedom and intellectual property law.* Upper Saddle River, NJ: Prentice Hall.

Satchwell, M. D. (2005). The tao of open source: Minimum action for maximum gain. *Berkeley Technology Law Journal, 20,* 1757-1798.

Stoltz, M. L. (2005). The penguin paradox: How the scope of derivative works in copyright affects the effectiveness of the GNU GPL. *Boston University Law Review, 85,* 1439-1477.

Suber, P. (2004) *Open access overview.* Retrieved July 10, 2006, from http://www.earlham.edu/~peters/fos/overview.htm

The Open Source Initiative. (2006a). *Open Source Initiative OSI-licensing.* Retrieved March 30, 2006, from http://www.opensource.org/licenses/

The Open Source Initiative. (2006b). *Open Source Initiative OSI—Certification Mark and Program.* Retrieved March 30, 2006, from http://www.opensource.org/docs/certification_mark.php

The Open Source Initiative. (2006c). *Open Source Initiative OSI—The Open Source Definition.* Retrieved March 30, 2006, from http://www.opensource.org/docs/definition.php

Valimaki, M. (2005). *The rise of open source licensing: A challenge to the use of intellectual property in the software industry.* Helsinki, Finland: Turre Publishing.

Walker, T. J. (2004). *Open access by the article: An idea whose time has come?* Retrieved July 10, 2006, from http://www.nature.com/nature/focus/accessdebate/13.html

Wikipedia. (2006). *Wikipedia: Copyrights—Wikipwdia, the free encyclopedia.* Retrieved March 30, 2006, from http://en.wikipedia.org/wiki/Wikipedia:Copyrights

KEY TERMS

Copyleft: Copyleft is a kind of licensing mechanism, with which licensees have to apply the same license the original works adopted to the derivative works.

Creative Commons Licenses: Creative Commons licenses are a kind of licensing model which applies to open content. Creative Commons licenses are composed by four elements (attribution, noncommercial, no derivatives, and share alike). Creative Commons licenses allow the licensees to make use of CC-licensed works with no need to get prior permission from the licensors as long as the licensees follow the conditions the licensors chose for the works.

License: It is a legal permission to commit some act.

License Compatibility: It is an abstract idea to illustrate whether two portions of content regulated by two different licenses can be combined within a work compatibly and produce the other resulting work.

Open Access Publishing: Open access publishing is a kind of publishing model, under which journals open access to the public immediate on publication and usually the authors of the journal articles do not need to pay the page fee for the publication.

Open Content: Open content describes the creative work which allows copying and modifying with no need to get extra permission from the licensors, such as works licensed under Creative Commons licenses.

Open Source Software Licenses: Open source software licenses apply to open source software. Open source software licenses feature that licensees can use, copy, distribute, and modify the regulated software on a royalty-free, worldwide basis.

ENDNOTES

1   Article 2(b) of the GPL stipulates that "You (licensee) must cause any work that you distribute or publish, that in whole or in part contains or is derived from the Program (GPL-applied program) or any part thereof, to be licensed as a whole at no charge to all third parties under the terms of this License (GPL)."

2   OSD # 1 states that an OSS license should not "restrict any party from selling or giving away the software as a component of an aggregate software distribution containing programs from several different sources."

3   OSD # 3 states that "The license must allow modifications and derived works, and must allow them to be distributed under the same terms as the license of the original software."

4   Open Access Law Project is established under Science Commons' publishing project to promote open access to legal scholarship. For more detailed information about Open Access Law Program, please see http://sciencecommons.org/literature/oalaw

5   Our "no endorsement" wordings are motivated by BBC's Creative Archive License. The detailed terms can be found on http://creativearchive.bbc.co.uk/licence/nc_sa_by_ne/uk/prov/.

6   BBC proposes five rules for Creative Archive Group License, which comprises "non-commercial," "share alike," "crediting" (attribution), "no endorsement and no derogatory use" and "UK." The first three rules are very similar to CC's; the last two are innovations created by BBC. For more details, please see http://creativearchive.bbc.co.uk/archives/2005/03/the_rules_in_br_1.html

7   "Modification record" in this chapter includes several possible meanings, e.g., the record about who did the modification; the record about when the modification was made; the record about which part of the original programs has been changed.

8   See article 2,3 of QPL.

9   According to Andrew M. St. Laurent's opinion, QPL's requirement that a licensee distributes modifications separately with the initial work can protect the reputation of the initial developers and make clear the primacy of the initial developers' works. See Andrew M. St. Laurent, "Understanding Open Source & Free Software Licensing" (2004, p. 87, O'Reilly). In this chapter, we further extend St. Laurent's viewpoints and come up with the new element "modification record" for OSS's licenses.

Chapter XXXI
FLOSS Legal and Engineering Terms and a License Taxonomy

Darren Skidmore
Monash University, Australia

ABSTRACT

This chapter introduces the reader to terms relevant to understanding free/libre and open source licenses, some of the relevant legal, and relevant software engineering terms that are useful in understanding the issues in FLOSS. Then a brief history of FLOSS licenses is given before introducing a taxonomy to help understand the types of licenses which are available in the FLOSS domain. A brief description to think about differing views of the usage and users of FLOSS is given in conclusion.

INTRODUCTION

The purpose of this chapter is to explain some of the issues which free/libre open source software (FLOSS) licenses are attempting to address, although it should be noted at the outset that these also apply to any type of software license. The chapter firstly discusses the legal terms applicable in intellectual property with an emphasis on FLOSS. To complement the legal issues, discussion turns to software terms and their definitions as part of software development and engineering. Having defined the two areas pertaining to the FLOSS licenses, a brief history is given before discussing a taxonomy of FLOSS Licenses. The chapter concludes with a brief discussion on how the view of user of the FLOSS may change the need for a type of license. The purpose of

explaining the legal and software engineering terms is because if a person does not have a background in these areas, then it is unclear as to why licenses, or debate about the outcome of licenses, are being made. The taxonomy is given to help readers understand that there a several license types, and to assist them in their choice of a license or in understanding the outcomes attached to a license.

As more organisations adopt or consider FLOSS, there is a greater need to understand at a more generic level the broad aims or outcomes of the effects of the licenses. The Commonwealth of Massachusetts compares 52 different licenses (Commonwealth of Massachusetts, 2004), the Open Source Initiative (OSI) compares over 58 (Open Source Initiative, 2004b), while the Free Software Foundation (FSF) lists and comments on almost 100 (Free Software Foundation, 2005a), and the ifrOSS lists over 180 different open source licenses (ifrOSS, 2005). Each independent license has different conditions and outcomes; legally it is important to understand the clauses in a specific license, but before that there is a need to understand the broad aims of the license, and to match that to the organisational needs and requirements, of an application. Certainly in terms of ICT governance a taxonomy of FLOSS licenses helps to match the organisational strategic and tactical aims with the operational choice of the specific license. There has been comment that there are too many open source licenses (Skidmore, 2006), with the OSI looking into the proliferation of licenses (Open Source Initiative, 2005), and comment that it is not that there should be less licenses, but that there should be a cleaning up of the terms used and agreement on how to word specific desired outcomes (Rosen, 2005).

The term FLOSS is used to describe free/libre and open source software. The word *libre* is specifically included to emphasis the concept is about freedom, rather than price. Also within this chapter, when referring to an application, the term can include a computer program, which could be a word processor, a Web browser, or an email program, but also an operating system such as Linux. Although there is a distinct difference in terms of what functions these different types of programs do, the issues in licensing are, in the main, similar. The term FLOSS means F/LOS-software, so when taking about the software just the acronym FLOSS is used.

Before discussing the licenses, it is important to firstly explain some of the legal issues which are trying to be addressed by the licenses, not all licenses with each of these issues, some of the legal issues did not really exist when some were created, or were not considered by the authors of the licenses. Complementary to the legal issues are the software engineering terms which also influence the licenses. Certainly debate about

new or updating FLOSS licenses are aimed at issues in software engineering or software usage which are being practiced now. In reading the chapter readers of course should be aware of the changes in the last four decades in both software engineering and in legal jurisprudence which, because they are constantly changing, do have implications for the FLOSS licenses.

BACKGROUND: LEGAL TERMS

The legal terms in Table 1 are not an exhaustive list of either the terms or of the scope of these terms, nor is the full complexity of the issues of applying to software treated, however the list does include the more important terms and states some of the more critical issues. Table 1 gives an overview of the terms that will be covered in this chapter. Although it is possible to sell an entire program, the most common form of transfer in software development is that of licensing (von Krogh & von Hippel, 2003), which is why the aspects pertaining to licensing are focussed on in this chapter. There are several works which are dedicated to the legal issues in software and explanations in depth of the issues (Rosen, 2004; St. Laurent, 2004; Välimäki, 2005).

Table 1. List of legal terms discussed

Intellectual Property terms
Author / Owner
Copyright / Patent
Trademark
Derivative Work
License / Contract
Jurisdiction terms
Choice of Forum
Choice of Law
Other Legal terms
Consumer Warranty
Export Control
Distribution
Written Language
Reasonable and Non Discriminatory Licensing (RAND)

Intellectual Property Terms

Author/Owner

Any intellectual property (IP) has both an author(s) and an owner(s), these can to be the same person, but often can be separate entities. With intellectual property except in the case of moral rights, it is the owner of the IP, who has control of that IP, not the author. Source code is covered under copyright law as with most written text. Although there is some jurisprudence specific to software, the same copyright laws apply to software as they do to the writing of a novel or scientific paper, or to a performance in the theatre.

Copyright/Patent

There is a vast difference between copyright and patent; which has important for implications for any software, no matter how it is licensed (Skidmore & Skelly, 2003). The critical difference to understand is that under copyright it is the authors' specific expression of the idea that is applicable and therefore controllable under copyright, whereas in patent, it is the idea itself which is controllable. Under copyright, as long as two or more authors have independently expressed the same thing, this is (simplistically) legal, or if an idea is expressed in one way, it may be possible to express it in a different way and not to infringe another's copyright. However, in patent law if an idea is held under a patent, even if the idea is expressed in a completely different language or even a different way, there is still infringement.

Copyright applies worldwide (or in any TRIPS nation [World Trade Organization, 1994]) and exists at the point of creation, whereas a patent must be granted in each separate jurisdiction, and the conditions under which a patent will be granted varies. In most jurisdictions the term for copyright is generally 70 years after death of author or 90 years for a corporation, whereas the patent term is for 20 years. Another difference is that it is possible

for a patent to be granted in the U.S. but not in the European Union, the patent holder may also be a different person, in different jurisdictions. This can lead to a situation where software which is freely usable in Australia violates a patent in the United States, and cannot be used freely in the U.S. A good example on the issue of patents in different jurisdictions is that of the Blackberry device, which is a personal data assistant (PDA) that is popular with executives to access their calendars and email while out of the office. The Blackberry network was under serious threat of being shut down in the U.S., which is a major market for the device, because of a patent dispute between RIM and the holder of a patent which claimed that the Blackberry infringed their patent, eventually settled after a long and drawn out legal battle. The patent, which was only valid for the U.S., meant that RIM could operate in any other jurisdiction in the world, but not in the U.S.

The issue of patents in software is becoming more important especially as now patents can be obtained for both business methods and for software (American Intellectual Property Law Association, 2000). Patents are written in a broad technical and legal manner, with business method patents can be checked by people without expertise in technology (Cohen & Lemley, 2001). It is difficult to search out and determine if the software is infringing a patent, if a search is ever done. There is also the risk of inadvertent additions of code which infringes a patent; if the code infringed copyright, then it could be rewritten, but when infringing a patent, the either the idea has to be removed or a royalty negotiated.

There has also been evidence of predatory actions by patent holders, both with ambit patent claims, preventing access to ideas or where companies participate in work on the development of industry standards, only revealing in the later stages that a critical patent is held (Kirk, 2002; Soat, 2002; United States Federal Trade Commission, 2003). To counter this some companies have declared that open source projects are able to use some of their patents and that they will also use their patent portfolios to respond to patent infringement demands (IBM, 2005a, 2005b; Novell, 2004; Open Invention Network, 2005; Red Hat, 2004).

Trademark

A trademark, in simple terms is where there is a recognisable brand. The protection of a trademark can be seen as important, in the case of FLOSS, some licenses, such as the Apache license state conditions that give the right to use the code base, for other purposes, but specifically disallow people for using the Apache name or similar branding.

Derivative Work

A derivative work within the context of software is a new or modified program, building on or using part of another work. One issue that arises is that some jurisdictions there is no legal definition of derivative work, so this definition needs to be written into the specific license or contract, it may in some cases be implied by the courts, but this is not certain. When some licenses were written, issues such as derivative work were not defined, because there was no need for this in that jurisdiction.

License/Contract

A license is permission by the owner of the intellectual property, for another to use that IP, if the subsequent user does not follow the conditions of the license, redress can be taken by the IP owner, under copyright law. A contract is an agreement between two parties on the conditions of use and the governing relationship between the property of the two parties. Redress can be sort under contract law. The classic conditions for a contract to be valid require that there must be an offer, an acceptance, and consideration. In many jurisdictions, consideration must be exchanged by both parties, in other words, both parties must give the

other party something of value, for a contract to be formed, which can be an issue in FLOSS, where there is not always value exchanged.

Jurisdiction Terms

When considering legal issues the jurisdiction of the country needs to be considered, especially in terms of software, which can easily be transferred from one location to another, which can easily be in a different legal jurisdiction.

Choice of Forum

Choice of forum, simply means, the jurisdiction in which the matter will be decided by the courts and law. This could be under the laws (and probably the courts) of the European Union, Victoria Australia, or New York in the U.S. The choice of forum can be written into the conditions so that there is legal certainty about which jurisdiction the case will be brought, or to ensure that the case is not considered under another forums law. The European Union discussed the creation of an EU FLOSS license partially for reasons of forum (Dusollier, Laurent, & Schmitz, 2004), a license has been written with governing law as being French (CeCILL, 2005a, 2005b), and NICTA in Australia has created a license for that jurisdiction (National ICT Australia, 2004).

Choice of Law

Choice of law, is a separate issue to choice of forum. The choice of law states under what type of law the case will be heard, such as the law of tort or the law of contract. A choice of law clause in software may also be used to say what it is not (Rosen, 2004), describes a FLOSS license, where the conditions specifically state that a law on the sale of goods is expressly excluded, because of the desire to treat the FLOSS as intellectual property rather than as goods.

Other Legal Terms

Consumer Warranty

In many jurisdictions, the consumer warranty laws, cannot be contracted away, and any clause which does this is void (Fitzgerald, 2003; Rosen, 2004; St. Laurent, 2004). Modern license, have added clauses to the licenses that allow for such consumer laws and situations.

Rosen (2004) raises several issues to do with the wordings used in FLOSS licenses. In licenses, conditions can be express (i.e., spelled out) or implied (i.e., not spelled out, but could be presumed), the older licenses for example, do not explicitly deal with the issues of patents, consumer warranty, and derivatives, which increases the reliance on implied terms.

Export Control

Export control has occasionally been a major issue in software. Encryption technology is considered by some countries, under the Wassenaar Arrangement[1], as munitions or a dual use good and technology (i.e., a bullet, or tank, or in this case a technology), which require an export permit to cross the national boundary. For many years, in some countries, software applications which contained encryption algorithms required a permit to be exported, the permit controlled what was allowed and what countries the application was permitted to be exported. This was the situation in the U.S., until the year 2000 (Electronic Privacy Information Center, 2000). Therefore some licenses contain clauses stating the software is required to comply with U.S. export laws.

Distribution

The term distribution in FLOSS has a specific meaning in that distribution occurs when the source code or application is given to an entity, outside the organisation who modified the code. If the modified application is created for a specific company, then generally that company can use the application for internal use. Only if the company gives the application or source code to an outside entity is the software then seen to have been distributed, depending on the terms of the license, the source code may then have to be shared with any who ask. This has never been tested in a court of law, nor is it certain if there is distribution when sharing between large multi-business units in organisations or with a federal government sharing with state government agencies.

Written Language

Depending on the jurisdiction there maybe a requirement that the legal documents, be in a specific language or that the parties have to specifically agree that the language of the document shall be in another language, such as the Squeak License (Apple Computer Inc., 2005) for the province of Quebec.

Reasonable and Non-Discriminatory Licensing (RAND)

Reasonable and non-discriminatory licensing (RAND) is a term used mainly in standards setting bodies, such as the W3C, where if an organisation wishes to have their IP used in a standard then they must make the IP available under reasonable and non-discriminatory terms. Although not perfect, this allows companies to participate in standards processes and not risk being locked into or out of the required adherence when a standard has been set.

Legal Test of Open Source Licenses

There has been one legal case, in Germany[2], which has upheld the FLOSS license conditions of the GNU GPL, and also an organisation, which

Table 2. List of software terms discussed

> Source Code
> Object / Binary Code
> Codebase
> Stack
> Documentation
> Software Bugs
> Linking / Software Libraries
> Remote Procedure Calls / Web Services
> Embedded System

tracks and enforces the conditions of the GNU GPL license, GPL Violations.org[3].

BACKGROUND: SOFTWARE TERMS

Source Code

Source code is the code which is written by programmers or software developers. The software developers will create the source code using a programming language, such as C++, Java, Bash, .NET, PHP, or SQL. Source code is human readable, and by convention contains comments to explain the code, and relevant variable names. Source code is complied into byte/object/binary code so it can be run on a computer. Depending on the programming language the compilation can be a completely separate computer file (application), or the code can be interpreted (compiled) at run time.

Object/Binary Code

The object code or binary code is the machine readable code, which has been complied into a form that can be executed or run by the computer. The process for converting from source code to object code is generally one way, although it is possible to recompile back to source code from object code, the recompiled source code will not have meaningful comments, or variable names,

making it harder to maintain and understand the workings of the program.

CodeBase

A larger software project might contain several separate applications or sections of code which have to be maintained. Some of these may be dependent on other sections or applications to run. As well in larger projects there might be different versions of the source code which are at various stages of testing or completeness this also includes proposed changes, and modifications. The term CodeBase is used for these collections of source code generally held in a central repository.

Stack

In software engineering there are many definitions of a stack, from the allocation of memory through to the collection of programs or subsystems required for a solution. In this chapter, it is the later definition which is taken. Many applications require support from other programs, applications, or libraries, which fit into the required software stack. If these applications or libraries are not available or the licensing fees make them uneconomical then this effects the dependent applications.

Documentation

Documentation does not mean the inText documentation or comments within the source code, but the user manuals, technical documentation, testing data, testing documentation, FAQs, and other documentation which can accompany software. Documentation has intellectual property considerations separate to those of software. The licenses applicable to the documentation of software documentation would generally be different to that of the software, due to different issues that need to be addressed.

Software Bugs

All software has bugs, some bugs are minor, some are major, a bug is a flaw in the program which prevents it from behaving either as it should, or in a manner that it should not. Among other reasons, the CodeBase of an application will change due to the maintenance to resolve bugs. In actively used programs thought must be given to the ongoing management of the CodeBase. In the cases of inactively used programs, a person or organisation might need access to the CodeBase to fix the errors themselves.

Linking/Software Libraries

Linking to software libraries is a common practice in software programming. Linking is done where there is a need to carry out a function or to return a result. A software library is efficient for this for several reasons, one is that the function or result maybe a commonly needed one, so doing it once rather than repeating it several times is worthwhile, this is also an efficient reuse strategy. Another reason to do this is that if there is a need to update the function then only one place needs to be updated. An example of this might be the calculation of pythagoras theorem, where rather than the code being separately put into several applications. The code could be put into a mathematical library, and then called upon by the applications, when needed. Linking to software libraries is generally done intra-computer, so the libraries and the application would be on the local hard drives of the machine running the application. In FLOSS, software libraries may have some issues, because if the functions are in a software library, the requesting application is using the source code (in the library) to produce a result but because this is not actually incorporating the actual source code into the requesting application and it is not clear what the implications are both philosophically and legally under some licenses.

Remote Procedure Call/Web Services

Similar to linking to a software library, a remote procedure call is a request by a software application for a task to be done, generally made to a machine or service that is remote to the requesting machine. Several standards have tried to address some of the complex issues to make this work, and make it easier for software programmers to create such calls, such as RMI, DCOM, and CORBA. A new protocol Web services is a further refinement of these standards and concepts. A Web service may provide information, for example, a stock price, or the result of a calculation, but could also carry out more complex procedures, such as Income tax calculations. Similar to software libraries, there may be an issue with some FLOSS philosophies in that it is possible to write a Web service application, which uses FLOSS source code, where the software services are going outside of the boundary of the organisation, but where the application itself is not distributed outside of the organisation. Therefore since the company has not distributed the application outside of the organisation, they are not required to share the new or improved source code with any who ask (Marson, 2005).

Embedded System

The definition of an embedded system can be used in different ways, but in general an embedded system, will be a system built for a specific purpose, as opposed to a system that can be used for many purposes or a general device. Examples of specific purpose devices are traffic light controllers, printers, routers, and lift controller. The personal computer is an example of a general device. The term embedded system is also used in the PDA and mobile phone devices although theses devices sometimes have functionality like PCs. Although in the past an embedded system used purpose-built components, there is a growing use of building an embedded system from a combination of hardware and software components, but limiting the functionality to that of the required device. For instance, some printers are using the operating system Linux (or a distribution of Linux) to carry out their functions.

BACKGROUND: HISTORY OF FLOSS

The classic traditional starting text for FLOSS is Raymond (2001). Another book as an alternative view, Wayner (2000), describes the personalities and schisms in the FLOSS community. A DVD, Moore, Wonderview Productions (Firm), and Seventh Art Releasing (Firm) (2003), is a documentary with an historic account of Linux and open source which includes interviews with many of the personalities in the FLOSS community.

For some people, open source is a philosophy and the licenses are merely the technical and legal method by which this philosophy is enacted. As part of the philosophy, there is a belief that any person should be able to access the source code, to learn, modify, and use as they wish. For others, FLOSS is the best choice in terms of engineering, and the use of a FLOSS license is a pragmatic decision.

The Free Software Foundation[4] (FSF), was created by Richard Stallman, who is the author of the signature open source license, the GNU General Public License[5] (GNU GPL) (Free Software Foundation, 1991). The FSF strongly believes that software should be free. They created the phrase "free as in speech, not as in beer." The FSF philosophy is governed by the four freedoms[6]:

- *The freedom to run the program, for any purpose (freedom 0).*
- *The freedom to study how the program works, and adapt it to your needs (freedom 1). Access to the source code is a precondition for this.*

- *The freedom to redistribute copies so you can help your neighbour (freedom 2).*
- *The freedom to improve the program, and release your improvements to the public, so that the whole community benefits (freedom 3). Access to the source code is a precondition for this.* (Free Software Foundation)

The GNU GPL contains a provision mandating that when source code is either modified or is taken from code which is licensed under the GNU GPL, that this new source code must be licensed under the GNU GPL, and if distributed, made available to others in source code form, upon request. This applies, theoretically, even if a single line of code is used. For this reason the GNU GPL is referred to by some as being viral or by others that the license will "propagate" (New Zealand State Services Commission, 2006). The term that was coined by the FSF and that is used to describe the effect of the GNU-GPL is copyleft. CopyLeft is the condition where a license requires that subsequent modifications and extensions to the application are made free as well (Free Software Foundation, 2005b).

Traditionally there were two licenses used in open source—the GNU GPL and the BSD license—although when referring to the BSD license, what is meant is a BSD style license, rather than the actual BSD license. The Berkley Software Distribution (BSD) was an operating system created at the University of Berkley in California. A BSD type license does not require that the modified application or applications that BSD licensed code is placed into be licensed under the BSD: the resulting application can be released under any license, including a CSS license. The best known example of this is Apple Mac OS X, which is based upon and borrows code from Mach, FreeBSD, and NetBSD. Apple took the code and created Darwin, which was the basis for OS X (Michaelson, 2004). The only requirement of the original license was to give attribution to the authors of the source code. The

Table 3. Open Source Initiative Open Source Definition (Open Source Initiative, 2004a)

```
1. Free Redistribution
2. Source Code
3. Derived Works
4. Integrity of the Author's Source Code
5. No Discrimination Against Persons or Groups
6. No Discrimination Against Fields of Endeavour
7. Distribution of License
8. License Must not be Specific to a Product
9. License Must not Restrict Other Software
10. License Must be Technology-Neutral
```

BSD community believe that code should be free, they view the GNU GPL as a contradiction in terms, in other words, that the GPL code is free except of the GPL (Wayner, 2000). There are also arguments that, because originally, the code was developed under public funding, that the public should have access to use the code in any way they wish, including the ability to exploit the work for commercial gain.

The Open Source Initiative (OSI) was formed in 1998 to market the concept of FLOSS to a wider community. The founders of the OSI believed that the attitude of the FSF was preventing organisations from adopting open source software. FLOSS was a pragmatic, software engineering view that was a valuable development methodology, not just a philosophy. The OSI could not trademark or control the term open source, therefore they created a trademark that could be used to certify that a license was compliant against a series of conditions. The OSI open source definition is briefly listed in Table 3 (Open Source Initiative, 2004a). All of the FSF licenses are OSI compatible licenses, as are the BSD style licenses. The OSI mark is only a certification of the OSI conditions; there are many licenses which claim to be open source that do not have an OSI certification.

In the domain of open source, the FSF is associated with the term *free* and the OSI with the term *open source*, with open source referring to licenses certified against the OSD. However, there are certainly licenses which refer to themselves as open source which do not conform to the OSD.

Analysing the licenses used in SourceForge, Weiss (2005) found the GNU GPL was used for 45% of the projects with 7% using the GNU LGPL, 5% BSD, with 22 other licenses sharing the remaining 43%. The GPL is the major FLOSS license and its aims are well liked, although the large figure can be partially explained because of the reciprocal obligations of the GNU GPL. Examining SourceForge project data from April 2006 using the FLOSSmole Query Tool[7] gave results of 66.9% GNU GPL, 10.3% GNU LGPL, and 7.0% BSD, with the remaining 57 licenses totalling 15.8% of the licenses used in the projects hosted on SourceForge.

MAIN FOCUS OF THE CHAPTER: SOFTWARE LICENSE TAXONOMIES

The term open source software has changed from what was once a reasonably simple term. Although FLOSS was used and exploited commercially, this was generally only in special sophisticated environments. However, the term has been appropriated and widened to include other meanings and agendas. This has been caused by many factors, including the growth and maturity in FLOSS, changes in jurisprudence, the need to address new issues in software engineering, changes in vendor strategy, and the need for companies to either protect their intellectual property or to market product.

The only aspect in common agreement between the conditions in the various licenses seems to be the ability to view the source code, after this the variation is almost endless. Therefore as a way of understanding the various types of licenses there is a need to organise them into rough. A taxonomy of the types of license has been developed to distinguish and place the license into various types, by outcomes of the licenses.

Table 4. Taxonomy of FLOSS licenses

Traditional Open Source Licenses
    Reciprocal Licenses
    Non-Reciprocal Licenses
    Linking Licenses
    Dual Licenses
Quasi Open Source Licenses
    Obligation Licenses
    Morality Licenses
    Viewable Source Licenses
    Membership Licenses
Open Source Support Licenses
    Content Licenses
    Open Standards Licenses
Public Domain
Closed Source / Proprietary Licenses

The taxonomy to consider licenses in can be seen in Table 4, the licenses are considered at two levels, this is because there is a need to separate the licenses out into four broad areas before discussing the classifications inside of those areas. The first broad area is traditional open source, which includes the licenses approved by the FSF and the OSI. Dual licenses are also included in the traditional open source. The next area is that of quasi open source licenses, which have an open source component, but generally have restrictions on the use or that place obligations on the consumers of the source code. An area for open source support licenses has been created because there is a need to specifically address issues and artefacts associated with FLOSS where the artefact is not source code. Although strictly not a FLOSS area, the public domain and closed source licenses complete the range of licenses.

Traditional Open Source Licenses

In traditional open source software, the source code is available for use by others. The source code is both visible and the source code can be added to, modified, or sections of the source code can be used in other programs. Limitations on

the resulting source code and application may exist, for example under the GNU GPL the new source code must be licensed under the GNU GPL. There may also be a dual license choice available, where a choice can be made between participating in the open source development or create a closed source application.

Reciprocal Licenses

The reciprocal licenses enforce that any resultant source code, which either borrows the code from another codebase, or adds to a CodeBase, must be licensed under the original source code license. Examples of these licenses are the GNU-GPL, the Common Public License[8] (CPL), and the European Union Public License[9] (EUPL). A reciprocal license does not mean that the license is the same as the GNU GPL or is considered by the FSF as a free license, but generally they have similar characteristics to the GNU GPL.

The GNU Lesser General Public License[10] (GNU LGPL) is a special license created by the FSF specifically to allow the linking by applications to software libraries. See the definition of linking licenses for more details on the GNU LGPL, however simply the GNU LGPL has characteristics of a reciprocal license for code that is used from GNU LGPL source, but can be linked to without creating the reciprocal conditions required by the license.

Not all reciprocal licenses are considered to be "free" by the FSF. For example, the CPL is not considered to be compatible with the GNU GPL license by the FSF because the license has some requirements, including in the way it deals with Patents, that make it incompatible with the GNU GPL. The conditions are not seen as bad by the FSF, just incompatible with the GNU GPL

Non-Reciprocal Licenses

A non-reciprocal license is similar to the original BSD license in that there is no requirement in the license that any derivative work must be licensed under the original license. Some other clauses may exist such as the need to give attribution, protection of trademark, or governing patent claims. In the context of this taxonomy a non-reciprocal license allows for the new source code to be used in anyway that is wished including making the software closed source. An example of a non-reciprocal license is the Academic Free License[11] (AFL). The AFL and the CPL are almost identical licenses, in terms of the clauses, the differences between the two is the conditions which make the CPL reciprocal and the AFL non-reciprocal.

Linking Licenses

Linking licenses are licenses where the terms of the license allow for other applications, or code to link to them, but do not require the linking application to be licensed under the license of the linked application. Linking licenses are useful in that the application licensed under the linking license can be used in the software application stack, and a vendor or programmer is able to use the application with other software applications, including proprietary licensed software, without risk of being forced to release their other source code under the linking license. Currently, the best example of this is the GNU Lesser GPL, or GNU LGPL (Free Software Foundation, 1999), and the license was created specifically so that applications could link to the GNU LGPL[12], because of the concern of some that if the GNU GPL was used then the linking program would be required to be then licensed under the GNU GPL.

To explain this better, take for example if Application Alpha was licensed under license XYZ and it linked to Application Beta, which was licensed under the GNU LGPL. This is perfect in keeping with the conditions of the GNU LGPL. If, however, Application Beta incorporated source code from Application Alpha, then under the conditions of the GNU LGPL, then Application Alpha would be required to then be licensed under the GNU

LGPL. This is because of the license conditions of the GNU LGPL: another type of linking license might have different conditions.

Dual Licenses

The dual license is not strictly speaking a specific license; this is where the owner of the intellectual property can license the IP under different licenses to different people or for different conditions. Therefore it is possible to give the FLOSS community a license that applies to FLOSS but also to license the same application under a commercial license to others. Some research has been done on the use of dual licenses (Välimäki, 2003). This is the case with database MySQL, where the company MySQL will allow an organisation that does not wish to use the MySQL database under the GNU GPL to pay for a separate license. A license known as the Sleepycat License[13] has similar properties, in that if the new application is open source, then it is treated as open source. If it is a commercial product, then a commercial negotiation must occur. The Mozilla Public License[14] is a reciprocal license but also has the option to be a dual license, in that the initial developer can stipulate a second license that may be used for licensing the application, although this is typically the GNU-GPL.

It is vitally important to understand that if a person or company wishes to use a dual licensing system, they must own the copyright in the source code, or have control of the intellectual property in the source and software so that they can create the dual license arrangement. Governance and management of the code base is critically important.

Quasi Open Source Licenses

The quasi open source licenses have taken on some of the characteristics of the traditional licenses, but have conditions that modify the levels of control or distribution of the licensed source code.

Obligation Licenses

Obligation licenses either impose restrictions on the modifications of the source code, such as how the modifications may be distributed, or give special privileges to the licensor. Generally, these are used by vendors where they may be trying to control the distributions for compatibility reasons, the Sun Community Source License[15] (SCSL) or for the vendor to be able to use any resulting innovation in their products, such as the Netscape Public License[16]. Microsoft has a license which is actually a copyleft style, but limits the software to being used only on Microsoft Windows Operating Systems the Microsoft Limited Permissive License[17] (Ms-LPL), thus creating the obligation, and also disallowing it from being an OSI certified license. The Ms-LPL is a subversion of the Ms-PL (Microsoft, 2005), which does not contain the obligation condition.

Morality Licenses

Morality licenses are licenses which include provisions that preclude users of the software, and/or those who wish to use sections of the source code, from using it for certain purposes. The source code is still available to be viewed, used and modified, but has limits on the uses. Because these licenses exclude fields of endeavour they are not able to be certified by the OSI, nor are they considered to be Free in terms of the FSF. The Hactivismo Enhanced-Source Software License Agreement (HESSLA) license specifically states that the program or code cannot be used to violate Human Rights (Hacktivismo, 2005). The Xineo freeware license states the program may not be used for commercial or military purposes (Xineo.net).

Viewable Source Licenses

Some licenses allow little more than the ability to view the source code, so a section was created called viewable source. The Microsoft Reference License[18] (MS-RL) permits only the viewing of the code for reference purposes.

Membership Licenses

Although there are no current examples of membership licenses in FLOSS, there are examples of the use of the concept. A membership license would allow the source code or other IP artefacts to be shared amongst a membership set. Organisations such as the Avalanche Corporate Technology Cooperative[19] and the Government Open Code Collaborative[20] have been setup to share IP between their members. The concept behind the collaboration is that these organisations are consumers of IP and wish to draw upon others experiences to implement and get value from ICT.

Open Source Support Licenses

Content Licenses

FLOSS is just source code, however, there is a need for documentation, and there may also be a need for the application to implement or comply to standards, therefore a category of open source support has been included. Content licenses are licenses which apply to content, generally to documentation, but can apply to other forms of content. The FSF have created a license for content to match the GNU GPL called the GNU Free Document License[21] (GFDL). There is also the creative commons license suite of licenses, which allows a copyright owner to choose how they share their IP. The creative commons has also created the infrastructure to support the IP owner and consumer in easy communication and control of the IP (Creative Commons, 2005).

Open Standards

Open standards are different to FLOSS in that standards are created as references so that software applications have interoperability when transferring data, such as EDI, XML, word processor documents, or even TCP/IP network traffic. An open standard is a standard which is accessible on reasonable and non-discriminatory (RAND) terms, compared to closed standards, where the vendor either does not release the details of the data standard or the information is only available within a closed consortium.

Public Domain

It is possible to place source code into the public domain, where anyone can take the source code and use it as they wish. Although sometimes confused with open source, the public domain is different: at the very least the owner may not place other restrictions as they wish on the code and still have it free.

Closed Source/Proprietary

Closed source licenses, are considered to be those which prevent access to the source code, although this is extremely simplistic, and only for the purposes of comparing in terms of being able to view the source code. There are many different types of closed source licenses which have a variety of conditions, this should not be forgotten or glossed over in the comparison with FLOSS.

FUTURE TRENDS: VIEWS OF FLOSS LICENSES

Traditionally, although FLOSS software was available to anyone, it was programmers and technically savvy people who participated and consumed the source code and software. However, as the number of users of FLOSS licensed software grows, including developers, vendors, end users, and applications, there is the need to consider that some types of FLOSS licenses have different advantages and disadvantages for different types of users or business strategies. Consideration of the type of license used, or adopted changes depending on the needs of the entity that wishes to use or develop the application.

There is a conventional view that the GNU GPL is the most free software license in the choices available, however if you are a developer who wishes to use code the best choice may not be the GNU GPL (Gacek & Arief, 2004; Michaelson, 2004). The developer's best choice may be to use BSD licensed source code as they can use the code for any purpose, with out obligation to contribute the code back to the community.

Consumers may not care to use the ability to use modify the code or so it may be irrelevant as to the type of FLOSS license governing the application. As a consumer or organisation, they might not care if others have the same software, but do require that they be able to continue to use the software. Rather than placing software into escrow, there is the choice of using FLOSS. There are risks in this scenario, in that a consumer maybe forced to participate in the FLOSS development to influence their software requirements (Edwards, 2005). However similar costs maybe applicable if the software is required to be maintained in any case. Software being used in a business process may require that it is not a copyleft style as they may not wish to return code to a competitor; alternatively an organisation may wish to gain from sharing ideas or business risk with the community. There has also been an increase in the use of open source software in embedded devices, or the use of an open source license to distribute the code in embedded devices, for which reciprocal licenses would force the disclosure of the code, but also that non-reciprocal licenses can be used to disseminate source code, ideas, and reputation.

There maybe reasons other than profit or costs in the selection of certain types of licenses. A European Union report discusses that the EU should use a copyleft style license because EU citizens or organisations should be allowed to use source code developed for the EU since their taxes have already paid for the software. This includes that others should not be able to take the source code and to create software and then sell it back to EU citizens as they have already paid for it. Therefore the EU is considering the use of a reciprocal style of license (Dusollier et al., 2004).

CONCLUSION

This chapter has listed a brief list of legal issues as well as an associated list of software engineering terms, and given a taxonomy which can be used to describe the broad characteristics of the FLOSS License. Understanding and consideration of all three are needed when using FLOSS, either for use internally, for just the application, or in expanding. It should also be remembered that non-FLOSS has similar issues associated with their use, and that in using any software, that there are licensing conditions applicable to that software which has to be taken into account. Just because it is FLOSS does not mean it is good nor just because it is not FLOSS does it mean that it is simple to comply with the license conditions.

Attention has also been drawn to the legal and software terms that are relevant in discussions about FLOSS. Some issues such as patents and linking to remote services will have an ongoing effect on the development of new licenses or conditions of licenses. Though the taxonomy only provides a guide, it is still extremely important that any specific choice be made after considering in depth the actual conditions of the license, as well as the business needs of the organisation or person from the ICT.

REFERENCES

American Intellectual Property Law Association. (2000). *Patenting business methods.* Retrieved 2003, from http://www.aipla.org/html/whatsnew/patentingbusiness2.pdf

Apple Computer Inc. (2005). *Squeak license.* Retrieved September 30, 2005, from http://squeak.org/download.license.html

CeCILL. (2005a). *CeCILL Free Software License Agreement version 2.0—English.* Retrieved November 11, 2005, from http://www.cecill.info/licences/Licence_CeCILL_V2-en.html

CeCILL. (2005b). *Contrat de licence de logiciel libre CeCILL version 2.0.* Retrieved November 11, 2005, from http://www.cecill.info/licences/Licence_CeCILL_V2-fr.html

Cohen, J. E., & Lemley, M. A. (2001). Patent scope and innovation in the software industry. *The California Law Review, 90*(1).

Commonwealth of Massachusetts. (2004). *Open source licenses: Quick reference chart.* In http://www.mass.gov/itd/legal/quickrefchart.xls (Vol. 44.5 Kb, pp. Chart in Spreadsheet format [.xls] of ~50 Open Source Licenses and their attributes). Commonwealth of Massachusetts.

Creative Commons. (2005). *Licenses explained: Creative commons.* Retrieved September 11, 2005, from http://creativecommons.org/about/licenses/

Dusollier, S., Laurent, P., & Schmitz, P-E. (2004). *Open source licensing of software developed by the European Commission* (Report). European Commission.

Edwards, K. (2005). An economic perspective on software licenses: Open source, maintainers and user-developers. *Telematics and Informatics, 22*(1-2), 97-110.

Electronic Privacy Information Center. (2000). *Cryptography and Liberty 2000: An International Survey of Encryption Policy.* Retrieved November 8, 2005, from http://www2.epic.org/reports/crypto2000/

Fitzgerald, B. (2003). *Legal issues relating to free and open source software* (Vol. 1). Brisbane,

Queesnland, Australia: Queensland University of Technology School of Law.

Free Software Foundation. (2004). *The free software definition*. Retrieved August 1, 2004, from http://www.gnu.org/philosophy/free-sw.html

Free Software Foundation. (1991). *GNU general public license Version 2*. Retrieved August 1, 2004, from http://www.gnu.org/copyleft/gpl.html#SEC1

Free Software Foundation. (1999). *GNU lesser general public license Version 2.1*. Retrieved August 7, 2006, from http://www.gnu.org/licenses/lgpl.txt

Free Software Foundation. (2005a). *Licenses*. Retrieved November 7, 2005, from http://www.fsf.org/licensing/licenses/

Free Software Foundation. (2005b). *What is copyleft?* Retrieved August 1, 2004, from https://www.fsf.org/licensing/essays/copyleft.html

Gacek, C., & Arief, B. (2004). The many meanings of open source. *Software, IEEE, 21*(1), 34-40.

Hacktivismo. (2005). *The hacktivismo enhanced-source software license agreement*. Retrieved September 30, 2005, from http://www.hacktivismo.com/about/hessla.php

IBM. (2005a). *IBM statement of non-assertion of named patents against OSS*. Retrieved January 20, 2005, from http://www.ibm.com/ibm/licensing/patents/pledgedpatents.pdf

IBM. (2005b). *New IBM initiative advances open software standards in healthcare and education*. Press Release.

ifrOSS. (2005). *License center*. Retrieved November 7, 2005, from http://www.ifross.de/ifross_html/lizenzcenter-en.html

Kirk, M. K. (2002). *Competing demands on public policy*. Paper presented at the Conference on the International Patent System.

Marson, I. (2005). *GPL 3 may tackle Web loophole*. Retrieved October 1, 2005, from http://www.zdnet.com.au/news/software/soa/GPL_3_may_tackle_Web_loophole/0,2000061733,39214742,00.htm

Michaelson, J. (2004). There's no such thing as a free (software) lunch. *ACM Queue, 2*(3).

Microsoft. (2005). *Microsoft Permissive License (Ms-PL)*. Retrieved November 8, 2005, from http://www.microsoft.com/resources/sharedsource/licensingbasics/permissivelicense.mspx

Moore, J. T. S., Wonderview Productions (Firm), & Seventh Art Releasing (Firm). (2003). *Revolution OS*. [S.l.] [Los Angeles, CA]: Wonderview Productions; Seventh Art Releasing [distributor].

National ICT Australia. (2004). *Australian Public Licence B Version 1-1*. Retrieved from http://nicta.com.au/director/commercialisation/open_source_licence.cfm

New Zealand State Services Commission. (2006). *Guide to legal issues in using open source software v2*. State Services Commission.

Novell. (2004). *Novell statement on patents and open source software*. Retrieved May 25, 2005, from http://www.novell.com/company/policies/patent/

Open Invention Network. (2005). *Open invention network formed to promote linux and spur innovation globally through access to key patents*. Retrieved November 15, 2005, from http://www.openinventionnetwork.com/press.html

Open Source Initiative. (2004a). *The open source definition*. Retrieved August 1, 2004, from http://www.opensource.org/docs/definition_plain.php

Open Source Initiative. (2004b). *Open source initiative OSI—Licensing*. Retrieved November 9, 2005, from http://www.opensource.org/licenses/

Open Source Initiative. (2005). *License proliferation*. Retrieved May 10, 2006, from http://opensource.org/docs/policy/licenseproliferation.php

Raymond, E. S. (2001). *The cathedral and the bazaar: Musings on Linux and open source by an accidental revolutionary* (rev. ed.). Beijing; Cambridge, MA: O'Reilly.

Red Hat. (2004). *Statement of position and our promise on software patents*. Retrieved May 10, 2005, from http://www.redhat.com/legal/patent_policy.html

Rosen, L. (2004). *Open source licensing software freedom and intellectual property law*. Upper Saddle River, NJ: Prentice Hall.

Rosen, L. (2005). *License proliferation*. Open Source Developers Lab. Retrieved from http://www.rosenlaw.com/LicenseProliferation.pdf

Skidmore, D. (2006). *Too many open source licenses! But do the existing licenses adequately encompass the diverse needs and concerns of particular stakeholders?* Paper presented at the Towards Open Source Software Adoption: Educational, Public, Legal, and Usability Practices. OSS 2006 tOSSad workshop proceedings, Como, Italy.

Skidmore, D., & Skelly, L. (2003). *Patents in information systems: International issues*. Paper presented at the the 4th International We-B Conference, Perth, Australia.

Soat, J. (2002). Small companies say they're being sued for employing common practices for doing business on the Net. *InformationWeek*.

St. Laurent, A. M. (2004). *Understanding open source and free software licensing*. O'Reilly.

United States Federal Trade Commission. (2003). *To promote innovation: The proper balance of competition and patent law and policy*. United States Federal Trade Commission,.

Välimäki, M. (2003). Dual licensing in open source software industry. *Systèmes d'Information et Management, 8*(1), 63-75.

Välimäki, M. (2005). *The rise of open source licensing: A challenge to the use of intellectual property in the software industry*. Turre Publishing.

von Krogh, G., & von Hippel, E. (2003). Special issue on open source software development. *Research Policy, 32*(7), 1149-1157.

Wayner, P. (2000). *Free for all: How Linux and the free software movement undercut the high-tech titans* (1ˢᵗ ed.). New York: Harper Business.

Weiss, D. (2005). *Quantitative analysis of open source projects on SourceForge*. Paper presented at the The First International Conference on Open Source Systems, Genova, Italy.

World Trade Organization. (1994). *TRIPS (trade-related aspects of intellectual property rights)*. Retrieved May 10, 2003, from http://www.wto.org/english/docs_e/legal_e/27-trips_01_e.htm

Xineo.net. *Xineo Freeware License*. Retrieved from http://software.xineo.net/flightsim/Licence.html

KEY TERMS

Free Software Foundation (FSF): A primary oganisation in the free/libre and open source space, created by Richard Stallman, and maintainer of the GNU software projects and the GNU GPL software license.

Open Source Initiative (OSI): Organisation which created the open source definition, a certification mark for open source software, also a primary organisation in the open source space.

GNU GPL: Primary open source license, GNU means *Gnu is Non-Unix*. The GNU is a recursive software programmers joke. The GPL is the general public license, which is the legal means of the FSF's philosophy of CopyLeft.

Traditional Open Source Licenses: These are the reciprocal, non-reciprocal, linking, and dual licenses.

Quasi Open Source Licenses: These are the obligation, morality, viewable source, and membership licenses.

Open Source Support Licenes: These are the content and open standards licenses.

ENDNOTES

[1] http://www.wassenaar.org/

[2] Unofficial English translation of the District Court of Munich "Harald Welte vs. Deutschland GmbH" 2004. http://www.jbb.de/judgment_dc_munich_gpl.pdf

[3] http://www.gpl-violations.org

[4] http://www.fsf.org

[5] The acronym GNU is "GNU is Not Unix", which is a recursive joke, the general public license is the GPL. Most people refer to the license as the GPL, but the FSF prefer to use the fuller term GNU GPL.

[6] Computers start counting from zero, this is why freedom 3 is the fourth freedom, and the first freedom is freedom zero.

[7] FLOSSMole, http://ossmole.sourceforge.net/ Data available at http://floss.syr.edu/OssMole/index.jsp. SQL query "SELECT count(*) as total_records, code FROM project_licenses group by code"

[8] Common public license, http://www-124.ibm.com/developerworks/oss/CPLv1.0.htm

[9] http://europa.eu.int/idabc/en/document/2623/5585#eupl

[10] GNU Lesser General Public License (v. 2.1), http://www.fsf.org/licensing/licenses/lgpl.html

[11] http://www.rosenlaw.com/AFL3.0.htm

[12] Previously to February 1999, the GNU LGPL was the GNU L[ibrary] GPL. With version 2.1, in February 1999 the name changed to the L[esser] GPL.

[13] http://www.sleepycat.com/company/licensing.html

[14] http://www.mozilla.org/MPL/MPL-1.1.html

[15] http://www.sun.com/software/jini/licensing/SCSL3_JiniTSA1.html

[16] http://www.mozilla.org/MPL/NPL-1.1.html

[17] http://www.microsoft.com/resources/sharedsource/licensingbasics/limitedpermissivelicense.mspx

[18] http://www.microsoft.com/resources/sharedsource/licensingbasics/referencelicense.mspx

[19] https://www.avalanchecorporatetechnology.net/

[20] http://www.gocc.gov

[21] http://www.fsf.org/licensing/licenses/fdl.html

Section V
Public Policy, the Public Sector, and Government Perspectives on Open Source Software

Chapter XXXII
On the Role of Public Policies Supporting Free/Open Source Software

Stefano Comino
University of Trento, Italy

Fabio M. Manenti
University of Padua, Italy

Alessandro Rossi
University of Trento, Italy

ABSTRACT

Governments' interest in free/open source software is steadily increasing. Several policies aimed at supporting free/open source software have been taken or are currently under discussion all around the world. In this chapter, we review the basic (economic) rationales for such policy interventions and we present some summary statistics on policies taken within the European countries. We claim that in order to evaluate correctly the consequences of such interventions one has to consider both the role and the administrative level at which such decisions are taken as well as the typology of software that is involved. Moreover, we argue that the level playing field cannot be taken for granted in software markets. Therefore, non-intrusive public policies that currently prevail at the European level in terms, for instance, of the promotion of open standards or in terms of campaigns aimed at informing IT decision-makers, are likely to be welfare enhancing.

INTRODUCTION

Governments' interest in free/open source (F/OS) software is steadily increasing. In Europe, this interest has become visible in the Lisbon Strategy and in the corresponding eEurope Action Plans 2002 and 2005 approved by the European Commission where it has been clearly stated the key role of open source software and open standards in pursuing the general objective of giving all citizens the opportunity to participate in the global information society.[1]

All over the world governments are considering various policies to support F/OS software; these policies go from the provision of "best practices" for the usage of open source to information campaigns aimed at making markets participants aware of all software alternatives, from simple expressions of preference towards F/OS software to large scale adoption of open source solutions in governments' offices and schools.

The role of the public sector in the software market is of primary importance. Governments not only set the legal and regulatory framework where economic agents interact, but they are also big software purchasers;[2] this double role makes governments key players in determining the future evolution of software markets and it is therefore of crucial interest to understand both the motivations and the effects of governments' interventions in this sector.

This chapter critically reviews the main arguments in favor or against public intervention supporting F/OS; we also provide some empirical evidence about the various public interventions that are already in place in Europe. The chapter is structured in three parts: in the first part, we provide a general analytical framework; public interventions may occur at different administrative levels (i.e., from municipalities to national or supra-national level), and they may have different motivations. These complexities have not received enough attention in the previous analyses on public interventions towards F/OS; the aim of this section is to offer a possible taxonomy for governmental policies in the software market and to discuss the many rationales for intervention but also the counterarguments that often have been put forward. In the following section, we present some evidence concerning the main public initiatives in Europe. Rather than focusing on any specific case study, we have collected information from the European IDABC, the program documenting the major initiatives supporting F/OS within the European Union. In this way, we have been able to draw some general considerations on the mo-

tivations and the characteristics of governments interventions implemented all across the EU. The subsequent section concludes by bridging the theoretical discussion with the empirical analysis. We claim that, if one considers that the largest share of the software market is represented by self-developed or customized products, the existing literature has placed too much emphasis on packaged software and arguments against public support of F/OS might be improperly grounded. Moreover, we believe that the level playing field cannot be taken for granted in software markets. Therefore, non-intrusive public policies that currently prevail at the European level in terms, for instance, of the promotion of open standards or in terms of campaigns aimed at informing IT decision makers, are likely to be welfare enhancing.

BACKGROUND: A GENERAL FRAMEWORK

It is useful to start our analysis by providing a general framework for discriminating the large heterogeneity of public interventions in the software market. In particular, we claim that, in order to judge correctly rationales, motivations, and consequences of public interventions, it is important to distinguish between the various roles played by policy makers and the various categories of software involved. We argue that many existing contributions, both in the scholarly and in the practitioners' debate, have not clearly taken into account these distinctions.

Public administrations, institutions, and governments play a double role in the software industry. On the one side, being big spenders for software licenses and software development, their adoption/use decisions represent a significant share of the demand thus having a major impact on market equilibrium. On the other side, by acting as legislators and regulators, governments do in various ways determine the evolution of the market; for instance, it is quite evident that the legislation towards intellectual property rights, either based on strong patent protection as in the U.S. or on weaker copyright legislations as it is within the EU, has a major influence on the functioning of the market and the diverging experiences on the two sides of the Atlantic stand as a clear example of this role. Similarly, as we discuss later in the chapter, governments frequently intervene mandating the adoption of open standards/interfaces; these policies are usually aimed at promoting compatibility and interoperability between different software platforms, thus creating a level playing field between different competitors; this kind of intervention clearly affects the efficiency of the market and therefore suggests a regulatory intention of the proponents.[3]

Obviously, it is often difficult to disentangle interventions of public authorities as adopters/users from those motivated on regulatory scopes; being large users, the decision to adopt a certain software package taken by public bodies affects the dynamic evolution of the industry and the equilibrium outcome, thus having regulatory consequences on the overall functioning of the market.

Irrespectively of the role played by a public administration, interventions may produce different consequences depending on the nature of the product involved. Software is not a commodity and the industry is extremely heterogeneous; indeed, the vast majority of software is either self-developed or custom while packaged software represents a minor share of the market.[4] The structure, the players, and the dynamics of mass-market and custom segments of the software industry are very different as well as different are likely to be the effects induced by the various public interventions.

In Table 1 we provide four examples of interventions distinguishing among different roles of public administrations and different typologies of software: three of these interventions are directly related to the promotion of F/OS, while the fourth refers to the well-known Microsoft European antitrust case. This last example relates to the F/OS world since, as a consequence of the antitrust action, Microsoft has recently announced its decision to allow access to some parts of the source code of its operating system.[5]

Rationales for Intervention: Review of the Literature

The literature on F/OS software in public administrations is quite substantial. Supporters of F/OS software have mainly focused on adoption of such technologies in the public sector and have based their arguments on technical, cost-efficiency or political-idealistic grounds. Regulatory scopes and therefore those rationales based on the consequences of F/OS public adoption on the overall functioning of the market have been receiving a

Table 1. Examples of public interventions supporting F/OS or regulating the market[10]

	Adoption/Development	**Market Regulation**
Custom	August 2005: the French Ministry of Foreign Affairs starts developing an open source architecture in order to integrate its computing system.[6]	October 2004: the Belgian administration published its white book concerning the use of open standards and open specifications for public sector purchased software.[7]
Packaged	September 2004: the Education Council of Castilla - La Mancha signed an agreement with Sun Microsystems to distribute Star Office 6.0 to the region's schools.[8]	EU's 2004 antitrust decision: Microsoft is required to disclose complete and accurate interface documentation which would allow non-Microsoft work group servers to achieve full interoperability with Windows PCs and servers. This will enable rival vendors to develop products that can compete on a level playing field in the work group server operating system market.[9]

much more limited attention by this stream of research.

Conversely, most of the critical voices in this debate have warned against detrimental consequences of both direct support/intervention and adoption of F/OS by public administrations on market performance.

In what follows, we briefly summarize the debate on F/OS software in the public sector; we devote the first subsection to provide a general overview of the most frequent motivations that have been proposed to justify public support towards open source. In the second subsection we look at the issue from a more critical viewpoint and we present the (often) skeptical view held by some economists and closed source practitioners.

Why Supporting F/OS?

Advocates of the F/OS movement put forward several rationales for public policies in the software market. Leaving aside pure idealistic-philosophical motives,[11] governments should support F/OS because of its intrinsic superiority with respect to closed source software. F/OS is considered to outperform proprietary software in terms of, for instance, higher reliability, security, flexibility, and maintainability of the code.[12] These superior features stem both from the organizational mode of F/OS which is characterized by the presence of a community of developers that continuously reviews the source code and fixes possible bugs, as well as from the fact that the availability of the source code makes it possible for the user to adapt the software to her/his own personal needs and to check every possible defect. Cost-efficiency is a second common rationale for policy interventions which is especially important for those public administrations that are pressured by budget concerns. The public sector would benefit from F/OS because of a number of reasons: net savings due to the reduced or non-existing licensing fees, the opportunity of freely contracting with software developers for subsequent code

maintenance/upgrade without being locked into the relationship with the initial provider, or the possibility of profiting from economies of reuse/collaborative development.[13] Similarly, a further beneficial effect would follow from a more efficient employment of public resources that would be shifted from license costs towards human capital investments.

With respect to the issue of innovation dynamics in the software industry, F/OS advocates also stress the importance and benefits of public intervention. Open source licenses guarantee the availability of the source code and the same legal rights as those of the original developer to every individual who is interested in a certain software product. This wide availability of the "updated state-of-art," within an industry characterized by cumulative generation of knowledge, is perceived to be of crucial importance to spur innovation. In this respect, Varian and Shapiro (2003) argue that, being typically based on open interfaces, F/OS encourages third-party innovation in terms of development of, for instance, adds-on and complementary products.[14] Similarly, Benkler (2002) considers self-organization in the distributed peer production model more efficient in "acquiring and processing information about human capital available to contribute to information production projects" than traditional institutions, such as markets and hierarchies. Henkel and von Hippel (2004) push this argument further, claiming that "user innovation," a fundamental trait in F/OS software development, is welfare enhancing.

From the national perspective, those countries, whose software industry is lagging behind or is not competitive in the international markets, may consider public support to F/OS a viable way to cultivate a domestic software industry, therefore reducing their dependency from foreign suppliers; this rationale for public intervention seems to be ranked particularly high in the agenda of both emerging[15] and developed[16] countries. Varian and Shapiro (2003) sponsor this opinion and emphasize that the GNU/Linux operating system is "an open platform on which commercial or open source applications can be built, thereby spurring the development of a robust domestic industry."[17]

Another common motivation for intervening in support of the F/OS movement is the stimulus of competition in the software market; this motive seems particularly relevant for those segments of the market characterized by the presence of dominant firms such as in the packaged software segment[18] and, more generally, in software procurement markets where dominant proprietary systems tie users to single suppliers, thus restricting competition.[19]

A More Critical View

During the last few years, several economists and other scholars have scrutinized the possible role of public policies in support of F/OS software. Apart from some relevant exceptions, the majority of authors seem to be rather skeptical about the welfare benefits that would accrue from governments directly stimulating F/OS.[20] One leading argument is that open source has emerged and, in many cases, has been extremely successful even without any intervention in place; therefore, there seems to be no need for public policies in order for F/OS to flourish. On top of that, focusing on closed source software, many authors claim that there is no clear evidence of significant failures in the software market and, consequently, there is no urge for governments' intervention. Evans (2002) and Evans and Reddy (2002) point out that the software industry is highly competitive[21] and also its performances in terms of growth, productivity, and R&D expenditures have been impressively high.[22] In other terms, software markets appear to be an example of well-functioning markets and, therefore, public funding to stimulate the emergence of alternatives to closed source software are prone to pick the "wrong winner." Moreover, a strong support to F/OS software may seriously undermine the incentives of commercial firms to

innovate or to improve the quality of their software (Schmidt & Schnitzer, 2003).

One of the main arguments in favor of F/OS is that it guarantees to public administrations significant reductions in software expenses; various authors point out that cost savings obtainable by adopting F/OS rather than proprietary software are by far smaller that those expected. The licensing fees represent only a minor part of software costs and a meaningful comparison between F/OS and commercial software has to be done in terms of the total cost of ownership (TCO) which also includes user training, technical support, maintenance, and possible upgrades of the software. On these grounds, the overall cost advantage of F/OS is less evident.[23]

The higher degree of innovativeness that, according to supporters, characterizes the F/OS development mode is also a strongly debated issue. Smith (2002) acknowledges the brilliant performances of proprietary software companies in terms of R&D expenditures and resulting innovation and declares himself rather skeptical about F/OS being able to replicate such figures.[24] Evans (2002) and Evans and Reddy (2002) go even further and claim that the theoretical argument according to which open source implies more innovation completely lacks of solid empirical evidence, given that many successful F/OS software projects draw strong inspiration from already existing closed source counterparts.

This discussion reveals a widespread skepticism among economists and closed source advocates about direct government policies in favor of F/OS software; nonetheless, there is a general consensus on the need of a broader set of interventions that somehow ensure the level playing field in the software market. In particular, various authors are making strong arguments against the current system of protection of intellectual property rights. A long series of decisions taken by U.S. courts during the last twenty years has extended software patent protection and has made it easier for applicants to obtain patents even for obvious inventions. These facts have induced large firms to accumulate sizable numbers of software patents, the so-called patent thickets, that can be strategically used in order to block competitors' innovation. As Bessen (2002, p. 13) points out, U.S. patent legislation may actually "sabotage the otherwise healthy open source movement" therefore potentially undermining competition from F/OS solutions.[25]

Finally, an issue that has drawn the attention of several contributors relates to the public funding of software R&D based on open source solutions. In this case, the non-rival and non-excludable nature of software goods, largely due to negligible replication costs, may induce policy makers to sponsor F/OS software projects as a means to increase social welfare.[26] While there is some consensus on the beneficial effects of this kind of interventions, the usage of restrictive licensing schemes (such as the GPL), is still very much debated: the software developed within publicly funded R&D projects should be made available to the widest possible audience but such restrictive licensing terms may undermine private appropriation of publicly funded basic science efforts.[27] In particular, closed source firms may be prevented from adopting and developing complementary applications for software distributed under GPL-like licensing schemes. Lessig (2002) suggests that governments should employ a non-discriminatory approach: publicly funded code should be released in the public domain or employing non-restrictive open source licenses (such as BSD-like ones).

MAIN FOCUS OF THE CHAPTER

Major Interventions in the EU

All across Europe, governments and public agencies are intervening in the software market in various ways; since September 2003, the major initiatives are registered on the Open Source

Observatory, a dedicated Web site compiled by the European Commission within the IDABC program.[28] For each intervention registered on this Web site a brief abstract and, usually, a series of official documents and press releases describing the content of the policy are available. In order to derive useful information, we have reviewed the existing documentation focusing on the most important interventions registered on the IDABC site, therefore disregarding public initiatives taken by very small municipalities. The dataset we have compiled starting from the IDABC documentation has been complemented with the information recovered from an independent investigation by the Center for Strategic and International Studies (see Lewis, 2004).

It should be noted that given the methodology used within the IDABC program, the information we have gathered does not represent the complete set of initiatives taken in the European public sector. Some typologies of policies or some countries might be underrepresented in the sample. However, we believe that our effort to summarize the existing policies in favor of F/OS software represents a useful starting point to analyze the major European initiatives within a unified setting.

Overall, we have collected information about 105 interventions, distributed across 14 European countries; France is by large the most active country with more than 28% of the interventions in our sample.[29] Around 8.5% of the policies have been taken at the EU level and therefore they should be common to all European countries.

To summarize the information derived from our dataset, we have grouped policies according to:

- **Type of software involved by the intervention:** We have distinguished between *custom*, *packaged* software, and broader interventions aimed at supporting the use of *open standards/interfaces*.
- **Political and administrative levels at which the intervention is taken:** We have

applied a two-tier classification distinguishing both between *government* and *public agencies*/bureaus (e.g., central government vs. postal services) and between *central* and *local*/regional level of intervention (e.g., central government vs. local municipality).

- **Type of intervention:** We have grouped interventions into three broad categories: *adoption* when the government/agency has decided to adopt a certain software, *advisory* when the policy consists of a general claim of preference towards open source and/or encourages the use of F/OS or it is aimed at informing potential adopters of the existence and characteristics of open source and, finally, *development* when the government actively promotes the creation of new software.

- **Rationale for intervention:** We have classified policies into seven non-exclusive broad categories: *cost-efficiency,* that pools together motivations such as savings in license fees, economies of reuse of the software, savings from collaborative development of projects, and more efficient employment of public resources (e.g., shift from license fees to investment in human capital); *code availability,* combining motivations connected to the technical advantages assured by transparency, security, robustness, and quality of the code; *interoperability,* in which the rationale for intervention lies in stimulating the diffusion of open standards and in promoting interoperability in the software market; *flexibility,* in which motivations are linked to flexibility advantages assured by, for instance, the possibility of tailoring the code to the user's needs, to assure integration and compatibility with existing systems, and so on; *enhanced competition,* combining interventions motivated by levelling the playing field, creating alternatives to proprietary companies, supporting domestic industries, stimulating technical independence from

dominant vendors, introducing competition in support, maintenance, and upgrade of systems and so forth; *efficiency in the public sector,* gathering motivations specifically related to the diffusion of best practices in public administration bodies; and, finally, *information diffusion,* a category representing those interventions motivated by the aim of increasing the available information and of raising consciousness about F/OS in the general public or, more specifically, in public administrations.

Table 2 shows the sample distribution of the various policies with respect to their type. F/OS adoption and advisory are the most common interventions in Europe: together they represent more of the 80% of the whole sample.

In Table 3 we go further into the detail and we present how the three types of policies are distributed between central and local decisional levels and between governmental authorities and public bureaus/agencies. More than 80% of the interventions in our sample are taken at the governmental level (both local and central) while agencies have played a much more limited role. Advisory policies aimed at suggesting and promoting F/OS prevail in central governments while at the other levels adoption is the most common type of intervention. This is not surprising once considered that central governments often provide "guidelines" for action while operative decisions are effectively endorsed at the local level and in agency bodies.

Table 2. Public policies classified in terms of type of intervention

Intervention	Freq.	%
Adoption	47	44.8
Advisory	39	37.1
Development	19	18.1
TOTAL	105	100

Table 3. Policies classified in terms of type of intervention and administrative level

Level	Intervention			
	Development	Adoption	Advisory	TOTAL
Central Gov.	8 (17.8%)	9 (20%)	28 (62.2%)	45 (100%)
Central Agency	1 (8.3%)	9 (75%)	2 (16.7%)	12 (100%)
Local Gov.	9 (21.4%)	24 (57.1%)	9 (21.4%)	42 (100%)
Local Agency	1 (16.7%)	5 (83.3%)	0 (0%)	6 (100%)
TOTAL	19 (18.1%)	47 (44.8%)	39 (37.1%)	105 (100%)

Table 4. Policies classified in terms of software type and administrative level

Level	Software		
	Custom	Packaged	Open Std.
Central Gov.	69%	73%	0%
Central Agencies	33%	66%	17%
Local Gov.	38%	78%	5%
Local Agencies	83%	33%	0%
TOTAL	53%	72%	8%

In Table 4 interventions are grouped according to the kind of software they are directed to: either software custom or packaged or towards the implementation of open standards. Note that in many cases, the intervention is not restricted to a unique type of software but it may involve two or all of them.[30] Table 4 suggests that local governments are more active towards packaged software while central governments do not seem to follow any particular pattern.

Restricting the analysis to central governments and central agencies, we have looked more closely at the motivations behind interventions. According to the available information, only in 37 out of 57 of the cases it was possible to collect official statements explicitly accounting for the rationales for intervention. The information we have gathered is presented in Table 5. Clearly, given the small number of observations, some caution has to be exerted when interpreting these data; however, it is worthwhile to highlight the major trends that characterize European policies.

Total figures in Table 5 show that cost-efficiency motivations are the most popular, followed by interoperability and code availability ones. Regarding specific policies, adoption policies are largely motivated by interoperability (viewed at the level of the single adopter) and cost-efficiency

Table 5. Public policies classified in terms of rationale for intervention (central government and agencies only)

Intervention	Total	Cost Efficiency	Code Availability	Interoperability	Flexibility	Enhanced Competition	Efficiency in Public Sector	Information Diffusion
				Type of Motivation				
Development	6 (100%)	1 (17%)	4 (67%)	1 (17%)	1 (17%)	1 (17%)	0 (0%)	1 (17%)
Adoption	11 (100%)	7 (64%)	2 (18%)	6 (55%)	3 (27%)	2 (18%)	1 (9%)	1 (9%)
Advisory	20 (100%)	11 (55%)	7 (35%)	11 (55%)	2 (10%)	8 (40%)	5 (25%)	7 (35%)
TOTAL	37 (100%)	19 (51%)	13 (35%)	18 (49%)	6 (16%)	11 (30%)	6 (16%)	9 (24%)

rationales (in particular, savings on license fees) while rationales regarding technical advantages of code availability and flexibility (all subcategories equally represented) are less cited, therefore suggesting that short-term advantages might be more salient than long-term ones in the stated motivations. On the other hand, pure regulatory motivations (such as stimulating market competition) are not explicitly accounted for. As far as advisory policies are concerned, interoperability (also considered at the market level) and cost-efficiency (all subcategories equally represented) are still fundamental rationales, but other regulatory motivations are popular as well (in particular, enhancing competition and raising awareness in markets). Finally, technical advantages of code availability represents the major rationale for R&D policies, while, surprisingly, motivation regarding cost-efficiency are rather infrequent.

FUTURE TRENDS

As we have briefly discussed in a prior section, economists are rather critical about intrusive public policies into the software market and, to some extent, we adhere to this skepticism.

Just to mention some arguments, the software industry has really proved to be extremely dynamic, characterized by high rates of growth and, while competition in some software segments might result in "winner-takes-all" outcomes, dominant positions have been frequently displaced by new comers (see Schmalensee, 2000); in a word, markets have performed reasonably well. Moreover, it is not yet clear if the production mode of open source is really more innovative than the proprietary one and empirical evidence on this issue is far from being clear-cut.

However, we believe that looking at the F/OS movement from an economic viewpoint, many relevant aspects have not received so far the attention that they should have deserved and the evidence on the EU experience reported above suggests some of the directions towards which the analysis should look at in order to better understand the actual effects of these policies.

For example, we believe that the distinction between custom and packaged software has not been properly taken into account in the literature. One of the main concerns against public support towards open source is based on the allegation that such policies would be detrimental for the incentives to innovate by commercial firms. We have already pointed out that almost two thirds of the market is represented by software that has been developed internally or that is customized and, as shown in Table 4, more than half of the interventions in our sample relates to this latter type of software. We are convinced that the above allegation cannot apply to this kind of software: customized software is by definition software "on demand" and the incentives to develop new lines of code arise at the moment of the call for tender, regardless of the open or close nature of the source code.

From the evidence presented in a previous section, it emerges that across the EU, together with cost saving reasons, public interventions in support of F/OS founded their motivations primarily on the desire of stimulating an open standard environment for software applications but also on the relevance of source code availability and on the intention to promote more competitive software markets.

It is recognized that proprietary software is likely to create important lock-in positions; the unavailability of the source code renders adopters dependant on the original software provider for further maintenance/development/upgrade of the code. Moreover, the use of closed standards, a typical solution employed by proprietary vendors, makes it more difficult for adopters to disengage themselves from software vendors. The absence of complete and public documentation regarding file and data storage formats and other communication standards might substantially increase

the switching costs thus rendering unprofitable the migration to other software packages. Lock-in is certainly a source of a relevant increase in life-cycle costs but these costs are extremely difficult to evaluate when one wants to compute correctly the total cost of ownership of a given software product.

On the contrary, a relevant feature of both open source code and open standards is that competition may be created in the aftermarket, and this may significantly reduce the cost of service, support, maintenance and interoperability.[31] Moreover, according to this view, fears of picking "wrong winners" through governmental advisory or adoption of F/OS solutions should be lessened if one takes into account that F/OS software is based on open formats that are commonly available and that might be employed by closed source vendors to develop compatible value-added proprietary solutions or interoperable adds-on and complementary products.[32]

While the above arguments apply to custom software in particular, a regulatory policy in support of open standards may found solid justifications also in the context of mass-market software; as a consequence of strong network effects, these segments of the software industry are often characterized by the presence of dominant players whose platforms have the typical features of "essential facilities." Controlling an interface (the key input) allows the dominant firm to protect its position and possibly to extend it to other complementary products. Similarly to the current practice in other industries, also for the case of software the provision of open access to the essential facility should be seriously considered in order to promote competition and to improve market efficiency.

CONCLUSION

The bottom line is to ensure that markets lead to efficient outcomes and therefore to exclude, based on economic grounds, that public interventions might be beneficial relates to the assumption that all potential adopters are properly informed about the alternatives that are available in the market. A recent empirical study on F/OS in the public sector shows that this is not necessarily the case. Ghosh and Glott (2005) show that a large share of IT administrators in the public sector ignore that in their agencies F/OS was actually employed.[33] More interestingly, the fact of being aware or not about the current usage of open source software has a major impact on the evaluation of the potential benefits of F/OS adoption. Nearly 70% of the "aware IT administrators" finds it useful to extend the use of open source in their agencies. This percentage shrinks to 30% among the IT administrators that were unaware that F/OS software was already employed in their institutions. Clearly, this evidence provides strong support for policies aimed at informing potential adopters about the characteristics and the availability of open source solutions.[34]

ACKNOWLEDGMENT

The authors would like to thank Bruno Caprile, Vincenzo D'Andrea, Sebastian Spaeth, Ruben van Wendel de Joode and two anonymous referees for their helpful comments on earlier drafts of this chapter. The usual disclaimer applies. Financial supports from Progetto di Ateneo 2006—University of Padua (for Stefano Comino and Fabio Manenti) and from MIUR under the projects FIRB03 and PRIN05 (for Alessandro Rossi) are gratefully acknowledged.

REFERENCES

Benkler, Y. (2002). Coase's penguin, or, Linux and the nature of the firm. *Yale Law Journal, 112*(3), 369-446.

Bessen, J. (2002). What good is free software? In R. Hahn (Ed.), *Government policy toward open source software* (pp. 12-33). Washington, DC: AEI-Brookings Joint Center for Regulatory Studies.

Bessen, J., & Hunt, R. M. (2004). *An empirical look at software patents* (Working Paper 03-17). Federal Reserve Bank of Philadelphia.

Comino, S., & Manenti, F. M. (2005). Government policies supporting open source software for the mass market. *Review of Industrial Organization, 26*, 217-240.

Danish Board of Technology. (2002). *Open source software in e-government.* Retrieved from http://www.tekno.dk/pdf/projekter/p03_open-source_paper_english.pdf

DeLong, J. B., & Froomkin, A. M. (2000). Speculative microeconomics for tomorrow's economy. *First Monday, 5*(2).

Evans, D. S. (2002). Politics and programming: Government preferences for promoting open source software. In R. Hahn (Ed.), *Government policy toward open source software* (pp. 34-49). Washington, DC: AEI-Brookings Joint Center for Regulatory Studies.

Evans, D. S., & Layne-Farrar, A. (2004). Software patents and open source: The battle over intellectual property rights. *Virginia journal of law and technology, 9*(10), 1-28.

Evans, D. S., & Reddy, B. (2002, May 21). *Government preferences for promoting open-source software: A solution in search of a problem.* NERA Economic Consulting report. Retrieved from http://papers.ssrn.com/sol3/papers.cfm?abstract_id=313202

Finnish Minister of Finance. (2003). *Recommendation on the openness of the code and interfaces of state information systems* (Working Paper 29/2003). Retrieved from http://www.vm.fi/vm/

en/04_publications_and_documents/01_pub-lications/03_working_group_memoranda/20031015Recomm/65051.pdf

Forge, S. (2005). Towards an EU policy for open-source software. In M. Wynants & J. Cornelis (Eds.), *How open is the future?* (pp. 489-503). Brussels, Belgium: VUB Brussels University Press.

Ghosh, R., & Glott, R. (2005). *Free/libre and open source software: policy support.* (Results and policy paper from survey of governments authorities.) Maastricht, The Netherlands: University of Maastricht, MERIT.

Ghosh, R., Krieger, B., Glott, R., & Robles, G. (2002, June). *Free/libre and open source software: Survey and study* (Deliverable D18, Final report, Part 2B: Open source software in the public sector: policy within the European Union). Maastricht, The Netherlands: University of Maastricht, International Institute of Infonomics.

Henkel, J. (2006). Selective revealing in open innovation processes: The case of embedded Linux. *Research Policy, 35*(7), 953-969.

Henkel, J., & von Hippel (2004). Welfare implications of user innovation. *The Journal of Technology Transfer, 30*(1-2), 73-87.4

Lessig, L. (2002). Open source baselines: Compare to what? In R. Hahn (Ed.). *Government policy toward open source software* (pp. 50-68). Washington, DC: AEI-Brookings Joint Center for Regulatory Studies.

Lewis, J. A. (2004, August 1). *Global policies on open source software.* Center for Strategic and International Studies report. Retrieved from http://www.csis.org/index.php?option+com_csis_pubs&task_view&id=3046

Schmalensee, R. (2000). Antitrust issues in schumpeterian industries, *American Economic Review, 90,* 192-196.

Schmidt, K., & Schnitzer, M. (2003). Public subsidies for open source? Some economic policy issues of the software market. *Harvard Journal of Law and Technology, 16*(2), 473-505.

Schmitz, P.E. (2001). *Use of open source in Europe, an IDA study.* European Commission, DG Enterprise. Retrieved from http://europa.eu.int/idabc/servlets/Doc?id=1973

Schmitz, P.E., & Castiaux, S. (2002). *Pooling open source software, An IDA feasibility study. Interchange of data between administrations.* European Commission, DG Enterprise. Retrieved from http://europa.eu.int/idabc/en/document/2623/5585#feasibility

Smith, B.L. (2002). The future of software: Enabling the marketplace to decide? In R. Hahn (Ed.), *Government policy toward open source software* (pp. 69-86). Washington, DC: AEI-Brookings Joint Center for Regulatory Studies.

U.S. Federal Trade Commission. (2003). *To promote innovation: the proper balance of competition and patent law and policy.* Retrieved from http://www.ftc.gov/os/2003/10/innovationrpt.pdf

Varian, H., & Shapiro, C. (2003). *Linux adoption in the public sector: an economic analysis,* mimeo. University of Berkeley, California.

von Hippel, E. (2005). *Democratizing innovation.* Cambridge, MA: MIT Press.

Wheeler, D. A. (2005). *Why open source software /free software (OSS/FS, FLOSS, or FOSS)? Look at the numbers!* Retrieved from http://www.dwheeler.com/oss_fs_why.html

KEY TERMS

Customers' Lock-In: A situation in which a customer is so dependent on a vendor for products and services that he/she cannot move to another vendor without substantial switching costs, real and/or perceived.

Economic Regulation: Set of restrictions promulgated by government administrative agencies through rulemaking supported by a threat of sanction or a fine. The main scope for government's regulation is to prevent markets' failures, in other words, situations in which markets do not efficiently organize production or allocate goods and services to consumers (as in the presence of a monopoly/dominant firm).

Essential Facility: In a vertically related market, it is defined as a facility, function, process, or service that meets three criteria: it is monopoly controlled; a potential competitor requires it as an input to provide services and to compete downstream with the monopoly supplier; and it cannot be economically or technically duplicated. Facilities that meet this definition shall be subject to mandatory unbundling and mandated pricing.

Intellectual Property Rights (IPRs): Intellectual property is a term used to refer to the object of a variety of laws, including patent law, copyright law, trademark law, trade secret law, and industrial design law. These laws provide exclusive rights to certain parties over intangible subject matter or over the product of intellectual or creative endeavor; many of them implement government-granted monopolies.

Proprietary Software (PS): Software products that are designed in such a way that others cannot access or view a product's source coding/the programming that allows the software to perform certain functions.

Source Code: The programming that allows software programs to perform certain actions or functions.

Total Cost of Ownership (TCO): Financial estimate aimed at helping consumers and enterprise managers to assess direct and indirect costs related to the purchase of any capital investment, such as (but not limited to) computer software or hardware.

ENDNOTES

[1] Further details are available at: http://europa.eu.int/information_society/eeurope/2005/index_en.htm. All the URLs provided in this chapter are active at the moment of writing the chapter (June 2006).

[2] Just to give a relevant example, the Dutch public sector spent around 400 million euros on software in 1997; see http://www.ososs.nl.

[3] For an example at the transnational level see the European Interoperability Framework for pan-European eGovernment services, mandating a series of policies, standards and guidelines aimed at "facilitating [...] the interoperability of services and systems between public administrations, as well as between administrations and the public" (http://europa.eu.int/idabc/en/document/2319/5644). For an application at the national level the reader may refer to the Dutch manual on open standards and open source software (OSOSS) in the procurement process, encouraging the adoption of open standards in the public sector (http://www.ososs.nl).

[4] According to Bessen (2002), packaged software has never accounted for more than a third of software expenses.

[5] See, for instance, Microsoft's Jan. 25, 2006 press release available at http://www.microsoft.com/presspass/press/2006/jan06/01-25EUSourceCodePR.mspx and the comments of Neelie Kroes (European Union's antitrust chief), stating that documentation enabling interoperability, rather than mere code disclosure, is at issue in order to meet EU's requirements (http://today.reuters.

com/business/newsArticle.aspx?type=tec hnology&storyID=nL26331447).

6   http://europa.eu.int/idabc/en/document/4549/469

7   http://europa.eu.int/idabc/en/document/3336/469.

8   http://europa.eu.int/idabc/en/document/1766/469.

9   http://europa.eu.int/rapid/pressReleasesAction.do?reference=IP/04/382&format=HTML&aged=l&language=EN&guiLanguage=en.

10   For a brief but comprehensive review of various national initiatives and policies on open source software see the links provided by the IDABC Open Source Observatory at http://europa.eu.int/idabc/en/document/1677/471.

11   A notable example of this kind of motivations can be found in the programs and activities of the Free Software Foundation, aimed at affirming the primacy of freedom ideals in the development and diffusion of software.

12   For a comprehensive survey on this topic see Wheeler (2005).

13   Reuse economies are savings due to recycling previously developed code as a basis for a new project; collaborative development economies are strategies of mutualization consisting in partnerships for joint development by the public sector, motivated by the needs of pooling efforts and sharing costs in building, maintaining and upgrading large software projects of common interest. See Schmitz and Castiaux (2002) for an assessment applied to FO/S software.

14   Bessen (2002) holds a similar view.

15   Support to domestic software industry lies at the core of the IT national policies of India and China. See, for instance, the remarks of the Indian President, A.P.J. Abdul Kalam, on the future challenges of information technology for developing countries (http://news.com.com/2100-1016-1011255.htmlnews.

com.com/2100-1016-1011255.html) or the speech of the Ministry of Science and Technology at the 2004 International Conference on Strategies for Building Software Industries in Developing Countries (http://www.iipi.org/Conferences/Hawaii_SW_Conference/Li%20Paper.pdf).

16   This occurs both at the national as well as at the local levels. See the statement by the Finnish Ministry Kyösti Karjula (http://www.linuxtoday.com/news_story.php3?ltsn=2002-06-17-011-26-NW-DP-PB) as an example of the first type and the deliberation of the autonomous province of Trento on the adoption of open standards and open source software (http://www.linuxtrent.it/Members/napo/deliberaPAT_n1492.pdf) as an instance of the second type.

17   Smith (2002) contrasts this view arguing that in a large number of countries, not only in the developed ones, a flourishing (proprietary) software industry already exists.

18   Among others, see the statement made by Boris Schwartz, deputy leader of the SPD parliamentary group, during the debate about the transition towards open source systems of the city of Munich (http://www.linuxtoday.com/infrastructure/2003052600126NWSWPB).

19   See, for instance, the recommendations of the Danish Board of Technology (2002) on supporting the emergence of alternatives in custom built software markets as means to foster competition and the recommendations of the Finnish Minister of Finance (2003), suggesting to include the possession of the source code in tender drafts in order to assure competitive bidding in future development and maintenance.

20   One notable exception is represented by Lessig (2002) who claims that government preference towards F/OS is justified by the presence of externalities that market forces do not internalize. For instance, software

developed for or adopted by some branches of the government could be employed usefully also by other branches if it is free or open source; the initial development/adoption decision should take into account also the potential benefits for future users.

[21] These authors provide several figures to support their argument. In the US the Herfindahl-Hirschman index (HHI) for the software industry is smaller that the average HHI computed for the US manufacturing industries; furthermore, during the period 1996-2000 there has been a decrease by 27% in the quality-adjusted prices for the packaged software.

[22] According to Evans (2002), in the year 2000 the R&D expenditure of software companies represented one tenth of the overall R&D undertaken within the industrial sectors while fifteen years before it accounted for only 1%.

[23] The empirical evidence comparing the TCO of open vs. close software solutions does not seem to be conclusive. For a comprehensive overview the reader may refer to the FlossPols report on policy support (Ghosh & Glott, 2005).

[24] Smith, Microsoft's senior vice president, also claims that often, in order to bring the software to the market, additional investments have to be done and these can not accrue from the F/OS world but can only come from the commercial one.

[25] For an empirical analysis on software patents see Bessen and Hunt (2004). According to these authors, the strategic accumulation of patent thickets seems to be the most convincing explanation for the large increase of software patenting in the US. Similarly, several panelists, according to a recent US Federal Trade Commission (2003) report, support the view that the patent protection system poses threats to innovation in the software industry. Lessig (2002) and von

Hippel (2005) argue in favor of lessening the extent of patent protection in the software industry. According to Evans (2002) and Evans and Layne-Farrar (2004), even though some (minor) reform of the patent legislation might be beneficial, software patents should not be banned altogether.

[26] See, for instance DeLong and Froomkin (2000) for an application to digital goods markets.

[27] Smith (2002) and Lessig (2002) hold the view that government should finance R&D activities but the resulting software should not be distributed under restrictive licensing schemes. On the contrary, Varian and Shapiro (2003) focusing on the Linux case argue that the adoption of GPL does not necessarily prevents the development of complementary applications. Henkel (2006) provides empirical evidence that, despite GPL's strict requirements in releasing derived works, firms can adopt several successful strategies in order to protect their own code enhancements.

[28] IDABC stands for Interoperable Delivery of European eGovernment Services to public Administrations, Businesses and Citizens; the information available on the Open source Observatory is collected by a special Web-team from staff members of the European public sector and also by searching the Internet for relevant information. The documentation we have collected is available at the following URL http://europa.eu.int/idabc/en/chapter/491.

[29] The large interest of public authorities in France has been documented also in a previous IDABC report, see Schmitz (2001).

[30] This fact explains why rows sum up to more than 100%.

[31] On these lines, Ghosh, Krieger, Glott, and Robles (2002) suggest that whenever it is feasible governments and public institutions should opt for software open source,

for example, by granting unlimited access to the source code, the right to modify the software and that to reproduce and distribute an unlimited amount of copies of the modified version under the same license restrictions. Forge (2005, p. 492) argues that policy-markers should mandate "backward compatibility, open access to program interfaces, and separation between operating systems and applications".

[32] Moreover, it is worth mentioning that in some cases policies supporting F/OS software are inspired by neutrality principles, therefore suggesting joint use rather than full substitution of closed source software by migrating to F/OS systems.

[33] According to the authors 30% of IT administrators were unaware of F/OS software usage and this figure increases in the case of small budget public agencies.

[34] A welfare analysis of the impact of various policies supporting F/OS in the presence of "unaware" potential adopters can be found in Comino and Manenti (2005).

428

Chapter XXXIII
Use of OSS by Local E-Administration:
The French Situation

Laurence Favier
University of Bourgogne (Dijon), France

Joël Mekhantar
University Jules Verne of Picardie (Amiens), France

Marie-Noëlle Terrasse
University of Bourgogne (Dijon), France

ABSTRACT

This chapter deals with the integration of OSS in local and territorial e-administration and its relations with the state level in France. France includes both many local collectivities: (36,568 local collectivities) on four levels (local, departmental, regional, and central) and a centralized State. The policies defined in France and promoted by initiatives from the European Union are leading to the definition of a normative framework intended to promote interoperability between information systems, the use of free software and open standards, public-private partnerships, development of know-how and abilities. These policies are applicable to State agencies but are not required for local and regional collectives because of the constitutional principle of administrative freedom. The chapter shows how the integration of all administrative levels can be operated in an e-administration framework OSS based, often coexisting with proprietary software. The legal, political, and technical (III) frameworks of such integration are presented.

INTRODUCTION

The last 2005 July 5th European parliament rejects the *attempts of the European Patent Office and its allies to impose software patentability on Europe. This vote promoted the diffusion of OSS, especially in e-government's applications.* In this background, we will focus on the effective use of OSS in French local e-administration. France includes many local and territorial collectivities:

Copyright © 2007, IGI Global, distributing in print or electronic forms without written permission of IGI Global is prohibited.

(36,568 local collectivities) Integration of electronic administration between the different levels (local, regional, national, international) has not yet truly been implemented in France, even less so has it been theorized. However, a key point in the success of electronic administration resides at the most local level, the town hall, where citizens use it to undertake their administrative requests. The users, businesses or citizens, wish to have efficient service without needing to bother with the differences in responsibilities or approaches for each of these levels. Local, uncoordinated initiatives may result in costly incompatibilities or redundant work. Furthermore, a paradoxical situation could occur since the new technologies, necessary for deployment of services, are a factor in increase of "digital fracture" (even that of its spreading in company environment).

The policies defined in France and promoted by initiatives from the European Union (IDABC networks, Government Online International Network, International Council for Information Technology in Government Administration) are leading to the definition of a normative framework intended to promote interoperability between information systems, the use of free software and open standards, *public-private partnerships*, development of know-how and abilities. In France, the *ADAE* (Agence pour le Développement de l'Administration électronique—the agency for development of electronic administration), in the framework of the ADELE program has performed this task by creating a strategic plan (PSAE) and a master plan for electronic administration (SDAE). These policies are applicable to State agencies but are not required for local and regional collectives because of the constitutional principle of administrative freedom.

This chapter deals with the integration of OSS in *local and territorial e-administration* and its relations with the state level. OSS often coexists with proprietary software: how their integration is operated? What are the legal (I), political (II), and technical (III) frameworks of such an integration?

BACKGROUND

The development of e-administrations within public organizations is a reality that has become progressively prevalent in the legal framework in France and, more generally, within the European Union (I). In this development, the problems of interoperability between the different levels of administration and the desire to be able to establish relations between the local, regional, national, and supranational levels, particularly between European nations, have raised the question of using open source software (Culnaert, 2004).[1] among administration specialists and decision-makers (I-2).

The Emergence of French and European Law on Local E-Administration

The development of e-administration in general and *local e-administration* in particular, with the transformation of procedures[2] and the explosion of local e-services, is a reality in France[3] and in Europe[4]. On the legal level, the French constitutional and administrative organization allows the prime minister to regulate the development of public services on the Internet[5] for the State and its public administration institutions.

However, the constitutional principle of free administration of public organizations leaves public organizations greater freedom in organizing themselves directly within the limits of their obligatory declaration to the National Commission for Information Technology and Civil Liberties (CNIL) in order to ensure the protection of personal data[6] by applying the Law of January 6, 1978. The CNIL publishes a practical guide, explaining the requirements it imposes on public organizations on this subject.[7] The Law of January 6, 1978 was amended by a new Law dated August 6, 2004. That established a distinction between the two types of requirements prior to the declarations, based on the nature and goal of the data processing: data processing subject to the general declaration procedure and that subject to the exceptional

429

authorization procedure. The declaration system makes up the general system established by the new law for data processing which does not risk extending to privacy and civil liberties. Based on this, the authorization system concerns very specific situations like data processing likely to infringe on privacy and civil liberties given its goals and characteristics as well as certain kinds of data processing done on behalf of the State. Very early on, this simple legal framework allowed pioneering towns to anticipate a digital future. The experience of the "digital town" Parthenay was a little like a laboratory. In effect, in this town, the first reflections on the subject "IT and local development" were carried out in 1993 and emerged from participation in several European programs: the METASA program, the MIND program[8], and the IMAGINE project, whose goal was to encourage a social appropriation of information and communication technologies.

A series of unpublished studies carried out at the University of Bourgogne[9] showed the abundance of ideas, which were demonstrated by *local e-administrations* in the United Kingdom[10] and in France, to offer new services to citizens.

In addition, a French ordinance of December 8, 2005[11] applicable both to state and local public organization services aims at simplifying the administrative requests by using electronic means. With this ordinance "an administrative authority can answer all information requests electronically that a user or another administrative authority sends to it by this method." Furthermore "When a user sends a request or information to an administrative authority electronically and receives confirmation of receipt (electronically), this administrative authority shall duly input and process the request or information without requesting confirmation from the user or asking him/her to resend it in another format."

Most importantly, this ordinance encourages the creation of e-administrations without restricting them to the single State services by indicating:

The administrative authorities[12] can create e-services within the limits of the measures of the aforementioned Law of January 6, 1978 and the rules on security and interoperability set forth in Chapters IV and V of this ordinance. When they implement such a service, in accordance with the former, the administrative authorities make their reason for creating it accessible as well as its method of use, particularly the possible communication methods. These methods are imposed on users.

For the development of *local e-administration*, despite the creation of the ADAE,[13] France is not yet set up for a true master plan comparable to that established in the United Kingdom. The e-Administration Plan of Action (P2AE) 2004-2007 within the framework of ADELE must be content with indicating:

The local public organizations are special and indispensable partners in the development of the e-administration. To this end, they participate in the study, development and creation of numerous services. Actors for change, they equally benefit from inter-departmental works that will be included in the framework for the e-administration plan of action.

Eventually, this plan only includes several services for local public organizations.[14] Under the framework of local interest related to their area of expertise, the former maintain control of the political options on whether to develop online services for the concerned citizens or not.

To develop online services, essential factors in the decision relate both to cost and certainty that the public organization will not become a prisoner of technology so that it may evolve its online public services to meet the needs of users. Consequently, the use of open source software seems to offer a satisfactory answer to this dual concern.

The Problem of Open Source Software within Local E-Administration Law

Before it was replaced by the ADAE (Agency for the Development of e-Administration), ATICA (Agency for Information and Communication Technologies) encouraged the use of open source software licenses in France for administrations starting in December 2002 with the publication of a guide.[15] The e-Administration Plan of Action 2004-2007 reiterated this approach and, for France, specified that the Government, through ADAE, also wishes to open a debate on processing shared add-on "open source" software based on the normalization model.

This debate will integrate the legal licensing aspects in order to evaluate the opportunity of defining licenses in accordance with these principles and in complete accordance with the law in European Union countries: The goal is therefore to bring about the success of a group of legal suggestions, allowing for the constitution of a core of freely reusable software.

This goal is the subject of a specific project file[16]. In addition, the same slant towards open source software is also taken to develop the diffusion of the AGORA tool[17] thanks to its licensing by GPL (general public license), as well as to migrate workplace software towards solutions based on open source software.[18] Similar suggestions exist in numerous European countries, like in Germany or in Spain, even if, in the last case, Parliament recently rejected the use of open source software. From a technical point of view, software is open when its source code is freely available, allowing the software to be duplicated, modified and redistributed. Access to the source code, and thus to all the instructions and program lines to modify the software, allows a community of developers and users to work together to constantly improve the software.

However, as ATICA specifies:

the availability of the source code is not the only requirement to define software as "open." In effect, "from a legal point of view, open source software is first and foremost software protected by copyright and subject to a license that regulates it and limits the rights and obligations related to it. Often compared with property systems, which usually only include user rights, open source software is distinguished by the most important rights being agreed upon by the software author and the license beneficiaries.[19]

Although it is still subject to copyright and intellectual property rights, open source software is of greater interest for local public organizations. In effect, their licenses not only allow the public organization-beneficiary to use the software but to study its functionality, change it for its own purposes, and redistribute the changes as well. License preserves the rights of the software author who remains free to distribute it under other licenses.

Legally, recourse to open source software particularly prevents the administration from having to pay royalties for software after the expiration of a contract. However, this risk exists in the case of using property systems.

The question of cost related to the right to use property systems which, after the computerization of the 1970s, was raised in a general fashion for all e-administration information systems is a major concern when it is a question of creating online public services. This is a strategic matter in the measure where the firm that holds the software rights can become an important limiting factor to the use of a service developed by the administration beginning with the said software. In other terms, from a legal point of view, the management of the license of use for property systems becomes more and more complex so that it is useful for administrations to switch to open source software if only to maintain control over developments in order to create and continue to develop online public services by preserving technical and financial expertise for future developments.

A dispute brought forth by the Bull Company against the National Health Insurance Agency for Wage Earners (CNAMTS) administration is very instructive in this respect. This agency, considered an administration, continued to use a software package for a period that was no longer covered by a transfer contract and without regard to the initial contract under which it was required to destroy the software package at the end of the transfer. The ruling handed down in this dispute was

that contrary to that argued by CNAMTS, the circumstances under which the Bull Company understood the execution of the initial contract, essentially, regarding royalties for the use of the software do not in any way prevent the Bull Company from requesting damages caused by a lack of gain; that from the instructions, the damages claimed by the Bull Company correspond to the market price amount calculated in proportion to the number of days in which the services were used outside of the contracted period, having been determined based on the amount of useful expenses shown by the co-contracting party for CNAMTS, increased, within the limits of the market price, by an amount corresponding to the recovery of damages suffered by the co-contracting party due to the wrongful behavior of the institution.[20]

This risk does not exist when the administration uses open source software. It remains free to continue to use the software and, especially, to change it to meet online public service needs without exposing itself to financial and legal limitations.

However, open source software remains protected by a license and, admittedly, it must be pointed out that until recently, French administrations only used licenses developed in the Anglophone world.[21] For this reason, in France,

to create better legal security while maintaining the spirit of these licenses, the CEA, CNRS and INRIA have launched a project to draft open source code software licenses under French law. In 2004, the CEA, CNRS and INRIA also drafted the *CeCILL*, the first license that defined the use and dissemination for open source software in accordance with French law, borrowing the principals of the GNU GPL.[22] The English text of version 2 of the license, which has the same legal value, is accessible online.[23]

In addition, it must be observed that, with these non-property systems developed in the private sector just like open source code software, there are fears regarding the use of e-Administration solutions due to the introduction of DRM (digital right management) and the development of a law on technical measures to protect digital contents. These questions that go above and beyond the framework of open source software pose specific legal problems regarding the protection of author's rights and neighboring rights[24] in the digital environment.[25]

These questions that go above and beyond the framework of open source software pose specific legal problems regarding the protection of author's rights and neighboring rights[26] in the digital environment.[27] To this regard, the Court of Cassation, with a ruling dated February 28, 2006 (Civ. 1, Appeal no. D 05-15.824 and E 05-16.2002, Decree no. 549 FS-P+B+R+I) overturned a ruling from the Paris Court of Appeals (Paris Court of Appeals, 4th Chamber, Section B, April 22, 2005) which ruled in favor of an individual on the impossibility of copying a DVD due to technical protection measures. The Court of Cassation considered

that by ruling thusly, whereas the scope of the normal use of the work, eliminating the exception of a private copy, takes into account the risks inherent to the new digital environment regarding the protection of author's rights and the economic importance whereas the use of the work, in DVD format, represents the depreciation of cinemato-graphic production costs, the court of appeals violated the aforementioned texts,

meaning that Articles L. 122-5 and L. 211-3 of the Intellectual Property Code were interpreted in light of the measures of Directive no. 2001/29/CE of May 22, 2001 on the harmonization of certain aspects of the author's right and neighboring rights for the information company, along with Article 9.2 of the Convention of Berne.

Moreover, the new French project of law named "*DAVDSI*"[28] could threaten open source software's diffusion. It allows for a sentence: edition, public access, public communication, knowingly done and whatever the form of such a publication, of a disposal designed for public access of non authorized works or protected objects. The distribution of software which allows information transfer (Web server, mailing ...) could be concerned.

MAIN FOCUS OF THE CHAPTER

The legal framework for the use of open source software in the administration attempts to respond to the growing use of open source software in the local administration as shown in all surveys of the past few years (II-1). This use is not only limited by the legislation on author's copyright. It is also limited by the political framework defined by the State: the interoperability of information systems must allow exchange of dematerialized information between the local level and the central administration (II-2). From this point of view, the use of open source software presents many advantages since it is based on open standards. But its effective implementation is accompanied by a new economic model limiting the independence of the State: the logic of outsourcing the management of information systems and strengthening the public-private partnerships by replacing or

finalizing the economics related to remuneration for intellectual property software (III).

Surveys' Results: The Effective Use of Open Source Software by Local French E-Governments

Different surveys' (APRONET 2004[29], MAZARS 2005, MARKESS 2005[30], FLOSSPOLS[31] 2005) confirm the European, and specifically French, interest in OSS. Whereas on 12/15/2005 the Spanish parliament rejected the proposed law aimed at imposing the use of open source software in the central administration under the pretext of supporting necessary competition between open source and property systems, the French State chose to stimulate the use of OSS. National agencies (ADAE—Agency for the Development of e-Administration and AIFE—Agency for State Financial Computerization's[32]) formulates recommendations, associations provide local e-governments with concrete help like the ADULLACT, the Association of Open Source Software for Developers and Users for the Administration and Local Public Organizations (http://www.adullact.org) or *AFUL*, the Francophone Association of Linux Users and Open Source Software for Education. ADULLACT is dedicated to support and coordinate the action of local public organizations, public administrations and hospital centers in order to promote, develop, share and maintain a common patrimony of useful open source software for public service missions. In addition, the part of OSS in the French administration information technology's budget is growing.

The information system for local public organizations today must overcome two difficulties: public organizations perform a number of jobs, 60 to 70 different jobs according to the MAZARS survey and their information system is often made up of groups of specialized software that do not communicate.

More over e-administration projects come to the same time: e-procedures, dematerialization of calls for tender, the new interactive services aimed at citizens (Internet sites, electronic administrative counters) and finally the productivity efforts required from all public functions. Local IT (information technology) specialists have never been confronted with as many similar works in progress.

Government purchasing, finances and accounting are the special information technology works in progress according to the MARKESS study. Then there are the intranets, human resources management (subject to the double effect of the "senior citizen boom" and decentralization) and extranets. Third, there are different work applications (management of consulting services, acts, legality audits, grants, social welfare, etc.) and other things related to citizen relations (Web sites, portals, online services, or other various tools) (geographic information system-SIG-, collaborative work, electronic document management-GED).

The reasons cited for using OSS are cost control (thank to the sharing of programs and knowledge) and the open standards needs in order to ensure the interoperability and upgradeability of the chosen solutions. The growing complexity of managing licenses becomes difficult to support in the public sector, which is characterized by the diversity of the jobs that it performs, the heterogeneous public institutions to which its services are aimed and the complexity of the cases that it handles (each individual is a peculiar case and can ask specific questions).

The open source alternative becomes a formidable tool in business negotiation. Its sole existence reduces bills from traditional editors by half. Now the OSS are used throughout the French Finance Department. The Copernic application for the general management of taxes is one example. Open source software is considered to now be the default choice for all fiscal applications. "According to our evaluations, our new JBoss application server using J2E cut costs by a fourth compared with our old property system tool" said Jean-Ma-

rie Lapeyre, technical director of Copernic, the new program that restructures the fiscal system (budget of 911 million Euro over nine years). According a report from DGME (Direction for the Modernization of the State[33]), the goal is that a third of the information projects are implemented with open source software, compared with the 10% from two years ago.

The rise in local public administrations expenses since 1982, following a notable decentralization of the State, budget control is the primary concern for 55% of respondents.

The French E-Government Scheme: Protecting the Public Treasury

The state does not limit the technical choices for local e-governments. It acts in two ways:

- Publication of a standard of interoperability named "Référentiel Général d'Interopérabilité"
- National experiments; these experiments consist in implementing an e-administration project in a region before offering it more generally to the entire nation once tested. Thus, since January 1, 2005, the e-Bourgogne[34] project was an experimental e-service platform for businesses and citizens

Nevertheless, public organizations are not completely free to choose the information system that they want to implement for e-administration's applications. It's the case of account which supposes the exchange of dematerialized data between the local e-government (for those who undertake the expense) and the public accountants who implement the budget. If the local e-governments are free to incur the expenses that they have voted on, the payment of the expenses is exclusively assumed by a public accountant, the Public Treasury (le Trésor Public). It ensures the legality of the account of local e-governments ... The Regional Accounting Offices control the legality of the operations carried out by public accountants under the accounting framework. Thus, there must be agreement between local e-government account and this of the municipal treasurer, a local agent of the public treasury.

We can absolutely imagine that more than 36,000 municipalities, nearly 100 departments, 25 regions, and thousands of other various local structures (township committees, EPCI,[35] etc.) each have their own information system. From the legal point of view, they have the right to do so because of the principle of free administration of local public organizations, a principle granted by the Constitution of 1958 and reinforced by the constitutional amendment of March 2003 (on which the Law of August 13, 2004 is based which deeply affects the role of Regional Accounting Offices). Some local governments give free rein their imagination. However, they always follow the rules of the public treasury accounting.

Today, the information technology application's "Helios" has been implemented to carry out the dematerialized exchanges between public organizations and the public treasury. All the accounting positions (3,400) will soon be equipped with this application. The format adopted by Helios will allow users to generate automatically and quickly accurate data regarding the budgetary and financial situation of public organizations from dematerialized budget documents which traced voted and performed expenses. Local information systems must send their dematerialized data under the framework of the standard exchange protocol (PES). Likewise, the regulation on project exchange "Acts" (help for assistance and safe electronic transmission) must be adopted by public organizations. This is available in the ADAE frame of recommendations: technical architecture, language of interoperability, protocols, and standard. The role and contents of the services requested for a tierce of e-transmissions[36] must be defined in relation with the technical architecture of the connection and the reception platform for acts from the Department of the Interior.

The local e-administration architecture is not determined in France without state protection, role assumed by the national agency, ADAE. We also understand why the French e-administration framework depends on the Finance Department within the General-Direction of the Modernization of the State whereas it was previously dealt with by the Ministry of Civil Service.

New Public-Private Relations

The need to be independent of property system editors often hides another dependency: that which it exercises in regard to information technology services' private companies. Open source software seems to be a factor promoting the dependence of the State on the private sector. The model of *tierce maintenance applicative,* an outsourced type of software maintenance for a company that uses an external service provider, is generalized with the use of open source software. Applicative maintenance consists in maintaining an information program in a state that allows it to fulfill its function: correction of errors, adaptation of operations to new hypothetical situations, and maintaining performance despite more and more users ... When this maintenance is performed with a third party, an outside service provider, this is a *tierce maintenance applicative.* In France, the Department of Finance (MINEFI: Ministère des Finances), entrusted a group of companies "Gemini, Linagora, Bull" with the support, maintenance, and creation of open source solutions for the tax management information system COPERNIC. The massive and large scale use of open source software, within the many MINEFI departments, since 2000, requires:

- Securing the choice of open source software, which has become a strategic axis in the construction of public information systems
- Registering the choice as an authentic lever to control and reduce costs

- Guaranteeing the evolution of the information system by ensuring an accrued interoperability, a continuous service and a high level of project maintenance through the updated training of public agents

MINEFI also has:

- An authentic and continuous service regarding the availability of used open source solutions (more than 100 supported and maintained software solutions)
- Know-how and expertise allowing for the creation of large migration works aimed at open source solutions in the following domains: file servers, mailing, workstations, public key infrastructure, and so forth.

This contract between administration and private sector represents the most important one in the field of open source software in 2005 in France, Europe and the world.

Local e-governments also use outsourcing more and more often. Seventy three percent of them have used or have planned to use external service providers, a proportion that is higher in the EPCI (84%) and General Councils (76%). The local administrations use a lot of consulting and support services before launching IT (information and communication technologies) projects. We also notice the importance of trusted third parties (36%) and the ASP[37] (25%), mainly for the dematerialized public markets, and in the emergence field of archiving. According to Markess International, this market is growing. This advisory's company estimated the French software and IT service market[38] for these administrations to be 3.3 billion Euro in 2005. This market should grow by 12.1% between 2005 and 2007 and reach nearly 4.2 billion Euro.

The new law on public-private partnerships[39] has a special field of application in e-administrations, whose development Jean Arthuis recommends.[40]

A new form of outsourcing, the public-private partnership, instituted by Ordinance no. 2004-559 of June 17, 2004, introduces a contract type into French law that is inspired by British law. The new public-private partnership contracts (PPP) can be extended over the long term, including an overall service starting from the inception of a building and its construction up to its maintenance and including the legal and financial assembling of the operations.

ADAE has been skeptical regarding the application of PPP to e-administration. According to its director, the intervention of a private service provider in a public information system requires both:

- The perennially and stability of businesses
- The independence of the used software tools
- The non-reuse of personal data for commercial means
- A draconian confidentiality clauses

We can see that the introduction of open source software in the administration is not harmless. Since it leads to a sharing dynamic between users, a development led by demand and not by supply, open standards to the detriment of property systems, it also generates a new economic model contributing to strengthening dependence on the public sector with regard to the private companies.

FUTURE TRENDS

Technical Integration of Open Source Software in French Local Administration

Legacy information systems of local administrations are getting increasingly complex and they cover increasingly wide and complex domains. This gives rise to three main issues:

- First, users (administrative staff members) tend to put high expectations on new information systems but, at the same time, they do not wish to change their working habits. Such an attitude can lead to a dead end in defining business patterns.

- Second, most administration information systems need to be considered as cooperating information systems since they relate to several business domains—typically encountered in the French society—which can be more or less intricate. Furthermore, it is difficult to take in charge the underlying vocabulary of such a cooperation with classical tools (e.g., thesaurus, ontologies) since they do not scale up easily.

- Third, in order to provide users with high quality services, new information systems use brand-new technologies (e.g., authentification, Web technologies, content-based information retrieval). This can lead to some form of digital divide among various information system users.

As long as information systems do not become too large and while they address a well-known domain, these three issues can be considered as orthogonal. Thus, they can be controlled by engineering staff. Nevertheless, when dealing with large-scale, complex, or innovative information systems, it can be difficult to separate issues and to build a meaningful information system proposal. Platforms for software engineering appear to be a promising approach in such a context. In the rest of this section, we argue on developing such platforms on an open source and open format basis.

New Services vs. Working Practices

Generally, development of a new information system is an opportunity to introduce new services. Such services can take various forms:

1. **Externalized services can be offered to citizens:** Providing e-services (information to be read, forms to be filled in, electronic requests to be sent) through a Web portal, providing clusters of services under a uniform access, and so forth

2. **Services relying on business expertise can be offered to citizens:** Providing external links to related services (which are offered by other organizations), offering access to a part of the business knowledge (technical explanations, documentations, etc.). Such externalized and expertise-related services do not change too much working practices

3. **A renewal of information systems can be an opportunity to interconnect separate administrations:** This can be a real improvement in offered services since citizens no longer need to repeatedly give basic records in each administrative office they go to. In such a case, there are two major tasks to perform: allow cooperation of services by making their bases of knowledge compatibles while guaranteeing the respect of legal protection.[41] A representative example in France is the DMP[42] project which aims at providing citizens with a unified electronic medical record (to be used by physicians, pharmacists, and hospitals). Health services use various identifiers for their own records of patients. Thus, it is necessary to produce a general identifier (which could be derived from the French national identifier since it was forbidden by the CNIL). The IdeoPass project has been carried out in order to create servers of patient identifications. Beyond the IdeoPass project, it is necessary to specify which type of access will be given to each actor of the healthcare system: physicians, nurses, pharmacists, social security staff members, and so forth.

4. **A new information system generally encompasses a workflow engine:** It is thus necessary to define precisely in which order and by whom each document will be treated.

In such a case, working practice may have to change in a significant way in order to comply with the workflow description. It may be a true challenge to obtain collaboration of administration staffs when such a workflow has to be used.

Open source software could be one of the ways to introduce more flexibility into working practices, in the sense that platforms based on OSS can be better tuned to users' requirements since the overall cost is diminished by license costs. At the same time, introducing OSS-based platforms in administration business implies an effort towards training of administrative staff members. Thus, acquired competencies make further evolutions easier.

Towards Shared Knowledge

As stated previously in this paper, the French administration is composed of a network of institutions working at various levels of responsibility and power. Most institutions have a substantial autonomy in organizing themselves to attain objectives that are fixed at the national level. Development of administration information systems has been carried out by local initiative of administration services. Yet, a trend of collaboration and uniformity is developing in various ways:

- Offer of domain-specific information systems (e.g., AMUE[43] is a national agency offering services for university management).
- Definition of national personal identification numbers which can be used by cooperative information systems (e.g., IdeoPass[44], which is an open source server of patient identification numbers which will be used for DMP, the unified electronic medical record).
- Norms for data collection in order to produce national statistics (e.g., DADS-U[45], which defines data to be used in statistical analyses for employee management).

As showed by Delmas-Marty[46] for national laws, major attempts to integrate administrative core businesses cannot be conducted hierarchically (from international to national and local). The horizontal integration of domain-related knowledge and procedures constitutes a much more reasonable objective. Such a selective integration must be conducted in such a way that no additional technical problems arise in sharing knowledge. Thus, open format bases of integrated knowledge, usable through open source interfaces make a lot of sense.

A Risk of Digital Divide

A widely discussed side-effect of computer dissemination is the digital divide among citizens depending on access to the internet world. Such a digital divide can also occur within administration staff members since not every employee is accustomed to sophisticated computer tools. An internal digital divide within administrative staff members can damage the balance between employees since some of them can be more or less unable to assure their part of work, and even lead to an upset of the hierarchical organization. Such an internal digital divide is a major issue in deploying new information systems or new working environment and needs to be taken into account from the very beginning of the project.[47]

Thus, it is rather important for e-administration platforms to enable integration of the basic tools (e.g., office automation tools) to which administration staff members are used to, while not propagating requirements for specific tools down to the core of platforms. Once again, open source and open format based platforms fully satisfy such requirements.

General Technical Perspectives

We believe that platforms for engineering of local administration domains should comply with the following three-fold statement: first, a major part of any administration platform should be released under one of the Open Source License and based on open formats; second, it is necessary to allow local administration staff to continue using tools they are used to; third, it is mandatory to offer a basic set of functionalities which can be fine-tuned for specific uses.

As we proposed[48], such platforms should be based on a core business description made available through open formats and plug-ins (either proprietary or open source plug-ins). As an illustration, we describe (Figure 1) a platform for education and job market surveys which encompasses:

- **Definition of vocabularies and specifications:** An integrated nomenclature[49]; a specification of data for statistical analysis of university teaching activities[50], which falls under the SISE project; a specification of data for statistical analysis of enterprise employee staffs[51], which is called DADS-U.
- **Open formats:** Including formats for basic

applications (e.g., RTF, OASIS[52]), as well as domain-specific open formats such as an official format for DADS-U and a format for SISE as published by the French educational department
- **Domain specific plug-ins:** Such as an Open Source DADS-viewer called DADS-U Vue[53], an open source statistical computing tool from R-Project[54], a proprietary software for student management called Apogee (at the present time, Apogee which was developed by the National Agency for Universities, cannot be plugged into such a platform)

CONCLUSION

The example of the French e-administration that we have just analysed allows readers to understand what are the difficulties and the advantages of OSS dissemination in a nation including both a centralized state and very numerous local e-governments. The promotion of OSS (by national agencies and associations) is more important than in many other national e-governments' master plan but it is also threatened by two kinds of obstacles:

Figure 1. An example platform for university management (with open source/format components in yellow and proprietary ones in green)

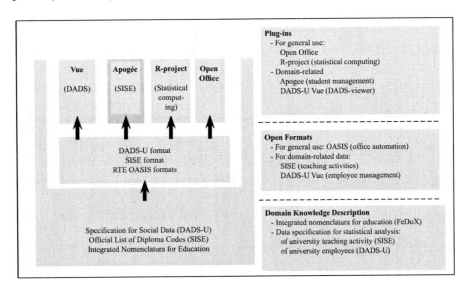

- The development of copyright law
- The knowledge management required to take advantage of OSS; human resources of administration don't have enough IT skills to operate the dematerialization of the system and to support its evolution because of the staff recruitment method

Copyright and knowledge management are not just a momentary step of the process of e-government implementation. They constitute a trend that central as well as local governments have to deal with.

REFERENCES

Arthuis, J. (2004, July 21). *L'informatisation de l'Etat. Pour un État en ligne avec tous les citoyens.* (Senate Information Report No. 422, 2003-2004). Retrieved from http://www.senat.fr/rap/r03-422/r03-422.html

Chantepie, P., Herbel, M., & Tarrier, F. (2003). *Mesures techniques de protection des œuvres et DRMS, Un état des lieux January 2003* (Report n° 2003-02, p. 156). Paris: Ministry of Culture and Communication.

Culnaert, E. (2004). *Le logiciel libre dans l'e-administration.* Retrieved from http://www.aecom.org/veille/008_libre_240904.htm

Chevallereau, F-X. (2005, June 27). *EU: eGovernment in the member states of the European Union.* (Independent Report, p. 555). European Communities.

DAVDSI (Droit d'Auteur et Droits Voisins dans la Société de l'Information). (2006). Journal Officiel, 178(3). Retrieved from http://assemblee-nationale.fr/12/dossiers/031206.asp

Delmas-Marty, M. (2006). *Le pluralisme ordonné* (Les forces imaginantes du droit - II), Seuil.

IRIS (Initiatives Régionales Innovations et Stratégies). (2005). *Les politiques en faveur du e-gouvernement en Europe.* Retrieved from http://www.egovinterop.net/Res/4/EGov%20in%20Europe-VF2.pdf

Eveno, E., & Jaeckle, L. (1998). Parthenay, modèle de *Ville Numérisée*? Report issued to DG III, Mind Project.

UK: The National Strategy for Local e-Government—Two years on. (2005, March 5). United Kingdom—Regional & Local. Retrieved from http://www.localegov.gov.uk/images/2%20years%20on%20-%20Realising%20the%20benefits%20from%20our%20investment%20in%20e-gov_227.pdf

Chantepie, P., Herube, M., & Tarrier, F. (2004). *Mesures techniques de protection des œuvres et DRMS, Un état des lieux January 2003* (Report n° 2003-02, p. 156). Paris : Ministry of Culture and Communication.

Maillard, T. (2004). La réception des mesures techniques de protection des œuvres en droit français: Commentaire du projet de loi relatif au droit d'auteur et aux droits voisins dans la société de l'information. *Légipresse, 208*(II), 8-15.

Maillard, T. (n.d.). Mesures techniques de protection, logiciels et acquis communautaire: Interfaces et interférences des directives 91/250/CEE et 2001/29/CE. *RLDI 2005/5*, 154.

Savonnet, M., Leclercq, E., Terrasse, M.-N., Grison, T., Becker, G., Farizy, A. S., & Denoyelle, L. (2006). Development platforms as a niche for software companies in open source software. In *Proceedings of the 2nd Open Source Software Workshop, OSS'06*, Italy.

SISE: Official report of the department of Education. (2000, June). Programme des opérations statistiques et de contrôle de gestion des directions d'administration centrale des ministères. *Bulletin Officiel de l'Education Nationale*, 5.

Zuliani, P., & Succi, G. (2004). An experiment of transition to open source software in local authorities. In *Proceedings of the 14th Conference E-Challenges*, Austria.

KEY TERMS

Copyright: Right to exercise ownership and control over a particular item.

Digital Right Management: Protection of online information.

French E-Administration: French national oversight administered through online media.

Interoperability: Ability for different computer systems at different administrative levels of government and in different regions to exchange information effectively.

Local E-Administration: Local government oversight administered through online media.

Public-Private Partnership (PPP): Cooperative agreements for public and private sector organizations to work together to achieve a common goal.

Territorial E-Administration: Federal/national government oversight administered through online media.

ENDNOTES

[1]   Éric Culnaërt, "Le logiciel libre dans l'e-administration," 2004: http://www.aecom.org/veille/008_libre_240904.htm

[2]   The most important example in France of local management is the transformation of public works contracts, obligatory since January 1, 2005.

[3]   For examples in France, see the site *service public-fr* which provides a list of links allowing for online requests on local sites: http://www.service-public.fr/teleservices/teleservices-local.html

[4]   For e-government politics in Europe, we will report the last summary report of François-Xavier Chevallerau, *EU: e-Government in the Member States of the European Union*, June 27, 2005, 555 p. as well as the studies on each member State, which can be downloaded at: http://europa.eu.int/idabc/en/document/4370/254. Also see the study carried out by the Regional Innovation Strategy Initiative (IRIS), *Les politiques en faveur du e-gouvernement en Europe*, June 2005, p. 49 http://iris.oten.fr/

[5]   This framework must come from the circular of the Prime Minister. See, for example, the circular of the Prime Minister dated March 12, 1993 regarding the protection of privacy in automatic data processing: Law no. 78-17 dated January 6, 1978, applied to the administrations and the entire public sector, regarding IT, files and civil liberties; the role of ministers and coordination by the Government commission at the National Commission for Information Technology and Civil Liberties (CNIL). Also see the circular of the Prime Minister dated October 7, 1999, regarding service and public institutions' Internet sites.

[6]   These obligations are shown on the French legal level with Law no. 78-17, amended on January 6, 1978 and the texts used for its application, particularly Decree no. 2005-1309 of October 22, 2005. In addition, at the level of the Council of Europe, France ratified the Convention to protect people with regard to automatic personal data processing dated January 28, 1981: http://conventions.coe.int/treaty/Commun/QueVoulezVous.asp?NT=108&CM=1&DF=09%2F01%2F01&CL=FRE. In addition, as regards European Union law, we must also consider the two directives of the European Parliament and Council: Directive 95/46/CE of the European Parliament and Council of October 24, 1995, regarding the protection

of physical people in processing personal information and the free circulation of data (Official Gazette no. L 281 dated 11/23/1995, p. 0031-0050): http://europa.eu.int/smartapi/cgi/sga_doc?smartapi!celexapi!prod!CELEXnumdoc&lg=fr&numdoc=31995L0046&model=guichett; Directive 2002/58/CE of July 12, 2002 regarding the treatment of personal information and the protection of personal information in the electronic communications sector (personal information and electronic communication directive), which can be downloaded at: http://europa.eu.int/eur-lex/pri/fr/oj/dat/2002/l_201/l_20120020731fr00370047.pdf

7   http://www.cnil.fr/index.php?id=1263

8   EVENO, Emmanuel and Luc JAECKLE (1998) "Parthenay, modèle de *Ville Numérisée* ?" Report issued to DG III, Mind Project.

9   Led by J. MEKHANTAR : 17 research files on local e-administration were created in 2004-2005 by Masters students in Local Public Organization Law at MÂCON and 13 research files in French and English were drafted by the 3rd year LLB students (10 from Manchester University and 3 from Queen's University in BELFAST). These studies analyzed local e-government in local French and British public organizations.

10   See: *UK: The National Strategy for Local e-Government—Two years on.* March 05, 2005 – United Kingdom – Regional & Local http://www.localegov.gov.uk/images/2%20years%20on%20-%20Realising%20the%20benefits%20from%20our%20investment%20in%20e-gov_227.pdf

11   Ordinance No. 2005-1516 of December 8, 2005 regarding the electronic exchange between users and administrative authorities and between the administrative authorities, Official Gazette of December 9, 2005: joe_20051209_0286_0009.pdf

12   By applying Article 1 of the aforementioned ordinance: "Considered as administrative authorities within this ordinance are the administrations of the State, the local public administration, public institutions that are administrative in nature, organizations managing social welfare systems related to the social security code and the farm laws listed in Article L. 223-16 and L. 351-21 of the Labor Laws and other organizations in charge of managing an administrative public service."

13   The Agency for the Development of e-Administration, created as an interdepartmental service available to the minister in charge of State reform, is currently integrated in the General-Directorate on State Modernization since its reorganization on January 3, 2006.

14   This principally deals with ADELE 69 Deployment of Co-Coverage with service-public.fr; ADELE 70 Local IT Systems (SIT); ADELE 71 Transformation of the Local Public Sector; ADELE 72 Civil Status Statistics and mayoral election results at the INSEE [French National Institution of Economic and Statistical Information]; ADELE 73 E-Burgundy; ADELE 74 Setting up an co-financing infrastructure allowing for the transformation of exchanges between public organizations and administrations. See the Plan of Action for e-Administration: http://www.adae.gouv.fr/article.php3?id_article=314

15   ATICA, Guide to Select and Use Open Source Software Licenses for Administrations, Paris, December 2002, p. 39: http://www.adae.gouv.fr/upload/documents/guide_LL.pdf

16   ADELE127. Set up of open source software and collaborative development. *Description*: To bring about the success of a group of legal suggestions, allowing for the constitution of a core of freely reusable software.

[17]   ADELE129. AGORA. *Description*: Functional architecture and techniques allowed rapid parameterization of Internet, intranet or extranet sites fulfilling a wide range of applications as well as evolved interfaces allowing for all the architecture to be simply managed, based on a very non-restrictive hosting architecture.

[18]   ADELE130. Workplace migration. *Description*: Experimenting with alternative open source software solutions.

[19]   ATICA, ibid, p. 4

[20]   CAA Douai, No. 03DA00786, May 3, 2005, CNAMTS: http://www.legifrance.gouv.fr/WAspad/UnDocument?base=JADE&nod=J7XCX2005X05X000000300786

[21]   Particularly: General Public License (GPL), LGPL (Lesser General Public Licence), QPL (Q Public Licence) and BSD (Berkeley Software Distribution).

[22]   http://www.cecill.info/

[23]   http://www.cecill.info/licences/Licence_CeCILL_V2-en.txt

[24]   See: Philippe Chantepie, Marc Herubel, Franck Tarrier, *Mesures techniques de protection des œuvres et DRMS, Un état des lieux January 2003*, Report n° 2003-02 – (I) Ministry of Culture and Communication, Paris, 2004, 156 p.

[25]   See the bibliography, accessible online at www.mtpo.org. Particularly: Thierry Maillard, "La réception des mesures techniques de protection des œuvres en droit français : Commentaire du projet de loi relatif au droit d'auteur et aux droits voisins dans la société de l'information," *Légipresse* 2004, no 208, II, pp. 8-15 ; Thierry Maillard, "Mesures techniques de protection, logiciels et acquis communautaire: Interfaces et interférences des directives 91/250/CEE et 2001/29/CE," RLDI 2005/5, no 154.

[26]   See: Philippe Chantepie, Marc Herubel, Franck Tarrier, *Mesures techniques de protection des œuvres et DRMS, Un état des lieux Janvier 2003*, Report n° 2003-02 – (I) Ministry of Culture and Communication, Paris, 2004, 156 p.

[27]   See the bibliography, accessible online at www.mtpo.org. Particularly: Thierry Maillard, "La réception des mesures techniques de protection des œuvres en droit français: Commentaire du projet de loi relatif au droit d'auteur et aux droits voisins dans la société de l'information," *Légipresse* 2004, no 208, II, pp. 8-15; Thierry Maillard, "Mesures techniques de protection, logiciels et acquis communautaire: Interfaces et interférences des directives 91/250/CEE et 2001/29/CE," RLDI 2005/5, no 154.

[28]   DAVDSI : (Droit d'Auteur et Droits Voisins dans la Société de l'Information): http://assemblee-nationale.fr/12/dossiers/031206.asp

[29]   Apronet (Association des professionnels de l'Internet des collectivités locales) www.anetville.com/public/article.tpl?id=9672&rub=8010

[30]   http://www.mazars.fr/ http://www.markess.fr. Markess is going to publish its latest analysis on *"Les Technologies de l'information en réponse aux enjeux des administrations publiques locales, 2005-2007,"* the results of a study carried out in mid-2005 on 300 local public administrations and 80 service providers.

[31]   http://www.flosspols.org/deliverables/FLOSSPOLS-D16-Gender_Integrated_Report_of_Findings.pdf

[32]   ADAE : Agence Nationale pour le Développement de l'Administration Electronique- AIFE: Agence pour l'Informatisation Financière de l'Etat.

[33]   http://www.modernisation.gouv.fr

[34]   https://www.e-bourgogne.fr/

[35]   Public undertakings for intercommunal cooperation (EPCI) are French administrative structure grouping municipalities that chose to develop a certain number of com-

36 The third of e-transmission ensures the transfer of the flow of information between administrations and local public organizations or between administrations by respecting a language and interoperability regulation.

37 Application service provider.

38 Source ibid. The results of this analysis have been issued to the elected members and managers of the administration at the November 23, 2005 conference organized by Markess International at the University of Paris VIII.

39 Ordinance No. 2004-559 of June 17, 2004 on partnership contracts (Official Gazette of June 19, 2004, p. 10994.

40 Jean Arthuis, on behalf of the Senate Finance Committee, July 21, 2004, Senate, Paris. *L'informatisation de l'Etat. Pour un État en ligne avec tous les citoyens.* Senate Information Report No. 422 (2003-2004). Senate, July 21, 2004. Available online: http://www.senat.fr/rap/r03-422/r03-422.html.

41 In France, such a legal protection us stated by specifications CNIL (Commission Nationale de l'Informatique et des Libertés).

42 DMP Dossier Medical Personnel, Personal medical record. URL http://www.d-m-p.org/.

43 AMUE National Agency for Universities Management. Agence de Mutualisation des Universités et Etablissements de l'Enseignement Supérieur. URL http://www.amue.fr.

44 IdeoPass. URL http://adullact.net/projects/ideopass.

45 DADS-U Déclaration automatisée des données sociales – Unifiée. http://www.travail.gouv.fr/dossiers.

46 M Delmas-Marty, Le pluralisme ordonné (Les forces imaginantes du droit - II), Seuil 2006.

47 P. Zuliani and G. Succi. *An Experiment of Transition to Open Source Software in Local Authorities*, Proceedings of the 14th Conference E-Challenges, Austria, 2004.

48 M. Savonnet, E. Leclercq, M.-N. Terrasse, T. Grison, G. Becker, A. S. Farizy, and L. Denoyelle. *Development Platforms as a Niche for Software Companies in Open Source Software.* Proceedings of the 2th Open Source Software Workshop, OSS'06, Italy, 2006.

49 Nomenclatures intégrées dans FeDoX. URL http://fedox.irisa.fr/Pages/nomenclature2.htm

50 SISE: Official report of the department of Education. "Programme des opérations statistiques et de contrôle de gestion des directions d'administration centrale des ministères." Bulletin Officiel de l'Education Nationale, Special No 5 du 1er juin 2000

51 Examples: Global view of a DADS-U file for a paid vacation fund. Exemples DADS-U V08R02 : Vue globale d'un fichier DADS-U pour une caisse de congés payés. URL www.cnsbtp.fr/caisses/doc/Exemples DADSU v08r02 v2.pdf

52 OASIS Open Document Format for Office Applications. URL http://www.oasis-open.org

53 DADS-U Vue. On Adullact Web Site, URL http://www.adullact.org/article.php3?id_article=316

54 The R-Project for Statistical Computing. The R Foundation for Statistical Computing, URL http://www.r-project.org//

Chapter XXXIV
Issues and Aspects of Open Source Software Usage and Adoption in the Public Sector

Gabor Laszlo
Budapest Tech, Hungary

ABSTRACT

This chapter introduces L-PEST model as the proposed tool for better understanding the fields are influenced by motivations and adaptation policy on FLOSS of public authorities and governments. Software usage in the public sector is a highly complex topic. In the confines of this chapter the selected case studies will show consideration to the vastly different needs and capacities and the different approaches and motivations towards the utilization of FLOSS by governments and/or local authorities. The primary objective of this chapter is to identify and describe the actors associated to the usage of FLOSS within and by the public sector. This chapter has made an attempt to fill this research gap and place the different actors into one complex model. It is hoped the proposed model assists better clarifying the intricate relationship between relevant factors. Nevertheless, much more research work is needed in the years to come. According to Michel Sapin, French Minister in charge of Public Administration and e-Government (2001), "The next generation e-government has two requirements: interoperability and transparency. These are the two strengths of open source software. Therefore, I am taking little risk when I predict that open source software will take a crucial part in the development of e-Government in the years to come."

INTRODUCTION

The digital economy transforms governments and governments took on new roles in those areas of the economy most affected by technological changes. Governments play important roles in creating the proper environment for ICT development, and also have a significant leading role as users of these technologies by creating new modes of public's behavior. Governmental functions and operations can be managed only by the extensive use of ICTs and by using software applications (Lanvin, 2003).

The world's largest consumers of computer software are usually governments and they thus can have considerable influence on the software market. Governmental usage of software can impact on virtually all aspects of civil life: the inclusion and participation of citizens in public life, the transparency and openness of decision making, and the elimination of the digital divide, digital persistence, and digital literacy. The question of which software is utilized by public administrations is, therefore, of fundamental importance. Free Software advocate Eben Moglen has said, "Who controls the software, controls life."[1]

In the early days of computing the common software model was based on the open source model. Software and hardware were often combined in a single package. The software was usually traded in the form of source code and computer users have shared their computer code. Many important early programs, also with government funding, were widely shared (Bessen, 2002).

Then, the late 1970s and early 1980s with the appearance the consumer computing saw the beginning of the commercialization of software products based on the proprietary model. The software that operates the hardware has become as important as the hardware itself.

A significant difference between open source and proprietary software is that the open source (as it is called) software source code is freely available to the user. In contrast, the proprietary software vendors release their product only in binary form and it is illegal for end users to decompile the binary machine code to usable source code.

Free/libre open source software (or FLOSS as it is commonly referred to) has gained enormous momentum all over the world. While this movement has been closely followed with attention by many advocates and practitioners, academic research on the subject has only started emerging. These research projects have focused mainly on individual motivations, knowledge sharing and the user communities themselves.

The primary objective of this chapter is to identify and describe the factors related to the usage of open source software within and by the public sector.

To achieve this objective, background is given on the discussion about government roles and policies towards open source software, as in the selected case studies.

One of the strengths of this chapter is that it presents a theoretical framework, a general model of software usage at large within the public sector and the identified factors assigned to global perspectives.

BACKGROUND

ICTs have the capacity to play a valuable role in improving the quality of life, particularly in health, education, agriculture, and the environment. To take one example, in the healthcare sector ICTs enable the implementation of tele-health programs in remote areas, allowing some health care to be provided remotely, independent of person-to-person contact. Further, improvements in medical equipment are also a result of advances in ICTs. In education, remote access to the knowledge bases, e-libraries and even e-learning systems and universities can deliver knowledge to rural areas, where such opportunities for learning would be unavailable without ICTs. Agriculture and environmental issues can be better managed by, for example, geographic information system (GIS) and weather forecasts.

However, at the same time, there exists the so-called digital divide, an umbrella term that is commonly understood to mean the gap between ICT *haves* and *have-nots*. Generally, the approach to the question of the capacity of ICT to increase standards of living and to that of the digital divide has focused on two main issues.[2] One focuses mainly on actual connectivity—infrastructure and access. Another approach beyond connectivity is to consider the level of the ICT literacy and skills of a particular population and as well, consequently take into consideration political and social cohesion aspects.

An improved economy can not alone eliminate the gap, so governmental "intervention" is a prerequisite for overcoming the digital gap. Today, governments, businesses, international groups, and nongovernmental organizations (NGOs) have undertaken numerous initiatives aimed at eliminating this digital divide (http://www.bridges.org/digital_divide). These initiatives have targeted not just the consequences of economic differences between countries and peoples and the relevant differences in access to technologies, but also the cultural capacity and political will necessary to apply these technologies for effective development. A nation's intellectual capital and capacity for innovation are based on its human capital, which is why it is so important for governments to make steps to strengthen the equality.

Wilson pointed to a four-sided social formation—a Quad—that has emerged at the heart of the still-inchoate knowledge society. "Conceptually 'quad' refers to persistent four-sided networked interactions of small groups of elites across four sectors of the political economy—government, private sector, research centers, and NGOs" (Wilson, 2003, p. 6).

The Quad theory predicts causal relationships between the architecture of the Quad and the subsequent performance of the ICT sector. The more robust the architecture of the Quad, the better performance of the ICT sector as a whole. The architecture and dynamics of the Quad relationships are different in every country and change time to time.

As a member and part of the Quad, the government has a special obligation to protect the integrity, confidentiality and accessibility of public information, to protect the privacy of its citizens, to educate the "next generation", to create jobs, and to preserve and make available the national heritage (also in electronic format)

Figure 1. The Quad (Source: Used with permission by E. J. Wilson)

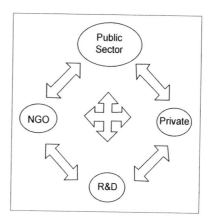

for the public and for the next generation. Other important roles for the governments are to make the country competitive in the globalized marketplace, and to carefully manage the budget (Stanco, 2003).

METHODOLOGY

This chapter provides an inductive general conceptual model—based on known and publicly available strategy documentation of various public sector and government initiatives for promoting or using FLOSS. The selection of key factors is grounded in available research literature on FLOSS and the above mentioned documentation and case studies.

OPEN GOVERNMENT

The average citizen has limited access to important government records, and what is available is often incomprehensible. An open government must be transparent and accountable and information related to the decisions an open government makes must be open to the public and freely available. Access to government and public information is regulated by law in many countries. Perrit states:

Freedom of information issues are centrally important in countries around the world, and the Internet's World Wide Web offers the potential to provide freedom of information at low cost. Achieving a sound information policy to promote open government requires constant vigilance by those who care about the goal. (Perrit, 1997, p. 397)

In the aftermath of 9/11 the relationship between IT, governments and their citizens has dramatically and radically changed. Security has become the most important factor. Yet, in the face of increased demands for security, for many within societies, the demands of privacy and trust remain paramount, thus giving rise to conflict between governments and their citizenry. Governments make greater efforts based on anti-terrorism legislation[3] to monitor their citizens' activities, while simultaneously citizens demand a greater ability to monitor the activities of his or her government.

INITIATIVES FOR FLOSS IN GOVERNMENT WORK

E-government work and what is commonly understood as general government work are now too closely intertwined to be realistically separated. At the same time, public administrations have special functions and operations which cannot be adequately handled with proprietary software applications on the market that are developed for multiple purposes (Stanco, 2003). The moderate opinions which stress that there is no need to make a choice between FLOSS and proprietary software vendors gather ground but the feasible solution is mixing these software.

It seems likely that all governments use FLOSS applications on some level, with or without open source label—though perhaps without deliberate

policy. Whereas many governments have policies or consideration towards FLOSS usage, the motivations may vary from cost reduction to security or dependency issues and within the broader context of policies to support such issues as equity or education. However, FLOSS policies and legislation as developed by national, regional or local governments around the World (USA, Canada, Australia, many countries in Africa or in Europe) are more often than not inadequate to support the viable realization of such policy goals.

The Center for Strategic & International Studies (2004) maintains—Government Open Source Policies—a list of such initiatives that were approved or proposed. This section highlights different approaches of adaptation and policy considerations for the implementation of FLOSS.

European Union

In the recent years many open source-related programs have been launched by the European Union. Fields of development of FLOSS within the EU include security, interoperability and e-participation. The software usage and the interaction between different systems is a complex approach. Interoperability is one of the key factors. One early Commission Working paper stressed the need for interoperability of program for public administration across the EU. It states that the proposed interoperability framework "will be based on open standards and encourage the use of open source software." "Interoperability, therefore, for both the public and enterprise sectors, is at the heart of the eEurope 2005 Action Plan and the achievement of the Lisbon goals" (Linking up Europe, 2003, p. 5).

In the European Union, the public sector were advised to avoid proprietary document formats, known as lock-in. Using the open standards would assure the desired interoperability and open standards would more greatly be supported by open source software. Using interoperable systems would guarantee equality among the citizens using different kind software applications (Promotion of Open Document Exchange Format, 2003).

On the other hand, notwithstanding the above-mentioned initiatives, the relationship between governments and open source is not unambiguous.

Extremadura

Extremadura was the poorest region of Spain, lagging behind the rest of the country in both the economic and technological field. Though short on financial resources, the region set very high goals for itself in its Regional Strategy on Information Society in 1997. The policy lay "in the application of technological innovation for the promotion of freedom and equal opportunities, taking advantage of and putting at the disposal of everyone, what is nobody's property: the knowledge gathered by Humanity all through History." Two formal strategic objectives were put forth: "Accessibility for all—the Internet as a public service" and "The stimulation of technological literacy."

Given the combination of Extremadura's strategic goals and the limited financial resources available, the use of FLOSS was a logical choice. The LinEx project, a combination of "Linux" and "Extremadura," was born of these strategic initiatives. The objective of the Linex project was to create a fully functional platform, based on FLOSS, providing universal access of IS tools to all citizens. While doing so, its aim was to provide adaptability, economic benefits, and security to as great a degree as possible, without losing sight of actual feasibility. LinEx is specifically designed for use in regional administration and schools. Early on in the project, it was decided that LinEx would not innovate the software itself, but rather concentrate on localization of the software and take care of the distribution. To avoid technical problems during the initial phase of the project, a Spanish company was hired. The region's gov-

ernment ships the resulting software for free to all of its citizens.

Extremadura was also simultaneously funding a development center whose task was to create accounting software, hospital applications and agricultural applications (IDABC, 2003).

Munich

Coming after the switch to Linux in the servers of the Bundestag in 2002, Germany's interior minister signed an agreement with IBM to offer the German government offices deep discounts on computer systems based on Linux (IBM signs Linux deal with Germany, 2002). Soon afterward, Germany's third largest city government, Munich, commissioned Client Study for the State Capital Munich (UNILOG Integrata, 2003) comparing the alternatives and assigning 6,218 (out of 10,000) points to a Linux/OpenOffice migration, versus 5,293 to an upgrade of Microsoft Windows. Based on this study the Munich municipal government made a decision to adopt for their computer systems open source software. The Council of Munich voted on May 2003 in favour of the adoption for its desktop and notebook computers an open source operating system and office applications. This move, unprecedented in scale in the European public sector, has been widely commented upon and discussed since then.

Following a test phase conducted in cooperation with SuSE Linux and IBM, the Council formally adopted on June 16, 2004 detailed plans to manage the migration process, which is expected to last until 2009. According to the plan the migration was to be gradual, starting in 2004 with office desktop applications (OpenOffice.org office suite and Mozilla Web browser running on the existing Windows NT desktops), and then moving to operating systems and more specialized applications over a period of five years. The municipal government of Munich released a statement in September 2005 that the completion of migration phase one, scheduled to be completed in 2005, had been pushed back to at least 2006. The reasons were that Novell Inc. announced in late 2003 the acquisition of SuSE and meanwhile legal problems regarding a proposed EU patent law. The chosen new Linux distribution was the Debian (Grassmuck, 2005).

USA

The most famous report concerning FLOSS usage within the Department of Defense (DOD) was released in 2003. "The goals of the study are to develop as complete a listing as possible of FOSS applications used in the DoD, and to collect representative examples of how those applications are being used." Over a two-week period the survey identified a total of 115 FLOSS (in the report named as FOSS) applications and 251 examples of their use. "The main conclusion of the analysis was that FOSS software plays a more critical role in the DoD than has generally been recognized. FOSS applications are most important in four broad areas: Infrastructure Support, Software Development, Security, and Research" (The MITRE Corporation, 2003, p. 2).

The Commonwealth of Massachusetts launched a new policy regarding the planning, development, and implementation of IT systems. "The goal of the Commonwealth's open initiatives is to ensure that investments in information technology result in systems that are sufficiently interoperable to meet the business requirements of its agencies and to effectively serve its constituencies" (Open Initiatives of Massachusetts, n.d.). The Massachusetts case illustrates the technology based considerations concerning software usage.

Brazil

The Government National Institute of Information Technology is charged with implementing open source software in Brazil. They released the first strategy planning document in 2003.

On the surface, the decision of the Brazilian government was a simple cost cutting measure. According to the National Information Technology Institute, Brazilians spend $1.1 billion every year on software licensing fees, and the federal government was the nation's biggest customer. The government is paying around $500 to Microsoft for license fees for every workstation. The government accounted for 6% of Microsoft's 2003 Brazilian revenues of $318 million. Switching to FLOSS would save millions of dollars (Kim, 2005). The decision to migrate to open source software on a national scale was not simply a matter of choosing one product over another. Although the Brazilian government identified economic reasons to migrate to open source software, it was a political decision that validated open source software as a movement. Through numerous open source projects, the government has tried to bridge the technology divide within the Brazilian population. While in the European Union the research experts recommend free software licenses for software deriving from public funds, Brazil has become the first country to require any company or research institute that receives government financing for the development of software to license it as open-source, meaning the underlying software code must be free to all (Benson, 2005).

Peru

Peru passed a law encouraging the procurement of free software by the government in September 2005. The bill was originally introduced in 2002. A Peruvian congressman stated in his letter to Microsoft: "The basic principles which inspire the bill are linked to the basic guarantees of a state of law, such as: the free access to public information by the citizen; the permanence of public data; the security of the state and citizens" (Greene, 2002). This bill has as its aim to establish measures and policies which will permit the acquisition of software licenses by the public administration under conditions of technology neutrality, and the free concurrence and equal treatment of suppliers. The technical evaluation of the software and hardware required by the public administration will be according regulations dictated by the National Informatic System governing body. The bill offers an excellent summary of the idea of neutral software usage: "The entity will ensure that the procurement answers to the principles of effectiveness and technological neutrality, transparency, efficiency, within the boundaries of austerity and economizing of public resources" (Peruvian bill translation, 2005). One essential item included in the bill also stress the need for the education of the employees and users of computer and IT technology.

South Africa

One of the best-known case studies concerns the South African government's official strategy for FLOSS. This was one of the first strategies, that officially recognized the legitimacy of the adoption of FLOSS within the public sector. The South African strategy highlights that "the government will implement OSS, where analysis shows it to be the appropriate option. The primary criteria for selecting software solutions will remain the improvement of efficiency, effectiveness and economy of service delivered by the government to its citizens" (Using Open Source Software in the South African Government, 2003, p. 24). One of the main strengthens of this strategy is the appreciation of the social benefits that could include, but are not limited to, better education, greater governmental transparency, more effective e-government services, and wider access to governmental information.

China

China has been very aggressively promoting Linux. The military has been one of the earliest adopters of Linux. The Red Flag Linux was de-

signed for use in government offices, schools and on home computers. Red Flag Linux, a Beijing-based provider of Linux software and services, is connected to the Chinese Academy of Sciences, the central government's top research institute. The main reasons for the adoption of Linux were political and the desire for independency from Microsoft (Einhorn, 2003). Membership in the World Trade Organization (WTO) and access to its benefits are strongly affected by the level of protection given to intellectual property rights in a country (Wong, 2004). According the Piracy Study (BSA, 2005), in the country there is a high frequency of pirated software. Since China became a full member of the World Trade Organization, the government has been trying to reduce software piracy within its country. This is another strong reason why government agencies and business are currently adopting the Linux operating system on their desktop workplaces.

L-PEST MODEL

As is shown in the above selected case studies there are many different approaches around the world to using FLOSS within the public sector. In this section a general model is introduced. The L-PEST model is a theoretical creation. The idea of the model reaches back to IDABC "The Many Aspects of Open Source" (n.d.) material. The original text summarized some of the various reasons for choosing different organizations of FLOSS. This idea was then extended, modified, and put into a model based on the research by the author. The author's proposed L-PEST model can give a broader picture as to the aspects of software usage in the public sector. With further and ongoing work, it can be applied to all kinds of software as a comprehensive tool.

The key factors were derived from motivations of governments within their environments, which were revealed in the case studies. The five key actors of the model—as shown on Figure 2—are:

Figure 2. L-PEST model

Legal environment Licensing Liability Piracy	
Political Privacy Digital persistence Digital heritage Open government Public procurement	**E**conomical Cost reduction Balance of the software market and transparency Innovation Job creation Dependency
Social Freedom and equality Education Behaviour of software use Digital divide	**T**echnological Quality Functionality Interoperability Transparency Support the standards Lock-in Security Localization

political, economic, social, and technical and all around these fields the legal environment can be founded. This structure maps real life.

Every actor has its own attributes, however in some cases there are attributes with different meanings. In this case, when an attribute could be assigned to more than one field it was put into the most characteristic actor (e.g., lock-in which is based on the technical elements may well also influence the economical aspect, or transparency has quite a different meaning in economics than in technology).

Legal Environment

The legal environment surrounds the model because it has an effect on the other four factors. It has own attributes as well. The constitution determines the framework activity of the country; the law determines the operation of society, while the economy is influenced by the law of economics. Acts regulate and ensure competition, and technology is also regulated by industrial law (including, patents, trademark, and copyright law).

Copyright is the most usual method of protection for software products. The copyright automatically and implicitly protects all intellectual creation, including computer software.[4] "The copyright laws, by default, do not allow for redistribution (nor even use) of software. The only way that redistribution can be done is by granting specific permission in a license (Working group on Libre Software, 2000, pp. 20-21). A license is a contract between the user and the licensor. The licensing model of about FLOSS differs from the proprietary software, but is based on same idea. In fact, open source licenses are also enforceable because they use, in one form or another, copyright law. Most open source licenses were designed according to the United States law. Open source (OSS) licenses are more permissive than free software (FS) licenses.[5]

One of the main threats for open source may be software patents—which are not currently common outside of the USA—but efforts to introduce them are in progress worldwide, usually lobbied for by large multinational corporations. The issue of software patents[6] divides even the governments of countries and the parliaments within them, as can be seen in many cases in Europe. On the other hand, many companies which have huge software patent portfolios in the USA, such as IBM and Novell, open numerous of their owned patents and put them at the OSS developing community's disposal. At the same time, many companies adopt and encourage OSS policy and business model.

Liability means that the software producers are responsible for their own products, warranties, and indemnifications. In reality almost all kinds of software, even the proprietary kind, are shipped on an "AS IS" basis, which means that the producer wriggles out of any kind of responsibility. In many countries, legislation does not allow the exclusion or limitation of this kind of liability.

Software piracy is a problem all around the globe and it can hurt a country in many ways. A country with poor protection for intellectual property rights is not as attractive to foreign investors. This is the reason why China, since joining the World Trade Organization (which strongly defends and pursues intellectual property rights protection), has made enormous efforts to reduce the prevalence of piracy within its borders. In a developing country, piracy is much more prevalent than in the industrialized nations; however, the greater dollar losses are incurred in the latter situation (BSA, 2005).

Political Aspects

The political aspect is related to government's function and roles. They may be distinguished between such roles as promoting social justice and functions such as tax collection (Lanvin, 2003). The government's role ensures the viable environment for ICT development and also the ICTs for Development. This can be summarized as a National Information Strategy that was well defined by the Library and Information Association of New Zealand:

A National Information Strategy addresses strategic issues to ensure that all citizens have the opportunity to access and utilize a nation's knowledge wealth in a way that will enhance the social, political and economic well-being of that country. It states the government position on the creation, management and use of information, and sets direction for government action in support of the strategic goals. (LIANZA, 2002, p. 7)

A national information strategy can be defined in terms of political planning or political action planning for development.

Privacy is a key factor in the interaction between governments and citizens. Whatever software is utilized by governments to control, manage and transmit the citizenry's personal data must be transparent in order to protect the citizen's right to privacy (Stanco, 2003). For example, an e-voting system without transparency leaves organizations and governments at the mercy of software providers.

The preservation of digital heritage and digital content has become a major challenge for society.

Digital persistence means continued accessibility to the stored content, even as the technology is changed—in this case for the governments' and public administrations' documentation. (It also preserves the original documents, in the case of national heritage.) It is in close relation with lock-in and dependency that it will be introduced. The secretary of administration and finance of the Commonwealth of Massachusetts, stated:

Our public policy focus is to insure that public records remain independent of underlying systems and applications, insuring their accessibility over very long periods of time. In the IT business a long period of time is about 18 months. In government it's over 300 years, so we have a slightly different perspective. (Kriss, 2005)

Economic Aspects

Within the scope of this chapter only several issues can be highlighted. Governments sometimes need to undertake intervention into the market on behalf of common good. A high degree of market transparency can result in disintermediation due to the buyer's increased knowledge of supply pricing. Transparency is important since it is one of the theoretical conditions, which reaches back to Adam Smith's invisible hands theory, required for a free market to be efficient. Consequently, it may well be true, that the government should not intervene in the free market except to assure neutrality and a level playing field for all types of software. The governments should to be assuring the neutral decision on software and public procurement and choices on software products should be made objectively, flexibly, and with a focus on a range of factors.

One of the primary economic concerns is the cost of software usage. Total cost of ownership (TCO) shows the real cost of software utilization. The purchased software will usually remain the property of the supplier; the consumer pays for the right to use the software. Total costs need be divided into two main categories, direct and indirect costs. The measurement is difficult because the indirect costs are extremely difficult to assess and measure (Wheeler, 2005). Another approach to the issue of value and cost can be to focus on the examination of return on investments (ROI). Both methods are extremely sensitive to the set of assumptions made by the individual or group taking the measure.

Research and development (R&D) and other innovation are more important than ever before. In their role as a member of the Quad (see Figure 1), governments should undertake to stimulate innovation. The economic benefits of such stimulation, as in the case of job creation, for example, are well known. There are numerous arguments that R&D that is financed through public funding should be released under FLOSS license. This kind of license supports the sharing of scientific results and dissemination of created information and value—and "there is not need to reinvent the wheel." Many FLOSS licenses are business friendly (Wong, 2004).

One of the major arguments in favor of FLOSS is concern over the issue of dependency; that is, the public becomes reliant on software suppliers. In many instances, there are painfully few options as to software vendors. Beyond the issue of economic costs incurred from near monopoly, the question of dependency also speaks to the issues of security and privacy protection.

Social Aspects

The ICTs have a huge potential to make life better, despite the consequences of the so-called digital divide. The dual societal pursuits of freedom and equality are furthered via the ability of citizens to access the information and services of national and municipal governments. The goal of open, transparent government is dependent upon ever-greater access that ICTs offer. Meanwhile, the choices governments make as regards open or proprietary software, and the value they place on either, act as an example to the public, as well as reflecting the governments' position vis-à-vis issues such as privacy and security.

Education, of course, also greatly impacts on the economic development and potential of a country. Governments, of course, play a major role in creating a proper environment for education. Digital literacy and elimination the digital divide are close correlation with education. In educational systems there are two major expenses related to software: in using proprietary software, schools must buy licenses for every single computer that uses the software, while at the same time, the school has to ensure the possibility that students do not abuse its use after the class.

Many NGOs are not able to afford commercial, non-pirated software. This is a compelling reason to seriously consider FLOSS as a viable option for NGOs. An excellent example of the benefits of open source is the "Human Rights Tool" open source software named Martus (http://www.martus.org/). The developer uses the new model of social entrepreneurship, which combines market forces with philanthropic capital and entrepreneurial drive. Social entrepreneurship as a focus of academic research has a relatively brief history and as yet no research has been made on the connection with FLOSS communities and businesses.

Technical Aspects

The measurable technical parameters are, among others, the reliability, performance and scalability of the systems. These parameters can be compared using the same technical analyses (Wheeler, 2005). The quality of a software product is a controversial field. Functionality of software means the software functions fit the users demand and requirements.

In technical context interoperability is used to describe the ability of different software and hardware form different vendors to exchange data and utilize the same protocols and to operate effectively together. If the competitors' products are not interoperable, the result may be monopoly. To avoid the vendor lock-in, it may be prudent for governments to take steps to encourage interoperability in various situations.

Transparency refers to the fact that, when software is developed, the original source code is available (or not) to public (or user) review. The government is responsible for storing a large amount of data in name of the public. *Lock-in* means that if the data is stored in closed format using proprietary software, the information will only with difficulty be available and retrievable for many decades to come. Since FLOSS and open standards make available the source code, the way in which information is stored is publicly known, or at least traceable. Lock-in can be only avoided by using open standards. Moreover, lock-in may

also refer to education where the brand-specific trainings confine the students and users.

Security is one of the main issues when software is used by governments and public administrations. Computing is crucial to the infrastructure of countries. Nowadays the information environment is extraordinary complex and fragile. Modern society is increasingly vulnerable in its technological and economical infrastructure, in its telecommunications, its energy sources, and its transportation. The infrastructure and information systems can be attacked, destroyed, disrupted, and corrupted by small groups or even single individuals. It is not necessary to destroy the infrastructure in its entirety, nor to attack it physically via traditional means: it can be crippled electronically, and virtually anonymously (Steele-Vivas, 1996). This vulnerability is a reason why the choice of software used is relevant and important. This refers also to the political actions.

Countries where English is not commonly spoken face a serious disadvantage when it comes to the uptake and dissemination of ICTs. However the translation is one of the major parts of localization, moreover, localization involves the task for adapting and customizing the products for local users' specific cultural and/or technical needs.

FUTURE TRENDS

The consideration and utilization of FLOSS by national and municipal governments will continue to grow in the coming years. One of the main fields where FLOSS can best be utilized is in the e-government services increasingly in demand.

A related area where FLOSS can be adopted is within Public Authorities, which are quite different in the each country and which therefore require the flexibility of localization, which FLOSS affords. Another main issue where FLOSS is already utilized with success regardless of cost consideration is in healthcare, which is one of the costliest segments within governmental services around the world. FLOSS can also improve the performance of healthcare services, whilst ensuring both interoperability and patient privacy.

Around the world, governments are developing e-voting systems, but the resistance to these systems by citizens link back to a lack of the trustworthiness of closed systems, which can be avoid by using freely available source code.

Countries in the developing world can gain the possibility to use high-quality free software as opposed to scaled-down versions of more costly proprietary software.

CONCLUSION

Information and communication technologies have drastically changed societies, influencing the everyday activities of both individuals and governments. The information society has become a reality and acted as a call to action by governments. Although much research has been done on the use and consequence of FLOSS in the public sector, not enough knowledge exists on public sector and government policy options and behaviour as regards the adoption of this software. In addition, there are numerous negative perceptions and misunderstanding about FLOSS.

This chapter has made an attempt to give a comprehensive overview of the different fields and aspects relevant to governments and peoples that are influenced by the choice of software that governments make. Another aim has been to delineate the relationship between these related issues and factors. FLOSS touches upon multiple areas as it was introduced in the paper, using the L-PEST model. Beyond a well-known cost consideration, the case studies and the proposed model showed the FLOSS and open standards could afford a workable social-economic-technological solution.

In using the model, it has become clear that the utilization of and decisions regarding software adoption, there are numerous factors that could, and should have impact on software decisions. Within networked societies, interconnectivity was the first step, and nowadays interoperability has gained emphasis. There are numerous arguments that software application that is financed through public funding should be released under FLOSS license. It is not enough that this software is freely available at no cost. With the freely available source code there is the opportunity for improved quality, while simultaneously avoiding lock-in and the development for only one platform. This can bolster the elimination of the digital divide and help foster participation and inclusion programs.

As it was introduced in the Quad theory, the relationships between the Quad's elements determine the performance of the information society and development as a whole. The members of the Quad are involved in the different categories of the L-PEST model.

Much empirical and theoretical work is still needed in this field and in reference to the presented model as well as a better graphical representation of the model. Future research will focus on a detailed examination of motivations and a more precisely defined analysis of every factor involved. In reference to Wilson's Quad model, it might be interesting to investigate stakeholder analysis in contrast of the Quad and L-PEST model.

Protagoras, Greek philosopher (c 485- c 410 BC) said: "There are two sides to every question." And this case there do exist disadvantages in the utilization of FLOSS. It should be noted that the advantages and disadvantages can be measured and evaluated in relation to those incurred by using proprietary software. This model has considered general recommendations focusing on FLOSS but also makes possible comparison between proprietary and FLOSS software.

ACKNOWLEDGMENT

The author wishes to thank Nora for her encouragement and patience throughout the duration of research and writing process, Matthew Strauss for his valuable feedback and improvement of wording of the manuscript, and the anonymous reviewers for their helpful comments and all those who provided advice and suggestions on earlier versions.

REFERENCES

Benson, T. (2005, March 29). Free software's biggest and best friend. *The New York Times*, p. C1.

Bessen, J. (2003). What good is free software? In W. R. Hahn (Ed.), *Government policy toward open source software* (pp. 12-33). Washington, DC: AEI-Brookings Joint Center for Regulatory Studies.

BSA. (2005). *Piracy study*. Retrieved January 8, 2006, from http://www.bsa.org/globalstudy/upload/2005-Global-Study-English.pdf

Einhorn, B. (2003). Why Gates opened windows in China. *Business Week Online*. Retrieved December 12, 2005, from http://www.businessweek.com/technology/content/mar2003/tc2003033_6406_tc058.htm?tc

Grassmuck, V. (2005). LiMux: Free software for Munich. In J. Karaganis & R. Latham (Eds.), *The politics of open source adoption (POSA) Version 1.0* (pp. 14-36). Social Science Research Council [Electronic document]. Retrieved June 3, 2005, from http://www.ssrc.org/wiki/POSA/index.php?title=Main_Page

Greene, T. C. (2002). MS in Peruvian open-source nightmare. *The Register*. Retrieved November 17, 2005, from http://www.theregister.co.uk/2002/05/19/ms_in_peruvian_opensource_nightmare/

IBM signs Linux deal with Germany. (2002). *BBC News*. Retrieved December 2, 2005, from http://news.bbc.co.uk/1/hi/business/2023127.stm

IDABC. (2003). *FLOSS deployment in Extremadura, Spain*. Retrieved December 12, 2005, from http://europa.eu.int/idabc/en/document/1637

Kim, E. (2005). F/OSS adoption in Brazil: The growth of a national strategy. In J. Karaganis & R. Latham (Eds.), *The politics of open source adoption (POSA) version 1.0* (pp. 53-59). Social Science Research Council [Electronic document]. Retrieved June 3, 2005, from http://www.ssrc.org/wiki/POSA/index.php?title=Main_Page

Kriss, E. (2005). Informal comments on open formats. *Mass.gov*. Retrieved December 21, 2005, from http://www.mass.gov/eoaf/open_formats_comments.html

Lanvin, B. (2003). Leaders and facilitators: The new roles of governments in digital economies. In S. Dutta, B. Lanvin, & F. Paua (Eds.), *The global information technology report 2002-2003—Readiness for the networked world* (pp. 74-83.). Oxford: Oxford University Press.

LIANZA (Library and Information Association of New Zealand). (2002). *Towards a national information strategy*. Retrieved October 10, 2004, from http://www.lianza.org.nz/text_files/nis_7nov02.rtf

Linking up Europe: the Importance of Interoperability for eGovernment Services. (2003). *Commission Staff Working Paper, Commission of the European Communities*. Retrieved October 22, 2005, from http://europa.eu.int/idabc/en/document/2036/5583

Open Initiatives of Massachusetts. (n.d.) *Mass.gov*. Retrieved December 6, 2005, from http://www.mass.gov/open_initiatives

Perritt, H. H., Jr. (1997). Open government. *Government Information Quarterly, 14*, 397-406.

Peruvian bill translation. (2005). Retrieved January 8, 2006, from http://www.apesol.org/news/199

Promotion of Open Document Exchange Format. (2003). *IDABC European eGovernment Services.* Retrieved November 28, 2005, from http://europa.eu.int/idabc/en/document/3428/5890

Stanco, T. (2003). *US: Testimony.* Retrieved October 1, 2005, from http://www.egovos.org/Resources/Testimony

Steele-Vivas, R. D. (1996). Creating a smart nation: Strategy, policy, intelligence, and information. *Government Information Quarterly, 13,* 159-173.

The Center for Strategic & International Studies. (2004). *Government open source policies.* Retrieved September 14, 2005, from http://www.csis.org/media/csis/pubs/040801_ospolicies.pdf

The Many Aspects of Open Source. (n.d.). *IDABC European eGovernment Services.* Retrieved January 19, 2005, from http://europa.eu.int/idabc/en/document/1744

The MITRE Corporation. (2003). *Use of free and open source software (FOSS) in the U.S. Department of Defense.* Retrieved October 22, 2005, from http://www.terrybollinger.com/dodfoss/dodfoss_pdf.pdf

UNILOG Integrata .(2003). *Client study for the state capital Munich.* Retrieved January 5, 2006, from http://hdl.handle.net/2038/490

Using open source software in the South African government. (2003). Retrieved September 19, 2005, from http://www.oss.gov.za/docs/OSS_Strategy_v3.pdf

Wheeler, D. A. (2005, November 14). *Why open source software / free software (OSS/FS, FLOSS, or FOSS)? Look at the numbers!* Retrieved December 12, 2005, from http://www.dwheeler.com/oss_fs_why.html

Wilson, E. J., III. (2003). *Forms and dynamics of leadership for a knowledge society: The Quad.* Retrieved June 14, 2005, from http://www.cidcm.umd.edu/wilson/leadership/quad2.pdf

Wong, K. (2004). Free/open source software. Government policy. *International Open Source Network, Elsevier.* Retrieved October 12, 2005, from http://www.iosn.net/government/foss-government-primer/foss-govt-policy.pdf

Working group on Libre Software. (2000). *Free software/open source: Information society opportunities for Europe?* Retrieved November 5, 2005, from http://eu.conecta.it/paper.pdf

KEY TERMS

Dependency: In this context, dependency means that the users are dependent on the software vendor for products and services so that he or she cannot move to another vendor without substantial cost.

Interoperability: Means the ability of systems to operate effectively together independently of different software or hardware vendors.

Localization: Means more than simply the translation of software; it refers to the customization of the software for local needs and demands.

Piracy/Copyright Infringement: The software piracy refers to the duplication, distribution, or use of software without the permission of the copyright holder.

Return on Investment (ROI): Generally, a ratio of the benefit or profit received from a given investment to the cost of the investment itself. This approach also focuses on the benefits and the measurement of the value of making an investment, not only the cost savings.

Total Cost of Ownership (TCO): A financial estimate for such things as (but not limited to) computer software or hardware. TCO is commonly used to support acquisition and planning decisions for a wide range of assets that bring significant maintenance or operating costs across a usable life of several years or more. TCO analysis is not a complete cost-benefit analysis. It pays no attention to business benefits other than cost savings.

Transparency: Transparency involves openness, communication, and accountability. In this context it refers to the fact that, when software is developed, the original source code is available (or not) to public (or user) review.

Lock-In, Vendor Lock-In: In technical terms it means that, if the data is stored in closed format using proprietary software, the information will only be available and retrievable with difficulty. The term also refers to dependency of different types of lock-in, such as when the users are 'locked-in' when trained for a specific technology or the dependency of the specific vendor.

ENDNOTES

[1] "Who controls the software, controls life. Well, it had better us. That's the real political meaning of the free software movement, " said Eben Moglen, professor of law, General Counsel, Free Software Foundation at Open Source Conference, May 2004, Toronto.

[2] The digital divide has a number of definitions and approaches. Examples can be found at: Bridge the Digital Divide (http://www.bridgethedigitaldivide.com/digital_divide.htm) Digital Divide Network (http://www.digitaldivide.net/)

[3] Many governments passed anti-terrorism laws, aimed at enhancing security and facilitated the capture of terrorists. Global Policy Forum Web page (http://www.globalpolicy.org/empire/terrorwar/liberties/libertindex.htm) looks at cases where the "War on Terrorism" threatens civil liberties. The European Union ratified controversial data retention legislation (Directive 2006/24/EC) on the retention of data generated or processed in connection with the provision of publicly available electronic communications services or of public communications networks and amending Directive 2002/58/EC). A week later on, EU and US representatives met for an informal high level meeting on freedom, security, and justice where the US expressed interest in the future storage of information.

[4] Creative Commons has built upon the traditional copyright law based on the all-rights-reserved concept to offer a voluntary some-rights-reserved approach. The Creative Commons licenses provide a flexible range of protections and freedoms for authors, artists, and educators. http://www.creativecommons.org

[5] FLOSS licensing approach based on differences between FS and OSS movement. The free software licenses do not allow closing"the source code while the permissive (OSS) licenses permit the creation of proprietary development. Philosophy on: "Why Free Software" is better than "Open Source" http://www.gnu.org/philosophy/free-software-for-freedom.html; Free Software licenses: http://www.fsf.org/licensing/; Open Source licenses: http://www.open-source.org/licenses/

[6] More detailed reading on software patents in the European Union and other involved issues can be found at: Software Patents in the EU (http://www.oreillynet.com/pub/a/network/2005/03/08/softwarepatents.html) and Software Patents vs. Parliamentary Democracy (http://swpat.ffii.org/).

Chapter XXXV
The Labor Politics of Scratching an Itch

Casey O'Donnell
Rensselaer Polytechnic Institute, USA

ABSTRACT

This chapter will focus on the economic and temporal/labor demands of creating free/libre and open source software (FLOSS). It begins by analyzing the economic and educational foundations of those countries most actively involved in FLOSS development, and how that affects the overall demographics of the FLOSS movement. Through examining the symbiotic relationship that the community has with commercial or closed software development, the educational and employment prerequisites, and overwhelming gendered makeup of the movement, we will come to see the movement in new ways. This is supplemented by an examination of how this economic structure could conceivably be exploited for increased economic gain at the expense of those individuals actually involved in the creation of the software. Finally, the chapter concludes by looking at possible ways in which FLOSS software could be opened up more broadly to non-technical software users.

INTRODUCTION

This chapter will focus on the economic and temporal/labor demands of creating free/libre and open source software (FLOSS). It begins by analyzing the economic and educational foundations of those countries most actively involved in FLOSS development, and how that affects the overall demographics of the FLOSS movement. Through examining the symbiotic relationship that the community has with commercial or closed software development, the educational and employment prerequisites, and overwhelming gendered makeup of the movement, we will come to see the movement in new ways. Expanding our understanding of who is actively involved in developing the software, will enable us to come to a better comprehension about what sorts of economic and temporal resources are necessary for its development and continued growth. This is supplemented by an examination of how this economic structure could conceivably be exploited for increased economic gain at the expense of those individuals actually involved in the creation of the software. Finally, the chapter concludes by looking at possible ways in which FLOSS software could be opened up more broadly to non-technical software users.

BACKGROUND

Recent quantitative studies of the FLOSS movement indicate that the overwhelming majority of FLOSS participants are from the United States and Western Europe. France and Germany lead the pack and the U.S. comes in third. When taken as a whole, Western Europe accounts for nearly 65% of the total number of developers active in the development of free software. When those developers from the U.S. are added the numbers become even more skewed. Nearly three quarters of FLOSS development occurs in these two regions (David, Waterman, & Arora, 2003; Ghosh, Glott, Krieger, & Robles, 2002).

While much has been said about the differing understandings of what precisely "freedom" refers to in the context of FLOSS, the focus is often on that of the source code. When the focus is not directly on the source code, there is a conflation between civil liberties and code liberties (Stallman, 2002). The freedom of developers typically only extends to their freedom to learn/modify that source code. The question of what economic, labor, and political demands precede this freedom is almost entirely neglected by leaders of the movement. While the liberatory promises of FLOSS are indeed admirable, the inability to see their relationship to other economic factors is problematic. While the software is indeed free in both senses of the word, it is difficult to assume that either kinds of freedom automatically indicate participation. Nor would this indicate the kind of discrepancies we see between Western Europe, the U.S., and the rest of the world. There is also the problematic extension of "user" status to that of developers (Karim & von Hippel, 2003). While some would site FLOSS as an exemplary example of participatory design (Schuler & Namioka, 1993), the fact remains that for the most part most current FLOSS

users are not typical users. The level of technical expertise and time required for altering the shape and direction of FLOSS projects is not typical. The future trends portion of the chapter looks at ways in which some FLOSS projects have made potential steps, and opportunities for continued pursuit of the more emancipatory claims made by FLOSS proponents.

If FLOSS is looked at as a social movement, as opposed to a development methodology or ideology, there are two important aspects to examine. While these are not the only aspects or approaches to understanding a social movement (Hess, 2005; Hess, Breyman, Campbell, & Martin, forthcoming), they are the most relevant to this chapter. Resource mobilization and frame analysis draw on what is often referred to as new social movement theory. Resource mobilization looks at how people and economic resources are drawn into a movement, it also scrutinizes the strategic connections which movements make in order to reach its goals (McAdam, Tarrow, & Tilly, 2001). Frame analysis on the other hand examines the kind of rhetoric being used by movement leaders to attract new followers (Benford & Snow, 2000). It is possible for there to be competing frames, or for frames to change over time. What becomes quickly apparent is that in the case of the FLOSS movement, by and large the reasons why people have become involved, and the resources necessary for them to do so departs rather dramatically from the primary frames being presented by the more vociferous leaders of the movement, Richard Stallman being the primary example.

MAIN FOCUS OF THE CHAPTER

The Labor Politics of the FLOSS Movement

The primary social and political-economic prerequisites of the FLOSS movement can be boiled down into three primary needs: higher education, software development employment opportunities, and a job market which supplies middle class or better wage rates. While it is certainly possible for FLOSS participants to have access to none of these resources, and still make significant contributions, the quantitative analyses of the FLOSS movement by and large indicate that this is not the case. While this chapter tends to highlight the political-economic demands of FLOSS development, other barriers to entry also beg further analysis, language being another reasonable starting place.

Looking closely at the quantitative data of other researchers (David et al., 2003; Ghosh et al., 2002), it quickly becomes apparent that there is a direct correlation between the three political-economic elements listed above, and those countries with the broadest participation in FLOSS development. All statistical numbers mentioned in the following section come from the most widely cited demographic numbers available from these studies. It is interesting that those countries providing the majority of FLOSS developers are also the leading developers and retailers of commercial software. In part one wonders if the lack of attention to this relationship is in part a denial of this correlation.

One of the most interesting pieces of the FLOSS story comes directly from colleges and universities, which seem to be a core component for successful involvement in the production of FLOSS software. Seventy percent of those who participate in the development of FLOSS software have at least a bachelor's degree, and almost half have graduate degrees. Many FLOSS projects are even started by academics during undergraduate or graduate careers. The Linux kernel and the GNU project itself are particularly good examples (Raymond, 2001). Many students even become acquainted with FLOSS projects while taking undergraduate courses that utilize FLOSS projects as teaching aids or tools for development. While students need not acquire degrees directly in IT related fields, a large percentage do. These same

trends would also indicate why by and large the number of women active in the development of FLOSS software is also low, because the number of women actively pursing IT degrees has continued to drop in recent years (Randall, Price, & Reichgelt, 2003). Without adequate educational training, very few people acquire the requisite technological expertise that enables them to participate. In both respects, educational programs seem to provide a kind of foundational level from which FLOSS participation comes from.

Directly linked to the involvement of academics and students in the production of FLOSS software is an existing buy in to what some would call a gift economy, but which is known by many other names such as "symbolic capital" or more simply "reputation" (Zeitlyn, 2003). These same ideals are often extend more broadly in speaking about hacker culture, of which the FLOSS movement is related to, but different in at least its public framing (Himanen, 2001). In part the shared history of FLOSS projects starting in academic institutions reinforces the idea that FLOSS economics has a great deal to do with reputation and a buy in to the notions of progressive science found in these institutions. This would also indicate why any kind of suggestion that there are barriers to entry or structural conditions which shape the landscape of the FLOSS movement to either be ignored or denied (Kelty, 2001). Others have also demonstrated that to a large extent FLOSS projects are in fact highly hierarchical at the level of practice, and that many projects are one person strong, while only the most active projects have five or more people. This suggests that FLOSS has more to do with itch scratching at a personal level than freedom (Healy & Schussman, 2003).

The need for IT employment opportunities is in many ways tightly tied to the first necessary component. In many respects, the market can place demands upon educational institutions to provide it with a labor pool meeting its needs. In this case however, because many students become interested in, and even vested into FLOSS during their educational years, it makes sense to leave it as the first demand. The relationships between the academy and labor markets aside, nearly 83% of those individuals working on FLOSS projects are employed in an IT related field. Put another way, only 17% of FLOSS developers are not employed in IT. Of those not actively employed in the IT sector, almost all of the remaining developers are actually students. Only 5% of the FLOSS movement is made up of people outside of the academy, or not employed in IT jobs. While this number is not insignificant, and perhaps worthy of more study, it makes it difficult to assume that this number will grow based simply on the nature of the movement. With this in mind, it quickly becomes apparent that IT employment opportunity provides both a motivating factor for pursing education, but also a demand for the kinds of expertise which are also required by FLOSS projects. Indeed, some have argued that the "reputation" mechanism built into FLOSS, which is supposed to link it into the meritocracy of the academic machine, might have more to do with a desire for economic gain or professional development (Watson, 2005).

In many contexts, free software work begins to occupy a kind of professional development space for software developers. It is a context in which they can work on larger software projects than could be done on their own, and begins to act as a kind of portfolio for job seekers. In places where commercial IT employment may be more competitive, FLOSS becomes an arena in which aspiring employees cut their teeth on real world projects in the hopes that it makes them more desirable job candidates. Nearly every survey respondent for the quantitative studies from which this chapter pulls its conclusions from noted that skill development and improved job opportunities were important motivators for why they were active in FLOSS development. These two motivators are directly tied to the availability of IT employment opportunities. This aspect too plugs into the reputation machine already mentioned;

many employees cite possible improvements in professional developemnt as a motivating factor for FLOSS work (Watson, 2005).

More than half of FLOSS developers receive some kind of compensation for their FLOSS work. This statistic brings us to our final political-economic demand which FLOSS development is based upon. Not only do FLOSS developers have IT job opportunities, most of the jobs that are available come with pay structures that place them firmly within a middle class or better lifestyle. Seventy percent of those involved with FLOSS development make at least one thousand Euros or better per month. Almost 50% make 2,000 or more Euros per month. Also, given the nearly 10% of those who are students, and likely having little or no income, this will bring down the average pay rates. Nearly 7% of FLOSS developers reported having no income, which means that their involvement is likely supported by other means.

Who is free to work on free software? While some will answer, "Everyone," the reality seems somewhat different based upon the information contained in survey data. There is a very specific demographic that dominates the development of FLOSS software. While the focus above has been on political-economic demands, there are others worth noting as well. One of the most under examined demands is temporal. Who is free to spend time working on FLOSS software? While some developers are employed to work on FLOSS projects, many do so on a volunteer basis. Nearly a quarter of developers spend only two or fewer hours per week developing FLOSS software. However, another 45% spend nearly two to ten hours working on FLOSS projects. It quickly becomes apparent that there is an implicit assumption that free time is available to be spent on software development. Nearly 40% of FLOSS developers are single. Only 20% are married. While it is not fair to say that FLOSS developers are by and large, bored and lonely. It is fair to say that the time demands of relationships and family life certainly have an impact on how much available time can be spent on the development of software (Hochschild & Machung, 1989).

Of course there are notable and interesting outliers for these demands, it is important that more work be done to better understand the ways in which these social and political-economic demands thread themselves through the FLOSS community. There may also be other means by which political-economic demands of FLOSS could be reduced, though one wonders what the consequence of such a movement would be. IT work/labor seems to flow through the core of the FLOSS software movement. Many of those involved in the development of these projects do so not only to scratch and itch, but also as a means of professional development, making money, educating themselves, and many others. This is done not out of a primary interest in freedom, but a knowledge of the market which they are a resource within, as well as out of a love or interest in the development of software systems. While others have attempted to characterize other motivating factors as "pivots" or forces that shape communities based upon "attainment" (Stewart, 2004), or the economics of OSS and beyond (Lerner & Tirole, 2002, 2004), external structural considerations are still absent. Can something be learned from acknowledging all of the motivating factors both internal and external that drive the movement?

It would appear based upon this survey data, that there are indeed prerequisites for participation in the FLOSS movement. However, these demands are never examined despite the pervasive use of the word *free* in numerous contexts and with different meanings. Even when utilized as a means of getting at freedom, the questions never probe any further about what other kinds of freedoms and opportunities must be operating for such a freedom to exist in the first place. This is in part because the FLOSS movement was born out of countries where these demands and prerequisites were already in place, there was never a need to re-examine them. Only now, as the FLOSS move-

ment has become broader and more global do we begin to see the need for re-examining these assertions. While in part this chapter is critical of this inattention, it does so in a spirit of renewed understanding and broader participation. Without critically examining these issues, the FLOSS movement will remain a predominantly the project of Americans and Western Europeans, when in so many ways the movement can conceivably offer so much more.

FUTURE TRENDS

Based upon this information, it seems reasonable to assume that as other countries find themselves more able to provide the political-economic foundations from which FLOSS development can spring, more global involvement will be found. However, it is concerning that the U.S., though recording the largest number of potential IT professionals, and some of the most well paid, actually contributes percentage wise the fewest developers to the FLOSS movement. Simultaneously, U.S. corporations are shifting operations to make more effective use of FLOSS software. The potential for economic exploitation are undeniable. While some companies are busy shifting software development operations offshore, others are busy reducing software teams and refocusing on FLOSS based initiatives. This kind of movement cuts away at the foundations, which make FLOSS development possible in the first place. If the economic foundations of this kind of development are examined and more widely acknowledged, it is possible that an improved relationship between free and commercial software could develop (Lancashire, 2001). If the focus remains simply on free however, without acknowledging the very real human costs associated with the development of free software, the potential for exploitation will remain.

By also taking seriously the demands of FLOSS involvement, commercial software organizations could find themselves able to gain more. If many software developers already spend significant amounts of non-work time involved in the development of free software projects, what could be done to ensure that the involvement was usable by the parent company? While it is true that others will stand to benefit from the same investment, so too does the company gain the possibility of earning from the investment of others. Though it does not translate directly into the language that guides most corporate organizations, that is that surprising, given that the FLOSS movement in many ways was a reaction against broader practices of software development throughout the industry.

In many ways the future trends associated with the politics and economics of the free software movement are tied up with the future of commercial software development. While IT workplace practices continue to change and adjust to the global economy, so too with the FLOSS landscape be shaped by these forces. Because of the demands which FLOSS makes upon education, employment, and capital, it only make sense that the future of software development capital, work, and education will continue to impact and shape this movement. Simultaneously, FLOSS has also had a significant impact on these three areas as well, and will continue to do so.

It is also possible that if the FLOSS community were to broadly adopt certain practices that enable new kinds of interaction with developers and projects, which even users without the available resources to contribute code can make new and innovative alterations to a project. Broadly speaking, this approach could be thought of as design for appropriation (Eglash, Crossiant, Di Chiro, & Fouche, 2004). It has already proven effective at encouraging new uses and expansive growth for applications that make the technological investment in such mechanisms; the Firefox Web browser for example currently has more than 5,000 available add-ins (Multiple, 2006). These kinds of design decisions could make significant alterations to the kinds of barriers to

entry that currently exist. This also asks FLOSS developers to take seriously the idea of a using user rather than consuming user, an idea, which pervades the commercial software industry (Gillespie, 2004).

CONCLUSION

The almost symbiotic relationship between commercial software development and the FLOSS movement needs to be acknowledged. With nearly 50% of FLOSS developers making their income from the development of other software packages, it is problematic to continue denying the social and political-economic factors that make these projects possible in the first place. The political-economic foundations of FLOSS software seem to lie in three primary categories: educational, employment, and work compensation. Each one of these is important to the involvement of software developers in FLOSS projects. Without these, broad participation in free software development would not occur. By not acknowledging these links, we open ourselves up to the possibility of exploiting IT workers in both established and emerging economies. While free software may indeed be free in the broadest senses of the words, the context in which free software labor occurs is not free of the realities of social and political-economic demands, and we must also keep those issues in view. It is also possible for FLOSS developers to make a conscious decision to alter these structural demands.

REFERENCES

Benford, R., & Snow, D. (2000). Framing process and social movements: An overview and assessment. *Annual Review of Sociology, 26*, 611-639.

David, P. A., Waterman, A., & Arora, S. (2003). *The free/libre/open source software survey for*

2003. Stanford, CA: Stanford Institute for Economic Policy Research.

Eglash, R., Crossiant, J., Di Chiro, G., & Fouche, R. (2004). *Appropriating technology: Vernacular science and social power.* Minneapolis: University of Minnesota Press.

Ghosh, R. A., Glott, R., Krieger, B., & Robles, G. (2002). *Free/libre and open source software: Survey and study.* Maastricht, The Netherlands: International Institute of Infonomics.

Gillespie, T. (2004). *Designing against user agency: A consideration of the FCC 'broadcast flag'.* Paper presented at the Society for the Social Studies of Science, Paris.

Healy, K., & Schussman, A. (2003). *The ecology of open-source software development.* Free/Open Source Research Community Working Paper. Retrieved from http://opensource.mit.edu/papers/healyschussman.pdf

Hess, D. J. (2005). Technology- and product-oriented movements: Approximating social movement studies and science and technology studies. *Science, Technology, & Human Values, 30*(4), 515-535.

Hess, D. J., Breyman, S., Campbell, N., & Martin, B. (forthcoming). Science, technology, and social movements. In E. Hackett & J. Wacjman (Eds.), *The handbook of science and technology* (3rd ed.). Cambridge, MA: MIT Press.

Himanen, P. (2001). *The hacker ethic, and the spirit of the information age.* New York: Random House.

Hochschild, A., & Machung, A. (1989). *The second shift: Working parents and the revolution at home.* New York: Penguin.

Karim, L., & von Hippel, E. (2003). How open source software works: 'Free' user-to-user assistance. *Research Policy, 32*(6), 923-943.

Kelty, C. (2001). Free software/free science. *First Monday, 6*(12). Retrieved from http://www.first-monday.org/issues/issue6_12/kelty/index.html

Lancashire, D. (2001). Code, culture, and cash: The fading altruism of open source development. *First Monday, 6*(12). Retrieved from http://www.firstmonday.org/issues/issue6_12/lancashire/index.html

Lerner, J., & Tirole, J. (2002). Some simple economics of open source. *Journal of Industrial Economics, 52,* 197-234.

Lerner, J., & Tirole, J. (2004). The economics of technology sharing: Open source and beyond. *National Bureau of Economic Research.* Retrieved from http://www.nber.org/papers/w10956

McAdam, D., Tarrow, S., & Tilly, C. (2001). *Dynamics of contention.* Cambridge, UK: Cambridge University Press.

Multiple. (2006). *Firefox add-ons.* Retrieved July 7, 2006, from https://addons.mozilla.org/search.php?app=firefox&appfilter=firefox&type=E

Ong, A. (1991). The gender and labor politics of postmodernity. *Annual Review of Anthropology, 20,* 279-309.

Randall, C., Price, B., & Reichgelt, H. (2003). Women in computing programs: Does the incredible shrinking pipeline apply to all computing programs? *Inroads—The SIGCSE Bulletin, 35*(4), 55-59.

Raymond, E. S. (2001). *The cathedral & the bazaar: Musings on linux and open source by an accidental revolutionary.* Cambridge, MA: O'Reilly and Associates.

Schuler, D., & Namioka, A. (1993). *Participatory design: Principles and practices.* Hillsdale, NJ: Lawrence Erlbaum Associates.

Stallman, R. M. (2002). *Free software, free society: Selected essays of Richard M. Stallman.* Boston: GNU Press.

Stewart, D. (2004). *Social forces and constraint in the attainment of community status.* Free/Open Source Research Community Working Paper, http://opensource.mit.edu/papers/stewart1.pdf

Watson, A. (2005). *Reputation in open source software.* Free/Open Source Research Community Working Paper. Retrieved from http://opensource.mit.edu/papers/watson.pdf

Zeitlyn, D. (2003). Gift economies in the development of open source software: Anthropological reflections. *Research Policy, 32*(7), 1287-1291.

KEY TERMS

Design for Appropriation: The idea that systems can be designed in such a way that they are more open to user manipulation or transformation.

Frame Analysis: Examining the rhetoric or presented meanings of social movement leaders as an insight into how a movement generates followers.

Labor Politics/Economics: The relationship between labor or work and broader social, political and economic aspects. This can also be the relationship between workers and those they work for. For more information, see Ong (1991).

Resource Mobilization: Examining the means by which social movements generate followers, connect with other organizations, and generate the resources necessary for its longevity and success.

Social Movements: "Social movements enhance public participation in scientific and technical decision-making, encourage inclusion of popular perspectives even in specialized fields, and contribute to changes in the policymaking process that favor greater participation from nongovernmental organizations and citizens generally" (Hess, Breyman, Campbell, & Martin, forthcoming, p. 1).

Structural Demands/Conditions: Refers to the relationship between different groups or entities and to a relatively enduring pattern of behavior or relation. Social systems, institutions, or norms become embedded in society in such a way that they are relatively unquestioned.

Symbiotic: Close relationship between two organisms, groups, or movements in close relation. These relationships are typically beneficial to both.

Symbolic Capital: The amount of prestige a person holds acting within a certain set of social structures. The use of the word capital implies its location as part of a system of exchange.

Users/Consumers: The distinction is made that there is a difference between users and consumers, that ones role is seen as more active and co-producing, and the other as passive and depleting.

Chapter XXXVI
Open Source Technology and Ideology in the Nonprofit Context

Jonathan Peizer
Internaut Consulting, USA

ABSTRACT

This chapter contextualizes open source development and deployment in the nonprofit sector and discusses issues of ideology that often accompany it. The chapter separates and defines the ideologies of application development, selection and use, describing the different issues and impacts each creates in the nonprofit context. The purpose of the article is to clearly articulate the unique dynamics of application development and deployment in the nonprofit or social value context and where to apply ideological considerations for best effect.

INTRODUCTION

This chapter contextualizes open source development and deployment in the nonprofit sector and discusses issues of ideology that often accompany it. Open source has intensified the ideological debate over what technology to deploy in a given circumstance. The nonprofit sector, always price sensitive to any technology solution, has embraced the idea of open source as a cheaper alternative to commercial applications. Open source is also viewed by some as embodying the humanistic and cooperative (vs. competitive) philosophy that defines the best practices of the sector.

Open source refers to a program in which the source code is available to the general public for use and/or modification from its original design free of charge. It is typically created as a collaborative effort in which programmers improve upon the code and share the changes within the community (Sacchi, 2002). Open source has come to mean different things to different constituencies. To software programmers it reflects a particular development methdology and philosophy. The more legally minded see it as licensing ideology that more easily allows sharing intellectual property. To users, especially nonprofit institutions with typically limited resources, it means free sofware and freedom from dependence on proprietary technology and related service models. In addition to these definitions, there is also a strong ideological lobby that sees open source as the alternative to commercial dominance by any one player in the software industry and as an equalizer with the potential of wresting control away from U.S. predominance in the software industry (Stewart & Gosain, 2001).

Because open source methodology and ideology have become so intertwined it is appropriate to ask if the right debate is taking place around it, particularly in the context of nonprofit implementations of this technology. Ideology and technology cohabit the same plane of existence on three distinct levels:

- **Development ideology:** How is the technology developed?
- **Selection ideology:** Why is the technology chosen?
- **Ideology of use:** What is the technology ultimately used for?

The most important and thorniest ideological consideration is the ideology of use. Unfortunately, far too much time is spent obsessing about the ideology of software selection to meet a particular need and far too little time considering the effects of its application. How software is deployed, particularly in a world that is hypersensitive to global security concerns, has much farther reaching implications and consequences than the ideologies used to create and select it (Kling, 1983).

BACKGROUND

Development Ideology: How is the Technology Developed?

Ideological considerations occur early in the development process. Is software developed for free, on a commercial basis or as a hybrid of the two (Lerner & Tirole, 2002)? Is an application designed to meet a social mission, a personal interest or a business requirement? On the legal front should applications be fully available to the public for the purposes of modification, or hidden behind proprietary legal constructs? From a standards point of view are considerations purely technical or are the needs of the disabled and disadvantaged taken into account when designing new technology specifications?

Developers ultimately decide why they build applications. They decide if they wish to generate profit, simply sustain ongoing development and maintenance costs or if contributing a piece of code to the world is payment enough for their efforts. In the current reality, lower price points, mass distribution networks, and a proliferation of useful toolsets have allowed software developers a far more significant range of ideological decisions to make when they create software. They have a plethora of commercial and open source languages, tools, operating systems, and even legal frameworks to choose from in order to develop and distribute their creations.

In this new environment it is also far easier to develop tools for the social sector than it ever has been. The advent of the PC in the 1980s made technology affordable for the first time to many nonprofits. The PC created a market for the social sector that in large part did not exist in the costlier mainframe context. In the '90s, the Internet once again lowered the barriers by providing a technology that allowed nonprofits to reach out and extend their constituencies at a far lower cost (Lee, 1997). Open source tools have unlocked even more development opportunities for this market. They have spurred commercial software developers to rethink their price structures in order not to lose this relatively new market consisting of literally millions of social purpose nonprofits, educational institutions and health facilities globally.

Developers of commercial software maintain a straightforward profit-based ideology for any market they sell to. However, that does not preclude them from doing pro bono work or developing applications for the social sector that are heavily discounted or distributed freely. Salesforce.com has a foundation and distributes discounted and free licenses of its products to nonprofits (Salesforce, 2005). Techsoup.org provides a variety of software aggregated from different vendors who are interested in providing discounted commercial applications to the nonprofit sector (Techsoup, 2005). Open source developers operate on a number of levels as well. Some have strong ideological convictions that tools should be developed free of charge for the social sector as well as for any non-commercial user (Stewart & Ammeter, 2002). Others are driven by a need to limit the dominance of a single, perceived commercial player. Still others simply wish to demonstrate their creativity to the world and to build a better mousetrap. There are even open source developers advocating the free distribution of source code while allowing an economic model based on distribution (Ellliott & Sacchi, 2004).

Both the commercial and open source developer community may operate on development ideologies that are purely technical, focusing on building software tools for other developers that allow them to in turn build end-user tools (Kuan, 2002). Developers may also choose to build generic end-user products that meet the needs of any sector. Word processors and spreadsheet products for example can be built using either commercial or open source tools, and following either commercial or open source principles of distribution. All sectors including the social sector have a need for these basic tools in whatever form they are built. The social sector also requires specialized mission-focused applications which are often not available as mass-produced shrink wrapped applications.

Some suggest that the promise of open source to the nonprofit sector lies in the open code base that allows developers around the world to collaborate on projects to produce or enhance new application (Kogut & Meitu, 2001). There is

an expectation of a whole slew of new mission critical applications to meet nonprofit needs at a reduced cost. This assumes a reasonable number of developers exist that are willing to devote time for little pay to work closely with nonprofits over significant periods, measured in years, to develop and upgrade these applications. It also assumes the problem has been that nonprofits have a hard time developing applications to meet their needs in the proprietary marketplace due to a slew of programmers not having access to code.

Experience indicates that nonprofits typically do not have the resources to implement basic technology right out of the box let alone supporting technical staff to develop and maintain their applications. Technology support organizations like NPower and the Circuit Rider movement work in the nonprofit context because technology in this environment requires that it be bundled with capacity and service (E-Riders, 2005; NPower, 2005). Capacity and service are what for-profits invest in internally so they can absorb and take advantage of the technology they implement. In the nonprofit environment only the largest nonprofits, (typically those with the capacity to generate income) invest in internal technology departments. The rest require low cost nonprofit technology service providers or consultants (McInerney, 2004). Making code accessible though an open source development ideology does not magically create a cadre of new and interested programmers willing to develop and maintain applications for the nonprofit environment. There must be an underlying economic model that provides resources to compensate them over years of development and maintenance (Lerner & Tirole, 2002).

It is clear why open source application efforts such as Mysql and Apache work (Kuan, 1999). These applications are about developers creating products for other developers in order to enhance their own efficiency and productivity (Dempsey, Weiss, Jones, & Greenberg, 1999). In the end, these products help anybody implementing a web server or database including nonprofits. The constituency for these applications is huge—much larger for example than for an application focused on case management for battered women. It is also clear why end user open source applications like Open Office developed for a mass audience (including for-profits and nonprofits) work as well. They have the benefit of well-paid technical staff employed by companies who may wish to work with the code to enhance internal needs or to experiment on their off time. Some governments, which are beginning to mandate open source usage, may contribute technical support to these endeavors as well.

Nonprofits certainly benefit from both the hard-core open source technical products like servers and databases and the mass-market open-source end-user products. However, they are not necessarily underwriting their development or enhancing the code themselves with phantom technical resources they cannot afford. The fundamental question an open source development ideology leaves unanswered is how one underwrites and sustains the development and continued maintenance of mission sensitive open source applications for the nonprofit sector (Franck & Jungwirth, 2001)? In the current environment, many of these applications are still subsidized by foundation underwriting—hardly a long-term solution for sustainability (Saint-Paul, 2003).

Finally there are also destructive software development ideologies. Some developers create viruses, worms, trojans and other harmful applications for no other purpose but to cause disruption. These development ideologies are distinct from *using* tools that are ideologically neutral or beneficial for destructive purposes, (for example using the ability to imbed hidden copyright or file information in images to pass terrorist messages along). In the latter case the technology itself is not designed to be destructive, but is used for that purpose. In the case of destructive development ideologies, both the development and use of the application are designed to be nefarious. Destructive developer

ideologies aside, commercial, non-commercial, or socially responsible development ideologies are all equally valid. They represent the product of their developer's creative interests in solving a particular problem. However, the fact that any technology, whatever the design intention, can be used for constructive or destructive puposes, underscores why the ideology of use often trumps the ideology of development.

Selection Ideology: Why is the Technology Chosen?

How should users choose a software application that best meets their particular requirements? Unfortunately, the questionable practice of applying software development ideology as the primary decision point to the software selection process is becoming far too common and has created an unnecessary complication for nonprofits trying to employ technology to meet their mission objectives. The idea that software selection choices should be made based primarily on the premise of free vs. commercial technology and open source vs. proprietary technology is entirely misguided.

Managing systems operations and satisfying the needs of real users meeting long term organizational objectives often produce professional IT managers that are agnostic pragmatists rather than ideologs. While it may be fashionable for some developers and users to equate open source with open society, most users trying to achieve their business objectives using IT as a process are interested in only one thing—that the software satisfies their need to get from point A to point B.

It benefits everyone when software developers make decisions to create a variety of free, commercial, proprietary, and open source solutions. The various ideologies chosen to develop these products, provides users the freedom to choose the best solution from a diversity of options to meet the objectives at hand. The ideology behind the application's development may be only one of many factors determining which software to choose—along with many other variables that must be prioritized. Solid operational prerequisites most often used to base software selection decisions on include:

- Do the application's functions meet the user specifications?
- Do the design considerations meet project requirements?
- Do the cost considerations meet project requirements?
- Do the security considerations meet project requirements?
- Do the networking considerations meet the project requirements?
- Are the necessary resources there to program or deploy the application?
- Are the necessary resources there to maintain the application?
- Are the necessary training and documentation resources available to satisfy project requirements?
- Is the hardware available and appropriate to meet the needs of the software application?
- Is there a facility to convert data?
- Are the necessary integration points there if the application must interface with other applications?
- What is the evolutionary trajectory of the software I choose?

Answering these questions may lead to selecting applications built on particular development ideologies. However, the selection process is based purely on an objective set of operational criteria to deliver the most effective solution satisfying a stated need.

Price is a very sensitive factor to the nonprofit community, often influencing the selection of applications. However, ease of installation and use and continued high touch support are also important factors to take into consideration when satisfying this sector. When applications don't work in this environment and there is no support around to provide basic assistance, users become very reticent to use the technology again; much more so than in the commercial context. A good project manager must weigh all these decisios before making a selection.

There are at least four reasons why selecting software based primarily on an ideological preference is not recommended:

- **Selection methodology should compliment the risk of any software implementation:** A software implementation is a costly and complex affair that involves a sophisticated behavioral interplay between people and technology. Often it means changing the way departments or whole institutions do things as they adapt to often less than intuitive automated processes. Most people are naturally resistant to these changes. Technologists who manage software implementations know that there are many pitfalls to watch out for even in the best of circumstances. Choosing an application for any reason other than how it meets specified business requirements is a tremendous gamble (Mosko, Jiang, Samanta, & Werner, 1999).
- **What criteria of selection actually make sense?** When building a house, is it best to select the tools to use based on the alloys they are built with? Their craftsmanship? Their cost? The method that went into forging them? The most logical and primary consideration would be to select the right tools necessary to complete the building project. Craftsmanship, cost, alloys, and method of creation might all be considerations, but these factors should be weighted based on how they contributed to the tool's success in helping complete the building project. As attractive as it might be, using a hammer forged on Thor's anvil is innapropriate if

what was really needs is a screwdriver from The Home Depot.

- **Defining the real cost of use and tangible benefits of social source for nonprofits:** Many organizations find open source vs. commercial applications more attractive because they are free to use. Often there is never any real plan to actually tinker with the application code to modify how it works—a major benefit of open source products. The limitation on technical resources are already major limitations in developing a product further in the nonprofit environment. If cost of purchase is the main motivation, let the buyer beware. The real costs of any application deployment outside of initial purchase relate to installation, training, data conversion, ongoing maintenance, support and new version upgrades. These must all be taken into consideration if using commercial or open source applications. What is free now may also have a cost later. The once free open source, the Star Office revision now has a price attached to it. This often happens as an application gains significant market share. The need arises to better support its continued development and maintenance for an increasing and more demanding end-user market in an organized and timely fashion.

In this sense, the nonprofit sector's use of open source may not be much different from their use of commercial applications. True, they are not paying retrogressive licensing schemes while the software is still free (Lerner & Tirole, 2005). However, they are not necessarily taking full advantage of the promise of open source either. They must still pay someone for long-term technical support for applications ideologically developed to meet a social good.

- **Comparing apples to apples in generating social value:** The "social value" case that some argue for open source software is not compelling enough to influence a selection decision (e.g., that because open source is free and open to redesign, nonprofits end up with access to richer, less costly, and more reliable applications, freeing themselves up to spend their limited resources elsewhere). In fact, there is just as valid an argument to support and opposite viewpoint. Consider this:

The social benefit of most open source applications is primarily in their free use and less so in their extensibility. The benefit of free, modifiable code would constitute a far more significant social benefit if most nonprofits took advantage of it, but most cannot because of resource constraints. There are also training and documentation costs associated with any new and significant software modification. Commercial software is typically closed and de facto has an expense connected with its purchase. However, it is often deeply discounted for the nonprofit and educational environments, although not all over the world as it should be. Software that is unaffordable but necessary is often pirated in developing countries that cannot afford it, nullfying the actual cost acquisition arguments of open source vs. proprietary software.

Commercial software developers that discount for their nonprofit customer base may create far more social value if they also convert some of their commercial sales revenue directly to philanthropic purposes. A number of philanthropic institutions and corporate social responsibility programs are funded by commercial software profits and are contributing to the global fight against aids, the reform of micro lending and economic development, training and education, library support, children's programs, media development, and a plethora of other social value activities. The Gates Foundation has the largest endowment of any U.S. foundation dwarfing the Ford, Rockefeller, and MacArthur endowments, and the Open Society

Institute's yearly allocations. It must allocate at least 5% of that endowment (about one billion dollars) of grant funding annually. One cannot separate the direct correlation between revenue generated from commercial software and the work of the Gates Foundation, the Microsoft Community Affairs Department, the Time-Warner AOL Foundation, The Real Foundation and Glaser Family Fund, The Paul Allen Foundation, and so forth. They do quality work and their people are just as dedicated as any other foundation staff to the value proposition of assisting civil society. It is disingenuous to compare the social value of both commercial and open source applications without recognizing this other dimension of social benefit that accrues from commercial application development.

Applying ideology to the selection process in either a commercial or open source context is a tricky business. The reality is the current IT environment is a hybrid technology environment. This has been for decades for decades. Many organizations currently support a mixed environment of proprietary and open source applications as the need dictates. Walk into many organizations today and you'll find internet servers running on open source linux, apache, and Mysql while the desktop environment supports Microsoft Windows and Office applications. While software developers choose the ideology they are most comfortable developing applications in, when it comes to selecting an application to meet a particular user need, its best to select applications based soley on operational criteria that best satisfies the need.

MAIN FOCUS OF THE CHAPTER

Ideology of Use: What is the Technology Ultimately Used for?

The deployment of any technology is by far the most interesting ideological concern but often the one least focused upon. Most software is

built to solve a particular problem or to create a new functionality. All technology development is informed by values. However, a technology tool, once developed, can be applied in many ways that reinforce the original intention, run counter to it or spur new possibilities never thought of by the developer. Ideological debates around technology development and selection are easier to have because the issues are far more limited, and revolve around technology choices and objective operational requirements. The genie is let out of the bottle only once a technology is deployed. The ideology of use poses far more serious ethical issues than the development and selection ideologies previously discussed. Following are five examples in the current global context.

Case #1: Ideology and Terms of Use

What if a technology allows encrypting hidden messages in a digital image to pass along to an intended recipient who has the key to unlock the message? This application can be used by the Otpor Student movement in Serbia in a bid to change an autocratic regime, or it can be used by Al Qaeda to communicate its next major terrorist attacks against a target. Should the usage of such tools be somehow regulated?

And what if they are regulated? Some years ago, then-Russian President Yeltsin issued a decree that the keys to all encryption designed into software and distributed in Russia must be provided to the FSB (the Russian successor of the KGB) (Anderson, 1995). That would cover the example above but it would also cover a securely encrypted, open source human rights application. A Chechen NGO in Russia using such an application to track human rights abuses would not necessarily be as fully protected by the laws in that country as a similar organization tracking abuses against Islamic citizens in the U.S. However what if this application did fall into the wrong hands and was used by a Chechen terrorist organization?

Here the ethical dilemma takes on an interesting twist. Reporting the encryption keys to the appropriate authorities could put a legitimate human rights organization in jeopardy given the anti-terrorist, anti-Chechen environment. However, not reporting the keys might allow the application to fall into the wrong hands, allowing secure encrypted communications in a country where it is clearly illegal without the government having a key. What is the responsibility of the developer who makes a secure application freely available in SourceForge (the online open source software repository)?

Hacktivismo has taken a crack at this type of ethical dilemma by developing an ideology-based licensing regime, the Hacktivismo enhanced-source software license (Hacktivismo, 2005).

This modified, open source license regime requires that applications be used for their intended purpose, to support Hacktivismo's political agenda: Assertions of liberty in support of an uncensored Internet. Martus, the secure, open source human rights monitoring application referred to above uses strengthened "anti-hacking" clauses in a standard open source software license to protect its application and users (Martus, 2005).

Making the application available with a license for intended use and clear instructions that it should be used legally in the environment in which it is deployed represents one viable solution for the developer to avoid both extremes. It creates a contract between the developer and the end user but leaves it up to the user in country to abide by both pre-requisites. Restricting the application's use in Russia altogether might turn out to be as ineffective as the PGP encryption software ban was in the United States. On the other hand providing pre-assigned keys is not really an option as neither the FSB or the developer have the processes and resources in place to track every user that could pull it off an open source application catalog like SourceForge.

This example is not as extreme as it sounds. Commercial vendors are making their software

code available to governments in order to meet their national security concerns in light of the global terrorist threat. In making the code available however, trust is being put in the various governments not to abuse or exploit this information.

Case #2: Ideology and Hacktivism

Denial of service attacks have brought down major Web sites like Yahoo and eBay causing millions of dollars in lost business and annoying service disruptions. They have even precipitated arrests for criminal mischief. However, the famous Chiapas denial of service (DoS) attack attributed to the Electronic Disturbance Theater was an act of civil disobedience, commonly referred to as hacktivism. Hacktivism promotes social causes online, in this case the plight of the indigenous people of Chiapas Mexico. In the current world context, what application of technology constitutes criminal behavior, terrorism or hacktivism/civil disobedience?

The originator of the Chiapas (DoS) attack argues that the Chiapas attack was technologically full of holes. It was acknowledged as easy to get around and obviously technologically flawed as DoS attacks go. It was designed as an act of civil disobedience to send a message clearly related to an issue of social importance. Finally, it was attributed to an organization with known credibility in the hacktivist community, a community driven to advocate for social justice through the creative use of technology. Given the new threats faced today, can the intent of these attacks be distinguished by the sophistication of the software involved, the nature of the cause, the amount of damage done or the entity from which it emanates?

Just as it is important to distinguish activism and civil disobedience from criminal behavior and environmental terrorism hacktivism must be distinguishable from cybercrime/cyberterrorism. Billions of dollars of national security technology R&D coupled with a push to standardize privacy and surveillance laws internationally have the potential to make the Internet a much less open and democratic place than it has been. It may be far easier to mislabel hacktivism cyberterrorism or at least criminal mischief in the future. Yet activism and civil disobedience are valid forms of protest and have been protected civil liberties (off-line). Even as very valid national and global security concerns are addressed, some provisions must be made for this form of speech that protects it online as it does offline. There may come a time when "traditional" hacktivists are included as arbiters of what constitutes hacktivism and what does not in a society that is more sensitive to national security concerns.

Case #3: Ideology and the Technical Fix

The Martus Human Rights application consists of two core parts that make up a secure client and server. The latter can sit in a different country to securely store human rights reports. The developer wishes to make Martus an open source application along with a modified open source license. However, doing so might open Martus up to dangerous hacking by those who would undermine the application and get to the human rights data it is designed to protect. Is the Hacktivismo modified licensing agreement the application's only protection [or enough] against people who would violate human rights? Does the nature of the application disqualify its submission as an open source product?

In this particular case, the design philosohpy of the application informs both its use and its security. The basic application can be modified as open source software. However, the security it uses to protect users against access to their records is the same strong encryption protocol employed by secure tools such as PGP. This encapsulated module within the Martus product cannot be modified. On the server side, the application designed to store information does nothing but authenticate users and store their data. It cannot even read the encrypted messages. There is not a whole lot of sophistication built into the server side outside of doing very discreet and simple tasks. The processing decisions are made on the client side. Hence there is far less reason to release the server side software as open source because it would not be particularly useful to build upon. The entire application speaks to both development and use ideologies focusing on two objectives: Making it secure enough for the human rights constituency to be able to trust it, and freely available as open source so they can afford to use it.

Case #4: Ideology and Destructive Technology

We assume viruses are all bad. But what if for national security purposes a democratic government creates a virus that infiltrates a terrorist's PC and captures his keystrokes so that important information is uncovered that prevents an attack and saves thousands of innocent lives?

It is technically feasible but how can it be assured that such a virus does not fall into the wrong hands or that is not used improperly in the right hands? Just as a socially responsible application can be used for destructive purposes, so can a typically destructive application be used for benevolent purposes. What is the intrinsic ideology of a gun for example? Protection or violence? The ideology of use and the user often determine the context. Using the gun as a good example, it is more logical to regulate applications typically used for destructive purposes than those purposed for benevolent use whatever the original design intention.

Case #5: Free Market Ideology and Technology

What is the responsibility of any commercial corporation that has developed its technology in a free and democratic society not to sell this same technology to repressive governments in order to censor, secretly monitor, or otherwise oppress its people? What is its obligation once in a repressive country not to use its software to help a government harass or detain its citizens in contravention on international conventions or treaties on human rights?

At this crucial intersection between social welfare and free enterprise we have not found the appropriate answer in many contexts. The debate around the publish-what-you-pay movement, conflict diamonds, generic drugs to the developing world, and breaking the technology filtering regimes of oppressive countries all have their roots in better defining the traffic lights for this intersection. Often governments are left to regulate business interests as a result of public outcry after the damage has already been done.

FUTURE TRENDS

Open source software continues to become more mainstream as greater numbers of developers contribute to the code base and the applications get better, more ubiquitous and user friendly. From the nonprofit's perspective it still remains to be seen how support of open source development efforts will coalesce around mission focused applications. Funders typically provide support to institutions with a official 501c3 nonprofit status working on social objectives, and not loose coalitions of developers. However, 501c3 entities like Aspiration (http://aspirationtech.org) and its Social Source Commons application are demonstarting alternative approaches by creating nonprofit developer and technology support communities around applications and issue areas. Unlike most corporations that can afford to employ technology support, nonprofits and funders alike are increasingly relying on external nonprofit technology support entities like E-Riders Npower

to provide both support for nonprofits who do not employ internal expertise (E-Riders, 2005; NPower, 2005). What this means is that unlike commercial entities who can afford their own technicians, nonprofits will prioritize solutions as much for the ability of their third party providers to support them as for any particular technology whether open source or proprietary. How these intermediary technology support organizations handle the open source and ideology question will have significant impact on what technology is actually employed in nonprofits.

In this new environment that seeks to strike a balance between civil liberties and national security the software ideology debate must focus on the on the more important issues of what software is developed and deployed for. At the same time the high software project failure rates must be taken into account and selection and implementation decisions applied realistically to the nonprofit context. Software selection must be based on criteria that allow for a higher probability of success precisely because of the low degree of resources and tolerance for failure in this sector. Software selection should be left to the same operational criteria that have always led to increased probability of successful application deployment—meeting a defined user need.

CONCLUSION

Technology is neither an enabler nor a facilitator of civil society in its own right. Nor is it a decider of its own ethical or non-ethical use. The mechanism that ultimately decides the ideology behind any given technology are the people and institutions applying it, regardless of the intent defined its original development. It should not be surprising that software development, an area of computer science, presents the same range of ethical dilemmas that most of the other sciences do.

REFERENCES

Anderson, R. J. (1995). Crypto in Europe: Markets, law and policy. In *Proceedings of the International Conference on Cryptography: Policy and Algorithms* (LNCS 1029, pp. 75-89). London: Springer-Verlag.

Dempsey, B. J., Weiss, D., Jones, P., & Greenberg, J. (1999). *A quantitative profile of a community of open source Linux developers*. School of Information and Library Science at the University of North Carolina at Chapel Hill.

Elliott, M., & Scacchi, W. (2004, August). Mobilization of software developers: The free software movement. Retrieved from http://www.ics.uci.edu/~wscacchi/Papers/New/Elliott-Scacchi-Free-Software-Movement.pdf

E-Riders. (2005). A standard definition of e-riding. *Eriders.net*. Retrieved October 17, 2005, from http://www.eriders.net/model/whatis/

Franck, E., & Jungwirth, C. (2001). *Reconciling investors and donators: The governance structure of open source*. [University of Zurich Working Paper]. Zurich, Switzerland: University of Zurich.

Hacktivismo. (2005). A description of the enhanced use license software. *kottke.org*. Retrieved October 17, 2005, from http://www.kottke.org/02/11/hacktivismo-enhancedsource-software-license-agreement

Kling, R. (1983). Value conflicts in the deployment of computing applications. *Telecommun. Policy*, 12-34.

Kogut, B., & Metiu, A. (2001). Open-Source software development and distributed innovation. *Oxford Review of Economic Policy, 17*(2), 248-264.

Kuan, J. (1999). *Understanding open source software: A nonprofit competitive threat*. Un-published manuscript, Haas School of Business, UC Berkeley.

Kuan, J. (2002). *Open source software as lead user's make or buy decision: A study of open and closed source quality*. Stanford Institute for Economic Policy ResearchMimeo.

Lee, E. (1997). *The labor movement and the Internet: The new internationalism*. Chicago: Pluto Press.

Lerner, J., &Tirole, J. (2002). Some simple economics of open source. *Journal of Industrial Economics, 52*, 197-234.

Lerner, J., & Tirole, J. (2005). The scope of open source licensing. *Journal of Law, Economics and Organization, 21*(1), 20-56.

Martus. (2005). Description of the application. *martus.org*. Retrieved October 17, 2005, from http://martus.org/

McInerney, P-B. (2004). *Ideological competition among organizations: Nonprofit technology assistance and the rise of a nascent organizational field*. Submission to the 2004 ASA Annual Meeting by the Department of Sociology, Columbia University.

Mosko, M., Jiang, H., Samanta, A., & Werner, L. (1999, December). *Software acquisiton metamodel*. UCSC-CRL-00-02. Retrieved from ftp://ftp.cse.ucsc.edu/pub/tr/ucsc-crl-00-02.ps.z

NPower. (2005). Description of its role as a technology service providers to nonprofits. *npower.org*. Retrieved October 17, 2005, from http://www.npower.org/about/index.htm

Saint-Paul G. (2003). Growth effects of nonproprietary innovation source. *Journal of the European Economic Association, 1*, 429-439.

Salesforce. (2005). Description of its role foundation giving activities. *salesforce.com*. Retrieved October 17, 2005, from http://www.salesforce.com/foundation/

Scacchi, W. (2002). Understanding the requirements for developing open source software systems. *IEEE Proceedings on Software, 149,* 24-39.

Stewart, K. J., & Ammeter, T. (2002). An exploratory study of factors influencing the level of vitality and popularity of open source projects. In *Proceedings of the 23rd International Conference on Information Systems*, Barcelona, Spain.

Stewart, K. J., & Gosain, S. (2001). An exploratory study of ideology and trust in open source development groups. In *Proceedings of the 22nd International Conference on Information Systems*, New Orleans, LA.

Techsoup. (2005). A description of Techsoup's Techsoup Stock providing free or discounted software and hardware to nonprofits from a variety of vendors. *techsoup.org.* Retrieved October 17, 2005, from http://www.techsoup.org/stock/default.asp?cg=header&sg=stock&visit=1

KEY TERMS

Application-Development Ideology: The context in which a software developer chooses to develop his application, it can be for gain, glory or to meet a social good.

Application-Selection Ideology: The context is which applications are selected for use. Historically, applications have been chosen to meet practical business requirements. However, the introduction of open source has ledd to a movement of some, particularly in the nonprofit community, advocating selection of software weighted more heavily on open source development ideology.

Application-Use Ideology: The context in which a user chooses to use a software application. It can be for constructive or destructive purposes, to meet a social need, a business requirement or any other utilitarian purpose.

Hacktivism: The use of technology in the context of civil disobedience, potentially breaking the law through technical means to protest perceived injustice.

Destructive Application Ideologies: Ideology that may occur in the process of development, selection or use creating applications dedicated to create disruption, or selecting and using applications specifically to cause disruption regardless of the reason for developing them.

Mission Sensitive Nonprofit Open Source Applications: Applications specifically designed to promote and further the mission objectives of a nonprofit such as case management for domestic violence or human rights monitoring applications. These applications are typically not mainstream, with a harder business case for supporting developers to create and maintain applications.

Open Source's Social Value Equation: An argument that through its collaborative development methodology and fee sharing of intellectual property among users, Open Source can be equated to the best principles of the nonprofit sector. While ideologically attractive, the notion fails to take into account that:

1. Nonprofits actually compete with each other for limited resources.
2. Open source development is often accomplished by a relatively small core team.
3. There is a cost of ownership that is somewhat hidden from nonprofits that focus on free applications without taking into account that technical support is still required to maintain it.
4. Revenue generated by major software vendors has been invested back into society (in the form of new foundations) to achieve high social value impact projects in a variety of issue areas.

Chapter XXXVII
Governance and the Open Source Repository

R. Todd Stephens
BellSouth Corporation, USA

ABSTRACT

This chapter examines the critical task of governing the open source environment with an open source repository. As organizations move to higher levels of maturity, the ability to manage and understand the open source environment is one of the most critical aspects of the architecture. Metadata can be defined as information pertaining to the open source environment that the organization defines as critical to the business. Successful open source governance requires a comprehensive strategy and framework which will be presented through historical, current-state, and future perspectives. The author expects that by understanding the role of open source metadata and the repository within, researchers will continue to expand the body of knowledge around asset management and overall architecture governance.

INTRODUCTION

Open source continues to make inroads into the corporate environment where it is now a standard embraced by most of the top tier corporations in America (Ferris, 2003). Applications like Apache and Linux have been phenomenally successful in providing real business value (Garvert, Gurbani, & Herbsleb, 2005). However, further research is needed in how organizations should govern the open source environment which requires more than the indemnification of the product. Open source governance requires the establishment of architectural standards that each and every group can adhere to in order to deliver bottom line business value. A centralized repository for downloading certified open source products ensures that the principles of asset management are implemented and managed effectively.

The driving purpose of the architecture community is to minimize the unintended effects on the business due to technology changes. Utilizing an open source repository for impact analysis will ensure that proposed changes will not create catastrophic events within the business itself. The repository provides the mechanism for inventory management which allows organizations to see what is already acquired, deployed, and supported within the environment. In addition, efforts like domain analysis, reuse, and release management are essential to the implementation of open source as an enterprise asset. When organizations embrace open source as a viable alternative to in-house or outsourced development, they must accept the responsibility and implications of transforming it from code to an enterprise asset.

BACKGROUND

The background section will review the core concepts that enable open source governance within a large scale deployment. Additionally, the historical precedent for a repository will set the stage for the introduction of the open source repository. Architecture governance is the practice and orientation where the technical architecture is managed and controlled at an enterprise-wide level. Maturity models provide a framework by which organizations can measure their progression of governance; software and open source maturity models will be reviewed. The higher levels of maturity define an environment where consistency, predictability, and ongoing optimization are the keys to success.

Architecture Governance

Information technology governance specifies accountabilities of technology related business outcomes and helps companies align their technology investments with their business priorities (Ross & Weill, 2005). Enterprise architecture is a set of frameworks, principles, guidelines, and standards created to guide the development and deployment of enterprise systems. The rate of change in the business is accelerating causing the cycle times allowed for implementing new systems to decrease. Existing technology infrastructure often gets in the way of rapid change and may inhibit the organization's ability to respond. By having an architectural governance program, large enterprises can respond quickly and effectively

to the demands of the business. One tool that can be used to determine the road map of governance is called a maturity model. A maturity model is a method for judging process maturity of an organization and for identifying the key practices required move to the higher levels.

Software Maturity Models

In 1986, the Software Engineering Institute (SEI) was asked by the U.S. Air Force to create a systematic method of evaluating software contractors. In conjunction with the MITRE Corporation, the study group produced a questionnaire that enabled the Air Force to judge a software provider as either successful or unsuccessful in its capabilities. The questions were divided in a number of groups (key process areas) and then assigned to specific levels within the model. The resulting model was called the capability maturity model (CMM). The levels describe the path a software provider must follow in order to move to the higher levels of maturity. These paths are actually a collection of key practices that must be mastered before moving to the next level (Baskerville & Pries-Heje, 1999). Maturity implies a potential for growth in capability and indicates both the richness of an organization's software process and the consistency with which it is applied in projects throughout the organization. In addition, productivity and quality resulting from an organization's software process can be improved over time through consistent gains in the discipline achieved by using its software process (Chrissis, Curtis, Paulk, & Weber, 1993). The CMM provides five levels of maturity: initial level, repeatable level, defined level, managed level, and optimized level.

Level 1: The Initial Level

At this level, the organization has a less stable software process and management practices. The process is ad-hoc and changes as work progresses.

All aspects of the process are unpredictable with no key process areas defined within the domain of the organization. When an organization lacks sound management practices, the benefits of good software engineering practices are undermined by ineffective planning and reaction-driven commitment systems.

Level 2: The Repeatable Level

At this level, the focus is on project planning, management, tracking, and the implementation of procedures and policies. The objective of this level is to establish an effective project management process that allows the organization to "repeat" successful practices and procedures used on earlier projects. Key process areas for this level include: requirements management, software project planning; software project tracking and oversight; software subcontract management; software quality assurance; and software configuration management.

Level 3: The Defined Level

This level focuses on the organization's defined standard software process, including software engineering and management processes. The activities are stable and repeatable and are implemented throughout the organization. Key process areas include: organization process focus, organization process definition, training programs, integrated software management, software product engineering, intergroup coordination, and peer reviews.

Level 4: The Managed Level

This level focuses on productivity, quality and the assessment of each defined process. Measurements are established for quantitative assessment and evaluation of software processes and products. At this level, the organization is capable of predicting quality trends within quantitative bounds. Key process areas include quantitative

process management and software quality management.

Level 5: The Optimized Level

This level focuses on continuous process improvement. At this level the organization has the ability to identify process weaknesses and product defects, and to improve both the process and product. Key process areas include defect prevention, technology change management, and process change management. Updates to the model are reviewed by a body of over 500 practitioners and approved by an advisory board of 14 senior software engineering professionals (Marshall & Mitchell, 2002). The CMM model has been adapted by several different disciplines including knowledge management, people capability, project management, and product development.

Open Source Maturity Models

Based on the success of the CMM, Golden (2005) defined an open source maturity model (OSMM) as a basic requirement for analyzing open source products. Each product is evaluated on six basic elements: product software, support, documentation, training, product integrations, and professional services. These six elements are scored against the basic requirements, ability to locate available resources, access element maturity, and the assignment of a maturity factor. The purpose of the OSMM is to provide a level of maturity for the open source product. Since there are 80,000 open source products, organizations will be faced with multiple options in deploying specific solutions. Additionally, Guliani and Woods (2005) defined an OSMM based on the products age, supported platforms, momentum, popularity, design quality, costs, and support associated with the open source product. The main issue with these models is that they only concern themselves with evaluating the product and not matching the level of maturity of the organization, architecture, and the client support required after the product is implemented. At the core of the long term value-add proposition for any enterprise asset, like open source, is the managing repository (Pereira & Sousa, 2004). The repository handles both the structured and unstructured information content required at the higher levels of maturity. Huang and Tilley (2003) describe the top two levels of maturity and the base requirement of a knowledge management system.

Traditional Roles of the Repository and Registry

A repository is basically a database application that contains information about an asset with the ability to attach unstructured documentation. Depending upon the object type, the repository may only store the metadata information or the actual object itself. Traditionally, metadata focused on database and the extraction transform and load (ETL) type metadata. The evolution from physical data structures to logical models, component descriptions and system definitions extends the metadata environment to a whole new world of possibilities. Blecher (2005) defines metadata as any information regarding the characteristics of any artifact, such as name, location, perceived importance, quality or value to the enterprise, and its relationships to the other artifacts that an enterprise has deemed worth managing. New technologies such as XML and Web services are also requiring new forms of repositories than can manage the asset in a design and production environment. Today, the vast majority of repositories are Web enabled which means they should follow the standards of design, usability, content management, and user centered design principles. This allows organizations to define the product, service, and an information framework in much of the same fashion that businesses build models for the online environment.

From the maturity model perspective, the repository plays a key role in several different areas

including governance, establishing information context, and reuse. A closer look at many of the maturity models show a distinct migration from a chaotic model of operation to a definable, repeatable, and electronic method of doing business which is the base criteria for level five maturity. The repository is the central part of this business model, just as the card catalog is the central point of information for any library. Organizations that implement open source repositories and then begin to expand the business functionality and integration are moving toward an open source transformation. Open source can then move away from a chaotic environment to a centralized point of service that is built from the business point of view. The maturity process for the repository begins with capturing the current inventory of open source products and then adds the services like version tracking, impact analysis, subscription services, information context, and measured reuse.

Producers, Consumers, and Librarian Responsibilities

Since open source is usually brought into an organization during the architecture or design time of a project, most resources do not think about reuse or change management with these applications. From the programmer's point of view, open source simply provides a starting point where a base set of functionality can be implemented fairly easily. With today's focus on cost savings and speed to market, development organizations will look for best practices, irregardless if they are formal or stealth. In one sense, the producer of the open source product is the collection of experts that worked together in order to produce the application. However, the organizational entity (Corporation or Educational Institute) looks toward a single point of contact for the product. This subject matter expert will have responsibility to evaluating, piloting, standardizing, and implementing the functionality into the organization. The open source producer is the architect, developer or development manager that decides that open source should be brought into the organization and integrated into the application environment. The responsibility of the producer or integrator is to ensure that the asset integrates seamlessly into the technology architecture and does not cause disruption of service or business functionality. As an asset, the producer must ensure that both structured and unstructured information is collected and loaded into a repository.

The consumer of open source products are directly related to the internal development communities. In addition, architects, designers, testers, and ongoing support may also be interested in the different dimensions of open source as it relates to the organizational deployment. The consumer of open source is any person or group that access the information describing the asset or the actual asset itself. Consumers can gain access to the information in a passive nature by reading and collecting the information for educational purposes. Consumers may access the information through an active delivery method where distinct consumer services are automated and built on the actual metadata information. The open source consumer is responsible for locating and accessing the reusable information, assessing the ability to reuse the asset, adapting to the asset environment and integrating the asset into the framework of technology. The greatest return on investment occurs when multiple consumers of open source work together as a community and ensure the application supports the business and evolves to a more agile technology environment.

The role of the librarian is to manage the information about the open source environment and act as a third party for the use and functionality of the asset. A portion of this functionality will be performed by the repository which provides the discovery, access and documentation services. The librarian is essential to the success of implementing open source into a large scale environment since they focus on providing value to both the producer and the consumer. As information about the open source environment flows into the repository, the librarian is responsible for ensuring the information is accurate and conforms to the defined domain of the meta-model. Data quality is essential in the long term success of the repository and the value to the business cannot be understated. Services offered to the open source producer include work flow, utilization metrics, content aging, and inventory statistics. The librarian also serves the consumer by ensuring the information required for implementation decisions is presented in a clear and concise manner. The consumer will also be interested in understanding the implementation environment; specifically reviewing the number of other implementations or capacity information. As a broker of information, the librarian works to ensure a solid relationship between the producer and consumer.

OPEN SOURCE REPOSITORY AND THE GOVERNANCE MODEL

Overview

Currently, open source components are described and packaged in environments that focus on the development community. Open source communities, like Source Forge, provide metadata elements such as administrator, developers, installation instructions, development status, licensing, along with several others. Like the majority of externally defined repositories, organizations are unable to add application or business functionality without replicating the data. Integration, which is essential in architecture governance, is rarely taken into account and many sources of open source software may not be as trustworthy as others.

Issues and Problems

By itself, the open source repository does not solve the issues around information technology governance. However, the repository does enable the organization to address these concerns from a maturity point of view and then add additional functionality as required by the various key process areas. As organizations move to the service-oriented architecture (SOA) and distributed agile solutions, these issues take on a more global impact to the organization as a whole. The following issues must be addressed as organizations begin to govern the open source environment.

Inventory Management

Perhaps the most important issue that must be dealt with early on is to understand the environment and the diverse technology parts that make up the application portfolio. Open source applications are only a small part of the infrastructure and each component must be subject to the rules and processes of inventory management. The basic question of what open source products do we have, who is using them, where are they installed, and what do they functional do are all important informational elements that must be captured. Even at the most basic level, executives cannot manage or govern the architecture without knowing what is actually in the environment.

Impact Analysis

While inventory management focuses on the what, impact analysis focuses on the relationships between the open source application and the other assets of the organization. Questions like who is using this product, what systems are using the open source component, who is the subject matter expert, what other products work with this application, if we replace this application what else is impacted and who will support the application. The importance of these questions becomes critical when looking at the technology environment as a competitive advantage and the lack of disaster recovery or high availability. Today's business environment cannot afford

downtown due to integration or security issues. Just knowing the impact of a change, an organization can save millions in revenue by avoiding a service interruption.

Reuse and Domain Analysis

Domain analysis can be described as the process of reviewing the business and process environments looking for commonalities and variability's that enable the creation of a domain model. This domain model is the core requirement for the development or use of reusable assets. This process of matching the business functionality to specific assets allows the architecture governance organization to ensure that only one version or one solution for a specific problem is implemented. For example, the domain analyst might take the Web server functionality and associate the Apache product which means that any group needing the basic functionality is required to utilize the same application. This ensures that reuse becomes a critical technology imperative and a requirement for any lifecycle project. Reuse, software reuse, and code reuse are three terms that are often misused and confused by the general practitioner. Reuse has been tossed around since 1994 where many organizations jumped all over the reuse bandwagon without much success. Software reuse goals and practices are not new, but full scale success has been hard to find. That being said, effective reuse of knowledge, processes, and software has been proven to increase productivity and quality of the IT organization. McIlroy (1969) published one of the earliest references to software reuse at the New York NATO Conference on Software Engineering. Early reuse efforts were primarily focused on reusing algorithms to ensure the consistency of calculations. Companies with scientific and engineering computing needs were early proponents of "function reuse" to ensure that a specific engineer calculation across all of their systems computed the same value. The initial goal was not to reduce costs to build systems nor quicker time-to-market. In 1995, Gamma, Helm, Johnson, and Vilissides (1995) published their unique approach to reuse in a book called *Design Patterns: Elements of Reusable Object-Oriented Software*. The result was a recognition that reuse must include much more than just code. Spanning from architecture to test-cases, reuse must be looked at in a holistic fashion in order to produce economic benefits. Organizations spend much more time on the architecture, design, analysis, and specification than actual coding.

Procurement and Version Control

How you bring open source into the organization is also critical for governance to operate effectively. *Freely available* software means that there is no single source for a particular application. Linux can be downloaded from thousands of sites, each of which has a wide variety of versions, support, and documentation. Ideally, the open source repository has only one version that is matched to the domain model and every implementation within the organization utilizes this single source. In doing so, the organization can manage the environment in a consistent manner that can deliver the functionality demanded by the business. In addition, by tracking who, when and how the organization downloads and implements open source, the architecture can ensure a secure and legal environment.

Metrics and Measurements

In order to move up the maturity model, you will eventually need to integrate metrics into the governance process. Some metrics, such as installations, versions, downloads, document views, and repository path analysis are inherent in the prior issues and concerns. As the organization matures their reuse program, other metrics may emerge as valuable instruments of governance. The metrics may include transaction volumes that pass through the open source application.

Transaction volume is a key determinate of which open source product should be implemented. Some products add functionality at a cost of transaction speed. In other cases, the speed of service is more important than additional functionality that may not be needed. Jeffrey Poulin (1997) was one of the first people to take an extensive look at measuring software reuse. In his 1997 book, he covered the principles, practices, and economic models for measuring component-based reuse within a corporation. Open source should implement an economic based reuse metric program in order to measure the core reuse of open source applications. Other metrics that could be implemented include industry support, return on investment (ROI), capacity, and performance. Metrics should be captured on a monthly basis and evaluated by utilizing trend analysis software which evaluates the information over an extended period of time. Ideally, the process of collection should be automated and have the ability to capture at any point in time. What growth percentage should be applied to the open source metrics? Again, long-term success is not defined by the explosion of growth in the first year but by the subsequent three to five years. The first few years may very well have triple digit growth but sustaining growth is the key to success and maturity.

Open Source Environment

Many people in the information technology field look at the open source repository as an application which provides a limited set of value points for the organization. The reality is that the repository itself is just one part of a much larger collection of products, services, tools, processes, and customer support components. Figure 1 provides one view of the open source repository environment that attempts to pull in some of these components into a single framework. This framework is based on the experience of the author in a Fortune 500 organization.

Figure 1. The open source portal and corresponding components

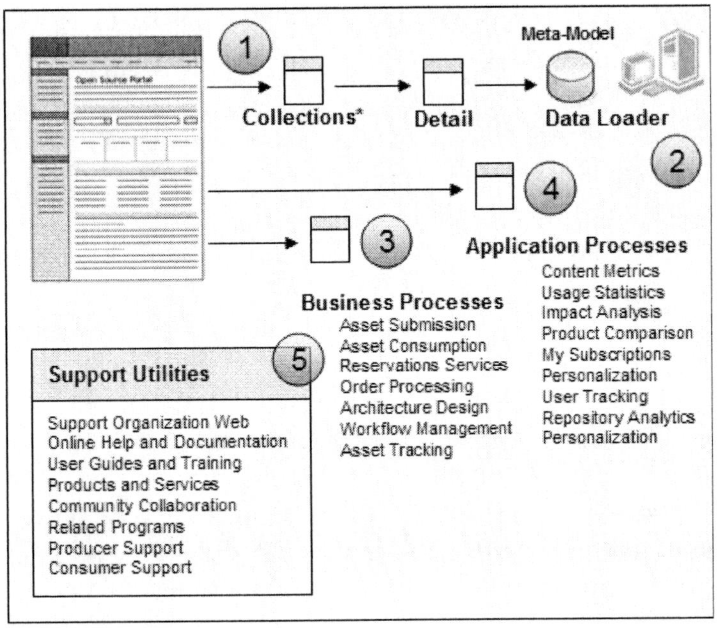

The structure of this diagram includes five basic components:

1. Open source portal
2. Traditional repository
3. Business processes
4. Application processes
5. Customer support environment

The Open Source Portal

Elena Varon (2002) indicates an enterprise portal gives end users access to multiple types of information and applications through a standard interface. The vertical portal addresses one aspect of a business, such as a human resources site that lets employees sign up for training classes and view pay stubs. Others define a portal as an interface for people to access and exchange information online. It is usually customizable and can be designed to provide employees, customers or trading partners with the information that they need, when they need it. Aiken and Finkelstein (2000) indicated that enterprise portals will be the primary method used by organizations to publish and access business intelligence and knowledge management resources. Similarly, the open source portal provides a single point of access for all open source products and services within the enterprise. The main portal page should contain some of the following functionality:

- Basic overview, user guide and online help for the repository
- Semantic and advanced Boolean search
- Multiple hierarchal structures for open source classification
- Usage based classifications: Latest additions, coming soon, top ten
- Key business functions for the repository
- Service provider support
- Personalization of the portal
- Related programs to the open source effort

Once a user has selected an option on the open source portal a collection of assets will be presented. A collection is a method of grouping assets based on the context of the selection. A collection may be presented via search engine or through a taxonomy classification system. In a large organization, the open source repository must be able to reduce the source set and place the components into the context requested by the end user. The detail page provides the metadata that describes the open source component itself. This metadata includes the generic (semantic) meta-model which is simple and straightforward metadata such as name, description, or keywords. The Dublin Core standard is one such generic meta-model standard. The context specific meta-model describes an asset within a specific context: Object Management Group (OMG), common warehouse model (CWM), reusable asset specification (RAS), and Web service definition language (WSDL) are just a few examples. These standards focus on specific types of resources or assets (structural metadata). Presenting this information in a single, usable, and functional page is critical to the success of the repository. While no open source metadata standard has been defined, a general meta-model can be used to capture the classification information.

The Data Loader

Located on the right side of Figure 1 is the data loader utility which actually loads the metadata information into the meta-model. Vendors provide a large collection of utilities that can harvest metadata from tools, databases, and a wide variety asset types. In addition to the automated loading utility, most applications provide librarian tools for versioning, data quality, integration, and data entry. Without metadata exchange standards within the open source community, the majority of the information will need to be loaded by hand.

Business Processes

The open source repository can be transformed from a passive store for open source information into an integrated solution for governing the environment. The key to this transformation are the business processes.

Asset Submission and Status Tracking

Metadata must be collected on each and every open source component submitted to the repository. Even when the majority of metadata is collected through an automated tool, basic information must be assembled in order to initiate the process of cataloging the open source components. An online form or series of forms can provide self service for collecting information from the open source provider. Ideally, this process could be automated with the use of integrated Web services. The librarian should have the open source information tagged as "pending" to indicate that the component still needs to be reviewed by the governance organization. Each open source component should be reviewed and scored based on support, maturity, functionality, and associated risk to the organization. Once the component is approved, then and only then should the functionality be exposed to the entire organization.

Asset Consumption

The repository can also provide services for the utilization of the open source components. Consumers work on the front end of projects to integrate reusable open source assets into the technology environment. One of the biggest problems with implementing enterprise architectures is understanding the environment from a usage point of view. Online forms can trigger the engagement process for utilizing assets as well as track the relationship between application and the asset. One of the challenges of enterprise application integration (EAI) is knowing what systems, applications, and data constructs are currently being used within the corporation. By integrating the concepts around consumption and implementing open source as a governed technology, the architecture will ensure a stable and robust environment that complements the agile organization.

Application Processes

Application processes are products and services that operate on the metadata information itself. While the business processes focused on the specific workflow of the environment, the application processes focuses on value-add from the repository data. One of the most basic application processes is measurement of the amount of content within the open source repository which would include components, documentation, and metadata elements. Another key metric is the actual usage of the open source information. Usage metrics can communicate the priority, reuse, as well as opportunities for the governance organization. Understanding the complete picture of open source usage is critical to building a long-term program. Impact analysis is the process of identifying or estimating the impact of a change in the environment. Impact information can be used when planning changes, making changes, or tracking the effect of changes implanted within the open source environment (Apiwattanapong, Harrold, & Orso, 2003). Other application processes include subscription services, reservation services, failed searches, and user tracking.

Customer Support Environment

The open source components as well as the overall architecture process needs to have a customer facing environment. The support group creates an environment of self-service and community support within the organization. Support groups approach the governance process from the service perspective as opposed to the technology view. Adding customer support utilities to the product mix is a positive step in creating a customer experience. Some of the basic components should include: user guides, online help, product and service overviews, frequently asked questions (FAQs), and training programs. In addition, producer and consumer communities can be created with a wide variety of collaboration tools in order to add value to the relationship. The open source repository environment is a complex collection of communications that are one way, collaborative, and interactive in nature.

Open Source Meta-Model

The key to any repository is to have a solid meta-model that allows the organization to catalog the open source component as well as provide services with the metadata information. The process should start with a basic set of elements and then begin to expand. Since the metadata will be stored externally to the open source package, effort should be made to reduce the complexity in order to ensure adherence to the standards. Table 1 provides a standard set of elements that should be associated to the open source package.

These elements provide the foundation of knowledge management around the open source asset which will enable basic functionality like search, taxonomies, and hierarchal classifications. The domain specifies a controlled set of values that can be applied to the field. Table 2 provides an example of the data elements applied to the open source package from Apache called Lenya.

Organizations should expand this core set of metadata elements to match the level of governance required by the executive community. Keeping in mind that the more metadata elements added requires additional investments in data quality, process management and analytical support. This meta-model can be expanded to include unstructured information, relationships with other components, and packaging information.

Table 1. Basic open source metadata e-elements

Field Name	Description	Domain
Title	Name of the open source package	No
Description	Detailed description of the package including value-add, utility and functionality	No
Class Fields	Classification fields like type or topic	Yes
Source	Source of the original package including URL, description and owning organization	No
Keywords	Key words and key phrases used to describe the application	Yes
Release	Version or release level	No
Usage	Describe how the application should be used or what business need will be addressed with this application	No
Technical Dependencies	List the technology requirements like operating system, database application, or Web servers	No
Contacts	Contacts information for owners, users, and subject matter experts	No
Dates	Dates like origination, valid through, release, and so forth.	No
License	Specific type of license	Yes
Status	Current internal status	Yes
Online Reference	Any online reference sites for documentation, support, or best practices	No

Table 2. Open source metadata elements for Apache Lenya

Field Name	Description
Title	Apache Lenya—Open source content management (Java/XML)
Description	Apache Lenya enables content management with the following features: content authoring, workflow, internalization, layout, site management and security
Class Fields—Context	End user interface
Class Fields—Class	Information worker class
Source	The Apache Software Foundation
Keywords	Content management, open source, apache, workflow, check-in, check-out, publishing, asset management
Release	1.2
Usage	Content management
Technical Dependencies	Cocoon, Ant
Contacts	John.doe@mycompany.com
Dates	06/15/2003
License	http://lenya.apache.org/license.html
Status	Active
Online Reference	http://lenya.apache.org/1_2_x/index.html

THE FUTURE OF OPEN SOURCE REPOSITORIES

Overview

Repository frameworks continue to make inroads into the information technology community. Advancements in SOA and the Web services are exposing the repository to a new community of developers and architects. Infrastructure maturity frameworks like information technology infrastructure library (ITIL) are also bringing the repository to the forefront as a central point of management and governance. The open source community will need to bring in the concepts of operational maturity and governance in order to extend the viability of the overall environment. This will create an opportunity for the repository to become the central point of control for the enterprise. While smaller organizations that only implement a few open source components may not need the repository, larger organizations will increasingly depend on the functionality as a competitive advantage. This progression of value reflects the maturity of open source within the corporation which must include the tools of governance. The convergence of enterprise architecture and governance will bring about more vendor support, tools, and standards which should create an environment for growth for the open source repository. As companies continue to out source development, integrate open source components, and deploy service architectures, the repository will become much more of a business requirement. Despite the fact that intangible assets, like open source, have been largely ignored by accounting, executives, and board of directors, most companies are increasingly reliant on them (McFarlan & Nolan, 2005). The controlling bodies must ensure that management knows what information or applications is being used, how it is being used, who is using it, and what value-add does it bring to the bottom line of the organization.

Business Trends

The world of business is evolving to a much more agile environment than in the past. One of the biggest trends today is the concept of out sourcing components of the business model to other organizations. Information technology has enabled the business to outsource their supply chain to companies like United Parcel Service (UPS) and their technology operations to EDS or Accenture. This trend allows organizations to focus on their core competencies. The impact of this on the open source environment is that organizations will continue to move toward standard technologies and business processes. While today the challenge of open source is functionality and support, tomorrow the challenge will be on integration and business agility. Business agility enables an organization to cope with the unpredictable changes, to survive unprecedented threats from the business environment, and to take advantage of changes as opportunities (Goldman, Nagel, & Preiss, 1996). The open source repository will evolve from a source of passive information to the point of integration for the business itself, irregardless if the open source component is an XML standard, application program, or open business process. Business agility will open the door for dynamic business models that create value only for a short period of time. Organizations must capitalize on these opportunities by deploying technologies that adapt to the changes in the business model; not in months but in days.

Another business trend that will impact the open source environment is the mobile workforce or information worker. Technology advancements have created an environment where work can be done around the world by anyone at anytime. This requires that information about the technology environment be available 24 hours a day. The repository allows mobile workers to access information and documentation from any location as long as the resource has access to the Intranet. New mobile devices, like the cell phone, personal data assistance (PDA), and laptop computers are changing the basic definition of work and the creation of value. Having the complete open environment documented and available online, the repository will become the single point of integration and the enabler of business value.

Technology Trends

Technology continues to evolve toward an environment that will remove all barriers to entry, barriers of geography, and barriers of time. The proliferation of computing into the physical world promises more than the ubiquitous availability of computing infrastructure; it suggests new paradigms of interaction inspired by constant access to information and computational capabilities. For the past decade, application-driven research in ubiquitous computing has pushed three interaction themes: natural interfaces, context-aware applications, and automated capture and access (Abowd & Mynatt, 2000). These movements are coming of age and the impact to the corporation cannot be understated. All of these advancements can be directly tied to the influence of the open source environment. Another trend is the move toward collaborative computing which is exactly how the open source community thrives. Collaboration will eventually move away from the development model to the utilization, standardization, and governance of the environment components.

Conclusion

In the current environment building an open source repository is more of a process issue than a technology one. The adoption of commercial software can be controlled in a straightforward manner through the procurement process. Open source adoption is far more difficult to manage because there is no single gateway to control how and when open source software is used. Companies that deploy open source must consider the same myriad issues they consider in any commer-

cial software deployment: security, governance, integration and lifecycle management. Central to this governance theme is the open source repository. The long term success of open source within the organization will be defined by the ability to govern the information technology environment as a core component of the infrastructure. In order to move to the highest levels of maturity, open source governance and the repository must be integrated into the core architecture.

REFERENCES

Abowd, G., & Mynatt, D. (2000). Charting past, present, and future research in ubiquitous computing. *ACM Transactions on Computer-Human Interaction, 7*(1), 29-58.

Aiken, P., & Finkelstein, C. (2000). *Building corporate portals with XML*. New York: Mc-Graw-Hill.

Apiwattanapong, T., Harrold, M., & Orso, A. (2003). Leveraging field data for impact analysis and regression testing. In *Proceedings of the 11th European Software Engineering Conference and 11th ACM SIGSOFT Symposium on the Foundations of Software Engineering*. Helsinki, Finland: The Association of Computing Machinery.

Baskerville, R., & Pries-Heje, J. (1999). Knowledge capability and maturity in software management. *The Database for Advances in Information Systems, 30*(2).

Blecher, M. (2005, December). What metadata is and why you should care. *Business Integration Journal, 12*.

Chrissis, M., Curtis, B., Paulk, M., & Weber, C. (1993). *Capability maturity model for Software*. Version 1.1, Software Engineering Institute, Pittsburgh, PA.

Ferris, P. (2003). The age of corporate open source enlightenment. *Queue, 1*(5), 34-44.

Gamma, E., Helm, R., Johnson, R., & Vilissides, J. (1995). *Design patterns: Elements of reusable object-oriented software.* Reading, MA: Addison-Wesley.

Garvert, V., Gurbani, A., & Herbsleb, J. (2005). A case study of open source tools and practices in a commercial setting. In *Proceedings of the Fifth Workshop on Open Source Software Engineering.* St. Louis: The Association of Computing Machinery.

Golden, B. (2005). *Succeeding with open source.* Reading, MA: Addison-Wesley.

Goldman, S., Nagel, R., & Preiss, K. (1996). *Co-operate to compete: Building agile business relationships.* New York: Van Nostrand Reinhold.

Guliani, G., & Woods, D. (2005). *Open source for the enterprise: Managing risks, reaping rewards.* Cambridge, MA: O'Reilly Media.

Huang, S., & Tilley, S. (2003). Towards a documentation maturity model. In *Proceedings of the 21st Annual International Conference on Documentation.* San Francisco: The Association of Computing Machinery.

Marshall, S., & Mitchell, G. (2002). An e-learning maturity model? In *Proceedings of the 19th Annual Conference of the Australian Society for Computers in Learning in Tertiary Education,* Auckland, New Zealand.

McFarlan, R., & Nolan, W. (2005). Information technology and the board of directors. *Harvard Business Review, 83*(10).

McIlroy, M. D. (1969). "Mass produced" software components. In P. Naur & Bl Randell (Eds.), *Software engineering,* Brussels (pp. 138-155). NATO Scientific Affairs Division.

Pereira, C., & Sousa, P. (2004). A Method to Define an Enterprise Architecture using the Zachman Framework. In *Proceedings of the 2004 ACM Symposium on Applied Computing,* Nicosia, Cyprus. The Association of Computing Machinery.

Poulin, J. (1997). *Measuring software reuse.* Reading, MA: Addison-Wesley.

Ross, J., & Weill, P. (2004). *IT governance: How top performers manage IT decision rights for superior results.* Watertown, MA: Harvard Business School Press.

Varon, E. (2002). Portals finally get down to business. *CIO Magazine,* 12.

KEY TERMS

Asset: An asset is any artifact that the organization defines as critical to the business.

Metadata: Metadata is information that describes the characteristics of an asset. This information may be in the form of structured metadata like title, description, and author; or unstructured metadata like implementation instructions or test cases.

Maturity Model: A maturity model describes an evolutionary path where activities/best practices are introduced in order to create a more stable, consistent, and definable environment.

Meta-Model: A meta-model defines the basic structure of the information being collected about an asset.

Governance: Governance is the act of managing the technical environment as a portfolio of assets. Managing the portfolio would include activities like domain analysis, inventory management, and reuse.

Repository: A repository is a software application that manages the asset information throughout the lifecycle including the acquisition, storage, publishing, and security rights of that asset.

Reuse: Reuse describes the activities of identification, generalization, development, and management which support practitioners utilizing existing assets vs. building from scratch.

Section VI
Business Approaches and Applications Involving Open Source Software

Chapter XXXVIII
Analyzing Firm Participation in Open Source Communities

Wouter Stam
Vrije Universiteit Amsterdam, The Netherlands

Ruben van Wendel de Joode
Delft University of Technology, The Netherlands
Twynstra Gudde Management Consultants, The Netherlands

ABSTRACT

Increasingly, firms participate in OSS communities. However, surprisingly little empirical research has been performed to understand firms' participation in OSS communities. This chapter aims to fill this gap in state-of-the-art research on OSS. We will discuss and analyze the results from a survey of 90 Dutch high-technology firms that are active in the market for OSS products and services. In the survey we asked the firms what activities in OSS communities they perform. One outcome is that firms' activities can be grouped into two distinct categories of activities, namely technical and social activities. This outcome is an important contribution to research on OSS that until now has viewed community participation as a uni-dimensional construct. The survey results also suggest that firms view their internal investments in R&D as a complement to their external product-development activities in OSS communities.

INTRODUCTION

The emergence of open source software (OSS) communities, in which individual contributors freely share their innovations, has presented organizations with new opportunities to sell their software products and services (Von Hippel & Von Krogh, 2003). Well-known examples of firms that use the communal resources of OSS communities are IBM, SUN, and Red Hat. These and other organizations hope to benefit from OSS because

they believe it constitutes a low-cost and high-quality knowledge resource that may spur new product development. Furthermore, they believe that characteristics of OSS communities, like the release of source code, may provide opportunities that lead to the early adoption of new products and hence lead to first-mover advantages (Dahlander & Magnusson, 2005). Engagement of commercial organizations in OSS communities may also provide various benefits to OSS communities, since firms may (a) enlarge the user base of the communities, (b) contribute scarce financial and human resources, and (c) perform a boundary-spanning function by linking the communities to various groups of non-technical users.

Despite the potential mutual benefits of community participation by firms, recent studies have suggested that commercial actors may have a tendency to demonstrate significant free-riding behavior and contribute little back to the joint effort that characterizes open source communities (Bonaccorsi & Rossi, 2004). Firms may focus only on their own benefits and as a consequence, exploit the communal resources while keeping their involvement in the community at a minimum. Although this behavior can harm both the firm and the community in the long run (Dahlander & Magnusson, 2005), surprisingly little empirical research has been carried out to examine if such free-ridership actually takes place. Little research has been performed to analyze the activities firms actually perform in OSS communities. As a result, an understanding of the conditions under which firms contribute to the development of OSS communities remains incomplete.

This chapter extends previous work on participation in OSS communities by firms. We achieve this by studying how such firms participate in OSS communities. Specifically, we are looking for factors that may explain any variation in the type and extent of participation across firms.

Based on survey data that was collected from 90 OSS firms in The Netherlands, our first aim is to show that the engagement of firms in OSS communities involves more than just technical activities such as contributing software code. We will show that organizations also perform social activities, such as organizing conferences and workshops that may facilitate knowledge sharing among community members and spur the wider adoption of OSS. By making a distinction between technical and social participation, we offer a more holistic perspective on the engagement of commercial actors in OSS communities.

Our second purpose is to explain what factors account for the observed differences between firms in the ways they participate in OSS communities. By demonstrating that the type and extent to which companies participate in OSS communities is logically connected to specific characteristics of these firms, such as their business models, we generate a better understanding of the conditions under which firms make certain types of contributions to OSS communities.

Our chapter proceeds as follows. First, an overview of state-of-art literature is given in which we will introduce the literature on community participation by individual developers and commercial firms. Next, we present our empirical study of Dutch OSS firms and discuss its main findings. We conclude with a discussion of future trends and present our overall conclusions.

BACKGROUND

Economic theory suggests that people only contribute to the production of a good if the benefits exceed the costs (e.g., Olson, 1965). Yet, people *participate* in open source communities without receiving direct tangible benefits for their efforts. In other words the efforts, or costs, involved in writing source code or solving other people's problems do not exceed the direct monetary benefits that can be gained from such activities. The reason for this lack of direct benefits is that the products and services created by active participants like the source code (the human-readable part of software) or the answers to questions can simply be downloaded for free. Thus, in OSS communities the costs of participation appear to outweigh the benefits. At the same time, however, research has shown that a surprisingly large number of individuals voluntarily *participate* in the communities (Hertel, Niedner, & Herrmann, 2003). This paradox has received much attention from researchers, who wondered: "Why do individuals *participate* in OSS communities?"

Participation in Open Source Communities by Individuals

Especially the earlier writings on OSS communities provide us with a number of potential answers to the question why individuals voluntarily *participate* in the communities. One of the dominant answers has been that OSS communities are *gift economies* in which individuals like to give (Bergquist & Ljungberg, 2001; Markus, Manville, & Agres, 2000; Zeitlyn, 2003). The concept of a gift economy can be traced back to Mauss (1990) who described a wide range of communities in which the giving of gifts laid the foundation for exchange. A *gift economy* relies on the principle of reciprocity and an implicit requirement to give (Mauss, 1990). In these systems "a gift is not so much a physical resource as a social and moral system by which sharing, collaboration, loyalty and trust are cultivated" (Bollier, 2001, p. 11).

Indeed, there are some indications that the principle of gift giving is important in OSS communities. "Open-source contributors have told us that they enjoy the sense of 'helping others out' and 'giving something back'" (Markus et al., 2000, p. 15). A respondent we interviewed argued: "It is nonsense to believe that in OSS you do not receive anything. If you do what you are good at, others will do the same. I receive a lot from others, which I could not have done myself. In the gift economy everybody is better off." As such, participants in the communities are said to create and sustain dynamic relationships with one another based on the exchange of gifts (Zeitlyn, 2003).

In an effort to better understand why individuals participate in OSS communities, researchers have adopted different techniques. One of the most frequently used techniques is survey research. One of the first large-scale and internationally-conducted surveys on OSS developers was by the University of Maastricht in the Netherlands and the company Barlecon Research from Germany. In the study called "Free/Libre and Open Source

Software: Survey and Study," Ghosh and Glott (2002) report on important findings regarding the participation of developers in OSS communities, which were derived from a large scale survey among 2,784 OSS developers who answered various questions about their participation in OSS communities. The use of surveys has provided better insight into the reasons why individuals participate in OSS communities. These motives can be related to the costs and the benefits of participation.

Concerning the costs of participation in OSS communities, researchers have argued that these costs are relatively low (Lakhani & Von Hippel, 2003). Low costs are important, because "when the costs of freely revealing an innovation are low, even a low level of benefit can be adequate reward" (Von Hippel, 2001, p. 85). Thus, due to the low costs the barrier for people to participate in the communities is also relatively low.

Although the costs are low, there must be an incentive for individuals to incur even such low costs. Therefore, researchers have focused most of their efforts to analyze and understand the benefits individuals enjoy as a result of their participation in OSS communities. These research efforts have shown that participation in OSS communities may offer a large amount of benefits to individuals, many of which are intangible. Some of the most-frequently identified benefits of participating in OSS communities are:

- Building a reputation in a community (e.g., Dalle & Jullien, 2003; Lakhani & Von Hippel, 2003)
- Learning and improving one's programming skills (e.g., Hertel et al., 2003; Von Hippel & Von Krogh, 2003; Lakhani & Wolf, 2003)
- Meeting a personal need with a software program that has a certain functionality (e.g., Edwards, 2001; Hars & Ou, 2002)
- Having fun (Lakhani & Von Hippel, 2003; Torvalds & Diamond, 2001)

A Logical Next Step: Why Do Firms *Participate*?

In recent years, open source software has received a lot of attention. Furthermore, or perhaps as a result of this attention, OSS is currently used by many private and public organizations. Good examples of such organizations are the New York Stock Exchange, Shell, the French daily Le Figaro, the U.S. army, national government in Brazil, and the city of Munich. Each of these organizations has adopted OSS to support one or more of its organizational processes.

Organizations do however not only adopt OSS for their internal processes; they may also participate in its development by contributing resources back to OSS communities. Firms like Yahoo and CNet, for instance, participate in the OSS community Apache and regularly contribute source code to the joint effort, solve bugs (mistakes) in the software and answer other people's questions on mailing lists. In light of this increasing participation of firms in OSS communities a logical next question is: "Why do firms participate in OSS communities?"

The answer to this question may be significantly related to the business model of firms, as firms will only invest their time and effort in the communities if they believe it will lead to additional benefits or reduce costs. Researchers have provided some arguments as to why and how firms make money or reduce costs from OSS (e.g., Goldman & Gabriel, 2005; McKelvey, 2001). Yet, company motivation has received far less attention in scientific research than individual motivation has (see also Bonaccorsi & Rossi, 2004), and hardly any surveys have been conducted to support the arguments as to why firms participate in OSS communities.

There is one notable exception. The exception is a survey conducted by Bonaccorsi and Rossi (2004). They conducted a survey among 146 Italian firms to understand why firms participate in OSS communities. Their most important conclusion is

that firms participate for different reasons than individuals do. Whereas individuals have many social motivations, firms will typically emphasize more on economic and technical reasons to participate. Bonaccorsi and Rossi found that the most important reasons for firms to participate are: (a) OSS communities allow small firms to be innovative, (b) contributions and feedback from the OSS communities are very useful to fix bugs and improve software, and (c) open source software is reliable and has a high quality. Another important conclusion they draw is that the more pragmatic motives of firms to participate are accepted by individual participants, which would mean that OSS communities are robust and can deal with differing motivations (Bonaccorsi & Rossi, 2004).

Another, in light of this chapter, relevant publication is by Grand, Von Krogh, Leonard, and Swap (2004). In their paper they propose a four-level model of company participation based on the level of resources a firm allocates to OSS. In level 1 a firm is primarily a user and therefore allocates a relatively low level of resources to OSS. This does not mean that open source software is free of costs to the firm since they do need to incur costs to install and run the software. In levels 2 and 3 the allocated resources to OSS steadily increase and in level 4 a firm's overall business model is based on OSS. Examples of level-4 firms are Red Hat and SuSE. Such firms will typically make significant contributions to a variety of OSS communities.

MAIN FOCUS OF THE CHAPTER

The Focus of this Chapter: Firm Participation

This chapter aims to contribute to the state of the art on firms' involvement in OSS communities. In particular, this chapter will focus on the question: "How do firms participate in OSS communities?" One of the primary reasons for focusing on this question is because we believe that the answer will provide crucial information about the business models of firms. We believe it will solve a crucial part of the puzzle as to how firms earn money from OSS. For instance, what type of activities do firms focus on? Does a software vendor perform different activities in OSS communities as compared to a hardware vendor? It could very well be that firms, depending on factors like their size or expertise, perform different types of activities in the communities. This observation would suggest (a) that they have a different business model and (b) that different activities are logically connected to different business models. Obviously, such information would be crucial for any firm that wants to earn money from OSS.

Our Empirical Study

To generate a better understanding of participation in OSS communities by private firms, we conducted a survey in 2005 among all Dutch high-technology firms that sell OSS-related products and services. The business owners were asked to fill out survey questions that covered their firms' business models and participation in open source communities. Since there were no comprehensive listings available of all OSS firms active in The Netherlands, we used several secondary-data sources to identify the research population. Relevant sources included (1) the membership list of the Dutch OSS branch organization, called "Vereniging Open Source Nederland," (2) the Web site of the governmental program "Open Standards and Open Source Software" (OSOSS) that contains a list of OSS firms, and (3) Internet searches by means of keywords such as "open source solutions," "open source products and services," and "Linux solutions." In total this resulted in an initial list of 127 firms. Interviews were then conducted with key informants who are knowledgeable about the industry (cf. Kumar,

Stern, & Anderson, 1993). Ventures that were not on the initial list, but were mentioned by more than one expert, were added to the initial list, which resulted in nine additional firms. During the data-collection process we encountered 11 ventures that either ceased operations or whose founders indicated that their firm was not active (anymore) in the OSS industry.

To maximize response rates, we followed several suggestions by Dillman (2000): firms were sent a letter stating the purpose and importance of the research project, followed by a phone call in which they were requested to participate. Whenever possible, appointments were made during which the questionnaires were personally delivered to the business owners. From the final population of 125 firms, 90 firms eventually returned a completely filled-out questionnaire, thereby yielding a response rate of 72%. We tested for non-response bias by comparing key attributes of respondents to those of non-respondents. For the variables of both firm size (as total number of employees) and firm age (as number of years since firm formation) t-tests indicated no significant differences.

Characteristics of Open Source Firms

Table 1 presents the descriptive statistics of the firms that are active in the Dutch market for OSS-related products and services. The industry is still in its infancy, with many young and small firms that are technically oriented. A typical firm has been in business only for five years, was founded by two entrepreneurs, and employs in total about six persons. These firms are managed by teams of entrepreneurs who already have more than 12 years of work experience in the IT industry, but who on the other hand have limited experience in the area of marketing and sales. Similarly, a significant share of firms' staff has a technical orientation with more than half the staff consisting of product developers. These findings are

Table 1. Descriptive statistics of OSS firms

	Min.	Max	Mean	S.D.
Firm age	0.50	29.25	5.24	4.18
Total sales in 2004 (x €1,000)	0	4,400	372.13	758.41
Number of company founders	1	8	1.94	1.27
Founding team IT industry experience[1]	0	60	12.18	10.91
Founding team marketing & sales experience[1]	0	24	2.68	5.00
Total staff (incl. founders) in 2005	1	50	5.87	8.27
Proportion of total staff with university degree	0	100	40.25	37.36
Proportion of total staff developing new products and services	0	100	52.79	39.58
R&D intensity[2]	0	150	28.00	28.08
Number of OSS projects involved in 2004	0	16	3.28	4.02
Number of OSS projects self-initiated	0	16	1.07	2.39

[1]   *Measured as the team's total number of years of work experience at time of firm formation*

[2]   *Measured as the proportion of sales in 2004 that is invested in the development of new products and services*

consistent with prior research showing that high-technology firms generally have a strong technical background and often lack sufficient marketing expertise that is necessary to successfully commercialize new products and services (Roberts, 1991). In contrast to a study of Italian OSS firms by Bonaccorsi and Rossi (2005), which reported that only very few employees were university graduates, we find that over 40% of firms' staff has in fact a university degree. This result supports earlier research that demonstrated that OSS developers are generally highly educated and confirms studies showing that high-technology firms generally employ more highly educated staff compared to less knowledge-intensive firms (Oakey, 1995). With regard to firms' engagement in open source projects, the data show that the average firm was involved in about three projects in 2004 of which one project was started by the firm itself. This finding suggests that firms are involved in multiple projects simultaneously and also shows that most firms not only take advantage of existing projects, but also contribute to the community by initiating new projects.

Firms' Business Models

As shown in Table 2, firms that offer OSS-related products and services in the Dutch market pursue a variety of *business models*. Most firms generate the majority of their revenues from open source solutions, but generally combine these with more traditional proprietary offerings. Interviews with the business founders suggested that many customers are still unaware of open source or perceive it as a risky alternative and as a consequence, many firms are more or less "forced" to also offer proprietary solutions. With regard to product offerings, we find that most firms sell little hardware solutions. Sales of a typical firm are based for about 41% on software development, while over 52% of revenues come from offering additional business services such as consultancy, support, and training. Interestingly, this pattern mirrors the business models of the more traditional Dutch IT firms that also predominantly generate revenues from selling IT services.

With respect to the distribution of firms' sales across three main customer groups, our data show

Table 2. Business models of OSS firms

	Min.	Max	Mean	S.D.
Division of Total Sales over Product Categories[1]:				
OSS-related products and services	0	100	71.62	33.49
Software	0	100	41.31	31.28
Hardware	0	60	5.65	10.63
Services	0	100	52.82	33.12
Division of Total Sales over Customer Groups[1]:				
Government and nonprofit	0	99	25.69	27.68
SMEs	0	100	43.16	35.34
Large firms (> 100 employees)	0	100	31.16	34.42
Foreign customers	0	50	3.81	8.85

[1]   *Measured as the proportion of total sales in 2004*

that small and medium-sized enterprises (SMEs) account for the largest proportion of sales (43%), followed by large firms (31%), and government and nonprofit organizations (26%). Given the fragmented nature of the SME market and the general tendency for governments and large firms to be early adopters of OSS, this finding is quite surprising. Finally, the limited share of sales to foreign clients (4%) suggests that most firms predominantly focus on serving the Dutch market. This finding can be explained by the observation that most firms rely for a large part on selling OSS-related services, which are generally more dependent on geography and more locally oriented than firms selling software packages that can be distributed across foreign markets.

Technical and Social Participation in Open Source Communities

Given our interest in the ways in which firms *participate* in open source communities, the business owners were asked to indicate the extent to which their firms performed a variety of activities in the open source community. We used a five-point Likert scale with individually labeled answer categories ranging from "never" to "very

often." In all cases, very often was coded as 5.0 while never was coded as 1.0 (i.e., larger values denote greater participation). Eleven items were included that were identified from previous empirical research (e.g., Ghosh & Glott, 2005; Von Krogh, Spaeth, & Lakhani, 2003) and conceptual work (e.g., Feller & Fitzgerald, 2002).

A principal-components analysis was performed to assess any interrelationships among the different activities and to look if these can be reduced to a smaller number of dimensions. A varimax rotation was performed on all factors satisfying Kaiser's criterion (i.e., eigenvalues of 1.0 or greater). This procedure produced two factors explaining 58.02% of the total variance. All items showed strong factor loadings of 0.69 or higher and cross-loadings below 0.30. Given the commonly used cut-off point of 0.30, our factor loadings demonstrate strong significance and are representative of the underlying components (Kim & Mueller, 1978). Our conclusion from this analysis is that we have empirically isolated two distinct factors that represent important dimensions of community participation by private firms. The first factor consists of six items and is labeled "technical participation" (Cronbach α = 0.87), while the second factor is made up of five items and called "social participation" (Cronbach $\alpha = 0.77$).

Technical participation by firms in open source communities refers to the activities firms undertake to make contributions to software development across a variety of open source projects. It involves activities that are directly or indirectly related to the development of new software such as contributing source code, writing software documentation, and participating in e-mail discussions. Active *technical participation* in OSS communities implies that firms not only use their access to communal resources to create and appropriate value for their own benefit, but also contribute to community development by sharing source code, technical know-how, and knowledge on end-user requirements with other community members. Compared to firms that view communal resources as a public good that is there for the taking, firms demonstrating extensive technical participation realize that a sustainable business model depends on their ability to become actively involved and deeply integrated in the developer community (Weber, 2004).

Social participation by firms in open source communities involves the activities companies initiate to facilitate knowledge sharing among developers, firms, end users, and other community members and that may promote the wider acceptance and adoption of open source software among individuals and organizations outside the OSS community. Examples of activities that reflect *social participation* include organizing workshops, conferences, and other events related to OSS, and participating in political activities to further the interests of the OSS communities. These events are settings in which representatives from various organizations and industries come together to share knowledge and experiences through face-to-face interactions, construct social networks, and learn about "best-practices" related to important technical and organizational aspects of OSS development and commercialization.

Active social participation by firms may help to overcome the relatively limited external legitimacy of the OSS movement, which refers to the problem that outsiders to the community may be reluctant to commit any resources to OSS-related business activities since they do not understand or acknowledge them (Aldrich & Fiol, 1994).

Determinants of Firms' Technical and Social Participation

The finding that the engagement of firms in open source communities can be subdivided into two distinct domains of activity, in other words, technical and social, generates the question what factors may explain any variation in the type and extent of participation across firms. In this part of our chapter we will analyze two important antecedents of technical and social community participation by firms: (1) the human capital characteristics of a firm's founding team and (2) the business model a firm pursues.

Drawing from the entrepreneurship and upper-echelons literatures that have shown the strong influence of the demographic characteristics of a firm's founding and top-management team on organizational structure and outcomes (Hambrick & Mason, 1984), our first prediction is that the work experience of a firm's founding team is related to the kind of activities it performs in OSS communities. The extent to which entrepreneurs already have worked in the IT industry and have experience in marketing and sales functions will affect their ability and willingness to engage in technical and/or social activities in OSS communities. Second, based on studies that have demonstrated the link between organizational strategy and structure (Miller, 1987), we expect to find a relationship between the kinds of business models that firms pursue and the way they participate in OSS communities. Variation in product offerings and customer groups across firms will influence the benefits that firms may

Table 3. Multiple regression analyses predicting technical- and social-community participation

Variables	Technical Community Participation	Social Community Participation
Firm age	-.08	-.11
Firm size	.17	.17
Founding team IT industry experience	-.22*	.37**
Founding team marketing and sales experience	.19†	-.29*
Staff developing new products and services	.22*	-.08
R&D intensity	.22*	-.14
OSS-related sales	.06	.22†
Government and nonprofit sales	-.04	.23*
Foreign sales	.32**	.00
Model F	3.49***	2.06*
R^2	.32	.22
Adjusted R^2	.23	.11

*Note: Standardized coefficients reported: †$p < .10$, * $p < .05$, ** $p < .01$, *** $p < .001$*

obtain from technical and social community participation, and hence affect the activities firms perform in OSS communities.

Table 3 shows the results of OLS regression analyses that respectively model firms' technical and social participation in OSS communities as a function of their human capital and business models. For both models, variance inflation factors (VIF) did not show any signs of multicollinearity (VIF < 1.67). We also checked for normality by conducting a Kolmogorov-Smirnov test, which supported the univariate normality assumption. Though both models are statistically significant, the first model with technical community participation as the dependent variable is both more significant ($p < .01$ vs. $p < 0.05$) and has a higher explanatory power (adjusted $R^2 = .23$ vs. .11) than the second model explaining firms' social community participation. Next, we describe our main findings with regard to the antecedents of technical and social community participation.

Founding Team Work Experience

Our results indicate that variation in the work experience of a firm's founding team members is significantly related to the type and extent of community participation activities displayed by that firm. Work experience in the IT industry is—a result we find highly surprising—significantly negatively associated with technical community participation (ß = -.22, p < .05), yet is significantly positively related to social community participation (ß = .37, p < .01). Firms founded by entrepreneurs with more years of experience in marketing and sales positions however, exhibit significant lower levels of social participation in open source communities (ß = -.29, p < .05). One explanation for these findings is that founders with more industry experience have larger social networks with other people that work in the same industry and social community participation may be a way to maintain these network relationships.

Furthermore, these business owners may share similar backgrounds making it more likely that they interact with each other.

The finding that firms with business founders who have a marketing background engage significantly less in social participation is also surprising, because we expected that these entrepreneurs are more inclined to undertake promotional activities. The explanation for this outcome may be that these business owners do not necessarily engage less in marketing activities, but that they put more effort in promoting their business activities *outside* the open source community. These firms may be more customer-oriented and therefore inclined to connect their business activities to individuals and organizations from outside OSS communities.

Commitment to Innovation

Our analysis with regard to the relationship between a firm's *business model* and community participation also produced a number of interesting findings. The results suggest that firms that demonstrate a commitment toward *innovation* engage significantly more in technical participation in OSS communities than firms that are less focused on the development of new products and services. Both R&D intensity ($\beta = .22$, $p < .05$) and the proportion of staff that is classified as product developers ($\beta = .22$, $p < .05$) have a significant positive relationship with technical participation. This suggests that firms in our sample view technical participation in OSS communities as a complement to their own R&D activities, rather than as a substitute to internal expenditures on *innovation* (cf. Chesbrough, 2003). It may also indicate that firms need a certain level of absorptive capacity (Cohen & Levinthal, 1990), which can be developed through internal R&D activities, before they can successfully engage in technical community participation.

Foreign Sales

Our findings indicate that firms that generate foreign sales also engage significantly more in technical-community participation than firms that are not active in foreign markets ($\beta = .32$, $p < .01$). Although causality between the two variables cannot be determined, it is a highly interesting outcome. It could for instance indicate that technical participation is necessary to serve international markets. Developer communities are by their very nature extremely internationally oriented, such that technical participation by firms with foreign sales may be required to access knowledge about world markets. An alternative explanation could be that technical activities like contributing source code and answering questions on mailing lists create international recognition for the firm. This recognition in turn may create international demand and thus foreign sales. Perhaps social participation does not result in international demand because this type of participation generally involves activities that are more locally oriented.

Focus on Open Source

Our results show that firms with a stronger focus on open source, in other words, they generate a larger percentage of revenues from OSS-related products and services, engage significantly more in social participation ($\beta = .22$, $p < .10$). Yet, no relationship was found with technical participation. These results indicate that social participation is a logical activity for firms that truly focus on open source. It may be that these firms have more to gain from social participation, as it provides them with new business opportunities and enhances their reputation in the community. Surprisingly, technical participation in OSS communities is less logically connected to the focus of firms on OSS. An explanation for this finding may be that even when firms sell many proprietary solutions, they still want to engage in technical community participation in order to take advantage of OSS as a complementary asset that can add value to internally developed products and services (e.g., Grand et al., 2004).

Sales from the Nonprofit Sector

Firms that generate relatively more sales from government and *nonprofit organizations* are significantly more involved in social participation ($\beta = .23$, $p < .05$). Important to note here is that no relationships were found between community participation and the percentage of revenues a firm receives from respectively small and large firms. These results suggest that in particular firms that target the nonprofit market segments recognize the importance of social-community participation, possibly because *nonprofit organizations* also engage extensively in social participation in OSS communities. Alternatively, firms that invest in social activities may generate additional revenue from the nonprofit sector. This could signal a tendency for *nonprofit organizations* to focus less on technical expertise of potential suppliers and much more on their reputation in the community, which is possibly better generated through social activities.

FUTURE TRENDS

There are a number of limitations to our research. The most important limitation is that we cannot explain some of the findings. Why do firms in which the founding members have experience in the IT industry perform fewer technical activities in OSS communities as compared to firms in which the founding members do have experience in marketing and sales? We would have expected a different relationship. The collection of qualitative data from interviews may help to further interpret our results.

Next, this study focused only on the Dutch market and solely included firms that operate in the Netherlands. However, many OSS communities are global in nature with firms from various parts of the world participating in them. Given that prior studies have shown that OSS adoption rates and participation in OSS communities may differ across countries (Ghosh & Glott, 2002), additional comparative studies are needed that contrast how differences in the economic, institutional, and cultural context in which firms operate, affect the way and extent to which they participate in OSS communities.

A third limitation is related to one of the previous limitations. Our data indicate that social and technical participation are positively correlated. Thus, firms that perform more technical activities are also more likely to perform social activities, and vice versa. Yet, firms do appear to make a well-balanced and purposeful choice between the two types of participation. This observation is supported by Table 3, which shows that quite a few variables correlate exactly opposite with the type of activities. This is most visible for the experience of the founding members. More IT experience implies less technical participation and more social participation. More marketing and sales experience, however, implies exactly the opposite. What explains these findings? Again, further research needs to be conducted to better understand these findings: Is it true that firms make a purposeful choice?

Another interesting strand of research would be to relate the findings of this research with the framework proposed by Grand et al. (2004). According to Grand et al. (2004) one would expect that firms in the first level of the framework, in other words, firms that predominantly use OSS, display hardly any type of participation except maybe some forms of social participation. Technical participation would typically be more appropriate for firms that engage and interact more frequent with OSS communities. One question we did not ask is: How long have you been an active participant in OSS communities? Neither did we ask: Do you consider yourself a user or an active participant? Future research should include such questions, as it might shed light on the validity of the framework proposed by Grand et al. (2004).

A final limitation is that we have not related our findings to the success or failure of firms in our sample. It would be truly fascinating if we could relate the types of participation in OSS communities to the level of innovation or profit a firm achieves. Future research efforts will need to research whether we can actually find such relationships.

CONCLUSION

In this chapter we have uncovered parts of an important gap that exists in the state-of-the-art literature on open source software communities. We have argued that little research, especially empirical research, has been performed to address the question why and how firms participate in OSS communities. In this chapter we have focused on the interface between firms and OSS communities, and in particular we examined ways in which firms participate in such communities. We reported important findings from a survey among 90 Dutch high-technology firms that are active in the market for OSS products and services. Based on this dataset we made a number of highly relevant and interesting observations.

Unquestionably, one of the most important outcomes of our research concerns the finding that the activities that firms undertake in OSS communities can be grouped into two distinct categories. A first group of activities, referred to as social participation, includes activities like organizing workshops and conferences. The second set of activities, which we label technical participation, includes actions such as contributing source code, bug fixes, and participating in mailing list discussions. Whereas prior work has conceptualized open source community partici-

pation as a uni-dimensional construct, our study suggests that it may be valuable to disentangle the concept into distinct dimensions that may have unique antecedents and consequences.

Next, based on our analysis of the correlates of social and technical community participation we were able to draw some highly interesting conclusions. One important result concerns the finding that firms seem to view their internal investments in R&D as a complement to their external product-development activities in OSS communities. This outcome supports the view that for firms that depend on open source business models, both internal as well as external participation in OSS communities are necessary conditions for innovation that possibly reinforce each other.

Our findings also seem to support a popular assumption about participation in OSS communities. Many researchers have assumed that users of open source software will first use the software and gradually perform more and more activities (Ye, Kishida, Nakakoji, & Yamamoto, 2002). Gradually, they will learn from their activities and will become more active participants. Assuming that social activities are more typical for users and technical activities are more typical for knowledgeable users and participants, our finding that social and technical participation are positively correlated would provide some first evidence to support this assumption.

Based on our outcomes we have shed insight in the differing ways in which firms behave in open source communities. We hope this insight helps firms to understand what their options are to become involved in OSS, and that they understand the ways that are available to them to make money from open source software. We do realize however, that further research is necessary to tie the insights in this chapter to important firm-level outcomes. It would be highly relevant for instance, to understand whether different types of participation have different effects on firms' innovative and financial performance. We hope this chapter has encouraged scholars to put further research efforts into this exciting new research domain.

REFERENCES

Aldrich, H. E., & Fiol, C. M. (1994). Fools rush in? The institutional context of industry creation. *Academy of Management Review, 19*(4), 645-670.

Bergquist, M., & Ljungberg, J. (2001). The power of gifts: Organizing social relationships in open source communities. *Information Systems Journal, 11*, 305-320.

Bollier, D. (2001). *Public assets, private profits: Reclaiming the American commons in an age of market enclosure*. Washington: New America Foundation.

Bonaccorsi, A., & Rossi, C. (2004). Altruistic individuals, selfish firms? The structure of motivation in open source software. *First Monday. Peer reviewed journal on the Internet, 9*(1).

Bonaccorsi, A., & Rossi, C. (2005). Comparing motivations of individual programmers and firms to take part in the open source movement: From community to business. *Knowledge Technology and Policy*, 55-77.

Chesbrough, H. W. (2003). *Open innovation: The new imperative for creating and profiting from technology*. Boston: Harvard Business School Press.

Cohen, W. M., & Levinthal, D. A. (1990). Absorptive capacity: A new perspective on learning and innovation. *Administrative Science Quarterly, 35*(1), 128-152.

Dahlander, L., & Magnusson, M. G. (2005). Relationships between open source software companies and communities: Observations from Nordic firms. *Research Policy, 34*, 481-493.

Dalle, J.-M., & Jullien, N. (2003). 'Libre' software: Turning fads into institutions? *Research Policy, 32*(1), 1-11.

Dillman, D. A. (2000). *Mail and Internet surveys: The tailored design method*. New York: John Wiley.

Edwards, K. (2001). *Epistemic communities, situated learning and open source software development*. Paper presented at the 'Epistemic Cultures and the Practice of Interdisciplinarity' Workshop at NTNU, Trondheim.

Feller, J., & Fitzgerald, B. (2002). *Understanding open source software development*. London: Addison-Wesley.

Ghosh, R., & Glott, R. (2002). *Free/libre and open source software: Survey and study*. Maastricht: MERIT, University of Maastricht.

Ghosh, R., & Glott, R. (2005). *FLOSSPOLS: Skills survey interim report*. Maastricht: MERIT, University of Maastricht.

Goldman, R., & Gabriel, R. P. (2005). *Innovation happens elsewhere. First edition: Open source as business strategy*. San Francisco: Morgan Kaufmann.

Grand, S., Von Krogh, G., Leonard, D., & Swap, W. (2004). Resource allocation beyond firm boundaries: A multilevel model of open source innovation. *Long Range Planning, 37*, 591-610.

Hambrick, D. C., & Mason, P. A. (1984). Upper echelons: The organization as a reflection of its top managers. *Academy of Management Review, 9*, 193-206.

Hars, A., & Ou, S. (2002). Working for free? Motivations for participating in open-source projects. *International Journal of Electronic Commerce, 6*(3), 25-39.

Hertel, G., Niedner, S., & Herrmann, S. (2003). Motivation of software developers in open source projects: An Internet-based survey of contributors to the Linux kernel. *Research Policy, 32*(7), 1159-1177.

Kim, J., & Mueller, C. W. (1978). *Factor analysis: Statistical methods and practical issues.* Beverly Hills; London: Sage.

Kumar, N., Stern, L. W., & Anderson, J. C. (1993). Conducting interorganizational research using key informants. *Academy of Management Journal, 36*(6), 1633-1651.

Lakhani, K., & Von Hippel, E. (2003). How open source software works: Free user-to-user assistance. *Research Policy, 32*(7), 922-943.

Lakhani, K., & Wolf, R. G. (2003). *Why hackers do what they do: Understanding motivation and effort in free/open source software projects* [Working Paper no.4425-03] Boston: MIT Sloan. Retrieved November 2004, from freesoftware. mit.edu/papers/lakhaniwolf.pdf

Markus, M. L., Manville, B., & Agres, C. E. (2000). What makes a virtual organization work? *Sloan Management Review, 42*(1), 13-26.

Mauss, M. (1990). *The gift: The form and reason for exchange in archaic societies* (Vol. Translation of Essai sur le Don (1950) Presses Universitaires de France). London: W. W. Norton.

McKelvey, M. (2001). The economic dynamics of software: Three competing business models exemplified through Microsoft, Netscape and Linux. *Economics of Innovation and New Technology, 10*, 199-236.

Miller, D. (1987). Strategy making and structure: Analysis and implications for performance. *Academy of Management Journal, 30*(1), 7-32.

Oakey, R. (1995). *High technology new firms.* London: Paul Chapman.

Olson, M. (1965). *The logic of collective action: Public goods and the theory of groups.* Cambridge, MA: Harvard University Press.

Roberts, E. B. (1991). *Entrepreneurs in high-technology: Lessons from MIT and beyond.* New York: Oxford University Press.

Torvalds, L., & Diamond, D. (2001). *Gewoon voor de fun (Just for Fun)* (C. Jongeneel, Trans.). Uithoorn: Karakter Uitgevers.

Von Hippel, E. (2001). Innovation by user communities: Learning from open-source software. *Sloan Management Review, 42*(4), 82-86.

Von Krogh, G., Spaeth, S., & Lakhani, K. R. (2003). Community, joining, and specialization in open source software innovation: A case study. *Research Policy, 32*, 1217-1241.

Von Hippel, E., & Von Krogh, G. (2003). Open source software and private-collective innovation model: Issues for organization science. *Organization Science, 14*(2), 209-223.

Weber, S. (2004). *The success of open source.* Cambridge, MA: Harvard University Press.

Ye, Y., Kishida, K., Nakakoji, K., & Yamamoto, Y. (2002). *Creating and maintaining sustainable open source software communities.* Paper presented at the International Symposium on Future Software Technology, Wuhan, China.

Zeitlyn, D. (2003). Gift economies in the development of open source software: Anthropological reflections. *Research Policy, 32*(7), 1287-1291.

KEY TERMS

Community Participation: Contributions to a community. They can be made by organizations and or individuals.

Firm: An organization that conducts business.

Social Participation in OSS Communities: Activities companies initiate to facilitate knowledge sharing among developers, firms, end users, and other community members and that may promote the wider acceptance and adoption of open source software among individuals and organizations outside the OSS community.

Technical Participation in OSS Communities: Activities firms undertake to make contributions to software development across a variety of open source projects. It involves activities that are directly or indirectly related to the development of new software such as contributing source code, writing software documentation, and participating in e-mail discussions.

Chapter XXXIX
Community Customers

Jeroen Hoppenbrouwers
Vrije Universiteit Brussel, Belgium

ABSTRACT

This chapter discusses the role of the project/product community in the open source product life cycle. It outlines how a community-driven approach affects not only the development process, but also (and more importantly) the marketing/sales process, the deployment, the operation, and in general the resulting software product. Participation in the community is essential for any organisation using the product, leading to the concept of a community customer. Specific community participation guidelines are given to organisations and individuals who deploy and use open source software, further develop it, or offer lifetime services on the product.

INTRODUCTION

Open source is not only about cost, or freedom to choose, learn, and modify. A very important aspect of open source projects is their *organisational* freedom. This freedom leads to both challenges and opportunities for organisations which intend to merely deploy open source products in their routine operations, not planning any development. Open source product procurement, deployment, and operational maintenance are different from those of traditional products, largely because of the organisation of the processes which breed and raise open source software.

We start from Evers' definition of an open source project, which is: "Any group of people developing software and providing their results to the public under an open source license" (Evers, 2000). However, we immediately want to add

that this definition, as many others, overemphasizes the importance of *development*. We would like to extend the definition by including *users* of the software, as will be argued in the rest of this section.

Bonaccorsi and Rossi (2003) analyze open source as a *process innovation*. Various economic questions have been raised on why such a process can produce anything at all, mostly concentrating on the traditional economic question: "Why do programmers write open source codes if no one pays them to do it?" The body of literature about this economic aspect is huge, and this chapter will not elaborate on this issue. Instead, we focus on the observation made by Bonaccorsi and Rossi that "There is a large group of individuals who are not capable of developing programmes but only of using them" (2003, p. 1244). They put this group next to the hobby developers and the members of the hacker culture, traditionally assumed to be the majority of open source contributors. For this chapter, we would like to further divide the first group into *individuals* and *organisations*. It is especially the organisational user participation in the open source process that is of interest to us.

This chapter focuses on the role of the community of stakeholders, usually simply called "the community," which forms around an open source product. Observations from various angles and theoretical background lead to concrete recommendations for organisations and individuals who consider adding an open source product to their ICT portfolio. The chapter does not aim at open source development, but explicitly addresses "end-using organisations" and explains why and how they have to consciously play a particular role in the community. When using open source products, they become a customer of the community, not of a vendor—but a customer they are, with associated real costs to pay and real benefits to enjoy. The term *community customer* will be introduced to define the role(s) such an end user, which may be an organisation, must play.

We can now rephrase our definition of an open source project: "Any group of people developing or deploying software common to the group and providing their development results and usage experiences to the public under an open source license."

BACKGROUND

Even after the formal founding of the Free Software Foundation (Stallman, 1985) and the subsequent translation of the principles of free software to business situations by the open source movement (Raymond, 1998a), it took a while before analysts worked out *why* the open source model works, and the issue still is not fully understood.

A popular insight, fielded by Raymond and many others, is that open source developers are mostly driven by "ego." They develop and show the world the results to boost their self-esteem. However, this analysis turns out to be over-simplified. A better analysis can be made by referring to existing (business) economic notions which got developed when studying non-profit economics, a relatively new field by itself (Hansmann, 1980). These insights also cover the non-developing community participants, often a much larger number than the actual developers (Craig & Beck, 1993). We will briefly summarize several known reasons why people may contribute to open source projects without being paid to do so, and place them in the context of their role in the community.

Rent-Seeker and Donator Approach

Two main aspects of open source community participation can be distinguished: *rent-seeking* and *donation*.

In *rent-seeking*, "emphasis is put on the fact that although no wages are paid to contributors, other pay-offs may turn the investment of labour into an open source project into a profitable decision" (Franck & Jungwirth, 2003, p. 402). This aim to mostly establish individual reputation is not only driven by ego, as Raymond states, but also can be used to improve credibility on secondary markets such as the job market or the market for venture capital (Lerner & Tirole, 2002). However, this only partially explains what happens.

An important remark must be made that the actual rent-seeking is not necessarily done by the individual open source community participant. Many examples exist where participants are paid by (for-profit or non-profit) institutions to work on an open source project. As O'Mahony (2003) states, "Contributors may be sponsored by firms, but they are not employees of the project and project relations are not guided by employment relations" (p. 1179). This group of contributors is likely not primarily motivated to contribute to the project due to its open source nature. In such a case, the participant's rent-seeking and his employer's rent-seeking are not of the same type.

While the individual community participant might be motivated by reasons which have nothing to do with open source, such as "just doing his job," his employer apparently is motivated by the open source nature. We want to emphasise that *contributors* are not necessarily *developers*, although most literature seems to quietly assume this.

Donators on the other hand are not driven by any immediate individual gain, either monetary or reputation, and therefore can be considered truly idealistic contributors that just want to improve something they value as a product (Rota, von Wartburg, & Osterloh, 2002). They do not contribute for nothing, they do have a goal, but the goal is not yet fully understood by mainstream economics and subsequently might receive less acknowledgment than classic rent-seeking motivations (Hansmann, 1980).

As Franck and Jungwirth (2003) argue, neither rent-seeking nor donation alone can sufficiently explain why the open source model works. It is the combination of both, the motivation mix, which makes the model successful. Therefore they consider one of the basic institutional innovations of open source projects the crafting of a governance structure which enables rent-seeking without crowding out donative behaviour, which is in line with Bonaccorsi and Rossi (2003). In particular, they explain why classical capitalistic firms based on the rent-seeking model struggle to not drive out idealistic donators who do want to help, but don't want to see their help turned into financial profit by and for the company. Open source projects avoid this problem, and subsequently may attract more community participants, potentially leading to more (community) customers.

Coordination Approach

Another clear distinction between open source and proprietary processes can be found in the type of the process itself. There is the *disclosure-feedback approach* used in open source projects, and the *secrecy-incorporation approach* used by traditional firms in the software industry (Franck & Jungwirth, 2003, p. 404). Not only do these two approaches require different communication paths between end-users and developers, they also influence the complete organisational culture of a software community.

It can be argued that institutional secrecy-incorporation culture might cause a built-in tendency for the vendor to focus on the code and the feature list from his own point of view only, growing a product that from the outside might be what customers think they want, but that from the inside slowly turns into a dinosaur. The open source culture on the other hand does not only drive the resulting product, but also the underlying code base and the road map. It has inherently less trouble with technical inbreeding and on top, people downstream of the developers still can get information about the used technology, coding style, future plans, and other internal issues. If the customers see developments that they do not favour, they have a direct communication path to the developers, without a sales and marketing organisation that tries to mediate. And they can always decide to take over the development, to fork off a branch, or to move to another product in time, depending on the costs and benefits associated with these options. Some customers may consciously select an open source product for these reasons, and dutifully accept the costs of community participation.

Motivation Structures

As the last part of the project domain, we want to discuss several motivation structures that can be used to get work done. The traditional prime motivation structure of capitalistic firms builds upon a complex mesh of trade secrets, information hiding, licenses, copyrights, patents, and all legal and economic institutions that are required to enforce these rules upon the market. With these instruments in place, firms can set up an incentive structure from the top down and assure proper activities by individuals via extensive monitoring. Monitoring includes observing input behaviour, apportioning rewards, giving assignments and instructions, terminating contracts, and so forth (Alchian & Demetz, 1972). Rent-seekers can be rewarded either by paying them proper wages for their work, or by granting them residual claims (profits) as a result of their monitoring (Franck & Jungwirth, 2003).

In the open source culture, both incentives (to have people perform good quality work and to reap the profits of investments) are present as well, yet differently implemented. Lerner and Tirole (2002) argue how individual reputation may make up for the work motivation without requiring immediate financial rewards, and the thoughtful keeping of maintainer and credit files which list individual contributions prevent shirking in a development group (Raymond, 1998b). Franck and Jungwirth further expand on this issue by introducing how experienced developers may gain further reputation by starting new (sub)projects and attracting good people to join in and make the project another success.

However, these incentives leave unexplained who is going to do the "non-sexy work." Typical volunteer contributions will be focussed on "sexy" activities, such as feature expansion. But, as Bonaccorsi and Rossi (2003) say, "The core development group does not carry out the bulk of the coordination effort" (p. 1247). They further add that "it is difficult to accept the idea that [mundane] low-gratification activities could be motivated by the same incentive structure than high-level, creative work." This opinion is shared by Lakhani and von Hippel (2003), who add that next to coordination and quality assurance work, much effort usually is required for documentation, translation, marketing, packaging and other "mundane" activities outside the typical developers' scope. The obvious need for these "mundane" activities may be a prime reason why organisations that are end-users of an open source product may decide to participate in the community. By mak-

ing sure these activities take place, they safeguard their own interest in the product.

Community Customers

From the driving forces listed above, it can be concluded that direct, individual participation in the community is not necessarily sufficient to make an open source project flourish. Many projects will need contributions outside the development core which are insufficiently rewarding to be taken up by volunteers (Bonaccorsi & Rossi, 2003, pp. 1246-1247; Lakhani & von Hippel, 2003).

The literature covers the case of so-called *hybrid business models* in which for-profit organisations shift their attention from development to providing services around an open source product. End-users, or end-using organisations, may purchase the services of such a vendor with the product and can ignore the underlying open source model. They talk primarily to the vendor in case of problems or questions.

An alternative approach is when the end-using organisation sees the open source community itself as the vendor, and becomes *a customer of the community.* This is distinctively different from free riding on the product—the organisation truly spends resources on the customership. In many cases, there is no monetary exchange between the organisation and the community, as often the community has no central representative which accepts money in exchange for services, such as with a traditional vendor (even if hybrid). But the customer certainly may spend resources on the open source product, by *donating effort around the product to the community.* A very common way of becoming a community customer is to pay employees or contractors to do some work around the product, which is not necessarily development, and then instructing them to donate the results to the community (Lakhani & von Hippel, 2003).

From the point of view of a product end-user who is not actively participating in the product's development, he has the choice of becoming a *vendor customer* or a *community customer*. In both cases, he needs to assure motivation by paying up. He can pay a vendor to get the product, or pay a person (or third party company) to assure the product becomes and remains available via the open source community. If an end-user decides to free-ride on an existing open source product and explicitly does not become a community customer, he runs the risk of being left in the dark when the product deviates away from his needs, which is exactly what would happen if he illegally duplicates commercial software. For individuals, this may be less of a problem, but for organisations it surely is, or at least should be considered a serious drawback.

The rest of this chapter will elaborate on the role that end-user organisations should play as they become a community customer.

PRODUCT vs. PROCESS

Related to the disclosure-feedback versus secrecy-incorporation approaches discussed previously, we can observe a key cultural aspect of open source projects which fundamentally differs from proprietary culture. Notice that we do not say *commercial* culture: It has been conclusively proven by now that commercial enterprise and open source can go hand in hand with hybrid business models (IBM, Sun Microsystems, Apple, Red Hat, Oracle, and others are spending considerable funds on open source product development, according to Bonaccorsi and Rossi), although just GPL-ing the sources of a previously proprietary product in the hope of increasing user involvement certainly is not sufficient (Dalle & Jullien, 2003).

The key difference may be that open source culture favours the *process* over the *product*. Typical proprietary culture favours the product over the process. Some open source products are fully comparable to their proprietary counterparts, yet the process which produced them is completely different. We already discussed how

the open source governance structure enables a sustainable process in a different way than proprietary governance; in the next sections, we discuss how this different process influences the result, the product. We will argue that community customers must focus on the process, not on the product, and that their donations should be aimed at the process.

Understanding how to use the community of an open source project may be the key to successful development, selection, deployment, and maintenance of the product. The nature of open source is such that ignoring the community may mean a significant increase in risk for open source product users. We will show what use should be made of the community in what phase of a project (where deployment of a product is a project in itself), and argue that (monetary) resources saved by acquiring an open source product instead of a commercial proprietary product should at least partially be invested into the community. Not because it is a noble thing to do, but because it is required for proper process management, which reflects back onto the product. Just as vendors need honest customers, open source communities need honest customers as well.

COMMUNITY CUSTOMER ROLES

We define a community customer as *an individual or organisation who wants to deploy an open source product, without having a direct aim to further develop the product, and who actively engages or instigates engagement in community participation to assure future suitability of the product for one's own purposes.* A community customer typically is not a developer. The term *customer* should be assumed to mean exactly that: offering resources, monetary or otherwise, to receive services or products in exchange.

From a practical point of view, any organisation that considers deploying open source software should be fully aware of the roles that the community needs to play in the process. Just as thinking of open source software as free beer misses the point, assuming that open source software can be treated as a shrink-wrapped box with a help desk phone number inside may lead to expensive mistakes as well. Becoming a community customer is not as easy as becoming a vendor customer—yet.

Development Process

This part of the open source process has been discussed in great length in various places and receives the most attention. For the purpose of our statement, we will not join the discussion but just observe that active participation in the development community may be a very attractive option for organisations that have operational feature requirements which are not yet fully satisfied by the current product (Green, 1999). Being open source, the introduced enhancements will of course find their way outside the organisation, so they cannot directly be used for competitive advantage. However, indirect advantages may be such that even considerable investments in development may pay off. IBM and Oracle are prime examples of companies which explicitly fund open source development of infrastructural projects, to reap the benefits of better infrastructure to build proprietary products and services on.

Organisations funded by tax payers, such as the Dutch SURFnet,[1] increasingly demand that any software developed with their funds must be open source. This does justice to the origin of the funds and prevents products from dying when their originating project terminates. Many examples exist of products, developed from government grants by a commercial party not bound to open source, that were shelved as soon as the project came to a conclusion, no matter how successful it was. Large public funding organisations such as the European Union have tried to assure product viability by stressing business plan development during or even before product development, and

also by rigorous matched funding requirements to force (commercial) partner tie-in. This approach has not been very successful. Making all publicly funded products open source by default could be a very attractive alternative.

Summarizing, active participation in the development process is true community customership for organisations, but by no means the only possibility.

Procurement Process

Software procurement has been a well-studied subject for many decades (Anderson, 1990), and open source procurement gets much attention as well. There appear to be a few basic differences between typical proprietary software procurement practices and open source software procurement practices. Some of these differences occur because open source software allows for much more information gathering from process details that are usually unavailable with proprietary software. Others occur because open source software typically does not have a commercial organisation with a marketing and sales budget behind it, which actively goes out to win new customers.

Although there are no licensing costs associated with open source products, their procurement costs are not zero. There is evidence that open source product assessment and selection might be significantly more expensive (for the customer) than proprietary assessment and selection, as much of the actual cost is shifted from the vendor to the customer. Using industry-average data, it was estimated that the sale of a proprietary learning system would cost the vendor over US$250,000 in proposal writing, large-scale demonstrations using detailed, prescribed scripts from the customer, expert presentations and so forth (Farmer, 2006). All this work needs to be done by other people (not from a vendor) when the procurement of an open source system is investigated. A part can come from documented community experiences, but it mostly is up to the customer to spend the resources. Although, obviously, the customer ends up paying the costs back to the vendor in case he decides to license the proprietary product, he does not need to pay in case he does not license the product. This no-purchase-no-pay option is unavailable when investigating open source software, as there is nobody else who takes the risk of spending the money in the hope of winning a sale and getting it back with profit.

The community already should play a very important role in open source procurement decision making. Van den Berg (2005) lists explicit community input for the decision process and indicators to measure them objectively. She suggests that visible community activity traces should determine whether the associated open source product should be shortlisted, and that in-depth reviews of (implicit) community test results and user experiences should play an important role in the final decision.

We would suggest that as standard part of an open source procurement process, the "customer" should donate its findings during the procurement process back to the community, no matter whether the product was eventually adopted or not. There is in no way any obligation to do so, but the open source culture as a whole favours this kind of contributions as in many cases they are not purely donative, but have serious rent-seeking components as well.

As Feldstein (2006) suggests, this might be one of the few ways to fundamentally alter the procurement process and gain significant financial advantage for nearly everybody. It may come at the expense of some proprietary vendors who use large marketing budgets to outsell less financially strong competitors, making use of the unavailability of free, objective assessments in the proprietary world. But in the end, the customers pay for these marketing efforts out of their own pockets, reason why they do have a long-term incentive to change the procurement process to a more open one. They can do so by contributing their findings to the community, which would satisfy the classic economic assumption of availability of full market information to all market parties.

As with most donations, the problem is that there is no immediate financial or other reward of donating experiences, especially not if they were negative and it was decided not to use the open source product. However, if there is a chance that the same organisation will in the future again review any open source product, the donation will be worth the effort, because it increases the likelihood that others will contribute their experiences as well, and it increases the respect that the donating organisation will gain among peer organisations. Having respect due to proven contributions is a significant asset in open source communities, and will usually lead to priority service in case the organisation needs something in return, such as concrete help. First-class tickets can be purchased, also in the open source world; see also Lakhani and von Hippel (2003).

So, to summarize, open source procurement must be approached differently than proprietary procurement. Initially an organisation needs to invest more of its time and resources, but the open source community will be inclined to help, and even more so if the organisation shows respect by donating its experiences back to the community straight away (preferably not after the whole procurement process has been completed, but much earlier). If the organisation decides to deploy the open source product, it will be immediately rewarded for its donation by not having to pay the marketing and sales efforts back to the vendor in licensing costs. Instead, it has invested soundly in its community reputation. It has become a true community customer.

Deployment Process

The community roles in the deployment process are for a large part the same as those in the procurement process. Documented previous experiences, best practices, how-to and other helpful guidelines are a valuable resource for any deployment project. These resources need to be built up by organisations that deployed the product in the past. Mature open source products usually have a significant body of this type of documents available. Lakhani and von Hippel (2003) summarize all these resources plus personal help as *field support*, and call it essential for open source project success.

Extra information may be gathered from the community by tapping into its people network. When specific questions or problems pop up during a deployment project, the community usually offers fast methods to get help from people (and organisations) that have been there. Here it pays off if the deploying organisation has established a level of visibility in the community, as there is a cultural priority mechanism in place which makes others more inclined to help if they feel that they have been helped as well, in other words, previous community contributions have been made. Note that it is not experience that counts in this process, but attitude. If somebody helps people that come after him, he will be helped by people that went ahead of him.

Therefore, an allowance in the project deployment budget should be made to document and file the experiences back to the community, and if possible to join the pool of active community members that can provide quick assistance to new users of the product. This active participation, beyond a one-time documentation donation, demarcates the line between treating the product as a stable entity and treating it as a living, growing being. It is natural that this demarcation coincides with the moment of deployment: Before deployment, the product was not actively used.

In practice, this active participation happens nearly unnoticed. Many open source deployers subscribe to the product's mailing list or become a regular visitor of the product's web site to stay informed of changes. According to Lakhani and von Hippel (2003), time spent on reading these resources may average 100 hours per year for active participants. Many of them then get into the habit of actually replying to cries for help from others "while they are there anyway." This is not time lost to charity; it is a sound investment in their organisation's visibility in the community. It will be noticed by other community members and when the organisation needs help itself, it will get it. As a community customer, they paid for it.

A way to disturb this process is by only asking and never returning the favour, or worse, by outright demanding something from the community as if it were a vendor. Therefore, deployment project management should explicitly favour active participation and visibility in the community, and not view it as idling on the Web instead of getting serious work done, or worse, helping the competition ahead. Postponing community contributions until community services are required means that the help will not come when it is needed most.

Operational Maintenance Process

Whereas during the deployment process the main community resource is the available documentation, the operational maintenance process is largely supported by quick responses to concrete (and often urgent) questions and the associated monitoring of other organisation's questions and answers. For many organisations that use open source products in their daily operations, such participation in the community has become a second nature.

Operational issues around software are not much different for both proprietary and open source products. Both need regular patching for bugs and for security problems. Both need to have feature development going on, as no environment stays the same for very long. The community provides these patching services as a natural part of the process, and the organisation using the open source product must be as committed to keep their installation up to date as with a proprietary product.

What may be different is that open source products tend to have a livelier patch cycle. Proprietary products typically have many months between releases and may provide an update service via the sales organisation, actively approaching their known customers with patches depending on the perceived urgency. Open source products may offer the same service, but usually do not actively approach customers. Instead, the community relies on being actively monitored. With the current trend towards online updates, many products from both proprietary and open source origin check for updates automatically, and even may apply the patches automatically without service interruption.

Open source products tend to have a quicker response than traditional vendors, with the community watching the product vigilantly and providing solutions to discovered problems often within hours. Especially when security problems are discovered, vendors may be tempted to keep the problem in-house and quietly solve it with the next patch release, hoping that nobody will produce an exploit in the mean time. Open source products usually do not tolerate this delay and rely on the community to provide a patch as soon as the problem is disclosed. This means that it is in the deploying organisation's own interest both to keep a keen eye on the reported problems and available patches, and to actively contribute in reporting perceived problems or even fixing them for the community.

Proprietary products tend to have a few major customers who are talking to the vendor's marketing organisation, while the small customers may be left in the dark. With open source products, in theory everybody can talk directly to the developers. However, some community customers will be larger than others, just as with major accounts in traditional commercial relationships. It is to be expected that the perceived account size in terms of customer contributions to the community also drives his influence on the developer core. But customer contributions are independent of actual organisation size in terms of number of licenses and other traditional indicators.

This open community steering means two things for organisations that deploy open source products: they have a heavier vote in the product's development if they are actively participating in the community, and if they are serious about some required feature which is not getting enough attention, they may develop it (or have it developed by a third party) and donate it to the community for further integration and maintenance. It is not uncommon for organisations using infrastructural open source products to see the funding of co-development of these products as a regular operational cost. Instead of hoping that the vendor steers the product towards a useful future, they get hold of the steering wheel themselves where required.

It will not come as a surprise that yet again the community role here is one of serve and be served. Free riding certainly is possible and takes place all the time, but organisations that are serious about their software have the option and nearly the obligation to take an active role and to invest real resources into their infrastructure. The paybacks are not necessarily immediate, but almost always guaranteed. Lakhani and von Hippel (2003) report that 98% of time spent on community communications is reserved for reading and learning about other people's problems, to improve one's own performance. Only the remaining 2% is actually spent on helping others. In the end, because of resource sharing and economies of scale, the result often will be obtained with less overall resource spending than with classical proprietary production where competition is the main driving force behind development. A side effect is a reduced chance that a critical product suddenly disappears from the market due to competition. It is much more likely that a timely course change takes place, or a friendly merger with another product that appears better designed or uses newer technology.

FUTURE TRENDS

With the growing importance of open source products in the world's ICT infrastructure and the proven viability of the business model underlying the process, it is to be expected that more organisations will deploy open source products in their daily operations. Care needs to be taken that these organisations become true community customers to assure continuity. Open source projects therefore will further develop the equivalent of sales and marketing efforts to explicitly build and foster their community of customers. It can already be observed that some of the larger communities spun off actual marketing departments with a real budget, often sponsored by a traditional commercial company which seeks rent from the after sale services.

A new market may emerge for companies which specialise in open source community customer relationship management, organising the community for a product without participating in it. They would get their funding from active community customers who outsource part of their involvement in order to concentrate on core product tasks while leaving the non-product-specific tasks to the specialised and more efficient company. Enterprises such as the Open Source Technology Group,[2] which exploits many Web-based systems that play an important role in the fabric of open source community building (such as SourceForge, Slashdot, Linux.com, Freshmeat, ThinkGeek, etc.), already move towards this market but are not fully there yet.

Other expected trends are the increasing commercialisation of all open source-related activities. Although there will be opposition from the fundamentalists who believe that any commercial activity should be rejected, commercial enterprise around an open source product or even a Free Software product in the most strict sense is totally in line with the basic assumptions of the model.

People will find creative ways to retain the important aspects of open source while adding known applications of capitalist economy to improve the overall efficiency. Existing proprietary vendors will increasingly participate in this process in order to survive, and several maturity models already position software vendors with a large interest in services higher up the maturity ladder than development-centric organisations (Farmer, 2006, p. 7). These services can be both around the product (deployment assistance, consultancy) but also directly add to the process, where the services typically are paid for by community customers and delivered to the community by specialized companies or individuals.

CONCLUSION

Any organisation planning to adopt an open source product must consider investing in the associated community, which is time-consuming and expensive, but usually well worth the effort. An indication of the investment required cannot be given yet, although licensing costs of comparable proprietary products obviously are the upper limit. With the scale advantages and organisational learning of an open source community, community participation costs should show a decreasing trend in time and stay under the licensing costs of a proprietary product in most cases. Further research should attempt to make this trend explicit and to develop theories to predict actual individual and total costs of open source projects.

What is evident in the process is that any attempt to consider open source or free software a bargain due to the absence of licensing costs is bound to cause a problem. Free lunches are rare. However, the freedom of choice what to eat for lunch, where, and when, is a benefit well worth the price.

REFERENCES

Alchian, A. A., & Demetz, H. (1972). Production, information costs, and economic organization. *The American Economic Review, 62*, 777-795.

Anderson, E. E. (1990). Choice models for the evaluation and selection of software packages. *Journal of Management Information Systems, 6*(4), 123-138.

Bonaccorsi, A., & Rossi, C. (2003). Why open source software can succeed. *Research Policy, 32*, 1243-1258.

Craig, J. S., & Beck, C. E. (1993). New look at documentation and training: Technical communicator as problem solver. *Information Systems Management, 10*(3), 47-55.

Dalle, J.-M., & Jullien, N. (2003). 'Libre' software: Turning fads into institutions? *Research Policy, 32*, 1-11.

Evers, S. (2000). *An introduction to open source software development.* Technische Universität Berlin, Fachbereich Informatik, Fachgebiet Formale Modelle, Logik und Programmierung (FLP).

Farmer, J. (2006). *On the cost of selling an enterprise learning system.* Retrieved February 2006, from http://www.immagic.com/eLibrary/GENERAL/IMM/I060108F.pdf

Feldstein, M. (2006). *More thoughts about Blackboard: "The fault, dear Brutus..."* Retrieved February 2006, from http://mfeldstein.com/index.php/weblog/permalink/more_thoughts_about_blackboard_the_fault_dear_brutus/

Franck, E., & Jungwirth, C. (2003). Reconciling rent-seekers and donators: The governance structure of open source. *Journal of Management & Governance, 7*(4), 401-421.

Green, L. (1999). *Economics of open source software.* Retrieved June 2006, from http://badtux.org/home/eric/editorial/economics.php

Hansmann, H.B. (1980). The role of nonprofit enterprise. *Yale Law Journal, 89*, 835-901.

Lakhani, K., & von Hippel, E. (2003). How open source software works: "Free" user-to-user assistance. *Research Policy, 32*(6), 923-943.

Lerner, J., & Tirole, J. (2002). Some simple economics of open source. *The Journal of Industrial Economics, 50*, 197-234.

O'Mahoney, S. (2003). Guarding the commons: How community managed software projects protect their work. *Research Policy, 32*, 1179-1198.

Raymond, E. S. (1998a). The cathedral and the bazaar. *First Monday, 3*(3). Retrieved November 17, 2005, from http://www.firstmonday.org/issues/issue3_3/raymond/

Raymond, E. S. (1998b). Homesteading the Noosphere. *First Monday, 3*(10). Retrieved November 17, 2005, from http://www.firstmonday.org/issues/issue3_10/raymond/index.html

Rota, S., von Wartburg, M., & Osterloh, M. (2002). *Trust and commerce in open source: A contradiction?* [Working Paper]. University of Zürich.

Stallman, R. (1985). *Free software foundation.* Currently active at http://www.fsf.org/

van den Berg, K. (2005). *Finding open options.* Master's thesis, Tilburg University. Retrieved February 2006, from http://www.karinvandenberg.nl/Thesis.pdf

KEY TERMS

Community Customer: An individual or organisation who wants to deploy an open source product, without having a direct aim to further develop the product, and who actively engages or instigates engagement in community participation to assure future suitability of the product for one's own purposes.

Development (Open Source): The act of writing program code to extend the functionality of an (open source) product. Explicitly limited to code writing to distinguish it from the many other productive activities around an open source product (reviewing, translations, packaging, end-user help, documentation writing, marketing, process management ...) which typically are not done by developers but also not by typical users.

Free Rider: An individual or organisation who acquires an open source product and actively uses it, without donating experiences or development results back to the community, but while still using the community resources.

Open Source Community: In this chapter the same as open source project, but community is better suitable, as it has more people semantics than project.

Open Source Community Customer Relationship Management: Task/process of actively following, facilitating, and fostering the community customers, so that they keep coming back to the community for help, and hopefully donate contributions to the community in return. CCRM may be outsourced to an organisation which does not itself participate in the community and may ask a fee for the work.

Open Source Project: Any group of people developing or deploying software common to the group and providing their development results and usage experiences to the public under an open source license.

ENDNOTES

[1]   http://www.surfnet.nl/
[2]   http://www.ostg.com/

Chapter XL
Open Source Software Business Models and Customer Involvement Economics

Christoph Schlueter Langdon
Center for Telecom Management, University of Southern California, USA

Alexander Hars
Inventivio GmbH, Bayreuth, Germany

ABSTRACT

This chapter is focused on the business economics of open source. From a strategic perspective, open source falls into a category of business models that generate advantages based on customer and user involvement (CUI). While open source has been a novel strategy in the software business, CUI-based strategies have been used elsewhere before. Since the success of e-commerce and e-business, CUI-based strategies have become far more prevalent for at least two reasons: Firstly, advances in information technology and systems have improved feasibility of implementation of CUI strategies and secondly, CUI-based economics appear to have often become a requirement for e-business profitability. This chapter presents a review of CUI-based competition, clearly delineates CUI antecedents and business value consequences, and concludes with a synopsis of managerial implications and a specific focus on open source.

INTRODUCTION

Open source software applications and source code are developed cooperatively in an Internet-based peer-to-peer network or community of programmers (Hars, 2002). Some call the open source development process, therefore, also as peer-to-peer production (Wikipedia.org, 2006). The open source model has caught the attention of business strategists and financial analysts (and executives and shareholders of software firms), because open source developers devolve most property rights to the public, including the right to use, redistribute and modify the software free of charge. Some industry observers argue that this approach will emerge as the prevalent way to design and write software; others have been more cautious seeing open source as a niche model (Hars & Ou, 2001; *The Economist*, 2006).

Open source is new and old at the same time. It is a new concept in the software industry. However, the attractiveness of open source is rooted in mechanisms and economics that have fueled business success in many other areas before. From a business strategy perspective open source fits into a broader category of business models based on customer and user involvement (CUI) that can provide superior economics.

A very visible example of this category of business models is Ikea, the Swedish furniture maker and retailer. Among consumers Ikea is known for its stylish yet affordable furniture. Among some business strategists and researchers Ikea is a prominent example of the economics of customer involvement, which has emerged as a key source of competitive advantage, particularly in the e-commerce area. Broadly speaking customer or user involvement describes a strategy that emphasizes engaging customers and user in business operations.

BACKGROUND

"Ikea Economics"

In the case of the Swedish furniture maker and retailer, Ikea, customer involvement is integral to doing business and creating economic advantage. Ikea customers are involved in business operations in that they pick their purchase off the Ikea warehouse shelf, drive it home and assemble it themselves.

Figure 1 depicts a two-tier industry system following Porter's value chain schematic (Porter, 1985). A product has to be developed, made, distributed, sold, and delivered. In the case of Ikea outbound logistics or delivery and final assembly are "outsourced" to the customer (see Figure 1). This saves Ikea cost compared to the competition that sells assembled pieces, which are bulky and, therefore, have to be home delivered. Furthermore, because Ikea furniture is assembled at the final

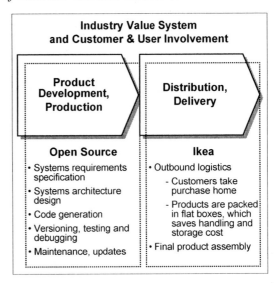

Figure 1. Open source and Ikea: Two examples of customer and user involvement

destination, a customer's home, products can be shipped in flat boxes without negative space, which further saves handling and storage cost throughout the entire supply chain and channel system. But the advantage of customer involvement doesn't stop here with merely lower cost. Customer involvement economics can be an enabler of other economic advantages. In the Ikea example, the cost advantage due to customer involvement is used or leveraged by splitting savings with the customer, effectively lowering product prices, often below the price of the competition. The lower sticker price makes stylish design affordable for a larger market, which increasing Ikeas market potential. This larger footprint, in turn, allows Ikea to benefit from another economic advantage, the one that has been the main economic engine of mass production, namely scale economies. In other words, at Ikea customer involvement has worked as a starter to ignite an economies of scale engine. This combination of customer involvement economics and scale economies have helped Ikea become the world's largest furniture maker and retailer with 221 stores in 34 countries as of Spring 2006 (http://franchisor.ikea.com,

Figure 2. Theoretic customer and user involvement model

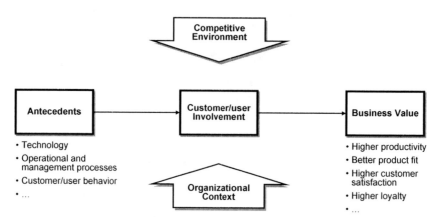

3/31/06). It also turned its founder, Ingvar Kamprad, into a multi-billionaire. *Forbes* magazine recently estimated Mr. Kamprad's fortune at $28 billion—trailing only Microsoft co-founder Bill Gates, U.S. investor Warren Buffett and Mexican industrialist Carlos Slim (Forbes, 2006).

Theory of Customer and User Involvement (CUI)

The literature defines customer involvement as the extent to which a customer is engaged as a participant in business operations, specifically in service production and delivery, including, for example, order processing and account management (Schlueter Langdon, 2003a, 2006). A first research construct has been developed and integrated into a broader theoretic model (see Figure 2; Schlueter Langdon, 2003a, 2006).

The customer involvement construct and its definition are rooted in several streams in the literature: "customer integration" and "customer relationship management" in marketing, "co-production" and "service encounter management" in service operations research, and "citizen participation" in the public policy literature.

In 1980, Whitaker introduced the notion of "co-production" in public service delivery in the field of public policy management (1980). At the

same time Hakansson appears to have introduced the notion of a "customer integration strategy" within the context of marketing strategies in industrial markets, defining it as the ability to adapt to specific customer needs to increase business benefits (Hakansson, 1980, p. 370).

Brown, Raymond, and Bitner (1994) have first explicitly used the phrase "customer involvement" in their categorization of research on service encounters. Brown et al. divided research on service encounters into three primary types, the second of which is focused "on customer involvement in service encounters and the customer's role in service production and delivery" (1994, p. 34). Chase (1978) first discussed a customer's role in the service delivery process. This perspective has been expanded in the service operations literature to also consider the customer as a partial employee (Czepiel, 1990; Bowen, 1986; Kelley, Donnelly, & Skinner, 1990; Mills & Moberg, 1982; Mills & Morris, 1986).

In the marketing literature Sheth and Parvatiyar posited that "relationship marketing attempts to involve and integrate customers, suppliers, and other infrastructural partners into a firm's development and marketing activities. Such involvement results in close interactive relationships with [...] customers [...]" (Sheth & Parvatiyar, 1995, p. 399). Furthermore, "consumers are increasingly becoming co-producers. […] In many instances, market participants jointly participate in design, development, production, and consumption of goods and services" (Sheth & Parvatiyar, 1995, p. 413). Gruen, Summers, and Acito (2000, p. 36) called this phenomenon "co-production."

The notion of customer integration is presented in the marketing literature as an extension of manufacturer-distributor relationships (Andersen & Narus, 1984, 1990). The theory base that underlies the marketing literature on manufacturer-distributor relationships and, therefore, the argument that customer involvement can enhance business value (see Figure 2) is a synthesis of exchange theory (Kelley & Thibaut, 1978; Thibaut & Kelley, 1959) and transaction cost economics (Williamson, 1975, 1985). Exchange theory states that parties transfer resources in relationships to enhance self-interest, while transaction cost economics reveals conditions under which certain organizational choices can maximize self-interest in the exchange relationship.

Specifically, the literature points to several consumer and seller benefits from tight customer integration. Lovelock and Young (1979) discussed the customer as source for increasing a service firm's productivity. Sheth and Parvatiyar (1995) indicate that consumers benefit from products and services that suit their needs better and sellers from higher customer satisfaction. Higher customer satisfaction in turn is positively related with customer loyalty and market share (Anderson, Fornell, & Lehmann, 1994; Anderson, Fornell, & Rust, 1997).

MAIN FOCUS OF THE CHAPTER

Customer and User Involvement and Business Value Categories

Since the success of the Internet in business, CUI-based strategies have become more prevalent. For one, advances in information technology and many open standards have increased information systems capabilities at lower cost to make CUI-based strategies feasible. For another, CUI economics are often required in the first place in order to make e-business operations profitable, because in electronic commerce companies have become expected to do more for less.

Analysis based on industrial organization theory clearly highlights this more-for-less dilemma. Tracking value chain activities, such as product search, reveals that the Internet-enabled change in the interaction between a consumer (demand side) and vendor (supply side) has led to an extension of the traditional value system (Schlueter Langdon & Shaw, 2000, 2002). In electronic commerce vendors are often doing more than in traditional commerce. Online vendors are supporting activities, which consumers have to perform manually in traditional channel systems. For example, instead of driving to multiple stores, walking up and down the aisles to search for a product and find a low price, shoppers can enter key words and at the push of a button, they can evaluate competing price quotes. Doing more is costly as online sellers resort to "softwarization," the wholesale automation of business transactions and processes using information systems (Schlueter Langdon, 2003b, 2003c). While labor cost may be saved, online vendors have to invest in the design, building and implementation of sophisticated information systems, and they continue to spend money on operations, maintenance and updates. The cost of selling goods online may be cheaper but consumers also expect lower sticker prices online. In order to turn a profit, online vendors often rely on CUI economics. Table 1 provides a systematic overview of major, generic CUI business value categories.

High customer involvement may allow for mass-customization of products and services using customer data or user profiles, which, in turn, may facilitate both—lower cost and higher revenue. To take a real-world example, Dell can leave customization of products (e.g., choice of

Table 1. Major, generic CUI business value categories

Cost	Revenue
Customer or user operates business process activities	

• Company saves employee time and expense
• Likely higher fixed cost for IS that can be operated by many customers instead of a few employees only
	• Goods can be purchased anytime and from anywhere
	  -> Higher quality, better product fit
	  -> Better customer data

Higher quality, better fit	
• Less inventory in entire channel system • Less slow moving and obsolete items • Less discounts	• Customer likes the feeling of being in control -> Higher customer satisfaction • Monopolistic competition pricing opportunities
Higher customer satisfaction	
• Lower churn saves customer acquisition cost • Positive word of mouth may save marketing expenses	• Higher loyalty • Higher lifetime customer value
Better customer data or "profiles" (behavior, wants and needs)	
• Data mining improves accuracy of targeting customers and saves marketing and sales cost • Lower marketing research cost	• Up-selling opportunities • Better next generation product • User lock-in and higher switching cost

microprocessor) and product bundling (e.g., PC with ink jet or laser printer) to individual preferences, which can increase up-/cross-selling opportunities and customer satisfaction (see Table 1). At the same time Dell can save inventory cost and write-offs, because customers trigger of manufacturing and assembly activities (see Table 1). Instead of Dell pushing products into the market, customers are pulling the product through the system, turning a made to stock system into a made-to-order flow.

CUI is not limited to specific industries, such as consumer products (Ikea and Dell). The auto industry has discovered CUI economics as a source of business advantage. For example, BMW, the German maker of luxury cars, has designed information systems so that European buyers can custom-design their own cars with any change possible until five days before production. As a result, 80% of European BMW buyers custom-design their vehicles and most last minute changes of orders are reportedly upgrades to bigger engines and more luxurious interiors, which tend to be more lucrative for the firm (*Business Week*, 2003). Another CUI example in the auto industry is the emerging area of vehicle relationship management (VRM). Automakers have begun to install black boxes into vehicles that often work similar to flight tracking devices in airplanes. The box is valuable in two ways. Firstly, it provides vehicle usage data, which is a function of vehicle model, the driver and its environment. Secondly, it provides a new, interactive channel system with every customer. The data and the channel can be exploited to better manage customer and vehicle relationships, hence VRM. Vehicle usage data can be exploited for diagnostics purposes to improve uptime. The new channel can be used to interact with customers to improve buyer satisfaction and loyalty. All it takes to unlock the value is user participation.

In the software industry open source software has emerged as an important implementation of CUI economics. Many applications are created by an open source community. Figure 1 illustrates that all essential software development activities of requirements specification, architecture design, code generation, debugging and testing as well as ongoing maintenance are left to a community of users.

FUTURE TRENDS

Discussion and Managerial Implications

In open source software advantages accrue along all three major dimensions of business performance: cost, time and quality. Figure 3 reveals how an average IT implementation project using open source software compares with a project that is being built traditionally. Results are based on a convenience sample of expert assessments. The business value parameters have been defined at a very high level: cost measures project cost including maintenance, time measures initial implementation as well as downstream modifications, and quality rates the degree of excellence and customer satisfaction.

In terms of cost, open source saves at least the profit or profit margin associated with a brand name product, brand name systems integration service and brand name maintenance contract. (The software business "has an exceedingly high gross [profit] margin of 90%, [...] a net profit margin of 27%. This shows that its marketing and administration costs are very high, while its cost of sales and operating costs are relatively low (McClure, 2004).)

Open source can save time, because documentation is public and exposed to public scrutiny, just like the source code itself (Hars, 2002). Furthermore, customer support is not limited to a vendor's office hours or a particular maintenance subscription level but open source documentation and expertise tends to be available online and anytime.

Quality can be better, firstly, because of transparency of the process and secondly, because of

transparency of qualification and achievements of contributors (Hars, 2002). This mirrors a key lesson of a free market system, namely that transparency tends to increase buyer value. Also, problems are fixed when a problem exceeds users' willingness to cope and not when decided by a vendor's corporate strategy or business policy.

Figure 3 summarizes the assessment of our convenience sample of experts, which includes senior developers and architects, and information technology executives of Fortune 500 companies (chief information officers, CIOs, and vice presidents).

Figure 3 compares an open source implementation with a traditional, branded solution along the aforementioned and defined business value categories of cost, time and quality. Results reflect a consensus among our experts that open source software beats a traditional solution in any category. The extent of this advantage can vary. First, there is variability within each category. Consider cost: some experts see an OSS implementation at 50% of the cost of a traditional solution. Others see it more at 75%. Second, there is variability across business value dimensions. Higher quality appears to be the most significant advantage, followed by lower cost.

Figure 3. CUI business value assessment: The open source example

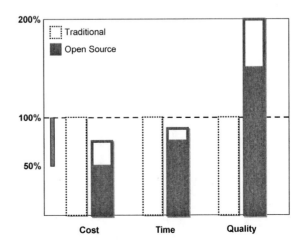

527

By nature, a high level comparison, such as the one presented in Figure 3, is constrained and implications are limited. First, this comparison is limited to an implementation that utilizes open source software instead of a commercial package (e.g., installing, configuring, integrating, testing, and maintenance). It does not include the writing of application source code. Furthermore, the comparison is focused on situations in which open source is a true alternative. Second, the business value parameters—cost, time, and quality—can be interdependent and, therefore, difficult to isolate. For example, in order to speed up a project the quality of the code may be compromised; to save money less qualified engineers are used who need more time to write the code, and so on. Expert interviews were conducted in a way that such effect would be additive to the assessments presented in Figure 3. Third, an average project is considered and, therefore, results aim to reflect a *central tendency*, which is useful as a guideline but it obscures the variance in size and complexity of information systems projects. Furthermore, the distribution may be skewed and in this case average values can be easily misinterpreted.

It is understood that a specific evaluation would require a dedicated analysis in order to properly compare alternatives quantitatively. In order to conduct such analysis, a multi-step approach would have to be devised. Figure 4 depicts an exemplary business intelligence analytics schematic derived from research theory (Schlueter Langdon, 2005, 2007).

Central to any business value assessment—and open source is no exception—is the identification of a causal model that underlies everything that follows (see Figure 4, phase two: qualitative assessment -> conceptual model). A causal model represents the most relevant variables and a set of logical relationships between them. It prevents confusing cause and consequences. The business practice of jumping straight into a spreadsheet to calculate a conclusion is a common mistake. No patient would accept treatment without prior diagnosis. By the same token, any reliable and robust business value assessment requires careful separation of independent and dependent variables as well as moderating effects grounded in theory and best practice. While medical doctors are trained extensively to administer diagnosis-based treatment many managers jump straight to actions, often based on gut instinct only. If key variables and cause-result relationships cannot be clearly identified and delineated on a single sheet of paper then it is not plausible that jumping to some spreadsheet-based calculation would suddenly solve the problem.

Figure 4. Business model evaluation method

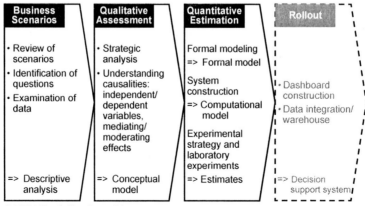

© 2005-06 Pacific Coast Research Inc.

At the conceptual modeling stage for assessing CUI benefits it is important to:

- Understand under which circumstances customer involvement can create benefits:
  - What are specific customer involvement antecedents?
  - What are key moderating effects (see Figure 2)?
  - Avoid reinventing the wheel and instead use existing theory, best practice and literature in information systems, management and marketing, for example.
- Understand how to leverage CUI economics:
  - Can CUI be leveraged to generate other advantages (see Table 1)?
  - Would it take partners to increase advantages?
- Understand how an incumbent business model may become vulnerable to CUI-based competition from either old rivals or new entrants or both.

Once a model has been designed it can be implemented. Typically, this means constructing a spreadsheet (see Figure 4, phase three: quantitative estimation -> estimates). This is also an opportunity to verify measurements concepts before collecting the required data. Finally, results would have to be evaluated.

CONCLUSION

Aforementioned issues can only be exemplary. Experience suggests that it is often misleading to suggest a generic solution. The business model evaluation method presented in Figure 4 distinguishes between major analytical phases. It would have to be adapted, modified and specified for a given decision problem. However, while actual outcomes may vary, Figure 3 suggests that an open source solution may in any case be an economical choice. This outcome coincidences with the observation that brand name software vendors increase the attractiveness of products that compete with open source packages and/or even offer products in an open source way.

ACKNOWLEDGMENT

The manuscript has benefited from thoughtful suggestions and comments of the many contributors and anonymous reviewers of the Special Interest Group on Agent-Based Information Systems (SIGABIS) of the Association for Information Systems (AIS, www.agentbasedis.org). The authors particularly acknowledge discussions with and advice from Steve Davis, Omar El Sawy, Mark Hayes, Jörg Heilig, Bob Josefek, Ann Majchrzak, Steffen Neumann, Kim Spenchian, and Ed Trainor.

REFERENCES

Anderson, E. W., Fornell, C., & Lehmann, D. R. (1994, July). Customer satisfaction, market share, and profitability. *Journal of Marketing, 56*, 53-66.

Anderson, E. W., Fornell, C., & Rust, R. T. (1997). Customer satisfaction, productivity, and profitability. *Marketing Science, 2*, 129-145.

Anderson, J. C., & Narus, J. A. (1984). A model of the distributor's perspective of distributor-manufacturer working relationships. *Journal of Marketing, 48*, 62-74.

Anderson, J. C., & Narus, J. A. (1990). A model of distributor firm and manufacturer firm working relationships. *Journal of Marketing, 54*, 42-58.

Bowen, D. E. (1986). Managing customers as human resources in service organizations. *Human Resource Management, 25*(3), 371-383.

Brown, S. W., Raymond, P. F., & Bitner, M. J. (1994). The development and emergence of service marketing thought. *International Journal of Service Industry Management, 5*(1), 21-48.

Business Week. (2003, June 9). *BMW's labor practices are cutting-edge, too.* Retrieved August 6, 2003, from http://www.businessweek.com

Chase, R. (1978, November-December). Where does the customer fit in a service operation? *Harvard Business Review,* 138-139.

Czepiel, J. A. (1990). Service encounter and service relationships: Implications for research. *Journal of Business Research, 20*(1), 13-21.

The Economist. (2006, March 18). Open, but not as usual. *Special Report: Open-source business,* 73-75.

Forbes. (2006). *The world's billionaires.* Retrieved March 31, 2006, from http://www.forbes.com/billionaires

Gruen, T. W., Summers, J. O., & Acito, F. (2000, July). Relationship marketing activities, commitment, and membership behavior in professional associations. *Journal of Marketing, 64,* 34-49.

Hakansson, H. (1980). Marketing strategies in industrial markets: A framework applied to a steel producer. *European Journal of Marketing, 14*(5,6), 365-378.

Hars, A. (2002). Open source software. *WISU, 4,* 542-551.

Hars, A., & Ou, S. (2001). Working for free? Motivations for participating in open source projects. *International Journal of Electronic Commerce, 6*(2), 25-39.

Kelley, H. H., & Thibaut, J. W. (1978). *Interpersonal relations: A theory of interdependence.* New York: John Wiley & Sons.

Kelley, S. W., Donnelly, J. H., & Skinner, S. K. (1990). Customer participation in service production and delivery. *Journal of Retailing, 66*(3), 315-335.

Lovelock, C. H., & Young, R. F. (1979, May-June). Look to consumers to increase productivity. *Harvard Business Review,* 168-178.

McClure, B. (2004, April 28). The bottom line on margins. *Investopedia.com.* Retrieved March 31, 2006, from http://www.investopedia.com

Mills, P. K., & Morris, J. H. (1986). Clients as partial employees of service organizations: Role development in client participation. *Academy of Management Review, 11*(4), 726-735.

Mills, P. K., & Moberg, D. J. (1982). Perspectives on the technology of service operations. *Academy of Management Review, 7*(3), 467-78.

Porter, M. E. (1985). *Competitive advantage: Creating and sustaining superior performance.* New York: The Free Press.

Schlueter Langdon, C. (2007). Instrument validation for strategic business simulation. In V. Sugumaran (Ed.), *Application of agent and intelligent information technologies* (pp. 108-120). Hershey, PA: Idea Group Publishing.

Schlueter Langdon, C. (2003a). Linking IS capabilities with IT business value in channel systems: A theoretical conceptualization of operational linkages and customer involvement. In *Proceedings of WeB December 2003,* Seattle, WA (pp. 259-270).

Schlueter Langdon, C. (2003b, June). IT matters. In Does IT Matter? An HBR Debate. *Harvard Business Review,* 16. Retrieved from www.hbr.org

Schlueter Langdon, C. (2003c). Information systems architecture styles and business interaction patterns: Toward theoretic correspondence. *Journal of Information Systems and E-Business, 1*(3), 283-304.

Schlueter Langdon, C. (2005). Assessing economic feasibility of e-business investments [White Paper Version 3.0]. Redondo Beach, CA: Pacific Coast Research.

Schlueter Langdon, C. (2006). Designing information systems capabilities to create business value: A theoretical conceptualization of the role of flexibility and integration. *Journal of Database Management, 17*(3), 1-18.

Schlueter Langdon, C., & Shaw, M. J. (2000). The online retailing challenge: Forward integration and e-backend development. In *Proceedings of ECIS July 2000 Conference*, Vienna, Austria (pp. 1025-1028).

Schlueter Langdon, C., & Shaw, M. J. (2002). Emergent patterns of integration in electronic channel systems. *Communications of the ACM, 45*(12), 50-55.

Sheth, J. N., & Parvatiyar, A. (1995). Relationship marketing in consumer markets: Antecedents and consequences. *Journal of the Academy of Marketing Science, 23*(4), 255-271.

Thibaut, J. W., & Kelley, H. H. (1959). *The social psychology of groups*. New York: John Wiley & Sons.

Williamson, O. E. (1975). *Markets and hierarchies: Analysis and antitrust implications*. New York: The Free Press.

Williamson, O. E. (1985). *The economic institutions of capitalism*. New York: The Free Press.

Whitaker, G. (1980, May-June). Co-production: Citizen participation in service delivery. *Public Administration Review*, 240-242.

KEY TERMS

Business Intelligence Analytics: Summarizes models and methods used to analyze data for the purpose of helping executives make better, more precise decisions.

Business Model: Describes how profit is generated; captures business logic by separating independent/dependent variables and mediating/moderating effects.

Co-Production: Has evolved to describe a situation in which people outside paid employment, such as customers, contribute to business value-added.

Customer and User Involvement: Describes the extent to which a customer is engaged as a participant in business operations, specifically in service production and delivery, including, for example, order processing and account management.

Customer Relationship Management (CRM): A broad term to cover concepts, methods, and procedures, and enabling information technology infrastructure that support an enterprise in managing customer relationships.

IT Business Value: Captures the business value derived from investments in information technology components and systems. Generic IT business value categories include cost, revenue, and quality.

Peer-to-Peer Production: Describes work performed and organized through the free co-operation of equals.

Chapter XLI
Investing in Open Source Software Companies:
Deal Making from a Venture Capitalist's Perspective

Mikko Puhakka
Helsinki University of Technology, Finland

Hannu Jungman
Tamlink Ltd., Finland

Marko Seppänen
Tampere University of Technology, Finland

ABSTRACT

This chapter studies how venture capitalists invest in open source-based companies. Evaluation and valuation of knowledge-intensive companies is a challenge to investors, and while many methods exist for evaluating traditional knowledge-intensive companies, the rise of open source companies with new hard-to-measure value propositions such as developer communities brings new complexity to deal-making. The chapter highlights some experiences that venture capitalists have had with open source companies. The authors hope that the overview of venture capital process and methodology as well as two case examples will provide both researchers and entrepreneurs new insights into how venture capitalists work and make investments.

INTRODUCTION

In the traditional view, the evolution of a technology-based new company is seen through separate consecutive stages. Business is based on creating tangible real assets; and in the end, the value of a company is also based on real assets. First, the technology is developed, which is foilowed by setting up the organization. Once the organization has reached a sufficient scale, internationalization is started. Finally, the value of the company is estimated with potential venture investment or through realization either through an initial public offering (IPO) or a trade sale.

However, due to the increased complexity of products and services, time-to-market tends to lengthen. In order to maintain sufficient resources until the company reaches profitability, external financing is needed. The time needed to turn a company's cash flow positive varies considerably. A long product development phase and slow market penetration prolong the period of negative cash flow. Simultaneous internationalization drains resources at an even higher rate. Since start-ups do not usually have collateral to secure bank loans, equity financing is the most evident form of financing. Venture capital funding is usually sought in order to get business development support in addition to plain financing.

New business ideas are increasingly more knowledge intensive, driven in part by the application of ICT as an enabling technology across industrial sectors. Also, the nature of business has changed: times-to-market are faster, development stages are no longer consecutive but can be simultaneous or even skipped, and companies are born global. Distinct from yesterday's industrial companies, today's knowledge-intensive companies' values are not based on their real assets but rather on their intangible assets such as knowledge, networks, and brand. Needless to say, intangible assets are considerably more challenging to value. The previous is even truer in the case of open source software (OSS) companies, since part of their business (and value) relies on open source (OS) communities in which people contribute their time and knowledge voluntarily into projects. Contributions are real but take place without formal contracts or incentive mechanisms, and people can easily abandon the community.

Furthermore, OSS companies that build their businesses on OS products (e.g., Google, JotSpot) have huge savings in time and licensing fees; they get to market faster and cheaper. This sets

even greater challenges for those valuing OSS companies. In theory, these free contributions should yield in higher valuations. On the other hand, the uncertainties involved should have the opposite effect.

The mission of this study is to compare traditional IT companies and their valuations and evaluations to those of OSS companies from the viewpoint of the venture capitalist. This is further divided into several subquestions:

- What are the special issues to be taken into account when evaluating OSS companies?
- Do venture capitalists assign a positive, negative, or no value to OSS companies and their communities when compared to traditional IT companies?
- Is there hype around OS?

Data for recent valuations of OSS and traditional IT companies were gathered from the VentureOne database. VentureOne (2005) is one of the leading venture capital research firms offering information on the venture capital industry. To better understand the investment decisions made and valuations paid for OS companies, and in order to get insights into what are the specialties in evaluation of OSS companies, two case studies were carried out. When designing the case study, based on the authors' initial understanding of the issues at hand, a pattern of interview questions was constructed. In addition to these semistructured interviews, data were gathered from publicly available sources. The interviewees were key managers of the case companies. Both of the cases present seed/early-stage venture capitalists that have been active in investing in OS companies. In addition, the case studies were backed up with several interviews with venture capitalists and entrepreneurs as well as feedback gathered from Internet online communities (for the questionnaire used, see Puhakka & Jungman, 2005).

533

BACKGROUND

Earlier Research on Evaluation and Valuation Theory

Venture capitalists evaluate their investment opportunities based on certain criteria. It is widely accepted that the three key investment decision criteria are management team, market projections, and product (Tyebjee & Bruno, 1981, 1984; MacMillan, Siegel, & Narasimha 1985).

In addition, venture capitalists have preferences, such as a venture's stage of development, its location, its industry or technology, and size of the investment required, which vary among one another (Seppä, 2000). These criteria and preferences are related to evaluation of an investment opportunity: does the venture have potential? Is it worth our time and money? Does it fit our investment strategy? Venture capitalists base their evaluation on business plans, meetings with the entrepreneurial team, and various researches.

Only after positive results from evaluation is it time to think about the value of the company. The process of valuation resembles business negotiation. Herein, "valuation means the process of placing a monetary value on an investment opportunity" (Seppä, 2003, p. 6). Venture capital valuations are not as straightforward as public market valuations or share prices. "Because of the fluctuations in the supply and demand of venture capital, investment valuations are not always determined according to the rules of efficient markets" (Seppä, 2003, p. 11). Valuation also can refer to venture capital funds' periodic valuations of investments (Association Française des Investisseurs en Capital [AFIC], British Venture Capital Association [BVCA], & European Private Equity and Venture Capital Association [EVCA], 2005).

Valuation of high-tech companies by venture capitalists theoretically has been studied extensively (e.g., Lockett, Wright, Sapienza, & Pruthi, 2002; Seppä, 2003). The value of a new venture is derived by discounting predicted future cash flows to the present. The discounting factor depends on the probability of returns. Even if a company has significant potential future cash flows, the risk of failure decreases its net present value.

Different methodologies exist in the valuation, but all aim at answering the same question: what is the present value of expected future earnings or the exit value of a company? The methods fall into the following four categories (Lockett et al., 2002):

1. Liquidation value-asset-based methods
2. Discounted cash-flow-based methods
3. Options-based valuation methods
4. Rule-of-thumb valuation methods (comparator valuations)

The concepts of present value and net present value (NPV) form the basis for the valuation of real assets and investment decision-making. Essentially, the method makes a comparison between the cost of an investment and the net present value of uncertain future cash flows generated by the venture. There are at least four major steps in a discounted cash flow for a proposed venture.

First, assuming that the venture is all equity financed (i.e., all necessary capital is provided by the shareholders), forecasts are needed for what the expected incremental cash flows would be to the shareholders if the venture were accepted.

Second, an appropriate discount rate should be established that reflects the time value and risks of the venture, which, therefore, can be used for the calculation of the present value of expected future cash flows. The concept of present value includes the notion of the opportunity cost of capital. The appropriate discount rate, or the cost of capital, first must compensate shareholders for the foregone return they could achieve on the capital market by investing in some risk-free assets. It also has to compensate them for the risk they are undertaking by investing in this project rather than in a risk-free financial asset. Thus,

534

the cost of capital is determined by the rate of return investors could expect from an alternative investment with a similar risk profile. Fortunately, the rich menu of traded financial assets provides venture fund managers with the opportunity to estimate the right price.

Third, based on the value additives of present values, the NPV of the venture is to be calculated. Once the cash flow forecasts are finalized and the appropriate discount rate is established, the calculation of the venture's NPV is a technical matter. When all future cash flows that need to be discounted arrive at their present values, and by adding them to the present value of the necessary capital outlay, the NPV of the venture is achieved.

Finally, a decision has to be made whether to go ahead with the venture or not. As the company proceeds toward profitability, the likelihood of success grows, and the value of the company grows. Thus, it can be argued that every step a company takes toward its goals increases its value.

Exit valuations of technology companies are dependent on the prevailing market situation. Because the presumed exit valuation is the most important measure when considering the value of a company at the last venture capital round before an IPO, it is obvious that exit valuations have significant effects on valuations at all investment rounds, although the effect diminishes toward the founding stage. Due to dramatic changes in exit valuations (e.g., during 1999-2000), there has been a wide variation in valuations at various venture capital rounds as well.

Hype and Uncertainties Vitiate the Theory

Every now and then, things get out of hand. In the 1990s, it was argued that revenues and earnings were neither sufficient nor relevant ways to put value to emerging e-businesses or dot-coms that had no revenues and actually no existing mecha-

nisms of extracting payments from customers. A way to assign value to a member in a Web community was proposed: a so-called "lifetime value of a customer" or a "price-to-eyeball multiple," an estimate of how much on average a customer would end up paying to a company (Valliere & Peterson, 2004).

Emerging OS companies face a similar challenge since part of their businesses (and values) relies on OS communities in which people contribute their time and knowledge voluntarily. Contributions are real but take place without formal contracts or incentive mechanisms, and people can easily abandon the community. The question rises how one should value community contributions like these. The International Private Equity and Venture Capital Valuation Guidelines (AFIC, BVCA, & EVCA, 2005) provide no aid on this. On the other hand, venture capitalists certainly have some views, since there are already several cases in which they have invested in OS companies.

Every venture capital (VC) investment is difficult to value due to the high degree of uncertainty in the performance. The valuation of OS companies is even more challenging, as there is yet neither history nor guidelines due to the uncertainties, for example, in the following:

- Profitability of business model
- Revenue streams
- Market acceptance
- Community commitment
- Competitive reactions
- Quality of software
- New General Public License (GPL) version in 2007

The list includes similar uncertainties that were involved in the dot-com bubble (Valliere & Peterson, 2004). Indeed, one can see the signs of hype in OS as well. Signs of hype surround certain companies (company hype), the OS market (market hype), and the activity of other investors as a group (investor hype) (Valliere & Peterson, 2004). In 2005, OS was getting increasing attention in the press, and venture capitalists were announcing OS strategies. However, it is too early to say whether this will lead to unreasonable valuations of OS companies.

MAIN FOCUS OF THE CHAPTER

Done Deals and Given Valuations

So far, the OS experience has not been a happy one for venture capitalists. According to the research firm VentureOne, some $714 million was invested in 71 OS companies in 1999-2000, and most of those projects collapsed (VentureOne, 2005). One of the biggest successes that is left of those experiences is RedHat Inc., which went public in 1999 and makes money selling enhancements and maintenance services to corporations using Linux OS operating systems. However, it still has some ways to go before reaching $200 million in revenues (RedHat, 2005) and is a relatively mild success with market value less than $5,000 million and earnings per share (EPS) of $0.33 (NASDAQ, 2005). So this is certainly no Google with market value just under $70,000 million (EPS $5.02) or eBay with market value just under $60,000 million (EPS $0.78) (NASDAQ, 2005) that aggressive venture capitalists often use as a reference as companies they want to fund as the "next big thing."

The biggest success so far with OS ventures, as they traditionally have been viewed, has been IBM's Linux service business that the company has grown as a separate emerging business opportunity unit and has managed to grow it from $0 to more than $2 billion in revenues in just 5 years. Still, there is no public record on how much IBM has invested in this venture to realize that growth (IBM, 2005).

Several studies have pointed out that Linux, Apache, and MySQL, for example, have reached

the maturity in which the technology or code is comparable or even superior to the existing proprietary ones. Furthermore, for example, Firefox has managed to take the market by storm extremely quickly without any significant marketing budget. In other words, early evidence seems to point to the OS approach, at least some cases, as an efficient way to develop technology and take that to market. However, at least the experiences from the first round financings of OS companies indicate that it is not necessarily the best way to do business.

After a few years of trying to figure out whether money can be made by OS companies, the answer from venture capitalists seems again to be a reluctant yes. Twenty OS businesses raised $149 million in venture money in 2004 in the United States alone (VentureOne, 2005). There are no numbers available for the rest of the world, but in Europe, several investments took place. Looking at that total, it would seem that most of the investments are still on a seed or first-round level (compared to an average level on various rounds of realized investments); if distributed evenly among companies, the amount would be $7.45 million.

Case: BlueRun Ventures

BlueRun Ventures was originally launched as Nokia Venture Partners in 1998 with $150 million initial invested capital from Nokia Corporation. Even with money from Nokia, it was designed right from the start to act independently of its only investor. It raised a second fund of $500 million in 2000, which then already included other investors besides Nokia, such as Goldman Sachs.

In 2005, Nokia Venture Partners raised its third fund of $350 million and changed its name to BlueRun Ventures. Today, BlueRun Ventures has offices in nine locations globally and manages $1 billion making investments into IT, mobile, and consumer technologies at seed- and early-stage levels (BlueRun Ventures, 2005). In looking at OS investment opportunities, the key issue identified by BlueRun Ventures is a strong community close to the company. In their view, OS is a transformation force that is forcing a unit price down and the only realistic counterforce to big incumbent companies such as Oracle or BEA systems.

Still, they consider the market being at an early stage of deployment since after the bubble there have been no notable initial public offerings by OS companies. From the investment point of view, they consider two uncertainties in OS: the size of the market and the fragile business attachment of dealing with a community.

BlueRun Ventures has quite a bit of experience dealing with OS companies and has looked at about a hundred companies from 2002 to 2005. However, in early 2006, it had just completed its first investment in this space. Seed and early-stage investments are tricky, as typically there is very little or no historical numbers to look at. As the company's partner noted, it really is not a science but rather a very subjective opinion of opportunities. The questions are typically, "Do I like this opportunity? How much money is needed to make it happen? Does it fit with the funds strategy?" After that, the actual valuation is actually based on negotiations, which rely more on people skills than anything else (A. Kokkinen, personal communication, November 11, 2005).

From BlueRun Ventures' perspective, valuations in the long run should be the same for both traditional and OS startups. However, the nature of seed-investments is different since communities in a way have taken care of development that is typically done with seed money, resulting in a technology but not in protected intellectual property rights (IPRs).

As the market is still developing, BlueRun Ventures has not been able to identify any OS-dedicated venture funds, even though it expects several of those to be formed. A prerequisite for an OS fund may be that first there should be four to five initial public offerings, which would

give enough evidence to the managers of funds in order to go to their investors and propose an OS fund (A. Kokkinen, personal communication, November 11, 2005). How this will turn out remains to be seen. Either OS will remain part of existing funds' investment targets, or OS-dedicated funds will be seen in the future. The latter would obviously result in more sophisticated ways of evaluating OS; otherwise, the competition for investors' money will continue to be played out between traditional software companies and OS companies in mutually accepted terms.

Case: Nexit Ventures

Nexit Ventures is a Finnish-based traditional venture capital company. It raised its Euro 100 million fund in 2000, which was later reduced to Euro 66.3 million. The investors are private institutions with 50% of their commitments outside Finland. The initial focus was seed and early-stage companies both in the Nordic and North America; later this was modified to early- and later-stage companies in the same geographical regions. The technology focus of mobile and wireless communication, from core components and enabling middleware to applications and services, has remained the same.

Nexit Ventures does not consider a pure OS company to be a viable investment opportunity. Rather, it sees the OS approach of collaborative effort to solve various issues to be an enabler for various things of potentially great value. For example, Apple's iPod makes it easy for consumers to utilize music downloaded from Web, whether the music is from legitimate sources or not. Still, the idea was that the closer one gets to the core of OS (the community), the harder it is to make money. Nexit Ventures considers OS be at every level of deployment from early adoption to maturity; it just is not always very visible, and there are legal uncertainties.

From Nexit Ventures' point of view, it is somewhat isolated in Finland about what is taking place globally (a bit paradoxical, as most things are said to take place on the Internet), and it has not yet seen a rise of OS-based businesses, often referred as Web 2.0 companies. There has been little discussion on the public media, especially compared to the United States. In the United States, where valuations are very high again, due diligence in follow-up rounds is quite weak, according to Nexit Ventures, as there is pressure to do hard sought deals.

Regardless, looking at opportunities and valuating them, Nexit Venture's comments corroborate those of BlueRun Ventures. The markets for venture capital investments are imperfect and always will be. Therefore, the valuations are not made with transparent scientific methods but rather are results of negotiations. In other words, it can be argued that the potential of one's business idea opens the door to negotiations with the venture capitalist, but the valuation that will take place with the investment is determined "by one's skills as a negotiator, that are impossible to quantify or to break down into a scientific model" (A. Tarjanne, personal communication, November 30, 2005).

FUTURE TRENDS

It might be that in the end, the biggest successes to financing community come from and to companies that are not really OS companies as such but rather use OS components to build new businesses; for example, Google, which, like most Web companies, was built on top of OS). From $1 million initial seed capital in 1998 and an injection of $25 million growth capital in 1999, the company realized the value to its investors in 2004 by going public, and by spring 2005, it had surpassed the Finnish pride Nokia with more than $80 billion in market capitalization, compared to just less than $80 billion for Nokia (Google, 2005).

As stated earlier, lareg amounts of money are invested into OS businesses, and we expect dedi-

cated OS funds to be formed in the near future. The key driver will be successful exits from OS investments. However, the first bets (i.e., seed round investments) to potential future successes have just been made, and how successful those will be can only be known in the coming years. Once we can get significant amounts of data, interesting quantitative comparisons can be made between investments in OSS and traditional IT companies.

CONCLUSION

Venture capitalists do not seem to put special value on OS companies. However, some of them recognize that there are distinctly different elements in evaluating OS companies. For instance, expected cash flows are likely to be bigger in businesses built on OS software than in similar traditional software companies, due to the savings, for instance, in licensing fees. Concurrently, the uncertainties in OS should increase the discount rate (see Figure 1).

Figure 1. Potential cash flow and risk measured by discount rate of the companies using OS or proprietary software (Source: Adapted from W. Cardwell, personal communication, November 15, 2005)

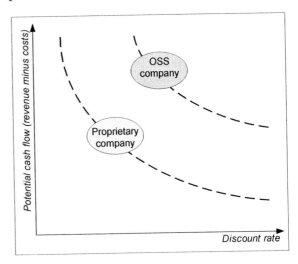

In interviewing the selected experts and looking at the selected cases, it seems that rather than putting effort into further understanding valuation methodologies, entrepreneurs should seek help in learning better negotiation skills. However, in the academic world, more complex approaches have been taken in valuating a company. It might be appropriate to ask whether the academics are really serving the industries if these methodologies are not actually used by the people in the venture capital industry.

The good news for entrepreneurs looking to launch new OS ventures is that money is available, and investors are making their bets again on OS. Still, the basic dilemma remains: while the venture capitalist is looking to become a shareholder as cheaply as possible, the entrepreneur, of course, is trying to retain as much ownership as possible. This would not be an issue if there were a transparent, objective way to estimate the value of the venture. However, as one interviewee said, this is not likely to happen, as the venture capital market remains imperfect. Unfortunately, there are many unknown factors affecting the present value of a startup that have to be estimated, and thus, objectivity is hard to maintain.

ACKNOWLEDGMENT

The authors wish to express their gratitude to interviewed investors and other persons who have expressed their valuable opinions in various forums. An earlier version of this chapter (Puhakka & Jungman, 2005) was presented at the eBRF 2005 Conference in Tampere, Finland, September 26-28, 2005, and published in the *Conference Proceedings, Frontiers of e-Business Research (FeBR 2005)*.

REFERENCES

Association Française des Investisseurs en Capital (AFIC), British Venture Capital Association (BVCA), & European Private Equity and Venture Capital Association (EVCA). (2005). *International private equity and venture capital valuation guidelines*. Retrieved September 16, 2005, from http://www.privateequityvaluation.com

Lockett, A., Wright, M., Sapienza, H., & Pruthi, S. (2002). Venture capital investors, valuation and information: A comparative study of the U.S., Hong Kong, India and Singapore. *Venture Capital, 4*(3), 237-252.

MacMillan, I. C., Siegel, R., & Narasimha, P. N. S. (1985). Criteria used by venture capitalists to evaluate new venture proposals. *Journal of Business Venturing, 1*, 119-128.

NASDAQ. (2005). *The NASDAQ stock market*. Retrieved February 14, 2006, from http://www.nasdaq.com

Nexit Ventures. (2005). *Nexit Ventures Web site*. Retrieved November 28, 2005, from http://www.nexitventures.com

Puhakka, M., & Jungman, H. (2005). Evaluation and valuation of open source software companies: A venture capitalist' perspective. In M. Seppä, M. Hannula, A.-M. Järvelin, J. Kujala, M. Ruohonen, & T. Tiainen (Eds.), *Frontiers of e-business research 2005* (pp. 855-865). Tampere, Finland: Tampere University of Technology & University of Tampere.

RedHat Inc. (2005). *RedHat Inc. Web site*. Retrieved December 3, 2005, from http://www.redhat.com

Seppä, M. (2000). Strategy logic of the venture capitalist. *Jyväskylä Studies in Business and Economics 3*. Jyväskylä: University of Jyväskylä.

Seppä, T. (2003). *Essays on the valuation and syndication of venture capital investments*. Doctoral dissertations. Helsinki University of Technology, Helsinki, Finland.

Tyebjee, T., & Bruno, A. (1984). A model of venture capitalist investment activity. *Management Science, 9*, 1051-1066.

Tyebjee, T., & Bruno, A. (1981). *Venture capital decision making: Preliminary results from three empirical studies*. Wellesley, MA: Babson College.

Valliere, D., & Peterson, R. (2004). Inflating the bubble: Examining dot-com investor behaviour. *Venture Capital, 6*(1), 1-22.

VentureOne. (2005). *VentureOne Web site*. Retrieved September 16, 2005, from http://www.ventureone.com

KEY TERMS

Evaluation: Subjective and qualitative assessment of an investment opportunity.

Proprietary: Belonging to or controlled by an individual or organization that has the ability to share that item (in this case, software code) with others.

Seed Company: Company in a stage of research, assessment, and development of an initial concept before reaching the start-up phase (FVCA Yearbook, 2004).

Startup Company: Company in a product development stage requiring further funds to initiate commercial manufacturing and sales (FVCA Yearbook, 2004).

Valuation: Process of placing a monetary value on an investment opportunity (Seppä, 2003).

Venture Capital: Equity investments made for the launch, early development, or expansion of a business (EVCA, 2005, www.evca.com).

Chapter XLII
Revenue Models in the Open Source Software Business

Risto Rajala
Helsinki School of Economics, Finland

Jussi Nissilä
University of Turku, Finland

Mika Westerlund
Helsinki School of Economics, Finland

ABSTRACT

Profit-oriented business behavior has increased within the open source software movement. However, it has proved to be a challenging and complex issue due to the fact that open source software (OSS) business models are based on software that typically is freely distributed or accessed by any interested party, usually free of charge. It should be noted, however, that like all traditional software businesses, the business models based on OSS ultimately aim at generating profits. The aim of this chapter is to explore the key considerations in designing profitable revenue models for businesses based on OSS. We approach the issue through two business cases: Red Hat and MySQL, both of which illustrate the complexity and heterogeneity of solutions and options in the field of OSS. We focus on the managerial implications derived from the cases, discussing how different business model elements should be managed when doing business with OSS.

INTRODUCTION

Whereas the business models of the traditional providers of proprietary software are grounded in one way or another on the distribution of access to the use of software-related intellectual property (IP) protected by copyrights, business models within the open source movement have to rely on other types of revenue models. This is due to the fact that open source software (OSS) business models are based on software that typically is freely distributed or accessed by any interested party, usually free of charge. OSS is often mistaken for shareware or freeware, but there are significant differences between the licensing models and the processes between and within these types of software. It should be noted, however, that like all traditional software businesses, the business models based on OSS ultimately aim at generating profits. However, profitability and business models of OSS are still poorly understood phenomena, and there is no single framework that would explain the potential determinants of firm-level revenue model choices.

In this chapter, we make an attempt to identify key considerations in designing successful revenue models in the OSS business. We explore the revenue models of two selected OSS business cases. Through these cases, we aim at identifying the firm-specific business model elements that guide, enable and constrain the choice of revenue model options in OSS business. As a limitation to the analysis presented in this chapter, we leave the exogenous factors (such as competition and other environmental factors) beyond the scope of our consideration.

BACKGROUND

In this chapter, we discuss the background of the OSS business, typical licence OSS choices, and the potential for conducting for-profit business with OSS.

Development of OSS Business

The history of the open source movement goes back to the early ages of computing. In the 1960s and 1970s, it was common for programmers in certain academic institutions (e.g., Berkeley, MIT) and corporate research centers (e.g., Bell Labs, Xerox's Palo Alto Research Center) to share computer program source codes with other programmers. It was not until the early 1980s that proprietary software became very popular, thus causing problems with cooperative software development (Lerner & Tirole, 2002). The predecessor of the open source movement, the Free Software Foundation (FSF), was founded in 1983

by MIT employee Richard Stallman in his attempt to formalize cooperative software development and create a complete free[1] operating system with necessary software development tools. This project was called the GNU Project. Stallman's general concept of free software possesses four essential freedoms (Stallman, 1999):

- Freedom to run the program
- Freedom to modify the program
- Freedom to redistribute the program
- Freedom to distribute modified versions of the program

Stallman didn't want to release software with restrictive copyright terms because it would prevent certain forms of valuable cooperation. On the other hand, releasing software to the public domain would leave it vulnerable to be copyrighted and included in proprietary packages. Thus, Stallman came up with the idea of copyleft, or protecting the freedom of software with the means of copyright laws. In addition, copyleft ensures that the modified works are also released under copyleft terms and, therefore, to the use of the community. Stallman, (2002) argues, "Proprietary software developers use copyright to take away the users' freedom; we use copyright to guarantee their freedom. That's why we reverse the name, changing 'copyright' into 'copyleft.'" To implement this idea, the FSF developed the GNU General Public License (GNU GPL), the first of the now extensive selection of copyleft licenses that are used to protect free/OSS. Meanwhile, the open anticommercialism of FSF led to a group of free software movement leaders deciding to find new ways to strengthen their cause, but with less radical means. They came up with the term "open source," which they thought would better describe the software ideals, and founded the Open Source Initiative (OSI). The idea of the organization was to promote the Open Source Definition (OSD), a set of terms for licences, which is more adaptable to commercial use than the approach FSF took.

OSI has since registered a certification mark, and there is a variety of OSI-certified licenses (including GNU GPL and other copyleft licenses).

What motivated the birth of OSI was the way free software was being developed in such projects as the Linux operating system since the beginning of the 1990s. The new development model introduced in the Linux project was first described in "The Cathedral and the Bazaar," an essay written by Eric Steven Raymond, one of the founders of the OSI (Raymond, 2001). The Linux development model was seen as a better way of software development that could lead to higher quality and rapid advancement. Cooperational software development was not only for the ideologists and community-spirited anymore, but rather something also to be used in more commercial projects. The new emphasis born with the OSI made it possible for the business world to intensively embrace OSS. Before 1998, relatively few people in the IT industry knew about free software; however, a couple years later, open source was on many people's lips. With the participation of big IT companies such as IBM, Hewlett Packard, and Nokia, open source has become a credible player in the IT field.

OSS Licensing

OSS, exactly defined, is software fulfilling the terms of distribution given in the OSD and adopting a license approved by the OSI (Open Source Initiative, 2004). Summarizing the ideas behind the terms in OSD, the software license must generate the following effects:

- Source code must be readable and available, either included with the binary code or publicly downloadable
- Free distribution of the software by any party, on any medium, to any party, gratis or for a fee

- Derivative works must be allowed, either under similar license or not, depending on the specific OSS license type
- No discrimination against persons, groups, or fields of endeavor

The nature of OSS is in the licensing terms and not just the accessible source code, which is just one part of the features the licensing terms generate. In addition, the licensing terms allow the free use, redistribution, and modification of the software. The copyright owner preserves the moral rights and some economic rights, such as the right to dual-license the software, but transfers many important rights to the users and developers of the software in order to enable the development of the software and to increase its adoption. It is important to understand that the OSD licensing terms allow the creation of many types of OSS licenses, each with different qualities. Välimäki (2005) categorizes OSS licenses into three functionality classes, ranging from the most liberal to the most restrictive. The categories are permissive licenses, licenses with standard reciprocity obligation, and licenses with strong reciprocity obligation. Standard reciprocity means that the distribution terms of the source code must be maintained in further developed versions, which is also called the "copyleft" effect. Strong reciprocity obligation means that in addition to standard reciprocity effects, derivative works and adaptations must keep the licensing terms intact, also called the "viral" effect.

Välimäki (2005) has studied the prevalence of different OSS license types. Table 1 presents the most popular licenses as surveyed in his study at SourceForge.net in 2004 (Välimäki 2005), together with their functionality and relative popularity in project licensing.

In Table 1, the popularity percentage refers to the occurrences of these license types among all OSS licenses (surveyed at the SourceForge. net in late 2004).

Table 1. Most popular OSS licenses and their functionality (Source: Adapted from Välimäki, 2005)

License	Functionality	Popularity
GNU GPL	strong reciprocity	66.50%
GNU LGPL	standard reciprocity	10.60%
BSD	permissive	6.90%
Public domain	permissive	2.70%
Artistic	permissive	2.00%
MIT	permissive	1.70%
Mozilla	standard reciprocity	1.50%
Common Public License	strong reciprocity	0.60%
Zlib	permissive	0.50%
QPL	strong reciprocity	0.40%
Open Software License	strong reciprocity	0.40%
Python License	permissive	0.40%
Academic Free License	permissive	0.30%

Special Characteristics of OSS Business

One of the most critical issues for OSS business is that the licensing terms allow free redistribution of the licensed software (i.e., the licenser doesn't necessarily gain any revenue from these copies of the software). In fact, charging a fee for OSS is usually not feasible, because (1) any buyer may start to resell the software or give it away and (2) fees could severely diminish the rate at which both developers and users adopt the software product (De Laat, 2005), which often is the motivation behind licensing a product as OSS. Therefore, it is usually not feasible to base the revenue logic on licensing fees. It is also possible to use OSS as part of a firm's other products; namely, software packages, hardware, and/or services. This approach is not free of challenges either, since the unique licensing of OSS may create risks as well as opportunities.

Many firms conducting business with OSS are in some way dependent on the OSS community for developing software in their product offerings, for support, or for customers. However, the OSS community is outside the hierarchical control of the firms since there are normally no contractual agreements between them. In addition, the idea of exploiting the financial value of a jointly developed community might go against the values of the community (Dahlander & Magnusson, 2005) in which the code is actively protected from being appropriated by commercial firms through the use of legal and normative mechanisms (O'Mahony, 2003). However, the attitudes and policies toward the commercial exploitation within the OSS community range from the critical attitudes of FSF and copyleft licensing to the more liberal attitude of OSI and permissive licenses.

Dahlander and Magnusson (2005) propose three approaches a firm can use to relate to the OSS community. In question is the parasitic approach in which the firm focuses on its own benefits without considering possible damages to the community. Since the firm doesn't share the norms, values, or rules of the community, the possibility to influence community development does not exist. The commensalistic approach is about benefiting from the community while leaving it otherwise indifferent. Since the firm isn't considered hostile, influencing

Table 2. Firm-community relationship (Source: Modified from Dahlander & Magnusson, 2005)

Approaches	Description	Nature of Relationship
Parasitic approach	Focuses on firm's own benefits without considering possible damages to the community	Search for useful input without obeying norms, values, and rules
Commensalistic approach	Firm aims to benefit from the community	Search for useful input from the community
Symbiotic approach	Firm tries to codevelop itself and the community	Give something to the community, often through a firm-established community

the community is possible but difficult. Also in question is the symbiotic approach in which the firm tries to co-develop itself and the community. This demands heavy involvement in community development and sharing of norms and values but also allows influencing community development in a desired direction. These approaches are illustrated in Table 2.

MAIN FOCUS OF THE CHAPTER

According to recent studies, the business-model concept includes some elements of business strategy and aims to describe the business as a manifestation derived from strategy (Rajala, Rossi, & Tuunainen, 2003; Osterwalder, 2004; Morris, Schindehutte, & Allen, 2005). It has also been defined as an abstraction of business (Seddon & Lewis, 2003), which characterizes revenue sources and specifies where the company is positioned in its value-creating network in a specific business. The essential elements of various business models are defined in different words by several researchers (Rajala et al., 2003; Hedman & Kalling, 2003; Osterwalder, 2004; Morris et al., 2005, Rajala & Westerlund, in press). Many of the studies identify a number of elements that are characteristic of various business models. These elements, expressed in different words by different authors, include the following: (1) offerings; (2) resources needed to develop and implement a business model; and (3) relationships with other actors (Timmers, 2003;

Osterwalder, 2004; Morris et al., 2005). Finally, these elements are interconnected with (4) the revenue model, including sources of revenue, price-quotation principles, and cost structures, which is characteristic of a particular business. Grounded on the previous review and summation of the prior research literature, we identify three business model elements in order to describe the revenue models in the OSS business. These business model elements are key considerations on which firms should focus after the decision to participate in an OSS business. In the following we discuss these elements in detail.

Offering

In the literature of business and management, the concepts of product strategies and product offerings are discussed widely (Cravens, 1987; Kotler, Armstrong, Saunders, & Wong, 1996). We see that offerings embody several aspects within the concept of a business model and, thus, affect the revenue model. Generally, type of offering, target market, product vs. service orientation, licensing model, and so forth, can be considered as aspects related to the product strategy. Likewise, the product offering includes aspects such as complexity, the essential benefit that the customer is really buying, and product features, styling, quality, brand name, and packaging of the product offered for sale (Kotler et al., 1996).

From the business model perspective, a defining characteristic of OSS as a product is that it is not a physical but rather an information product.

Information, or digital, products have unique characteristics that differ largely from physical product characteristics. However, certain open source business models, such as widget-frosting and accessorizing (see the following), consist also of physical products. In addition, OSS revenue models such as support selling, service enabling, and software franchising, are comprised mostly of service components, which also have a very different nature.

In addition to the type of offering, license types are considered part of the offering element in our conceptual model as a determinant of revenue model choices. Indeed, the licensing issues and commitment to the principles of OSS licenses (GPL, etc.) are key issues related to information products such as OSS solutions (Lee, 1999).

Resources

The development of resources in the industrial-network perspective is linked to its strategy (Håkansson & Snehota, 1995; Gadde & Håkansson, 2001; Sallinen, 2002). According to this view, resources vary according to the business and product strategy. The resources and capabilities of a firm are among the central issues in understanding and analyzing its business. This accentuates the essence of resources in core competencies (Selznick, 1957; Prahalad & Hamel, 1990), as they are generally seen as firm-specific property that is subordinate to the core competencies. The resource-based view of the firm originated from the work of Penrose (1959) and was further developed by Wernerfelt (1984). According to Penrose (1959), bundles of resources that are activated in different ways lead to incoherent performance and heterogeneous outputs in various organizational settings.

In our analysis of the resources in the OSS business, we share the view of Metcalfe and James (2000), who define tangible and intangible assets as physical and nonphysical resources, and capabilities as intangible knowledge resources.

Furthermore, we see that the increasing complexity of OSS markets makes it difficult for firms to have all the necessary resources in their possession to compete effectively. This view is consistent with the research of Ariño and de la Torre (1998). These resource-related approaches provide us with a basis on which to identify key resources in various types of OSS business models. They deepen our understanding, especially of how resources are applied and combined by a firm, and take inimitable resources as a basis for the creation of sustainable capabilities as described in other technology-intensive industries such as those by Hart (1995) and Gabrielsson (2004).

Relationships

We see that the elements in our conceptual model are interrelated with each other and, therefore, are consistent with Håkansson and Snehota (1995) and Rosenbröijer (1998) that capabilities of a company reflect its success in combining resources to perform activities through internal and external relationships.

As pointed out in the previous discussion, we need to consider the interaction of companies with other actors as an inseparable part of a business model, similar to offerings and resources. Timmers (2003) points out that in the context of business models, the focus shifts from creating value through internal activities to creating value through external relations. He identifies these relationships within the value-creating network as an important element in the development and distribution of offerings. In addition to being an important intangible company asset, a firm's network offers access to the resources of other network actors (Foss, 1999; Gulati, Nohria, & Zaheer, 2000; Chetty & Wilson, 2003, Möller & Svahn, 2003).

Revenue Model

Discussion of the revenue models in the context of OSS has traditionally been problematic since the

OSS movement emphasizes free distribution of intellectual property. However, since the emergence of the OSS movement, there has also coexisted a favorable attitude toward earning money and, more generally, toward profit-oriented behavior based on the OSS (Raymond, 2001).

Concerning open source as an economic phenomenon, De Laat (2005) argues that whether an enterprise involved in the open source business chooses to license its own software product as open source or tries to benefit from existing OSS products, the ways of making money with open source are basically the same. These ways include selling services to facilitate OSS use, selling connected hardware, and selling commercial closed applications to use with OSS. However, Hecker (1999) has identified eight possible revenue models to be applied in conjunction with OSS. These models are described in Table 3.

Although Hecker's list of OSS revenue models (summarized in Table 3) was published as early as 1999, it still remains one of the most comprehensive classifications of OSS revenue models. It clearly points out that a company has a multitude of options to capture revenue with OSS.

CASE EXAMPLES

In our literature review, we identified three endogenous business model elements (i.e., offering, resources, and relationships) that affect

Table 3. Summary of OSS revenue models (Source: Modified from Hecker, 1999; Välimäki, 2005)

Revenue Model	Description	License Types	Revenue Sources
Support selling	A for-profit company provides support for a software that is distributed free of charge.	Any	Revenue comes from media distribution, branding, training, consulting, custom development, and post-sales support for physical goods and services.
Loss-leader	A no-charge open source product is used as a loss leader for traditional commercial software (i.e., the software is made free by hoping that it will stimulate demand for a related offering of the company).	Varies	Complementary offerings (e.g., other software products)
Widget-frosting	Companies that are in business primarily to sell hardware can use this model to enable software such as driver and interface code. By making the needed drivers open, the vendor can ensure that they are debugged and kept up to date.	Any	The company's main business is hardware. This is quite similar to the loss-leader model.
Accessorizing	Companies that distribute books, computer hardware, and other physical items associated with and supportive of OSS.	Any	Supplementary offerings
Service enabler	OSS is created and distributed primarily to support access to generating revenue from consulting services and online services.	Any	Service fees
Brand licensing	A company charges other companies for the right to use its brand names and trademarks in creating derivative products.	Strong reciprocity	Copyright compensations
Sell it, Free it	A company's software products start out their product life cycle as traditional commercial products and then are converted to open source products when appropriate.	Alteration of license type	Initial revenue from software product offerings converted into other models (e.g., the loss-leader model)
Software franchising	A combination of several of the preceding models (in particular, brand licensing and support sellers) in which a company authorizes others to use its brand names and trademarks in creating associated organizations doing custom software development; in particular, geographic areas or vertical markets.	Strong reciprocity	The franchiser supplies franchisees with training and related services in exchange for franchising fees of some sort

the revenue models in the OSS business. In this chapter, we illustrate these determinants and their interconnectedness with the revenue model in two empirical examples: MySQL and RedHat. We see that these case examples improve understanding of the interrelatedness of these business model elements, and especially their role as determinants in setting up the revenue model. Furthermore, the cases illustrate the complexity and heterogeneity of solutions and options related to revenue models in the field of OSS business.

MySQL

The MySQL trademark and copyright are owned by the Swedish company MySQL AB. Two Swedes, David Axmark and Allan Larsson, founded MySQL AB, together with Michael "Monty" Widenius, a Finn who is broadly appreciated as the chief designer and developer of the system. The company develops and maintains its key product offering, the MySQL open source database system, in close collaboration with the OSS community over the Internet. Unlike projects such as Apache, MySQL is owned and sponsored by a single for-profit firm, MySQL AB. In addition to providing the database product under the GPL license, the company sells support through service contracts as well as commercially-licensed copies of the MySQL database software, and employs people all over the world to communicate about the use and development of the product.

Offering

The offering of MySQL AB is a multithreaded, multiuser SQL (structured query language) relational database server (RDBS) software. The software is available either under the GNU GPL or under other licenses when the GPL is inapplicable to the intended use. MySQL provides database products for integrating software vendors and original component manufacturing (OCM) partners, enterprise organizations, and private

users in the OSS community. To distribute its offering to a large number of users worldwide, MySQL AB has applied a dual licensing principle by making the MySQL database software available for free on the Internet under the GPL and selling it under proprietary licenses when the GPL is not an ideal option and in situations such as inclusion of MySQL technology in closed source products. In summary, the core offering of MySQL AB embodies an in-house developed software product and related services.

Resources

As a symbol of the key resources of MySQL AB, chief technology officer Widenius began programming databases in 1981. He worked previously in Tapio Laakso Oy developing systems that needed data storage. Similarly, Axmark and Larsson, his two colleagues and later cofounders of MySQL, collaborated in programming projects from 1983 to 1995 and accumulated knowledge about database systems. By licensing the MySQL product under an OSS license, the company transferred some of its internal intellectual property resources to the open source community, thus gaining possible future clients as well as developers and enthusiasts to support its offering. The internal programming resources can still be considered the key element in the MySQL business model. Currently, 80% of the source code in the MySQL core database product (version 4.0) is programmed by in-house programming resources; the company has systematically invested in professional management resources to successfully manage its growing for-profit business.

Relationships

As already described, the collaboration based on personal relationships between key individuals can be seen as the key determinant of success in the early phases of the MySQL product development. This open atmosphere and knowledge-sharing culture between the cofounders of MySQL AB provided a sound base for enlarging the network to OSS-oriented Internet communities. At present, partners in the business network of MySQL include companies such as suppliers, distributors, outsourcing service providers, other key companies in the OSS field, commercial research institutions, and other strategic partners. Relationships with these actors are based on commercial multi- or bilateral activity. Furthermore, relationships in the business network include collaboration with public (government) organizations, research institutes, and so forth.

Relationships within the OSS community are a multifaceted phenomenon. According to the company CEO, the community of 5 million MySQL users includes several groups that produce MySQL books and articles as well as conduct courses and presentations. Furthermore, these ecosystems develop applications in different OSS projects. Currently, MySQL AB is balancing between the OSS community and commercial business networks that have somewhat disparate needs and values. We see that MySQL AB depends on the OSS community for its ecosystems and even more for the customer base, but they mostly conduct the product development in-house. However, the company also has made a significant contribution to the OSS community by licensing the database as an OSS. Therefore, we define MySQL's approach toward the OSS community as a symbiotic one.

The Revenue Model

MySQL AB is often cited as the champion of the second generation of open source projects. These projects are open source but are directed by for-profit companies. The revenues of these corporations derive from selling consulting services for their products. MySQL AB makes MySQL available under the GPL for free and sells it under proprietary licenses for clients when the GPL is not an ideal option (e.g., inclusion of MySQL technology in a closed source product).

Currently, MySQL AB receives more income from proprietary license sales than from its other income sources, branding, and services. Its main income seems to come from embedded commercial users (Välimäki, 2003). In terms of Hecker (1999), the revenue models of MySQL AB include features from support selling and dual licensing, both of which can be considered incarnations of the loss-leader model.

Red Hat

The U.S.-based Red Hat is one of the world's leading Linux software provider and one of the highest profile companies employing OSS in its business model. Red Hat's offerings resemble those of a classical software vendor: software distributed on CDs or over the Internet, deployment support, add-on products, and so forth. The unique aspect of the business model is that, for the most part, Red Hat has neither developed the software offering itself nor paid the development for suppliers. The role of Red Hat in its value network is related to its main activities in packaging, branding, and distributing the open source Linux operating system, thus making it usable for those who are not familiar with the ins and outs of the constantly evolving project.

Offering

Red Hat offers Linux and open source solutions into the mainstream by making high-quality, low-cost technology accessible (Rappa, 2005). In particular, Red Hat provides operating system software along with middleware, applications, and management solutions. In recent years, the target market has shifted mainly to corporate customers, thus influencing the heavy emphasis on enterprise Linux and network tools. Major parts of the software offering are provided under the GPL, which governs the redistribution of source code as well as monetary licensing rights for the binaries (Microsoft, 2005). In addition, Red Hat offers support, training, and consulting services to its customers worldwide and through top-tier partnerships. These services range from complete Linux migration to client-directed engineering to custom software development, especially in industry-specific solutions.

Resources

From the perspective of Red Hat's business model, it is obvious that key resources are related to brands and their development and management, as well as to marketing and business management. The funding provided by investors has enabled Red Hat to systematically develop these resources. In addition to marketing and management capabilities, relationships with OSS communities as the supplier network form a key resource in Red Hat's business model. Indeed, the company makes an extensive use of external resources for developing the software in its offering. The internal production resources include personnel and technology aimed at producing services.

Relationships

Red Hat has succeeded in establishing strong ties with large enterprise and academic customers such as Amazon.com, AOL, Merrill Lynch, Credit Suisse First Boston, DreamWorks, Veri-Sign, Reuters, and Morgan Stanley. In addition, its customer portfolio includes local, state, and federal governments in various countries. The company also maintains key industry relationships with hardware and middleware suppliers. In June 2002, Red Hat, Oracle, and Dell formally launched a combined Linux effort that includes joint development, support, and hardware and software certification. It was considered as an emphatic declaration in the strategy of Red Hat to focus on enterprise customers. Due to the inherent sharing nature of OSS, Red Hat considers balance as a key aspect in building a successful

business without sacrificing customer trust, and in creating shareholder value without severing ties to the open source community.

Red Hat is gaining significantly from the software produced in the OSS community. It participates in OSS and Linux development by collaborating in standards creation as well as sponsoring the Fedora Project. According to the classification of Dahlander and Magnusson (2005) presented in the theoretical part of the study, the company's approach toward the OSS community could be defined as a symbiotic relationship, although the emphasis on enterprise customers embodies commensialistic elements.

The Revenue Model

Despite the release of software under the GPL-license mode, the services employed by Red Hat for commercial viability places a layer of restriction upon the binary and source code usage based on support contracts. This hybrid approach enables the company to provide OSS solutions in a commercial way (Microsoft, 2005). Thus, the primary revenue model is currently what Red Hat calls "subscriptions," which allows the company to effectively develop and deliver its technology based on customer feedback, as well as to provide support to customers over the life of an agreement. In terms of Hecker (1999), we identify this revenue model as support selling.

It has been claimed that this is a high-margin activity demanding only a little investment (Mantarov, 1999). On the other hand, little investment means lower entry barriers, and support offers a very weak basis for differentiation to gain sustainable competitive advantage. Microsoft clearly has nearly a monopoly on desktop operating systems, but its market share in services related to desktop operating systems is much smaller. Thus, there is potential for revenue models based on service provisioning, as in some OSS-based businesses.

Table 4. Summary of the cases

Business Model Elements	MySQL	Red Hat
Offering	Core offering embodies an in-house developed data-base software product and related services.	Operating system software maintenance and services along with operating system software.
Resources	The internal programming resources and professional management resources.	Resources related to the development and management of brands, as well as to marketing and business management.
Relationships	Balancing between the OSS community and commercial business networks that have somewhat disparate needs and values. Dependence on the OSS community mainly as a user community.	OSS community as significant product developer.
Revenue model	A majority of revenue originates from proprietary license sales, and a smaller proportion stems from other sources such as services. The main income seems to come from business users. The revenue model includes features from support selling and dual licensing, both of which can be considered incarnations of the loss-leader model.	The primary revenue model is currently what Red Hat calls "subscriptions," which allows the company to effectively develop and deliver its technology based on customer feedback, as well as to provide support to customers over the life of an agreement. The revenue model is identified as support selling.

CONCLUSION

This chapter aims at identifying the key determinants of OSS revenue model choices. On the basis of our literature review and through our case studies, we see that there are several motives for firms to participate and contribute to the OSS movement.

In this chapter, we identify three business model elements that affect firms' revenue model choices. These identified elements are offering, resources, and relationships. The type of offering in terms of the user environment and, thus, the target market of the software (private vs. enterprise applications and desktop vs. server applications) constrain the possibilities to form a revenue model. Furthermore, the licensing model affects the revenue model choice through defining the free and commercial components, as well as the use and further development terms and conditions of the software.

In addition to the type of offering, we argue that a firm's resources are an important factor affecting the revenue model. We see that the internal resources and capabilities of firms are essential determinants of the actor-driven devel-

opment activity in the collection and integration of divergent OSS components into commercial offerings. Our cases illustrate that they strongly enable and constrain the possibilities to collect revenue based on OSS. Furthermore, relationships between business actors and the OSS community form the essential external resource and capability base of the firm. The importance of relationship management is emphasized in balancing between the noncommercial culture of OSS communities and the for-profit business networks. The objectives and characteristics of these two networks differ in terms of the development of loyalty, trust, and motivation of actors into activities in which some actors may benefit economically.

The managerial implications of this chapter suggest that profit-seeking firms in the field of OSS must maintain a balance between their profit-oriented business objectives and the noncommercial principles of the OSS community. This is consistent with Dahlander and Magnusson (2005), who argue that an intention to control the community development may allow a firm to manipulate the development toward its strategic goals, but might also diminish the creativity and general interest of the community toward the project.

Our empirical observations from the two case examples indicate that the selection of the revenue model is dependent on other business model elements. The case of MySQL illustrates that the need to maintain relationships with both the OSS community and the business network has led to a revenue model based on dual licensing. In this model, the community has access to the software for free, but business users may buy a software license for their commercial purposes. Furthermore, the dual-licensing model used by MySQL illustrates that a change in any of the elements of the identified key determinants may affect the revenue model choice. In this model, the company owns all copyrights to the software and, therefore, can license the software with two licenses, one allowing gathering of revenue from sold copies of the software and the other based on the principles of the loss-leader model.

The lesson learned from the Red Hat case is that internal resources (e.g., well-known brands) and superior commercialization capabilities allow a company to benefit from the development efforts of the OSS community. The business model of Red Hat is based on the ecosystem developing the core product collaboratively. The role of Red Hat in this collaboration is to deliver the results of the development work commercially added with service elements essential for the users of software.

REFERENCES

Ariño, A., & de la Torre, J. (1998). Learning from failure: Towards an evolutionary model of collaborative ventures. *Organization Science, 9,* 306-325.

Chetty, S. K., & Wilson, H. I. M. (2003). Collaborating with competitors to acquire resources. *International Business Review, 12,* 61-81.

Cravens, D. W. (1987). *Strategic marketing.* Homewood, IL: Richard D. Irwin, Inc.

Dahlander, L., & Magnusson, M.G. (2005). Relationships between open source software companies and communities: Observations from Nordic firms. *Research Policy, 34*(4), 481-493.

de Laat, P. B. (2005). Copyright or copyleft? An analysis of property regimes for software development. *Research Policy, 34*(10), 1511-1532.

Feller, J., & Fitzgerald, B. (2002). *Understanding open source software development.* Pearson Education Limited.

Foss, N. J. (1999). Networks, capabilities, and competitive advantage. *Scandinavian Journal of Management, 15,* 1-16.

Gabrielsson, P. (2004). *Globalizing internationals: Product strategies of ICT companies.* Series A: 229, Helsinki: Helsinki School of Economics.

Gadde, L.-E., & Håkansson, H. (2001). *Supply network strategies.* Chichester, UK: Wiley.

Gulati, R., Nohria, N., & Zaheer, A. (2000). Strategic networks. *Strategic Management Journal, 21,* 203-215.

Håkansson, H., & Snehota, I. (1995). Analyzing business relationships. In D. Ford (Ed.), *Understanding business marketing and purchasing* (3rd ed., pp. 162-182). London: Thomson Learning.

Hart, S.L. (1995). A natural-resource-based view of the firm. *Academy of Management Review, 20,* 986-1014.

Hecker, F. (1999). Setting up shop: The business of open source software. *IEEE Software, 16*(1), 45-51. Retrieved August 24, 2000, from http://www.hecker.org/writings/setting-up-shop.html

Hedman, J., & Kalling, T. (2003). The business model concept: Theoretical underpinnings and empirical illustrations. *European Journal of Information Systems, 12,* 49-59.

Kotler, P., Armstrong, G., Saunders, J., & Wong, V. (1996). *Principles of marketing*. Hertfordshire, UK: Prentice Hall.

Lee, S. H. (1999). *Open source software licensing*. Retrieved February 14, 2006, from http://eon.law.harvard.edu/openlaw/gpl.pdf

Lerner, J., & Tirole, J. (2002). Some simple economics of open source. *Journal of Industrial Economics, 50*(2), 197-234.

Mantarov, B. (1999). *Open source software as a new business model: The entry of Red Hats Software, Inc. on the operating system market with Linux*. Retrieved August 24, 2000, from http://www.lochnet.net/bozweb/academic/dissert.htm

Metcalfe, J. S., & James, A. (2000). Knowledge and capabilities: A new view of the firm. In N. J. Foss & P. L. Robertson (Eds.), *Resources, technology and strategy: Explorations in the resource-based perspective*. New York: Routledge.

Microsoft. (2005). Software licensing models. Retrieved March 23, 2006, from http://www.microsoft.com/resources/sharedsource/licensingbasics/licensingmodels.mspx

Möller, K., & Svahn, S. (2003). Managing strategic nets: A capability perspective. *Marketing Theory, 3*, 201-226.

Morris, M., Schindehutte, M, & Allen, J. (2005). The entrepreneur's business model: Toward a unified perspective. *Journal of Business Research, 58*, 726-735.

O'Mahony, S. (2003). Guarding the commons: How community managed software projects protect their work. *Research Policy, 32*(7), 1179-1198.

Open Source Initiative. (2004). *The open source definition* (Version 1.9). Retrieved March 5, 2004, from http://www.opensource.org/docs/definition.php

Osterwalder, A. (2004). *The business-model ontology: A proposition in design science approach*. Academic dissertation, Universite de Lausanne, Ecole des Hautes Etudes Commerciales, Lausanne, France.

Pateli, A. G., & Giaglis, G. M. (2004). A research framework for analysing eBusiness models. *European Journal of Information Systems, 13*(4).

Penrose, E. (1959). *The theory of the growth of the firm*. New York: Oxford University Press.

Perens, B. (1999). The open source definition. In C. DiBona, S. Ockman, & M. Stone (Eds.), *Open sources: Voices from the open source revolution*. Sebastobol, CA: O'Reilly & Associates, Inc.

Prahalad, C. K., & Hamel, G. (1990). The core competence of the corporation. *Harvard Business Review, 32*, 79-91.

Rajala, R., Rossi, M.., & Tuunainen, V.K. (2003). Software vendor's business model dynamics case: TradeSys. *Annals of Cases on Information Technology, 5*, 538-549.

Rajala, R., & Westerlund, M. (In press). A business model perspective on knowledge-intensive services in the software industry. *International Journal of Technoentrepreneurship*.

Rappa, M. (2005). *Case study: Red Hat. Managing the digital enterprise*. Retrieved November 20, 2002, from http://digitalenterprise.org/cases/redhat.html

Raymond, E. S. (2001). *The cathedral and the bazaar: Musings on Linux and Open Source by an accidental revolutionary*. Sebastopol, CA: O'Reilly & Associates, Inc.

Rosenbröijer, C.-J. (1998). *Capability development in business networks*. Doctoral dissertation, Swedish School of Economics and Business Administration, Helsinki.

Rossi, M. A. (2004). *Decoding the "free/open source (F/OSS) software puzzle" a survey of theoretical and empirical contributions.* Quaderni, Siena: Dipartimento di Economica Politica, Università di Siena.

Sallinen, S. (2002). *Development of industrial software supplier firms in the ICT cluster.* Doctoral dissertation, University of Oulu, Oulu.

Seddon, P. B., & Lewis, G. P. (2003). Strategy and business models: What's the difference. In *Proceedings from the 7ʰ Pacific Asia Conference on Information Systems*, Adelaide, South Australia (pp. 1-30).

Selznik, P. (1957). *Leadership in administration: A sociological interpretation.* New York: Harper Row.

Stallman, R. M. (1999). The GNU operating system and the free software movement. In C. DiBona, S. Ockman, & M. Stone (Eds.), *Open sources: Voices from the open source revolution.* Sebastobol, CA: O'Reilly & Associates, Inc.

Stallman, R. M. (2002). *What is copyleft?* Retrieved November 20, 2002, from http://www.gnu.org/licenses/licenses.html

Timmers, P. (2003). Lessons from e-business models. *ZfB—Die Zukunft des Electronic Business, 1,* 121-140.

Välimäki, M. (2003). Dual licensing in open source software industry. *Systemes d´Information et Management, 8*(1), 63-75.

Välimäki, M. (2005). *The rise of open source licensing. A challenge to the use of intellectual property in the software industry.* Helsinki: Helsinki University of Technology.

Wernerfelt, B. (1984). A resource-based view of the firm. *Strategic Management Journal, 5,* 171-180.

KEY TERMS

Business Model: An abstraction of business, or the manifestation of strategy, that characterizes the business and specifies in which the company is positioned in its value-creating network.

Offering: An inseparable part of a business model that includes aspects such as complexity; the essential benefit that the customer is really buying; and product features, styling, quality, brand name, and packaging of the product offered for sale.

Relationships: The ties and interaction of companies with other actors.

Resources: Specific properties that are subordinate to the core competencies of companies.

Revenue Model: The method of value capturing that includes the description of the sources of revenue, price-quotation principle, and cost structure.

Software Licensing: The definition and agreement of rights to use, redistribute, or modify software.

Source Code: The programming that allows software to perform a particular function or operation.

ENDNOTE

[1] The adjective "free" refers to freedom, not price.

Chapter XLIII
Open Source for Accounting and Enterprise Systems

Thomas Tribunella
State University of New York at Oswego, USA

James Baroody
Rochester Institute of Technology, USA

ABSTRACT

This chapter introduces open source software (OSS) for accounting and enterprise information systems. It covers the background, functions, maturity models, adoption issues, strategic considerations, and future trends for small accounting systems as well as large-scale enterprise systems. The authors hope that understanding OSS for financial applications will not only inform readers of how to better analyze accounting and enterprise information systems but will also assist in the understanding of relationships among the various functions.

INTRODUCTION

This chapter will inform the readers about the feasibility and potential applicability of open source software (OSS) to the functional areas of accounting and finance. Small and enterprise-scale systems will be examined. The chapter will review background information and frameworks for analyzing the business case related to financial applications of OSS.

OSS systems can provide support to individual business functions or integrated suites of functions. For example, open source enterprise systems provide an integrated set of business functions that are organized around business processes.

In this chapter, we will address the concerns of managers and educators who are interested in learning more about open source business systems. We studied available OSS accounting and financial applications by reviewing available documentation on Web sites. For a number of enterprise applications we reviewed, the system functionality and market positioning, downloaded the systems and studied system requirements, installed and set up the systems, and reviewed the license agreements. Initially, the chapter will review the current state of OSS business systems with a focus on definitions and functional applications of small accounting systems and larger enterprise systems. We will then address the critical factors and decision frameworks relevant to the adoption OSS for accounting and financial applications. In addition, we will explore future trends in OSS financial reporting systems.

BACKGROUND OF OSS ACCOUNTING AND FINANCE APPLICATIONS

In this section, we will discuss the business issues that are a required background in order to have a general understanding of accounting and financial applications with OSS. Open source is used to describe "a software program or set of software technologies that are made widely available by an individual or group in source code form for use, modification, and redistribution under a license agreement with having very few restrictions" (American Bar Association, 2006). The logic behind the open source philosophy is that users must be able to read, redistribute, and modify the source code for a piece of open source software. In contrast, a traditional software license is designed to protect the intellectual property of the software developer and severely restricts reading, redistributing, and modifying source code. Since an open source license gives broad rights to read, redistribute, and modify the source code for a piece of OSS, users constantly improve the OSS by adapting it to various applications and fixing bugs.

The intellectual and legal origin of most open source license agreements can be traced to two sources: the GNU General Public License (GPL) and the University of California BSD Unix license agreements (McGowan, 2001). These agreements reflect the goal of creating a community environment in which innovation and quality improvements are rapidly shared and distributed through common ownership of intellectual property rather than through individual or organizational ownerships through copyrights (Kennedy, 2001). Improvements made by individuals are made publicly available back to the community.

Statistics available from www.freshmeat.net (Freshmeat, 2006), a Web site described as one of the largest indexes of Unix and cross-platform OSS, indicates that these two license forms (or close revisions of them) account for almost 80% of the license agreements used by projects tracked on the site. Since 20% of the projects utilize different types of agreements, users must examine carefully the license agreement of the system they want to use.

We downloaded and reviewed the license agreements for a number of enterprise, accounting, and financial applications (see Tables 1 and 2). For this sample, The GNU General Public License was the most common agreement. It is important to note that as the target market for these systems moves toward large enterprises, commercial licenses and hosted licenses emerge (Tustena CRM, 2006). Given the variations of licensing agreements demonstrated in this sample, users must carefully compare the license agreement with the requirements of their organizations.

The trade press and other publications emphasize that OSS is about back-office technology such as servers and operating platforms. The relevance of OSS to functional areas, including accounting, finance and enterprise systems, is not well understood. Historically, OSS has focused on technology components such as the Linux operating system and the Apache Web server. Open source business applications are beginning to emerge, the most familiar being OpenOffice, an OSS application office suite supporting word processing, presentation, and spreadsheet applications. Now available as OSS is a variety of accounting, financial, and enterprise systems applications. Reflecting the potential these offerings have in the marketplace, venture capital is flowing into open source business applications, which be an indicator that these OSS business applications will play a significant role in the future (Cook, 2004; Marshall, 2005; Stein, 2005).

Finding operating systems and servers to support the various open source accounting and

Table 1. OSS small accounting systems

Name of Product	Operating System	Database (if any)
Appx-BANG	Windows, OS-Independent, Linux	MySQL, Oracle
BestBooks	OS-Independent	JDBC
CentraView	Windows, Linux, Mac OSX	MySQL
EzyBiz	Linux, OS-Independent	
GRISBI	Windows	XML-Based
Lazy8 Ledger	OS-Independent	
Muhasebeci	Windows, Linux	MySQL
NetAccounts	OS-Independent	
NOLA	Linux, Windows	
OpenAccounting	OS-Independent	
OSAS	Windows	
PHPBalanceSheet	OS-Independent	
Quasar	Linux, Windows	
SQL-Ledger	UNIX, Mac OSX, Windows	Perl DBI/DBD
Tiny ERP	Linux	
TinyBA	Windows, Linux	
Traverse	Windows	SQL
TurboCash	Windows	
XIWA	Linux	

finance (A&F) systems (except venture capital) is not a problem. Microsoft (MS) Windows, Linux (Red Hat), Solaris (Sun), BSD, Mac OSX, and UNIX all have accounting applications that run on their platforms. OSS systems will run on most popular servers such as Microsoft, Apache (the most popular server with more than 50% of all installations), Sun, and so forth. Table 1 displays some of the more well-known OSS small accounting systems and the operating systems they require (Sourceforge, 2006).

To compile Table 1, we went to SourceForge and checked all software under the Topic menu listed as office/business -> financial -> accounting. This identified more than 500 systems as of June 2006. We then identified all systems rated as mature, stable, or production. Many systems display the term *accounting* under topics, but we question whether some of these systems are true accounting systems. Therefore, we examined company Web sites to determine through product information if the systems were capable of most traditional accounting functions. In addition, we checked to determine if the company was still actively operating and if the software was current. We also reviewed recent journal articles on the subject to see if the companies were cited. We understand that this market is in a continual state of change and many new products are appearing. We feel that we have highlighted most of the more well-known systems, but we may have missed a few. Certainly, the OSS accounting market would benefit from some consolidation.

In addition to small accounting applications, there are other business applications available

Table 2. OSS enterprise scale accounting systems

Name of Product	Target Market Segment	Database	Open Source License Agreement
Compiere	ERP and CRM, small to medium enterprises and large corporations	Oracle, MySQL	GNU Public License
GnuCash	Desktop financial manager	PostgreSQL	GNU Public License
Tiny ERP	Small to medium business ERP	PostgresQL	GNU Public License
Tustena CRM	CRM, including large enterprises	SQL	Mozilla Public License
WebERP	ERP for small to medium enterprises	MySQL	GNU Public License

in the form of OSS systems. For example, functions such as personal financial management, e-commerce, office suites (e.g., spreadsheet, word processing, database, graphical presentation), Web browsers, RDBM (relational database management), e-mailing clients, and strategic planning are all available in OSS. The enterprise scale accounting systems usually interface with a relational database as well as with other applications. Table 2 displays some of the more well-known OSS enterprise scale accounting systems and summarizes the target market applications, database, and license requirements of each system (Compiere, 2006; GNUCash, 2006; Sourceforge 2006; TinyERP, 2006; Tustena CRM, 2006; WebERP, 2006).

Most open source accounting and finance (A&F) systems work very well with relational database systems. As a matter of fact, some open source systems run on open source RDBM systems. For example, Compiere runs on MySQL, a mature and stable open source RDBM system. Even though there are many related applications that work with A&F systems, this chapter will focus mainly on the financial applications of these systems. However, we will briefly discuss closely related applications since they interface with A&F applications.

MAIN FOCUS OF FINANCIAL OSS SYSTEMS: FUNCTIONS AND ADOPTION CONSIDERATIONS

In this section, we first will describe the functions of small and enterprise-scale A&F systems, and then we will discuss the frameworks as well as the considerations used to evaluate these systems. Evaluation criteria will include quantitative financial models and qualitative maturity and strategic considerations.

Small-Scale Business Systems

A&F systems can be viewed as a functional set of application modules that can be mixed and combined. Traditional OSS modules support functions such as general ledger, accounts receivable, accounts payable, purchase orders, sale orders, inventory management, and fixed assets. OSS modules that are based on the same operating systems are often mixed and combined with the help of consultants who program the patches and make the code publicly available. Most proprietary systems do not operate well with OSS modules at the functional application level. Table 3 displays the typical accounting modules that come with a standard small business OSS package (Romney, 2006).

Table 3. Module application features: Small to mid-size accounting systems

Application	Description
General Ledger	The chart of all accounts and balances that supports the double entry system of accounting.
Accounts Receivable	Amounts due from customers for credit sales.
Accounts Payable	Amounts due to suppliers for credit purchases.
Purchase Order	Used to place and record orders with suppliers.
Sales Order	Used to place and record sale orders with customers as well as completing sales tax returns.
Inventory Management	Used to keep track of goods purchased from suppliers and merchandise available for sale to customers.
Fixed Assets	Used to keep track of the purchase, depreciation, and disposition of long-term productive assets such as property, plant, and equipment.
Payroll	Used to record and pay employees as well as file payroll tax returns.
Project Management	Used to track revenue and costs related to specific jobs or projects.
Financial Statements	Compiles financial statements such as balance sheets and income statements from general ledger accounts.

Table 4. Core application and business process features: Enterprise scale systems

Application Module	Description
Financial	Includes general ledger, accounts receivable, accounts payable, legal consolidation, cost center accounting, product cost controlling, and activity-based costing.
Operations and Logistics	Includes inventory management, materials requirements planning, materials management, plant maintenance, production planning, project management, purchasing, quality management, routing management, shipping, and vendor evaluation.
Sales and Marketing	Includes order management, pricing, sales management, and sales planning.
Human Resource Management	Includes human resource time accounting, payroll, personnel planning, and travel expenses. Also includes vacation and sick time tracking..
System Administration and Management	Includes tools to support ERP system installation and management, such as security management.

Enterprise-Scale Business Systems

There are many small business OSS packages, but one can ask if open source systems support enterprise-level applications such as support for inventory and manufacturing operations. The answer is clearly yes, as a few OSS systems operate at the enterprise level. For example, Compiere can handle enterprise-level applications with integrated business processes such as quote-to-cash, requisition-to-pay, CRM (customer relationship management), PRM (partner relationship management), supply chain management, performance measurement, and a Web store. Table 4 displays the typical business process modules that should come with an enterprise-level OSS package (Davenport, 1998; The ERP Fan Club and User Forum, 2006; O'Leary, 2000; Stein, 2006; Sumner, 2005).

In order for OSS business systems to have a significant impact on the market in the future, they will have to reach enterprise scale. The term *enterprise system* has a broad definition. One definition focuses on the capacity, robustness, and scalability of the underlying technology: An enterprise system is an information system that offers a high quality of service and can support the large volumes of processing and data typical of a large organization. Such systems typically require independent server hardware and a dedicated administration. When the term enterprise system is applied to business applications, it has a more restricted definition. An enterprise system not only provides the quality and capacity to support a large organization as described previously, but its business functionality is broader than what is required for a specific workgroup, department, or small business. An enterprise system provides cross-functional capability to support multiple business operations such as accounting, finance, production, sales, and marketing (Davenport, 1998). Table 5 shows examples of the advanced application features of some enterprise systems that achieve

cross-functional integration among various departments in a large organization (e-consultancy, 1999; Swanton, 2004; Techtarget, 2006).

As can be seen in Table 5, open source A&F systems usually support and interface with Web-based applications such as online transactions, CRM systems, Internet catalogs, and electronic banking. The underlying infrastructure of open source operating systems, open source database systems, and open source Web servers has a demonstrated record of stability and reliability. For example, a high percentage of Web servers are based on open source platforms such as Apache servers.

What are the implications for open source A&F systems? Open source A&F systems must be stable, reliable, and able to process transactions with multiple simultaneous users. The larger enterprise systems must handle large groups of users. One view, as discussed by Wheeler (2005), is that the open source process with many developers having access to and contributing to the source code inherently produces system software that runs with higher stability and has less down time

Table 5. Advanced application features: Enterprise scale systems

Application Module	Description
Supply Chain Management (SCM)	The management of information between partners in the supply chain to enable the control of goods, services, and money from the acquisition of raw materials to the final customer product.
Partner Relationship Management (PRM)	Supports communication among companies and their partners, which enables shipping schedules and real-time information to be available to all.
Auditing Information System	Includes tools for auditing businesses and systems, documenting the progress of an audit, and preparing reports.
Customer Relationship Management (CRM)	Includes one-on-one marketing, telemarketing, sales force management, call center automation, e-selling, data warehousing, and customer service.
Internal Controls Management	Includes tools to plan and manage enterprise systems audits and verify internal controls.
E-Business	Technology to enable employees, customers, suppliers, and business partners to collaborate. Includes business-to-business and business-to-consumer capabilities such as Web stores and Internet catalogs.
Strategic Enterprise Management	Includes tools to manage and integrate strategic planning, budgeting, forecasting, and performance management.

than proprietary systems such as Windows-based systems (Wheeler, 2005).

There is other research and testing that demonstrates different conclusions. Zhao and Elbaum (2003) compare the development process for OSS and proprietary software. While the open source process does involve more developers and should have more testing performed in parallel, this results in improved detection of defects, not necessarily improved debugging and correction of defects (Zhao & Elbaum, 2003). Paulson (2004) and Stamelos (2002) also present findings that question the superior reliability of open source software.

Collectively, this research is ambiguous and does not provide clear-cut direction to decision-makers regarding the quality and reliability of open source A&F applications. The open source model does enable organizations considering the adoption of open source A&F applications the ability to download and rigorously evaluate the application. However, to fully test the application using samples of an organization's data and processes involves much of the work to actually implement the system, so this benefit may not be as positive as it first seems.

Another advanced application of enterprise systems are built-in security and audit modules. In theory, the audit risk associated with A&F systems is not greater with OSS than with proprietary systems. Since the code is freely available, the OSS community finds weaknesses in the program that could violate system security. Since the systems are open, they are patched by a wide variety of users at a very rapid rate. Those patches become open and available to other users in the community. Therefore, a well-maintained OSS should be a low-risk system from a security point of view. There are very few reports of open source A&F systems having been hacked. The more widely used proprietary systems have audit modules that support the work of external auditors such as CPA firms and governmental regulators. OSS systems are lagging in this area. However, Tiny ERP does have an audit module.

Maturity and Stability Frameworks for Understanding OSS Systems

There are several maturity models, such as the generally recognized as mature (GRAM) and generally recognized as safe (GRAS) models (Wheeler, 2006). Maturity and safety are important considerations in OSS because immature systems will not have a critical mass of support, consulting, training, vendors, and users. The GRAM and GRAS models are conceptual. More quantitative models are the business readiness rating (BRR) developed at Carnegie Mellon (Center for Open Source Investigation, 2005), open source maturity model (OSMM), and CapGemini OSMM (CapGemini, 2006). The quantitative models employ a rating system in which important attributes and goals of the system are rated and then weighted proportionately. Then a score is calculated, and the OSS systems are ranked in order of maturity and acceptability. Table 6 displays the framework for the open source maturity model (OSMM) (Golden, 2006).

Notice in Table 6 that software elements are listed on the left side of the model. Then each element is scored based on its maturity. Next, each element is given a weighting factor based on its importance to the organization. In the next phase of the analysis, the element maturity scores are multiplied by the weighting factor to generate element-weighted scores. In the final phase of the analysis, the element-weighted scores are totaled to produce a product maturity score. The various product maturity scores are then compared to determine the best OSS system. A well-designed maturity model should help managers understand the development and stability level of an OSS product.

Sourceforge.net identified more than 500 OSS accounting projects as of early 2006. The vast majority of these products is neither mature nore ready for commercial applications. Sourceforge.net ranks OSS in the following seven categories:

Table 6. Open source maturity model (OSMM)

Element	Potential Score	Actual Score	Weighting Factor (default weights)	Element Weighted Score (actual score x weighting factor)
(1) Product Software	0 to 10		4	
(2) Support	0 to 10		2	
(3) Documentation	0 to 10		1	
(4) Training	0 to 10		1	
(5) Product Integrations	0 to 10		1	
(6) Professional Services	0 to 10		1	
Total of Weighting Factors			10	
Product Maturity Score (max. = 100)				100
Type of User →	Early Adaptor	Pragmatist		
Purpose of Use:				
Experimentation	25	40		
Pilot	40	60		
Production	60	70		

- 7: Inactive
- 6: Mature
- 5: Production/Stable
- 4: Beta
- 3: Alpha
- 2: Pre-Alpha
- 1: Planning

Examples of major OSS producers with mature or stable systems are Compiere, GnuCash, TurboCash, Traverse (MS compatible), OSAS (NT, Linux, Unix), Lazy8 Ledger, NOLA, NetAccounts, SQL-Ledger, PHPBalanceSheet, WebERP, OpenAccounting, Quasar, CentraView, and TinyBA. There are also many systems under development (Sourceforge, 2006).

Getting support and training for an open source A&F system can be difficult if the system is not mature. Very mature systems that are industry leaders have a support infrastructure equivalent to the infrastructure provided by mature proprietary software vendors, including the following:

- Call center and help-desk support
- Online assistance such as demonstrations, documentation, and forums
- Consulting services
- Training seminars and conferences

However, there is a contrast between these services for open source A&F and the offerings from proprietary software vendors. The services from a proprietary software application vendor focus on providing support to customers while they are making their purchase decisions and then providing support while customers install, configure, and operate the application. The consultants for open source A&F applications expand their support to address the needs of software developers with technical documentation and recommendations during the development process.

An organization considering the use of open source A&F applications must look carefully at the vendors' services and support, determine the balance between support for development and operational usage, and evaluate whether the offerings of the vendor meet its needs. In addition, the revenue model for support vendors in open source is different from the proprietary software model. An implication of this is that an organization using open source may receive its support from a network of suppliers. These suppliers may include support from the core organization leading the open source A&F development, operational support from independent consultants, patches and improvements from partners, and value-added resellers (VAR) who address various aspects of the support value chain.

Accordingly, organizations adopting open source A&F most likely do not have the one-stop alternatives that exist for proprietary software. Adopting organizations will need to assess their capabilities to be actively engaged in selecting consultants and managing their support.

Financial Frameworks for Understanding OSS Systems

There are several frameworks that can help us understand the cost and value of OSS information systems and that can be applied to accounting applications. For example, the total cost of ownership (TCO) as well as capital budgeting models such as return on investment (ROI), net present value (NPV), payback period (PB), and internal rate of return (IRR) can shed light on the value of these systems.

Costs for an information system fall into a number of categories: purchase cost for new software, hardware, and networking technologies; resource costs to install, set up, and configure the hardware and software; and ongoing administration and maintenance costs. The primary savings for an organization adopting open source A&F will be the purchase cost. Generally, an OSS is available free or at a very low cost. Therefore, the initial software (SW) cost is very low. Hardware requirements for Linux- and Unix-based operating systems are very low since these systems run very efficiently and can operate very well on used equipment. The most common model for categorizing costs related to information systems is the TCO (David, 2002). The TCO model is displayed in Table 7.

TCO includes all expenses associated with owning and maintaining work stations within an organization. It is a holistic view of IT-related

Table 7. Total cost of ownership

Measure	Calculation	Percent of Use
ROI: Return on Investment	Income from Investment / Average Investment	40.7%
TCO: Total Cost of Ownership	See Table 7	29.1%
IRR: Internal Rate of Return	Rate of Return when NPV = 0	13.6%
ROA: Return on Assets	Net Income / Average Total Assets	08.2%
Other Measures	**See Below**	**08.4%**
NPV: Net Present Value	Investment – Present Value of Net Cash Flow at the Desired Rate of Return	
PBP: Payback Period	Investment / Net Cash Flow	
ARR: Accounting Rate of Return	(Net Cash Flow – Depreciation) / Initial Investment	
SLD: Straight Line Depreciation	(Cost – Salvage) / Useful Life	
BEU: Break Even Units	Fixed Costs / Contribution Margin Per Unit	
RI: Residual Income	Net Project Income – ([Cost of Capital] [Capital Investment])	

Table 8. Profitability and capital budgeting methods for measuring IT investments

Acquisition	Administrative Costs	
	Control	**Operations**
Hardware Costs	Centralization: Control of software and network administration from one department.	**Installation and upgrades:** Installing updates and new systems.
Software Costs	Standardization: Similar hard and software configurations throughout the end-user community.	**Evaluation:** Analyzing the latest technology that becomes available.
		Power consumption: Costs of energy per work station.
		Training: Cost of helping end users understand system features.
		Downtime: Cost of system failures and repairs.
		Fuzz: Personal use of company systems.
		Auditing: Cost of monitoring systems.
		Viruses: Cost of repairing software and data from intrusions.
		Support: Cost of services to address user problems.

costs at an enterprise level. TCO includes acquisition costs, control costs, and operation costs. Acquisition costs account for approximately 20% of the total costs. It has been posited that investing in control will reduce many operational costs (David, 2002).

Consulting, training, and change management are usually very expensive when an organization converts to an OSS or to any other system. Ongoing administrative costs will likely be the same for various open source A&F systems. Finally, maintenance costs are also impacted since frequent updates, corrections, extensions, and patches are frequently released by the open source community.

An additional dimension of cost and time affected by open source systems is the request for proposal (RFP) cycle. Much of the RFP process is invested in analyzing the licensing proposals from each potential vendor, assessing the payback from the investment, and negotiating the terms and conditions of acquiring the software. Organizations utilizing OSS business applications should see this process shortened and reduced in complexity, since the software is usually free and license agreements follow standard models.

Table 8 displays the most common financial models for judging information technology (IT) projects. The percentage of use was reported by a *CIO* magazine research report in which 256 IT professionals reported the metric they used to measure IT initiatives (CIO, 2001).

Following is a summary of the IT budgeting and measurement tools in Table 8 (Romney & Steinbart, 2006; Williams, Haka, Bettner, & Meigs, 2005):

- **ROI:** Compares the annual cash flow with the initial investment to produce a return on investment percentage. If the cash flow is unequal over the life of the project, managers can use the average annual cash flow. ROI does not consider the time value of money, which is a significant flaw in the method for long-term projects.
- **IRR:** Calculates the effective interest rate that would result, assuming a net present value of zero for the project. In other words, IRR is the discount rate that makes the NPV of an investment (or project) equal to zero. Using this method, managers will select projects with higher IRRs.

- **ROA:** Determines the return on the book value of the average assets related to the system.
- **NPV:** Calculates and sums the discounted future cash flows of the benefits minus the costs. NPV discounts all cash flows on an investment back to present value using a required ROI. Accordingly, the analyst tries to determine if the present value of future cash flows (revenues or savings) from the system is greater than the current investment required to finance the system. Under this method, managers will select projects with higher positive NPV.
- **PBP:** Calculates the number of years before the new savings from the project equal the initial cost of the investment. The method calculates the time it will take to recoup an investment in terms of nominal dollars. It does not consider the time value of money, which is a significant flaw in the method. Under this method, managers will select projects with shorter payback periods.
- **ARR:** Calculates the percentage increase in operating income from an investment in nominal dollars. ARR does not consider the time value of money, which is a significant flaw in the method. It does consider depreciation, which is a method to allocate the cost of an asset to accounting periods in a systematic and rational manner required by generally accepted accounting principles.
- **BEU:** With break-even (BE) analysis, we can determine how many units we need to sell (or savings we need to gain) in order to break even on an IT project. Furthermore, we can calculate forecasted and projected levels of profits analysis. The analysis generates understandable income statements and graphical presentations of potential IT project results and is a popular technique in the MIS industry. It explains how cost drivers affect cost behavior and allows for sensitivity analysis. It shows how changes in cost-driver activity levels affect variable and fixed costs. The tool is easy to quantify and calculate break-even sales volume in total dollars and total units. BEU provides a visual representation of project performance by creating a cost volume-profit graph and helps to supply information for forecasts and projections. It also calculates sales volume in total dollars and total units to reach a target profit.
- **Residual Income (RI):** Determines the net income of a division less the cost of capital on the division's capital investment.

These measures should be improved with the implementation of OSS systems if those systems are less expensive and have the same functionality. Accordingly, chief financial officers (CFOs) should look favorably on OSS systems.

Strategic Factors Related to OSS Financial Systems

The decision regarding implementing an OSS enterprise system includes additional factors. In order to address the business processes within an individual organization, enterprise systems require a significant amount of customization to the specific requirements of the organization. Before adopting an OSS enterprise system, an organization must analyze the capability of the candidate software to support this customization. Does the candidate software meet the needs of the organization as delivered? If not, then what tools and processes are defined to support customizing the enterprise system? Proprietary systems such as SAP R/3 and Oracle Applications enable configuration, which customizes the application to the business requirements without writing software. If configuration capability is not supported, then the organization must assess the adoption and extension the software needs to meet the organizations requirements.

Closely aligned with customization is a strategic question. The open source model generally requires that changes to the software be shared with the open source development community. Frequently, the specialization of business processes within an organization is a source of competitive advantage. Typically, OSS license agreements enable organizations to modify the source code and freely use the modifications internally as long as the software containing the modifications is not distributed publicly. Organizations using OSS systems must carefully review the license agreements and be sure that their plans are supported.

FUTURE TRENDS

In the future, most A&F systems will support markup standards such as XML (extensible markup language) and XBRL (extensible business reporting language). The hypertext markup language (HTML) is a standard that defines the format of information exchanged between Web browsers and Web servers. However, it has a fixed set of information types that it can exchange. XML language is a standard created to overcome the restrictions of HTML by providing mechanisms to extend, in an application-specific manner, the types of information that are exchanged.

XBRL is a specific standardized set of extensions created using XML for financial applications. It is the markup language used to tag financial information for the U.S. Securities and Exchange Commission's (SEC's) EDGAR (Electronic Data Gathering and Retrieval) database. In the short run, XBRL will soon be used by the SEC to accept financial reports that contain data in XBRL-compliant form. XBRL is voluntary for now, but in the future, given the current trend of government regulation of the financial markets, it may become required (Debreceny, 2005). The creation of languages such as XBRL will allow

for the rapid communication of data among organizations, systems, and networks.

XBRL allows users to increase the speed of the financial reporting process and may lead to a continuous reporting process in the future. In an environment of continuous reporting, accountants and auditors will have to transition from periodic reviews of batches of financial information to a constant monitoring of a flow of financial information. The security and control of these online, real-time systems will create new challenges and opportunities for accountants and auditors (Debreceny, 2005). Furthermore, since XBRL-based tags can be used to identify grains of data for financial applications, new ways of understanding the financial reporting process and testing data quality will have to be developed (Tribunella, 2005).

Organizations and companies in Europe and Asia have emphasized de facto standards more than U.S. companies, which tend to focus more on innovation and rapid technology migration. How will these factors affect OSS financial systems? Standards are emphasized because of the vendor independence they provide. In the technology domain, OSS has exploited standards to offer technologies that support standards at a lower cost. The role of standards in A&F systems domains is much smaller. It is too early to tell whether a trend for standardization will be driven by European and Asian governments and organizations.

CONCLUSION

The adoption of OSS accounting and financial systems is not widespread. Proprietary systems such as QuickBooks, Peachtree, and Cougar Mountain have a majority of the market share of small business systems. Closed source systems such as SAP, Oracle, and Microsoft dominate the enterprise systems market. However, Compiere (an OSS) reports 930,000 downloads of its enterprise system and is supported by 44 partners with worldwide locations.

Open source A&F systems are in a state of rapid evolution. They are not well developed at the enterprise level, but there are many small business accounting systems with a complete set of standard modules. However, only a few of the systems are mature with a network of vendors that support and train users as well as provide consulting and installation. The lack of support should change as more venture capital flows into the OSS industry. Given the low cost and stability of an OSS system for A&F applications, one can make a strong business case for its implementation. Accordingly, we believe these systems will gain greater acceptance in the business community in the future.

The bottom-line questions that an organization must answer are whether the applications meet its needs and whether the costs are affordable. Since open source A&F systems are in their infancy the current answer is yes in a minority of cases. But for those organizations that choose to employ OSS systems, the next question is whether the structure of pricing, support, and maintenance cost is less than the proprietary alternative. Given the business criticality of these applications and the potential need to customize them, the jury is still out.

REFERENCES

American Bar Association. (2006). *An overview of "open source" software licenses.* Retrieved February 9, 2006, from http://www.abanet.org/intelprop/opensource.html

CapGemini's OSMM. (2006). Accessed July 2, 2006, from www.seriouslyopen.org

Center for Open Source Investigation at Carnegie Mellon West. (2005). *Business readiness rating for open source: A proposed open standard to facilitate assessment and adoption of open source software.* Retrieved July 2, 2006, from http://www.openbrr.org

CIO. (2001). *CIO research reports: Measuring IT value.* Retrieved July 5, 2006, from http://www.cio.com

Compiere ERP and CRM homepage. (2005). Retrieved July 7, 2006, from http://www.compiere.org

Cook, J. (2005). Venture capital: Investors see open source software potential. *Seattle Post-Intelligencer.* Retrieved November 11, 2005, from http://seattlepi.nwsource.com

Davenport, T. H. (1998). Putting the enterprise into the enterprise system. *Harvard Business Review, 76*(4), 121-131.

David, J. S., Schuff, D., & St. Louis, R. (2002). Managing your IT total cost of ownership. *Communications of the ACM, 45*(1), 101-106.

Debreceny, R. S., Chandra, A., Cheh, J. J., Guithues-Amrhein, D., Hannon, N. J., Hutchison, P. D., et al. (2005). Financial reporting in XBRL on the SEC's EDGAR system: A critique and evaluation, *Journal of Information Systems, 19*(2), 191-210.

e-consultancy. (1999). Oracle announces strategic enterprise management; major components available now. *ZDNet UK.* Retrieved February 9, 2006, from http://www.e-consultancy.com/newsfeatures/19803/oracle-announces-strategic-enterprise-management-major-components-available-now.html

The ERP Fan Club and User Forum. (n.d.). *Enterprise resource planning.* Retrieved February 9, 2006, from http://www.erpfans.com/erpfans/erpca.htm

Extensible Markup Language (XML). (2006). Retrieved February 9, 2006, from http://www.w3c.org\xml

Freshmeat. (2006). Retrieved July 6, 2006, from http://www.freshmeat.net

GnuCash. (2005). Retrieved July 6, 2006, from http://www.gnucash.org

Golden, B. (2005). *Creating your open source ERP strategy.* Retrieved June 8, 2005, from http://SearchOpenSource.com

Golden, B. (2006, September-October). The open source maturity model. *Enterprise Open Source Journal,* 22-25.

HyperText Markup Language (HTML). (2006). Retrieved February 9, 2006, from http://www.w3c.org\markup

Kennedy, D. M. (2001). *A Primer on open source licensing legal issues: Copyright, copyleft and copyfuture.* St. Louis, MO: Saint Louis University Public Law Review. Retrieved July 5, 2006, from http://www.denniskennedy.com/opensourcedmk.pdf

Marshall, M. (2005). *Net start-ups face odd problem: More VC cash than they need.* Retrieved October 14, 2005, from http://www.siliconvalley.com

McGowan, D. (2001). *Legal implications of open source software.* Retrieved July 5, 2006, from http://www.law.umn.edu/uploads/images/254/McGowanD-OpenSourceFinal.pdf

O'Leary, D. E. (2000). *Enterprise resource planning systems: Systems, life cycle, electronic commerce and risk.* London: Cambridge University Press.

OpenOffice. (2006). Retrieved February 9, 2006, from http://www.openoffice.org

Paulson, J., Succi, G., & Eberlein, A. (2004). An empirical study of open source and closed source software products. *IEEE Transactions of Software Engineering, 30*(4), 246-256.

Romney, M., & Steinbart, P. (2006). *Accounting information systems* (10th ed.). Upper Saddle River, NJ: Pearson Prentice-Hall.

Sourceforge. (2006). Retrieved January 31, 2006, from http://www.sourceforge.net

Stamelos, I. L., Angelis, A., & Oikonomou, G.B. (2002). Code quality analysis in open source software development. *Information Systems Journal, 12*(1), 43-60.

Stein, S. (2006). *EDI impact on ERP system implementation.* Retrieved February 9, 2006, from http://www.msc-inc.net/ERP_Implementation.htm

Stein, T. (2005, September). Has the free software paradox been solved? *Venture Capital Journal.*

Sumner, M. (2005). *Enterprise resource planning.* Upper Saddle River, NJ: Pearson Prentice Hall.

Swanton, B. (2004). Oracle internal controls manager keeping ahead of SOA projects. *ZDNet.* Retrieved February 9, 2006, from http://techupdate.zdnet.com/techupdate/stories/main/Oracle_Internal_Controls_Manager.html

TechTarget. (2006). Retrieved February 9, 2006, from http://searchcrm.techtarget.com/sDefinition/0,,sid11_gci214321,00.html

TinyERP. (2006). Retrieved February 5, 2006, from http://tinyerp.com.

Tribunella, T. J., Neely, M. P., & Triubunella, H. R. (2005). Academic and practitioner interests regarding emerging technologies in accounting. *Journal of College Teaching and Learning, 2*(5), 31-41.

Tustena CRM. (2006). Retrieved February 9, 2006, from http://www.tustena.com/crm/default.aspx

WebERP. (2006). Retrieved February 9, 2006, from http://www.weberp.org/index.php

Wheeler, D. A. Homepage. (2005). *Why OSS/FS? Look at the numbers.* Retrieved February 9, 2006, from http://www.dwheeler.com

Williams, J. R., Haka, S. F., Bettner, M. S., & Meigs, R. F. (2005). *Financial and managerial accounting: The basis for business decisions* (13th ed.). New York: McGraw Hill Irwin.

Zhao, L., & Elbaum, S. (2003). Quality assurance under the open source development model. *Journal of Systems and Software*, 66-75.

KEY TERMS

Accounting Information System (AIS): A subset of the management information systems composed of the people, processes, and assets that are responsible for the financial information of an organization. The AIS collects transaction data, monitors internal controls, and produces accounting information such as financial statements and budgets.

Auditing: An independent objective review and assessment of an organization's financial processes and information to validate that appropriate internal control processes are followed, that the information resulting from these processes is valid, and that risks are being monitored and responded to appropriately.

Cross-Functional Integration: The process of combining the various functional business activities within an organization by bridging the boundaries and enabling the flow of information among the various organizational functions.

Database Management System (DBMS): A specialized software package that serves as the repository of an organization's data. The DBMS organizes and manages the data so they are available to applications programs such as the accounting information system.

eXtensible Business Reporting Language (XBRL): XBRL is a specification for the reporting and communication of financial information. XBRL is an extension of extended markup language (XML). Financial information is de-scribed by a set of tags that is standardized for representing financial information and enabling its communication between information systems using the Internet.

Internal Control: The set of management procedures, either manually performed or automated by information systems, that are utilized to assure that an organization's management policies and procedures are adhered to and that the objectives of the organization are being achieved. Internal controls processes include monitoring risks and monitoring the reliability and quality of information within the organization.

Relational Data Model: A model in which data are viewed by users as two-dimensional tables. Each table represents an entity type such as a customer. Rows are instances of an entity, and columns are attributes of the entity. The tables relate or link to each other through shared attributes.

Request for Proposal (RFP): An RFP is a document utilized in the acquisition process for the purchase of software and services. An RFP documents the needs of the acquiring organization and defines all specific requirements that the acquiring organization has related to functionality, delivery time, post-acquisition support, additional services, and so forth. The RFP also defines specific requirements for vendors responding to the request, including information that is required in their requests, the deadline for responses, financial disclosure, security and intellectual property rights, and so forth.

Strategic Planning: A plan created by top management to achieve the general long-range vision and mission of the organization. This plan may include multi-year goals related to technology infrastructure, large capital projects, governance policies, financial budgets, and market share objectives.

Chapter XLIV
Open Source Software and the Corporate World

Sigrid Kelsey
Louisiana State University, USA

ABSTRACT

This chapter discusses various ways that open source software (OSS) methods of software development interact with the corporate world. The success achieved by many OSS products has produced a range of effects on the corporate world, and likewise, the corporate world influences the success of OSS. Many times, OSS products provide a quality product with strong support, providing competition to the corporate model of proprietary software. OSS has presented the corporate world with opportunities and ideas, prompting some companies to implement components from the OSS business model. Others have formed companies to support and distribute OSS products. The corporate world, in turn, affects OSS, from funding labs where OSS is developed to engaging in intellectual property disputes with OSS entities. The consumer of software is sometimes baffled by the differences in the two, often lacking understanding about the two models and how they interact. This chapter clarifies common misconceptions about the relationship between OSS and the corporate world and explains facets of the business models of software design to better inform potential consumers.

INTRODUCTION

Open source software (OSS) is impacting the corporate world in numerous ways, from providing software and competing with its proprietary software companies to changing the direction of the software industry. While some corporate giants are embracing the OSS business model, launching OSS projects of their own, and supporting existing OSS projects, others are vigorously competing with the OSS movement and

its products. Still others are capitalizing on successful OSS products by packaging, distributing, and providing support for them. Sharma et al. (2002) assert that the success of OSS is turning the software industry from a manufacturing to a service industry in which customers are paying more for support and service than for the product itself. In addition, the OSS model of production has gained recognition as an "important organizational innovation" (Lerner & Tirole, 2002, p. 1). Without a doubt, the OSS movement has had a substantial influence on the software industry and the corporate world.

BACKGROUND

Both the OSS and proprietary models of software productions have existed since the early days of software development. Unix, for example, was developed at Bell Laboratories in the late 1960s and early 1970s and distributed freely to universities during the 1970s. Unlike the altruistic motivations of many OSS products, the reason for Bell Laboratories' free distribution was to keep the "consent decree" that resulted from a 1956 antitrust litigation that prevented AT&T from marketing computing products (Vahalia, 1996). In fact, AT&T's 1979 announcement that it would commercialize UNIX prompted the University of California Berkeley to develop its own version, BSD UNIX (Lerner & Tirole, 2002). AT&T's move to make the cooperatively developed UNIX into a proprietary product came four years before Stallman's decision to develop GNU and General Public License.

By 1980, a business model for software had emerged, restricting the copying and redistribution of software by copyright. Bill Gates had already established himself as a supporter of this proprietary model, stating in his February 3, 1976, "An Open Letter to Hobbyists":

As the majority of hobbyists must be aware, most of you steal your software. Hardware must be paid for, but software is something to share. Who cares if the people who worked on it get paid? ... Is this fair? ... One thing you do do is prevent good software from being written. Who can afford to do professional work for nothing? (Gates, 1979)

Gates' letter indicates the differences in philosophy between proprietary and free software proponents that have existed since the early days of software development.

In 1984, computer scientist Richard Stallman, frustrated that all available operating systems were proprietary, quit his job at MIT to develop the GNU (pronounced guh-noo, a recursive acronym for GNU's Not Unix) system. His goal, in addition to developing a new operating system, was to change the way software was created and shared, giving users freedom to modify or add to programs, redistribute the programs with their changes, cooperate with each other, and form communities. Stallman also developed the concept of "copyleft" and the GNU General Public License (GPL) in 1989, publishing all of his work under that license. Copyleft gives software a copyright and users permission to change the software, add to it, and redistribute it, as long as it remains under the GPL terms. By preventing the software from entering the public domain, the GPL prevents users from turning free software into a proprietary derivative. Thus, the beginnings of the OSS movement were a reaction to the proprietary corporate model. In 1990, University of Helsinki student Linus Torvalds wrote the Linux kernel, releasing it under GPL, and filling the gap for a piece of Stallman's system still under development. Soon after, the Apache Web server was developed, providing an OSS application for Linux. This combination of software offered a new option to Internet service providers and e-commerce companies, which, until then, had only proprietary options.

Stallmans's Free Software Foundation Web page, reminding readers that free software means "free" as in "free speech," not as in "free beer" (Free Software Foundation, 2005), echoes a concept brought forth perhaps more eloquently by Thomas Jefferson and widely-quoted by OSS advocates that "ideas should freely spread from one to another over the globe, for the moral and mutual instruction of man, and improvement of his condition ... ." With the growth of the OSS movement, some of the values of the OSS culture have diversified, but freedom and sharing remain integral to its success and completely dissimilar to the proprietary model of development and distribution.

The corporate and OSS models and philosophies continued to influence one another and develop throughout the 1990s. The "open source" label came out of a 1998 meeting, and shortly thereafter, the Open Source Initiative was created. Also in 1998, the Digital Millennium Copyright Act (DMCA), criminalizing the production of software for the purpose of evading copyright, and the Sonny Bono Copyright Term Extension Act, extending U.S. copyright terms by 10 years, both passed. Despite the divergent directions the two movements were taking, the difference between free software and proprietary software has never been reducible to a battle between anti-corporate OSS proponents and the profiteering corporate world, as many people perceive. Corporate companies with a stake in the software industry have, in fact, navigated various approaches to succeed in an industry in which the motivations for developing software go beyond the commercial value of the product or ownership of intellectual property. While the two models of software production and distribution are competitive in many ways, it was also in the 1990s that it became common for commercial companies to interact with the OSS community (Lerner & Tirole, 2002).

During 1998, Torvalds appeared on the cover of *Forbes*, Netscape announced a decision to make the next version of its Web browser an OSS product, and IBM adopted the Apache Web server as the core of its Websphere line of products (O'Reilly, 1999). Like IBM, some corporate giants have chosen to use and support OSS voluntarily. Others have found it necessary to contribute in order to market products to Linux users; still others have fought intellectual property battles with OSS constituents. The relationship between the OSS and corporate cultures is complex, but it is clear that the OSS culture is making an impression on the corporate world, and vice versa.

As the OSS community has grown to include professionals, students, hobbyists, corporate giants, universities, and others, the freedom of ideas and sharing knowledge remains the crux of the OSS ideology. To integrate into the OSS culture, therefore, the corporate world must be willing to share its developments. This chapter summarizes some ways in which the OSS movement has motivated change in the software industry and corporate world, citing some specific examples of corporations reacting to OSS software and strategies in different ways, which serves to illustrate the larger picture.

MAIN FOCUS OF THE CHAPTER

Corporate Culture and Motivations of OSS Developers

A misconception often associated with OSS developers is that they are volunteer programmers, willing to "dedicate their time, skills, and knowledge to the OSS systems with no monetary benefits" (Ye & Kishida, 2003, p. 1). In fact, there are many money-making opportunities for open source developers, from providing software support to programming for companies or institutions using the software. While it is true that many OSS developers are paid to make the developments, Eric Raymond is quick to point out that while OSS developers may be paid for their contributions to the software, their salaries rarely depend on the sales value of their software (Raymond, 2001). OSS contributors may work for support companies, universities, and other organizations with motivations not attached to selling the software. This is a key difference between the OSS and proprietary software business models. Stallman's 1985 GNU Manifesto not only outlines his reasons for creating GNU but also offers some suggestions for how programmers can make money in an OSS environment.

While too often OSS advocacy is reduced to an anti-Microsoft position, the challenge that the OSS community has posed to the software giant does provide an illustration of the extent of the movement's success. For example, throughout the Microsoft Corporation antitrust case, Linux was a named threat to Windows domination, with Microsoft CEO Steve Ballmer referring to open source as a "cancer" (Microsoft Exec, 2006). In the midst of the antitrust trial, Eric Raymond became the recipient of two leaked internal Microsoft memos, posting them on a Web site and naming them the "Halloween Documents." The documents, acknowledged by Microsoft to be authentic but according to Raymond dismissed as an engineering study not defining company policy, discuss the success of Linux, acknowledging the achievements of the OSS movement and outlining strategies to "beat" Linux (Raymond, n.d.). Microsoft is not the only corporation combating the success of Linux and other OSS products.

The SCO Group is a software company currently involved in a number of disputes regarding intellectual property, including lawsuits with IBM, Red Hat, and Novell. SCO filed a complaint against IBM in March 2003 claiming that IBM has misappropriated SCO's proprietary knowledge by contributing to the GNU/Linux systems with code SCO claims to own, alleging damages of at least $1 billion. The result of the ongoing litigation will set a precedent for future cases.

In 2005, Columbia University law professor Eben Moglin formed the Software Freedom Law Center to help protect OSS development from similar litigation. The center provides pro bono legal services to FOSS projects and developers; its mission to help provide FOSS developers with "an environment in which liability and other legal issues do not impede their important public service work. The Software Freedom Law Center (SFLC) provides legal representation and other law-related services to protect and advance FOSS." His foundation is one of several helping to defray legal costs for litigation against FOSS

developments. The Open Source Development Labs (OSDL) Linux Legal Defense Fund has raised more than $10 million to provide legal support for Linus Torvalds and others subject to SCO litigation (Goth, 2005, p. 3).

While Microsoft and SCO are resisting the OSS model of business, others in the corporate world have come to see the benefit of working with OSS producers and products. Silicon Valley's NetApp, for example, became involved in Linux because its Linux-using customers were experiencing difficulty moving files between their computers and NetApp filers. Although it was a Linux problem, customers complained to NetApp, and with a vested interest in fixing it, NetApp cofounder and chief of engineering talked to Linus Torvalds. Mistrustful of companies like NetApp, Torvalds declined NetApp's offer to fix the problem, naming instead his choice programmer for the job, Trond Myklebust. NetApp, along with Linux developers worldwide, could submit suggestions to Myklebust in hopes that he would accept them. Therefore, if NetApp was to market its product to Linux users, it was obliged to join the OSS culture (Lyons, 2004). The NetApp circumstances demonstrate that any company wishing to make its product compatible on a Linux platform has a stake in the OSS world. Yet while the OSS culture is able to influence the actions of the corporate world, so the corporate world is able to do likewise.

In 1999, NetApp began funding the University of Michigan's Center for Information Technology Integration (CITI), home to a lot of Linux NFS development. By 2002, NetApp was paying Myklebust a stipend and providing him office space in the lab and a company-paid apartment in Ann Arbor. Peter Honeyman, scientific director of the lab where Myklebust works who receives $192,000 a year from NetApp, notes, "What's in it for [NetApp] is sales; it can sell into the Linux market. This is not about philanthropy. There is plenty of mutual benefit going on here" (quoted in Lyons, 2004). Torvalds, who was mistrustful of NetApps' offer to help, works at a Beaverton,

Oregon, lab funded in part by Hewlett-Packard (Lyons, 2004). In response to the apparent conflict between picking up a salary from a revenue-hungry corporation and developing OSS, Torvalds compares himself to an athlete with a corporate sponsor (Lyons, 2004).

The OSS culture, therefore, is not separate from the proprietary model of development; rather, the two models interact with and influence each other in many ways. Indeed, the OSS, corporate, and academic worlds have a complex relationship, each able to control, to some extent, the others' directions.

A 2004 *Forbes* article notes that many top technical firms hire Linux programmers in hopes of manipulating the direction of Linux development (Lyons, 2004). Hewlett Packard Vice President Martin Fink acknowledges that the closer he can get to Torvalds, the more influence he can have on Linux, saying "I try to keep it under two hops. ... The way to get stuff done in the Linux community is to hire the right people." In 2003, Hewlett Packard generated $2.5 billion in Linux-related revenue; IBM $2 billion; and Red Hat, which distributes a version of Linux, $125 million in revenues. Linux runs in datacenters of places like Charles Schwab & Co. and Sabre Holdings (Lyons, 2004). These corporations have recognized the benefits of the OSS culture, and many of them have become sponsors of its research.

IBM has been a powerful corporate advocate of OSS development for years. In 2005, IBM promised free use of 500 of its U.S. patents to open source developers, stating, "The open source community has been at the forefront of innovation and we are taking this action to encourage additional innovation of open platforms" (IBM, 2005). IBM's Bob Sutor, vice president of standards, says that this move was made in hopes of starting a "patent commons" for companies to contribute intellectual property for open source developers to use freely without fear of litigation (Goth, 2005, p.4) from companies like SCO. OSS supporters

generally believe that software patents hinder advancements in software research; in Europe, efforts are underway to prevent laws that would allow the patenting of software (Carver, 2005).

With IBM and other such corporations designing their products with OSS platforms in mind, and contributing to the furthering of OSS research, it is plain that the two cultures have learned to work together.

Reliable Code, Reliable Support

The success of OSS projects like Linux, Apache, and Perl evince the success of the bazaar model on the code itself. The traditional paradigm of collaborative development follows Brooks' Law, which ascertains that only a select circle of experts should be allowed to collaborate, with little or no feedback, to improve a product before it is finished. Brooks' Law states, "Conceptual integrity in turn dictates that the design must proceed from one mind, or a very small number of agreeing resonant minds" (Neus & Scherf, 2005, p. 216). Eric Raymond dubs the bazaar approach Linus' Law in which software is released early and often, evolving as users around the world use it and contribute to it. Making the code freely available and open to review by one's peers makes the quality better (Bergquist & Ljungberg, 2001). Open source has proved itself to be a formidable model for creating quality software, and as OSS projects become even more widely adopted, the culture and communities grow larger.

Customers often question the availability or longevity of support available for OSS. Without a revenue-generating company supporting it, it is difficult for OSS newcomers to imagine that any support will exist. But the success of OSS projects like Linux, GIMP, and Apache provide examples of the bazaar model's success. In fact, with proprietary software, the support is proprietary as well, where anyone who is able to provide support for OSS is free to do so. O'Reilly Media,

Inc., a strong supporter and early advocate of OSS, points out on its Web site that its success came in part because it was not "afraid to say in print that a vendor's technology didn't work as advertised" (O'Reilly, n.d.). Besides publishing numerous support books, O'Reilly provides online services and hosts OSS summits and conferences. O'Reilly is not the only company providing support for OSS. The OSS culture of sharing and helping gives assurance that with any successful OSS product, adequate support is available.

FUTURE TRENDS

A July 2005 article reports that 70% of Web servers on the Internet use Apache compared to roughly 25% using Microsoft's Internet Information Server (Bradbury, 2005). Already, European governments have adopted OSS for their computing needs, and California has started a U.S. trend toward the same, making a 2004 recommendation for the use of OSS in its performance reviews. Products like OpenOffice, named by Developer.com as a 2006 Open Source Product of the year, offer products that are able to compete with Microsoft's Office Suite. The state of Massachusetts is currently deciding whether to go forward with a decision made by the former CIO to use OpenOffice, with Harvard Law School Professor John Palfrey predicting, "If Massachusetts gets this right, others will follow" (McMillan, 2005).

OpenOffice is already common in Israel, in part because OpenOffice works well with the Hebrew language and because Microsoft software is expensive. China, where software theft has discouraged proprietary companies from marketing, has embraced OSS, creating the China Standard Software Company (CSSC) and the China Open Source Software Promotion Alliance (Bradbury, 2005). The growing trend to adopt OSS has spread worldwide. With such support, the OSS culture and movement will continue to grow. Market re-

searcher IDC predicts that by 2008, Linux server sales could approach $10 billion.

The GPL has yet to be ruled enforceable in a U.S. court of law; until now, it has only been enforced in private negotiation or settlement agreements (Carver, 2005). In Germany, however, a Munich district court has ruled it valid and enforceable (Carver, 2005). The result of ongoing litigation between SCO and IBM will set a precedent for how the GPL is interpreted in the United States.

Lerner and Tirole acknowledge that the future of the open source development process is difficult to predict with existing economic models and that further research is needed from an economic perspective.

CONCLUSION

The predominating shared norms, values, attitudes, and behavior that characterize OSS culture are deeply rooted in valuing freedom and sharing. As OSS has grown to offer software options for large entities like governments, companies, and universities, reasons for joining the OSS movement diversify. While the movement has grown and the culture has shifted, the basic values have remained in tact. Its success has impacted other cultures and traditions worldwide, from academic publishing and research to government to the corporate world. Clearly, the initial ideas and philosophies set forth by Jefferson and echoed by Stallman are affecting the culture of research worldwide, with the OSS movement proof that a culture of sharing is beneficial to everyone involved.

REFERENCES

Association of College and Research Libraries (ACRL), Association of Research Libraries (ARL), SPARC, & SPARC Europe. 2003. *Create: New systems of scholarly communications; change: old systems of scholarly communication.* Retrieved from http://www.createchange.org/resources/CreateChange2003.pdf

Bergquist, M., & Ljungberg, J. (2001). The power of gifts: Organizing social relationships in open source communities. *Information Systems Journal, 11*(4), 305.

Bradbury, D. (2005). *The future is open source: Should Microsoft be watching its back?* Retrieved January 18, 2006, from http://www.silicon.com/research/specialreports/opensource/0,3800004943,39150625,00.htm

Carver, B. W. (2005). Share and share alike: Understanding and enforcing open source and free software licenses. *Berkeley Technology Law Journal, 20*(1), 443.

Free Software Foundation. (2005). *The free software definition.* Retrieved December 1, 2005, from http://www.fsf.org/licensing/essays/free-sw.html

Gates, W. H., III. (1979). *An open letter to hobbyists.* Retrieved July 7, 2006, from http://www.digibarn.com/collections/newsletters/homebrew/V2_01/homebrew_V2_01_p2.jpg

Goth, G. (2005). Open source infrastructure solidifying quickly. *IEEE Distributed Systems Online, 6*(3), 1.

Hars, A., & Shaosong, O. (2001). *Working for free? Motivations of participating in open source projects.*

IBM. (2005). *IBM statement of non-assertion of named patents against OSS.* Retrieved January 29, 2006, from http://www.ibm.com/ibm/licensing/patents/pledgedpatents.pdf

Lattemann, C., & Stieglitz, S. (2005). Framework for governance in open source communities. In *Proceedings of the 38ᵗʰ Hawaii International Conference on System Science.*

Lerner, J., & Tirole, J. (2002). Some simple economics of open source. *Journal of Industrial Economics, 50*(2), 197.

Lyons, D. (2004). Peace, love and paychecks. *Forbes, 174*(5), 180. Retrieved July 11, 2006, from http://www.msnbc.msn.com/id/5907194/

McMillan, R. (2005). CIO who brought OpenOffice to Massachusetts resigns: Whether Peter Quinn's departure helps or hinders state's move away from Microsoft remains to be seen. *InfoWorld.* Retrieved January 29, 2006, from http://ww6.infoworld.com/products/print_friendly.jsp?link=/article/05/12/28/HNmasscio_1.html

Microsoft exec leaves suddenly: Windows live official helped company fight Linux threat (2006, June 21). *The Seattle Post-Intelligencer*, p. E1.

Moore, J. (2003). *Revolution OS: Hackers, programmers & rebels unite.* In W. Productions (Producer): Seventh Art Releasing.

Neus, A., & Scherf, P. (2005). Opening minds: Cultural change with the introduction of open source collaboration methods. *IBM Systems Journal, 44*(2), 215.

Open Source Initiative. (n.d.). *The open source definition.* Retrieved December 1, 2005, from http://www.opensource.org/docs/definition.php

O'Reilly, T. (n.d.). *O'Reilly media: History.* Retrieved July 11, 2006, from http://www.oreilly.com/history.html

O'Reilly, T. (1999). Lessons from open source software development. *Communications of the ACM, 42*(4), 32-37.

Raymond, E. S. (n.d.). *The halloween documents.* Retrieved January 30, 2006, from http://www.catb.org/~esr/halloween/index.html

Raymond, E. S. (2001a). *The cathedral & the bazaar.* Beijing: O'Reilly.

Raymond, E. S. (2001b). *How to become a hacker.* Retrieved January 24, 2006, from http://www.catb.org/~esr/faqs/hacker-howto.html#what_is

Rheingold, H. (1994). *Virtual community.* London: Minerva.

Sharma, S., Sugumaran, V., & Rajagopalan, R. (2002). A framework for creating hybrid-open source software communities. *Information Systems Journal, 12,* 7.

Stallman, R. (1985). *The GNU manifesto.* Retrieved from http://www.gnu.org/gnu/manifesto.html

Unsworth, J. M. (2004). The next wave: Liberation technology. *The Chronicle of Higher Education, 50*(21), B16-B20.

Vahalia, U. (1996). *UNIX internals: The new frontiers.* Upper Saddle River, NJ: Prentice-Hall, Inc.

Ye, Y., & Kishida, K. (2003). Toward an understanding of the motivation of open source software developers. In *Proceedings of the 25th International Conference on Software Engineering (ICSE'03)* (pp. 419-429).

KEY TERMS

Free Software (FS): Software that users have the freedom to alter, use, and redistribute, usually under the terms of the General Public License. Closely related to Open Source Software, the two terms are sometimes used interchangeably. "Free" is not associated with cost but with the freedom associated with it. However, free software is often cost-free as well.

General Public License (GPL): A license created by Richard Stallman that protects free software from being turned into proprietary software.

Open Source Software (OSS): Software that allows the user to see and alter the source code; closely related to free software.

Proprietary Software (PS): Software that does not allow the user to see or alter the source code.

Chapter XLV
Business Models in Open Source Software Value Creation

Marko Seppänen
Tampere University of Technology, Finland

Nina Helander
Tampere University of Technology, Finland

Saku Mäkinen
Tampere University of Technology, Finland

ABSTRACT

This chapter explores how the use of a business model enables value creation in the open source software (OSS) environment. We argue that this value can be attained by analyzing the value creation logic and the elements of potential business models emerging in the OSS environment, since profitable business is all about creating value and capturing it properly. Open source (OS) offers one possibility for firms that are continuously finding new opportunities to organize their business activities and increase the amount of value they appropriate according to their capabilities. Furthermore, the concept of a business model is considered a tool for exploring new business ideas and capturing the essential elements of each alternative. We propose that a general business model is also applicable in the context of OSS, and we provide a list of questions that may help managers deal with OSS in their businesses.

INTRODUCTION

Firms have recognized an increasing need to improve their abilities to change the way their business operations are organized. Thus, they assess new business opportunities and evaluate them in terms of whether they would suit the firm's business portfolio. A business model is considered a tool for exploring new business ideas and capturing the essential elements of each alternative. It is a construct for mediating technologies' development and economic value creation; in other words, it is an abstract representation of the business logic of a company. OS is a phenomenon that almost every company has encountered in the last couple of years. Obviously, it offers opportunities for the creation of new business, and thus, exploring the types of alternatives it may offer for value creation is a subject of growing interest.

We begin the chapter with a brief discussion of value creation and business models, which are applied and analyzed in the special context of the OSS environment. We argue that a general business model typical of proprietary software business is also applicable in the context of OSS. However, the elements of such a business model appear and are implemented in the OSS context in a different way than in the proprietary software business. One reason for this is that the value created in an OSS project often cannot be owned by single companies. This argument of the differences between OSS and proprietary software business forms the starting point of our analysis and is taken into account throughout the chapter.

The objective of this chapter is to explore how use of a business model enables value creation within the OSS environment. We argue that this value can be attained by analyzing the value creation logic and the elements of potential business models emerging in the OSS environment, since profitable business is all about creating value and capturing it properly. Firms are continuously finding new opportunities to organize their business activities and increase the amount of value they appropriate according to their capabilities. OS may offer one possibility for this.

BACKGROUND

Differences between Business Based on Proprietary and Open Source Software

In our examination, we have distinguished the three most salient points separating proprietary and OS software as (1) OS and licenses, (2) networks and their actors, and (3) the customer.

The main differences emerge from the openness of source code and licenses. OS code enables anyone to further develop the original code, and the license ensures that the will of the original developer holds. With proprietary software, the source code is not available, and typical licenses restrict utilization of the source code to only the commercial supplier of the software. Woods and Guliani (2005) stated, "The most important difference between software created by the OS communities and commercial software sold by vendors is that OSS is published under licenses that ensure that the source code is available to everyone to inspect, change, download, and explore as they wish. This is the essential meaning of open source: the source code ... can be obtained and improved by anyone with the right skills."

The openness and availability of the source code further mean that the value in OS projects is created for the network, not for individual companies or other entities or individuals. As it is, the business models of the companies involved in OSS projects must be linked to the business models of other network actors and perhaps include components outside the network. Thus, management of network relationships has a key role in OS business operations (Dahlander & Magnusson, 2005).

The idea is that by openly sharing the software code with others, each actor can do the part it does best and the cooperative effort's outcome is characterized by high quality. Additionally, when all actors have had the opportunity to do those parts of the development work that are nearest their respective core competencies, the development work usually feels easy, fun, and rewarding (Torvalds, 2001). A noteworthy feature of OSS is that the knowledge to create the product is not in the hands of firms but resides within various actors in the firm. Posing a challenge for utilization of this knowledge is that actors involved in OSS networks sometimes have very contradictory intentions and expectations. For example, firms usually are more focused on the issue of monetary value, while many of the coders participating in the OSS community find that money is not the first or even sometimes the last, motivator.

Additionally, when we consider the issue of creating value for the end customer, the role of the customer in the OSS environment is not always clear. In principle, all of the software coders can be understood as customers since they develop software for their own use. It is often claimed that a good OS project starts "by scratching a developer's personal itch." Apart from that, the coders seldom think in terms of specific customers for their projects; instead, all who want to utilize their software are free to do so. Thus, customer segmentation, while a typical consideration in proprietary software business networks, is not considered in OSS communities. More detailed analyses of the differences can be found in the works of Kooths et al. (2003, pp. 74-79) and Lerner and Tirole (2004), who reviewed the multidimensional nature of differences between the proprietary and OS approach to the software business.

Perspectives on Value Creation

In this section, value creation is discussed from the monetary and nonmonetary standpoint and in terms of various value creation functions and evaluation criteria; finally, it is discussed as something related to both the object of exchange and the interactive relationship between customer and supplier. During the interaction, the value is perceived by both parties.

While both academics and actors in the field commonly make use of the concept of value, it is often unclear what is actually meant by it in different contexts (Ford & McDowell, 1999; Lindgreen & Wynstra, 2005; Ramsay, 2005; Woodall, 2003). From a rather broad perspective, the concept of value can be regarded as the trade-off between benefits and sacrifices (Berry & Yadav, 1996; Lapierre, 2000; Parolini, 1999; Ravald & Grönroos, 1996; Slater, 1997; Walter, Ritter, &

Gemünden, 2001). These costs and benefits can be understood in monetary terms, but they can also be seen as including nonmonetary rewards such as competence, market position, and social rewards (Walter et al., 2001). Nonmonetary costs might include time, effort, energy, and conflict invested by the customer to obtain the product or service.

Both monetary and nonmonetary viewpoints are also visible in the analysis of direct and indirect value creation functions. According to Walter et al. (2001), the direct value creation functions are volume, profit, and safeguarding functions, while innovation, marketing, scouting, and access functions are indirect functions. Volume and profit functions are usually easier to measure in monetary terms, whereas the other value functions are basically nonmonetary in nature, although in the end, they should liquidate to money.

These monetary and nonmonetary costs and benefits, however, are eventually evaluated in the mind of the customer. Parolini (1999) discusses absolute and differential value, the latter of which should be understood as dependent on the customer's own expectations and evaluations. Thus, value is something the customer perceives. Furthermore, the customer always perceives the value of a certain product or service in relation to other possible solutions (Parolini, 1999).

Additionally, Parolini (1999) discusses the various criteria the customer can use in considering the value of products and services. These criteria for products and services are whether they improve the performance of the customer or reduce costs; whether they are hygienic or motivating; whether they are under the control of whomever is performing the analysis; and whether the costs are borne before, during, or after the purchase.

Value before (i.e., potential value), during (i.e., exchange value) and after (i.e., use value) purchase are important elements to take into account when discussing value creation. Value creation should be understood as a process during which the customer and supplier interact. During the interaction, the product or service is exchanged between the parties, and the benefits and sacrifices are thus realized. However, there is also a great amount of interaction between the parties in the relationship that is not directly related to the object of exchange. This interaction, however, usually does influence how the customer perceives the total value gained.

To be more precise, the benefits and sacrifices, whether understood in monetary or nonmonetary terms, are related naturally to the product or service that is exchanged between the supplier and the customer, as Reidenbach, Reginald, and McClung (2000) suggest when they define value as "the interaction between the benefits that customers want from a particular product/service and the price they are willing to pay to acquire the benefits provided by that product/service." However, Thomas and Wilson (2003) suggest that consideration of benefits and sacrifices should not be limited only to something related to the object of exchange; instead, they say, value should be considered also in relation to the benefits and sacrifices that occur in/from the relationship between the supplier and the customer. In other words, customers do not perceive the value merely through the object of exchange; they also take into account the whole relationship with the supplier as an influence on the amount of perceived net value (Lindgreen & Wynstra, 2005).

Kothandaraman and Wilson (2001) also address the issue of understanding value creation related to the product as well as to the overall process through which the product is developed, marketed, and delivered to the customer. Understanding value creation as a process between the supplier and the customer makes visible the relevant roles of both the customer and the supplier. We argue that just as it is not enough to study a relationship from the viewpoint of one party alone, the analysis of value creation also should not focus on only the customer's perspective, the latter being, unfortunately, the main area of concentration in recent literature (for refreshing

Figure 1. Perceived value and effort of the firm with proprietary vs. open source software

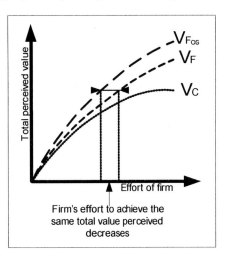

exceptions, see Möller & Törrönen, 2003; Walter et al., 2001).

The customer and the supplier both have their own views and influences on the value that is created, and both parties also want to capture their own share of the value. Figure 1 illustrates the viewpoint of both the customer and the supplier in value creation.

In order for a business to be profitable, the value captured by the supplier (denoted by V_F) should be higher than that created for the customer (V_C). We argue that the value perceived by a customer may change due to use of OS but not in every case. The value may change; for instance, the customer may perceive the utilization of OS components as more valuable than proprietary components for ideological reasons. We argue that the effects of utilizing OS components, nevertheless, may result in higher perceived value for the supplier firm. If, for example, a firm uses OS components in its product development, it may achieve either more value with the same effort or the same value with less effort (see in Figure 1 the difference in effort level between V_{Fos} and V_F).

The successfulness of the interaction between the supplier and the customer influences the net value perceived by the counterparts also in the OS environment. Thus, the supplier needs to keep in mind that it is not only the functionality of the actual object of exchange (i.e., the software), but it is the services offered around the software and the whole relationship with the supplier that influence the value perception of the customer. In many OS cases, the supplier, in fact, is offering services to the customer, while the source code as the actual object of exchange could be acquired by the customer directly from the specific OS project. The customer is actually acquiring the software from the specific supplier because he or she trusts the ability of the supplier to create more value in the form of smooth cooperation, upgrading, and maintenance services.

MAIN FOCUS OF THE CHAPTER: PURPOSE AND ELEMENTS OF BUSINESS MODELS

The Purpose of a Business Model

A business model is seen as a tool for exploring new business ideas and capturing the essence of each alternative. It is an abstract representation for mediating the development of technology and economic value creation. The business model concept often is discussed at only a superficial level (Porter, 2001). The model could be a tool allowing different strategic alternatives to be examined and developed before actions are taken, such as a shift in strategy or other change. There has been much confusion about the division of tasks between a strategy and a business model. Some have even considered the two concepts to be synonymous, while others have strongly argued that business models should include strategic aspects. By definition, a business model should encompass the business logic of a company. Still other authors have not seen any use for the latter concept, viewing it as *the emperor's new clothes*.

Although the concepts of a business model and strategy are highly complementary, they are not the same. A strategy focuses on value appropriation, while a business model explains how value is created for all stakeholders. Chesbrough and Rosenbloom (2002) made three clear distinctions between the two. First, a business model is based on value creation for the customer, but emphasis on capturing that value and sustaining it is part of the scope of a strategy. Second, financing of the value creation is implicitly assumed in business models, whereas a strategy explicitly considers the financing issues of value creation because of the underlying assumptions of shareholder value creation. Finally, there is a difference in the assumptions about the state of knowledge held by the firm and its stakeholders. Business models consciously assume limited and distorted information and knowledge, while a strategy is built on analysis and refinements in knowledge and, therefore, assumes the existence of a plentitude of reliable information to be transformed into knowledge. A practical distinction describes business models as a system that shows how the pieces of a business fit together, while strategy also includes competition (Magretta, 2002).

We specify the purposes of a business model in accordance with the view of Chesbrough and Rosenbloom (2002), who argue that the functions of a business model are as follows:

- To articulate the value proposition
- To identify a market segment
- To define the structure of the value chain within the firm
- To estimate the cost structure and profit potential
- To describe the position of the firm within the value network
- To formulate the competitive strategy

An explicit business model makes visible at least some of the invisible assumptions made during the design of a model. We are able to visualize what boundaries guide our thinking processes and may also restrict applicability. Moreover, the prerequisites for success of the business model in question may become clearer.

The Elements of a Business Model

An important consideration is the context-specificity of a business model. Is there need for specific models that are targeted to a particular industry? Regardless of the several industry-related papers devoted to these (i.e., business models for e-business) (Rappa, 2003; Rayport & Jaworski, 2001; Weill & Vitale, 2001), we propose that there is no need for a context-specific business model. A generic business model should involve the same elements, regardless of the industry in which the model is used. A context-specific model (perhaps with a prefix) should be seen as a local application of a general business model.

OSS as a phenomenon does not require any special business model as such. A generic business model could act as well in that environment

as in any other. Indeed, the requirements for such a business model are the same as in the general case. Yet, of course, the application of a general model reflects the characteristics of this particular business environment. For example, Timmers (2003) pointed out that the focus shifts from creating value through internal activities to creating value through external relations, and the number of relationships multiplies. He proposed that these relationships within the value-creating network are an inseparable part of the business model of a firm.

We propose that the generic business model of Osterwalder, Pigneur, and Tucci (2005) has all of the elements needed to fulfill the aforementioned purposes for a business model (see Figure 2).

For some readers, these elements may ring a bell. Indeed, "business plan" is used sometimes as a synonym for "business model." Business plans are useful tools for developing new businesses. However, they are a bit too heavy and inflexible for considering new business practices, and, as practice has many times demonstrated, they cannot show how the whole business should function

Figure 2. The elements and structure of a business model (Source: Modified from Osterwalder et al., 2005)

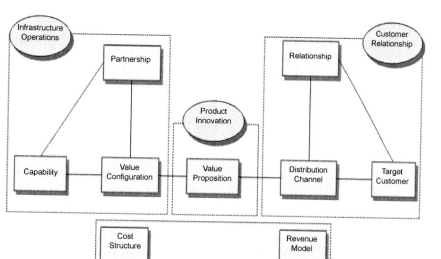

Table 1. Business model elements and example questions in OSS

Element	Description	Questions
Value Proposition	Gives an overall view of a company's bundle of products and services	Does the utilization of OSS affect the way the customer perceives our value offering? How do we take OSS into account in customer marketing?
Target Customer	Describes the market segments to which a company wants to offer value	Who are our target customers? Does OSS impose restrictions or provide wider access for certain market segments?
Distribution Channel	Describes the company's the various means of getting in touch with its customers	Could we use open distribution (Sourceforge or other), our own web site, or other servers? How are potential utilizers going to find us or our product?
Relationship	Explains the kind of links a company establishes between itself and its different customer groups	What is an appropriate OS license to use? What kind of relationship are we going to create with the community?
Value Configuration	Describes the arrangement of activities and resources	What is our role in the community? How do we share resources and carry out activities with the other actors and community players?
Core Competency	Outlines the competencies necessary to actualize the company's business model	What are the competencies we especially seek from and can offer to the community? How are we to manage relationships and maintain sustainable development?
Partner network	Portrays the network of cooperative agreements with other companies that are necessary for efficiently offering and commercializing value	What kinds of agreements are we going to make with various participants? How does utilization of OSS affect our partners outside the community?
Cost Structure	Sums up the monetary consequences of the means employed in the business model	What kind of cost structure do the aforementioned choices involve? Are we able to cope with the economic consequences?
Revenue Model	Describes the way a company makes money through a variety of revenue flows	What revenue models should we choose? Do we prepare revenue and risk sharing models?

(i.e., indicating what mechanism makes the business idea in question tick). Table 1 explains how the suggested business model approach functions. When a firm is considering utilization of OSS, it should answer certain questions. The example questions are presented separately for each element of a generic business model, and the elements therein are also described in more detail.

Product Innovation

The value proposition explains the overall bundle of products and services offered by the company. It typically starts from an innovation or product idea, and the aforementioned considerations of value creation are tightly linked to this element. The firm's ability to interpret its own intentions correctly is crucial. If the value proposition remains vague or even misleading, this may predict difficulties in the implementation of the rest of the business model, and in the worst case, desired success in the market does not occur.

Customer Relationships

First, selection of target customer(s) is closely linked to the value proposition element. If, for example, a firm chooses a developer group or a community as its primary target customer, it must understand who the secondary customers (the primary customer's customers) are and the nature of their needs. In other words, selection of target customer demarcates what kinds of value propositions are within range. Second, decisions concerning distribution channels mainly specify how the targeted market segment is to be reached. These decisions involve marketing, communication, and advertising. For example, Red Hat has put a lot of effort into advertising in order to reach its potential customers. Third, license policy and attitudes toward OS communities are the main items falling under the relationships element. Choosing an appropriate license may reflect the possibility a certain business idea has of success.

Infrastructure Operations

The first thing that must be understood in planning and starting to negotiate with suitable communities is the "onion model of communities" (Nakakoji, Yamamoto, Nishinaka, Kishida, & Ye, 2002). There may be a multitude of people linked loosely to a particular community, but not all are equal in importance. A firm, if it is to understand what kind of decision-making mechanism a community utilizes, must find a successful way to cooperate (Crowston & Howison, 2005; Mockus, Fielding, & Herbsleb, 2000; Raymond, 2000). For instance, IBM's way to cooperate with Eclipse community seems to work very well. IBM supports the community's development by donations and by hiring experts to work for community purposes; these actions also mutually support IBM's business purposes. Second, to benefit from what OS may offer, a firm must be able to carry out operations such as searches, evaluations, and negotiations. Thus, the core competencies of such a firm must lie in these areas. It has been recognized that it is not very easy to assess the maturity of an OS project (Comino, Manenti, & Parisi, 2005; Woods & Guliani, 2005). Finally, a partnership network actually is created on the basis of decisions made about earlier elements. It is in considering this element that decisions concerning agreements between parties are made. Agreements typically include descriptions of responsibilities and documents outlining the sharing of revenues and risks, for example.

Financial Aspects

Cost structure, as well, is based on choices made earlier. The efficiency of the firm's operations (activities such as search and evaluation) determines the company's internal cost structure. External cost structure is based mainly on the selection of licenses, partners, and customers. Finally, revenue model is based on the value proposition, the choice of target customer, license model, and other environmental elements. A famous example of successful licensing model is MySQL and its dual licensing. To develop and distribute OS applications under a GPL license, it is free to use MySQL, whereas a commercial license is offered for business purposes. There is no consensus as to whether some revenue models are better than others; one can only go by examples of what may create a successful business.

Appearances of the elements of the generic business model presented here differ not only between proprietary and OSS contexts but also among types of OS intensive firms. Some firms are more involved with OSS than others. For example, some utilize OSS tools in their own software development; others use OSS components as part of a system solution sold to end customers, and some firms are built entirely on OSS. Furthermore, the skills required can vary widely, depending on the maturity of the OS project (Woods & Guliani, 2005). In an OS business, a firm should recognize its own position desired and skill level needed. When the OS world evolves, OS expertise is becoming increasingly for sale, and thus, firms have more opportunities to buy expertise that they do not possess by themselves.

We argue that the generic business model presented in this chapter is applicable in considering each of these uses of OSS. It is just the way in which OSS affects the elements of the business model that varies. For example, a firm that utilizes OSS as a software development tool does not need to think about most of the questions presented in Table 1; whereas a firm that has built all of its operations on OSS needs to consider all of the suggested questions.

DISCUSSION AND FUTURE TRENDS

Academics and businesspeople often speak of business models when they really mean only parts of one (Linder & Cantrell, 2000). The basic message of this chapter is that the elements of a business model remain the same regardless of industry. These elements are necessary if the model is to cover all aspects of business. Choosing a license is a very important part of a business model creation, but the license on its own does not dictate the business model. We have already demonstrated how these elements can be considered in the OS environment from a business perspective.

One of the premises for a successful business is that the value perceived by the customer must be higher than the monetary counterpart—price. Traditionally, only when this is the case may a monetary transaction occur. The very essence of a business model is that it is a construct mediating the creation of value from a technological

potential. Thus, the concept of value must be regarded as multidimensional; and perceived value, in particular, as seen from both customer and supplier perspectives, is important when one considers value proposition. As Raymond (2000) pointed out in his seminal book, the developmental work for an OS project should be executed according to the top-down principle, not bottom-up. Therefore, managers can remember as a basic guideline that the firm should first be very clear as to what needs it hopes to address by taking part in or even simply utilizing OSS. When these basic questions have been answered, the process may proceed further.

The suggested business model is, in its current form, still somewhat abstract. However, it nonetheless may aid in structuring, thinking, visualizing, and further developing the planned mode of operations. Some authors have already developed computer-aided tools to assist in the implementation phase (Gordijn, 2004; Osterwalder et al., 2005). Another deficiency in the model is the weak link to operations. The nine elements should be further developed and grounded soundly in existing theoretical frameworks. Although some authors (Fogel, 2006; Woods & Guliani, 2005) have already offered practical viewpoints and guidelines for managers, academic research should take a more normative direction. Existing business models are typically only descriptive, whereas managers call for normative guidelines that could help in daily work. Linking elements more closely to the operational level of a firm would be of use in finding ways to operationalize and finally implement a particular business model. At the moment, this can give us only a static picture of business.

When the number of firms involved in OS increases, interest in the issue of value creation with OSS will grow as well. Business model discussion then will be a key area of interest, since a business model is a tool for value creation. One avenue for future research involves not limiting the discussion to critical analysis of successful business models for single firms but, instead, addressing the matter of how the elements of business models of several firms acting in the OSS network will engage and codevelop. In the network context, the question of firms and their ways of operating is only a starting point. The more interesting and potentially fruitful question is that of interaction and relationship management among the commercially oriented firms and the individual coders involved in the community.

CONCLUSION

In this chapter, we have addressed value creation and business models in the context of OSS. We defined the elements of value and a business model, and additionally provided a list of questions that should help managers deal with OSS in their businesses. We also have made comparisons between businesses with operations based on proprietary software and OSS. Although there are clear differences between the two, there is no need for a new kind of generic business model.

The generic business model suggested in this chapter is also applicable in the OSS field; only the emphasis and appearance of the elements of the model may vary. All in all, nonmonetary value and voluntary value division between network actors is typical of OSS business, meaning further that there are differences in value creation logic between businesses based on OSS and ones centered on proprietary code. This also causes variations within the elements of the business model.

REFERENCES

Berry, L. L., & Yadav, M. S. (1996, Summer). Capture and communicate value in the pricing of services. *Sloan Management Review*, 41-51.

Chesbrough, H., & Rosenbloom, R. S. (2002). The role of the business model in capturing value from innovation: Evidence from Xerox Corporation's technology spin-off companies. *Industrial and Corporate Change, 11*(3), 529-555.

Comino, S., Manenti, F. M., & Parisi, M. L. (2005). *From planning to mature: On the determinants of open source take off.* Retrieved December 13, 2005, from http://opensource.mit.edu/

Crowston, K., & Howison, J. (2005). The social structure of free and open source software development. *First Monday, 10*(2). Retrieved from http://www.firstmonday.org/issues/issue10_2/crowston/index.html

Dahlander, L., & Magnusson, M.G. (2005). Relationships between open source software companies and communities: Observations from Nordic firms. *Research Policy, 34*, 481-493.

Fogel, K. (2006). *Producing open source software: How to run a successful free software project.* O'Reilly.

Ford, D., & McDowell, R. (1999). Managing business relationships by analyzing the effects and value of different actions. *Industrial Marketing Management, 28*, 429-442.

Gordijn, J. (2004). E-business model ontologies. In W. Curry (Ed.), *E-business modelling using the e3-value ontology* (pp. 98-128). Oxford, UK: Elsevier Butterworth-Heinemann.

Kooths, S., Langenfurth, M., & Kalwey, N. (2003). Open source software. An economic assessment. *MICE Economic Research Studies, 4*, 95.

Kothandaraman, P., & Wilson, D. T. (2001). The future of competition: Value-creating networks. *Industrial Marketing Management, 30*, 379-389.

Lapierre, J. (2000). Customer-perceived value in industrial contexts. *Journal of Business & Industrial Marketing, 15*(2/3), 122-140.

Lerner, J., & Tirole, J. (2004). *The economics of technology sharing: Open source and beyond*

(No. 10956). Retrieved February 22, 2005, from http://opensource.mit.edu/

Linder, J., & Cantrell, S. (2000). *Changing business models: Surveying the landscape.* Accenture Institute for Strategic Change. Retrieved March 14, 2004, from http://www.accenture.com/global/research_and_insights/

Lindgreen, A., & Wynstra, F. (2005). Value in business markets: What do we know? Where are we going? *Industrial Marketing Management, 34,* 732-748.

Magretta, J. (2002, May). Why business models matter. *Harvard Business Review,* 86-92.

Mockus, A., Fielding, R. T., & Herbsleb, J. (2000). A case study of open source software development: The Apache server. In *Proceedings of the ICSE.* Retrieved from http://opensource.mit.edu/

Möller, K. E. K., & Törrönen, P. (2003). Business suppliers' value creation potential: A conceptual analysis. *Industrial Marketing Management, 32,* 109-118.

Nakakoji, K., Yamamoto, Y., Nishinaka, Y., Kishida, K., & Ye, Y. (2002). Evolution patterns of open source software systems and communities. In *Proceedings of the International Workshop on Principles of Software Evolution,* Orlando, Florida. Retrieved from www.kid.rcast.u-tokyo.ac.jp/~kumiyo/mypapers/IWPSE2002.pdf

Osterwalder, A., Pigneur, Y., & Tucci, C. L. (2005). Clarifying business models: Origins, present, and future of the concept. *Communications of the Association for Information Systems, 16*(1). Retrieved from http://cais.aisnet.org/articles/default.asp?vol=16&art=1

Parolini, C. (1999). *The value net: A tool for competitive strategy.* West Sussex, UK: John Wiley & Sons.

Porter, M. E. (2001). Strategy and Internet. *Harvard Business Review, 79*(3), 62-78.

Ramsay, J. (2005). The real meaning of value in trading relationships. *International Journal of Operations & Production Management, 25*(6), 549-565.

Rappa, M. (2003). *Business models on the Web.* Retrieved November 9, 2004, from http://digital-enterprise.org/models/models.html

Ravald, A., & Grönroos, C. (1996). The value concept and relationship marketing. *European Journal of Marketing, 30*(2), 19-30.

Raymond, E. S. (2000). *The cathedral and the bazaar.* Sebastopol, CA: O'Reilly.

Rayport, J. F., & Jaworski, B. J. (2001). *E-commerce.* Boston: McGraw-Hill.

Reidenbach, R. E., Reginald, W. G., & McClung, G. W. (2000). *Dominating markets with value: Advances in customer management.* Morgantown, WV: Rhumb Line Publishing.

Slater, S. F. (1997). Developing a customer value-based theory of the firm. *Journal of the Academy of Marketing Science, 25*(2), 162-167.

Thomas, S., & Wilson, D. T. (2003). Creating and dividing value in a value creating network. In *Proceedings of the 2003 IMP conference.* Retrieved October 3, 2005, from http://www.impgroup.org/

Torvalds, L. (2001). *Just for fun.* Keuruu, Finland: Harper Collins.

Walter, A., Ritter, T., & Gemünden, H. G. (2001). Value creation in buyer-seller relationships: Theoretical considerations and empirical results from a supplier's perspective. *Industrial Marketing Management, 30*(4), 365-377.

Weill, P., & Vitale, M. R. (2001). *Place to space. Migrating to eBusiness models.* Boston: Harvard Business School Press.

Woodall, T. (2003). Conceptualising 'value for the customer': An attributional, structural and dispositional analysis. *Academy of Marketing Science Review, 12.* Retrieved from http://oxygen. vancouver.wsu.edu/amsrev/theory/woodall12-2003.html

Woods, D., & Guliani, G. (2005). *Open source for the enterprise. Managing risks, reaping rewards.* Sepastopol, CA: O'Reilly.

KEY TERMS

Business Model: A tool for exploring new business ideas and capturing the essence of each alternative.

Competitive Strategy: How a firm attracts customers, withstands competitive pressures, and strengthens the firm's market position.

Core Competency: The set of skills that an organization must perform well in order for the organization to be successful in comparison with its rivals.

Value Chain: The generic value-adding activities of an organization that provide an analysis tool for strategic planning.

Value Network: Three or more organizations strategically collaborate to create superior value to the end-customer.

Value Proposition: How an organization will differentiate itself to customers, and what particular set of values it will deliver.

Chapter XLVI
Novell's Open Source Evolution

Jacobus Andries du Preez
University of Pretoria, South Africa
Yocto Linux & OSS Business Solutions, South Africa

ABSTRACT

Novell, Inc. was a leading network operating system provider in the 1980s and early 1990s. However, in the mid-1990s, Novell lost market share in the network operating system market. To counter this loss of market share, Novell made a strategic decision to go open (i.e., to make use of open standards and open source business strategies). Novell employs a subscription strategy, selling subscriptions to its Linux desktop operating system called SuSE. Novell has subsequently successfully handled the changeover from being a proprietary network operating system provider to being a leader in Linux and open source solutions. For example, a comparison of the financial results of Novell's fourth quarters of 2004 and 2005 shows an increase of 418% in Linux revenue to US$61 million. Novell has demonstrated that open source business strategies are feasible and profitable.

INTRODUCTION

Novell, Inc. is one of only a few multinational organizations that originally produced proprietary software and is now driving and successfully implementing a free/libre open source software (FLOSS) business strategy. Since 1994, Novell has been actively making use of open standards and open source software (OSS) from both a technical and a business point of view. Today, Novell uses open source standards and software in its business strategy.

In recent years, researchers have taken a keen interest in the open source sphere and how it can be applied to business strategies and business models (Koenig, 2004; Raymond, 2000, August; Raymond, 2000, September).

A concern exists within the academic world that in this arena there is no substantial evidence on whether the processes and practices are effective within the business environment and whether the theories are not prematurely adopted in an enthusiastic manner (Bitzer & Schröder, 2004; Scacchi, 2004). Goode and Golden (2004; 2004) suggest that organizations are reluctant to be initial adopters of open source strategies without knowing whether or not OSS can bring substantial financial benefit to their organizations' business.

Raymond (2000, August) points out that by studying this question, one will gain valuable insight into the economics of open source use. Therefore, there is a need for studies to be done on organizations that have successfully implemented an OSS strategy (Raymond, 2000, August). Not enough practical core studies have been done based on any successful use and implementation of effective open source strategies and business models (Bitzer & Schröder, 2004; Scacchi, 2004).

This chapter is an attempt to fulfill the need for such a study and will hopefully prove that an open source business strategy is a feasible and profitable option. The outcome is based on a practical case study.

BACKGROUND

Apart from any studies done, Raymond (2000, August) suggests that organizations releasing their products as open source compel information technology organizations to focus on the service industry rather than on the product manufacturing industry.

Specifically, he suggests that Linux distributors should compete with each other in a manner that would benefit us all. They are required to compete on service and support rather than product and price. Legally and ethically, Linux distributors can only sell service, administration, support, distribution, media, training, and its brand to consumers and clients who are willing to comply with the terms and conditions of the GPL license under which the Linux kernel is licensed (Lerner & Tirole, 2005; Raymond, 2000, August).

Novell has followed a route that has allowed it to enter the open source market more effectively by providing Linux distribution and Linux support, and by selling proprietary software along with open source Linux distribution. This allowed Novell to profit from selected proprietary products as well as to enter the service industry. Novell is believed to have effectively entered the service

market and is considered a successful open source provider, having followed a systematic rather than a "big-bang" approach.

The intention in this study is to look at several factors to determine whether or not Novell has made a success of its One Net strategy (a world without information boundaries), which is mainly driven by OSS.

To do this, I will show that Novell actively changed from being a proprietary software provider to being mainly a service provider of open sources in particular, changing its strategy to deliver a business solution by making use of Linux and OSS. This study will examine Novell's corporate history, its public financial statements (10K filings), and interviews with Novell personnel to show that open source is a viable and alternative to proprietary software.

OSS Business Models

According to Young (1999), making money with OSS is very similar to making money with proprietary software. This is achieved by producing a good product, properly marketing it, taking care of one's customers' needs, and building a brand that represents excellent service and quality.

Hendry (2002) maintains that open source use enables companies to make money, save money, and form better business partnerships with greater compatibility by means of various credible business models.

Similarly, Dahlander (2004) contends that although contributions to the OSS process are public, this should not be misconstrued as meaning that innovators are prohibited from capturing private returns from their contributions. In other words, an enterprise can make money from open source use.

The benefits of using Linux, according to Young (1999), are not its ease of use, the operating system's robustness, its high reliability, or the OSS tools with which Linux is distributed, but rather the benefit of control it provides to use, change, and redistribute the source code, as well as the freedom it represents in allowing access to the source code for understanding and modification or customization.

Gacek, Lawrie, and Arief (2004) describe the primary way to obtain private returns from OSS as providing service and distribution packages for OSS. Another means of commercializing OSS is by using open sources as a basis upon which other proprietary software can be built.

Hawkins (2004) asserts that open source business models can be subdivided into two categories: business models for the software consumer and business models for the software producer. When referring to the models for consumers, this signifies the total cost of ownership (TCO) of the chosen software solution. When referring to the models for software producers and, in particular, the revenues of the company, there are a few prospective sources of revenues, such as sale of software, support of software, increased hardware sales, training, consulting, customization, distribution, and the value of internal use.

McKelvey (2001) maintains that there are three idealistic business models that assess advances in knowledge-intensive products and services; to wit, firm-based control, network-based model, and a hybrid model. Each of these can then be subdivided into the two facets of innovation; namely, economic value and creation of novelty.

Hecker (1999) suggests that in order to implement an effective open source strategy, an organization should consider the implications and manage the following factors: code sharing, third-party technology, source code sanitization, export control, and a new software development process.

By providing solutions on time to the business' customers, according to Raymond (2000, August), a business can make money using any one or more of the seven open source business models he describes.

Koenig (2004) highlights seven business strategies that can give hardware or software vendors a competitive advantage. These strategies are the optimization strategy, the dual license strategy, the consulting strategy, the subscription strategy, the patronage strategy, the hosted strategy and, finally, the embedded strategy.

The subscription strategy, also known as the revenues-for-services strategy, is one in which a provider charges a license fee for software mainly to provide maintenance and consultation services. Novell uses this strategy. Novell acquired SuSE (Software- und System-Entwicklung) in an attempt to supplement its declining NetWare maintenance revenue and to enter the Linux desktop market in which the adoption rate is very promising (Koenig, 2004).

A particular approach, as described by Covey (2000), highlights a way to sell and make money with OSS. Covey specifies that the trick is not to sell a support contract but rather an administration contract. He explains that users of systems do not need support all that often but do require their systems to be administered on a regular basis. Users of computing systems require their computers to be updated with the latest security patches and application updates, something users do not want to do or do not have the relevant experience or knowledge to do.

On the other hand, Hohensohn and Hang (2003) maintain that open source service providers can be subdivided into five categories: distributors as OSS service providers, large hardware producers, large software firms, global system integrators, and specialized open source service providers.

Mantarov (1999) illustrates how a small firm (in 1999) such as Red Hat Software Inc. could enter a mature market by implementing an innovative strategy and turning threats and barriers into opportunities.

Novell traditionally made use of proprietary software and business strategies that coincided with the proprietary software. By going open, Novell made use of OSS and open source business strategies. The strategic decision to go open is explained in the next section, which summarizes Novell's corporate and open source history.

Figure 1. Novell's timeline of major events

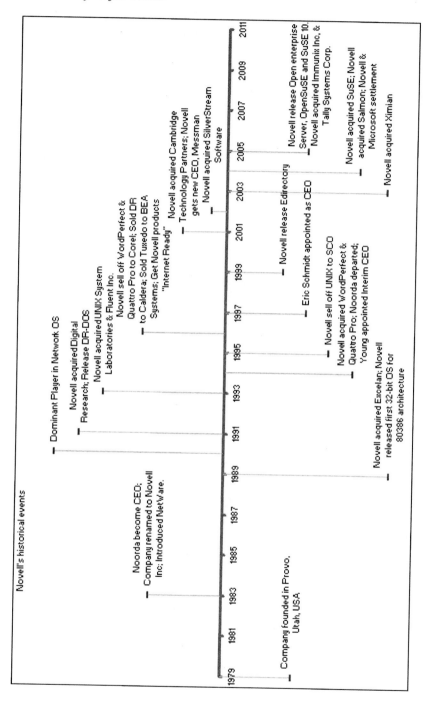

Novell's History

See Figure 1, Novell's timeline of major events, for a graphical representation of Novell's history.

The company was founded in 1979 in Provo, Utah, as Novell Data Systems Inc. At the time, Novell was a computer hardware manufacturer producing CP/M-based systems. The company was cofounded by Jack Davis and George Canova. The name Novell, suggested by Canova's wife, was a misinterpretation and was originally thought to mean "new" in French. Safeguard Scientific provided the seed capital for the company startup. The company did not do well initially, and both founders left the company soon afterwards. Victor Vurpillat, who originally organized the seed capital for the company, did not want the company to liquidate and persuaded Raymond Noorda to join the company as president (Novell Pressroom, 2004; Wikipedia, 2005).

During January 1983, the firm was renamed Novell, Inc., and Raymond Noorda was subsequently appointed CEO in May 1983. Under Noorda's guidance, Novell helped to establish the corporate network market with the introduction of the local area network (LAN). That year, Novell introduced a multiplatform network operating system (NOS) called Novell NetWare, the first LAN software based on file-server technology. The NOS made use of proprietary standards developed by Novell called IPX (Internet Packet eXchange) and SPX (Sequenced Packet eXchange), which were based on XNS and created the standards from IDP and SPP (Novell Pressroom, 2005; Wikipedia, 2005).

In the 1980s, network software began sharing files and printers within the LAN and expanded to include the management of wide area networks (WANs), which made enterprise-class computing possible. During the 1980s, Novell did extremely well, aggressively increasing market share by selling costly Ethernet network cards at a reduced price. In 1989, Novell acquired Excelan to gain valuable experience and TCP/IP-related software technologies. That year, Novell also released the very first commercially available 32-bit operating system for the 80386 CPU series processors (Novell Facts, 2004; Novell Pressroom, 2005; Wikipedia, 2005).

By 1990, Novell was the dominant player in providing NOS for any businesses that required a computer network. In 1991, Novell acquired Digital Research and released Novell DOS (also known as DR-DOS). This was done in order to break the Microsoft monopoly in the operating system market. Novell moved further away from its original market (smaller companies) to target larger corporations. Unfortunately, at the same time, Novell underinvested in research and development, which resulted in its key products being too complex to administer and control properly (Wikipedia, 2005).

In May 1991 Microsoft announced that it would be discontinuing the OS/2 partnership and would focus its time and resources on the Windows platform. This included the Windows NT kernel. This allowed them to enter the local area network market. During June 1993, Novell acquired Unix System Laboratories from AT&T, which gave it the rights to the UNIX kernel as well as Tuxedo (Transactions for UNIX, Extended for Distributed Operations), a transaction-orientated middleware platform used to manage distributed transaction processing. This was apparently done to compete directly with Microsoft in the enterprise networking and distributed transaction area. A month later, in July 1993, Novell acquired Fluent Inc. (a multimedia software company) for US$17.5 million. In the 1990s, Novell's NetWare operating system was updated to include key features for distributed enterprises.

During February 1994, Novell released the first commercially available, distributed, secure authentication system and enterprisewide directory service. That same year, in June, Novell acquired WordPerfect and Quattro Pro from Borland Inc. to gain entry into the office suites, workgroup, and standalone desktop applications market.

However, Novell was losing the network operating system market to Microsoft. With Novell losing market share and strained by the new competition, Noorda left Novell in 1994. John Young was appointed interim CEO. Novell was subsequently forced to sell UNIX to Santa Cruz Operation (SCO) in 1995. By 1996, Novell had sold WordPerfect and Quattro Pro as a package deal to Corel. DR was sold to Caldera Systems, and Tuxedo was subsequently sold to BEA systems in 1996.

In 1996, Novell interim CEO John Young realized that the Internet would make a tremendous impact on the traditional network market. He took a strategic decision to make all the company's products Internet-ready by supporting standard Internet protocols such as the TCP/IP protocol stack.

Eric Schmidt was appointed CEO in March 1997. He continued to drive the current strategy to get Novell's products portfolio Internet-ready. The result was NetWare 5 and Novell Directory Services.

In the last months of 1999, Novell released a high-availability cluster system as well as an e-directory. Novell's e-directory, a cross platform directory service, was a key requirement to ensure true interoperability, allowing effortless exchange and use of data across the Internet.

In a strategic move, Novell acquired consulting firm Cambridge Technology Partners in July 2001 in an effort to deliver both products as well as quality services to its customers. This partnership allowed Novell to deliver networking solutions that assisted companies with their business challenges. In 2001, CEO Eric Schmidt moved to Google Inc. and was subsequently replaced by Jack Messman, then CEO of Cambridge Technologies Partners.

July 2002 saw another bold step by Novell with the acquisition of SilverStream Software, a Web services-oriented applications development firm. With the acquisition of SilverStream Software, Novell acquired the expertise to convert business processes to Web services. The business area, called Novell exteNd, contains XML and Web service tools based on J2EE (Java 2 Enterprise Edition).

In August 2003, Novell acquired Ximian, an open source Linux desktop management solution. With this acquisition, Novell gained two open source visionaries, Nat Friedman and Miguel de lcaza, and two key open source projects, Mono (an open source Microsoft .NET implementa-

tion), and Gnome (a Linux desktop management platform). This gave Novell tremendous exposure in the open source community.

In January 2004, Novell acquired SuSE, Europe's leading commercial Linux distribution. With this acquisition, IBM invested a bold US$50 million in Novell to show its support for the acquisition.

Novell acquired another firm, Salmon, a UK-based IT consultancy firm, in July 2004, in order to strengthen its consultancy delivery. In November 2004, based on the SuSE distribution, Novell released the enterprise desktop, Novell Linux Desktop 9. Also in November, Novell and Microsoft settled a legal antitrust case for US $536 million based on Microsoft's efforts in the mid-1990s to eliminate competition in the office productivity applications market.

Later, in February 2005, Novell released e-directory developer interfaces to the open source community. At that time, Novell also launched the open source collaboration server initiative (an open source project providing calendar and mail functionality). In March 2005, Novell released the Open Enterprise Server, a secure suite of services that provides networking, communication, and application services. A month later, in April 2005, Novell acquired Tally Systems Corporation, an IT asset management solutions company. In May, Novell announced the acquisition of Immunix Inc., a host-based application security solutions provider. Later that year, in August 2005, Novell released SuSE as an open source project and named this project openSuSE (Novell Facts, 2005). Shortly afterwards, Novell released SuSE 10.

Novell's Strategy

"Novell will accelerate the adoption of Linux by working with its partners to remove barriers to Linux adoption" (Novell Keynote Presentation, 2005, May, p. 9).

It is apparent from the published works on open source business strategies that Novell acquired

SuSE in an attempt to increase its diminishing NetWare maintenance revenue, and aims to get the fast adoption rate of the Linux operating system on board (Koenig, 2004).

According to an internal McKinsey consultancy study, 30% of the income from enterprise solutions comes from license fees and about 70% from implementation of the solution (Koenig, 2004). In addition, a 2000 U.S. Department of Commerce report states that not since 1962 has software package cost exceeded 30% of the total software investment (Hoch, Roeding, Purkert, Kindner, & Muller, 1999). In line with this, Novell's software license net revenue for 2004 and 2005 was 25% and 22%, respectively (Novell Press, 2005). This confirms that the other 70+% of the software investment goes toward consultation, maintenance, and other related services.

Novell was forced in the mid-1990s to radically change the way it operated as well as to change the direction of the business due to Microsoft's entrance into the market. They accomplished this by making sure their software products were Internet-ready by guaranteeing that the products supported the IP protocol and other related Internet protocols. Since then, Novell has invested readily in acquisitions to make sure its diminishing services income could be boosted.

With the rapid adoption of the increasingly popular Linux operating system, Novell made a firm decision to supplement its Netware income with that of Linux. This became apparent when Novell acquired SuSE Linux. This move allowed Novell to make use of an open source subscription strategy, entering the desktop operating system market, which it is believed will become a lucrative market.

Careful study of the yearly and quarterly reports over the last seven years shows that open standards and open sources did assist Novell to slow down the decrease in its income from the Netware operating system and related services. It becomes apparent that Novell's main goal is not to derive its primary income from Linux server licenses or related services but rather to invest in the desktop market and try to acquire a fair share of the lucrative desktop operating system market. This is noticeable when one looks at the acquisitions of Ximian and SuSE in 2003 and 2004, respectively.

The acquisition of Ximian allowed Novell entrance into the desktop management solution arena, and with that acquisition, it acquired two mainstream desktop projects the Mono project (a .NET framework for the Linux desktop) and the GNOME project for managing the Linux desktop.

In 2005, Novell was actively driving the Mono development, seemingly in an effort to convince the Microsoft Windows' developers later on that all of their software development efforts can easily run on a Linux desktop solution. If Novell implements this strategy well, they might be able to convince a large enough developer base to convert to Linux and open source. On the other hand, they might only succeed in creating a second limited adoption solution similar to the situation of Microsoft Office and Open Office.org.

Later in 2005, Novell released SuSE as an open source project (openSuSE) in the hope that it will gain widespread support and adoption by the open source community as well as capitalize on the development of SuSE by the community. Naturally, Novell wants SuSE to become the Linux desktop of choice.

Close to the end of the study, there were complaints by a minority shareholder that Novell was not focusing on its core business and that expenditure was too high. The shareholder suggested to Novell that it should cut back on its spending on R&D, its Netware expertise, and noncore business areas, and invest more in Linux and open source projects (*Computer Business Review*, 2005).

Just a few weeks later, Novell released a press statement stating that it could reduce annual run rate expenses by more than US$110 million by attending to the concerns of that minority shareholder (Novell Press, 2005).

MAIN FOCUS OF THE CHAPTER

Financial Data

See Appendix for financial graphs.

Looking at the financial data, it is easy to see that during the dot-com boom, Novell did well in terms of earnings per share vs. free cash flow. After the crash of the market, it is noticeable that Novell suffered and had to actively change its strategy, which led to its pursuit of the open source option.

Although Novell's gross profit margin was above 60% from 2000 to 2005, it is obvious that the operating profit margin went below 0% to a minimum of -26.3% due to the number of acquisitions it made after 2001.

In terms of liquidity ratios, since 1999 Novell has always kept the current ratio above 1.50 and the quick ratio above 1.40. In 2005, the ratios rocketed to 166.8 and 158.8, respectively.

After the first acquisitions made by Novell in 2001, it is clear that the long-term debt was significantly affected, dropping from about 128% in 1999 and 69% in 2001, to -2324% in 2002 and -7100% in 2003, only to return to -56% in 2004 and 55.0% in 2005.

The return on equity (ROE), the return on invested capital (ROIC), and the return on assets (ROA) figures support the debt ratios, the liquidity ratios, and the operating profit margins. The ROE, ROIC, and ROA took a significant dip in 2001 from about 10% in 1999 and 3.5% in 2000 to about -20% in 2001, 2002, and 2003. Novell did see a recovery in these figures in 2004 with an ROE ratio of 3.2% and 26.9% in 2005.

Again the same pattern is noticeable by comparing sustainable growth from 12% in 1999 to -23% in 2002, only to return to 3.2% in 2004 and 26.9% in 2005.

Looking at net income, Novell's went down from US$190 million in 1999 to –US$272 million in 2001, only to recover from –US$246 million in 2002 to US$31 million in 2004. In 2005, the net income rose by 1210% to US$372.6 million.

The financial results for the full fiscal year of 2005 were released in early December 2005. Novell showed the Linux revenue going up by 418% to US$61 million for the fourth quarter in October 2005 compared to the fourth quarter in October 2004 (Novell Press, 2005).

It is clear that although Novell initially lost income from its traditional sources, the company did, indeed, manage to change its strategy, which in the long term has resulted in it successfully replenishing its diminishing NetWare income from that of Linux, OSS, and related services. Novell managed to build yet again a profitable enterprise, this time by making use of open source business models (an open source subscription strategy) rather than by making use of traditional proprietary software business models.

Findings

It is understandable that Novell makes use of open source strategies as part of its One Net strategy.

By looking at the major events in Novell's history between 1994 and 2005, it is apparent that the change to OSS has recently brought success. For two years, 1995 and 1996, after Novell lost the battle with Microsoft, Novell struggled to keep solvent. That was so until Novell switched to open standards and included the open source option as part of its main business strategy.

In the late 1990s, the information and communication technology (ICT) sector was booming; it was the era of the dot-com boom, which was later followed by a market crash. Eric Schmidt was CEO of Novell at the time, and Novell did well. Trying to determine the effect the ICT boom had on Novell is problematic, but what is unmistakable is that without OSS and an open source business strategy, Novell would probably not be here today. What is certain is that despite the market crash, Novell kept on doing well.

With a steady decline in the Netware income, Novell had to do something to replace that income. Knowing that the ICT market is changing, Novell made a strategic decision to focus more on service delivery than on the selling of product licenses. This is apparent from the acquisitions made shortly afterwards. According to Koenig (2004), the strategy Novell is following is an open source subscription strategy. This will allow Novell to sell SuSE Linux subscriptions and provide a support service bundled with the package, thus supplementing the decreasing maintenance income from Netware.

The author does not believe that Novell, in its business strategy, is making use of the traditional strengths of Linux as a server. Novell appears to be focusing primarily on the lucrative Linux desktop market by applying the open source subscription strategy. It seems that Novell is making use of Linux servers to get a foot in the door with the Linux desktop.

Earnings per Share

EPS, calculated in U.S. dollars, measures the return made on behalf of each issued ordinary share. For example, a company that made US$100 million last year and has 10 million shares out-standing would state earnings of US$10 per share. This value is calculated after paying preferred shareholders and bondholders as well as taxes.

From 2001 to 2004, Novell's earnings per share took a bit of a plunge from previous years. Between 2001 and 2004, Novell made significant investments in acquisitions that could have contributed to the fall in earnings per share. In 2005, however, Novell improved on the results of 2004 and showed a slight profit.

Free Cash Flow per Share Leveraged

Free cash flow is defined as the amount of cash a company makes after all deductions (taxes, dividends, interest). Free cash flow is used to allow all companies to be evaluated on a cash basis. In many countries around the world and in the U.S., interest expense is tax-deductible at the business level. Leveraged cash flow includes this tax benefit.

From analyzing the earnings per share and free cash flow per share, it is clear that Novell struggled to show pure profits, but this is understandable since a large amount of the profits was reinvested in acquisitions. The 2005 results

Table 1. Earnings per share vs. free cash flow 2000-2005 (Source: Novell Company Information, 2006)

Free cash flow/Year	2000	2001	2002	2003	2004	2005
Earnings per share	0.13	-0.70	-0.63	-0.42	0.08	0.98
Free CF per share leveraged	0.12	-0.07	-0.14	-0.69	-1.44	0.90
Free CF per share un-leveraged	0.12	-0.07	-0.14	-0.69	-1.44	0.90

Table 2. Profitability ratios 2000-2005 (Source: Novell Company Information, 2006)

Profitability/Year	2000	2001	2002	2003	2004	2005
Gross margin	71.8%	67.8%	60.4%	60.3%	64.3%	63.0%
Operating margin	6.1%	-26.3%	-8.1%	-5.0%	6.4%	38.9%
After-tax margin	4.3%	-26.0%	-21.8%	-14.6%	2.7%	31.1%

showed a positive cash flow and an improvement from the last four years.

Free Cash Flow per Share Unleveraged

Unleveraged cash flow is similar to leveraged cash flow except that it does not include the tax benefit. Looking at Novell's free cash flow from 2000 to 2005, the leverage was not large enough to influence the free cash flow per share.

Gross Margin

The gross margin, also called the gross profit margin, specifies the contribution from the company's core business toward covering the company's operating expenses. In many industries, the higher this is, the better.

Between 2000 and 2005, Novell kept the gross margin comfortable and acceptable, showing good profits.

Operating Margin

The operating margin is used to measure the performance and profitability of the company.

Novell took an initial dip in operating margin in the year of the dot-com bust (2001). Also that year, Novell made a significant investment in Cambridge technology. After that, Novell started to improve its profitability, showing a small profit in 2004. In 2005, Novell showed a significant improvement from the 2004 figures.

After-Tax Margin

The after-tax margin is similar to the profit margin, except that it takes taxes into account. This is also a good indicator of the company's profitability and performance.

Novell's figures for after-tax margin follow a similar trend with those of the operating margin, which implies that a lot of profit was absorbed by the operating expenses from 2000 through 2005.

Inventory Turnover

The inventory turnover ratio determines a company's activity or liquidity. The inventory turnover can be compared to industry averages. This ratio indicates how many times an inventory has been sold and replaced; the higher the value, the better the inventory is being managed.

With reference to Novell, there was an almost exponential growth in the inventory ratio between 2000 and 2002. It appears Novell did manage its inventory exceptionally well.

Current Ratio

The current ratio indicates the degree to which assets cover the claims of short-term creditors. A value of more than 1 is desirable since it allows the company to meet its short-term debt obligations. A high value may also indicate that assets are not being used effectively to generate new revenue.

Table 3. Liquidity ratios 2000-2005 (Source: Novell Company Information, 2006)

Liquidity/Year	2000	2001	2002	2003	2004	2005
Inventory turnover	125.92	376.22	4494.00	-	-	-
Current ratio	2.214	1.681	1.555	1.645	2.215	266.8
Quick ratio	1.966	1.526	1.438	1.571	2.137	258.8

Table 4. Debt ratios 2000-2005 (Source: Novell Company Information, 2006)

Debt/Year	2000	2001	2002	2003	2004	2005
Debt to total invested capital	1.0%	1.7%	0.7%	0.7%	38.2%	30.5%
Operating cash flow to long-term debt	32.2%	69.2%	-2324%	-7109%	-55.9%	55.0%

Table 5. Earnings 2000-2005 (Source: Novell Company Information, 2006)

Earnings/Year	2000	2001	2002	2003	2004	2005
ROE	4.0%	-21.5%	-23.2%	-17.3%	3.2%	26.9%
ROC/ROIC	3.9%	-21.1%	-23.0%	-17.2%	1.9%	18.5%
ROA	2.9%	-14.3%	-14.8%	-10.3%	1.4%	13.5%

Novell appears to have had the current ratio well under control since 2000, with exceptional results in 2005 compared to those of previous years.

Quick Ratio

The quick ratio, also referred to as the acid test, is similar to the current ratio except that it excludes inventory from current assets. The value can indicate whether or not the company can meet its obligations in difficult times. A value of greater than 1 is normally to be expected, but it should be compared to industry averages.

Here again, quick ratios over the years are well within a comfortable range, similar to that of the current ratios. Again, the 2005 results are well above those of previous years.

Debt per Total Invested Capital

The ratio indicates the level of financial leverage a company has, which is the total amount of external investments used to finance a company's business. The debt used in the ratio is the total debt obligations of the company. The ratio provides a better insight into the company's long-term leverage and risk.

Novell showed low figures between 2000 and 2003, which implies that it did not use its debt effectively to generate new returns. In 2004 and 2005, the debt was used more effectively, which produced better results.

Operating Cash Flow per Long-Term Debt

This ratio is calculated by using the previous four quarters of operating cash flow (rolling cash flow) divided by long-term debt. This ratio indicates how well operating cash flow covers debt. A low ratio suggests a potential solvency problem.

After Novell made significant investments in acquisitions in the period from 2001 through 2004, the operating cash flow per long-term debt was significantly affected. Novell had a serious solvency problem in 2003 but began rectifying it in 2004. By 2005, the problem had been overcome, and the operating cash flow showed a significant improvement from that of 2004.

Return on Average Common Equity

The ROE percentage shows the rate of return on the investment for the company's common shareholders. This ratio can be used to determine how well an organization reinvested income to generate additional income.

Most of the financial figures for Novell between 2001 and 2003 attest to the fact that the company was having a difficult time then. The ROE was no exception, its figures being unacceptable for that period of time. The year 2004 showed a slight positive swing in this regard with a great improvement in 2005.

Return on Investment Capital

The ROC percentage shows how effectively a company is utilizing its capital to generate profits. The indicator can be used to evaluate companies in terms of viability of products and management efficiency. It is also widely used to evaluate financial institutions but is not limited to the financial sector.

Again, the ROC figures are alarming for the period between 2001 and 2003, which can be attributed to the decrease in Netware sales and to Novell investing heavily in new acquisitions to replace the diminishing Netware income with income from Linux. There were signs of improvement in 2004, with a good ROC figure in 2005.

Return on Assets

The ROA percentage, also sometimes referred to as ROI (return on investment), is used to determine how profitable a company's assets are in generating revenue. In essence, it defines how many dollars in profit can be made from each dollar of assets the company controls.

Although the figures for 2001 to 2003 are unacceptable, a noticeable change is evident from the figures for 2004 and 2005. What emerges is that Novell made the appropriate changes to ensure the company would be solvent and showed some good profits.

Retention Ratio

The retention ratio is the exact opposite of the dividend payout ratio. The ratio indicates the proportion of net income that is not paid out as dividends to shareholders.

It is clear that as a result of making use of all the profits to reinvest in acquisitions, Novell has not paid out any dividends to its shareholders for the past six years.

Sustainable Growth Rate

The ratio defines the rate at which a company can grow without having to increase financial leverage. If the growth of the company surpasses this rate, it needs to finance its growth through external means.

Table 6. Turnover 2000-2005 (Source: Novell Company Information, 2006)

Turnover/Year	2000	2001	2002	2003	2004	2005
Retention Ratio	100.0%	100.0%	100.0%	100.0%	100.0%	100.0%
Sustainable Growth	4.0%	-21.5%	-23.2%	-17.3%	3.2%	26.9%
Asset turnover	93.3%	82.7%	106.4%	118.3%	121.0%	86.4%

The figures indicate that only in 2004 did Novell start showing that the company could grow before the growth needed to be financed. There was a significant improvement in 2005, showing that the company can grow by 26.9% before it requires external funding for expansion.

Asset Turnover

The ratio indicates the amount of sales generated from each dollar of assets. By using the company's assets, the ratio can be used to determine a company's efficiency in making sales. The ratio is indirectly proportional to the profit margin.

Taking into account the high asset turnovers, it can easily be deduced that Novell has low after-tax-profit margins. This is also an indication of the fact that Novell is focusing on a service model rather than a product model.

Novell in the Republic of South Africa

The author's geographic location is in the Republic of South Africa.

In the Republic of South Africa, Novell RSA is following a similar strategy to one that failed for Microsoft when Microsoft entered the network operating system market. In the mid-1990s, Novell catered mainly to the SME market and began to focus on large enterprises. Microsoft introduced its desktop operating system and office productivity suite for the home and SME market, which resulted in Novell losing most of its market to Microsoft. The strategy in South Africa is similar, except that Novell RSA is focusing on local government and large public enterprises rather than on large private enterprises.

So far, this appears to be working well in South Africa, where Novell has won several key government tenders. It remains to be seen whether Novell will be able to maintain its dominance in South African government contracts with the inception of a local Linux distribution called Ubuntu (a Zulu and Xhosa word roughly translating to "humanity toward others"). It seems that the project owner, Mark Shuttleworth, is following the exact opposite strategy to Novell South Africa, focusing on small to medium enterprises (SME), schools, and nongovernmental organizations (NGOs) rather than on large government contracts. This is very similar to the strategy Microsoft followed in the mid-1990s, which allowed it to gain market share against Novell.

Shuttleworth is entering into schools and community-based projects very successfully and appears to be gaining wide support in developing countries such as South Africa and Brazil. The Ubuntu Linux distribution, at the time of the study, was the number-one Linux distribution for several months running, according to distrowatch. org. Ubuntu is actively competing against larger distributions such as Red Hat and SuSE. eWeek (2005) rated both Ubuntu 5.10 and SuSE 10.0 high in terms of maturity, polish, and innovation and as being ready for the organization's desktop.

FUTURE TRENDS

At the end of the study, it became apparent that the ICT sector is experiencing what could easily be interpreted as *déjà vu*. Instead of having a repeat of the dot-com boom, there appears to be an increased interest and speculation in open source business. It is likely that venture capital companies are investing millions of dollars in open source startups because of the widespread belief that the open source service model is the one that will replace the current proprietary product license model (ZDNet UK Insight, 2005). A boom is highly unlikely since fewer than 20 companies secured venture capital in 2005 as open source companies, compared to the hundreds of thousands of companies that are developing proprietary software for commercial and internal use.

During 2004 and 2005, Novell made significant investments in acquisitions, the exact same strategy it followed about 10 years earlier when it embarked on a spending spree. Novell's acquisitions during 1993 and 1994 were short-lived, and today, the same scenario should ring warning bells. Novell should carefully monitor this pattern since it could lead to another selling spree similar to that of the mid-1990s.

Future research is required into the study of open source business models, particularly service-based business models and case studies. Research is required to determine how open source business models are implemented and how successful the business models and companies are that choose to implement it. Further research is also required in determining whether utilizing OSS and strategy will become the new way of doing business.

CONCLUSION

After careful study of Novell's corporate and open source history as well as its financial statements of the last seven years, it is evident that OSS is a viable alternative to proprietary software. Novell was able to rebuild the company after being at a low ebb in the mid-1990s and has grown into one of the biggest contenders in Linux and OSS today, or as eWeek (2005) stated, "Novell is pulling itself out of its NetWare grave with SuSE Linux sales and support" (p. 2).

OSS and open source business strategies not only assisted Novell in supplementing its diminishing NetWare income but also allowed it to replace its proprietary software income with that of Linux and OSS.

Today, Novell is making Linux and SuSE an alternative and attractive option for business. Novell has also shown the world that switching from a proprietary-based model to an OSS model is viable, feasible, and, indeed, profitable.

NOTE

The author (J. A. du Preez) has permission from Novel, Inc. to publish this research on matters regarding the company.

ACKNOWLEDGMENT

The author would like to thank Dr. A. B. Boake for his helpful comments and suggestions. Any errors are the author's.

REFERENCES

Bitzer, J., & Schröder, P. J. (2004). *Call for papers: The economics of open source software development.* Elsevier. Retrieved November 30, 2005, from http://opensource.mit.edu/papers/bookcallforchapters.pdf

Computer Business Review. (2005). *Novell under pressure from investors.* Retrieved September 16, 2005, from http://www.cbronline.com/article_news.asp?guid=0142D6B9-0B2B-4ACC-8C0E-F4F5A2CB4497

Covey, J. (2000*). A new business plan for free software.* Freshmeat Editorials. Retrieved December 3, 2004, from https://freshmeat.net/articles/view/143

Dahlander, L. (2004). Appropriation and appropriability in open source software. *International Journal of Innovation Management, 8*(4), 1-25.

eWeek. (2005). Going broke with free software. *eWeek Enterprise News & Reviews.* Retrieved July 15, 2005, from http://www.eweek.com/article2/0,1895,1833612,00.asp

eWeek. (2005). Upgrades lift Ubuntu and SuSE. *eWeek Labs Reviews.* Retrieved November 3, 2005, from http://www.eweek.com/print_article2/0,1217,a=163715,00.asp

Gacek, C., Lawrie, T., & Arief, B. (2004). *The many meanings of open source IEEE software*, *21*(1), 34-40.

Golden, B. (2004). *Succeeding with open source.* Boston: Addison-Wesley Professional.

Goode, S. (2004). Something for nothing: Management rejection of open source software in Australia's top firms. *Information & Management*, *42*(5), 669-681.

Hawkins, R. E. (2004). The economics of open source software for a competitive firm: Why give it away for free? *Netnomics*, *6*(2), 103-117.

Hecker, F. (1999). Setting up shop: The business of open source software. *IEEE Software*, *16*(1), 45-51.

Hendry, K. (2002). *Making money with open source.* Retrieved November 11, 2004, from http://www.cs.helsinki.fi/u/campa/teaching/oss/papers/hendry.pdf

Hoch, D. J., Roeding, C. R., Purkert, G., Kindner, S. K., & Muller, R. (1999). *Secrets of software success: Management insights from 100 software firms around the world.* Boston.

Hohensohn, H., & Hang, J. (2003). *Product- and service-related business models for open source software.* Siemens Business Services GmbH. Retrieved November 11, 2005, from http://mysite.fh-coburg.de/~wielandt/OSSIE03/ossie03-HohensohnHang.pdf

Koenig, J. (2004). Seven open source business strategies for competitive advantage. *IT Managers Journal.* Retrieved November 30, 2005, from http://www.itmanagersjournal.com/articles/314?tid=85

Lerner, J., & Tirole, J. (2005). The scope of open source licensing. *Journal of Law, Economics, and Organization*, *21*(1), 20-56.

Mantarov, B. (1999). *Open source software as a new business model.* Graduate Centre for International Business, University of Reading. Retrieved November 30, 2004, from http://bmantarov.free.fr//bojidar/essays/OSS_as_a_new_business_model.pdf

McKelvey, M. (2001). The economic dynamics of software: Three competing business models exemplified though Microsoft, Netscape and Linux. *Economics of innovation and New Technology*, *10*, 199-236.

Novell Company Information. (2006). *Investor relations; Annual reports.* Retrieved February 3, 2006, from http://www.novell.com/company/ir/annrpts.html

Novell facts. (2005). Retrieved December 12, 2005, from http://www.novell.com/company/fastfacts.html

Novell Keynote Presentation. (2005). In *Proceedings of the Novell Linux Infrastructure Event.* Retrieved November 15, 2005, from http://partnerweb.novell.com/partners/events/nlie_partners_keynote_script_english.doc

Novell Pressroom. (2005). *Novell announces restructuring to more closely align expenses with core business strategy.* Retrieved November 15, 2005, from http://www.novell.com/news/press/item.jsp?contentid=ded62eaedb847010VgnVCM10000024f64189____&sourceidint=hp_a3_aligns

Novell Pressroom. (2006). *Novell corporate history.* Retrieved February 1, 2006, from http://www.novell.com/news/press/pressroom/history.html

Novell Pressroom. (2005). *Novell reports financial results for fourth fiscal quarter and full fiscal year 2005.* Retrieved December 15, 2005, from http://www.novell.com/news/press/item.jsp?contentid=1eed0300713e7010VgnVCM10000024f64189

Raymond, E. S. (2000). *The cathedral and the bazaar.* Revision 1.57. Retrieved December 15, 2005, from http://catb.org/~esr/writings/cathedral-bazaar/

Raymond, E. S. (2000). *The magic cauldron.* Revision 1.19. Retrieved December 21, 2005, from http://catb.org/~esr/writings/magic-cauldron/

Scacchi, W. (2004). *Call for papers: Free/open source software development processes.* John Wiley & Sons, Ltd. Retrieved November 30, 2005, from http://serl.cs.colorado.edu/~seworld/database/4515.html

Spredgar® Software. (2006). *Spredgar software product, Financial ratios and graphs.* Retrieved January 6, 2006, from http://www.spredgar.com/download2.htm

Wikipedia. (2005). *Wikipedia the free encyclopedia.* Retrieved July 16, 2005, from http://en.wikipedia.org/wiki/

Young R. (1999). Giving it away: How red hat software stumbled across a new economic model and helped improve an industry. In C. DiBona, S. Ockman, & M. Stone (Eds.), *Open sources: Voices from the open source revolution* (1st ed., p. 114). Sebastopol, CA: O'Reilly and Associates.

ZDNet UK Insight. (2005). *Is open source a bubble ready to burst?* Retrieved December 15, 2005, from http://insight.zdnet.co.uk/software/0,39020463,39235813,00.htm

KEY TERMS

Business Model: A business model (also called a business design) is the mechanism by which a business intends to generate revenue and profits (http://en.wikipedia.org/wiki/Business_model).

Business Strategy: Business strategy or strategic management is the process of specifying an organization's objectives, developing policies, and plans to achieve these objectives, and allocating resources in order to implement the plans (http://en.wikipedia.org/wiki/Business_strategy).

Earnings: Income, generally defined, is the money that is received as a result of the normal business activities of an individual or a business (http://en.wikipedia.org/wiki/Earnings).

Earnings Per Share (EPS): The earnings returned on the amount invested initially (http://en.wikipedia.org/wiki/Earnings_per_share).

Free Cash Flow: Measures a firm's cash flow remaining after all expenditures required to maintain or expand the business have been paid off (http://en.wikipedia.org/wiki/Free_cash_flow).

Liquidity Ratios: Ratios that show the relationship of a firm's cash and other current assets to its current liabilities (http://dwc.hct.ac.ae/courses/badm300/glossary/glosl.htm).

Profitability Ratios: A group of ratios that shows the combined effects of liquidity, asset management, and debts on operating results (http://dwc.hct.ac.ae/courses/badm300/glossary/glosp.htm).

Turnover: In accounting, the number of times an asset is replaced during a financial period (http://www.investopedia.com/terms/t/turnover.asp).

Ubuntu: A South African ethic or ideology focusing on people's allegiances and relations with each other. The word comes from the Zulu and Xhosa languages (http://en.wikipedia.org/wiki/Ubuntu).

APPENDIX

Summary: Novell Annual Financial Reports 2000–2005

Figure 2. 2000-2005 earnings per share vs. free cash flow (Source: Spredgar® Software, 2006)

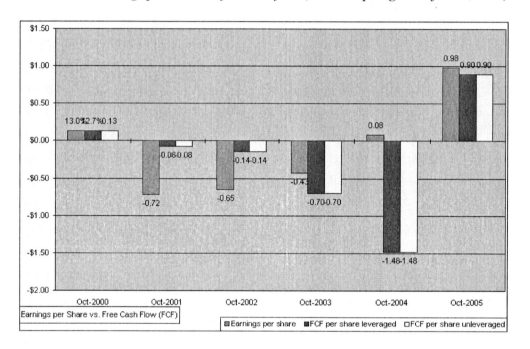

Figure 3. 2000-2005 profitability ratios (Source: Spredgar® Software, 2006)

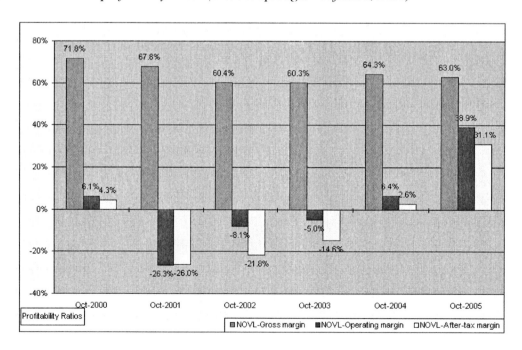

Figure 4. 2000-2005 liquidity ratios (Source: Spredgar® Software, 2006)

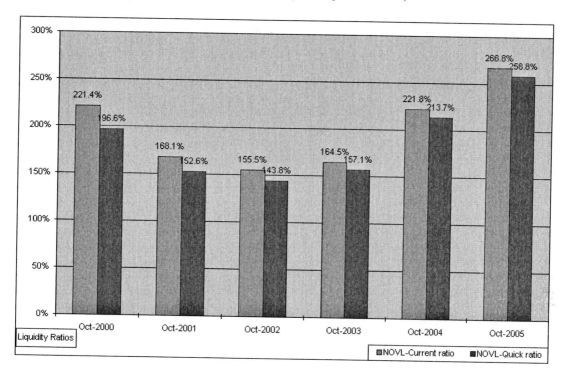

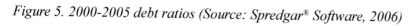

Figure 5. 2000-2005 debt ratios (Source: Spredgar® Software, 2006)

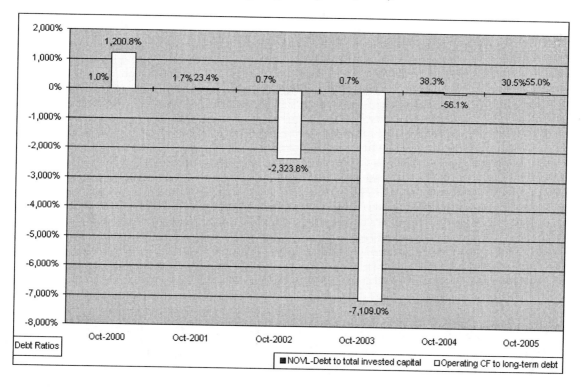

Figure 6. 2000-2005 earnings (ROE, ROIC, & ROA) (Source: Spredgar® Software, 2006)

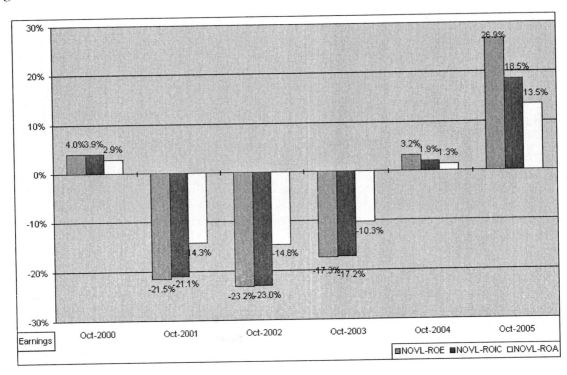

Figure 7. 2000-2005 turnover (Source: Spredgar® Software, 2006)

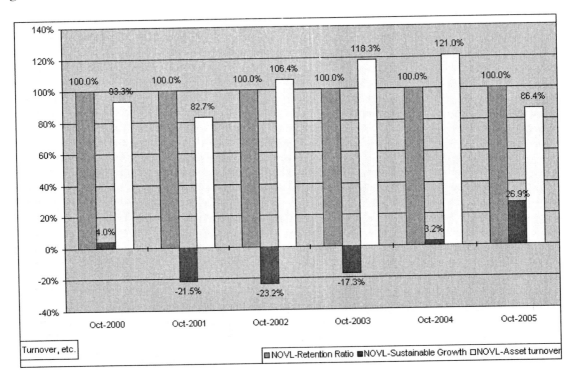

Section VII
Educational Perspectives and Practices Related to Open Source Software

Chapter XLVII
Communities of Practice for Open Source Software

Leila Lage Humes
University of São Paulo, Brazil

Nicolau Reinhard
University of São Paulo, Brazil

ABSTRACT

This chapter studies the use of communities of practice in the process of disseminating open source software (OSS) in the University of São Paulo. The change management process included establishing an OSS support service and developing a skills-building training program for its professional IT staff, supplemented by a community of practice supported by an Internet-based discussion list. After using the resource extensively during the early phases of the adoption process, users replaced their participation in this local community by a mostly peripheral involvement in global OSS communities of practice. As a result of growing knowledge and experience with OSS, users' beliefs and attitudes toward this technology became more favorable. These results, consistent with the theory of planned behavior constructs, provide useful guidance for managing the change process.

INTRODUCTION

Strategic and economic considerations led the University of São Paulo to institute a program to promote the use of OSS, also driven by a national OSS dissemination policy that draws a lot of mass media attention. The major drivers were software license cost reductions and independence from single vendors, as well as the perception that OSS allows increased user control over systems and interoperability.

OSS technology has many adherents among Brazilian faculty members and students. However, the university's professional IT staff, responsible for the infrastructure and administrative systems and a significant share of the total IT budget, has been more conservative regarding technological innovation.

These professionals are largely autonomous in their technological decision-making, and therefore, they had to be motivated to adopt the new technology. The university, besides providing all necessary support services to implement OSS, offered them special courses in order to make the process of adopting and implementing the technology easier. This chapter presents the case study of the change management program for the dissemination of OSS at the university, having as its main target the computer professionals of the various institutes and schools in charge of local IT infrastructure and support to end users.

The research approach was action-research, the authors being the sponsors and managers responsible for the dissemination process and the establishment of the community of practice (CoP). This chapter is structured as follows: the first section presents the motivation for OSS, its concepts and the theoretical framework for the case study. The case is described in the second section, whereas the third section analyzes the case and presents the conclusions of the study, with emphasis on the use and evolution of the CoP.

BACKGROUND: THE APPROACH TO THEORETICAL CONCEPTS AND RESEARCH

Open Source Software

Open source software (OSS) is based on the principle that computer programs should be shared freely among users, giving them the possibility of introducing improvements and modifications.

The Free Software Foundation (FSF), founded in 1984 by Richard Stallman, aimed at recreating the "open" environment of computers' early days, replaced by the establishment of the for-profit software industry. OSS users and developers engage in intense voluntary worldwide cooperation leading to community-based continuously evolving systems that can safely be used in critical applications and infrastructure (Nuvolari, 2004). The use of OSS is growing steadily. The Campus Computing 2003 survey (Green, 2003) found that 11.1% of all network servers in American higher

education institutions run on Linux. Another survey conducted by the authors in 2003 found that 20% of corporate low-platform servers in Brazil are based on the Linux operating system (Reinhard & Foresti, 2003).

Cooperation among OSS users and developers is maintained through an elaborate infrastructure for sharing knowledge and communication, including issue-reporting/tracking repositories, discussion lists, chat rooms, forums, electronic journals, specialized media, and meetings. A strong culture and group behavior have been developed in connection with it, enabled by the Internet (Scacchi, Gasser, Ripoche, & Penne, 2003).

OSS is developed as distributed work, with ample freedom for the creation and distribution of nonstable versions of systems, but with special governance mechanisms for the establishment of standards, verification, and distribution of so-called stable software versions.

Theory of Planned Behavior (TPB)

Ajzen (1991) proposed the theory of planned behavior to explain and predict individuals' intentions to exhibit a given behavior. Intention is seen as a function of the beliefs related to the following:

- Attitude toward the behavior (evaluation of the behavior)
- Subjective norm (perceived social pressure to conform)
- Perceived behavioral control (perceived ease or difficulty to perform)

TPB can be considered a suitable model for studying the behavior of computer professionals deciding on the adoption of OSS, since their behavior is largely under their volitional control (i.e., it is essentially their own decision whether or not OSS will be adopted in their departmental computing environments).

Figure 1. Model of the theory of planned behavior (Source: Ajzen, 1991)

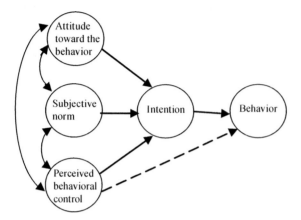

The change management process aims to increase the motivation of the university's IT staff to try the new technology (OSS) by:

- Creating a more favorable attitude toward the adoption of open software by demonstrating its benefits
- Promoting the perception of increased behavioral control through the provision of technical resources, knowledge, skills, and central support (including the CoP)

Subjective norms in favor of OSS (induced by the administration, academic community, and society at large) also played an important role in the process.

Communities of Practice (CoPs)

A CoP is a group of people with common purposes, experiences, and interests, who are willing to provide and share information, devoting time to collaborate with the group in solving problems beyond organizational structures and boundaries. There is a well established tradition of using CoPs for the development and dissemination of OSS knowledge.

CoPs are a way of capturing, documenting, and sharing explicit and tacit knowledge among its members. The interaction of members in these communities allows them to learn from each other by observing how they act in emerging situations and solve real problems, and how they generate new knowledge during this interaction (Orlikowski, 2002). According to Wenger, McDermott, and Snyder (2002), group members learn by working together, developing together a common sense of how work should be done and what it takes to accomplish tasks. Wenger (1998a) defines "practice" in this context as follows:

A concept of practice includes both the explicit and the tacit. It includes what is said and what is left unsaid; what is represented and what is assumed. It includes language, tools, documents, images, symbols, well-defined roles, specified criteria, codified procedures, regulations and contracts that various practices make explicit for a variety of purposes. (Wenger, 1998a, p. 47)

Vygotsky (1978, cited in Borthik, Jones, & Wakai, 2003, p.111) emphasizes the importance of learning experiences that support the gradual development of the learners' capabilities so they learn to do by themselves things that initially they could do only with assistance. Regardless of the source or form of the assistance, the goal is for learners to develop the capabilities they first experienced in assisted or collaborative learning situations (Bereiter & Scardamalia, 1985, cited in Borthik et al., 2003, p. 109). Making use of each other's expertise depends, however, on learners' recognizing expertise asymmetries and on their willingness to collaborate with each other in order to benefit from the expertise distributed among them (Borthick et al., 2003).

CoPs are an effective way of creating and organizing this knowledge. Group knowledge, both explicit and tacit, is created when a member presents a problem and the solution emerges through collaboration. CoPs may have members acting as coordinators or moderators and may develop formal organizational forms. Participation can be voluntary or mandatory. An individual can easily migrate from one CoP to another or participate simultaneously in several CoPs.

Members of a CoP must trust each other, and their activities must be perceived as adding value. Members can have different levels of participation (Wenger, 2002):

1. Active participation in discussions (usually 10-15% of the members)
2. Occasional participation in discussions (about 15-20%)
3. Peripheral participation (members who usually read others' messages but rarely ask questions)

According to Gongla and Rizzuto (2001), the evolution of CoPs is influenced by a dynamic balance of people, processes, and technology. These authors identify the following five stages of CoP evolution:

1. **Potential stage:** The community forms itself around a nucleus and is comprised of people with some common characteristics related to work or personal interests.
2. **Building stage:** The community grows. Founding members define its characteristics, how it will be built, and how it will present itself. Processes and structures are created. During this stage, the community defines its identity and reason for existence.
3. **Engagement stage:** The community is fully operational, growing in size and complexity, and learning more about itself and its environment. This learning helps to improve its structure and builds its capacity for leveraging tacit and explicit knowledge.
4. **Active stage:** Community members analyze, understand, define, and evaluate their contributions to the community and its environment.

5. **Adaptive stage:** The community starts to perceive and adapt to external conditions. Continuous adjustments and innovations create new solutions, processes, and groups, influencing and creating new tendencies in the community's area of expertise. Few CoPs manage to reach this stage and maintain this level because their perceived importance encourages migration to other forms of governance and institutionalization.

MAIN FOCUS OF THE CHAPTER: CASE STUDY OBJECTIVES AND METHODOLOGY

The study's first objective was to identify the beliefs and attitudes of the potential adopters and actual users of OSS, understand the relation between these factors and management action, and evaluate the users' decisions to adopt the new technology or increase its usage. For this first objective, the authors chose to use the theory of planned behavior (TPB) model for analysis, since, according to Taylor and Todd (1995), it can provide more effective guidance to IT managers and researchers interested in the study of system implementation. Taylor and Todd (1995) compared three technology acceptance models: the TPB model, the technology acceptance model (TAM) proposed by Davis (1989), and a decomposed version of TPB. They concluded that TPB, which adds subjective norms and perceived behavioral control as key determinants for both intention and IT usage provides a fuller explanation of behavioral intention and IT usage behavior.

The second objective was to understand the role of CoPs and other sources of support in this process and their evolution over time. Gongla and Rizzuto's (2001) model for the evolution of CoPs was adopted to study the strength of interaction among the members of the community. According to this model, the evolution of a CoP is influenced by a dynamic balance of people, processes, and technology.

The research tools used in this study consisted of a series of surveys applied to all participants of the training program during the first seven months of the research, analysis of secondary data from the University's Statistical Yearbook, USP (2005), monitoring the discussion list during two and a half years, semi-structured interviews, and participant observation (one of the authors was the corporate manager of the university's OSS program).

The survey was conducted in 2003 and its questionnaires were answered by a total of 147 course participants (all of whom were computer professionals employed by the university) with various levels of knowledge regarding the adoption and use of OSS. The training program included the lead instructor setting up an experimental CoP to provide users with support after the course and evaluate the evolution of their knowledge over time. After the establishment of the CoP, its evolution was evaluated for two and a half years. Semistructured interviews were used as a research tool for evaluating the instructors' and course participants' perceptions of the usefulness of the CoP.

The Case Study Context

The University of São Paulo (USP) is the largest research university in Brazil, with 70,000 students and 5,000 faculty, 35 Units (Schools and Institutes) spread over seven campuses in the State of São Paulo, and course offerings in all major fields of science, technology, and the arts.

The university's IT infrastructure is managed by a corporate steering committee that oversees a central IT center, three regional facilities (responsible for their campus) and local units (in all institutes and schools), with a total staff of 600 IT professionals. University administration is highly decentralized. Units have their own IT staff and are autonomous in managing their budgets and grants obtained from external agencies, which leads to a significant diversity of resources, technologies, and organizational approaches.

For this study, units were classified according to their OSS usage stage in the following categories:

* **Initial:** OSS in initial stage of implementation
* **Intermediate:** Few OSS-based services in regular use
* **Advanced:** Consolidated use, a large number of applications based on OSS

The OSS Innovation Program

Budget restrictions and strategic considerations (adoption of OSS is a goal heavily promoted by the Brazilian federal government) led the university to institute an OSS adoption promotion program, starting with the Linux operating system, which included the creation of an OSS repository, a support service, and a series of weeklong courses for staff computer professionals. One of the authors was the corporate manager of this initiative who decided to use innovation management concepts and tools (TPB, CoPs, support for technology adoption, courses, measurements, etc.).

The courses were offered in various locations, and participants could choose among the following courses, depending on their prior knowledge of OSS: Introductory—Basic Linux Installation (77 participants); Intermediate—PHP and Applications (43 participants); and Advanced—Security (78 participants). These courses were taught by a total of 11 instructors from the Central and Regional IT Centers and led by one head instructor. There was a fairly uniform distribution of the university's IT staff between beginners and advanced users of OSS.

The head instructor responsible for the courses created a CoP devoted to OSS, starting with face-to-face meetings (mostly during the courses), followed by discussion lists, forums, and so forth. Many participants, especially those taking the introductory course, were not familiar with these communication resources, but language and cultural uniformity in addition to the high credibility of the central IT center staff led to the acceptance of these structures.

OSS issues are highly visible in local mass media, generating a lot of folklore but also conveying objective knowledge about its characteristics, benefits, challenges, and available solutions. As one instructor put it:

People (the professional staff, systems analysts and technicians) read a lot about Linux in magazines and newspapers, but know very few actual users who can answer their questions. Linux is believed to be for hackers, requiring extensive knowledge of IT. Those who lack this knowledge shouldn't even try using Linux.

On the other hand, the academic community at USP led by some influential professors is, in general, in favor of this trend. OSS is used extensively for teaching and research by both faculty and students. In general, adherents see themselves as more innovative and competent, and tend to develop strong group behavior.

The university's corporate administrative systems and network infrastructure management, however, depend largely on proprietary software. At the unit and department level, there are many applications and operating systems based on proprietary software.

Analysis of Results: Evolution of Beliefs Related to OSS Characteristics

The introductory course participants answered the same questionnaire twice, before and after the course, whereas those who took advanced courses and were already familiar with the technology were asked to answer the questions only at the start of their courses. The goal was to identify the differences between the two groups' beliefs about OSS characteristics and the changes induced by the courses.

Since participation was not completely optional (central administration had urged unit deans to send their IT staff to these courses), it cannot be said that the instructors were "preaching to the converted," and therefore, the answers of the participants at the beginning of the courses reasonably reflected the community's beliefs. Being able to give anonymous answers also encouraged the candid expression of individual beliefs.

The survey shows that some of the participants' beliefs changed during the introductory course, particularly regarding OSS security and overall quality. The other beliefs evaluated were good cost/benefit relationship, features, reliability, ease of use, technical support, documentation, and warranty/services.

The TPB model used to study the individual adoption of OSS is based on the relationship between three constructs (attitude, behavioral control, and subjective norms) and their influence on the intention of adopting OSS, which was represented in this research by the variable "interest in OSS."

The TPB constructs consisted of the following variables:

1. **Attitude:** This is composed of security, a good cost-benefit relationship, relevant properties, reliability, and overall quality.
2. **Behavioral control:** This is composed of technical support, documentation, warranty/service, and ease of use.
3. **Subjective norms:** The survey questions for building this construct were (1) Do you consider the university's OSS adequate? (2) Do you take other units' software usage into account? (3) Do you take your colleagues' software usage practices into account? and (4) Do you take governmental recommendations into account?

A factor analysis (principal components with varimax rotation) was performed in order to evaluate the TPB constructs. The outcome was a single factor for attitude and behavioral control and two factors for construct subjective norms. The correlation between the attitude and behavioral control of constructs was significant at the 1% level. The interest in OSS was significant with attitude and behavioral control at the 1% and 5% levels, respectively, while subjective norms were significantly correlated with attitude at the 5% level. These correlations are consistent with the relationship proposed by the TPB model (except for the direct influence of the behavioral control construct on the subjective norms construct) (i.e., the interest in OSS is positively correlated with attitude and behavioral control).

If these relations can be posited as causal relationships, then the results can be interpreted as confirmation of the effectiveness of the university's strategy for promoting the adoption of OSS among its professionals through an effort to induce attitudes (beliefs) favorable to the technology and an increase in the perception of behavioral control over the adoption process by these professionals.

The survey results also showed the following:

1. The perception of the overall quality of OSS improves with the increase of both the IT staff's experience with the technology and the unit's stage of OSS adoption.
2. This perception of OSS quality is also related to the IT staff's willingness to implement OSS.
3. Users' favorable perceptions of OSS quality are related more to the stage of adoption in their units than to the chronological dimension of their experiences with OSS. Advanced users become leaders and references in their communities, reinforcing the adoption process.
4. The intention to adopt OSS is related directly to a positive attitude toward the

technology, an empirical finding that is consistent with TPB.

Communities of Practice

Initially, we will analyze the CoP that was established during the introductory course, which was offered several times in different locations over the course of four months, always with the same positive result.

Use of Internal Discussion List by Course Participants

IT staff members enrolled in the one-week introductory course were encouraged to join the CoP and its discussion list created for them by the head instructor. In order to motivate them to join, the instructor used the list during the courses to distribute lecture notes, technical information on Linux versions and security bugs, new applications, practical hints for installation, and so forth. This stage of a CoP can be identified with the potential stage of Gongla and Rizzuto's model, the community being formed around a nucleus by people with certain common interests.

Given the challenge of establishing the trust atmosphere needed for the satisfactory evolution of the CoP, the list was set up on a list server (yahoogroups) that had no connection with the university.

The head instructor who created the CoP is a Linux enthusiast and had been working with the system for more than four years. He also maintains the university's Linux site and helps users with their problems. His charisma and communication skills helped to build a trust relationship with and among the participants.

Although widely accepted through the mutual trust developed during the course, not all participants became active users of the list. As the instructor said:

Some people are afraid of showing that they don't know that much; in other words, they are afraid of asking elementary questions in the list. They would rather ask them in person or through personal e-mail.

Or in the words of a participant:

My doubts are much more basic than those put to the list. I still don't know much about Linux. If someone's question to the list interests me, I contact him directly by phone or e-mail, instead of using the list.

This participant, of course, was also an active member of the CoP, only using different communication channels. One incident demonstrates the emergence of the CoP's governance structure; due to the informality of the list, one participant started using inappropriate wording in his messages. Another member immediately rebuked him, and the situation did not occur again. Therefore, at this stage, the community was defining its rules and the form of presenting itself to the world, characterizing the building stage of the community, according to Gongla and Rizzuto.

The percentage of active list users is similar to what is reported in the literature: 26% made some sort of contribution, either asking or answering questions through the list. The 7% most active members accounted for 30% of the messages, and the instructor himself generated 38% of all messages. Other members sent the remaining 32% of the messages.

When the list started, answers were provided mostly by the instructor. Later on, other more experienced members began giving advice as well. The CoP also had a large number of lurkers, who remained on the list but only read messages.

The questions participants posed to the list were rated by the authors according to their level of difficulty as a proxy for the users' increasing levels of competence. The evolution of question complexity for the first seven months is presented in Figure 2, indicating that as members became more knowledgeable, difficult questions replaced simple ones. In fact, after seven months of use, most of the questions on the list were complex ones.

Figure 2. Evolution of question complexity over time

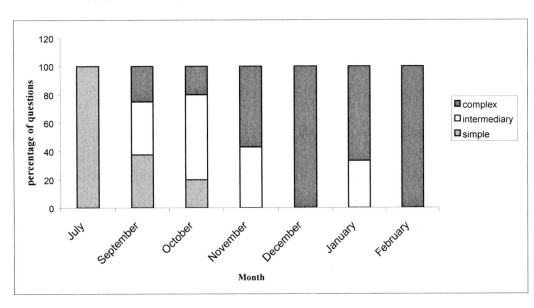

Figure 3. Evolution of the total number of questions posed to the CoP during the two-and-a-half-year period

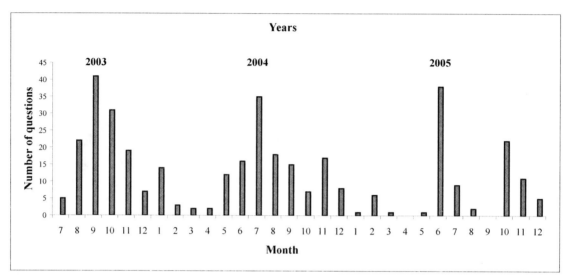

Figure 3 presents the total number of messages posted in the community during the period of two and a half years, showing a significant variation in list activity. The silent periods coincided with a lack of courses or of new releases. However, the announcement of a new Linux release or a new course with the arrival of newcomers would trigger an increase in community activity.

The instructor added information on the interaction among community members:

Members of this community also call me on my office phone. The number of questions answered by phone is at least twice the number of questions that I regularly answer in the discussion list. I also answer questions through my personal MSN messenger address.

This comment also indicates that users do not limit themselves to any single communication channel but rather adopt them according to personal values and perceptions of their social networks. According to Wenger et al. (2002):

Building trust, exploring the domain and discovering the kind of ideas, methods, and mutual support that are genuinely helpful take time. Most of all, community members need to develop the habit of consulting each other for help. ... The trust community members need is not simply the result of a decision to trust each other personally. It emerges from understanding each other. (Wenger et al., 2002, p. 84)

The instructor analyzes the building of trust among members of the community:

I believe that people feel free to share their opinions with other members of the community. This can be stated by the way people express themselves. However, there are some people that are still afraid of submitting simple questions and being judged as someone that is not that knowledgeable. The lack of knowledge sometimes inhibits interaction. In my opinion, people need to trust each other in order to freely interact within a community.

Nevertheless, users would still rate the list as a valuable learning resource. Considering only the list interactions, Figures 2 and 3 indicate that as members became more knowledgeable, difficult questions replaced the simple ones, and the total number of messages fell over time during certain periods of time.

The instructor also tried another form of interaction. He invited members of the CoP to join his Orkut community because there one could get to know people better as well as their preferences, personal skills, and other interesting characteristics. He assumed that the most active members of the community would join his Orkut community, but only 18% did so. He also mentioned that some who joined the Orkut community would not participate in the CoP list. Information on the Orkut community was provided in the course material available on the university's Linux site.

Some members of the community also became instructors of courses outside the university and used the same material in these courses. During the courses, the students were taught how to interact in a discussion list. The instructor always recommended that before submitting a question to a list, students should look for the information available on the Internet and in other discussion lists. Some of them had no experience with interaction in CoPs or discussion lists. The instructor also gave them advice on sites and tutorials available on the Internet. In his opinion, combining the information dispersed over the Internet is a difficult step for a beginner; the student is only able to find incomplete and scattered information and is incapable of organizing it in such a way to produce a solution to a problem.

In Brazil, a great challenge Linux beginners face is related to language. Most tutorials are available only in English, which is a problem for many users. Therefore, the discussion list in the country's language (Portuguese) is very useful to them.

Some people attending the course had no chance to install and work with Linux in their units. They would read the messages posted to the list but would not participate in discussions.

As far as the authors know and as stated by the instructor, there were no face-to-face meetings with members of the community after the training program. It should also be mentioned that according to the instructor, the more advanced members of the community quickly learned to find help in resources such as general discussion lists and tutorials available on the Internet. They would then use the CoP only as a last resort. As a result, the questions posed on the list became very complex, and sometimes the instructor would not know the answers. The users' attitudes toward the CoP and their recognition of its importance indicate that the community had reached the engagement stage. Its continuity would depend on new members joining it and an ongoing demand for new information. Other structures developed with some of the members becoming focal points for OSS support in their unit and its environment.

The discussion list was only one of the communication channels used by participants. CoPs build various channels of communication. Intermediate and advanced users still consider personal contact with individual colleagues to be an important source of information, in addition to the structured archival information repositories.

According to our research, the CoP has not progressed to the more advanced active and adaptive stages. These data are consistent with the opinion of the more advanced participants: CoPs are more important in the early stages, losing their value to the increasingly competent user over time. According to Wenger (1998b), CoPs preserve the tacit aspects of knowledge that formal systems cannot capture. For this reason, they are ideal for initiating newcomers into a practice.

CoPs for Advanced Users

Participants in the advanced course, who were more experienced users of OSS and providers

Table 1. Frequency (percentage) of mentions of CoP benefits: gains in learning, socialization and knowledge sharing (multiple responses allowed)

	Adoption Stage		
Benefits of CoP	*Initial (%)*	*Intermediate (%)*	*Advanced (%)*
Reduction in learning time	50	60	58
Improvement in quality and dissemination of best practices	100	46	58
Knowledge retention	0	23	25
Incentive to disseminate knowledge	50	64	75
Organizing and storing knowledge	0	23	50
Greater integration of members	50	46	67

of the university's support services, were asked about their attitudes and behavior toward support structures and the use of CoPs. These courses had a total of 78 participants, with 38 of them stating that they were members of at least one CoP. It is important to note that these answers relate to any OSS CoP in which a respondent participated.

One can see that there is great diversity in the levels of professional expertise and adoption of OSS, as shown by the choice of training courses. This diversity is important for the dynamics of the CoP. In general, list members have a favorable perception of the CoP's organization, effectiveness, and trust, an important condition for its performance and continuity.

Although there is a high degree of trust and collaboration among CoP members, the frequent lack of a strong structure and leadership can impact the CoP's effectiveness. Nevertheless, members perceive it as very significant. (1) For units in the initial stage of OSS adoption, all members considered it important; (2) for units in the intermediate stage of OSS adoption, 81% considered it important; and (3) for units in the advanced stage of adoption, 9% considered it very important, 82% considered it important, and only 9% considered it unimportant for the use of OSS.

Table 1 focuses on learning, socialization, and knowledge-sharing through the CoP. Members of units in the advanced stage value the list for its dissemination of knowledge and organization facilities.

The perceived benefits of CoPs to members in the initial, intermediate, and advanced stages are, respectively, (1) cost reduction, mentioned by 50%, 41%, and 42%; (2) productivity gains, mentioned by 100%, 68%, and 67%; (3) increased innovation through collaboration, mentioned by 25%, 50%, and 58%; and (4) incentive for collaboration between members, mentioned by 25%, 59%, and 75%. Therefore, whereas members in the initial stages look more for immediate productivity gains, members in the advanced stages value the collaboration aspects of the CoP more.

Participants from the intermediate and advanced courses were also asked about their use of information sources for solving problems. Table 2 shows the sources most frequently cited by participants, classified by their unit's stage of OSS adoption.

Table 2 also shows that there are distinct patterns of source usage by stage of adoption, with more advanced users being able to use more structured sources and their own personal relationships network. Advanced users are also less likely to actively seek help through CoPs and discussion lists.

Some of the members built very secure environments to be used in their units by faculty members, researchers, and so forth, employing

Table 2. Frequency of mentions of information sources for problem solving by unit adoption stage (multiple responses allowed)

	Adoption Stage		
Information Source	*Initial (%)*	*Intermediate (%)*	*Advanced (%)*
Tutorials	12	42	22
Support through CoP, sending messages to discussion lists	6	19	7
Searching discussion lists and forums	10	26	12
Asking colleagues	10	34	23
Internet search	13	41	24
Reading FAQs	11	29	16
Other	3	4	-

advanced Linux resources, and shared this knowledge with colleagues (Humes, 2004). It is important to emphasize that the university has various other discussion lists about Linux maintained by researchers, but there is no central coordination of these communities. Therefore, there is also no communication among them, and the university does not map the knowledge of those groups.

CONCLUSION

The process of OSS diffusion at USP has been successful in creating and promoting positive attitudes to the technology. Mechanisms such as establishing support centers, training courses, and CoPs were instrumental in this process. The survey shows that the training courses produced a positive change in beliefs regarding OSS and that they had a favorable impact on intention to adopt OSS. Professionals working for units at more advanced levels of adoption have more positive beliefs regarding OSS and a willingness to use it, a possible measure of the program's success.

An internal CoP created by the university demonstrated its usefulness to users in the early stages of adoption. This community reached the engagement stage, helping members solve their initial problems, supporting them in their learning processes, paving the way for the strengthening of their social networks, and enabling them to take part in global communities. Having fulfilled this purpose for the initial group of members, the community became less useful and was ultimately discontinued.

This chapter also provides empirical evidence to support the usefulness of the TPB for OSS adoption by demonstrating the connections among beliefs, attitudes, intentions, and decisions to adopt and increase the use of OSS. Additionally, it provides empirical data on the perceived usefulness and actual usage patterns of a CoP created for the specific purpose of promoting the use of OSS. For the practitioner engaged in promoting the dissemination of OSS in complex organizations in which innovations are mainly dependent on voluntary adoption, this chapter can also provide useful managerial guidelines.

REFERENCES

Ajzen, I. (1991). The theory of planned behavior. *Organizational Behavior and Human Decision Processes, 50*(2), 179-211.

Borthick, A. F., Jones, D. R., & Wakai, S. (2003). Designing learning experiences within learners' zones of proximal development (ZPDs): Enabling collaborative learning on-site and online. *Journal of Information Systems, 17*(1), 107-134.

Gongla, P., & Rizzuto, C. R. (2001). Evolving communities of practice: IBM global services experience. *IBM Systems Journal, 40*(4), 842-862.

Green, K. C. (2003). *Campus computing 2003.* Encinco, CA: Campus Computing.

Humes, L. L. (2004). *A adoção de software livre na USP: Um estudo de caso.* Uunpublished master's thesis, School of Economics, Business Administration and Accounting, University of São Paulo, São Paulo, Brazil.

Markus, M. L., Manville, B., & Agres, C. (2004). *Virtual organization design: Lessons from the open source movement.* Retrieved February 23, 2005, from http://web.bentley.edu/empl/m/lmarkus/Markus_Web_Documents_(pdf)/Virtual_Organization_Open_Source.pdf

McDermott, R. (2004). *Knowing in community: 10 critical success factors in building communities of practice.* Retrieved February 10, 2004, from http://www.co-i-l.com/coil/knowledge-garden/cop/knowing.shtml

Nuvolari, A. (2004) *Open source software development: Some historical perspectives.* Retrieved February 15, 2005, from http://opensource.mit.edu/papers/nuvolari.pdf

Orlikowski, W. J. (2002). Knowing in practice: Enacting a collective capability in distributed organizing. *Organization Science, 13*(3), 249-273.

Reinhard, N., & Foresti, N. (2003). Fogo cruzado. *Informationweek (Brasil), 5*(104), 20-22.

Scacchi, W., Gasser, L., Ripoche, G., & Penne, B. (2003). Understanding continuous design in F/OSS Projects. In *Proceedings of the 16th International Conference on Software and Systems Engineering and its Applications (ICSSEA-03),* Paris. Retrieved February 10, 2005, from http://www.ics.uci.edu/%7Ewscacchi/Papers/New/ICS-SEA03.pdf

Taylor, S., & Todd, P. A. (1995). Understanding information technology usage: A test of competing models. *Information Systems Research, 6*(2), 144-176.

USP. (2005). *University of São Paulo statistical yearbook.* Retrieved March 15, 2005, from http://sistemas.usp.br/anuario/

Wenger, E. (1998a). *Communities of practice: Learning, meaning and identity.* Cambridge: Cambridge University Press.

Wenger, E. (1998b). Communities of practice: Learning as a social system. *The Systems Thinker, 9*(5). Retrieved February 20, 2005, from http://www.ewenger.com/pub/pub_systems_thinker_wrd.doc

Wenger, E., McDermott, R., & Snyder, W. M. (2002). *Cultivating communities of practice: A guide to managing knowledge.* Boston: Harvard Business School Press.

KEY TERMS

Adoption: The adoption of an innovation may be conceptualized as a temporal sequence of steps through which an individual passes from initial knowledge of an innovation to a decision to adopt or reject it, to put the innovation to use, or finally, to seek reinforcement of the adoption decision made.

Change Process Management: Activities involved in defining and instilling new values, attitudes, norms, and behaviors within an organization that supports new ways of doing work and overcomes resistance to change; building consensus among customers and stakeholders on specific changes designed to better meet its needs; and planning, testing, and implementing all aspects of the transition from one organizational structure or business process to another.

Communities of Practice (CoPs): Groups of people with common purposes, experiences, and interests, who are willing to provide and share information, devoting time to collaborate with the group in solving problems beyond organizational structures and boundaries.

Linux: An operating system very similar to Unix that is suitable for use on a wide range of computers. It consists of a kernel that is the core of the operating system and a wide range of free utilities and application programs that are available in coordinated packages named Versions.

Linux Distribution: A version of a Unix-like operating system for computers comprising most of an operating system, the Linux kernel, and other application programs. There are currently more than 300 Linux distribution projects in active development that are constantly revised and improved by their respective developers.

Open Source Software (OSS): The principle that computer programs should be shared freely among users, with the possibility of introducing improvements and modifications. Therefore, users can make changes, build new versions, and incorporate changes.

PHP: PHP hypertext preprocessor is a scripting language used to create dynamic Web pages.

Chapter XLVIII
Selecting Open Source Software for Use in Schools

Kathryn Moyle
University of Canberra, Australia

ABSTRACT

Schools are places where the choices made about computing technologies not only reflect their technical requirements but also reflect the philosophical priorities directing those choices. Schools can deploy a startling range of software (i.e., operating systems, databases, office productivity software, and applications software) for specific teaching and learning purposes. Applications software deployed in schools must be suitable for use by students who are young and often have limited reading and fine motor skills. Back-end software must be robust enough to handle hundreds and sometimes thousands of users concurrently. One issue that faces schools interested in deploying open source software is the number of choices available; there is a wide variety of open source software that is suitable for use in schools. It is intended that this chapter provide readers with entry points to selecting open source software by identifying criteria that can be used by schools to shortlist potential open source software appropriate for their local environments.

INTRODUCTION

Schools are characterized by their diversity, complexity, and multidisciplinary nature; they are unique and complex organizations in which students are, for the most part, legal minors. Societies invest heavily in education since it is the way in which societies reproduce themselves (Berger & Luckmann, 1979). Schools, therefore, are dynamic and inherently social, political, and cultural places (Johnson & Christensen, 2004) in which values and philosophies are on show in practical and concrete ways. Indeed, the choices a school makes about computing technologies can operate as indicators of the values and philosophies that school endorses.

In the 21ˢᵗ century, including computing technologies into education is occurring throughout the world. Countries are at different stages in this process, but in general, the deployment of technologies is moving from individual, stand-alone computers to integrated technologies that are networked and, when possible, connected to the Internet (Hepp, Hinostroza, Laval, & Rehbein,

2004). Since models of open source software (OSS) development are based on contributing to the public good through online networked activities (Bessen, 2004), the paradigm shift away from personal to networked computers linked to the Internet makes OSS viable both technically and philosophically for the education sector. Some of the technical and philosophical contributions that OSS can make to education are discussed here in order to report the criteria proposed for identifying suitable OSS for use in schools.

TECHNOLOGIES IN SCHOOLS

Schools vary in the way computers are deployed for student use. Some schools may not have any computers in classrooms but may have them in a computer laboratory or library resource center. Other schools may have no computer laboratories but may have computers in classrooms or on portable carts; other schools may have computers in classrooms as well as in computer laboratories. Still other schools use thin-client or terminal service solutions (Moyle, 2005). Wireless, portable, and handheld technologies are also finding places in schools (Preparing Tomorrow's Teachers to Use Technology, 2002). In countries where basic access to computers is approaching universal, there are pedagogical moves away from teaching computing skills per se to integrating technologies into the teaching and learning (Guttman, 2003; Hepp, Hinostroza, Laval, & Rehbein 2004).

Integrating technologies into school education depends upon a robust information technology (IT) infrastructure: the hardware, software, and telecommunications (where it is available). Schools are becoming increasingly sophisticated IT environments in which hundreds of users can be logged on at any one time, but not all schools use software in the same way or to the same extent. A challenge for schools is to determine what infrastructure is appropriate for their contexts.

In education, traditionally the term *infrastructure* has referred to the physical attributes of schools; it now includes an IT infrastructure. Schools use a variety of IT infrastructure models. The choice of model depends upon an array of educational, social, and economic factors. If we accept that the main role of schools is teaching and learning, however, and if we accept that teaching and learning should include the integration of technologies, then we also must accept that the infrastructure of schools must emerge from what we want to happen in classrooms with our students, irrespective of whether those classrooms are physical, in an online environment, or are a mixture of both. Over the past several years in both developing and developed countries, there have been sustained efforts to put in place both school-based and systemic networked IT infrastructures (Farrell & Wachholz, 2004; Guttman, 2003; Programme for International Student Assessment (PISA), 2005). Schools and education departments, however, continue to grapple with the best way to organize themselves in order to ensure technologies are usefully and meaningfully deployed (United Nations Educational, Scientific and Cultural Organization (UNESCO) Bangkok, n.d.). It is timely, therefore, for schools to consider whether OSS has a place in their IT infrastructures.

Schools use various sorts of software for the respective pedagogical and administrative purposes they undertake (British Educational and Communications Technology Agency (BECTA), 2005; Hepp et al., 2004; Moyle, 2003). Some schools deploy only proprietary software, while others deploy only OSS; still others deploy a mixture of both. Some schools and school systems commission software development (e.g., student reporting, human resource, and payroll systems) as well as purchase off-the-shelf products (e.g., Microsoft Office). Over the past decade, however, the inclusion of OSS into schools' IT portfolios has been an emerging phenomenon around the world.

When considering the use of OSS in schools, it is important to recognize that creating and using this software is as much a social process as it is a technical one; it is underpinned with a different development model to that used to develop closed, proprietary software. The differences between the two models are eloquently captured by Eric Raymond in his book, *The Cathedral and the Bazaar* (2001). Indeed, understanding that there are differences between the way open and closed software is developed raises questions for schools about the nature of the control and use of software, which, in turn, raises questions about what human, organizational, and physical models are adopted; the nature of decision-making exercised; and who has access to the skills, facilities, and knowledge essential in order to design, implement, and sustain technologies in schools. There is no ultimate truth, however, about how software ought to be constructed or used in schools. As such, it is important to understand the philosophy and processes underpinning the development of OSS in order to understand how to select suitable software for schools.

SOFTWARE IS SOCIALLY CONSTRUCTED

Software is created by people. Irrespective of the nature of software and its purpose, software is socially constructed. All technologies, including the infrastructure or architecture established to support software deployment, are socially constructed; they are designed and built by people who have their own views about what problems require solving and how a particular problem can be solved with software. Closed proprietary software is developed in secret, and the source code cannot be viewed by anyone other than the developers.

People within open source communities contribute to a software project in ways in which they are able. People write software programs to enable themselves and other people to use computers to communicate with each other in a variety of ways; through document and presentation development, via e-mail, and through the use of rich multimedia on CD, DVD, and over the web. Programmers working over the Internet develop OSS in a devolved manner. Those people with sufficient programming knowledge contribute to the development of the software code. Programming requires the use of a language, which are known as programming languages, the languages in which authors write the commands required to make computers work the way they want them to (Raymond, 2001). Communities of developers communicate with each other via the Internet to create software. Anyone with the skills to understand the languages of programming can contribute.

Others can aid in the development processes by testing and debugging software, writing user documentation, and helping others use the software. This work is conducted through mailing lists. This devolved model is used for open source product development, testing, and maintenance. Those accessing OSS from the Internet can get help desk support through user groups. People contribute their ideas and experiences so the collective is able to develop greater wisdom. These groups can be considered to be akin to communities of learners (Whyte, 2000).

Members of OSS projects share systems of beliefs and values about software development and accessibility. There is the belief that software should be freely redistributable. It is considered a good thing that it can be modified to suit the social and cultural requirements to which the software can be put, which justifies the contribution of considerable collective effort. It is, therefore, a culture that encourages code sharing. The capacity to redistribute source code gives users of the software control over the technologies instead of vendors controlling customers by restricting access to the software code. The ability to participate in an open source community requires higher skill levels than simply maintaining the operation of software. The development of these skills, however, is supported by working in open source communities of learners and assists the process of continuous improvement of both the software and the skills of the people involved. Open source communities, therefore, are educative in themselves.

Understanding that software is the artifact of people collaborating either in secret or in public is fundamental to being able to make informed decisions about which OSS is suitable for a given school context. Accepting that software development is a social phenomenon that produces software assists in understanding how OSS communities operate, which, in turn, assists in evaluating whether certain pieces of OSS have a place in a school's IT portfolio.

SOFTWARE AND SCHOOLS

Schools use software for a variety of purposes. At the back end of a school's IT architecture, software is used to run servers, intranets, and proxy caches, and to provide printing, file serving, e-mail, and Internet access, and to run desktop computers. Common open source operating systems software deployed in schools include versions of Linux, Debian, Mandrake, SUSE, and Redhat Fedora (K-12 Linux 2006; Schoolforge-UK 2006).

Some schools use terminal services by utilizing the Linux Terminal Server Project (LTSP), an add-on piece of software for Linux that allows many computers to be used simultaneously. Applications run on the server with a thin client terminal handling input and output. Computers used in terminal services configurations tend to be low-powered, have no hard disk, and are quieter than desktop computers. In classrooms in which there are many computers operating, quieter options are very attractive. LTSP is also becoming popular in schools since it allows students to access computers without the purchase of expensive

desktop machines. Examples of distributions using LTSP include Skolelinux, AbulEdu, Edubuntu (an Ubuntu derivative), Deworks, and K12LTSP, which works with Fedora, (K-12 Linux Terminal Server Project, 2006).

At the front end, school software can be conceptually divided into two parts: the requirements for running the administrative functions of a school and the software used for enhancing teaching and learning, including specific curriculum software applications to achieve identified learning outcomes. Both the administrative and curriculum sides of a school network deploy office productivity software. Open source office software used in schools includes OpenOffice, which provides word processing, spreadsheets, and presentation software; and KOffice, an integrated office suite for KDE, the desktop environment used on Linux (Open Source Victoria, 2005). Curriculum software such as GIMP (or GNU Image Manipulation Project), OpenOffice Draw, and Blender 3D are used for manipulating graphics; Audacity is used for manipulating sound. Open source online games such as Lin City NG and NASA World Wind are available for teaching and learning purposes (Open Source Victoria, 2005).

MAKING CHOICES ABOUT OPEN SOURCE SOFTWARE IN SCHOOLS

Evaluating suitable OSS solutions alongside proprietary software must now be part of any responsible school's considerations concerning which software should constitute its IT portfolios. Making choices about what is the most suitable software in any given school environment, however, can be an interesting but time-consuming exercise. Questions concerning whether to include OSS within the IT portfolio of a school tend to revolve around balancing the following five demands:

- School context
- Educational, ethical, and social requirements
- Technical demands
- Business case
- Administrative and legal requirements

To balance these demands, some criteria are proposed to assist schools in their selection of suitable OSS. It is intended that this approach will enable schools to make informed choices about which OSS matches their particular environments.

SOFTWARE PROCESSES AND ARTIFACTS

It is important for school sector decision-makers to understand that OSS developments focus on processes for making the software as well as on the artifact itself. Although many OSS communities indicate which version of a software development is the most stable, the communities tend to operate on the principle that the software is always in beta release. As such, selecting OSS for deployment in schools requires an evaluation of the quality of the software and of the community that develops and sustains it. Understanding the depth and maturity of an OSS project can assist in making decisions about whether to deploy it in a school.

In the business sector, Bernard Golden (2005) has documented the importance of understanding the maturity of an OSS project in his book, *Succeeding with Open Source Software*. Golden (2005) outlines an approach to evaluating OSS called the open source maturity model (OSMM), which is structured to enable businesses to make comparisons between software alternatives and to check the match between the business requirements and the software under consideration. He highlights for those working in the commercial sector, the importance of the following:

- Assessing open source business models and determining how they align with those of the business in question
- Managing risk, including the licensing issues associated with OSS
- Locating and assessing technical support, training, and documentation resources

While schools have different motives from that of the business sector, it can be seen in the following that there is the capacity to translate some of Golden's work to the school sector.

SOME SELECTION CRITERIA

Being able to make judgments about the maturity of open source projects is a necessary consideration for the viability of such software deployments in schools. To consider any specific piece of OSS for inclusion in a school's IT portfolio, it must have the following characteristics:

1. Be appropriate for deployment in K-12 school environments
2. Have leadership and a dedicated core developer group
3. Have an active community around the software
4. Provide reports of developments and plans for features development
5. Be able to run on multiple hardware and software platforms
6. Have well-documented license conditions
7. Have third-party support and/or other strategic alliances
8. Provide rapid turnaround processes for supporting requests and bug fixes
9. Provide well-documented technical information and quality assurance processes
10. Have professional development of both teaching and technical staff easily available

A brief exploration of these criteria follows. The priorities placed on each of these criteria by the reader will vary, depending on the school context and the expectations of the functionality of the software.

Appropriate in K-12 School Environments

An overriding criterion all software has to meet for deployment in schools is that it is appropriate to the school sector. Server and other back-end software have to be sufficiently robust in order to enable hundreds of users to be logged on concurrently without the quality and functionality of the software degrading or ceasing to work. Front-end software has to be suitable for use by students who are young and often have varying levels of reading abilities and fine motor skills. Schools, therefore, require software that is robust, durable, and interoperable with other major pieces of software that it deploys.

The more software is interoperable with other pieces of software being used in a school, both technically and in its ability to share content, the more streamlined the integration of technologies into schools can be. Two OSS learning environments developed specifically for the education sector are Moodle (stands for modular object-oriented dynamic learning environment) and the learning activity management system (LAMS). Martin Dougiamas, the leader of Moodle, and James Dalziel, the leader of LAMS, communicate regularly to ensure software compatibility, interoperability, and collaboration between their two projects and with proprietary software vendors.

The degree of success a piece of OSS may enjoy within the open source community also depends in part upon its interoperability. Increasingly, then, open standards are being seen as fundamental to both the work of schools and to the open source community at large. Open standards are recognized as important in the future developments of OSS. Open standards can be considered to be commonly agreed, publicly available specifications for achieving a specific task (Krechmer, 2005). Software that adheres, for example, to the sharable courseware object reference model (SCORM) and is compliant with the World Wide Web Consortium (W3C) (http://www.w3.org/) guidelines for open Internet standards are not only technically desirable but increasingly are being seen as attractive to the work of schools.

There is a range of organizations that aims to provide standards to the open source community, including the Free Standards Group (FSG) (http://www.freestandards.org). The FSG has emerged from the open source community to develop open international standards that enable portability of software within the Linux environment. The aim of open source standards is to write once, run everywhere (FSG, n.d.). The not-for-profit OpenStandards.net provides a portal (http://www.openstandards.net) that has links to a wide range of IT standards bodies such as the W3C.

Front-end or application software for use in schools requires sufficiently simple yet rich functionality to enable children to use it. Complex sets of keystrokes to log on, for example, can make the software difficult and even impossible for use by school students. Images, audio, and graphics have to be clear and synchronized and load rapidly in order to maintain student interest. Font sizes have to be large enough for young children to read, and the content has to be verifiable and factual. Given that most of the users of software in schools are students, the software must also have adequate in-built security measures between the school and the outside world.

While each of these characteristics may seem self-evident, not all software has these characteristics; indeed, much commercial educational software designed for deployment in the university sector and then marketed to schools does not meet these criteria, so there is room for OSS projects.

Leadership and a Dedicated Core Developer Group

Schools purchase certain pieces of proprietary software because their branding is recognized, trusted, and perceived to be of high quality. Making choices between recognized proprietary pieces of software and open source alternatives, however, requires schools to feel secure about the alternatives. The leadership of a software project and the longevity of the core developer group is one indicator schools can look at in order to determine whether an OSS project has veracity and, therefore, is likely to be viable for deployment in a school. As such, understanding how OSS projects are led and managed can contribute to building trust in that project.

Many schools, however, do not understand that identifiable communities contribute to the development of a particular piece of OSS. Each community has a recognized leader who has the last say about which developments to accept and which to reject. The ability to provide leadership to the core developer group is what makes an OSS development community viable. Communities that tend to make the most successful and enduring OSS have explicit philosophical objectives; robust and rigorous development, testing and approval processes for improvements; and clear decision-making processes. For example, the Debian community elects its leader through a vote of its members.

The leader of an OSS development is critical for getting a project up and running. An open source development is typically commenced by the leader instigating a software project, writing a code that shows some possibility, and inviting others to join in the work of the project (Weber, 2004). Software projects that are durable, such as the Linux and Moodle developments, have identifiable, respected, and decisive leaders. Linus Torvalds, the instigator of the Linux software, wrote the first code and then opened it up for others to view and contribute to solving the programming problems. Torvalds remains the final arbiter on adopting a code contributed by members of the community or not (Weber, 2004). Similarly, Martin Dougiamas, who originally developed Moodle, continues to lead that project, enabling it to mature. The ability of a leader to moderate between the members of a community and maintain momentum for ongoing development and maintenance of a software is fundamental to its durability and success and, therefore, its applicability within the education context.

An Active Community around the Software

Schools have to make choices about various options for software that will have longevity. They require software that will have an ongoing life beyond the initial startup phase. The size of the community contributing to an OSS project can be used by schools as an indicator of that software's viability. OSS communities are made up of people who identify themselves with the development of a particular piece of software. Members of an OSS project tend to behave in ways in which trust; building a valued reputation among peers; and being generous with time, expertise, and source code are highly regarded traits of a community's participants (Pavlicek, 2000). These characteristics are also similar to those that schools traditionally value. While size does not necessarily translate into quality software, the size of a community contributing to its development is indicative of the value placed on the development and of the enthusiasm with which the open source programming community views the project. For example, there is a community of more than 1,000 active developers working on the software practical extraction and reporting language (PERL) (http://www.perl.org), a language often used for programming Web applications such as creating CGI programs.

Reports of Developments and Plans for Features Development

In order to plan for the future, schools require software that has a clearly identifiable life cycle. Knowing who is the open source community and what its plans for future software developments are assists schools in planning their own deployments and upgrades. An indicator of the health and viability of an OSS project for a school, therefore, is whether there are ongoing feature developments planned. Mature OSS development projects maintain plans for future developments and provide regular reports back to their community of developers concerning progress toward achieving the planned developments. The learning environment software Moodle, for example, provides a roadmap for forthcoming features (http://docs.moodle.org/en/Roadmap) and provides documented plans for the future (http://docs.moodle.org/en/Future). The PERL community maintains a Web page that provides both weekly progress updates concerning work that has been undertaken and summaries of the status of the projects being undertaken by PERL developers. Chief technology officers in schools can review these reports in order to be informed about whether a piece of OSS is suitable for their contexts.

The Software can be Run on Multiple Hardware and Software Platforms

Since various schools run different hardware platforms and different configurations of those platforms, the ability of OSS to function well on a range of hardware provides greater flexibility for the uptake of that software. The compatibility of the software with hardware commonly used in schools is, therefore, important. A detailed source of information about hardware compatibility with the operating system Linux can be found at http://www.linuxcompatible.org/. The Web sites for individual pieces of OSS also provide direc-tions concerning which hardware platforms will support that particular piece of software.

Some schools run both proprietary and non-proprietary software operating systems. Checking the Web site of an OSS project should indicate its compatibility with various operating systems. If it does not, then the software should be avoided for school use. When an OSS Web site indicates that the software will run on both open and proprietary operating systems, schools have the greatest flexibility for deployments. Furthermore, BECTA (2005) has published a suite of eight case studies that schools can use to assist them in decisions about OSS operating systems and hardware compatibility questions.

Well-Documented License Conditions

Software that provides the capacity for enhancement and modification is valuable to schools so they can customize and badge or brand it to their own requirements. But proprietary licenses that allow such changes are not necessarily financially viable or easily accessible to schools. License management then is an emerging and time-consuming problem for schools, and thus, the more simple and straightforward a software license is, the easier it is for schools to manage. OSS licenses work within copyright laws with the community making the software available under specified license terms. While licenses for most proprietary software are designed to reduce or prevent copies of the software from being made and to prevent changes to the software, OSS licenses are designed to guarantee people's abilities to share and modify the software.

OSS licenses can also be very attractive to schools since they allow the software to be distributed without limits to any number of machines or number of users accessing the software, which means that schools can make many copies of a piece of OSS and distribute it freely to staff and students to use at school and at home. This characteristic of unlimited distribution of OSS is important for schools since it reduces the amount of administrative time to monitor software licenses into which a school has to enter and avoids piracy conflicts.

While there are several different OSS licenses, the two main categories are copyleft and non-copyleft. Copyleft licenses leave the right to copy the software in place. The GNU General Public Licence (GPL) is the most common copyleft license and requires that all modified versions of the software must also be OSS. Non-copyleft licenses do not insist on the right to freely redistribute the software. The most common non-copyleft license is the Berkeley Software Distribution (BSD) license. The Open Source Initiative (OSI) credentials OSS licenses. Once accredited, the software can use the phrase Certified OSS to accompany the software. Schools interested in investigating the license associated with a piece of OSS can find detailed information about that license at http://opensource.org/licenses/.

Third-Party Support and/or Other Strategic Alliances

Knowing that a piece of software is going to have longevity is important for schools' planning and maintenance schedules. It is, therefore, reassuring to schools that as a piece of OSS matures and develops a reputation as a piece of high-quality software, it also tends to attract international interest and industry support through third-party publications, conferences, support documentation, and the provision of expertise. Red Hat Fedora and Moodle provide ready illustrations of this point.

Red Hat is a U.S. commercial company that has outlets around the world. It was one of the first Linux distributors and offers enterprise-level support services. Red Hat was one of the first OSS distributors to establish partnerships with companies such as Oracle and Sun. In late 2003, Red Hat split its corporate distributions from its desktops and renamed the Red Hat desktop operating system Red Hat Fedora. While the Fedora Project is a Red-Hat-sponsored open source project, it is also supported by the open source community and has the goal of building a complete, general-purpose operating system from OSS. Schools around the world use Red Hat publications and support services to assist their OSS deployments.

When OSS is developed for a specific audience such as the education sector, the software benefits from those industry connections. The development of the learning environment Moodle (http://moodle.org), for example, has been designed specifically to support social constructionist frameworks of online teaching and learning (http://docs.moodle.org/en/About_Moodle). Schools around the globe are taking up the use of Moodle, and Moodle Moots were conducted at Oxford University in England and in Adelaide and Sydney, Australia, in 2005 and 2006, with more planned in the foreseeable future. In 2006, the National Educational Computing Conference (NECC), the largest school education technology conference in the world, hosted several sessions and workshops about how to deploy Moodle in schools.

Rapid Turnaround Processes for Support Requests and Bug Fixes

Schools must maintain robust IT infrastructures. Teachers and students require the technology to work when they require it. There can be significant adverse ramifications to IT system crashes and virus invasions in schools, especially during test and examination times. Rapid turnarounds for support requests and bug fixes are, therefore, essential for schools. User forums and mailing lists can provide insights into how a community responds to requests for support. Chief technology officers investigating OSS options can look at the project's changelogs to see lists of new features, bug fixes, improvements, and other known issues for each release of the software. Open source projects that quickly address and fix bugs demonstrate that a healthy community is sitting behind the software, and for schools, software with a community supporting bug fixes is important in order to ensure that the software remains robust and stable. Some projects may use bug-tracking software that enables the status of each bug to be tracked. Schools can use portals such as Sourceforge.org and Freshmeat.net to gain data about OSS projects, including information such as activity, bug fixes, and user rankings.

Well-Documented Technical Information and Quality Assurance Processes

The increasing complexity of school IT environments is challenging the human resources models used in schools. As the number of computers in a school increases, it becomes necessary for technical support to be either on site or easily and rapidly accessible. An interested teacher may have performed such a function previously, but there is now recognition that the role of a teacher is to teach, not to maintain, upgrade, and troubleshoot a school's computing network (BECTA, 2002). In the 21st century, the provision of expert technical support is a mission-critical component in being able to efficiently deploy both proprietary software and OSS in schools.

Well-documented technical information, then, is necessary so that in-school technical officers can download and install software without difficulty. Since OSS is created over the Internet by a devolved group of developers, the maintenance of high-quality documentation is vital to the ongoing health of an OSS project. Well-documented technical information, therefore, can be used by schools as an indicator of the maturity and health of an OSS project. The quality of a piece of OSS also can be seen in the quality assurances processes put in place and managed to validate and verify the quality of that software. Furthermore,

design and code reviews of the software and documentation of test cases (Golden, 2005) can be used as indicators to the quality and robustness of the software.

Technical documentation associated with software can be accessed by a school via the Internet or can be obtained from service vendors. Technical documentation also can be presented in changelogs, release notes, and installation instructions, as well as in manuals. Important to schools is that the technical documentation is easy to access and understand; especially the installation and user documentation for the software distributions. The provision of accessible and easily understood technical documentation can provide school support staff with the necessary assistance to enable OSS to be deployed.

While the quality and availability of the technical documentation associated with a piece of OSS can be used as an indicator of the maturity of the software, not all OSS projects are good at maintaining their technical documentation in up-to-date and easy-to-read formats. Schools wishing to use OSS must check the quality of that particular software's documentation before deploying it. To assist schools in this process, a brief review of the technical documentation associated with five OSS projects is provided. The documentation accompanying these pieces of software is suitable for use in schools and is easy to locate on the Web.

Operating System

Debian is a stable Linux distribution and arguably the most widely used OSS distribution in the world. The Debian community adopts licenses that are OSI approved. Documentation for Debian is available electronically in several languages and in various formats, including PDF and HTML (http://www.debian.org/releases/stable/install-manual). The installation manual is written for a technically competent user and guides him or her through each installation step.

Office Productivity Software

OpenOffice is compatible with many operating systems and hardware platforms. It comes with easily accessible technical documentation to assist installation and troubleshooting. Many of its features and interfaces are similar to other proprietary office productivity software brands. The technical documentation and online help documentation associated with OpenOffice is written for end users and, as such, is nontechnical in its language.

Learning Environments

Moodle (www.Moodle.org) provides easily accessible and simple documentation that can be located on the Moodle home page. The technical documentation is regularly updated and is supported by frequently asked questions (FAQs) and online forums. The documentation has been tailored for teachers, administrators, and developers. Recent changes to the software are also documented on http://docs.moodle.org/.

Graphics Manipulation

GIMP is a graphics manipulation program that runs on multiple operating systems. The documentation includes books, tutorials, and mailing lists. Support documentation is provided in several languages and covers the tools and options found in the GIMP software.

Sound Editing

Audacity is a cross platform audio editing program. The Audacity home page (http://audacity.sourceforge.net/) provides information for both developers and the user community. Release notes, online help, FAQs, and tutorials are available online to support its use.

These preceding five software projects are used in schools because they are easy to install and use, and each has well-documented technical information.

Professional Development of Both Teaching and Technical Staff Is Available

The provision of training and professional development of technical, administrative, and teaching staff is required if the deployment of any piece of software is to be successful in schools. Indeed, the importance of professional development to support the integration of technologies into school environments has long been recognized (Bosco, 2003; Yee, 2000) by both government and nongovernment agencies alike. Golden (2005) also highlights the importance of training and development in the implementation of OSS in the business sector. Yet the development of staff in the use of technologies is often poorly executed by organizations, including schools. Many third-party organizations, however, provide support, including training and professional development services, to schools to enable them to easily and efficiently include and maintain OSS in their IT portfolios. A quick search of the Internet provides names of such third-party support services that specialize in training and professional development associated with the deployment of OSS and that schools can access.

CONCLUSION

There is a wide array of OSS available to schools and, as such, it is now prudent for decision-makers deploying software in schools to consider OSS. Schools face challenges in determining whether OSS is suitable for them and in determining which software projects are viable and sustainable. These challenges include understanding how OSS is developed and how OSS can be deployed in schools' IT infrastructures. To determine the viability of OSS for deployment in schools, however, the following characteristics must be examined concurrently: (1) the quality of the software, (2) the community of developers, and (3) the maturity or status of the software development as a project.

To assist decision-makers in schools, this chapter has presented some criteria to assist in the selection or not of OSS. It is intended that the outline provided here can contribute to developing understandings about whether OSS has a place in a school's IT portfolio. The selections of software that various schools make, however, are unlikely to be uniform. Schools will balance differently the demands of their respective school contexts; their educational, ethical, and social requirements; technical demands; business cases; and administrative and legal requirements in order to make decisions about whether to include OSS in their IT architectures.

The choices that schools make about the software they deploy, however, will not only be an indication of the technical requirements associated with their particular contexts but will also be a concrete reflection of the philosophical priorities they endorse. In the 21st century, schools have to ask themselves whether they are truly communities of learners in which they examine and implement what they value in every aspect of their schools. They now have to question whether they value open or secret processes of software development and examine how their choices translate into the decisions they make about their schools' IT infrastructures, because these structures are on show for all to see.

REFERENCES

Berger, P., & Luckmann, T. (1979). *The social construction of reality. A treatise in the sociology of knowledge.* London: Peregrine Books.

Bessen, J. (2004). *Open source software: Free provision of complex public goods.* Social Science Research Network (SSRN), USA. Retrieved February 8, 2006, from http://papers.ssrn.com/sol3/papers.cfm?abstract_id=588763

Bosco, J. (2003). Toward a balanced appraisal of educational technology in US schools and recognition of seven leadership challenges. In *Proceedings of the Consortium for School Networking (CoSN) Conference*, Washington, DC (pp. 2-8).

British Educational and Communications Technology Agency (BECTA). (2002). *ICT subject leaders: Outline ICT job description.* Retrieved February 8, 2006, from http://www.teachict.com/teacher/ict_subjectleader.doc

British Educational and Communications Technology Agency (BECTA). (2005). *Open source software in schools. A case study report.* Retrieved February 8, 2006, from http://www.becta.org.uk/corporate/publications/documents/BEC5606_Case_Study_16.pdf

Computer Economics. (2005). *Help desk staffing metrics, executive summary.* Retrieved February 8, 2006, from http://www.computereconomics.com/article.cfm?id=1076

Farrell, G., & Wachholz, C. (2004). *ICT in education: Meta-survey on the use of technologies in education.* United Nations Educational, Scientific and Cultural Organization (UNESCO) Asia Pacific Regional Bureau for Education. Retrieved February 8, 2006, from http://www.eldis.org/static/DOC14840.htm

Free Standards Group (FSG). (n.d.). *The imperative for Linux standards. A recommendation for the future. A white paper prepared by the Free Standards Group.* Retrieved February 8, 2006, from http://www.freestandards.org/docs/FSG_Imperative_WP_Public.pdf

Goldman, B. (2005). *Succeeding with open source.* Boston: Addison-Wesley Professional.

Guttman, C. (2003). *Education in and for the information society.* United Nations Educational, Scientific and Cultural Organization (UNESCO), Publications for the World Summit on the Information Society (WSIS) UNESCO, Paris. Retrieved February 8, 2006, from http://www.unescobkk. org/education/ict/v2/detail.asp?id=15685

Hepp, P., Hinostroza, E., Laval, E., & Rehbein, L. (2004). *Technology in schools: Education, ICT and the knowledge society.* World Bank. Retrieved February 8, 2006, from http://www1.worldbank. org/education/pdf/ICT_report_oct04a.pdf

Johnson, B., & Christensen, L. (2004). *Educational research: Quantitative, qualitative, and mixed approaches* (2ⁿᵈ ed.). Boston: Allyn and Bacon.

K-12 Linux. (2006). *Case studies.* K-12 Linux. Retrieved February 8, 2006, from http://k12ltsp. org/casestudy.html

Krechmer, K. (2005). The meaning of open standards. In *Proceedings of the Hawaii International Conference on System Sciences.* Retrieved February 10, 2006, from http://www.csrstds. com/openstds.html

Moyle, K. (2003). *Report of the trial of open source software conducted at Grant High School.* Department of Education and Children's Services (SA) South Australia, Australia. http://www. educationau.edu.au/...report_trial_open_source_ GHS.pdf

Moyle, K. (2005). An infrastructure for what? What infrastructure? *New technologies online conference,* International Networking for Educational Transformation, Specialist Schools Trust. http://www.sst-inet.net/olc/papers.aspx?id=4

Open Source Victoria. (2005). *Free software for schools, catalog of open source software for education.* Open Source Victoria. Retrieved December 1, 2005, from http://www.cybersource. com.au/about/education_FOSS_catalog.pdf

Pavlicek, D. (2000). *Embracing insanity: Open source software development.* Indianapolis: SAMS Publishing.

Preparing Tomorrow's Teachers to Use Technology. (2002). *Digital handhelds: The future of connected teaching and learning?* Retrieved February 8, 2006, from http://pt3.altec.org/stories/digital_handhelds.html

Programme for International Student Assessment (PISA). (2005). *Are students ready for a technology-rich world? What PISA studies tells us.* Organization for Economic Co-Operation and Development. Retrieved December 8, 2005, from http://72.14.207.104/ search?q=cache:FYC6xe3Pl1IJ:www.pisa.oecd. org/dataoecd/28/4/35995145.pdf+Are+students +ready+for+a+technology-rich+world&hl=en& gl=au&ct=clnk&cd=2

Raymond, E. (2001). *The cathedral and the bazaar. Musings on Linux and open source code by an accidental revolutionary* (2ⁿᵈ ed.). O'Reilly.

Schoolforge-UK. (2006). *Case studies.* Retrieved December 8, 2005, from http://www.schoolforge. org.uk/index.php/Case_Studies

United Nations Educational, Scientific and Cultural Organization (UNESCO) Bangkok. (n.d.). *Issues and rationale.* Retrieved February 8, 2006, from http://www.unescobkk.org/index. php?id=794

Weber, S. (2004). *The success of open source.* Cambridge, MA: Harvard University Press.

Whyte, B. (2000). "Upgrading": Co-constructing a community of learners. In *Proceedings of the Australian Association for Research in Education Annual Conference 2000,* Australia. Retrieved December 8, 2001, from http://www.aare.edu. au/00pap/why00394.htm

Yee, D. L. (2000). Images of school principals' information and communications technology leadership. *Journal of Information Technology for Teacher Education, 9*(3), 287-302.

KEY TERMS

Beta Release: The stage of software development in which all the features in their initial form have been implemented. Only bugs are fixed at this stage. In the OSS development cycle, beta releases of software are released widely in order that bugs can be identified and fixed rapidly.

Infrastructure: The structural components that together contribute to a full structure or organization. The term *infrastructure* often is used to refer to the physical elements of an entity such as a school but also refers to an information technology infrastructure that includes the hardware and software to create the system or structure.

Interoperability: The ability of products (in this case, software) to work together seamlessly.

License Management: The process of ensuring that the legal requirements specified in any one software license are met by the users in an organization.

Pedagogy: The processes of teaching children.

Professional Development: The process of learning undertaken to build the capacity of people working in a particular occupation or organization.

Public Good: Goods or services provided in the public interest and in which the processes undertaken to provide a public good do not inhibit other people's freedoms.

Schools: Institutions organized by groups within a society to educate younger members of that society. School buildings are the traditional places in which such learning occurs; however, the necessity to physically attend school is starting to change with the advent of the Internet.

Socially Constructed: The process used by entities, agencies, organizations, or other groups of people that enables goods and services to be created, invented, or produced through understood social processes created by the members of that group of people.

Values: Principles to which an individual or organization subscribes.

Chapter XLIX
Open Source E-Learning Systems:
Evaluation of Features and Functionality

Phillip Olla
Madonna University, USA

ABSTRACT

E-learning applications are becoming commonplace in most higher education institutions, and some institutions have implemented open source applications such as course management systems and electronic portfolios. These e-learning applications initiatives are the first step to moving away from proprietary software such as Blackboard and WEBCT toward open source. With open source, higher education institutions can easily and freely audit their systems. This chapter presents evaluation criteria that was used by a higher education institution to evaluate an open source e-learning system.

INTRODUCTION

Techniques for delivering educational material are constantly evolving to keep pace with new technologies and society habits. Educational content can be created in a variety of formats, such as video, online courses, telecourses, and podcasts, which are just a few of the alternatives to the traditional brick-and-mortar classroom environment. These alternative formats are creating a paradigm shift that is exemplified by the term *e-learning*, which is sometimes called online education or distance learning. The growth in e-learning is compounded by the confluence of Web-based technologies, advances in digital storage, processing and media, and the ongoing boutique approach to software development. This convergence of technologies facilitates education and learning that become ubiquitous and more engaging for both students and educators (Koohang & Harman 2005). E-learning relates to all activities relevant to instructing, teaching, and learning using various types of electronic media. The electronic delivery conduit could be the Internet, intranets, extranets, satellite TV, video/audiotape, and/or CD-ROM.

There is a variety of software applications and platforms that can be used for e-learning. They are defined using a variety of terms, including educational knowledge portal (EKP), learning management systems (LMS), virtual learning environments (VLE), education via computer-mediated communication (CMC) or online education. They might also be called a managed learning environment (MLE), learning support system (LSS), or learning platform (LP). This chapter presents a list of criteria that need to be considered when an organization is considering the implementation of an e-learning system.

E-learning applications are expected to reduce institutional expenses and increase institutional revenues (Harvey, 2004; Moallem, 2004; Porter, 2003). Some higher education institutions are considering the use of open source e-learning applications. Open source software products are freely available for delivering education online (Coppola & Neelley, 2004). Siemens (2003) proposes that the benefits of using an open source model are increased quality, greater stability, superior performance, improved functionality, reduced vendor reliance, reusability, reduced costs, auditability, reliability, and quick bug fixes.

This chapter is structured as follows: The first section provides an introduction to open source software (OSS), followed by an overview of the features and functionality that can be incorporated in any e-learning system. This is followed by evaluation criteria that can be used to evaluate open source e-learning systems.

BACKGROUND

Open Source E-Learning Software

There are various interpretations of OSS (Fuggetta, 2003); however, generally open source refers to a software's source code that is freely available to anyone who wishes to extend, modify, and improve the code. Examples of open source projects include Linux (http://www.linux.org), Apache (http://www.apache.org), Mozilla (http://www.mozilla.org), and OpenOffice (http://www.openoffice.org) (Koohang & Harman, 2005). The GNU project (http://www.gnu.org) defines free software as "a matter of the users' freedom to run, copy, distribute, study, change and improve the software." There are four elements that are emphasized by the GNU: (1) the freedom to run the program for any purpose, (2) the freedom to study how the program works and adapt it to your needs, (3) the freedom to redistribute copies, (4) the freedom to improve the program and release your improvements to the public so the whole community benefits (freedom 3). The open source model encompasses a set of principles and values that ensures the integrity of OSS. One of the prominent organizations that advocates open source projects is the Open Source Initiative (OSI) (http://www.opensource.org). OSI is a not-for-profit organization that recommends the following 10 guiding rules that are widely accepted by the open source community:

1. Free redistribution
2. Source code must be included
3. Derived works; allow modifications
4. Integrity of the author's source code
5. No discrimination against persons or groups
6. No discrimination against fields of endeavor
7. Distribution of license
8. License must not be specific to a product
9. License must not restrict other software
10. License must be technology-neutral

E-learning applications are becoming commonplace in most higher education institutions, and some have implemented open source applications such as course management systems and electronic portfolios. These e-learning applications initiatives are the first step to moving away from proprietary software toward open source. With open source, higher education institutions can easily and freely audit their systems. There is a view that open source systems are open and transparent and reduce the vendor lock-in. The system becomes flexible. There will be ultimate access/control, ownership, and freedom. The open system encourages increased exchange of ideas that advances innovation (Koohang & Harman, 2005). Young (2004) proposes that successful implementation of an open source model depends on (1) community building, (2) agreeing on a common definition of open source, (3) allocating and securing budget for free software, (4) encouraging institutions to switch to open source, and (5) have a positive working relationship with companies. Coppola and Neelley (2004) delineated several benefits of OSS for open learning. They are as follows:

1. The software evolves more rapidly and organically.
2. Users' needs are rapidly met as the OSS model harnesses their collective expertise and contribution.
3. New versions are released often and rely on the community of users and developers to test it, resulting in superior quality software tested on more platforms and in more environments than most commercial software.
4. The development team is often largely volunteers, distributed, many in numbers, and diverse. Often, paid members of the development team will manage the project and organize the work of the volunteers.
5. Security is enhanced because the code is exposed to the world.

The open source model promotes collaboration and sharing of resources. It creates a community of people that work together to achieve common goals (Koohang & Harman, 2005), especially in the open learning environment. Coppola and Neelley (2004) also suggest that an open source model promotes freedom to choose, increases user access/control, encourages a link to a global community, promotes quality, and enhances innovation in teaching and learning. The following

section describes some of the features of e-learning systems.

MAIN FOCUS OF THE CHAPTER

Features and Functionality of E-Learning Systems

E-learning applications comprise different features and functionalities that support the online learning environment. One of the key features expected from any open source system is the ability to facilitate communication between students and the tutor. It is also important to have a system that has capabilities such as creation of announcements, calendar entries, discussions, links, syllabi, course descriptions, and other course content using templates. Students should have the ability to e-mail other students, professors, or predefined distribution groups, along with access to a searchable e-mail address book.

The discussion board is a virtual space used to promote dialogue between students and the instructor. Typically, instructor-led discussions can be viewed by date and thread. Discussion posts may include attachments and URLs. Posts can be plain text, formatted text, or html. Discussion threads tend to be expandable and collapsible in order to view an entire conversation on one screen.

It is important that groups have their own shared file area and a private group discussion board in order to facilitate collaborative learning, since the ability to form cohorts is critical for distance learning courses. Groups can be defined either at the course level and apply across all activities that support them or at the individual activity level. In some systems, group work is managed through the use of project sites that are separate from the main course site. Each project site can have its own shared file exchange, discussion tool, calendar, announcements, chat, and group e-mail list.

It is important that the e-learning system allows real-time communication, also known as instant messaging, among students enrolled in the specified course as well as between the students and the instructor. This allows for quick interaction between students, which is unlike a typical discussion board in which it can take days for students or professors to respond. Real-time communication is crucial in maintaining dynamic conversations and debates, similar to what might take place in a traditional college classroom. This instantaneous communication also enhances student teamwork.

Some e-learning systems contain features that push academic information to user cell phones, PDAs, or external e-mail addresses. Another important feature is a course calendar, a flexible tool for both instructors and students. Instructors can post course-related events and announcements and can assign tasks. This allows the instructor to plan lessons and balance workload across several courses. For students, calendars provide the ability to monitor important deadlines.

One of the most important features of an e-learning system is the ability to effectively manage assessments and students grading. An e-learning testing engine allows the creation, randomization, and scoring of the most common test formats, including true/false, multiple-choice, multiple-answer, matching, fill-in-the-blank, and short-answer/essay type questions. Some systems will also allow test questions to contain images and audio files. Test security features should include the ability to set specific times when students are permitted to take tests and to set a specific time limit on a test. The systems should support a fully functional grade book that categorizes grades by assessment and by student, and should provide the capability to export scores to an external spreadsheet. Most systems provide the functionality to allow students to securely submit their work to their tutors, a feature called a digital drop box for submitting to professors completed

assignments that are time and date stamped. Students have the ability to view their individual grades as well as to compare themselves against the overall performance of the class by viewing overall percentages.

Most e-learning systems allow the tutor some level of customization; the instructor should be able to easily change the appearance of a course by changing the order and name of menu items and the location and width of the navigation menu. Custom tools can be created and quickly added and removed from course or student home pages. Students can customize the sounds, colors, font sizes, and layout of the tools within the interface. All registered students should have access to their own home page, which provides access to each of the classes in which they are enrolled as well as any groups of which they may be a part. The individual home page also lists any events that are linked to classes in which the student is enrolled as well as system-wide events from the student's personal calendar.

The systems normally provide templates to choose from when designing an online course. The templates can contain specific university logos and colors schemes, and users can change navigational options according to their preferences.

It is important that any chosen system possess adequate help and support. This feature provides tutors with access to supporting material such as an online instructor training guide, help files, and context-sensitive help; online groups to share documents, course components, schedules, and other collaborative tools and learning objects

An important functionality is the ability to allow instructors to post online lectures in several popular video and audio formats, including MPEG, WAV, MP3, QMOV, and others. This flexibility allows multiple pedagogical methods to be used in presenting course material. The next section will present evaluation criteria that can be used to evaluate a course management system.

FUTURE TRENDS

Evaluation of Existing Course Management Systems

The evaluation criteria of any e-learning system must start first with an understanding of the goals of the institution. There are potential trade-offs to consider when assigning weights to these criteria, which should be determined by the university's vision and strategy. For example, a university may want an e-learning package that can meet its future requirements and is easy to implement. However, putting an emphasis on meeting future requirements may require a package that uses state-of-the-art component technology, even though that technology has not been successfully implemented by other universities and may contain bugs, making it harder to implement initially. Such factors should be considered carefully and weighted in order of importance to the institution. These evaluation criteria would include the following:

- **Known requirements:** Ability of the package to meet the university's current academic and administrative requirements and future requirements that are currently known to exist
- **Unknown future requirements:** Ability to modify the package to meet the university's new requirements as they become known
- **Implementability:** Ability to implement the package easily; this might include an analysis of the vendor's background, software maturity, technology maturity, modifications, third-party implementer considerations, implementation assistance provided by the e-learning vendor, quality, documentation, and training
- **Support:** Ability of the vendor to support both the package and the university in the future; factors include vendor responsiveness, quality, development methodology, modifications, financial stability, warranty, user groups, and support functions
- **Cost:** Total cost to purchase and implement the package as well as ongoing maintenance and support costs. These costs include the following:
  o **Annual software license fee**
  o **Software purchase price**, including discounts
  o **Cost of additional hardware**
  o **Cost of customizing the package** to individual specifications
  o **Cost of installing the software and integrating it with other systems**
  o **Cost of converting data** (e.g., course Web sites on the system not chosen)
  o **Cost of training** the system administrators and those who will become e-learning faculty trainers
  o **Cost of additional products**, such as software tools needed to run the system and hardware needed to run the system
  o **Annual base maintenance package**, cost of modifications and maintenance of required hardware (including depreciation)
  o **Ongoing costs for training**, help desk support, system administrators, and application programmer costs for ongoing customizations, installations, and support

Once the university has established its overall goals and weighted them in terms of importance, it can then move on to an evaluation of the features of specific products and how those might best meet the needs of its constituents. Standard features of contemporary e-learning management systems include the following:

- Course scheduling and organization
- Student enrollment and administration
- Course content delivery capabilities

- Management of online class transactions
- Tracking and reporting of learner progress
- Assessment and measurement of outcomes
- Reporting of achievement and completion
- Student records management
- Hosting capabilities
- Virtual classroom and live collaboration tools
- Content assembly and authoring tools

The next generation integrated e-learning will likely include the following additional features:

- Object-oriented and Web-based architecture
- Skills gaps analysis/pretest and test-out features
- Profiling and mapping of personalized learning paths
- Employee competency and performance management
- Seamless integration with other enterprise systems

- E-commerce and wireless (mobile e-learning) capabilities
- Compliance with industry standards

The new generation of e-learning systems is increasingly browser-based and does not require many downloads or plug-ins on the user's desktop. While the emergence of completely Web-based applications is not a revolutionary technological shift, it is a major evolutionary process that provides a number of benefits to vendors, customers, and end users. The most important advantages of these are shorter implementation times, increased scalability, easier systems maintenance, enhanced deployment and data management, improved software control, and fewer memory problems on the user's desktop.

In addition to supporting the university's vision, mission, and goals, the evaluation of a specific e-learning tool must take into consideration the learner, the faculty, and the administration. Listed in Table 1 is an evaluation of 10 of the most popular open course e-learning systems. Each was evaluated on a scale of one to five, with five representing a particular strength of the product

Table 1. Open source course management systems comparison copyright (Source: Olla Crider, 2006)

	.LRN	Bodington	Claroline 1.4	ClassWeb 2.0	KEWL 1.2	Moodle 1.5.3	Sakai 2.0	ATutor 1.5	CHEF	Course manager
Communication										
Discussion Forums	5	3	1	2	4	5	3	4	2	3
File Exchange	4	4	2	0	3	3	4	4	3	0
Internal E-Mail	5	0	5	0	5	2	4	3	0	3
Online Journal/Notes	5	2	0	0	2	0	0	5	0	2
Real-Time Chat	0	0	3	0	5	4	3	5	3	2
Audio/Video Services	0	0	0	0	5	0	0	0	0	5
Whiteboard	0	0	0	0	5	0	0	5	0	0
Subtotal Communication	**19**	**9**	**11**	**2**	**29**	**14**	**14**	**26**	**8**	15

continued on following page

Table 1. continued

Productivity										
Bookmarks	0	0	0	0	5	0	0	0	5	**0**
Calendar/Progress Review	5	0	2	0	3	3	5	3	4	**3**
Orientation/Help	3	0	0	0	4	3	5	4	2	**2**
Searching Within Course	4	0	0	0	5	3	3	3	1	**0**
Work Off-line/Synchronize	5	0	0	0	3	0	0	4	0	**0**
Subtotal Productivity	**17**	**0**	**2**	**0**	**20**	**9**	**13**	**14**	**12**	5
Student Involvement										
Groupwork	5	4	3	0	4	4	4	4	0	**2**
Self-Assessment	3	5	5	0	3	3	4	3	0	**2**
Student Community Building	5	0	0	0	3	0	3	4	0	**0**
Student Portfolios	5	0	3	0	2	4	4	3	0	**0**
Subtotal Student Involvement	**18**	**9**	**11**	**0**	**12**	**11**	**15**	**14**	**0**	4
Administration										
Authentication	5	3	3	0	3	5	5	4	4	**3**
Course Authorization	3	5	0	0	3	4	4	5	3	**3**
Hosted Services	5	0	0	0	0	5	4	4	0	**3**
Registration Integration	5	0	3	0	3	4	3	5	0	**3**
Subtotal Administration	**18**	**8**	**6**	**0**	**9**	**18**	**16**	**18**	**7**	12
Course Delivery Tools										
Automated Testing and Scoring	1	3	3	2	5	5	5	4	1	**2**
Course Management	5	3	0	0	2	4	3	3	0	**0**
Instructor Helpdesk	4	4	4	0	3	3	5	3	0	**0**
Online Grading Tools	0	3	0	0	5	5	4	3	2	**3**
Student Tracking	0	0	3	0	0	5	0	3	0	**2**
Subtotal Course Delivery	**10**	**13**	**10**	**2**	**15**	**22**	**17**	**16**	**3**	7
Curriculum Design										
Accessibility Compliance	3	4	0	0	4	5	3	5	2	**5**
Content Sharing/Reuse	0	0	0	0	0	0	0	4	0	**0**
Course Templates	4	4	4	0	0	4	3	4	0	**2**
Curriculum Management	0	0	0	0	0	0	0	0	0	**5**
Customized Look and Feel	5	4	0	3	3	5	5	5	0	**4**
Instructional Design Tools	0	0	0	0	4	5	3	4	0	**2**

continued on following page

644

Table 1. continued

Instructional Standards Compliance	5	5	0	0	0	5	5	5	0	0
Subtotal Curriculum Design	**17**	**17**	**4**	**3**	**11**	**24**	**19**	**27**	**2**	18
Technical Specifications										
Hardware/Software										
Client Browser Required	4	0	4	0	2	4	5	4	0	**3**
Database Required	MySql others	,	MySql	MySql	MSSql	MySql	MySql Oracle	MySql	,	MSSql
Server Software Allowed	4	3	2	3	3	5	4	3	3	**3**
UNIX Server	4	4	4	0	0	4	4	4	4	**0**
Windows Server	1	4	3	3	2	3	4	3	2	**3**
Subtotal Hardware/ Software	**13**	**11**	**13**	**6**	**7**	**16**	**17**	**17**	**9**	9
Pricing/Licensing										
University or Private	U	U	U	U	U	U	U	U	U	**P**
Costs	Free	Free	Free	Free	Free	Free	Free	Free	Free	$$$
Open Source	Yes	Yes	Yes	Yes	Yes	Yes	Yes	Yes	Yes	Yes
Variety of Optional Extras	1	2	2	0	3	4	5	4	2	**3**
Software Version	2.1.3	2.6	1.71	2.0.3	1.2	1.5.3	2.1	1.5.2		**2.4**
TOTAL SCORE	95	69	59	13	106	118	116	136	43	73

being evaluated and one offering the least desirable functionality of those reviewed. Features evaluated included communication, productivity, student involvement, administration support tools, course delivery tools, curriculum design, technical specifications, and pricing/licensing. An overall rating was assigned by tallying the scores of each product in each of the functional areas. It should be noted that some systems were particularly strong in some functional areas and had a decided edge in a particular category.

It is the author's opinion that such evaluations should include both technical and nontechnical considerations in order to build the strongest level of support and capabilities. It likewise should include feedback from faculty, students, departments, and administrators in order to optimize input on the components most likely to be used, appreciated, and anticipated by each group.

CONCLUSION

Higher education leaders must find a way to reduce the cost and complexity of system integration work while ensuring that their electronic learning systems are built on a reliable and scalable architecture that allows them the flexibility to meet the needs of diverse teaching and learning styles. The educational technology systems of the future must be built from the perspective of enterprise infrastructure. They must be based on an open and modular framework that can be used by software vendors, and they must meet the needs of entire campuses, individual departments, and even single courses. In addition, they must take advantage of international standards that are being used by formal educational systems around the world.

The design of open source e-learning systems are now flexible enough to adapt to a wide range of instructional requirements and styles yet stable enough to allow faculty and students to concentrate on teaching and learning and not on the technology itself. They are robust enough to successfully scale up to support an ever-increasing workload, to adapt to new technologies over time, and to integrate with the existing campus infrastructure.

Recent years have seen strong growth in the availability of open source course management systems, and many tools now exist for the evaluation of these systems. An educational institution wanting to take advantage of these new tools may find that it is a laborious task to identify those systems that best align with its mission, values, and goals. Using the categories presented in this chapter may assist in the evaluation criteria.

REFERENCES

American Bar Association (n.d.) *An overview of "open source" software licenses: A report of the Software Licensing Committee of the American Bar Association's Intellectual Property Section.* Retrieved October 12, 2005, from http://www.abanet.org/intelprop/opensource.html

The Centre for Educational Technology Interoperability Standards (CETIS). (2004). *Open source e-learning technology hits prime time.* Retrieved from http://www.cetis.ac.uk/content2/20040724101134

Collier, G., & Robson, R. (2002). *What is the open knowledge initiative? A white paper prepared by Eduworks Corporation for O.K.I.* September 20, 2002. Retrieved October 30, 2005, from http://web.mit.edu/oki/learn/whtpapers/OKI_white_paper_120902.pdf

Coppola, C., & Neelley, E. (2004). *Open source open learning: Why open source makes sense for education.* Retrieved October 27, 2004, from http://www.rsmart.com/assets/OpenSourceOpensLearningJuly2004.pdf

EdTechPost (n.d.) *EdTech post open course management systems.* Retrieved September, 2006, from http://www.edtechpost.ca/pmwiki/pmwiki.php/EdTechPost/OpenSourceCourseManagementSystems

EduTools (n.d.). *Course management systems comparison tool by the Western Cooperative for Educational Telecommunications.* Retrieved November 27, 2005, from http://www.edutools.info/course/compare

E-learning systems information and services. (n.d.). Retrieved October 29, 2005, from http://www.e-/learningcentre.co.uk/eclipse/vendors/opensource.htm

Free & Open Source Software Portal, United Nations Educational, Scientific and Cultural Organization (UNESCO). (n.d.). Retrieved November 5, 2005, from http://www.unesco.org/cgibin/webworld/portal_freesoftware/cgi/page.cgi?g=Software%2FCourseware_Tools%2Findex.shtml&d=1

Fugetta, A. (2003). Open source software: An evaluation. *The Journal of Systems and Software, 66*, 77-90.

Harvey, A. (2004). Building a learning library. *The British Journal of Administrative Management, 3/4*, 26-27.

Jansen, C. M., Bach, V., & Osterle, H. (n.d.). *Knowledge portals: Using the Internet to enable business transformation*. Institute for Information Management at the University of St. Gallen, Switzerland. Retrieved October 27, 2005, from http://www.isoc.org/inet2000/cdproceedings/7d/7d_2.htm

Koohang, A., & Harman, K. (2005). Open source: A metaphor for e-learning. *Informing Science Journal, 8*.

Kumar, V. M. S., Merriman, J., & Long, P. D. (2001). Building open frameworks for education. Retrieved October 20, 2005, from http://www.educause.edu/ir/library/pdf/erm0169.pdf

Moallem, M. (2004). Distance learning and university effectiveness. Review of C. Howard et al. (n.d.). Distance learning and university effectiveness: Changing educational paradigms for online learning. *Information Management, 17*(3/4), 29-30.

Moodle Software homepage. (n.d.). Retrieved October 28, 2005, from http://moodle.org/doc/

The open knowledge initiative. (n.d.) Retrieved August 29, 2006, from http://www.okiproject.org/

Open source e-learning ratings. (n.d.). Retrieved November 20, 2005, from http://www.opensourcee-learning.com/index.php?option=content&task=view&id=388&Itemid=143

Paulsen, M. F. (2003). *Online education and learning management systems: Global e-learning in a Scandinavian perspective*. Oslo: NKI Forlaget.

Porter, L. (2003). ABCs of e-learning: Reaping the benefits and avoiding the pitfalls. *Research Library, 50*(2), 273.

Sakai Project homepage. Retrieved October 28, 2005, from http://www.sakaiproject.org/

Siemens, G. (2003). *Open source content in education: Part 2. Developing, sharing, expanding resources*. Retrieved October 27, 2004, from http://www.elearnspace.org/Articles/open_source_part_2.htm

Sun Microsytems. (n.d.) *Open knowledge initiative*. Retrieved November 2, 2005, from www.sun.com/edu

Virtual learning environment. (n.d.) Retrieved October 18, 2005, from http://en.wikipedia.org

Young, J. (2004). Five challenges for open source. *Chronicle of Higher Education*.

KEY TERMS

E-Learning: Education delivered electronically, typically over the Internet but also via a network or stand-alone computer. E-learning is computer-enabled transfer of skills and knowledge. E-learning applications and processes include Web-based learning, computer-based learning, virtual classrooms, and digital collaboration. Content is delivered via the Internet, intranet/extranet, audiotape, videotape, satellite TV, and CD-ROM.

Evaluation: To assess the effectiveness of something according to pre-existing criteria.

Evaluation Criteria: The factors that individuals track/follow in order to determine the effectiveness of an item being assessed for quality.

Free Software (FS): Software that can be used, studied, copied, modified, and redistributed without any restrictions, as defined by the Free Software Foundation (FSF).

Functionality: Degree to which an item operates or can be operated as intended by its designers/creators.

GNU: A project sponsored by the Free Software Foundation; a complete operating system based on the Linux kernel. GNU is an acronym for Gnus not UNIX. The project has developed its own kernel called HURD and maintains a library that will link to both free and proprietary software (GNU Project, 2004)].

Open Source Software (OSS): Software distributed both as source code and in compiled form. It cannot discriminate against any field of endeavor, group, or individual. It must come with a license that does not restrict derivative works and must not restrict any party from selling or giving away the code. Further, rights to use the code cannot be tied to a specific program and cannot restrict any other software or program to be of a certain origin or type (Open Source, 2005).

Chapter L
The Role of Open Source Software in Open Access Publishing

David J. Solomon
Michigan State University, USA

ABSTRACT

This chapter discusses the rapid transition from paper to electronic distribution of scholarly journals and how this has led to open-access journals that make their content freely available over the Internet. It presents the practical and ethical arguments for providing open access to publicly funded research and scholarship and outlines a variety of economic models for operating these journals. There are hundreds of journals that are run on volunteer effort by a few people or even a single person. Journal management software that can streamline the peer-review process as well as other aspects of operating a journal can dramatically reduce the effort of operating these journals and allow them to flourish. The availability of high-quality, open source journal management software is playing an important role in facilitating the success of small volunteer-run, open-access journals.

INTRODUCTION

This chapter discusses the Open-Access Initiative (OAI) in scholarly publishing and how open source journal management software can be a critical resource for small open-access journals published by volunteers. The issues that will be covered include the following:

- Rapid transition from paper to electronic distribution of scholarly journals and its economic implications
- Practical and ethical arguments for open access to research and scholarship
- Alternative models for funding the dissemination of scholarship and the key role open source software can play in facilitating open access to scholarship
- Future trends in the organization and funding of scholarly publications

BACKGROUND

Although a few scholarly journals[1] were distributed electronically prior to the World Wide Web (the Web), the development of the Web made electronic distribution of journals practical. Today, the majority of scholarly journals are available via the Internet, and electronic dissemination is quickly becoming the dominant means by which these journals are distributed (Van Orsdel & Born, 2002).

This rapid transition and the inherent differences between paper and electronic distribution have thrown the 340-year-old, multibillion-dollar scholarly publication system into turmoil. With electronic dissemination, many of the most resource-intensive roles that have traditionally been played by both publishers and librarians are disappearing, and it is not yet clear who will perform the roles that remain and how the evolving system will be organized and financed (Solomon, 1999).

Along with speed and convenience, electronic distribution has significantly reduced the cost and effort required to publish a journal. While these efficiencies are evident throughout the publication process, the most striking difference is that electronic dissemination essentially has removed the cost of distribution.

Since the incremental cost of distributing each copy of a paper journal is significant, the only practical means of funding these journals is through subscription fees. With electronic publication, funding a journal by other means and disseminating the content of these journals at no charge is both feasible and, in the view of many people, highly desirable. The calls for free and open access to scholarly journals started almost as soon as they began appearing in digital form (Harnad, 1990). By 2002, the movement organized itself into what is commonly called the open-access initiative (Budapest Open Access Initiative, 2002). There are compelling reasons for open access to scholarship that involve both practical and ethical issues.[2]

As noted by Willinski (2006), open access is not an all-or-nothing phenomena, but rather a continuum with many forms. He defines 10 styles of open access that provide different types of access that largely reflect how the cost of publication is funded. At the most basic level, there

are two general approaches to open access: the development of open-access journals and authors archiving their own manuscripts in open-access archives. These have been termed the "gold" and "green" roads to open access (Guedon, 2004).

At its most limited form, there are journals that make abstracts freely available. At the other end of the continuum are what Willinsky (2006) terms subsidized journals, which provide immediate open access to their full content with the cost of operating the journal subsidized by other means. Other models include partial open access, in which some material is freely available and the rest is available only by paid subscriptions; delayed open access, in which material is restricted to paid subscribers initially and at some point is made freely available; and author-paid models, in which the material is made freely available but authors must pay a fee to publish in the journal.

All these models provide some level of access over the traditional subscription fee model; however, all but subsidized journals limit open access to some extent or charge authors as a means of funding the publication process. Unfortunately, any restriction on access, including charging authors for publication, places barriers to the dissemination of research and scholarship that reduces the value of the information.

ARGUMENTS FOR OPEN ACCESS TO RESEARCH AND SCHOLARSHIP

There are compelling ethical and practical reasons for providing complete unrestricted access to scholarly literature. From an ethical standpoint, much of the cost of scientific research and other forms of scholarship is funded though public sources. The National Institutes of Health (NIH), for example, is spent approximately $29 billion on biomedical research in fiscal year 2006 (National Institutes of Health, 2005). Willinsky (2000) has called the product of this research public knowledge and argues that since the research is publicly funded, the knowledge it produces is public property and should be made freely available to anyone who wishes to access it.

For more than 340 years, scholarly journals have formed the most comprehensive, accurate, and up-to-date repository of knowledge. In a very real sense, these journals in aggregate form are our archive of scientific and scholarly knowledge (Guédon, 2001). Journal publishers have traditionally required authors to sign over copyrights in exchange for publishing their works. Publishers have argued that this is necessary to allow them to fund the publication process. With the cost reductions of electronic publication, it is getting harder to justify this assertion. Although our tax dollars largely pay for basic research, by requiring the copyright, journal publishers end up owning the embodiment of the knowledge generated by the research. This state of affairs has become almost ludicrous. For example, the public pays $60,000 to $80,000 to fund the research that results in an article from an NIH-funded grant, while the cost of publishing the article is only $2,000 to $3,000; yet the journal publisher ends up owning the copyright to the article (Willinsky, 2006).[3] Furthermore, in the United States, indirect payments from research grants to universities and other research organizations provide a significant portion of the funding for research libraries. It is these same libraries that purchase the majority of the subscriptions to scholarly journals. In essence, not only is the research publicly funded but so is the cost of disseminating the research.

Fifty years ago, virtually all scholarly journals were owned by professional societies. They operated these journals for the benefit of their profession, and in most cases, they operated the journals as a loss subsidized by the society. As the scientific enterprise began to expand rapidly after World War II, the need for journal space grew, and the budgets of research libraries increased. By the 1970s, publishing scientific journals became profitable. This, along with the need for additional journal space as the scientific enterprise continued to expand,

resulted in the rapid growth of the commercial scholarly publishing industry. This industry has become extremely profitable and is publishing a growing percentage of the scholarly journals.

The cost of commercially published journals tends to be much higher than journals published by societies. According to one set of studies done in the area of physics, the cost of commercially published journals was as much as an order of magnitude higher than society-published journals (Barschall, 1988). The cost of journal subscriptions has skyrocketed over the last 40 years to the point that it is significantly limiting the abilities of even well-funded research libraries in the United States to maintain their journal collections. This has prompted what librarians have called the serial pricing crisis.[4] Furthermore, the commercial publishing industry is rapidly consolidating into a few huge publishing companies that each owns hundreds of scientific journals.

The high cost of journal subscription fees has resulted in scientists and scholars in developing countries largely being cut off from the literature in their fields.[5] This has also resulted in the general public in the United States and other developed countries being cut off from the scientific and scholarly literature they fund through their tax dollars. While most scientists and other scholars in the developed world have reasonable access to the literature in their fields via their university libraries, the access is far less convenient than it could or should be. Rather than seamlessly moving from hyperlinked references to the full text of an article, scholars must either pay a fee in the range of $30 per full-text article via the publisher's Web site or work through their university library's electronic journal portal to access the article via the library's subscription. In my experience, working through these library portals to a specific article is a tedious process that takes about five minutes. This does not seem like much time; however, since scholars often review dozens of articles in researching a topic, it adds up to a great deal of wasted time compared with clicking on a single hyperlink to access an article in an open-access journal or archive. This wasted time reduces the efficiency of the research process and is a drain on the resources the public invests in research.

The advantages of moving to open-access journals and funding publication through means other than subscription fees seem obvious. However, the issues involved in radically changing a huge and well-established publication system are complex. As one might expect, there is a great deal of resistance to the OAI from the commercial publishers as well as from many scientific societies that are concerned about losing or having a significant reduction in the financial support for their journals. Although a growing number of university faculty members is becoming aware of the economic and ethical issues surrounding scholarly publishing, the majority has not yet embraced the OAI. This can be seen in the lack of success of the NIH's recent initiative encouraging its grantees to archive publications from their grant-funded projects in PubMed Central, the National Library of Medicine's open-access archive. During the first eight months of the program, less than 4% of the eligible articles was archived (Zerhouni, 2006).

Despite the fact that there is still limited support for the OAI among scientists and scholars, a growing number of open access journals and open access archives is appearing. Most of the focus on the topic in the literature has been on large, well-organized initiatives such as the Public Library of Science (2006), BioMed Central (2006), and author self-archiving in PubMed Central (2006). What is often forgotten is that there are hundreds of open-access journals that have been created by individuals and small groups of colleagues. In many cases, these are excellent journals that have become well respected and are having a significant impact on disseminating scholarship. They generally fall into the category of subsidized journals that neither charge for access nor charge authors for publication. They tend to be subsidized by the people who have created the journals and, to some extent, by their employers or societies. In this sense, these journals are the purest form of open access and, as such, the most efficient and effective means of disseminating scholarship. Interestingly my experience with several of these journals shows that they tend to share many of the characteristics of open source projects described in Jill Coffin's insightful article on open source cultures (2006).

As I have argued elsewhere (Solomon, 2006), these subsidized journals often face a dilemma. As they become successful, both access and submissions increase, as does the workload of operating these journals. Unlike other funding models, with subsidized journals there is no direct link between the success of the journal and the resources available for publishing the journal. The people operating these journals can become overwhelmed by the rapidly expanding workload as the journal becomes successful. This was the case for an open-access journal in medical education, *Medical Education Online* (MEO), which I founded in April 1996. Open-access journal management software can be a key resource in helping these journals survive and continue to allow their content to be freely available even as they become successful.

ROLE OF OPEN SOURCE SOFTWARE IN MAINTAINING SUBSIDIZED OPEN-ACCESS JOURNALS

Publishing an electronic journal can be done with virtually no funding. All that is needed is an e-mail account for communication and a Web hosting site for disseminating manuscripts. Both of these are available to most university faculty members through their institutions. If not, these resources can be purchased from an Internet service provider for as little as a few dollars a month.

Publishing an electronic journal requires effort, including the effort of conducting the peer-review process and providing feedback to authors; copy editing and typesetting manuscripts; indexing; handling correspondence concerning the journal; and updating and maintaining the journal Web site. These are tasks that can be done by an individual or a small group of scholars interested in creating an electronic journal in their field of scholarship. There are hundreds of journals operating in this fashion.[6] From my own experience with MEO, operating a subsidized electronic journal is quite manageable for even an individual when the journal receives only a small number of submissions. When these journals grow and prosper, as they often do, the workload can grow very quickly and become unmanageable.

Web-based journal management software provides a straightforward and effective means of significantly reducing the workload necessary to operate a journal by automating much of the clerical and administrative aspects of the process. The workload of managing the peer-review process, in particular, can be reduced by Web-based software. Software can also streamline the process of indexing journals and maintaining the journal Web site and can provide tools that facilitate accessing the material in the journal more effectively. While journal management software cannot solve the resource issues of operating a successful subsidized journal, it can be extremely helpful in reducing the workload as these journals become successful. I believe this can often be the difference between the success and failure of these journals.

The co-editors of MEO and I have found journal management software to be essential for implementing another strategy for maintaining a subsidized open-access journal as it grows. The most time-consuming aspects of operating a scholarly journal are the editorial tasks of conducting manuscript review. This requires carefully reading each manuscript submitted to the journal, assigning reviewers, aggregating the feedback from reviewers in making a publication decision, and providing constructive feedback to authors. In most cases, articles are accepted with revisions, and the editor must work with the author(s) to complete these revisions.

At MEO, we are addressing the workload of conducting the peer-review process by implementing a system with multiple review editors. By distributing the effort of this critical and time-consuming aspect of operating a journal among a number of different volunteer review editors, we are keeping the workload of operating a successful journal manageable. The strategy is also flexible; additional editors can be added to meet an increasing number of submissions while keeping the editorial tasks of the volunteer editors reasonable.

MEO currently has two managing editors; one performs an initial review of each new manuscript submission and assigns the manuscript, if suitable, for peer review to one of six review editors who manage the peer-review process until a manuscript is either ready for publication or is rejected. Seven of us are at various locations spread throughout the continental United States, and one of the review editors is located in Singapore. We generally have about 20 manuscripts in some stage of the peer-review or revision process and receive about 80 to 90 submissions a year. Without a Web-based peer-review management system, it would be nearly impossible to manage the review process with the editorial board distributed at various locations throughout the world.

JOURNAL MANAGEMENT SOFTWARE

MEO began using a rudimentary version of our journal management software about five years ago. It allowed manuscripts to be submitted electronically via a Web form system and maintained a database of both submissions and reviewers. The software automated the process of sending requests to review and tracked the requests to review sent to each of our more than 250 reviewers. Even this minimal system saved a significant amount of time.

A much more automated system was put in place about 6 months ago in order to allow us to move to the distributed review system described previously. The system tracks assignments of manuscripts to review editors as well as each manuscript through the various stages of the review process. Review assignment and feedback to authors is done by e-mails generated by the system; reviewers enter their ratings and feedback via Web forms. The submission process continues to be Web-based, and reviewers download review copies of the manuscript from our server. All correspondence and review feedback is stored in the database. Each reviewer's areas of interest/expertise as well as a detailed review history with their acceptance/completion of reviews, ratings, and written feedback are stored in the database and are available to the review editors when assigning manuscripts to reviewers. Review editors can also track the completion of reviews and view an individual reviewer's feedback or an aggregated summary of the reviews via the Web-based review system. The review system has significantly reduced the workload of managing the review process and has allowed us to keep track of 20 to 30 manuscripts at various stages of the review and revision process as well as the review activity of more than 200 volunteer peer reviewers; it has also made it possible to implement a system of multiple geographically distributed review editors.

I wrote the journal management software in PHP, a widely used open source scripting language, and used MySQL, an open source relational database system, to store and manage the data. We are still in a pilot phase of implementing the software, and at this point, I am not comfortable making it widely available. Once it is stable, I plan to make the software available for use by other journals through a general public license.

There is a number of journal management software systems that are currently available (McKirnan, 2002); most of them are quite expensive and probably not affordable for the editors of small subsidized open-access journals. Fortunately, there is an excellent open source journal management system, Open Journal System (OJS) (2002), developed by the Public Knowledge Project (2005), a federally funded project at the University of British Columbia and Simon Fraser University. OJS is currently being used by more than 550 journals around the world and offers a comprehensive software system that goes well beyond automating the peer-review process. It automates virtually the whole process of creating and managing a journal and a journal Web site and is highly customizable. It is written in PHP and can interface with MySQL as well as other database software.

We chose to develop our own software because of the legacy issues of transferring a journal that had been in operation for nearly a decade to a new format and to ensure that we could implement our distributed review process. For new subsidized, open-access journals, OJS offers an excellent means to reduce the workload of operating a journal as well as the need for technical expertise in Web development. OJS has strong user support and is a good example of a successful open source project. OJS is available at no charge, and the installation process is fairly straightforward. Since it is written in PHP and works well with MySQL, both widely available open source packages that run on a variety of operating systems, use of OJS can be feasible even for open-access journals with very limited resources.

FUTURE TRENDS IN OPEN-ACCESS SCHOLARLY PUBLICATION

The transition of scholarly journals from paper to electronic publication has been very rapid, and it is still not clear how these journals will be organized and funded in the future. It is quite likely that there will be several competing models rather than a single dominant model for disseminating research and scholarship.

At the present time, the traditional subscription-based funding model continues to dominate scholarly publishing, though I suspect this will change fairly rapidly. There is a growing awareness and acceptance of the rationale for open access publishing among scientists and scholars. However, what appears to be the prime motivating factor for change is government funding agencies. At the time of this writing, there are two bills under consideration in the United States Senate that will mandate that articles funded through federal granting agencies be made publicly available within a set period of time after publication. These include a bill to amend the Public Health Service Act to establish the American Center for Cures (THOMAS, 2006a), which would also require all research publications funded by the Public Health Service to be made available within six months after publication through PubMed Central, the National Library of Medicine's open-access archive. The second bill is the Federal Research Public Access Act of 2006 (FRPAA), a more general bill to provide for federal agencies to develop public access policies relating to research conducted by employees of that agency or from funds administered by that agency that would provide open access no later than one year after publication (THOMAS, 2006b). While there is a great deal of opposition to these bills from both the commercial publishing industry and some scientific societies, there seems to be strong public support for the concept. In a recent Harris poll, 82% of those surveyed wanted open access to publicly funded research for everyone (SPARC Open Access News Letter, 2006). Similar initiatives to require and/or support open access to government-funded research results are also being pursued in Europe.

I am not sure that the type of delayed access strategy mandated by CURES and FRPAA is the most sensible approach for providing open access. The societies and publishers opposing the legislation may be correct in that providing a grace period during which publishers can charge for access prior to complete open access may not ensure adequate funding for operating their journals. It also appears that the cost of electronic publication may be higher than first estimated (Butler, 2006). Additionally, the six- to 12-month delay in open access mandated by these bills unnecessarily impedes the flow of scientific information.

In my view, a more prudent approach would be to develop a system that allows journals publishing federally funded scholarship to apply through a grant or contract process for funding to cover the cost of publication. The grant or contract program should also support the continued development of open source software such as OJS for increasing the efficiency of the publication process as well as training material and other resources to help scholars form their own journals. While there would be significant costs involved in such a program, it is the public that currently largely funds the publication process through indirect and direct funding of research libraries that pay journal subscription fees. As open-access journals replace journals funded by subscription fees and as electronic journals replace paper journals that require a significant amount of resources to distribute and then warehouse, the cost of operating our research libraries will be significantly reduced. It is my belief that it will be possible to reduce the indirect payments from federal grants to universities and other research institutions that are used to fund these research libraries and, instead, use those funds to directly support the development and maintenance of open-access journals.[7] This would be a much more efficient system that would provide seamless access to scholarly journal articles not only for university faculty in developed countries but anyone else interested in accessing the material. It is quite conceivable that such a funding system could be cost neutral or even result in cost savings over our current system. This is particularly true if one factors in the cost of the time wasted by scientists and scholars to access our current journal system.

However, the funding models for scholarly publishing evolve, and subsidized open-access journals operated by an individual or a small group of colleagues mainly through their own efforts are likely to remain a limited but valuable niche for disseminating scholarship. Open-access journal management software that can significantly reduce the workload of operating a journal is essential for maintaining this valuable and highly efficient means of disseminating research and scholarship.

REFERENCES

Barschall, H. H. (1988, July). The cost-effectiveness of physics journals. *Physics Today*, 56-59.

BioMed Central. (2006). Retrieved July 22, 2006, from http://www.biomedcentral.com/

Butler, D. (2006). Open-access journal hits rocky times. *Nature, 441*(22), 914. Retrieved July 6, 2006, from http://www.nature.com/nature/journal/v441/n7096/full/441914a.html

Coffin, J. (June 2006). Analysis of open source principles in diverse collaborative communities. *First Monday, 11*(6). Retrieved July 21, 2006, from http://firstmonday.org/issues/issue11_6/coffin/index.html

Guédon, J. (2001). In Oldenburg's long shadow: Librarians, research scientists, publishers, and the control of scientific publishing. In *Proceedings of the Association of Research Libraries (ARL)*, Toronto. Retrieved April 5, 2006, from http://www.arl.org/arl/proceedings/138/guedon.html

Harnad, S. (1990). On-line journals and financial fire-walls. *Nature, 395*, 127-128. Retrieved June 23, 2006, from http://www.princeton.edu/~harnad/nature.html

McKirnan, G. (2002). Web-based journal manuscript management and peer-review software and system. *Library Hi Tech News, 19*(7).

National Institutes of Health. (2005). *Summary of the FY 2006 president's budget.* Retrieved July 1, 2006, from http://www.nih.gov/news/budget/FY2006presbudget.pdf

Open Access Initiative. (2002). *Budapest open access initiative.* Retrieved April 27, 2006, from http://www.soros.org/openaccess/

Open Journal System. (2006). *PKP@SFU.* Retrieved July 13, 2006, from http://pkp.sfu.ca/?q=ojs

Public Knowledge Project. (2005). *The public knowledge project of the University of British Columbia.* Retrieved July 13, 2006, from http://www.pkp.ubc.ca/

Public Library of Science. (2006). *PLoS public library of science.* Retrieved July 20, 2006, from http://www.plos.org/

PubMed Central. (2006). *U.S. National Institutes of Health (NIH) free digital archive of biomedical and life sciences journal literature.* Retrieved July 22, 2006, from http://www.pubmedcentral.nih.gov/

Solomon, D. J. (1999). Is it time to take the paper out of serial publication? *Medical Education Online, 4*(7). Retrieved July 22, 2006, from http://www.msu.edu/~dsolomon//f0000016.pdf

Solomon, D. J. (2006). Strategies for developing sustainable open access scholarly journals. *First Monday, 11*(6). Retrieved July 22, 2006, from http://www.firstmonday.org/issues/issue11_6/solomon/index.html

Suber, P. (2006). *Welcome to the SPARC open access newsletter, issue #98.* Retrieved July 17, 2006, from http://www.earlham.edu/~peters/fos/newsletter/06-02-06.htm#frpaa

THOMAS. (2006a). *Library of Congress THOMAS.* Retrieved July 17, 2006, from http://thomas.loc.gov/ [search term: Center for Cures].

THOMAS. (2006b). *Library of Congress THOMAS.* Retrieved July 17, 2006, from http://thomas.loc.gov/ [search term: Federal Research Public Access Act].

Van Orsdel, L., & Born, K. (2002) Periodicals price survey 2002: Doing the digital flip. *Library Journal.* Retrieved April 12, 2006, from http://libraryjournal.reviewsnews.com/index.asp?layout=article&articleid=CA206383&publication=libraryjournal

Willinsky, J. (2000). Proposing a knowledge exchange model for scholarly publishing. *Current Issues in Education, 3*(6). Retrieved July 22, 2006, from http://cie.ed.asu.edu/volume3/number6/

Willinsky, J. (2006). *The access principle: The case for open access to research and scholarship.* Cambridge, MA: MIT Press.

Zerhouni, E. A. (2006). *Report on the NIH public access policy.* Rockville, MD: National Institutes of Health. Retrieved July 14, 2006, from http://publicaccess.nih.gov/Final_Report_20060201.pdf

KEY TERMS

Electronic Dissemination: The dissemination of digital material via the Internet. For the purposes of this chapter, the term refers to the dissemination of journal articles in digital form via the World Wide Web.

"Gold" and "Green" Roads to Open Access: Two general strategies for achieving open access to scholarship. The gold road is via open-access journals that make their material freely available via the Internet. The green road is via authors archiving articles published in traditional subscription fee journals in archives that allow the content to be freely available via the Internet.

Journal Management Software: Helps manage and track manuscripts through the peer-review and publication process. It can automate a significant amount of the work required to operate a journal, but far from all of it. A central argument of this chapter is that open source journal management software can be a key asset in allowing small subsidized open-access journals with few resources to continue to operate and thrive as they become established and as their submissions grow.

Open-Access Initiative (OAI): Sometimes called the Budapest Open Access Initiative or OAI, the term was coined at a meeting in Budapest in December 2001 of the Open Society Institute. The initiative strives to promote the free and unrestricted online availability of scientific and other scholarly journal articles.

Scholarly Journals: Generally peer-reviewed journals that publish original research or scholarship by the researchers or scholars who performed the research or scholarship. They originated in the 17th century and up until about 50 years ago were largely owned and operated by scientific and scholarly societies. Since then, an increasing number is owned and operated at a profit by commercial publishers.

Serial Pricing Crisis: A term commonly used by librarians to describe the dramatic increase in the cost of journal subscription fees, particularly among scientific, technical, and medical (STM) journals, that has been occurring over the last 30 to 40 years. These price increases are limiting the ability of even well-funded research libraries in the United States to maintain their journal collections.

Subsidized Open-Access Journals: Open-access journals in which all material in the journal is made freely available to all readers via the Internet from the time it is published, and there is no charge to authors for publication. Hence, these journals derive no income from their operations. The cost and effort of publication is funded by some type of subsidy. In many cases, these journals are operated via volunteer effort that is the sole support of the journal.

ENDNOTES

[1] The term *scholarly journal* is used to refer to peer-reviewed journals used as a means of disseminating scholarship throughout most academic fields. The issues discussed in this chapter refer most acutely to what librarians term the STM (scientific, technical, and medical) journals.

[2] For a good overview of OAI, see http://www.earlham.edu/~peters/fos/overview.htm, retrieved April 14, 2006.

[3] Willinsky based this estimate on dividing the yearly NIH budget by the number of articles generated during a year, which came out to $60,000. He estimates the researchers' institutions; usually publicly funded universities, contribute another $20,000 per article. The $2,000 to $3,000 estimated cost for publication is based on data from the Public Library of Science (Butler, 2006).

[4] For example, see http://www.arl.org/stats/arlstat/graphs/2002/2002t2.html

[5] It should be noted that this problem is being abated somewhat by some publishers making the electronic versions of their journals freely available to libraries in developing countries.

[6] For examples, see the Directory of Open Access Journals (http://www.doaj.org/), which, as of July 11, 2006, contained 2,303 journals, a significant portion of which fall into this category.

[7] I am not suggesting that we cut off support for society or even commercially published journals. They should also have the ability to apply for funding for publishing federally funded research.

Chapter LI
An Innovative Desktop OSS Implementation in a School

James Weller
University of Cape Town, South Africa

Jean-Paul Van Belle
University of Cape Town, South Africa

ABSTRACT

This chapter presents a case study of a migration to open source software (OSS) in a South African school. The innovative aspect of the case study lies in how the entire implementation was motivated by the collapse of the school's public address system. It was found that an OSS-based message system provided a more cost-effective replacement option whereby the speakers in the school were replaced with low-cost workstations (i.e., legacy systems) in each classroom. Interestingly, this OSS implementation happened despite the fact that, in South Africa, Microsoft Windows and MS-Office are available free of charge to schools under Microsoft's Academic Alliance initiative. The chapter also analyzes some critical themes for adoption of OSS in the educational environment.

INTRODUCTION

There has been an increased interest and awareness of OSS in South Africa (RSA) for various reasons. The work of the Shuttleworth Foundation (TSF) is one reason. In addition, OSS is increasingly becoming a practical alternative to support efforts to cross the digital divide in developing countries. OSS is stable and, arguably, more reliable than its mainstream proprietary competitors (Wheeler, 2005; Whittle, 2002). The availability of OSS support for the development community (GITOC, 2003) is, indeed, an added advantage.

OSS source code can be modified to solve scalability issues (Hughes, 2003; Wheeler, 2005), and some research suggests that OSS may be more secure than proprietary software (Arendse, Colledge, & Dismore, 2002; Wheeler, 2005a). It is also cost effective in that it is capable of running on older hardware, prolonging the hardware's useful lifetime (GITOC, 2002).

While OSS has been accepted for some time as a viable alternative to proprietary software (PS) in the network server market, desktop usage of OSS still remains fairly limited (Prentice & Gammage, 2005). The high PS licensing and computer hardware costs in South Africa relative to the developed countries in combination with the several other perceived advantages of OSS have prompted several OSS on the desktop pilot projects in the education, public, and private sectors.

RESEARCH METHODOLOGY

The aim of this research is to explore a deeper understanding of issues that arise out of and inform migration into desktop OSS. It is an inductive, qualitative, and exploratory study. Research design was followed by data collection, analysis, interpretation, and drawing of conclusions that, in turn, informed the migration model.

A case study research method was considered relevant to the purposes of the study. This method has already garnered significant acceptance in the field due to its ability to provide subtle yet deep insights into social phenomena surrounding information systems (Klein & Myers, 1999; Walsham, 1995). The case study method enables investigations of social phenomena in their natural, real-world context and attempts to extract a deep, rich understanding of these phenomena (Benbasat, Goldstein, & Mead, 1987; Broadbent et al., 1998). A set of qualitative questionnaires were used to collect data through interviews. Data from existing documents and field observations were used as a support framework to the case study.

Thematic analysis was utilized to analyze data obtained in the case study. This involved extracting the common experiences/phenomena mentioned in multiple interviews and grouping together all specific talk related to these experiences. Themes were then identified by bringing together these fragments of conversation to form a comprehensive picture of the experience or phenomenon (Aronson, 1994).

Data for the case study were collected by conducting semistructured interviews with the school's IT manager, network administrator, staff members, and pupils. In addition, several docu-

ments provided to the researchers were analyzed, including a proposal to introduce a computer-based announcements system at the school, as well as basic internal training documentation for the Red Hat desktop environment.

BACKGROUND

Arguments Supporting OSS Usage in Education

Information and communication technologies (ICTs) are a key resource required in the field of education, especially in countries affected by the digital divide (Kotschy, 2002). Most of the affected countries are in Africa, where digital divide studies reflect wider ICT access and developmental inequalities. The existing ICT infrastructure within SADC countries, for example, is more developed in urban than in rural areas (bridges.org, 2002). Countries such as the DRC have outdated and costly telecommunications infrastructure inherited from colonial times, while landmines as a result of civil wars render most areas unusable in countries such as Angola (Bridges.org, 2003). ICTs have the potential to improve the quality of education as well as the quality of life for the people exposed to the technology (Tong, 2004). One of the largest barriers to utilizing ICTs in education is the cost of proprietary software (Tong, 2004). Additional barriers include the security risks associated with proprietary software, the trend of increasing proprietary software license costs, and the cost of hardware required to run proprietary software, especially as newer versions are released and support for older versions is discontinued (Glance, Kerr, & Reid, 2004). There is evidence that educational institutions, particularly tertiary educational institutions (TEIs), are showing significant interest in desktop OSS owing to the aforementioned factors (Conlon, 2004). Additional factors identified by a group of surveyed TEIs included the potential for reduced dependence on proprietary software vendors for support and upgrades, the ability to customize the software in-house, reduced total cost of ownership (TCO), and the ability to use the available source code for teaching purposes (Glance et al, 2004).

South African Examples of Desktop OSS in Education

A number of cases of OSS on the desktop use in the South African education section has been publicized. Two early successful projects conducted in Grahamstown involved Nathaniel Nyaluza Secondary School and Nombulelo Senior Secondary School (Halse & Terzoli, 2002). Through the help of members of the Rhodes University Computer Science Department, both schools have functioning computer labs with high-quality software. The research concluded that even with minimal computing resources, a satisfactory solution can be reached; however, better hardware does facilitate the provision of better, more advanced services.

Another successful desktop OSS project was completed at Alexander Sinton High School in Cape Town (Bardien, 2002). This was a migration from Microsoft to Mandrake Linux and OpenOffice.org. The initial solution used obsolete computers as thin clients, and the learners at the school transitioned quickly and fearlessly. The low cost of the client machines allowed the school to deploy a decentralized infrastructure with PCs at various locations in the school, and ensured that each learner could have his or her own computer when visiting the computer lab.

Perhaps the most widespread and, arguably, most successful implementation of OSS in the educational sector is the Shuttleworth Foundation's (TSF's) tuXlabs initiative, which endeavors to provide disadvantaged South African schools with basic computing facilities. These facilities are based on a 100% pure OSS architecture. As of December 2005, there were 154 successfully implemented tuXlabs in operation (TSF, 2005).

THE PINELANDS HIGH SCHOOL CASE STUDY

This section provides an analysis of the implementation of OSS on the desktop at Pinelands High School (hereafter referred to as "the school" or "Pinelands"), a secondary education school in Western Cape Province, South Africa. The school is located in the suburb of Pinelands in Cape Town. This area is not considered a previously disadvantaged or low-income area, so the fact that the school was utilizing OSS on the desktop and the reasons surrounding this were of particular interest.

Background to the Case Study

Pinelands' first implementation of OSS came about as a result of the increasing burden of software license fees for the Novell operating system that was used to run its main server. The current IT manager joined the school at a time when the school was running version 3 of Novell and when version 5 was about to be released. Management realized that a Novell upgrade was needed and that the school needed to make a long-term decision about its future server operating system. After some investigation into costs and feasibility, the Novell server software was replaced with Linux in 2002. An outside company was paid a fairly large sum of money to perform the migration; however, this once-off payment resulted in approximately R36, 000 annual savings on license fees.

At this point, with its limited knowledge of OSS, the IT department began to investigate the feasibility of further utilizing OSS at the school, including investigating a possible relationship with TSF and its tuXlabs project. In May 2002, Microsoft announced its agreement with the South

African Government (the Microsoft Academic Alliance) to provide free software to state schools (Microsoft South Africa, 2002). This would have included Pinelands High School. Indeed, the school's annual Microsoft licenses expenditure decreased from R40,000 (≈US$6,000 in 2002) to R5,000 (< US$1,000) in 2003. Thus, it is interesting to note that Pinelands still decided to go ahead with its tuXlab project, since software license costs could no longer be considered a driver. The reason was that the IT manager realized that a relationship with TSF opened up a wide range of new possibilities, particularly the free Linux training offered. This was seen as a potentially significant benefit and cost-saving in terms of being able to better maintain the existing Linux server, as well as the thin client machines that would form a tuXlab in house. This training is explained and further investigated later in the analysis of the case.

Pinelands does not run a traditional tuXlab in the sense that there is no lab of 20 to 30 thin client machines running desktop OSS at the school. Instead, a number of old Pentium 1 PCs obtained by the school were converted into thin clients for use in the library to help students perform and write research. TSF provided a server, and volunteers from TSF came into the school on a weekend and performed all of the necessary setup and testing tasks for the network, thin clients, server, and required software. This library, tuXlab, was completed in November 2003.

The catalyst for getting OSS software onto more desktops at Pinelands was the failure of the school's PA (intercom and announcements) system in January 2004. A quote for R40,000 to repair the system was obtained soon afterwards. However, no guarantee of the repairs could be assured by the repair company. Furthermore, the amount required to completely replace the existing system was considered exorbitant by school management. At this point, the IT manager came up with the idea to replace the announcements system with a computer-based one. She described

how she presented the motivation for the idea and sold it to the school's governing body:

I went to them with my feasibility [analysis], as best as I could understand it at that stage, saying, for R40, 000 this is what I could do: I could bring in the old machines, [fix up and extend our network infrastructure] to put a copper network in and also a wireless network across to the buildings that are just too far away for copper, and for R40, 000 this is what I could do for you: a computer in every room, which will cover the announcements system requirement, but which will also give you a whole bunch of other things. And that was really the catalyst that got [desktop OSS] into every classroom.

The new computer-based announcements system, affectionately known as IntraCom, is a Web-based application that runs on the school's intranet. The system allows staff to log on, view, and read announcements to the class at a predefined time in the day. Staff can also post their own new announcements for viewing by all other users. The nature of the system also means that there is no invasion of teaching time; staff members read the announcements to their pupils at their own convenience.

While this solution effectively satisfies the requirements for a school announcements system, it also provides many additional benefits. Old announcements can be archived and searched for later retrieval. The Web application is accessible from any networked computer in the school, which allows pupils to log on and review announcements, something not possible with a traditional intercom system. Furthermore, staff can access the Internet and e-mail, as well as produce text documents, spreadsheets, and presentations using OpenOffice.org version 1, all from the comfort of their own classrooms.

Although the thin clients in the classrooms are only used by school staff at this stage, the additional benefits arising from having a computer in every classroom as well as a working announcements system, all for the cost of the old announcements system being repaired with no guarantee, is a positive aspect of utilizing desktop OSS at the school. Furthermore, the computers in the library running desktop OSS provide an affordable solution to satisfy pupils' needs for Internet research stations.

Pinelands still actively maintains ts Microsoft desktop software (i.e., Microsoft Windows, MS-Office, etc.). Nevertheless, the process of getting OSS onto desktops at the school should still be seen as a migration rather than an implementation from scratch. This is because the process involved many changes at the school, including changes to the network infrastructure and the hardware being used, as well as many changes for both the staff and pupils using the new OSS. This is especially true in terms of a completely new look and feel to become accustomed to, exposure to new difficulties and frustrations, new training materials to absorb, and new ways (or software used) to perform daily tasks.

The following section describes some of the themes surrounding the implementation of the desktop OSS at Pinelands High School, which emerged from a thematic analysis of interview transcripts and written source documentation.

Some Themes that Emerged from the Case Study Analysis

Financial Benefits as the Main Driver for Migration

The main motivating factor for Pinelands' migration to OSS was found to be financially related. Several facets of this theme and the differing perceptions of their relative importance to various stakeholders were uncovered during analysis. Although the Microsoft Academic Alliance was already in effect in 2002, Pinelands still went ahead with their migration to desktop OSS. The financial reason for the migration, therefore, was

not primarily related to software licensing costs. This was confirmed by the IT manager.

The first financial benefit of having desktop OSS for Pinelands was the vast reduction in hardware costs. The IT manager illustrated the degree to which the school can save money on hardware by using desktop OSS instead of Microsoft products:

Where we already had a Windows machine in a classroom, that's great, but where I simply couldn't afford it, to put a R3, 000 massive box with 256MB [RAM] minimum to run Windows, I could put a thin client in at R400, or literally scrap that I'd built up ... we've been able to use any machine that I can get my hands on that I can convert into a working thin client, and that saves me R3, 000 every time [as opposed to] if I needed to get Microsoft working.

In addition, according to the school's IT manager, the older hardware currently being utilized as thin client machines does not incur any significant maintenance costs.

Another financially related perceived benefit of using desktop OSS encountered in the case was the ability to redistribute funds, which would otherwise have been allocated to proprietary software licenses and/or a new intercom system, into other areas of the school. This point was raised by almost all of the interviewees. One of the teachers that was interviewed gave a good illustration of this point when asked for her view on the benefits of utilizing desktop OSS:

Researcher: *Would you recommend other organisations like yours migrate to desktop OSS? If so, why? If not, why not?*

Teacher 1: *I think for our school we should go the open source route because it does save money and we can then use the money for other things, like more computers, facilities, books for my English department, etcetera. I think that while comput-*

ers are useful, we need to spend the money on other more practical things, rather than buying expensive computer programs.

Another teacher, who claimed to know very little about OSS, realized the potential financial savings that utilizing desktop OSS could bring. She commented that in her view, the facts that new desktop OSS could potentially be customized in-house to suit the needs of the school and that a lot of OSS was free were very positive aspects of using it.

The IT manager further went on to emphasize that the financial benefits of using desktop OSS are not always tangible. She pointed out that there is no difference on the financial "bottom line," but that a lot more can be done with IT using the same funds allocated when using OSS instead of proprietary software.

Ultimately, it would appear that financial factors were the main driving force behind the migration decision.

Partial Migration

Both the school's IT manager and network administrator stressed that it was not practical to migrate the entire school to OSS on the desktop. This was primarily due to the fact that the school was running several legacy applications, particularly its student records system. This critical application, along with other important legacy packages, runs only on Microsoft Windows, and the vendor has no immediate plans to rewrite it to run on Linux. The IT manager has no plans to do away with the Microsoft infrastructure for as long as it takes to keep the legacy packages running and stressed that a pragmatic phased approach to migration is better than an all-out crusade.

Problems with Training

One benefit of TSF's school tuXlabs program is that every school that is donated a tuXlab receives some free training on basic computer literacy, OpenOffice.org, and some educational applications. In the case of Pinelands, the training process did not work very well. The idea was that the three hours of initial training given to a select group of teachers would cause a ripple effect, and those teachers would then transfer their skills on to other teachers. This did not work at the time, because TSF had not received the proper training process and did not have knowledgeable enough persons employed to perform the training. (This was subsequently realized and fixed by TSF.) It was also pointed out by more than one interviewee that the training materials received at the time were somewhat basic in nature. The school's IT manager commented on the ineffectiveness of the training and the reasons for this:

The quality of training being given at that stage was disastrous and the guy eventually left [TSF], because he really didn't know [OpenOffice.org] and when we went to him and said 'This is how it works in Microsoft, how does it work in [OpenOffice.org]?' he really didn't know and sort of said 'Oh I'll come back to you' or something.

The school's IT manager had some previous business background and experiences related to training and emphasized the importance of ongoing training, not simply a once-off engagement. When asked about the merits of attempting to train a small staff that then creates a ripple effect throughout the organization, she responded by explaining:

My experience, in business as well as here, is that that won't happen. If you don't have an ongoing commitment from the organisation, you will lose your training, through staff turnover or whatever; you won't have a second generation of people trained from the inside.

The IT manager described how the school's IT department took what TSF had provided and built on it by offering voluntary internal training sessions to staff members who had a desktop OSS machine in their classrooms, or anyone else who was interested. The network administrator also emphasized that the IT department had tried to organize voluntary training sessions for the teachers, but the teachers did not have time to attend. For this reason (or possibly through lack of communication), neither the network administrator nor any of the staff or pupils interviewed knew of any planned training sessions in the near future.

All of the staff and pupils interviewed said that they had not received any formal instruction on the usage of the desktop OSS; some were not even aware that there were training materials available to them and had taught themselves. This implies either a lack of communication within the school about training or, more likely, a problem with making the training sessions voluntary. Those staff members who complained about problems with using the desktop OSS were the same staff members who said they did not receive training. It is possible that had the staff attended the offered training (assuming they knew about it), they would not be experiencing the problems they described. Thus, it would appear that a very strong emphasis on desktop OSS training needs to be recognized by management and, more importantly, its importance communicated to new users.

The IT manager and network administrator pointed out that the IT department at Pinelands, specifically the network administrator, was well prepared for the desktop OSS migration in terms of their own skills and training. This, however, was because the network administrator was able to volunteer at several installs and attend training sessions in advance. The IT manager pointed out that this may not always be the case, however, especially for most of the previously disadvantaged schools being given a tuXlab. She summed up very well a distinct weakness of the volunteering approach:

Any school would be entitled to [the training that the network administrator received], but we're in the privileged position where we have 2 IT people and can afford to take up those [volunteering] opportunities, whereas an underprivileged school that's been given 20 machines and their maths teacher is also running the computer lab; he just physically doesn't have the time to do this; get out and do the volunteering and do the training ... so although it looks good on paper, coming from a teacher's perspective, it's actually not practical because they just don't have enough hours in the day.

IT Manager

User Apathy, Resistance, and Acceptance

Inevitably, with change comes resistance, and Pinelands' migration of certain teachers from Microsoft to desktop OSS was no different. While the network administrator and IT manager both mentioned that there were several naysayers throughout the migration, from the proposal to the actual implementation, it was generally accepted by the staff that the move was going to happen because it was the best solution to the announcements system problem.

Interestingly, the IT manager, when asked about staff involvement in the migration process, highlighted the generally apathetic attitude toward IT at the school, rather than user resistance.

Researcher: *Was any resistance encountered during the migration, and if so, where did this resistance come from?*

IT Manager: *No, again, apathy. I don't think people really knew what we were trying to do.*

This seemingly apathetic attitude of the school's staff should not be viewed in a negative light. It is most likely the result of having been exposed to IT for an extended period of time (unlike many of the schools receiving a tuXlab donation) and the fact that, as the IT manager explained, most staff probably had very little idea of what implementing desktop OSS in the classroom to access a Web-based announcements system actually meant.

It was discovered that, following the migration, user resistance was not nearly as strong as anticipated. Although several staff members that ended up having an OSS machine in their classroom were not initially comfortable with using it, the only thing they had to use it for was to access the announcements system via a Web browser. Thus, although OpenOffice.org and several other packages useful to the teachers are available from the classroom thin clients, the staff members who are uncomfortable with OSS simply do not use them, preferring instead to use the Microsoft machines elsewhere in the school. The IT manager cites this as one of the benefits of not completely migrating to desktop OSS; the users that prefer to use Microsoft products have not been faced with drastic change. However, this could be seen to negate the entire point of migrating to desktop OSS in the first place.

In fact, all of the users interviewed, both pupils and staff members, mentioned that the desktop OSS installed, particularly OpenOffice.org and Mozilla Internet browser, were both easy to learn and use and that they did not care whether it was Microsoft or not. As one teacher commented:

I would use any program that was user friendly and catered for those who didn't know anything about it. As long as it's easy to use, and [OpenOffice.org] is, I would use it.

It would appear that the resistance to change that did occur in the school concerning the usage of OSS was and still is due to a loss of familiarity or the comfort zone with existing software, as opposed to a dislike or fear for all things open

source. This user resistance to change can occur even when it seems to go against economic sense, as illustrated in this example given by the network administrator:

We use a program called SketchPad for drawing, which had a problem with it, so I found a very nice open source replacement, but [name omitted for confidentiality] overruled me because he said that he knows SketchPad, so that's the one he wants to keep.

Finally, the IT manager made an interesting assertion, that the reason most pupils were not fazed by using the desktop OSS was probably because of the more adventurous nature of children and the differences in the way in which children and adults learn. She illustrated the point very well:

Kids will just experiment, move around and be self taught, whereas folk of my generation will want a checklist, and as long as they can do something sequentially, monolithically and the way they want, they're happy. The moment something happens that isn't on the piece of paper, it throws them, or if they have to do something differently, it throws them. So kids are far better at adapting.

Support Costs and Problems

While users at Pinelands are generally happy with the usability of the desktop OSS, certain support issues have been and continue to be encountered. The school's IT manager emphasized the fact that while OSS does save the school money when it comes to hardware, the software is far from free; support costs are a major consideration. Maintaining the actual thin clients, who are used to accessing the desktop OSS, was not perceived to be a major cost or highly demanding in terms of technical expertise; the real cost was seen to be paying specialists to maintain the Linux servers. While (only) the desktop OSS server is maintained by TSF and the local OSS community, as part of the tuXlab deal, the IT manager emphasized that for any OSS implementation, support availability and costs are a big problem, as illustrated by the following quote:

If you don't have the expertise in-house to maintain a Linux system, you have to go out there and find specialists, and Linux specialists are few and far between and expensive. And that's why our relationship with [TSF] has helped a bit—to get around some of those big issues of [support costs]—but at this stage, until there is a broader base of Linux specialists out there, it is a problem.

Although no longer tied down by heavy Novell licensing fees, Pinelands' IT department now has a budget of between R10,000 and R15,000 (≈ US$1,500 to US$2,000) annually for Linux server support and development.

CONCLUSION

This chapter described an interesting case study of the introduction of OSS-on-the-desktop in a South African school. The main driver for the installation was the replacement of the school's public address system. In addition, however, OSS offers many advantages toward bridging the digital divide at the school level in developing countries such as South Africa. Minimal license costs associated with freely or minimal cost distribution makes OSS a favorable option. The OSS access code can also be modified to suit the needs of the user, making it a more flexible option for innovative uses by schools. It was found that an OSS-based message provides the most cost-effective replacement option, whereby the speakers in the school were replaced with very low-cost workstations in each classroom.

An interesting aspect is that this OSS implementation happened despite the fact that, in South Africa, Microsoft Windows and MS-Office are available free of charge to schools under Microsoft's Academic Alliance initiative.

The chapter analyzed some critical themes for adoption of OSS in the educational environment. These were found to be related to financial considerations being the initial driver for OSS, the importance of user training, although a project can succeed without it, the fact that user apathy may be more common than user resistance, and the criticality of ongoing support.

To end the analysis of the themes emerging from this case study, it is pertinent to quote the Pinelands High School IT manager, who succinctly expressed the approach toward OSS projects at the school:

It's always been fairly pragmatic; what's going to be the best for us in this environment, as opposed to a crusade saying 'thou shalt go open source' and bite the bullet, whether that means I can't find the drivers or it costs me more than I intended.

It is hoped that this case study will shed some light on some of the issues other educational sector implementers of desktop OSS may face. However, since social and contextual issues differ markedly between organizations, not all of the aforementioned themes are expected to apply, and new ones are likely to surface.

REFERENCES

Aronson, J. (1994). A pragmatic view of thematic analysis. *The Qualitative Report, 2*(1).

Bardien, R. (2003). *Linux case study: Alexander Sinton High School.* Retrieved December 10, 2005, from http://casestudy.seul.org/cgi-bin/caseview1.pl?recnum=82

Benbasat, I., Goldstein, D., & Mead, M. (1987). The case research strategy in studies of information systems. *MIS Quarterly, 11*(3), 369-386.

Bridges.org. (2002). *SADC-WEF consultation report on e-readiness: Better, faster, cheaper: developing and leveraging world class ICT networks for social and economic advancement.* Retrieved January 3, 2006, from http://www.bridges.org/e-readiness

Bridges.org. (2003). *World Economic Forum-NEPAD e-readiness policy programme: Building capacity to narrow the digital divide in Africa from within.* Retrieved January 3, 2006, from http://www.bridges.org/e-readiness

Broadbent, M., Darke, P., & Shanks, G. (1998). Successfully completing case study research: Combining rigour, relevance and pragmatism. *Information Systems Journal, 8,* 273-289.

Conlon, J. (2004). I did it with Linux. *Journal of Applied Educational Technology, 2*(2), 3-4.

Glance, D., Kerr, J., & Reid, A. (2004). Factors affecting the use of open source software in tertiary education institutions. *First Monday, 9*(2). Retrieved from March 20, 2006, from http://www.firstmonday.org/issues/issue9_9/glance

Halse, G., & Terzoli, A. (2002). *Open source in South African schools: Two case studies* (Working paper). Highway Africa 2002, Johannesburg.

Klein, H., & Myers, M. (1999). A set of principles for conducting and evaluating interpretive field studies in information systems. *MIS Quarterly, 23*(1), 67-94.

Kotschy, P. (2002). The African digital divide, Linux and open source. In *Proceedings of the Conference on the African Digital Divide.* Retrieved December 10, 2005, from http://www.linuxafrica.co.za/conf_african_digital_divide.html

Microsoft South Africa. (2002). *Microsoft's software donation to South African government schools.* Retrieved December 10, 2005, from http://www.microsoft.com/southafrica/education

Prentice, S., & Gammage, B. (2005). *Enterprise Linux: Will adolescence yield to maturity.* Gartner Symposium/ITxpo.

Tong, T. (2004). *Free/open source software: Education.* Retrieved December 10, 2005, from http://www.iosn.net/education/foss-education-primer/fossPrimer-Education.pdf

TSF (The Shuttleworth Foundation). (2005). *Shuttleworth tuXlab Program.* Retrieved December 10, 2005, from http://www.tuxlab.org.za/index.htm

Walsham, G. (1995). Interpretive case studies in IS research: Nature and method. *European Journal of Information Systems, 4*(2), 74-81.

Wheeler, D. (2005). *Why open source software/free software (OSS/FS, FLOSS, or FOSS)? Look at the numbers!* Retrieved April 16, 2005, from http://www.dwheeler.com/oss_fs_why.html

Whittle, S. (2002). Secure, reliable, flexible … and free. IT training. Retrieved April 16, 2005, from http://search.epnet.com/direct.asp?an=7339794&db=buh

KEY TERMS

Linux: An open source version of the UNIX operation system originally developed by Torvalds Linus. It has many distributions such as Ubuntu, Red Hat, SUSE, Knoppix, and so forth (also known as distros). Linux versions have been developed for an extremely wide variety of hardware platforms ranging from handheld devices such as cell phones and PDAs to massive super-computer clusters. The term Linux actually refers to the kernel around which the distros are built. Most of the software tools and applications included with the distros were developed under the GNU project of he Free Software Foundation; hence, a more accurate description is GNU/Linux.

Microsoft Academic Alliance (MSDN AA): An initiative by Microsoft to promote the use of Microsoft's developer tools, platforms, and servers for instruction and research by significantly reducing their price to educational institutions. In South Africa, government (i.e., nonprivate) schools can apply for free use of, inter alia, the Windows operating system and MS-Office software by teachers and learners (pupils).

Mono: An open source implementation of the common language infrastructure, based on the .NET Framework specification (www.mono-project.com)

Open Source Software (OSS): Software distributed under a license that allows users to copy, modify, and redistribute the software.

Operating System (OS): Software that controls the execution of computer programs and may provide various services such as hardware control, file storage, input-output functionality, and user interface. It acts as the interface between the hardware and the applications.

OSS on Desktop or Desktop OSS: OSS applications that are utilized by everyday users to perform daily work tasks. This is in contrast to Server OSS, which are applications running on the server side. OSS on the desktop usually refers to a combination of an OSS operating system—usually a Linux distribution—and OSS productivity software such as OpenOffice, FireFox, or similar applications.

tuXlab: As a joint initiative of (partnership between) the Shuttleworth Foundation (www.tuxlab.org.za) and South African Schools tuXlab are computer centers installed with OSS as an economical and sustainable way to bring the power of computing to the learners in South Africa.

Chapter LII
Rapid Insertion of Leading Edge Industrial Strength Software into University Classrooms

Dick B. Simmons
Texas A&M University, USA

William Lively
Texas A&M University, USA

Chris Nelson
IBM Corporation, USA

Joseph E. Urban
Arizona State University, USA

ABSTRACT

Within the United States, the greatest job growth is in software engineering and information management. Open source software (OSS) is a major technology base for enterprise application development. The complexity of technologies used by industry is often an obstacle to their use in the classroom. In this chapter, a major software development paradigm change that occurred in about the year 2000 is explained. CS education programs have been slow to adapt to the paradigm change due to problems such as the tenure system, inexperienced student laboratory assistants, lack of leading-edge software tool support, lack of software team project servers, unavailability of help and mentoring services, and software unavailability. This chapter explains how these problems can be solved by creating an open source-based shared software infrastructure program (SSIP) sponsored by industry, but planned and implemented by SSIP member universities at no cost to member universities.

INTRODUCTION

Today students are saying no to computer science (Frauenheim, 2004). CS faculty members have panicked in what David Patterson (2005) calls Chicken Little rumor mongering. He tells everyone to stop whining about outsourcing. In our opinion, CS faculty should panic and adapt to a new software development paradigm. Patterson makes an invalid implied assumption for his article in that CS in some way is related to information technology (IT) jobs in U.S. industry (or, for that matter, that CS is useful to a software engineer). He is correct to say that U.S. IT jobs are increasing. Also, software engineering (SE) degree programs and jobs are increasing. His domino theory of job migration is not correct. We agree with Patterson that U.S. programmers should worry about both India and China. We do not agree that either India or China will have to worry much about the Czech Republic. Both India and China have such large populations and low wages that major CS job migration will mainly be to these two countries. The middle processes of a software product development software life cycle (DSLC) may completely migrate from the United States.

Every Fortune 1000 company with which we are familiar takes advantage of low labor costs in India and/or China. Unfortunately for CS, approximately 80% of high-paying CS jobs in the past have been with Fortune 1000 companies. Jobs that will remain in the United States will go to students that are familiar with open standards, a wide variety of solutions including open source solutions, software development tools that support open standard visualization design models and open source integrated development environments. In this chapter, open standards will be defined as standards that are publicly available. The Object Management Group (OMG) (2006) is an example of an organization that was created to produce open standards. OMG is an open membership, not-for-profit consortium that produces and maintains computer industry open standards for interoperable enterprise applications. OMG membership includes virtually every large company in the computer industry and hundreds of smaller ones. OMG's most widely used standard is described by the unified modeling language (UML) specification. UML is used worldwide to model application structure, behavior, architecture, business process, and data structure. We use the term open source software (OSS) to refer to software that has Open Source Initiative (OSI) (2006) licenses. Examples of OSS are Linux, Apache, Eclipse, and Derby. We also include open-standard compliant software that is provided free for classroom use to universities. An example is IBM Rational Software Architect (RSA).

The objective of this chapter is to explain how leading-edge industrial-strength software can be introduced into the university classroom by using OSS, open standards, distance learning, and infrastructure shared among cooperating universities. In this chapter, we will describe the evolution of software development during the 20th century, the paradigm change at the beginning of the 21st century, and the problems with existing university information technology education. Then we will describe a shared software infrastructure program (SSIP) to rapidly introduce leading-edge industrial software solutions into university classrooms at no cost to SSIP member universities.

BACKGROUND

Software education emerged during the last 50 years of the 20th century. During the mid-1900s, computers were applied to create firing tables for the military. Scientists programmed these computers using computational algorithms. Computer memories were small and expensive, and successful software depended on efficient algorithms. As computer use grew, universities began to offer programming courses based on algorithm methodology. The application of mathematical science of algorithms to computers led to a new field called computer science. As demand for computer programmers grew, computer science programs at U.S. universities grew in number. U.S. universities had the computers, while universities outside the United States and Europe did not have access to computers. As the size and complexity of computer systems continued to grow, one could not rely on the theory of algorithms to provide acceptable solutions. At a 1968 NATO conference in Europe (Naur & Randell, 1968), computer professionals realized that the software for major systems would have to be engineered based on engineering science and practice. That is when the term *software engineering* was introduced. In 1984, the U.S. Department of Defense created a Software Engineering Institute at Carnegie Mellon University (2006) to advance the practice of software engineering.

Throughout the 1990s, the cost of computers continued to decline, and the capabilities of computers increased. Computer cost was no longer a barrier to the spread of computer-related education programs to universities throughout the world. This has been aided by the creation and expansion of the World Wide Web (WWW) over the Internet.

Education Programs at the Beginning of the 21st Century

Computer-related educational programs at the beginning of the 21st century fall under the umbrella term information technology, which includes computer science, computer engineering, information management, and software engineering. Overall, the demand for information technology knowledge workers worldwide is increasing. The U.S. information technology education programs hit hardest by use of off-shore contractors are the science-based computer science programs. Enrollment in U.S. computer science programs is on the decline. These programs emphasize the middle or coding process of the DSLC. The

coding process is the easiest to outsource from high labor cost regions to off-shore low-labor cost regions. All indications are that the computer science down trend will continue. Demand for computer engineering graduates remains strong. The fastest growing demand is for information management and software engineering graduates. Software engineering and information management programs teach students about all phases of the DSLC. The U.S. Department of Labor Statistics (2005) projects software engineers to be one of the fastest growing occupations through at least 2014. Major universities are beginning to offer software engineering certificates, bachelor degrees, and master degrees.

At the beginning of the new millennium, the outlook for software engineers is strong. Distance learning technology is becoming common. Software development is being practiced with project members distributed around the globe. Open source, open standards, and interoperable software are being demanded by customers. Software knowledge will continue to change and expand at a very rapid rate.

MAIN FOCUS OF THE CHAPTER

University Environments

At present, all students that come to the university are computer literate and have their own computers. Many already know how to program and are connected to the Internet. They are looking for software knowledge that will qualify them to find jobs in which they can create complex software products. They are not finding this software knowledge in most computer science (CS) departments.

We now return to David Patterson's problems. As the president of the Association for Computing Machinery (ACM), he has to be a cheerleader for CS faculty who run around yelling that the sky is falling. For most of the CS degree programs, the sky is falling. Many of the CS degree programs are in small liberal arts colleges and former teacher colleges. These schools do not have the technical background or faculty required to give students the software knowledge that will be provided by the increasing number of IT and SE degree programs. As CS job demands decline, CS programs will continue to shrink with the faculty having to face problems of dying programs unless they adopt new methodologies to quickly introduce into classrooms the latest software knowledge required by industry

Woodie Flowers (2000), a mechanical engineering professor at MIT, recently asked, "Why should education change, we have been doing it this way for 4,000 years?" He said that over the next decade, educators will have to restructure their curricula in order to accommodate the World Wide Web. Change in the university moves at glacial speed. Software knowledge is continually expanding and growing much faster than the current education process can adapt to in order to meet the needs of industry. Ways must be found to upgrade software knowledge that is taught to IT professionals graduating from universities today.

The problem with education of the IT professional can be traced back to the first programmers. The first IT professionals were scientists and engineers who knew how to build and operate the first computers. Many programmers were mathematicians. When universities began to use computers in the 1950s and 1960s, engineering schools emphasized teaching hardware, science schools emphasized teaching programming languages, and business schools emphasized business applications. The science schools originally placed programming languages courses in mathematics departments. As time progressed, mathematicians teaching programming languages separated from the mathematics departments to create CS departments. Instead of teaching the latest software knowledge, the mathematicians began to teach what they knew best: computational mathemat-

ics and theory of algorithms. An algorithm is a procedure for solving a mathematical problem in a finite number of steps that frequently involves repetition of an operation (*Webster's new Collegiate Dictionary*, 1981). It can be shown that for any problems other than toy problems, it is impossible to prove that complex software products terminate in a finite number of steps. Thus, the time spent teaching theory of algorithms is time wasted. Tenured faculty members hired to teach mathematical algorithms will probably continue teaching algorithms until they retire. Since faculty members decide which young faculty members are hired and later tenured, they will probably continue to hire computational mathematics and algorithm specialists.

During the 1960s and 1970s, computers came into common use in industry. People with almost any background could be trained to operate the computer applications in industry. With the advent of the personal computer in the 1980s, the people familiar with computers continued to expand. With the commercialization of the Internet in the 1990s and the introduction of Internet browsers and computer games, virtually everyone below middle age used computers. Essentially every high school graduate that enters a university today has a computer that can be connected to the Internet. They learn how to program in high school and are usually familiar with some form of database management system. They learn how to access Internet servers through the use of browsers. If they would like to become an IT professional, they expect to learn the latest software knowledge that industry demands. CS departments that hire specialists in computational mathematics, theory of algorithms, and computational complexity theory will continue to lose students. Unless they change, CS departments probably will be absorbed eventually back into mathematics departments. By working with industry, university IT programs can teach the latest software knowledge to their students who will then be in high demand when they seek jobs in industry.

Year 2000 Productivity Paradigm Change

During the latter part of the 20th century, the demand for U.S. software developers continued to exceed the supply. During the 1950s, 1960s, 1970s, and 1980s, computer hardware was expensive and many developing countries could not afford computers. Thus, if you wanted to become an IT professional, you almost had to study in the United States. As a result, virtually every university created CS degree programs that were advertised as the correct degree program for the IT professional. The result is a huge oversupply of PhD and Master's degree graduates who learned mathematics theory but obtained very little software knowledge. During the 1990s, computer costs continued to decline to where students and universities even in the poorest countries could afford computers. The *Wall Street Journal* pointed out that the auto worker salary in Germany was $33 per hour, while an auto worker in China earned $0.98 cents per hour. The salary differential for knowledge workers such as software developers is similar. Leading up to the year 2000 was the conversion of all legacy software in the United States to handle a four-digit year instead of a two-digit year built into existing software products. There were not enough experienced programmers in the United States to handle the demand for COBOL programmers. Companies turned to the software houses in India to help with the conversion. The large Fortune 1000 companies were very pleased with the results, and after 2000, they began to out source computer coding to offshore companies in India. Recently, the Chinese commercial software industry, although lagging behind India's, has been undergoing major structural shifts that could make it the Asian industry leader (Kshetri, 2005). Chinese developers are making a major commitment to OSS. Large U.S. companies are out-sourcing the middle processes of the DSLC, while the upstream requirements elicitation, requirements specification, and soft-

ware architecture processes and downstream acceptance testing and software product installation will remain in the United States. CS educational programs emphasize the middle DSLC processes, while SE and information management programs emphasize the upstream and downstream DSLC processes. While the demand for IT workers in the United States is increasing, the demand for CS professionals is decreasing.

As mentioned earlier, CS faculty members have begun to panic and grasp for schemes to restore CS popularity. Former ACM President David Patterson (2005) suggests expanding student recruiting in high schools by ACM's new CS Teachers Association. He recognizes that software knowledge continually changes and places emphasis on keeping job skills up to date. Former ACM President Peter Deming along with Andrew McGettrick (Denning & McGettrick, 2005) point out that CS places too much emphasis on coding and not enough emphasis on other DSLC processes, including the use of advanced software tools to support requirements gathering, defect tracking, configuration management, middleware services, advanced software solutions, and software process visualization tools. They recognize that emphasis placed on analysis of algorithms and complexity theory as the heart and soul of computing is a mistake. Their solution is wrong. Recruiting of students is not the solution to declining enrollments. What is needed is emphasis on software knowledge that today's computer professional needs in order to be competitive in the global marketplace. A solution must be found to overcome the current problems with CS programs.

CS Education Program Problems

Major problems that must be corrected in order for CS graduates to be attractive to employers include tenure system, inexperienced laboratory assistants, software tool support, software team project servers, inadequate department support personnel, help and mentoring services, and available software.

The university tenure system is a type of union for university faculty. It is almost impossible to remove a tenured faculty member once tenure has been granted. Tenure empowers faculty but does not make them accountable. Tenured CS faculty members who are specialists in computational mathematics and algorithms will remain faculty members for approximate 30 years between the time tenure is granted and retirement. In most cases, they will continue to teach computational mathematics and algorithm theory even though there is very little industrial or student interest in these areas. Also, these subject areas are unnecessary for an understanding of the software knowledge required by the IT professional.

In universities with major research programs, many of the funded graduate student teaching assistants that oversee laboratories for software team projects have never used any advanced software tools used in industry to create software products. Even though companies may provide these tools to universities at no cost, the laboratory assistants must understand them and be able to help and mentor student teams in project courses.

Many software products can be used directly out of the box. Software users expect to be able to load a new software system and then begin to immediately start using the system. Heavy-duty software tools are not out-of-the-box. To set up a software tool environment, a software tool administrator must create directories and security as well as initialize parameters in which the tools in an environment work together. The typical CS department does not have enough software support staff or funds to hire additional staff to administer a suite of advanced tools.

Student teams working on a capstone project to create a software product must have access to a server for testing the software product. Often student projects crash servers during testing and interfere with other people trying to use that same server. Software projects need a server as a type of sand box for operating their software product. Normally, CS departments do not have the resources to dedicate servers to student projects.

Many software vendors provide free training and use of software tools to universities. But they provide minimal help facilities to answer specific questions that arise while trying to use the tools. Students must be able to contact knowledgeable people who will answer their questions in a timely manner. When students are learning to use complex software design and testing tools, they would like access to a mentor to guide them. Ideally, help and mentoring services should be available 24 hours a day, seven days a week (24/7).

As part of students gaining software knowledge required by industry, students must have easy access to software products and tools. Many vendors will provide free software and licenses to universities. Acquiring the software and licenses to use the software can be a problem when universities are not paying for the software. There needs to be a service to expedite the process of acquiring software for classroom use at no cost to universities.

By helping universities to quickly introduce best software development practices, improved processes, and advanced software tools, students that graduate from these programs will gain software knowledge that is in high demand by industry. The goal of the SSIP is to set up an infrastructure shared among universities in which universities can easily introduce the latest leading software knowledge into both undergraduate and graduate classrooms without building a costly infrastructure at each university. Member universities will contribute software knowledge infrastructure to the SSIP and will use the SSIP as a resource to support their classes. Operation of the SSIP is supported by industry sponsors at no cost to universities. Eventually, the SSIP would like to provide infrastructure to every interested university in order to teach the latest leading-edge software knowledge to their students.

Shared Software Infrastructure Program (SSIP)

SSIP was created in spring 2005. Current industrial sponsors of the program are AVNET, IBM, and Intel. The program is sponsored by companies who share a vision of integrated information flow within and among enterprises based on OSS, open standards, and global interoperability. The SSIP will support tools compliant with the OMG computer industry specifications for interoperable enterprise applications. Services provided by the SSIP will be determined by the member universities that use the SSIP Web site. Services and software will be provided to member universities at no cost to the universities. Costs of operating the SSIP and developing the infrastructure will be borne by sponsors.

Initial courses supported by the SSIP were capstone software engineering courses that had a software project in which teams of students develop a software product starting with the customer requirements and finishing with a demonstration of a working product. Students are introduced to a full set of computer-aided software engineering (CASE) tools. CASE tools were introduced across all phases of the DSLC. Each week a new tool with open was introduced. For each tool, the SSIP staff provided an overview, tool use examples, and online tutorials, and suggested assignments and a tool Web site. SSIP 34 Member Universities for fall 2006 include the following:

- Arizona State University, Tempe, Arizona
- Auburn University, Auburn, Alabama
- California State University, Los Angeles, California
- DePaul University, Chicago, Illinois
- Iowa State University, Ames, Iowa
- Louisiana State University, Baton Rouge, Louisiana
- Marquette University, Milwaukee, Wisconsin
- Mississippi State University, Mississippi State, Mississippi
- Neumont University, South Jordan, Utah
- North Carolina State University, Raleigh, North Carolina
- Pace University, New York City, New York
- Purdue University, West Lafayette, Indiana
- Queens University, Kingston, Canada
- Rutgers University, New Brunswick/Piscataway, New Jersey
- San Jose State University, San Jose, California
- Sacramento State University, Sacramento, California
- Southern Methodist University, Dallas, Texas
- Texas A&M International University, Laredo, Texas
- Texas A&M University, College Station, Texas
- Texas A&M-Corpus Christi, Corpus Christi, Texas
- Texas State University, San Marcos, Texas
- Texas Tech University, Lubbock, Texas
- University of Arizona, Tucson, Arizona
- University of Arkansas, Fayetteville, Arkansas
- University of California – San Diego, San Diego, California
- University of Houston -- Clear Lake, Houston, Texas
- University of Kentucky, Lexington, Kentucky
- University of Missouri – Rolla, Rolla, Missouri
- University of North Texas, Denton, Texas
- University of Oklahoma, Norman, Oklahoma
- University of Tennessee at Chattanooga, Chattanooga, Tennessee
- University of Tennessee at Knoxville, Knoxville, Tennessee

- University of Texas at Arlington, Arlington, Texas
- University of Texas at Dallas, Dallas, Texas

The current OSS tools supported by the SSIP for the software engineering capstone courses are the following:

- **Apache:** HTTP server and application server
- **CVS:** Configuration management system
- **Derby:** Database management system
- **Eclipse:** Platform for building an integrated development environment with plug-ins for tools
- **FireFox:** Web browser
- **Gantt Project:** Project planning software
- **Java:** With supporting tools
- **JRequire:** Requirements engineering tools
- **Linux:** Operating system
- **Tomcat:** Application server

At no cost to SSIP member universities, SSIP sponsors are very helpful in closing the information technology gap between software used by industry and software used in classrooms at universities. Avnet has agreed to provide computer server hardware, and Intel has agreed to support software and provide access to the Intel Software College (2006) where students can learn how to optimize and accelerate applications and to enhance software design, anticipate and address potential issues, and improve application performance. They also provide online courses as well as live and recorded Webcasts. IBM and IBM Rational provide computer servers, operational support, and the following software tools:

- **ClearCase:** Configuration management system
- **ClearQuest:** Defect tracking and change management system

- **DB2:** Database management system
- **ProjectConsole:** Visual project monitoring tool
- **PureCoverage:** Code coverage tool
- **Purify:** Automatic error detection tool for finding runtime errors and memory leaks
- **Quantify:** Performance analysis tool
- **RequisitePro:** Requirements tracking tool
- **Robot:** Automated functional regression testing tool
- **RSA:** Rational Software Architect visual modeling tool
- **SoDA:** Report generation tool that supports day-to-day reporting and formal documentation requirements
- **Test Manager:** Test management tool
- **Websphere:** Web server technologies
- **SSIP distributes content using the SSIP Web site located at the following URL:** http://ssi7.cs.tamu.edu/ssi/

For each software tool, the SSIP Web site contains a short tool overview describing the tool in terms easily understood by a student. There are also online tutorials for how to use the tool as part of the student team project. Use-cases are used to describe the relationship of the user to the sample application of the tool. Where available, a WWW link points to the tool Web site. New SSIP member universities are provided WWW linkages to course Web sites for courses that use SSIP content and services. The SSIP provides a user help service to answer questions about any of the tools. Where required, mentors are made available to provide one-on-one tool use help. SSIP servers are available for student project teams to test their project software products. SSIP user forums can be set up for member schools to discuss all aspects of introducing the latest software technology into classrooms.

Many of the CS education program problems are solved by using the SSIP. Students that come to the university today have their own personal client computer that can connect to the Internet. They can connect to computer servers for everything needed in a university curriculum. The SSIP has servers available to SSIP member universities to support member university courses. By using OSS and free software tools provided by industry, the latest software solutions used by industry can be introduced into classrooms at very little cost to the SSIP member university.

Tenured university faculty members who control the courses taught can reduce the time it takes to introduce into the classrooms the latest advanced software solutions used by industry. In the lecture part of a course, the faculty member introduces software development theory, practice, and processes. In the software project laboratory, tools used to support software development are introduced by distance learning through an SSIP Web site. Examples showing the use of each tool are provided. When additional assistance is required, the SSIP operates a help desk and can supply mentors.

The problem of inexperienced laboratory assistance is solved by SSIP supplying services in which students in the laboratory obtain all of the knowledge that is necessary to learn and effectively use software tools. As a result, laboratory assistants spend most of their time managing the student laboratory assignments and activities.

Software tools and support of the tools are provided through the SSIP Web site. The goal is to minimize the support staff that must be provided at the local university. The cost of development tools is minimized by use of OSS and by free software provided by industrial sponsors. Interoperability of applications developed by student software development teams is assured by emphasizing open standards.

Industry today is looking to hire students who know how to be a productive team member. Often universities are reluctant to let student projects use department computer servers shared with other applications for fear that the students will cause the servers to fail. Student project teams need a type of sand-box server on which the student team can build a software product. Sand-box computer servers are provided by the SSIP for use by SSIP member universities.

Help desks and mentoring services are expensive. Industry provides extensive help desk and mentoring services to their customers at great costs. Individual universities cannot afford to provide these infrastructure services to students for the wide variety of software tools needed to support team software development projects in order to create the latest software solutions. By member universities sharing these services, the SSIP can provide services to a large number of universities at a low cost to SSIP sponsors and at no cost to the universities. The SSIP can make these services available 24/7.

Without outside help, universities have difficulty making the latest software solutions available to the students in the classroom. Three barriers to availability are cost, training, and licensing. An SSIP goal is to solve the availability problem by providing open source or free software tools at no cost to SSIP member universities, software tool training classes by distance learning through the SSIP Web site, and free licensing to SSIP member universities.

The SSIP has been well received by member schools. Leading-edge software knowledge is being introduced into university classes at no cost to the university. We are very encouraged with the SSIP success to date. We plan to continue to take advantage of the existing environment in which every student has his or her own Internet-connected client computer on which the student can access the latest software knowledge content from Internet-connected SSIP servers.

CONCLUSION

With the beginning of the new millennium, software development is in a state of change. Low-cost client computers that can be interconnected by the

Internet are available worldwide. Software development teams can be globally distributed around the world. OSS tools can be used to create infrastructures to help introduce industrial strength software into university classrooms. The latest software development process and practices along with open standards can help university students learn how to create enterprise-level interoperable software solutions. The SSIP is an example of how universities working with industry can cooperate to share infrastructure to rapidly close the gap between advanced software technology used by industry and the software knowledge and skills taught in the classroom.

REFERENCES

Denning, P. J., & McGetrick, A. (2005). Recentering computer science. *Communications of the ACM, 48*(11), 15-19.

Flowers, W. (2000). Why change? Been doin' it this way for 4,000 years! In *Proceedings of the ASME Mechanical Engineering Education Conference*, Fort Lauderdale, Florida. Retrieved July 6, 2006, from http://www.asmenews.org/archives/backissues/may/features/educonf.html

Frauenheim, E. (2004). Students saying no to computer science. *ZDNet News.* Retrieved July 6, 2006, from http://news.com.com/Students+saying+no+to+computer+science/2100-1022_3-5306096.html

Intel Software College. (2006). Retrieved July 6, 2006, from http://or1cedar.cps.intel.com/softwarecollege/HomePage.aspx

Kshetri, N. (2005). Structural shifts in the Chinese software industry, *IEEE Software, 22*(4), 86-93.

Naur, N., & Randell, B. (1968). *Report on a conference sponsored by the NATO SCIENCE COMMITTEE*, Garmisch, Germany.

Object Management Group. (2006). Retrieved July 6, 2006, from http://www.omg.org

Open Source Initiative. (2006). Retrieved July 6, 2006, from http://www.opensource.org/

Patterson, D. (2005a). Stop whining about outsourcing! *ACM Queue, 3*(9), 63-64.

Patterson, D. A. (2005b). Restoring the popularity of computer science. *Communications of the ACM, 48*(9), 25-26.

Software Engineering Institute at Carnegie Mellon University. (2006). Retrieved July 6, 2006, from http://www.sei.cmu.edu/

U.S. Department of Labor Statistics. (2005). *Occupational outlook handbook, 2006-2007 edition.* Retrieved July 6, 2006, from http://www.bls.gov/oco/ococ267.htm

Webster's new collegiate dictionary. (1981). Springfield, MA: C. & C. Merriam Company.

KEY TERMS

Capstone Project: Designed for students to synthesize and integrate knowledge acquired through course work and other learning experiences.

Computer-Aided Software Engineering (CASE) Tools: Software tools used to assist in the development and maintenance of software.

Development Software Life Cycle (DSLC): Includes the multiple phases during which defined information technology work products are created or modified as part of the software development process. The last phase of development occurs when the software product is placed into operation.

Interoperable Software: Software that operates with various kinds of software applications and systems by agreeing on a common method with which to communicate and exchange data with one another.

Open Source: Refers to software that has Open Source Initiative (OSI) (2006) licenses. Examples of open source software are Linux, Apache, Eclipse, Derby, and so forth. Also included is open standard compliant software that is provided free to universities for classroom use. An example is IBM Rational Software Architect (RSA).

Open Standard: Refers to standards that are publicly available. The Object Management Group (OMG) (2006) is an example of an organization that was created to produce open standards.

Outsource: To send work that would normally be done by employees in a company to workers that are employed by an outside company.

Productivity Paradigm Change: The improvement of productivity by use of the Internet, clients and servers connected to the Internet, improved communication technologies, advanced software tools, and outsourcing to low-cost labor regions.

Shared Software Infrastructure Program (SSIP): The goal of SSIP is to set up an infrastructure shared among universities in which universities can easily introduce the latest leading software knowledge into both undergraduate and graduate classrooms without building costly infrastructure at each university.

Software Tool: A software product that software developers use to create, debug, or maintain software.

Chapter LIII
Wikis as an Exemplary Model of Open Source Learning

Robert Fitzgerald
University of Canberra, Australia

ABSTRACT

In their simplest form, Wikis are Web pages that allow people to collaboratively create and edit documents online. Key principles of simplicity, robustness, and accessibility underlie the wiki publication system. It is the open and free spirit of Wikis fundamental to open source software (OSS) that offers new contexts for learning and knowledge creation with technology. This chapter will briefly consider the role of technology in learning before discussing Wikis and their development. The emerging literature on the application of Wikis to education will be reviewed and discussed. It will be argued that Wikis embody an exemplary model of open source learning that has the potential to transform the use of information communication technologies in education.

INTRODUCTION

Wikis are an instance of what is known as a read/write technology. They allow groups of users, many of whom are anonymous, to create, view, and edit Web pages. In many cases, these pages are online, but there are instances of Wikis used as personal notebooks (e.g., Tiddlywiki, http://www.tiddlywiki.com/). All wiki systems use a simplified html markup language, but as their use spreads, so does the appeal of more user-friendly java-based WYSIWYG editors. It will be argued that the simplicity, accessibility, and openness of wikis support a model of collaboration and knowledge building that represents an exemplary model of learning with technology. This model is congruent with many of the key principles embodied in free and open source software (FOSS) and sociocultural theories of learning (Lave & Wenger, 1991; Vygotsky, 1978; Wenger, 1998). Many Internet-based communities and groups are already finding ways to embrace these forms of learning as a part of their ongoing process of community capacity building. In contrast, formal places of learning such as schools and universities have been slow to explore the potential of this technology. This chapter will briefly consider the role of technology in learning before discussing Wikis and their development. The chapter argues that FOSS and Wikis in particular offer education far more than just low-cost software or even sound principles of practice; they open up a space for new models of learning and knowledge creation with technology. The emerging literature on the application of Wikis to education will be reviewed before considering Wikis as an exemplary model of open source learning.

TECHNOLOGY AND LEARNING

While the application of computing technology to teaching and learning has at least a 30-year history, there is a large body of literature that suggests education is still struggling to use technology effectively (Cuban, 2001; Healy, 1998; Oppenheimer, 2003; Postman, 1993; Stoll, 1999). Results from large international studies show that the dominant use of technology tends to focus on skills and involves learners as content users and not content creators (Kozma, 2003). Part of the problem is that formal places of learning by their very nature are highly structured contexts. The role of context is important because there is a direct relationship between form and quality of the pedagogy and the form and quality of the resultant learning. Different teaching approaches and learning contexts result in different outcomes for students. In Boaler's (1997) study of mathematics classrooms, she showed that teacher-centered and rule-based teaching approaches not only produce low levels of student engagement but work to effectively limit the scope of the learning outcomes. There is a strong suggestion from her work that routine-style classrooms generate routine knowledge and that this is neither of the quality nor quantity required for real-world mathematical problem solving. Her key finding is that context matters. The Russian neuropsychologist Alexandra Luria understood this relationship well when he argued that cognition is a function of context. "Cognitive processes ... are not independent and unchanging 'abilities' ... they are processes occurring in concrete, practical activities and are

formed within the limits of this activity" (Luria, 1971, p. 266). In effect, Luria was saying that cognition is plastic, a finding that has subsequently been confirmed by contemporary neuroscience (Goldberg, 1990, 2001). The activities and tasks we set for learners not only determine the type and quality of knowledge that is produced but, more importantly, set the parameters for the development of their cognitive processes. Therefore, from a philosophical and practical design point of view, the contexts or settings of learning should be as open and free as possible.

There is also a growing body of literature suggesting that young people learn in different ways to past students and, therefore, require (and even demand) different teaching approaches (Gee, 2003; Oblinger, 2004; Prensky, 2001). Chris Dede (2005) has written extensively in this area, and in his assessment, these learners seek to co-design their learning experiences and prefer communal learning over individual learning. Anyone who has recently studied in schools or universities will know that despite institutional rhetoric to the contrary, these new modes of teaching and learning are not widespread. The pedagogical challenge is to use technology in ways that build upon learners' existing experiences and foster the creation of what von Krogh, Ichijo, and Nonaka (2000) refer to as communities of knowledge. In education, there is widespread recognition of the need to explore more collaborative approaches to learning (Jonassen, Peck & Wilson, 1999; Kozma, 2003; Laurillard, 2002; Sefton-Green, 2004; Somekh, 2004). Wikis offer one such tool, which is already a part of many learners' everyday lives as are a wide variety of other social software such as blogs and social networking applications like MySpace (http://www.myspace.com/).

BACKGROUND TO WIKIS

The founding developer of the World Wide Web (WWW), Sir Tim Berners Lee, first conceived of the Internet as a way for people to both read and write data. The reality of nonintuitive operating systems, html-based coding, clunky file transfer tools, and security restrictions guaranteed that while we could all read the Web, very few of us were able to easily write and publish material. To address some of these issues, Ward Cunningham developed the first Web-based Wiki (http://en.wikipedia.org/wiki/WikiWikiWeb) in 1995 as part of the Portland Pattern Repository (PPR). The origin of wikis, however, goes back much earlier to the 1980s and Cunningham's work on hypertext using Hypercard™ (eWEEK, 2006). In his original formulation (see Table 1), Cunningham developed wiki software to adhere to a number of core design principles (http://c2.com/cgi/wiki?WikiDesignPrinciples).

These basic principles result in a Web publication system that is simple, robust, and accessible. The system is designed to position users differently from traditional publishing models—from a Web reader to a Web author. This can be seen particularly in the universal principle in which any writer is automatically both a space organizer and an editor.

FREE AND OPEN LEARNING

Free and open are key principles underpinning FOSS. This chapter argues that these also underpin learning with Wiki technology. For a piece of software to be open source, it must also adhere to four freedoms summarized in Table 2 (http://www.gnu.org/philosophy/free-sw.html).

While these freedoms relate to OSS, they are enacted within a community of software developers in which community learning and knowledge construction is central. In this sense, open can be used to refer to something that is visible and without barriers. In relation to FOSS, it also refers to a principle of practice that embodies an attitude of generosity. This is perhaps best thought about in terms of an individual (or group) that makes himself or herself available to others and is free and willing to think about new ideas. The notions of free and open have much to offer our conceptions of learning and the underpinning processes of collaboration and knowledge construction with Wikis.

Table 1. Wiki core design principles

Design Principle	Core functionality
Open	Should a page be found to be incomplete or poorly organized, any reader can edit it as they see fit
Incremental	Pages can cite other pages, including pages that have not been written yet
Organic	The structure and text content of the site are open to editing and evolution
Mundane	A small number of (irregular) text conventions will provide access to the most useful page markup
Universal	The mechanisms of editing and organizing are the same as those of writing so that any writer is automatically an editor and organizer
Overt	The formatted (and printed) output will suggest the input required to reproduce it
Unified	Page names will be drawn from a flat space, so that no additional context is required to interpret them
Precise	Pages will be titled with sufficient precision to avoid most name clashes, typically by forming noun phrases
Tolerant	Interpretable (even if undesirable) behaviour is preferred to error messages
Observable	Activity within the site can be watched and reviewed by any other visitor to the site
Convergent	Duplication can be discouraged or removed by finding and citing similar or related content

Table 2. Four freedoms of open source software

Freedom	Description
0	The freedom to run the program, for any purpose
1	The freedom to study how the program works, and adapt it to your needs Access to the source code is a precondition for this
2	The freedom to redistribute copies so you can help your neighbour
3	The freedom to improve the program, and release your improvements to the public, so that the whole community benefits. Access to the source code is a precondition for this

EXAMPLES OF WIKI PROJECTS

The use of Wikis had a significant development when Jim Wales launched Wikipedia (http://en.wikipedia.org) and the Wikimedia Foundation (http://wikimediafoundation.org) in 2001. For the first time, this project made Wiki technology and tools freely and readily available to the public while also offering a series of projects to which they could contribute. The most recognized Wiki is the free-content encyclopedia Wikipedia (http://en.wikipedia.org/wiki/), which is maintained by a group of volunteers from around the world and is generally regarded as a rich information resource (LeLoup & Ponerio, 2006; Lih, 2004).

Jim Wales explained:

Wikimedia's mission is to give the world's knowledge to every single person on the planet in their own language. As part of that mission, Wikipedia is first and foremost an effort to create and distribute a free encyclopedia of the highest possible quality. Asking whether the community comes before or after this goal is really asking the wrong question: the entire purpose of the community is this goal. (Wales, 2005)

Wikipedia currently comprises 1,315,437 English language articles and nearly two million registered accounts (http://en.wikipedia.org/wiki/Special:Statistics). Roughly 25% of these articles have been translated into other languages (nine at last count). Since July 2002, Wikipedians have made more than 70 million edits. Web traffic statistics rank Wikipedia as the 17th most popular Web site (out of 250 million) on the Internet with more than 2.9 billion page views in the month of August 2006 (http://www.alexa.com). Currently there is a number of related Wikimedia Foundation projects that are in development, including Wikipedia (http://www.wikipedia.org/); Wiktionary (http://wiktionary.org/); Wikibooks; Wikinews (http://www.wikinews.org/); Wikiquote (http://www.wikiquote.org/); Wikicommons (http://commons.wikimedia.org); Wikisource (http://wikisource.org/); Wikispecies (http://species.wikimedia.org) and the soon to be officially established Wikiversity (http://en.wikibooks.org/wiki/Wikiversity). Each of these projects takes the basic wiki model and extends it into a specific area with a specific goal. For example, Wikibooks (http://en.wikibooks.org) is a project designed to produce open-content textbook modules to create global curricula. To date 21,019 book modules have been developed for more than 1,000 books. Some of these books are available in PDF format.

The large-scale application of wikis to community knowledge building can also be seen in Wikia (http://www.wikia.com) and Wikispaces (http://www.wikispaces.com/). The aim of these initiatives is to provide individuals and communities with a Web site they can use to create open content around their areas of interest. For example, establishing a Wikia site requires the topic to appeal to a large number of people and that its content will have some longevity. Computer game players have been particularly active, creating communities around their games. One example of how this is used can be seen in the ways that players of the massively multiplayer online game Runescape (http://www.runescape.com/) have built encyclopedic knowledge about all aspects of the game (http://www.wikia.com/wiki/Runescape).

LITERATURE REVIEW ON WIKIS

Many of the applications of Wikis are entirely congruent with the so-called Web2.0 and social software models that attempt to offer simple and robust technologies to non-expert users so they are able to create content and build communities. The last few years have seen a proliferation of simple tools for authoring Web content, particularly in the area of personal blogging. Wikis are still without peer when it comes to the large-scale collaborative authoring of Web content (Lamb, 2004; Wagner, 2004). A literature based on the application of Wikis is beginning to emerge, although to date it is primarily conceptual and descriptive. Generally, there is widespread agreement that Wikis represent an innovative and potentially powerful tool for collaborative content creation and sharing (Bold, 2006; Engstrom & Jewett, 2005; Godwin-Jones, 2003; Lamb, 2004; Wegner, 2004). There have been few studies that have tested empirical propositions, although many of these have been exploratory in nature.

Wikis have been studied in language learning (Godwin-Jones, 2003; LeLoup & Ponerio, 2006; Wang et al., 2005; Wei et al., 2005) as tools in higher education (Augar, Raitman, & Zhou, 2004; Bold, 2006) to promote forms of participatory journalism (Lih, 2004), as a tool for story-telling in primary schools (Désilets & Paquet, 2005), and examined for their potential role in increasing citizen participation in e-government initiatives (Wagner, Cheung, & Ip, 2006).

In summary, the empirical work finds the following:

- The effective use of Wikis appears dependent on a clear goal matched to a group of committed uses (Godwin-Jones, 2003).
- Highly structured environments that rely on top-down approaches (as opposed to bottom-up) limit the potential of Wikis as a tool for learning (Engstrom & Jewett, 2005; Wagner, 2004).
- Wikis such as Wikipedia are a rich source of information that can promote content creation, sharing, and discussion (LeLoup & Ponerio, 2006, Lih, 2004).
- It is important to augment students, Wiki work with strategies to promote deep and critical thinking to ensure high quality work emerges (Engstrom & Jewett, 2005).
- Wikis support a short edit-review cycle that ensures the rapid development of content (Lih, 2004).
- Employing the user as organizer and editor (many "eyeballs") is a highly effective strategy for ensuring quality (Lih, 2004).

There have been widespread calls for more research on Wikis (Lih, 2004; Wagner, 2004; Wei et al., 2005). It will be particularly important to develop research methods that are sensitive to both the quality of content produced in concert with how this content emerges within a community of learners. There is encouraging work on the development of metrics to assess the quality of Wiki (see Wikipedia) articles based on edit histories (Lih, 2004), but we also need to examine and assess the quality of the articles. Wiki edits are easily quantified, but what they relate to is not always clear. For example, a recent study found an inverse relationship between the quantity of Wiki edits and final exam scores (Wang et al., 2005). The authors advised caution against interpreting these findings as evidence that wikis are counter-productive to learning, but it does highlight the need for more nuanced and in-depth empirical studies on Wikis.

Wikis have generated considerable interest in education because they appear to support more collaborative models of teaching and learning. It is fair to say that there is considerable anecdotal evidence that Wikis can and should play a key role in e-learning in support of a more conversational and dialogic approach to knowledge creation and sharing.

AN EXEMPLARY MODEL OF OPEN SOURCE LEARNING

Wikis offer a different model of creating, editing, and sharing knowledge that is consistent with the educational push toward what have become known as sociocultural or constructivist approaches to learning. A founding thinker in these areas, Lev Vygotsky (1978), contended that learners neither receive knowledge nor simply discover it. They learn in social contexts in interaction with both humans and tools. A key concept for Vygotsky was the zone of proximal development (ZPD) in which he said all learning takes place. In Vygotsky's basic model, it is adults who scaffold young learners, helping to extend their thinking and learning. However, as the technological tools develop and evolve, we are beginning to see ways that both humans and their tools can scaffold learning. The technological spaces that make up wikis enable new forms of sociotechnological ZPDs that support both individual and community knowledge creation.

This focus on community and the power of joint construction is taken up in The Wisdom of Crowds (Surowiecki, 2004). Surowiecki argues that the collective knowledge of large groups is often unrecognized and almost always undervalued by society. He explains that many everyday activities, from voting in elections and the operation of the stock market to the way Google locates Web pages, depend on the collective input and knowl-

edge of large groups. Of course, not all crowds are smart, but Surowiecki believes that under the right conditions, crowds can act more wisely than an expert individual. To achieve the best results, crowds must be able to support diversity of opinion, relative independence in an individual's thinking, a model of decentralization that allows individuals to draw on their local knowledge and aggregation; that is, embody a process whereby individual knowledge can be combined into an integrative whole. The ways technology might be used to support the development of smart crowds is a relatively unexplored area; however, applications such as Wikis, blogs, and multiplayer games certainly show how large groups of people can productively interact online. While there is a huge qualitative difference between group interaction and wisdom, there is a relationship to be explored that highlights the importance of developing large-scale social technologies such as Wikis.

The previous discussion suggested that collaborative knowledge creation should be an important feature of formal learning; however, to date, and particularly with reference to the use of technology, it has not been. Some of the uses of technology by young people in their everyday lives seem to get closer to this goal. The way they work in their communities around multiplayer games to talk to each other and build knowledge is one example. The next section considers the potential that Wikis offer to achieve this goal.

One of the strengths of Wiki software is its capability to document and record aspects of the knowledge creation process. From an educational point of view, this can provide valuable insights to the knowledge construction process. In most Wikis, an article features a number of views: the article page, a discussion page, article editor, and history. A rich edit history features full revision history permitting comparison between current and last entry. Edits can further be identified via flags and commenting, helping others understand the changes that have been made. To help assure quality, edits also appear on a recent-changes page.

686

Rules can be set up to show pages that have been changed since the last visit. A list of contributions by users offers various analyses, often with full history and comparison tools. In combination, these tools open up the possibility of exploring the relationship (and tension) between individual and group constructions.

Surowiecki said:

Any "crowd"—whether it be a market, a corporation, or an intelligence agency—needs to find the right balance between the two imperatives: making individual knowledge globally and collectively useful (as we know it can be), while still allowing it to remain resolutely specific and local. (Surowiecki, 2004, p. 140)

Wikis allow both individual contributions and the evolving group product to sit alongside each other. The examples of Wikis outlined here seem able to create new forms of sociotechnological ZPDs for learners. These zones support both individual and community knowledge creation in ways that are consistent with the notion of communities of practice (Lave & Wenger, 1991; Wenger, 1998).

Wikis appear well suited to building knowledge in which the representation of balanced opinion is valued. While there is no guarantee that this prevents wikiwars, it does seem that strong opinion is better suited to other spaces such as blogs. It is clear from the work of the Wikimedia Foundation that there are no hard and fast rules to using Wikis. What is apparent, though, is that many of the more successful projects embody a spirit of community characterized by openness and freedom.

CONCLUSION

In the end, the success of innovations in learning such as Wikis will be seen in the increased capacity of individuals and their communities to create and apply new knowledge. Incorporating tools that not only facilitate but also document the effective management of information and creation of knowledge is now essential for an innovative and productive 21st-century society. Wikis are significant and innovative because they attempt to position learners as knowledge creators rather than simply content users. They also represent the application of new collaborative technologies in ways that are free and open. In terms of education, the Wiki model locates the challenge of improving information, data management, and knowledge creation processes within a community model. It also builds on strategies for increasing the capacity of all community members to expand their ways of thinking creatively and working collaboratively. A key feature of this approach is that knowledge and practice are shared in a spirit of generosity. The extensive use of Wikis in education begs the question: Are we really prepared to engage in this form of open source learning?

REFERENCES

Augar, N., Raitman, R., & Zhou, W. (2004). Teaching and learning online with wikis. In R. Atkinson, C. McBeath, D. Jonas-Dwyer, & R. Phillips (Eds.), *Beyond the comfort zone: Proceedings of the 21st ASCILITE Conference* (pp. 95-104). Retrieved June 17, 2006, from http://www.ascilite.org.au/conferences/perth04/procs/augar.html

Boaler, J. (1997). *Experiencing school mathematics: Teaching styles, sex and setting.* Buckingham, UK: Open University Press.

Bold, M. (2006). Use of Wikis in graduate course work. *Journal of Interactive Learning Research, 17*(1), 5-14.

Cuban, L. (2001). *Oversold and underused: Computers in the classroom.* Cambridge, MA: Harvard University Press.

Dede, C. (2005). Planning for neomillennial learning styles. In D. Oblinger & J. Oblinger (Eds.), *Educating the net generation* (pp. 224-249), Retrieved July 4, 2006, from http://www.educause.edu/educatingthenetgen

Désilets, A., & Paquet, S. (2005). Wiki as a tool for Web-based collaborative story telling in primary school: A case study. In P. Kommers & G. Richards (Eds.), *Proceedings of World Conference on Educational Multimedia, Hypermedia and Telecommunications 2005* (pp. 770-777). Chesapeake, VA: AACE. Retrieved June 17, 2006, from http://iit-iti.nrc-cnrc.gc.ca/iit-publications-iti/docs/NRC-48234.pdf

Engstrom, M. E., & Jewett, D. (2005). Collaborative learning the wiki way. *TechTrends, 49*(6), 12-16.

eWEEK.com. (2006). *Father of Wiki speaks out on community and collaborative development.* Retrieved July 14, 2006, from http://www.eweek.com/article2/0,1895,1939982,00.asp

Godwin-Jones, R. (2003). Blogs and Wikis: Environments for on-line collaboration. *Language Learning & Technology, 7*(2), 12-16. Retrieved June 20, 2006, from http://llt.msu.edu/vol7num2/emerging/

Goldberg, M. E. (1990). *Contemporary neuropsychology and the legacy of Luria.* Hillsdale, NJ: Lawrence Erlbaum.

Goldberg, M. E. (2001). *The executive brain: Frontal lobes and the civilized mind.* New York: Oxford Press.

Healy, J. (1998). *Failure to connect: How computers affect our children's minds.* New York: Simon & Schuster.

Jonassen, D., Peck, K. L., & Wilson, B. G. (1999). *Learning with technology: A constructivist perspective.* Upper Saddle River, NJ: Prentice Hall.

Kozma, R.B. (2003). Technology and classroom practices: An international study. *Journal of Research on Technology in Education, 36*(1), 1-14.

Lamb, B. (2004, September/October). Wide open spaces: Wikis, ready or not. *Educause,* 36-48.

Laurillard, D. (2002). *Rethinking university teaching.* London: Routledge.

Lave, J., & Wenger, E. (1991). *Situated learning: Legitimate peripheral participation.* New York: Cambridge University Press.

LeLoup, J. W., & Ponerio, R. (2006). Wikipedia: A multilingual treasure trove. *Language Learning & Technology, 10*(2), 12-16. Retrieved June 20, 2006, from http://llt.msu.edu/vol10num2/net/

Lih, A. (2004). Wikipedia as participatory journalism: Reliable sources? Metrics for evaluating collaborative media as a news resource. In *Proceedings of the 5th International Symposium on Online Journalism.* Retrieved July 7, 2006, from http://jmsc.hku.hk/faculty/alih/publications/utaustin-2004-wikipedia-rc2.pdf

Luria, A. R. (1971). Towards the problem of the historical nature of psychological processes. *International Journal of Psychology, 6,* 259-272.

Oppenheimer, T. (2003). *The flickering mind: The false promise of technology in the classroom and how learning can be saved.* New York: Random House.

Postman, N. (1993). *Technopoly: The surrender of culture to technology.* New York: Vintage Books.

Stoll, C. (1999). *High tech heretic: Why computers don't belong in the classroom and other reflections by a computer contrarian.* New York: Doubleday.

Surowiecki, J. (2004). *The wisdom of crowds.* New York: Doubleday.

von Krogh, G., Ichijo, K., & Nonaka, I. (2000). *Enabling knowledge creation.* New York: Oxford University Press.

Vygotsky, L. (1978). *Mind in society.* Cambridge, MA: Harvard University Press.

Wagner, C. (2004). Wiki: A technology for conversational knowledge management and group collaboration. *Communications of the Association for Information Systems, 13,* 265-289.

Wagner,C. Cheung, S. K., & Ip, K. F. (2006). Building semantic Webs for e-government with wiki technology. *Electronic Government, 3,* 36-55.

Wales, J. (2005). *Letter from the founder.* Retrieved July 14, 2006, from http://wikimediafoundation. org/wiki/Founder_letter

Wang, H. C., Lu, C. H., Yang, J. Y., Hu, H. W., Chiou, G. F., Chiang, Y. T., & Hsu, W. L. (2005). An empirical exploration of using wiki in an English as a second language course. In *Proceedings of 5th IEEE International Conference on Advanced Learning Technologies* (pp. 155-157). Retrieved July 17, 2006, from http://www.iis.sinica.edu.tw/IASL/webpdf/paper-2005-An_Empirical_Exploration_on_Using_Wiki_in_an_English_as_Second_Language_Course.pdf

Wei, C., Maust, B., Barrick, J., Cuddihy, E., & Spyridakis, J. H. (2005). Wikis for supporting distributed collaborative writing. In *Proceedings of the Society for Technical Communication 52nd Annual Conference.* Retrieved July 7, 2006, from http://www.uwtc.washington.edu/research/pubs/jspyridakis/STC_Wiki_2005_STC_Attribution.pdf

Wenger, E. (1998). *Communities of practice: Learning, meaning, and identity.* New York: Cambridge University Press.

KEY TERMS

Constructivist: An approach based on the work of Lev Vygotsky, who contended that learners neither receive knowledge nor simply discover it. They learn in social contexts in interaction with both humans and tools.

Free and Open Source Software (FOSS): A term first described by Richard Stallman referring to a software development process in which the software source code is made freely available for subsequent modification and development.

Hypercard™: A hypermedia program developed by Apple Computer in the 1980s.

Open Source Learning: A model of learning inspired by the key principles or freedoms embodied in the FOSS movement.

Web2.0: A term coined by Tim O'Reilly (http://tim.oreilly.com/) referring to a range of second-generation Web publishing and social networking technologies.

Wiki: A form of read/write technology that allows groups of users, many of whom are anonymous, to create, view, and edit Web pages.

Wikia: A project to provide communities with Wiki-type Web sites (see http://www.wikia.com).

Wikimedia Foundation: An international nonprofit organization run by Jim Wales, using wiki technology to promote free and open large-scale collaborative content creation projects (http://wikimediafoundation.org).

Wikipedia: A free-content encyclopedia (http://en.wikipedia.org/wiki/).

Zone of Proximal Development (ZPD): The difference between what learners can do by themselves and with the assistance of more capable adults or peers.

Chapter LIV
A Perspective on Software Engineering Education with Open Source Software

Pankaj Kamthan
Concordia University, Canada

ABSTRACT

As the development and use of open source software (OSS) becomes prominent, the issue of its outreach in an educational context arises. The practices fundamental to software engineering, including those related to management, process, and workflow deliverables, are examined in light of OSS. Based on a pragmatic framework, the prospects of integrating OSS in a traditional software engineering curriculum are outlined, and concerns in realizing them are given. In doing so, the cases of the adoption of an OSS process model, the use of OSS as a computer-aided software engineering (CASE) tool, OSS as a standalone subsystem, and open source code reuse are considered. The role of openly accessible content in general is discussed briefly.

INTRODUCTION

The steady rise of OSS (Raymond, 1999) over the last few decades has made a noticeable impact on many sectors of society in which software has a role to play. As reflected from the frequency of media articles, traffic on mailing lists, and growing research literature, OSS has garnered much support in the software community. Indeed, from the early days of GNU software to the X Window System to Linux and its utilities, and more recently the Apache Software Project, to name a few, OSS has changed the way software is developed and used.

Software engineering (Ghezzi, Jazayeri, & Mandrioli, 2003) advocates a disciplined and systematic approach to the development of high-quality software within budget, schedule, and other organizational constraints. This chapter discusses the symbiosis between traditional software engineering and open source software development (OSSD) from an educational standpoint.

The organization of the chapter is as follows. We first outline the background necessary for the discussion that follows and state our position. This is followed by a detailed treatment of key software engineering practices that are addressed in light of OSS. We then discuss the use of OSS in software engineering education (SEE). Next, challenges and directions for future research are outlined, and finally, concluding remarks are given.

BACKGROUND

The concept of open source can mean different things in different contexts (Gacek & Arief, 2004; Perens, 1999). For the purposes of this chapter, we will use "open source" as a single encompassing term that subsumes all of the following: free/freely available or libre/liberated software whose source is available without cost to the user, imposes minimal nonrestrictive licensing conditions, and is based upon nonproprietary technologies. Software that does not fall into this category is termed non-OSS. For example, commercial software is one class of non-OSS.

As the use of OSS in various sectors of society increases, the question of how they are actually engineered garners interest. A software engineering perspective toward OSS is necessary for a variety of reasons: OSS may be adopted and used in critical areas of an organization and thus needs to be carefully examined with respect to non-OSS alternatives; OSS installed in an organization may need to be maintained over time and, therefore, needs to be well understood by maintenance engineers; and current OSS practices could be of interest from an academic (teaching, learning, research) standpoint.

Although OSS itself has a long, rich history, it is only in recent years that a software engineering viewpoint toward it has been taken (Spinellis & Szyperski, 2004; Vixie, 1999). Annual workshops in recent years under the label of Open Source Software Engineering have also created an awareness of this important area.

As OSS becomes prominent, the issue of its outreach in an educational context arises. In this chapter, we take the position that students studying software development should be exposed early to this rapidly growing area. In fact, the use of OSS in computer science education has been emphasized in recent years (Attwell, 2005; González-Barahona et al., 2000; Liu, 2003). It has also been suggested (Cusumano, 2004) that developing OSS could also help students in their future career paths.

However, the current studies of OSS-based education are limited in one or more of the following ways: the discussion is often confined to the case study of a specific OSS, does not highlight the problems associated with introducing OSS, does not address software engineering exclusively, or ignores aspects of software engineering that OSS does not address. One of the purposes of this chapter is to address these concerns.

ELEMENTS OF SOFTWARE ENGINEERING AND ITS EDUCATION AND THEIR MANIFESTATIONS IN OPEN SOURCE CONTEXTS

This section looks at six broadly classified aspects; namely, management, process, modeling/specification, standards, documentation, and quality/measurement, which are common in most SEE contexts, and examines the extent to which they are realized (or not) in an OSS environment. In doing so, we inherently set the limits of the use of OSS in SEE, which is discussed in the following section.

Management

Managing a software project is important for its eventual success. We shall limit our discussion largely to measuring success and team, time, and configuration management.

The goals of developing software in educational and OSS contexts are different. In software engineering, the software product is a means to an end, not an end in itself. It has been reported (Cusumano, 2004) that OSS often lacks precise specification of goals and, as a result, fails to define success. The reason for abandoning an OSS project often are not given or made public. In SEE, there is a price for not performing up to the expectations or not working to full potential, which is often exhibited in a grading differential.

Although software engineers are often bound by an organizational or professional code of ethics, this is not the case is OSS, which is carried out on an honor system. Specifically, there are little or no repercussions for not following up on work or on schedule, or stalling the project altogether. This flexibility may be attractive in a professional context but does not scale well in an educational setting. In lieu of mimicking real-world software projects as well as due to natural limitations of schedules at educational institutions, there are inevitable time constraints associated with course projects. However, there is little sense of urgency in OSS projects.

There are differences between the social structure of a team of students in a software engineering environment and participants in the OSSD. In general, software engineers working on a software project in a professional or learning context are collocated, while those in OSS developers form a distributed community (Crowston & Howison, 2005; Thomas & Hunt, 2004). There is also a notable difference with respect to social bonding. The students most likely belong to the same institution and may take multiple courses together. The students also may be related on a personal level (roommates, siblings, friends), while that is not the norm in an OSS development in which the participants are loosely related. There is no inherently hierarchical team structure in OSS. There is usually a core group that contributes the most with a sporadic participation by others (Michlmayr, Hunt, & Probert, 2005). On the other hand, assuming responsibility and accountability individually and as a team are at the heart of software engineering.

The distributed nature of contribution as well as the desire of the developers to be able to disseminate up-to-the-minute code has led to a

usually strong support for configuration management (version control, bug tracking, or build management) (Asklund & Bendix, 2001) in OSSD. Posting nightly builds for tryout is quite common in an OSS environment. However, in the author's experience with the practice of SEE, configuration management is not as pervasive in educational software projects as it is in OSS and is usually limited to version control and backups.

Process and Workflows

In software engineering, students are normally introduced to both prescriptive and agile process models. The former are often rigid/bureaucratic and involve heavy use of documentation. The latter allow flexibility by virtue of sensitivity to the social and organizational environment in which software is being created and involve lightweight documentation. Each is useful in its own right with respect to the characteristics of various application domains and in different team environments.

The OSSD process, known as the Bazaar model (Vixie, 1999), is not subsumed by any of these conventional software process models, although it is much closer to the latter than it is to the former. For example, many of the practices of extreme programming (XP) (Beck & Andres, 2005), an agile process model, are applicable to OSS (Nishinaka, 2001). However, two of the key practices of XP (namely, Onsite Customer and Pair Programming) do not scale well in the distributed, nonproximal environment of the OSS. The Bazaar model also differs from other iterative process model frameworks such as the Unified Process (UP) (Jacobson, Booch, & Rumbaugh, 1999) that embrace certain aspects of agility. For example, UP has a strong emphasis on customer involvement and is model-driven, both of which are not a commonplace in OSSD.

Traditional software process workflows typically include software requirements (problem definition), software design (high-level view of the solution), implementation (low-level working solution), and testing (verifying whether the solution, in fact, matches the problem). In an OSSD process, software requirements are usually absent, the focus on design is informal, and there is much attention on implementation and, in some cases, on testing. Indeed, several OSS utilities (notably for properly structuring source code and for unit testing) have been created just to support the last two phases.

It is a commonly held belief in the software engineering community that the quality of a software process directly impacts the quality of the software product, and therefore, much research in the last two decades has focused on means for software process improvement. Indeed, process maturity is an integral topic in many courses related to software process engineering. Based on project retrospectives, organizations continually strive to improve their software processes in order to make them more effective while remaining cost-friendly. However, unlike the case of traditional software process environments in which organizations can make use of the capability maturity model (CMM) (Paulk, Weber, Curtis, & Chrissis, 1995), there seems to be little systematic effort toward addressing the maturity of the OSSD process.

Modeling and Specification

Modeling, particularly during early phases of software development, is playing an increasingly important role in activities and deliverables in software engineering (Beydeda, Book, & Gruhn, 2005). Early modeling is crucial from the point of view of understanding the problem and solution domains in an implementation neutral manner and control and prevention of problems that can propagate into later stages. Modeling in its different degrees of formality plays a central role in both XP and UP and is a determinant of the process maturity of an organization. Some form of modeling is introduced in most practical software engineering courses.

The unified modeling language (UML) (Booch, Jacobson, & Rumbaugh, 2005) has emerged as a standard language for modeling the structure and behavior of object-oriented systems, and its use in the last few years in SEE has increased dramatically. The author has recommended a proper use of UML (Kamthan, 2004) for domain and use case modeling in several courses. However, there is little evidence of use of UML and, in general, of any form of systematic modeling in OSSD.

Formal specifications are also integral to many courses in software engineering (Alagar & Periyasamy, 1998) in which the safety requirements or design of a critical system need to be precisely (mathematically) expressed. However, once again, there is little evidence to support the use of mathematics in OSS problem or solution domains for system analysis or synthesis, respectively. This evidently limits the use of OSS, even in part, in safety-critical software. A similar argument holds as the definition and design of real-time or embedded systems also gradually begin to depend on formal specifications.

Standards

There is a variety of reasons for introducing and adhering to standards in software engineering. Standards provide a common ground for a team, streamline efforts, and when applied well, are known to contribute to quality improvement (Schneidewind & Fenton, 1996). Lack of standardization often can lead to communicability problems (among humans) and interoperability problems (among machines). The author has been a strong proponent of the use of standards throughout SEE, has made mandatory use of IEEE and/or ISO/IEC standards in process documents, and strongly encouraged standardized (ANSI, ECMA) definitions of programming languages and corresponding compilers/interpreters.

The use of standards in OSSD is usually limited to implementation-level concerns. The OSS approach serves as a platform for trying out new technologies and developing proof-of-concept implementations. In doing so, the use of standards is limited to data formats such as the hypertext markup language (HTML) or the extensible markup language (XML) and presentation languages such as the cascading style sheets (CSS).

Documentation

The role of communication is central to any software development. The documentation forms the message carrier within the communication infrastructure of a software project.

The role of documentation is usually accentuated in software engineering. The courses related to technical communication and programming methodology early in the curriculum form the basis of internal documentation of software developed in later courses. In some cases, creating external documentation (user manual or a help system) may also be required.

In contrast, it has been the author's experience that often OSS is apparently weak with respect to both internal and external documentation. Any documentation, if at all, tends to focus more on the implementation rather than early stages (of requirements or design). Process documentation is not always adopted and followed. The documentation at times may not be complete or may only be sketchy. At times, help or tutorial documents are not updated to synchronize with the latest code releases. The OSS style of writing currently in place at times tends to be informal rather than technically inclined to the issue at hand. In other words, in general, OSS is not a hallmark of how documentation should be written. That these issues be pointed out to the students early is critical, especially if it is their first contact with a systematic use of documentation in software; otherwise, the perceptions and

habits tend to coagulate and are harder to change with the passage of time.

Quality and Measurement

In software engineering and its education, there is much emphasis on quality in all aspects of software (project, process, product, and occasionally people).

The issue of OSS quality in general, and concerns of performance, security, and usability in particular have been addressed (Schmidt & Porter, 2001; Halloran & Scherlis, 2002; Seidel & Niedermeier, 2003; Michlmayr, Hunt, & Probert, 2005).

There are many OSSs that exhibit high quality. However, the approach to quality assurance and assessment is not systematic (Fenton & Pfleeger, 1997), and therefore, the results do not seem to be repeatable. In OSS, peer reviews are used as a technique for an informal evaluation, whereas formal inspections are apparently nonexistent. In general, comprehensive collections of test cases, test suites, or test harnesses are rare, and broad testing is even rarer. More importantly, participation is voluntary, and monitoring is almost nonexistent. The linear relation of the number of bugs found to improve the quality proposed by the OSS development process (Raymond, 1999) is a bit simplistic and, indeed, has been termed as a fallacy from a software engineering perspective (Glass, 2003).

The view of quality that is usually taken in SEE is the following: to improve an aspect of a given entity, we must be able to quantify that aspect. Therefore, the issue of quality is closely related to that of measurement (Fenton & Pfleeger, 1997). For example, if we wish to improve space-efficiency, we could measure the source program file size and, in turn, the lines of code (or number of characters); to improve structural complexity of a program, we could measure the number of decision structures, parent-child classes, method calls, and so forth. Once again, there is little

Table 1. A high-level view of the framework for deploying OSS in SEE

Legality	
Feasibility	
Teaching and Learning Goals	
Application	OSS Process Adoption, OSS as a Software Development Tool, OSS as a Sub-System, OSS for Reuse
Theory	OSS for Pedagogy, OSS for Learning

evidence to support rigorous measurement in OSS contexts.

Having compared OSS to traditional software engineering and its education, we now turn our attention to realizing OSS in SEE.

IMPLICATIONS OF OPEN SOURCE SOFTWARE IN SOFTWARE ENGINEERING EDUCATION

We have previously advocated different (but not necessarily mutually exclusive) ways in which OSS can be used in SEE (Kamthan, 2006): for pedagogy and learning, which are theoretical in nature; and adopting the OSS process as tools that support software production, as one of the subsystems, or for the purpose of source code reuse, which are practical in nature. We note that the applied aspects can all occur within the same software project. These approaches need to be aligned with teaching and learning goals to which the contributing factors include the pedagogical aims of the institution (that will likely vary between, for instance, a polytechnic school and a university), alignment with respect to overall program curriculum, and student background. Since software engineering is a practical discipline, all the aims and activities from its initiation to its completion should be feasible. To help achieve that, analytical hierarchy process (AHP) and quality function deployment (QFD) are two commonly used project management techniques. Finally,

laws regarding OSS vary across jurisdictions (e.g., Canada, Germany, and Russia), and therefore, any use and/or development of OSS must be legally acceptable where it is carried out.

The precise articulation of the teaching and learning goals, of the criteria and techniques to be adopted for carrying out a feasibility study, or of legal issues is beyond the scope of this chapter. Table 1 summarizes our approach for integrating OSS in SEE.

We now discuss the theoretical and application-specific elements of the framework in more detail.

Open Source Software for Pedagogy

OSS could be deployed for the purposes of teaching in a classroom. The availability of source code in OSS provides a unique opportunity for the teacher to experiment.

Source code internals of software (i.e., usually larger in scale than those accompanying the commonly used textbooks) can be shown, and aspects of its design and quality can be debated in the classroom. Educators, for example, could point out both successful and failed OSS efforts and reasons for being so. As compared to toy theoretical examples in textbooks, the OSS real-world contexts often can provide better opportunities for teaching intricate concepts.

The openness of OSS in contrast to non-OSS becomes all the more valuable when a deep knowledge of system internals is necessary for understanding. This is particularly the case in systems software courses in which, for example, the design of an operating system kernel of an OSS such as Linux can be discussed.

Educators can also use OSS as a basis for assigning course projects on similar topics. The openness of the source code helps them judge the feasibility of a software project for a given team size and the time allowed. OSS also could be used as a basis for reverse engineering in which, given a certain OSS, students can be asked to create a high-level model or visualization for it or to refactor (Fowler et al., 1999) it to improve some of its quality attributes while still preserving its functionality.

It has been the author's experience that OSS also can serve as a starting point for discussing social aspects of software engineering such as software ethics (Qureshi, 2001) and licensing issues. For example, how well a given OSS follows the principles put forward by the Software Engineering Code of Ethics and Professional Practice (SECEPP) of the ACM/IEEE-CS Joint Task Force on Software Engineering Ethics and Professional Practices are worthy of examination and class discussion.

Open Source Software for Learning

OSS provides a useful workbench for learning. OSS can be used for self-learning purposes outside the classroom (e.g., at home). The ascent of affordable personal computers, high-speed Internet connectivity, and the use of the Web as an information base are having a major impact on the way students study and learn at home. The constructivist theories of learning have emphasized learning by doing, and the availability of OSS source code provides a unique opportunity for students to experiment and thereby enhance their skills.

We note, however, that the lack of sufficient documentation and timely technical support, if

at all, can pose obstacles for putting this into practice.

Adopting the Open Source Software Process

As part of a course project, students could be made to simulate an OSS environment for developing software by adopting the OSS process and the practices in it. The resulting software will then be an OSS whose development will be open to the public. As an example, SourceForge could provide a medium for development, collaboration, and distribution.

However, this may be the most challenging of all the applications of OSS in SEE. First, this will require extra effort on the part of the educator that may not be in line with the requirements of mainstream courses. The Bazaar model requires a different mindset from traditional approaches and may need to be tailored for an educational use. For example, instilling the sense of teamwork in physical proximity and collectively experiencing the issues that go with it are an important part of learning. Some institutions discourage coursework outside their confines and expect ownership of the final product.

Fairness in evaluation is also an issue. For example, once a team has set up a place on Source-Forge, should it be allowed to solicit help and feedback from those not registered in the course? What is the impact of openness of source across teams? These questions need to be addressed and satisfactorily answered prior to any OSS initiative in education.

Open Source Software as a Software Development Tool

We need software to develop software, and OSS utilities could prove to be quite useful in that regard. Examples are Apache Maven for project management, MediaWiki for fostering teamwide communication, ArgoUML as a UML modeler, IBM Eclipse as a multipurpose authoring environment, CCDoc for C++ documentation, Bugzilla for issue tracking, Apache Ant for building, and JUnit for unit testing, to name a few.

However, some of the hindrances one faces are the following: OSS is not always feature-rich in comparison to its non-OSS counterparts; the OSS utilities used may not be interoperable with each other; or students may find that all-in-one multi-utility packaged commercial integrated development environments (IDE) are more convenient to use for programming purposes than individual isolated pieces of software.

Open Source Software as a Subsystem

Reinventing the wheel is considered inertia in software development and, at times, is not practical. For example, it is not always realistic to develop everything that is required from scratch for a software project.

OSS can be used as auxiliary software and thereby supports the system under development. In that regard, OSS support in general has been exemplary. A systematic approach for creating Web applications has been termed Web engineering (Ginige & Murugesan, 2001), and OSS has played a crucial role in advancing this discipline. Indeed, the author's experience with the support of OSS in Web engineering for applications such as Course Registration System, Distributed Battle Ship Game, Fine Art Auction System, Patient Medical Record System, and Student Personal Information Portal, has in general been quite encouraging. For example, a project involving a Web application could use Amaya as the user agent on the client side and Apache Web Server along with Apache Tomcat or MySQL/PHP application server, as deemed necessary, for a dynamic delivery of resources on the server side. This can be supplemented by other software for quality assurance, including the use of information rep-

resentation language conformance checkers (for CSS, for HTML or for markup languages based on XML) and tools for checking Web accessibility such as A-Prompt.

One of the obstacles faced in the use of OSS as a subsystem is that due to security considerations, certain educational institutions do not allow arbitrary installations of network software by students. In cases in which they do, system administrators may consider it beyond their domains of responsibility and may not be willing to provide any technical support whatsoever.

Open Source Software for Reuse

This approach to OSS in software engineering advocates reuse portions of OSS code in assignments or as part of the system under development as for the course project. Examples include OSS libraries or frameworks. It ameliorates the tedium of writing the entire code from scratch, particularly for routine primitive functions such as creating a menu bar, finding the inverse of a matrix, or drawing an ellipse.

However, students treating reused code as a black box without really understanding the internals, the degree to which reuse should be allowed, and appropriate acknowledgement are some of the issues that remain a challenge. There is also the issue of evaluating work based on reuse, particularly when it has to be balanced against originality. For example, if the usability of software A (40% original, 60% reuse) is deemed much better than software B (60% original, 40% reuse), should A be graded higher than B if it is known that it was the reuse in A that made the difference? Similarly, should a team be penalized for choosing a software library that they did not know at the time of use had subtle floating-point errors that only became explicit in specific use cases?

We note here that reuse is neither truly free nor automatic. Efforts of reuse that are not an integral part of planning at the outset of a software project can be detrimental to productivity and maintainability. Also, according to the COCOMO II cost estimation model (Boehm et al., 2001), it comes at a price of learning and adapting to new situations.

Guidelines for Open Source Software in Software Engineering Education

Based on the previous discussion and our experience, we present the following broadly classified guidelines for the use of OSS in SEE:

- **OSS use planning:** Educators planning to adopt OSS could look into the usefulness of it for future careers of students since, for example, some OSS for the same domains are more broadly used in industry than others; check the history of the OSS and see if the evolution has been stable; verify claims particularly related to quality, if any, and look into the amount of testing; check the availability of any nontrivial (representative) examples and how well they work; check whether the OSS is sufficiently documented before recommending its use; and go through the licensing conditions. Indeed, close collaboration with systems administrators

of the corresponding departments can be quite useful in such decision-making. An incremental approach starting from a minimal and well-defined list of OSS is highly recommended.

- **OSS reuse:** Educators should set criteria for the degree for reuse of OSS and make it known to students. Students could be asked to formally declare any OSS code reuse and a precise articulation for doing so. Given more than one option for the use of an OSS as a subsystem, students could be asked how and why they chose one over the other. Possible criteria for choice of a subsystem could be availability, ease of installation, interoperability with the system being built, portability, and past experience. Finally, in order to minimize reuse of OSS without reflection, students could also be questioned to reflect understanding of any reused code.

- **OSS in perspective:** Outlining the benefits as well as pitfalls/shortcomings to students can be useful in placing the scope of OSS into context. In Tables 2 and 3, we summarize some of the trade-offs that could help in decision-making toward the use of OSS in SEE contexts.

Table 2. Advantages of the use of OSS in SEE

General/Administrative	• The possibility for educational institutions to be able to make available a broad collection of software without incurring heavy costs as well as be able to provide OSS utilities for which there are no commercial parallels.
	• The flexibility of trying out different OSSs and examining them at any level of desirable detail prior to making a commitment.
Teaching/Learning	• The opportunity for both teachers and students to experiment (e.g., with source code internals) more freely, which is in agreement with the spirit of teaching and learning.
	• The prospect for students to develop their own personal collection of tools specialized for various tasks (modelers, compilers, debuggers, etc.) in a software project within minimal cost.
	• The opportunity for students to contribute to an existing OSS in various directions (e.g., reengineering, reverse engineering, discovery of software design patterns, or extensions via implementation of further modules).
	• The opportunity for students to participate in the development of an OSS for their own software projects.

Table 3. Disadvantages of the use of OSS in SEE

Project	• Usually, the absence of precise estimates of schedules and details of other aspects that provide the overall picture of the software project plan.
Process	• The traditional OSS process model does not explicitly support any customer involvement in its phases, an aspect that is important for today's interactive systems. • Little or no evidence of early modeling of software that could be used as inspiration for similar domain contexts. • Minimal trace between phases of a process and that from phases to process artifacts. • Sporadic rationale for design decisions, including the use of algorithms and data structures, which led to implementation. • It can be difficult to make objective assessments of software projects that make broad reuse of open source code. • In many cases, there are no explicit guarantees for technical support when needed or at all. • The use of OSS, particularly those whose breadth of testing is not known, in safety-critical contexts.

FUTURE TRENDS

Among the possible domains that OSS addresses (Nakakoji & Yamamoto, 2001), it would be of interest to examine the ones more congruent to software engineering. OSS has already had a major impact on Web applications and Web services, but their broad use in real-time and embedded systems is yet to be seen.

Taking into consideration the human factor is important to both teaching and learning. In feedback to the author over the years, students find it important that the subject being communicated is fun to learn, and OSS can provide that avenue (Luthiger, 2005). Computer games offer a variety of technical challenges related to user interface/interaction design and incorporation of 3D graphics. They can also introduce many of the software metaphors (Boyd, 1999) without resorting to unnecessary terminology. Introducing such games as part of software projects (Rucker, 2002) and the use of OSS libraries to realize that would be of interest.

Among the open source possibilities, this chapter focuses mostly on OSS; a natural extension of this work would be to look into the use of open content (excluding source code) in soft-

ware engineering. The aim of open content is to facilitate the prolific creation of freely available, high-quality, well-maintained content (not including software). The significance of open content for education in general has been highlighted in Attwell (2005). The continually increasing price of textbooks, none of which may be suitable as-is to a given course, is one motivation for open content in SEE.

To that regard, there are a few promising initiatives on the horizon. The Open Sources Education (Tadeusz & Ostrowska, 2006) is a platform for e-learning that has been applied to management courses in universities in Poland, and adaptation of its didactic and communicative aspects to SEE would be of interest. The MIT OpenCourseWare and Rice Connexions are two commonly cited examples of institution-initiated efforts of making course content open to the public-at-large. The participation of other institutions will enable a wide range of choices and will be crucial for the success of open-course content. The Directory of Open Access Journals (DOAJ) is an Internet-based service that covers free, full-text, quality-controlled scientific and scholarly journals in various disciplines, including those related to software engineering, and in various natural languages. Such services could help level the playing field and open new vistas in research-oriented higher educational contexts in the software engineering discipline, particularly where affordability is an issue.

CONCLUSION

Today, OSS has reached the level of maturity that it could be embraced as well as criticized, but not ignored. If the predictions of software business models (Cusumano, 2004; Feller et al., 2005) are correct, OSS and non-OSS will continue to co-exist. Both OSS and non-OSS have their own share of strengths and weaknesses, are most likely to co-exist, and any approach to software development should take them into consideration. There is much that software engineering and commercial OSSD can learn from each other (Asundi, 2001), and indeed, recent industrial support of OSS efforts has led to mutual benefits.

If one of the goals of SEE is to prepare students for their future careers, we must look at OSS objectively. OSS has much to offer SEE; however, the transition from one to the other is hardly straightforward. However, the adoption of OSS in education need not be seen with skepticism but rather with cautious optimism.

In conclusion, OSS is bringing about change in the way software is being developed and used. To embrace this change requires a reflection and reexamination of the current state of the curriculum. For that to come to realization, the current software engineering culture (Wiegers, 1996) in educational institutions will need to evolve.

ACKNOWLEDGMENT

The author would like to thank CUPFA (Concordia University, Montreal, Canada) for its support via a professional development grant and the anonymous reviewers for their detailed feedback and suggestions for improvement.

REFERENCES

Alagar, V. S., & Periyasamy, K. (1998). *Specification of software systems*. Springer-Verlag.

Asklund, U., & Bendix, L. (2001). *Configuration management for open source software*. Paper presented at the First Workshop on Open Source Software Engineering, Toronto, Canada.

Asundi, J. (2001). *Software engineering lessons from open source projects*. Paper presented at the First Workshop on Open Source Software Engineering, Toronto, Canada.

Attwell, G. (2005). *What is the significance of open source software for the education and training community?* Paper presented at The First International Conference on Open Source Systems (OSS 2005), Genova, Italy.

Beck, K., & Andres, C. (2005). *Extreme programming explained: Embrace change* (2nd ed.). Addison-Wesley.

Beydeda, S., Book, M., & Gruhn, V. (2005). *Model-driven software development*. Springer-Verlag.

Boehm, B. W., Abts, C., Brown, A. W., Chulani, S, Clark, B. K, Horowitz, E., et al. (2001). *Software cost estimation with COCOMO II*. Prentice Hall.

Booch, G., Jacobson, I., & Rumbaugh, J. (2005). *The unified modeling language reference manual* (2nd ed.). Addison-Wesley.

Boyd, N. S. (1999). Using natural language in software development. *Journal of Object-Oriented Programming, 11*(9), 45-55.

Crowston, K., & Howison, J. (2005). The social structure of free and open source software development. *First Monday, 10*(2).

Cusumano, M. A. (2004). Reflections on free and open software. *Communications of the ACM, 47*(10), 25-27.

Feller, J., Fitzgerald, B., Hissam, S. A., & Lakhani, K. R. (2005). *Perspectives on free and open source software*. MIT Press.

Fenton, N. E., & Pfleeger, S. L. (1997). *Software metrics: A rigorous & practical approach*. International Thomson Computer Press.

Fowler, M., Beck, K., Brant, J., Opdyke, W., & Roberts, D. (1999). *Refactoring: Improving the design of existing code*. Addison-Wesley.

Gacek, C., & Arief, B. (2004). The many meanings of open source. *IEEE Software, 21*(1), 34-40.

Ghezzi, C., Jazayeri, M., & Mandrioli, D. (2003). *Fundamentals of software engineering* (2nd ed.) Prentice-Hall.

Ginige, A., & Murugesan, S. (2001). Web engineering: An introduction. *IEEE Multimedia, 8*(1), 14-18.

Glass, R. L. (2003). *Facts and fallacies of software engineering*. Addison Wesley.

González-Barahona, J. M., Heras-Quirós, P. D. L., Centeno-González, J., Matellán-Olivera, & Ballesteros-Cámara, F. (2000). Libre software for computer science classes. *IEEE Software, 17*(3), 76-79.

Halloran, T. J., & Scherlis, W. L. (2002). *High quality and open source software practices*. Paper presented at the Second Workshop on Open Source Software Engineering, Orlando, Florida.

Jacobson, I., Booch, G., & Rumbaugh, J. (1999). *The unified software development process*. Addison-Wesley.

Kamthan, P. (2004). A framework for addressing the quality of UML artifacts. *Studies in Communication Sciences, 4*(2), 85-114.

Kamthan, P. (2006). *Open source software in software engineering education: No free lunch*. Paper presented at the The 2006 Canadian University Software Engineering Conference (CUSEC 2006), Montreal, Canada.

Liu, C. (2003). *Adopting open source software engineering in computer science education*. Paper presented at the The Third Workshop on Open Source Software Engineering, Portland, Oregon.

Luthiger, B. (2005). *Fun and software development*. Paper presented at the The First International Conference on Open Source Systems (OSS 2005), Genova, Italy.

Michlmayr, M., Hunt, F., & Probert, D. R. (2005). *Quality practices and problems in free software projects*. Paper presented at the The First International Conference on Open Source Systems (OSS 2005), Genova, Italy.

Nakakoji, K., & Yamamoto, Y. (2001). *Taxonomy of open source software development*. Paper presented at the First Workshop on Open Source Software Engineering, Toronto, Canada.

Nishinaka, Y. (2001). *Open source software developments in XP style*. Paper presented at the First Workshop on Open Source Software Engineering, Toronto, Canada.

Paulk, M. C., Weber, C. V., Curtis, B., & Chrissis, M. B. (1995). *The capability maturity model: Guidelines for improving the software process*. Sebastopol, CA: Addison-Wesley.

Perens, B. (1999). The open source definition. In C. DiBona, S. Ockman, & M. Stone (Eds.), *Open sources: Voices from the open source revolution*. O'Reilly & Associates.

Raymond, E. S. (1999). *The cathedral & the bazaar*. Sebastopol, CA: O'Reilly & Associates.

Rucker, R. (2002). *Software engineering and computer games*. Addison-Wesley.

Qureshi, S. (2001). How practical is a code of ethics for software engineers interested in quality? *Software Quality Journal, 9*(3), 153-159.

Schmidt, D. C., & Porter, A. (2001). *Leveraging open source communities to improve the quality and performance of open source software*. Paper presented at the First Workshop on Open Source Software Engineering, Toronto, Canada.

Schneidewind, N. F., & Fenton, N. E. (1996). Do standards improve product quality? *IEEE Software, 13*(1), 22-24.

Seidel, W., & Niedermeier, C. (2003). *Open source software: Leveraging software quality in the industrial context*. Paper presented at the First Workshop on Open Source Software in an Industrial Environment (OSSIE 2003), Erfurt, Germany.

Spinellis, D., & Szyperski, C. (2004). How is open source affecting software development? *IEEE Software, 21*(1), 28-33.

Tadeusz, K., & Ostrowska, T. (2006). *The open sources education: A real time education*. Paper presented at the The 17th Annual Information Resources Management Association International Conference (IRMA 2006), Washington, DC.

Thomas, D., & Hunt, A. (2004). Open source ecosystems. *IEEE Software, 21*(4), 89-91.

Vixie, P. (1999). Software engineering. In C. DiBona, S. Ockman, & M. Stone (Eds.), *Open sources: Voices from the open source revolution.* Sebastopol, CA: O'Reilly & Associates.

Wiegers, K. (1996). *Creating a software engineering culture*. New York: Dorset House,

KEY TERMS

Agile Development: A philosophy that embraces uncertainty, encourages team communication, values customer satisfaction, vies for early delivery, and promotes sustainable development.

Coding Standard: A documented agreement that addresses the use of a formal (such as markup or programming) language.

Domain Model: A simplified abstraction from a certain viewpoint of an area of software interest.

Formal Specification: A software representation with well-defined syntax and semantics that is usually used to express software requirements or detailed software design.

Pair Programming: A practice that involves two people such that one person (the primary person or the pilot) works on the artifact while the other (the secondary person or the copilot) provides support in decision-making and provides input and critical feedback on all aspects of the artifact as it evolves.

Quality: The totality of features and characteristics of a product or a service that bear on its ability to satisfy stated or implied needs.

Software Engineering: A discipline that advocates a systematic approach of developing high-quality software on a large scale while taking into account the factors of sustainability and longevity as well as organizational constraints of time and resources.

Software Pattern: A reusable entity representing knowledge and experience aggregated by an expert in solving a recurring problem in a domain.

About the Contributors

Kirk St.Amant, PhD, is an assistant professor of technical communication and rhetoric at Texas Tech University, USA. He has a background in anthropology, international government, and technical communication and rhetoric, and his research focuses on intercultural communication—particularly in relation to online media and outsourcing relationships. He has taught face-to-face and online courses in intercultural communication, rhetoric, and technical communication for Texas Tech University, James Madison University, Mercer University, and the University of Minnesota. He has also taught courses in international business, e-commerce, and distance/online education for the USAID-sponsored Consortium for the Enhancement of Ukrainian Management Education (CEUME) and the Kyiv Mohyla Business School.

Brian Still, PhD, is an assistant professor teaching technical communication at Texas Tech University, USA. He has more than a decade of experience in information technology, including work as an application developer and Internet services manager.

*   *   *

Ray Agostinelli is founder and president of Kaivo Software, Inc., a consulting firm based in Boulder, CO, USA that specializes in open source software development. His interest in non-proprietary technologies dates to the mid 1980s when he served as a UNIX trainer and development manager for corporate IT and sales personnel. Before founding Kaivo in 1999, he held executive management positions in several technology companies serving both corporate and nonprofit clients. Ray is a 1985 graduate of Dartmouth College with degrees in physics and philosophy.

Noor Al-Nahas graduated with an MIS from the American University of Sharjah (AUS) in the UAE. She is currently completing her MBA at AUS. Her research interests are in e-business strategy formulation, knowledge management, and open source software. She has published in the *SAM Advanced Management Journal*, the *Proceedings of the Information Resource Management Association*, and the *Proceedings of the 8th BIS Conference*.

Reuven Aviv is an associate professor in the Department of Computer Science in Tel Hai Academic College, Israel and previously served as the chairperson of the department. He is also on the academic staff of the Open University of Israel and a member of Chais Research Center. He holds a PhD degree in mathematical physics from Tel Aviv University. He specializes in complex network analysis—computer communication networks, security, and asynchronous learning networks, and he has published numerous papers and given conference presentations on this topic. He and his colleagues were recently awarded grants from Sloan Foundation and the Israel Science Foundation to pursue their research in this area.

Brian D. Ballentine is an assistant professor and the professional writing and editing coordinator for the English Department at West Virginia University, USA. He holds degrees from John Carroll University, the University of Rochester, and Case Western Reserve University. Before joining the English Department, he was a senior software engineer for Philips Medical Systems, where he designed user interfaces for Web-based radiology applications and specialized in human-computer interaction. Among other projects, he is currently researching and writing a textbook, *Technical Communication for Engineers*, due out next year.

James Baroody is a distinguished lecturer and chair of the Decision Sciences and MIS Department in the E. Philip Saunders College of Business at RIT, USA. He holds a BS from the University of Richmond, an MS from the College of William and Mary, and a PhD from the University of Wisconsin-Madison. Dr. Baroody has worked extensively as a software and systems architect, software developer, and project manager in government, education, and industry. His most frequent role has been unifying business requirements with technology selection and system architecture. His research interests include information systems strategy, enterprise information systems, and service-oriented architectures.

Beatrice A. Boateng is an instructional technologist who recently completed her PhD at Ohio University, USA. She has interests in the use of instructional technology in the development and delivery of online (distance education) courses and free open source software as alternatives to proprietary software.

Kwasi Boateng is an assistant professor of new media at the School of Mass Communication at the University of Arkansas at Little Rock, USA. He has interests in online/Web journalism, media policy, and regulation and free/open source software.

Andrea Bosin is a researcher in the Mathematics and Computer Science Department, University of Cagliari, Italy. His research interests include data mining and knowledge discovery, distributed and service-oriented architectures, and grid computing.

Vanessa P. Braganholo is a professor in the Department of Computer Science at the Mathematics Institute, Federal University of Rio de Janeiro, Brazil. She has received the Doctor of Science degree from Federal University of Rio Grande do Sul in 2004. She has several publications in the database area. She has participated as a member of program committees, on conference organization committees, and as a reviewer of journals. She is a member of the Brazilian Computer Society. Her research interests are on semi-structured data and query processing.

Daniel Brink graduated from the University of Cape Town, South Africa, with an Bachelor of Commerce (Honors) degree in information system in 2005. He is currently an entrepreneur and avid supporter of the open source movement, having co-written a number of articles to promote the use of open source software.

Ralf Carbon is a researcher at the Fraunhofer IESE, Germany. His work focuses on agility in product-line engineering and service-oriented computing. He is involved in open source evaluation projects together with the University of Kaiserslautern since 2002. Before joining Fraunhofer IESE in 2005, he received a diploma in computer science from the University of Kaiserslautern and worked with their Software Engineering Research Group.

Marcus Ciolkowski is a researcher and project manager at Fraunhofer IESE, Germany, and at the University of Kaiserslautern, Germany. He received an MS in computer science from the University of Kaiserslautern. His research interests include empirical methods for software engineering and quality management. He has been involved in teaching open source courses since 2001 and has organized several international workshops on empirical software engineering.

Stefano Comino is lecturer in economics at the Department of Economics of Trento. His main fields of interest are industrial organization, innovation, software markets, regulation, and contract theory.

Megan Conklin is an assistant professor in the Department of Computing Sciences at Elon University, USA. Her primary research focus is on data mining and large database systems, particularly for software engineering data. She has a PhD in computer science from Nova Southeastern University.

Robert Cunningham is an associate lecturer at the School of Law & Justice, Southern Cross University, Australia. He has scholarly publications relating to a broad range of subject matter including sustainability, biotechnology and agriculture, and international trade. He occasionally contributes to legal information, advocacy, and education at the Northern Rivers Community Legal Centre in his capacity as solicitor.

Francesca da Rimini is a practising artist, video maker, writer, and occasional curator. After co-managing two major research and publishing projects documenting creative applications of new technologies, she became the founding executive officer of the Australian Network for Art and Technology in 1989. A member of the art collectives VNS Matrix (1991-1997) and identity_runners (1999-2006), she has helped create distinctive media art and Internet projects. In 1999 she received a prestigious Australia Council New Media Fellowship. Currently, she is a PhD candidate at the University of Technology, Sydney, Australia, where she researches social software, cultural activism, and the digital commons.

Bruno de Vuyst is a graduate of Antwerp and Columbia law schools and is an associate professor at Vesalius College, Vrije Universiteit Brussel (VUB); Advisor Industrial Policy, VUB; and secretary-general, Brussels I³ Fund (the incubation and spin-off fund of the VUB) and a director of VUB spin-offs; he is an elected representative of the Brussels Bar at the General Assembly of the Flemish Bars (OVB). He is on the board of editors of Intellectuele Rechten – Droits Intellectuels and Ad Rem, the member publication of OVB. He is on counsel at Lawfort Brussels, specializing in intellectual protection law.

Nicoletta Dessì is a professor of database systems and director of the Mathematics and Computer Science Department, University of Cagliari, Italy. Her major fields of study and research are distributed and service-oriented architectures, data mining and knowledge discovery, and e-learning.

Jacobus Andries (Jaco) du Preez currently calls Centurion in South Africa home. He holds a master's degree in information technology (MIT) at the University of Pretoria in South Africa. For the last three years he has been the technical director of an open source consulting company called Yocto Linux & OSS Business Solutions. His speciality areas are the use of open source software in the enterprise, enterprise architecture, business process management, and service-oriented architecture.

Alfreda Dudley-Sponaugle is a lecturer in the Department of Computer and Information Sciences at Towson University, USA. She currently teaches the computer ethics courses in the curriculum. Her research focus is technology, ethics, and culture. Her interests include information technology, systems analysis, management information systems, databases, and computer science education. Professor Dudley-Sponaugle's recent publications involved Web accessibility, ethics, and diversity issues in information technology. She is advisor to the National Society of Black Engineers chapter at Towson.

Zippy Erlich is senior lecturer in the faculty of the Computer Science Department at the Open University of Israel, and she served as the head of the department for four years. She has developed curricula for undergraduate and graduate programs of study in computer science. She received her BSc degree in mathematics and statistics, MSc in applied mathematics (both from Tel-Aviv University), and PhD in Computer science from the University of California, Los-Angeles. Before joining the Open University, she headed the Data Processing Department of the Israeli Navy Computer Center. Her research interests include measurement of information systems success, data mining, social networks, computer systems security, and e-learning.

Theodoros Evdoridis received a 5-year BEng (2003) in information and communication systems engineering and a MSc in information and communication systems security—from the University of the Aegean. He is currently a laboratory tutor and a doctoral student at the same university. His main research interests are information and database systems security.

Alea Fairchild is an associate professor of management and computer science at Vesalius College, as well as faculty in the Economics Department of Vrije Universiteit Brussel, Belgium. Her technical expertise lies in open architectures and interoperability. Her recent areas of research have included knowledge management and productivity metrics for technology. Dr. Fairchild received her doctorate in applied economics from Limburgs Universitair Centrum (now Universiteit Hasselt) in Belgium, in the area of banking and technology.

Laurence Favier is an associate professor in information and communication sciences at the University of Bourgogne (Dijon), France. He is a member of the research laboratory CRIS of the University of Paris X (Nanterre) and a member of Centre Georges Chevrier in the Human Sciences Home (Maison des Sciences de l'Homme) of Dijon. His research interests focus on e-government, e-governance, and knowledge management. He is also director of the program PRATSIC (for Research Program on Technology for Information and Knowledge Society) in Maison des Sciences de l'Homme of Dijon.

Robert Fitzgerald is a research fellow in the Learning Communities Research Area at the University of Canberra (Australia). His main interests are in the application of information and communication technologies (ICTs) to learning, collaboration, and problem solving. His current research focuses on the so-called Web2.0 technologies, and he leads a project, funded by the Australian Carrick Institute for Learning and Teaching in Higher Education, on the use of social software to support peer learning. Robert has held academic positions in Hong Kong and Australia and is a regular reviewer for international journals and conferences.

Ingbert R. Floyd is a doctoral student in the Graduate School of Library and Information Science at the University of Illinois at Urbana-Champaign. His research interests focus around the design of sociotechnical systems, particularly systems which are continuously evolving and thus require continuous design. In particular, he is interested in developing participatory design methodologies which take the rich descriptive and observational data from studies in social informatics and socio-technical systems theory, and utilize their results to inform the design methodology and design process.

Maria Grazia Fugini is a professor at Politecnico di Milano, Italy. Her research interests are in information system security and development. She is involved in the WS-Diamond and SEEMP UE Projects on Web-based information systems for e-government

Jochen Gläser received his PhD in sociology of science at the Humboldt-University Berlin in 1990 and completed his habilitation in sociology by the Free University Berlin in 2005. He is currently a Fellow at the Research School of Social Sciences of the Australian National University, Australia. His major research interests are the sociology of science, sociological theory, economic sociology, and qualitative methodology.

Alexander Hars (alexander.hars@inventivio.com) is founder and CEO of Inventivio, Germany, an innovative software company specializing in Java-based knowledge applications. After many years of research and teaching in information systems in the United States and Europe, Alexander Hars has founded a software company to bridge the gap between theory and practice and turn innovative ideas into marketable software products. Dr. Hars obtained a PhD degree and the habilitation degree (German tenure) from the University of the Saarland. He taught for several years at the University of Southern California. He has published two books on knowledge management and reference models for information systems. He has published several articles on the open source phenomenon. His research interests are knowledge management, applications of speech technology, and applications of speech technology.

Jens Heidrich received the BS degree (Vordiplom) and the MS degree (Diplom) in computer science with a minor in mathematics from the University of Kaiserslautern, Germany. Since July 2001, he is a researcher at the Software Engineering Research Group (AGSE) at the University of Kaiserslautern, Germany. His current research interests include software project management, software development methodologies, quantitative analysis and evaluation of software development processes and products.

Nina Helander, PhD, is a senior researcher at The Institute of Business Information Management, Tampere University of Technology, Finland. Her research interests are related to software business, in particular open source software and software components from a managerial point of view. She is especially interested in studying the relationships between different OSS network actors and the logic of value creation of these actors. Helander has publications on value creation, business networks, software component business, and OSS business models. In her doctoral dissertation, Helander studied software business from value-creating network perspective.

Sungchul Hong is an assistant professor in the Department of Computer and Information Science at Towson University, Maryland, USA. He received his PhD in management science from the University of Texas at Dallas. His major research interests are E-Commerce related technologies including automated algorithms in various markets, XML-related applications, and intelligent computer agent systems.

Jeroen Hoppenbrouwers holds a PhD in information science since 1997 and has worked as a research engineer at Tilburg University and Vrije Universiteit Brussels, mostly on industrial research projects. He specialises in language-related information science and systems architecture. His main interests are on the boundary between language and modelling of both information systems and semantic domains, often combined in ontology management. As a sidetrack, Jeroen values open source software for many more reasons than just *free beer* and tries to integrate professional processes in typical open source development projects.

Leila Lage Humes is a PhD student at the School of Economics, Business, Administration and Accounting, University of São Paulo (FEA-USP), Brazil, and also the manager of IT staff training programs at the University's Central Information Technology Coordination Department. Her previous positions include working as a network manager and as a support manager at the University of São Paulo.

Juha Järvensivu (MSc 2005) is studying towards a PhD at Tampere University of Technology, Tampere, Finland. He works as a researcher at the Institute of Software Systems, Tampere University of Technology. His primary research interests focus on open source software development and programming of graphical user interfaces.

Isabel John is a researcher at the Fraunhofer IESE, Germany. She works in several research and industrial projects in the context of software product line engineering and open source. Her work focuses on product line modeling, domain analysis and scoping. She has organized several workshops in the area of product line engineering and held several talks and tutorials on the topic. She received a diploma in computer Science from the University of Kaiserslautern.

M. Cameron Jones is a doctoral candidate in the Graduate School of Library and Information Science at the University of Illinois at Urbana-Champaign, USA. His research interests include human-centered computing, collaborative systems design, rapid prototyping and evaluation methods, and end-user appropriation and innovation. He is interested in how the widespread availability and use of open-source software and open-access APIs are impacting information systems design and development practices. He also researches and teaches approaches to Web programming which explore the creative appropriation, combination, and reuse of Web technologies and content.

Hannu Jungman, MSc (Eng.), is a project manager at Innovation Research Development Tamlink Ltd. and PhD student at Tampere University of Technology (TUT), Finland. Mr. Jungman's research and business interests are in finding more efficient, ownership-based operating models for pushing growth-oriented university spin-offs and other ventures to the radar scan of the venture capital industry, or, in other words, pushing them across the capital gap and the knowledge gap. Before, he worked in management accounting and consulting of growth-driven companies in Finland and abroad.

Pankaj Kamthan has been teaching in academia and industry for several years. He has also been a technical editor and participated in standards development. His professional interests and experience include software quality, markup languages, and knowledge representation.

Sigrid Kelsey is an associate librarian at Louisiana State University Libraries, USA. Her publications cover topics including Web design and usability, e-mail communication, online teaching, and information-seeking behavior. She has taught a graduate course in information technology and is currently co-editing a book on computer-mediated communication.

Mathias Klang (klang@ituniv.se) is a researcher in legal informatics at the University of Göteborg, Sweden. His research interests and publications lie primarily in the areas of the law in connection with topics such as technology, democracy, human rights, free expression, censorship, open access, and ethics. He has published several articles in these topics. He completed his PhD, "Disruptive Technology: Effects of Technology Regulation on Democracy," in 2006. In his free time, Mathias is also project lead for Creative Commons Sweden and a member of the Swedish Team of the Free Software Foundation Europe.

Tung-Mei Ko has been the project manager of Open Source Software Foundry (OSSF), Institute of Information Science, Academia Sinica, Taiwan since 2005. Her Research area focuses on the legal topics about the Free/Open Source Software. She received her Master of Law degree (LLM) from the Institute of Tax Law, Münster University, Germany in 2002 and her BL degree from Chung Yuan Christian University, Taiwan. During 2003-2005, she worked in the Science and Technology Law Center (STLC) of Institute of Information Industry (III) in Taiwan.

Gabor Laszlo currently teaches at Budapest Tech, Hungary and also manages the Information Society Research and Education Group. He is a PhD candidate at the Budapest University of Technology and Economics. His research interests focus on open source software and its applications in the public sector, and also e-learning. He has background in both economics and technology and has been a presenter at numerous scientific conferences. He has participated as co-author in publications of working progress booklets for the Hungarian Information Society and also wrote as co-author a position paper on strategy-planning document for the National Open Source Strategy in Hungary.

Pierre-Paul Lemyre is a lawyer of the Quebec Bar and is in charge of the conception and development of LexUM International Projects. He is highly interested in the challenges that lasting development poses, as well as in syndication technologies for access to law and in the issues related to free and open source software.

Diego Liberati is the chief scientist for information for the Control and Biomedical Engineering with the Electronic, Information and Communication Institute of the Italian National Research Council at Milano Institute of Technology, Italy.

Kwei-Jay Lin received a PhD in computer science from the University of Maryland, College Park. He is a professor in the Department of Electrical Engineering and Computer Science at the University of California, Irvine, USA. His research interests include service-oriented systems, e-commerce, real-time systems, scheduling theory, and Web-based systems. Dr. Lin is the editor-in-chief of the *International Journal of Service Oriented Computing and Applications* (published by Springer), and the editor-in-chief of the *Software Publication Track, Journal of Information Science and Engineering* (published by Academia Sinica, Taiwan). He has been a co-chair of the IEEE Technical Committee on E-Commerce since 2004.

Yi-Hsuan Lin is the legal lead of Creative Commons Taiwan. She received her Master of Law degree (LLM) from Southern Methodist University, Dallas, Texas in 2001 and her BL degree from Fu Jen Catholic University, Taiwan. During 2002-2003, she worked in MediaTek Incorporation of Taiwan, which is the world's leading digital media solution provider. She has worked for Institute of Information Science, Academia Sinica, Taiwan since 2004.

Yu-Wei Lin, Taiwanese, holds a PhD in sociology from the University of York (UK). Her PhD research investigates the heterogeneity and contingency in the free/libre open source software (FLOSS) social world. Challenging existing writings that depict an idealistic and harmonious hacker culture dominantly residing in FLOSS communities, Lin seeks to explore the socio-technical dynamics in FLOSS development and examine the diverse articulations and performances in which hacker culture and hacker identity are both reflected and constructed. Lin works as research associate at the ESRC National Centre of E-Social Science at the University of Manchester, UK.

William Lively received his PhD from Southern Methodist University in Computer Science and Electrical Engineering. Dr. Lively was a member of the technical staff of the Advanced Scientific Computer Group at Texas Instruments. He is a professor of computer science at Texas A&M University, USA and co-principal investigator of the Shared Software Infrastructure Program. He has been a consultant to many industrial firms and has served as principal investigator and/or co-principal investigator for numerous projects sponsored by government and industry.

Saku Mäkinen, PhD, is a professor of technology management at the Institute of Industrial Management, Tampere University of Technology (TUT), Finland. Dr. Mäkinen has been previously with the Department of Marketing, Faculty of Business Administration, National University of Singapore (NUS). He received his PhD in technology strategy from TUT, Finland. His research interests include international business, technology and innovation strategy, and management and industry evolution. He is the director of Center for Innovation and Technology Research (CITER, http://www.tut.fi/citer) at TUT.

Fabio M. Manenti is an associate professor in economics at the Department of Economics of Padua. His main research fields are industrial organization, networks and Internet economics, telecommunications, software markets, regulation, and antitrust and competition policy.

Marta Mattoso is a professor of the Department of Computer Science at the COPPE Institute from Federal University of Rio de Janeiro since 1994, where she co-leads the Database Research Group. She received a doctor's degree in science from Federal University of Rio de Janeiro. Dr. Mattoso has been active in the database research community for more than ten years, and her current research interests include distributed and parallel databases, data management aspects of Web services composition and genome data management. She is the principal investigator in research projects in those areas, with fundings from several Brazilian government agencies, including CNPq, CAPES, FINEP and FAPERJ. She has published over 60 refereed international journal articles and conference papers. She has served on the program committees of international conferences, and she is a reviewer of several journals. She is a member of the Brazilian Computer Society, where she is the Society Publication's director and the ACM, where she is an editor of the ACM-SIGMOD Digital Symposium Collection.

Joël Mekhantar is a professor of law at University Jules Verne of Picardie (Amiens). He is a member of CURAPP, Research Center in Law and Sociology in University Jules Verne of Picardie. Previously, he was an associate professor at University of Burgundy in Dijon (1990-2005). His principal areas of publishing and teaching are in constitutional law, public finance, civil service, e-administration, and e-democracy (legal and political aspects). He has written Droit politique et constitutionnel, (éd. Eska, Paris, 1998, p. 732); Finances publiques—Le Budget de l'État, (éd. Hachette, Paris, 2006, p. 160).

Tommi Mikkonen (MSc 1992, Lic. Tech. 1995, Dr. Tech 1999, all from Tampere University of Technology, Tampere, Finland) works on software engineering related topics at the Institute of Software Systems at Tampere U of Technology, Tampere, Finland. Professor Mikkonen's primary research interests include open source software development, software architecting, mobile devices programming, and their relations.

Bernardo Miranda is a senior technology developer at World Travel Holdings, an Internet-based travel agency. He has experience in many technologies, including Java, XML, databases and parallel processing. He obtained a Master of Science degree in systems and computer engineering at Federal University of Rio de Janeiro, Brazil in 2006. During the research at the university, he developed an open source middleware for parallel query processing using commodity databases. He has also participated as technology developer of the ParGRES project. ParGRES is an open source project with governmental funding that aims to provide parallel processing capabilities to the PostgreSQL database system.

Andrew Mowbray is a professor of law and information technology at the University of Technology, Sydney (UTS), Australia. He is the co-director of the Australasian Legal Information Institute (AustLII). He has written various pieces of software (including the Sino search engine) and has played some administrative roles in the law faculty at UTS.

Kathryn Moyle is one of Australia's leaders on open source software in schools. She lives in Australia's national capital city: Canberra, where she is an associate professor at the University of Canberra in the School of Education and Community Studies and is the director of the Learning Communities Research Area. Her primary responsibilities include undertaking research into information and communication technologies (ICT) in K-12 education. Dr. Moyle researches and publishes on a diversity of topics, focusing on practical educational, social and technical issues in schools. Politicians through to educational practitioners refer to her work.

Emmanuel Mulo is the webmaster at Uganda Martyrs University and an assistant lecturer in the Department of Computer Science and Information Systems of Uganda Martyrs University in Uganda. He provided technical assistance in the migration project. He is specialized in programming and FOSS tools for Web development. Currently, he is pursuing his MSc in computer science at Delft University of Technology in The Netherlands.

Dirk Muthig heads the Product Line Architectures Department at the Fraunhofer IESE. He has been involved in the definition, development, and transfer of Fraunhofer's PuLSE (Product Line Software Engineering) methodology since 1997. Dirk Muthig teaches product line engineering at the University of Kaiserslautern and has organized and participated in numerous workshops. He received a diploma in computer science, as well as a PhD, from the University of Kaiserslautern.

Chris Nelson, is the senior software engineer of the IBM Developer Skills Program. He has over 20 years of software development experience as both a programmer and an architect. He is also the owner of a small company. As part of the IBM Developer Skills Program, Chris works to increase developer skills on IBM's standards-based software through education and university curriculum programs.

Jussi Nissilä is a researcher and project manager at the Department of Information Technology at University of Turku, Finland. His work includes teaching and research in the areas of software entrepreneurship and productization. His research interests include business models and strategies in ICT business, virtual communities and network organizations and ICT in developing economies. He has published articles in top international conferences as well as in books on business and information systems.

Alessandro Nuvolari completed his first degree in economics and social sciences at Bocconi University in Milan (Italy), before obtaining his PhD at the Eindhoven Centre for Innovation Studies (ECIS). He is currently an assistant professor in economics of science and technology at the Eindhoven University of Technology, The Netherlands. His research interests include the nature of innovation and technical change during the British industrial revolution and the relationships between IPR regimes and the innovative performance of industries.

Casey O'Donnell is a PhD candidate in the Science and Technology Studies Department at Rensselaer Polytechnic Institute, USA and has performed fieldwork at game studios over two years. His dissertation, "Playing the New Economy: Video Game Development in India and the United States," is funded by an NSF grant. His work examines the diverse forces and activities that shape software development and makes it tenable in today's globalized economy. His research questions, "What can the worlds of software developers teach us about the "new" economy?" and "How do worlds differ across national and cultural boundaries?" links software development to global processes.

Phillip Olla is an associate professor at the school of business at Madonna University in Michigan His research interests include knowledge management, space Internet connectivity, mobile telecommunication, and health informatics. Over the last decade, in addition to university level teaching, Phillip has worked as an independent information technology consultant. His experience is primarily in the Internet and telecommunication and space industry. Olla has worked on a wide variety of pioneering projects in conjunction with mobile network operators and mobile service providers including British Telecom, Hutcinson 3G, T-Mobile, and IBM Global Services. He received his PhD from the Department of Information Systems and Computing at Brunel University, UK. He is a member of the editorial board for the *Industrial Management & Data Systems Journal,* the book review and software review editor for the *International Journal of Healthcare Information Systems,* and a member of the Editorial Advisory & Review Board for the *Journal of Knowledge Management Practice.*

Jennifer Papin-Ramcharan is the engineering and physical sciences librarian at the University of the West Indies – St. Augustine Campus, Trinidad and Tobago. She has several years experience as a lecturer in electronics, mathematics, physics, and computer science and also as an engineer. She is committed to user-centered service and the empowerment of individuals by connecting them to information and its effective use.

Jonathan Peizer has over two decades of experience in strategic planning, development, management, and successful execution of projects employing a broad variety of technologies. Peizer is a social entrepreneur with significant experience working in international, cross-cultural environments in over 75 countries. He has worked for Citicorp, AFS, Cheyenne Software, and the Soros Foundations, and he is the author of the book *The Dynamics of Technology for Social Change* (http://technologyforsocialchange.org). He currently manages his own consulting firm (Internautconsulting.com), a socially responsible eCommerce Enterprise (Greentealovers.com), and he developed the nonprofit capacity resource, Capaciteria.org. He is founder and board chair of the NGO Aspiration.

Barbara Pes is a researcher at the Mathematics and Computer Science Department, University of Cagliari. Her research interests include service-oriented architectures, data mining, and bio-informatics.

Daniel Poulin is professor at the Faculty of Law of the University of Montreal. Professor Poulin teaches information technologies in the Cyberspace Law Program. He is also director of LexUM, the foremost laboratory in Canada working on the computerization of law. His current research interests relate to legal information system design. Systems designed and implemented by LexUM include CanLII, Juris International and Portail du droit francophone Web sites.

Mikko Puhakka is researcher of open source business models at Helsinki University of Technology, Finland. Mikko is a founder of Holtron in 1994. Since Holtron's establishment Mikko has carried out numerous advisory assignments to technology companies as well as special assignments to government on innovation policy as well as on challenges of the information economy. Further he has managed several venture capital investments including an investment into MySQL, the leading open source database company in 2001. Besides the research work, Mikko continues to advise select clients from both private and public sector. Further, more specifically related to open source, Mikko is an advisor in the initiative for the Finnish Centre for Open Source Software (COSS).

Risto Rajala is a research fellow in information systems science and coordinator of the Business Networks Research Programme at the Helsinki School of Economics, Finland. He has published several articles on resources, services, networks and revenue models of software companies in international refereed publications. His current research activity is focused on the analysis of business models in the software industry.

Nicolau Reinhard is an associate professor of business administration at the School of Economics, Business Administration and Accounting of the University of São Paulo, Brazil, a Member of the University's Information Technology Steering Committee, and coordinator of the IT Management MBA Program of FIA (Administration Institute Foundation). Previous activities include positions in academic and government organizations (technical, management, and governance). He is a member of ACM, AIS, and the IEEE Computer Society and is an associate editor of the *Journal of Global Information Management, e-Service Qu@rterly*, and a member of the Board of Editors of Information and Management.

Llewellyn Roos is an analyst programmer at Open Box Software in Cape Town. He obtained his business science honours degree specializing in information systems at the University of Cape Town, where he focused on the private sector adoption of OSS. He continues to work in the field of enterprise applications and Web-based technologies.

Alessandro Rossi is lecturer of management at the Department of Management and Computer Science in Trento, Italy. His main fields of interest focus on modularity, software project management, design of complex artefacts, and free/open source software.

Bruno Rossi received a Laurea (MSc) in economics at the University of Trento, Italy in 2000 and a Laurea (BSc) in applied computer science from the University of Bozen-Bolzano (Italy) in 2004. From 2000 to 2003, he worked in two different software companies. From 2004 to 2005 he was a research assistant at the Faculty of Computer Science, and in January 2005 started his PhD program in computer science under the supervision of Prof. Giancarlo Succi. His research interests include the different models of software development, the open source model, technology adoption models and software engineering in general.

Francesco Rullani, after his graduation (2002), has worked as a junior researcher at the Fondazione ENI Enrico Mattei studying the impact of ICT on firms' location choices. During his Ph.D. at Sant'Anna School of Advanced Studies, Pisa, Italy, he has been visiting researcher at Stanford University (2004). Recently, he has been granted a Fondazione IRI scholarship (2006) to spend a year as a post-doc fellow at Copenhagen Business School. His Ph.D. thesis is focused on open source software, and in particular on developers' motivations, and on the organization and sustainability of this model of innovation.

Barbara Russo received her master's and PhD degrees in mathematics from the University of Trento, Italy, respectively in 1991 and 1996. Since 1996, she has been a post-doc fellow and research assistant in mathematics at the University of Trento, Italy. In 2003 she was full time research assistant at the Free University of Bolzano-Bozen, Italy. She is currently a research associate at the Faculty of Computer Science of the Free University of Bolzano-Bozen. She is a member of the Center for Applied Software Engineering of the Free University of Bolzano-Bozen. Her current research interests include statistical analysis of software metrics, models in software reliability (special focus on open source projects), and evaluation of agile software methodologies.

Sofiane Sahraoui is an associate professor of MIS at the American University of Sharjah in the UAE. He received his PhD in MIS from the University of Pittsburgh in the USA in 1994. His research interests are in e-government as a cornerstone of the information society and knowledge economy, open source software, IT planning, enterprise modeling, and the management of IT change in general. He has published in leading academic publications such as the *Journal of Information Technology Management*, the *Journal of End-User Computing, Behaviour & Information Technology, Human Systems Management*, the *Journal of Global Information Technology Management*, the *Journal of Computer Information Systems*, and a variety of other IT publications, both refereed and non-refereed.

Christoph Schlueter Langdon (csl@ebizstrategy.org) co-founded and chairs the Special Interest Group on Agent-Based Information Systems (SIGABIS, http://www.agentbasedis.org) of the Association for Information Systems (AIS). He is affiliated with the Center for Telecom Management of the University of Southern California (USC), USA after having been a full-time professor of USC's Marshall School of Business for five years. Prior to joining USC, Langdon was a scientist in the Artificial Intelligence Group of the Beckman Institute for Advanced Science and Technology at the University of Illinois at Urbana-Champaign. His research is focused on IS capabilities and their implications for business strategy using next generation analytical tools, such as agent-based modeling and strategic simulation. Results and insights have appeared in leading publications, including *Communications of the ACM*, IEEE journals, and *Harvard Business Review*. Chris Langdon has also been an advisor to Global Fortune 500 companies and governments on digital interactive channel development and business intelligence analytics, first as a consultant with Accenture (formerly Andersen Consulting), and then with Pacific Coast Research Inc., a boutique management advisory firm. Recent clients include DaimlerChrysler, Deutsche Telekom and Nissan. He has been educated at the Darmstadt University of Technology, Germany, and University of Illinois at Urbana-Champaign and received graduate degrees in Engineering and Finance & Business Economics, and a PhD degree in economics, all summa cum laude.

Marko Seppänen is a senior researcher and holds MSc in industrial engineering and management. Presently, he works for the Center for Innovation and Technology Research (CITER, http://www.tut.fi/citer) at Tampere University of Technology, Finland. He has several years of experience in research and teaching in the areas of technology management, management accounting and project management and has published actively on these topics.

Dick B. Simmons received his PhD in computer and information sciences from the University of Pennsylvania, USA. He served as an officer in the U.S. Army Signal Corps, worked as a design engineer for Radio Corporation of America, and as a technical supervisor at Bell Telephone Laboratories. He is a fellow of the IEEE. He currently is a professor of computer science at Texas A&M University and co-principal investigator and director of the Shared Software Infrastructure Program. He has served as principal investigator and/or co-principal investigator for over 38 Pprojects sponsored by government and industry.

Darren Skidmore has over 15 years experience in the ICT industry, including over six years as an academic at the University of Melbourne. He has conducted and presented research on free/libre and open source software at international, academic, and industry conferences. Research areas include jurisprudence, fiduciary usage, business value, and history of ICT. He holds an honours degree in information systems, a Master of Commercial Law degree, and a graduate diploma in cross cultural communication from the University of Melbourne, with a graduate diploma in applied finance and investment with FINSIA. He is currently completing a PhD in evaluation of software and software environments with Monash University, looking at issues such as the metrics used, the effects of architecture, best-practice frameworks, total cost of ownership, and risk. The research will also look at the understanding of the limitations of the evaluation process and the effects of reaction to favorable and adverse events on the process.

Marcus Vinicius Brandão Soares holds an MSc degree in computing and systems engineering from COPPE/UFRJ, Rio de Janeiro, Brazil. He is a member of the NECSO Science and Technology Studies research group at UFRJ (www.necso.ufrj.br), the Brazilian Institute for Electronic Law (www.ibde.org.br), and the Foundation for Free Information Infrastructures (www.ffii.org). He presented lectures in many universities in Brazil (UNIRIO, PUC-Rio, COPPE-UFRJ), France (École de Mines de Paris), and the USA (University of Maryland). His interests include new institutional economics, intellectual property rights, and actor-network theory. He can be found at oares@marcusvinicius.eti.br.

David J. Solomon is an associate professor in the Department of Medicine and the Office of Medical Education Research and Development at Michigan State University, USA. He is editor and founder of *Medical Education Online* (http://www.med-ed-online.org) an open access electronic journal that has been covering all aspects of health professions education since 1996. He has a PhD in educational psychology and works mainly in the area of program evaluation and performance assessment. His other major scholarly interest is the communication of research.

Frank Soodeen has been a librarian for sixteen years. For most of this period he was responsible for the development and management of the national standards information centre in Trinidad and Tobago. For the past six years he has been working as a systems librarian at The University of the West Indies – St. Augustine Campus, Trinidad and Tobago, where his focus has been on Web and multimedia development, digital library initiatives, and online course delivery. He now manages the UWI Libraries Systems Unit.

Wouter Stam is a doctoral candidate in strategy and entrepreneurship at the Vrije Universiteit Amsterdam, The Netherlands. He was also a visiting scholar at the Wharton School of the University of Pennsylvania. His current research interests include the antecedents and consequences of social networks, entrepreneurship, innovation, and the emergence of OSS firms. His dissertation is financed by a grant received from the Netherlands Organization for Scientific Research (NWO) and focuses on how internal resources and external social networks interactively shape the innovative and financial performance of OSS firms.

R. Todd Stephens is the director of the Collaboration and Online Services Group for the BellSouth Corporation, an Atlanta-based telecommunications organization. Todd is responsible for setting the corporate strategy and architecture for the development and implementation of the enterprise collaborative and metadata solutions. Todd writes a monthly online column in *Data Management Review* and has delivered keynotes, tutorials, and educational sessions for a wide variety of professional and academic conferences around the world. Todd holds degrees in mathematics and computer science from

Columbus State University, an MBA degree from Georgia State University, and a PhD in information systems from Nova Southeastern University.

Giancarlo Succi received a Laurea degree in electrical engineering (Genova, 1988), an MSc in computer science (SUNY Buffalo, 1991), and a PhD degree in computer and electrical engineering (Genova, 1993). He is a tenured professor at the Free University of Bolzano-Bozen, Italy, where he directs the Center for Applied Software Engineering. He has been a principal investigator for projects amounting more than 5 million dollars in cash and, overall, he has received more than 10 million dollars in research support from private and public granting bodies. Dr. Succi is a Fulbright Scholar and a member of the IEEE Computer Society.

Rania Suleiman is a graduate of MIS from the American University of Sharjah (AUS) in the UAE. Her research interests are in knowledge management and open source software. She has published in the *Proceedings of the 8th BIS Conference.*

Marie-Noëlle Terrasse is an associate professor in computer science at the University of Burgundy, (Dijon) France and a member of the research laboratory LE2I. Her research interests include UML-based metamodeling and its application to domain related frameworks (e.g., e-government frameworks), and her enterprise-directed applications include Web site modeling and open source-based developments. She has participated in various program comittees and conference organizing comittees (e.g., Libre Software Meeting in 2005) and is a poject leader for Apogee (a university application for the management of students).

Thomas Tribunella earned his PhD at the State University of New York at Albany, his MBA at the Rochester Institute of Technology, and his BBA at Niagara University, and he is a certified public accountant. He has been teaching, writing, and consulting for over 20 years. Dr. Tribunella has published over 35 journal articles, case studies, book chapters, and conference proceedings and has also won four best-paper awards for his research.

Michael B. Twidale is an associate professor in the Graduate School of Library and Information Science at the University of Illinois at Urbana-Champaign, USA. His research interests include the usability of open-source software, and how open-source models can be applied to participatory usability and data quality management. Dr. Twidale is interested in the development of novel information technologies through rapid prototyping and evaluation methods, and how end users collaboratively learn and appropriate technologies to develop creative new uses. He also researches and teaches on human-computer interaction, CSCW, information systems design, and the collaborative aspects of data-quality management.

Theodoros Tzouramanis received his 5-year BEng (1996) in electrical and computer engineering and his PhD (2002) in informatics from the Aristotle University of Thessaloniki. Currently, he is lecturer at the Department of Information and Communication Systems Engineering of the University of the Aegean, Greece. His research interests include access methods and query processing for databases; database security and privacy; and geographical information systems.

Joseph E. Urban received his PhD from the University of Louisiana at Lafayette. He worked at the University of Miami, the University of Southwestern Louisiana, and part-time at the University of South Carolina, while with the U.S. Army Signal Center before joining Arizona State University. He is currently a professor of computer science and engineering and directs the Inclusive Learning Communities program. Urban leads the Software Process, Environment and Automation Research Group. He is currently on leave to the National Science Foundation. He has authored more than 90 technical papers and has supervised the development of seven software specification languages.

Tere Vadén, is an assistant professor of philosophy and hypermedia at the University of Tampere, Finland and the University of Skövde, Sweden. He has published articles on the philosophy of language and art, and theories of information society. He is currently directing the Open Source Research Group at the Hypermedia laboratory in the University of Tampere.

Niklas Vainio is a researcher at the Hypermedia laboratory in University of Tampere, Finland, in its Open Source Research Group. He is interested in philosophical issues related to copyright, information society, and free/open source software communities. He has a background both in philosophy and software development and has previously published on free software philosophy and open source community structures.

Jean-Paul Van Belle is an associate professor in information systems at the University of Cape Town, South Africa. Before joining UCT, he set up the Information Systems Department at the University of the Western Cape. He obtained his MBA in the field of financial modelling and his doctorate in the field of enterprise modelling. Currently he is researching various aspects of e-commerce and m-commerce and supervises a number of doctoral and masters students in this area. He is a family man with three children and a keen outdoors person.

Karin van den Berg, MSc, is a recent graduate of Tilburg University, The Netherlands, in the field of information science and has written her master's thesis on the subject of open source software evaluation. Currently, Ms. van den Berg works full time as a freelance programmer specializing in the Web scripting languages of PHP in The Netherlands.

Victor van Reijswoud is a professor of information systems in the Department of Computer Science and Information Systems, Uganda Martyrs University in Uganda and the architect of the FOSS migration presented in this chapter. He is also chairman of the East African Center of Open Source Software based in Uganda. Before moving to Uganda, he was engaged in research and lecturing at several universities in Europe, Africa, and the USA.

Ruben van Wendel de Joode is an assistant professor in the School of Technology, Policy and Management, Delft University of Technology, The Netherlands. His research focuses on OSS communities. He has published his work in journals like *IBM Systems Journal, Computer Standards and Interfaces, Knowledge, Technology and Policy,* and *Electronic Markets.* His research on OSS communities has been financed by two grants received from the Netherlands Organization for Scientific Research (NWO). He is also the lead author of the book *Protecting the Virtual Commons: Self-Organizing Open Source and Free Software Communities and Innovative Intellectual Property Regimes* (2003).

Yuanqiong Wang is an assistant professor in the Department of Computer and Information Sciences at Towson University, Maryland, USA. She received her PhD in computer and information systems from New Jersey Institute of Technology. Her major research interests involve decision support systems (DSS), asynchronous learning network, knowledge management, and HCI.

James Weller is a graduate of the University of Cape Town and holds BSc (computer science) and BCom(Hons) (information systems). He is currently working as a business and systems integration analyst for a global management consulting firm in the mobile telecommunications industry. He maintains a keen interest in open source technologies, particularly operating systems and Web-based content delivery systems.

Mika Westerlund is a research fellow in the Finnish Graduate School of Marketing (FINNMARK) at Helsinki School of Economics, Department of Marketing and Management and coordinator of the ValueNet research consortium. His research interests address business networks, relationship value, and business models of companies, especially in the software industry.

Dave Yeats (dave.yeats@auburn.edu) is an assistant professor in the technical and professional communication program at Auburn University, USA, where he studies the communication practices of the open source software community. In addition to open source software development, his research interests include usability, documentation management, and technical communication pedagogy.

Index